3rd International Symposium on New Energy Technology Innovation and Low Carbon Development (NET-LC 2024)

E3S Web of Conferences Volume 528 (2024)

Guangzhou, China
29 – 31 March 2024

Editors:

S. Yao
M. Zhang
O. Mahian
L. Boon Han

ISBN: 978-1-7138-9803-0

Printed from e-media with permission by:

Curran Associates, Inc.
57 Morehouse Lane
Red Hook, NY 12571

Some format issues inherent in the e-media version may also appear in this print version.

For additional information, please contact EDP Sciences – Web of Conferences
at the address below.

EDP Sciences – Web of Conferences
17, Avenue du Hoggar
Parc d'Activité de Courtabœuf
BP 112
F-91944 Les Ulis Cedex A
France

Phone: +33 (0) 1 69 18 75 75
Fax: +33 (0) 1 69 28 84 91

contact-edps@webofconferences.org

Additional copies of this publication are available from:

Curran Associates, Inc.
57 Morehouse Lane
Red Hook, NY 12571 USA
Phone: 845-758-0400
Fax: 845-758-2633
Email: curran@proceedings.com
Web: www.proceedings.com

TABLE OF CONTENTS

RENEWABLE ENERGY UTILIZATION AND SMELTING TECHNOLOGY

CARBON EMISSION RESEARCH AND SUSTAINABLE DEVELOPMENT

Preface

In the context of the accelerating process of globalization, the continuous growth of energy demand and the increasingly severe environmental problems, new energy technology innovation and low-carbon development have become the focus of global attention. In order to promote the innovation and development of new energy technologies and promote international exchanges and cooperation, we successfully held the 2024 3rd International Symposium on New Energy Technology Innovation and Low Carbon Development (NET-LC 2024).

The Symposium was held in Guangzhou, China from March 29th to 31st, 2024, attracting experts, scholars, and business representatives from all over the world. The conference mainly focused on the latest research field of "new energy technology innovation and low-carbon development", aiming to provide an international cooperation and exchange platform for practitioners of new energy technology innovation and low-carbon development to share research results, discuss problems and challenges, and explore cutting-edge technologies.

During the conference, participants conducted extensive and in-depth discussions on a number of topics such as the innovation of new energy technologies, strategies and practices of low-carbon development, policies and market mechanisms, and environmental impact assessment. Many experts and scholars delivered wonderful speeches, sharing their latest research results and practical experience in new energy technology innovation and low-carbon development. At the same time, participants also carried out heated discussions and exchanges on how to promote the wide application of new energy technologies and promote the sustainable development of low-carbon economy.

In order to fully showcase the results of this Symposium, we have specially compiled this collection of proceedings. This collection includes excellent academic papers submitted by the participants, covering many aspects such as Photovoltaic Power Generation, New Energy Power Generation System and Grid Connection Technology, New Energy Vehicles, Electric Vehicles, etc. These papers not only represent the latest research results of the current new energy technology innovation and low-carbon development, but also provide valuable reference and inspiration for our future work.

In the process of soliciting manuscripts, we have received positive responses and strong support from the experts, scholars and business people around the world. They submitted papers and shared their research results and practical experience, contributing wisdom and strength to the success of this Symposium. Here, we would like to express our heartfelt thanks and high respect to all the contributors.

Finally, we would like to thank all those who participated in NET-LC 2024, including the organizers, speakers, delegates and staff. It is your joint efforts and hard work that have made this Symposium a complete success. At the same time, we would like to thank all the experts, scholars and business people for their attention and support to this Symposium, and look forward to meeting again in the future Symposium.

The Committee of NET-LC 2024

Committee Member

Conference General Chair
Prof. Tianshou Zhao, Southern University of Science and Technology, China
Prof. Liejin Guo, Sun Yat-sen University, China & Sun Yat-sen University, China
Prof. Junfu Lv, Tsinghua University, China
Prof. Huihe Qiu, Hong Kong University of Science and Technology, China
Prof. Jidong Lu, South China University of Technology, China
Prof. Lizhi Zhang, South China University of Technology, China

Publication Chair
Prof. Shuhuai Yao, The Hong Kong University of Science and Technology, China
Prof. Mingming Zhang, Harbin Institute of Technology (Shenzhen), China
Prof. Omid Mahian, Ningbo University, China
Prof. Lim Boon Han, Universiti Tunku Abdul Rahman, Malaysia

Program Committee Chair
Prof. Ronghui Qi, South China University of Technology, China
Prof. Lin Zeng, Southern University of Science and Technology, China
Prof. Yun Wang, University of California, Irvine, USA
Prof. Haslenda Hashim, Universiti Teknologi Malaysia, Malaysia

Local Committee Chair
Prof. Weiliang Wang, Jinan University, China
Prof. Shunchun Yao, South China University of Technology, China
Prof. Simin Huang, Dongguan University of Technology, China
Prof. Jinghong Zhou, Changchun Institute of Technology, China

Committee Member
Prof. Defilla Steivan, Tianjin University, China
Prof. Muhyiddine Jradi, University of Southern Denmark, Denmark
Prof. I. Neethi Manickam, Francis Xavier Engineering College, India
Prof. Masoud Taghavi, Chung-Ang University (CAU), South Korea
Dr. Mahajan Sagar Bhaskar, Prince Sultan University, Saudi Arabia
Dr. Somnath Ganguly, Bankura Unnayanyi Institute of Engineering, India
Dr. Shilong Huang, North China Electric Power University, China
Prof. Lishuai Jiang, Shangdong University, China
Prof. Fei Wang, Qingdao University of Science and Technology, China
Prof. Kamil Arslan, Karabuk University, Turkey
Prof. Qing Fu, Sun Yat-sen University, China
Dr. Wenyang Deng, South China University of Technology, China
Prof. Li Liu, Guangxi University, China
Prof. Kanghua Zhong, South China University of Technology, China
Prof. Jiashen The, Universiti Sains Malaysia, Malaysia
Prof. Nor Asiah Muhamad, Universiti Teknologi Brunei, Brunei

Statement of Peer review

In submitting conference proceedings to *Web of Conferences*, the editors of the proceedings certify to the Publisher that

1. They adhere to its **Policy on Publishing Integrity** in order to safeguard good scientific practice in publishing.
2. All articles have been subjected to peer review administered by the proceedings editors.
3. Reviews have been conducted by expert referees, who have been requested to provide unbiased and constructive comments aimed, whenever possible, at improving the work.
4. Proceedings editors have taken all reasonable steps to ensure the quality of the materials they publish and their decision to accept or reject a paper for publication has been based only on the merits of the work and the relevance to the journal.

Title, date and place of the conference

2024 3rd International Symposium on New Energy Technology Innovation and Low Carbon Development (NET-LC 2024)
March 29-31, 2024
Guangzhou, China

Proceedings editor(s):
Shuhuai Yao, The Hong Kong University of Science and Technology, China
Mingming Zhang, Harbin Institute of Technology (Shenzhen), China
Omid Mahian, Ningbo University, China
Lim Boon Han, Universiti Tunku Abdul Rahman, Malaysia

Date and editor's signature

Shuhuai Yao
Mingming Zhang
Omid Mahian
Lim Boon Han

Remaining oil potential and adjustment countermeasures in fault X well area of Lamadian Oilfield

Wenjie Jing

Dynamic room, Geological Research Institute, No.6 Oil Production Plant, Daqing Oilfield, Daqing Heilongjiang province, 163100, China

Abstract. The distribution characteristics of remaining oil in fault well area X of Lamadian oilfield are studied by numerical simulation, and adjustment countermeasures are formulated. Through the establishment of numerical model, the geological reserves and residual reserves of each small layer are analyzed, and the oil saturation distribution diagram output by the model is calculated, and the distribution characteristics of different types of remaining oil are summarized. The results are as follows :(1) the residual reserves are mainly distributed in the channel, accounting for 57.0 % of the total residual reserves, followed by the main sheet sand; (2) The remaining oil is mainly affected by imperfect injection-production relationship, fault occlusion and plane interference, accounting for 95 % of the total reserves of the type. Imperfect injection-production relationship accounted for 73.58 % of remaining oil types. The second is the enrichment of remaining oil from plane disturbance, accounting for 12.3 %. Residual oil of fault occlusion accounted for 9.16 %. The optimal well spacing was 200m, and the final recovery was 2.84 % higher than the original well pattern.

1. Introduction

Lamadian oil field is a tectonically controlled multi-layer non-homogeneous and severe sandstone oil field, in which the X fault is a fault-block tectonic reservoir, with the development of river, main body mat sand and non-main body mat sand, and 73 faults are developed in the area. Through nearly 40 years of water-driven development, through self-injection, water injection, multiple encryption, injection and polymerization adjustments, the variable well production system as well as operations have resulted in a complex distribution of residual oil in the subsurface[1-3]. Therefore, this study utilizes geological, seismic, logging, and numerical simulation to predict the distribution of residual oil within the reservoir, especially the distribution condition of its enriched parts[4-5]. As of 2018, the oilfield has entered the late stage of extra-high water content development, and is in the stage of water-polymerization and two-drive development, with scattered distribution of residual oil and great difficulty in dredging. However, due to the small density of well network and imperfect injection and recovery in the fault edge, some residual oil exists in some well areas in the fault edge, indicating that the fault edge is a potential area for secondary development [6]. According to the recent years of the oilfield's fault edge dredging situation, the X fault zone (area of 4.78km^2), which has not been implemented for secondary development, is taken as the study area to carry out the study of secondary development potential. Due to its complex structure, large scale of faults and many sedimentary units, it is necessary to carry out fine research on the reservoir development, tectonic features and utilization status of the study area, accurately deal with the complex spatial intersection and contact relationships between faults such as intersection and truncation, improve the quality of the tectonic model, and construct a fine geologic model; and then determine the potential of the study area for adjusting the secondary development.

This paper combines the actual development needs of the oilfield to formulate reasonable and effective adjustment countermeasures. It provides the basis for improving the development effect of X-fault zone, and it is also of great significance for the transformation of oilfield from fine to precise development.

2. Geological model of X-fault well zone in Lamadian Oilfield

Lamadian oil field X crude oil properties are good, according to the analysis results of the surface crude oil of the test wells within the study area, the average surface crude oil density is 860 kg/m^3, the crude oil viscosity is 22.9 mPa·s, the freezing point is 26.5 °C, the wax content is 23.00 %, and the gum content is 17.90 %. According to the statistics of high-pressure physical properties sampling and analysis, the original formation pressure is 11.27 MPa, the average saturation pressure is 10.7 MPa, the volume coefficient is 1.03, the original gas-oil ratio is 48.5 m^3/t, and the average formation temperature is 45 °C. Therefore, the crude oil has medium density, medium saturation pressure, and medium solidification point is 26.5 °C. Therefore, the crude oil is characterized by medium density and medium viscosity, and belongs to

Corresponding author's e-mail: 2804576644@qq.com

https://doi.org/10.1051/e3sconf/202452801001

normal temperature and normal pressure black oil reservoir (shown in figure1) .

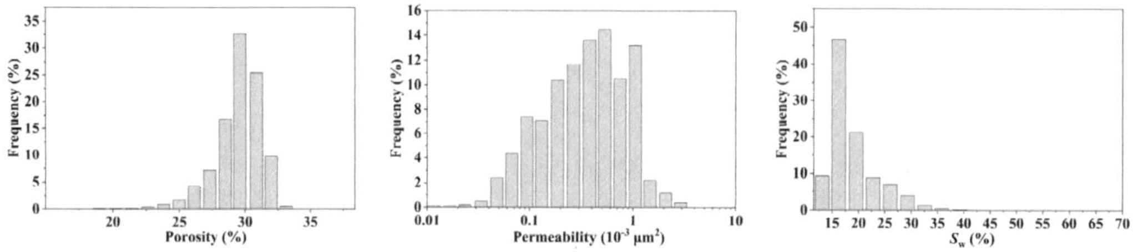

(a) Porosity histogram (b) Permeability histogram (c) Water saturation histogram

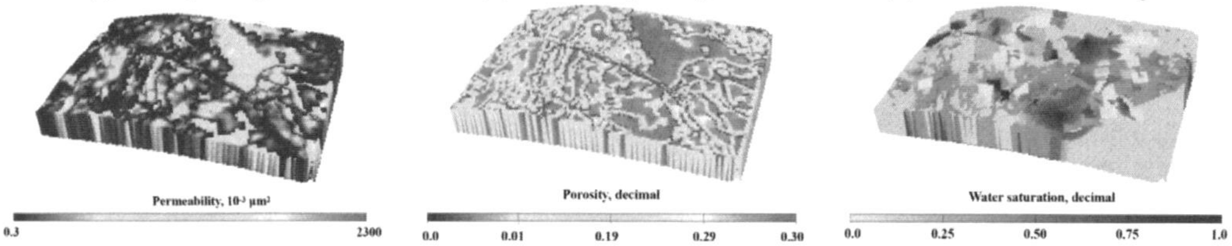

(d) Porosity model (e) Permeability model (f) Water saturation model

Figure 1. Stratigraphic parameters and their mathematical modeling

The fitted geological reserve of X fault is 6227.67×10^4 t, comparing with the actual reserve of 6258.14×10^4 t, with an error of 0.48 %, which is a good fit. The fitted reserves can meet the requirements of residual oil analysis (shown in table 1).

Table 1. Comparison of X-fault number model reservoir fits

Reservoir Layer Number	Physical reserves (10^4 t)	modeled reserve (10^4 t)	fitted reserves (10^4 t)	Relative error of fit (%)
R1	2355.1	2374.45	2342	-0.56
R2	2215.3	2234.95	2208	-0.33
R3	1687.74	1688.583	1677.67	-0.6
Total	6258.14	6297.98	6227.67	-0.48

3. Characterization of X residual oil distribution in Lamadian oilfield

The factors affecting the distribution of residual oil are many and complex, but they can be divided into two main aspects: geological factors and development factors. Geological factors mainly include reservoir inhomogeneity, tectonics and faults. The most important development factor is the degree of perfection of the injection and recovery system and the configuration relationship with the geological factors (shown in figure 2). Using the oil saturation distribution map output from the model calculation, layer-by-layer residual oil analysis is carried out for 114 sedimentary units in the block, based on which, the distribution characteristics of different types of residual oil are summarized.

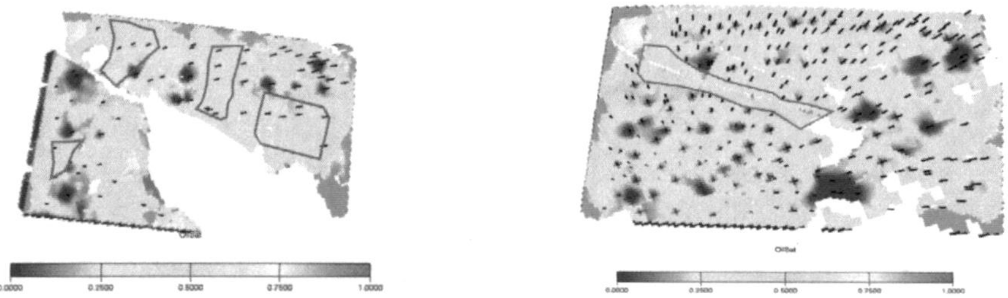

(a) Imperfect injection and recovery control type residual oil (b) Fault-masked residual oil

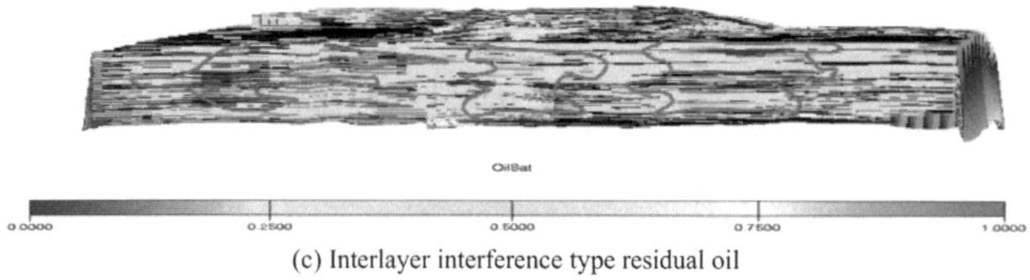

(c) Interlayer interference type residual oil

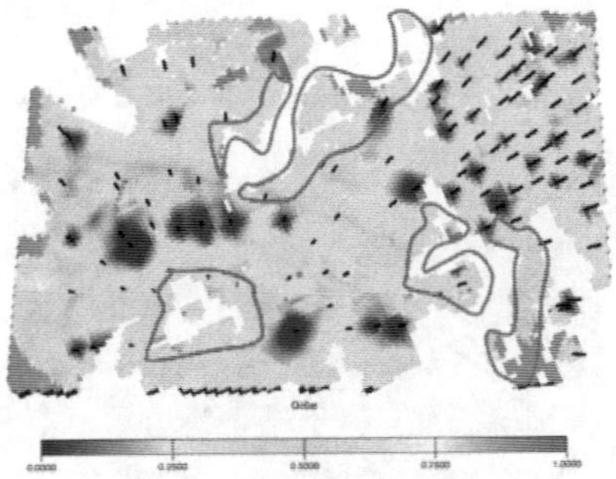

(d) Sand body edge type residual oil

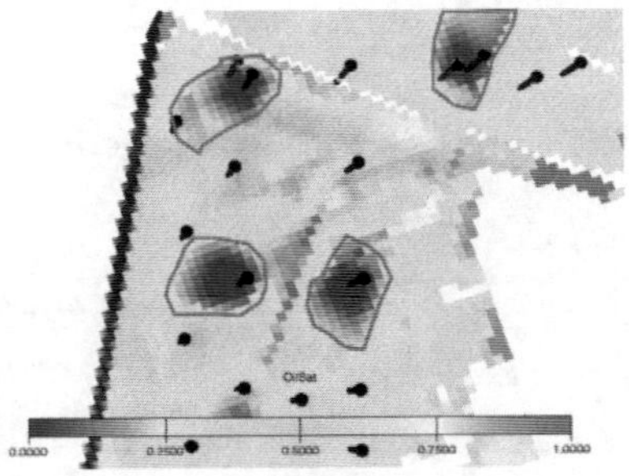

(e) Planar interference type residual oil

Figure 2. Schematic diagram of different types of residual oil

Residual oil statistics show that the distribution of residual oil is mainly dominated by imperfect injection and extraction relationship, fault blocking and plane interference, which overall accounts for 95 % of the total reserves of the type, and imperfect injection and extraction relationship occupies the main type of residual oil, accounting for 73.58 %. Secondly, plane interference residual oil is enriched, accounting for 12.3 %, and fault blocking type residual oil accounts for 9.16 %.

4. adjustment countermeasures based on the distribution of residual oil in complex fracture block oil reservoirs

For the remaining oil type of X fault, the main secondary development measures of X fault are local encryption adjustment and high water content well profiling. The results of residual oil research show that there are mainly five types of residual oil in X fault, for the distribution of residual oil, combined with the well seismic combined with tectonics, reservoir and numerical simulation results, the preferred wells, deducting the flooded layer, high water-cut layer, and at the same time, combined with the distribution of the existing network of wells, make a good adjustment of the injection system of the well area, and improve the relationship between the injection and production.

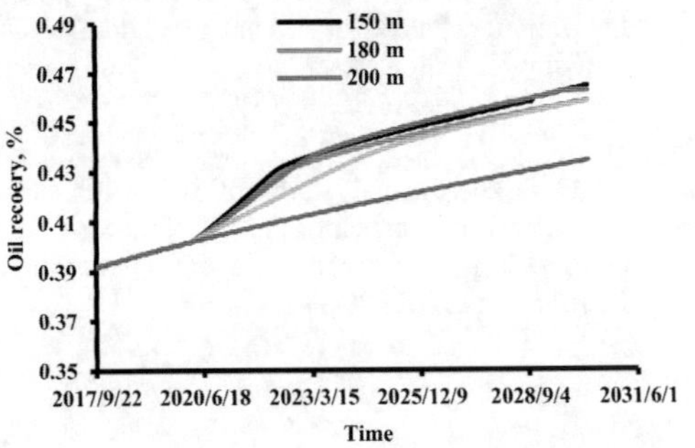

Figure 3. Recovery curves under different schemes

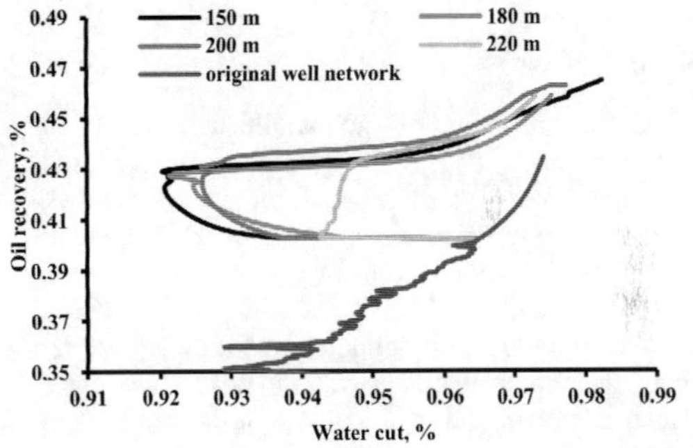

Figure 4. Integrated water cut versus degree of recovery under different schemes

The X-fault numerical model area predicts that by 2030 (shown in figure 3 and figure 4), the extraction degree of the original scheme is 43.44 %, the comprehensive water content is 97.40 %, and the extraction degree of each encrypted scheme is 46.47 %, 46.28 %, 45.88 %, and 45.85 %, which is 3.03 %, 2.84 %, 2.44 %, and 2.41 % higher than that of the original well network respectively. The increase in extraction degree of Scheme I and II with 150m well spacing is higher compared with other schemes, and Scheme I with 200m well spacing is the second highest. Comprehensive water content comparison, the water content of each encryption scheme is 98.24 %, 97.74 %, 97.53 %, and 97.30 %, which is 0.84 %, 0.34 %, 0.13 %, and -0.1 % higher than that of the original scheme, respectively. The comprehensive water content of Scheme 1 150m well spacing encryption method is higher than other schemes. In the comprehensive comparison, Scheme 1200 m well spacing encryption scheme with higher extraction degree, fewer encrypted wells and slower rise of water content is preferred as the optimal scheme.

When laying wells along the fault strike, the distance of the laying location from the fault is crucial, because too close a distance may cause the drilling process to cross the fault, and if the distance is too far away it is not possible to ensure the extraction of the remaining oil on the edge of the fault, so it is necessary to set an appropriate distance for extraction.

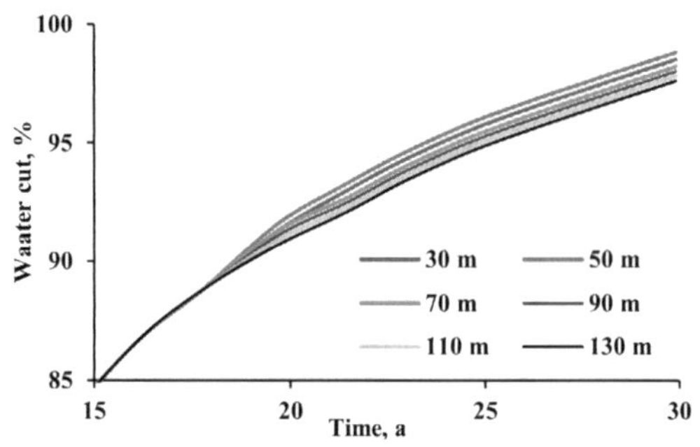

Figure 5. Comparison of water content of encrypted wells at different distances from faults

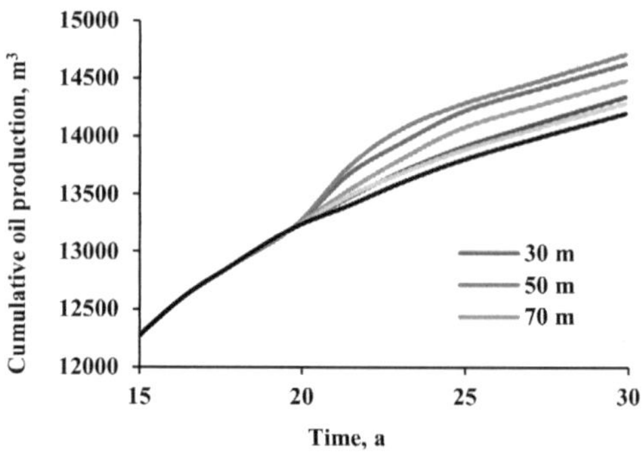

Figure 6. Comparison of cumulative oil production from encrypted wells at different distances from faults

It can be seen from the production index and the planar residual oil distribution law that (figure 5 and figure 6), within a certain range, the cumulative oil production gradually decreases as the distance between the encrypted wells and the fault increases; however, the cumulative oil production of the encrypted wells 50 m from the fault is higher than that of 30 m. This is due to the fact that the distance from the fault is too close to the fault, which will lead to the reduction of the control range of a single well, resulting in the decrease of the production capacity. At the same time, too close to the fault will increase the drilling risk. In the end, option 2 is chosen, i.e., the optimal distance between the encrypted wells and the fault is 50 m.

5. Conclusion

(1) From the viewpoint of the mobilization status of different sedimentary phases, the remaining oil reserves of the formation are mainly distributed in the river channel, and its remaining reserves account for 57.0 % of the total remaining oil reserves of the formation, followed by the main body mat sand.

(2) According to the distribution law of remaining oil in X fault, the remaining oil can be divided into five types: imperfect injection and extraction control type, fault blocking type, plane interference type, interlayer interference type, and sand body edge type, in which the remaining oil is mainly dominated by imperfect injection and extraction relationship, fault blocking and plane interference, and the whole accounted for 95% of the total remaining reserves in the type. Imperfect injection and extraction relationship occupies the main residual oil type, 73.58 %; followed by plane interference residual oil enrichment, accounting for 12.3 %; fault blocking residual oil accounts for 9.16 %.

References

1. Liu, T, Fawad, N, Li, C, Li, H, He, R, Xu, J, Ahmad, Q A (2023) Physical simulation of remaining oil distribution in the 3rd-order architecture unit in beach sand reservoir. Frontiers in Earth Science, 10, 1108525 .
2. Razavian M, Fatemi S (2021) Catalytic evaluation of metal azolate framework-6 in pristine and metal doped modes in upgrading heavy residual fuel oil[J]. Journal of Analytical and Applied Pyrolysis, 2: 105093.
3. Boesen T, Haber E, Hoversten G M (2021) Data-driven semi-supervised clustering for oil prediction[J]. Computers & Geosciences, 148(7623):104684.
4. Melzer, L, Stephen (2017) Residual oil zones changing commercial reservoir models[J]. Oil and Gas Journal, 115(4):52-55.
5. Fu, H, Bai, Z, Guo, H, Yang, K, Guo, C, Liu, M, Song, K (2023) Remaining Oil Distribution Law and Development Potential Analysis after Polymer Flooding Based on Reservoir Architecture in Daqing Oilfield, China. Polymers, 15(9): 2137
6. Wang, K, You, Q, Long, Q M, Zhou, B Wang, P (2023) Experimental study of the mechanism of nanofluid in enhancing the oil recovery in low permeability reservoirs using microfluidics. Petroleum Science, 20(1): 382-395 .

https://doi.org/10.1051/e3sconf/202452801002

Evaluation of the development effect of polydynamic drive in Class IIB oil reservoir of Lamadian Oilfield

Yanli Sun

The seventh operation area of the sixth oil production plant of Daqing Oilfield Co., LTD, Daqing Oilfield, Daqing Heilongjiang province, 163100, China

Abstract. With the continuous deepening of oilfield development, the target of oilfield development is gradually shifting towards reservoirs with poor physical properties, and polymer flooding can better utilize such reservoirs.The heterogeneity of Class II B oil layer in Lamadian Oilfield is severe, and polymer flooding takes effect later, with a small decrease in watercut.A study was conducted on the evaluation of polymer flooding development effectiveness for Class II B oil layers through reservoir numerical simulation . Research has shown that when the initial watercut is constant, a lower injection pressure can increase the utilization degree of remaining oil in the middle and high permeability layers of Class II B oil layers. With the increase of injection pressure, the response time is shortened, and the decrease in watercut is reduced; As the watercut increases, the injection pressure should gradually increase in order to improve the recovery of low permeability reservoirs;As the watercut increases, the injection pressure should gradually increase in order to improve the recovery of low permeability oil layers;At a watercut of 95% and injection pressure of 7MPa, a field experiment was conducted with a molecular weight polymer of 12 million , and the response time was 6 months. The cumulative oil increase was 14675m^3, achieving good polymer flooding effect.

1. Introduction

At present, Most oilfields in China are dominated by terrestrial deposits, and after a long period of water injection development, the water drive development effect deteriorates and the oil production decreases, so the polymer enhanced recovery technology becomes the preferred choice[1-2], and it has been widely applied in the mines[3]. The final recovery of water-driven in type II reservoirs can reach 45.3%, and the recovery of polymer-driven can be increased by 13.32%[4-6]. For some blocks, the development of Class II B oil formation using the production conditions of Class II A, there is a phenomenon of poor results. So the polydynamic recovery conditions of extensive Class II oil reservoirs are not applicable to Class IIB oil reservoirs.

The depth of the Sa III4-10 oil layer in the North-North Block I of the Lamadian Oilfield is 990 m~1100 m. After the boundary adjustment of the subdivided sedimentary unit, the upper and lower units show a better distribution of the compartments. Class I compartments are widely distributed in patches, and class II, III, IV, and V compartments are set in class I compartments in the form of stars and dots .

Analyzing the stratigraphy of North North Block I, it is found that its oil layer thickness is thin, permeability is low, the proportion of the thickness of Class I oil layer connection is low, and the proportion of the thickness of Class I connection is 13.5% lower than that of Class I oil layer . According to the field data, it was found that the proportion of Class II oil layers accounted for about 23.8% of the overall oil layer distribution, of which the Class IIB oil layer accounted for about 40% of the Class II oil layers. The permeability planar abruptness coefficient is larger, the horizontal non-homogeneity is stronger, and the planar coefficient of variation is also higher reaching 0.79.

Statistical data on the formation permeability of the block found that the permeability distribution of wells with ineffective injection polymerization was concentrated in three permeability intervals of 0.2~0.4 μm^2, 0.4~0.6 μm^2, and 0.6~0.8 μm^2, as shown in Figure 1.

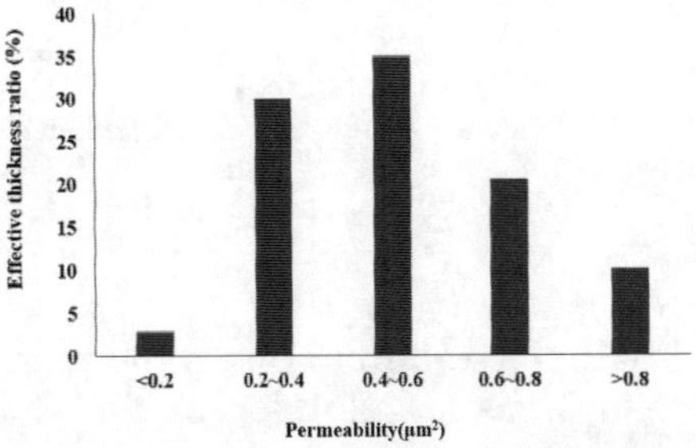

Figure 1. Permeability distribution in inresponse areas

In this paper, based on the research block, numerical simulation research is carried out to optimize and preferably select the poly injection parameters of Class II B oil formation, and clarify the characteristics of the

Corresponding author's e-mail: sunyanli0809@163.com

adjustment of the development parameters of Class II B oil formation.

2. Characterization of oil reservoirs in Class II-B reservoirs

After years of water injection and development in Lamadian oilfield, the reservoir physical properties, pore structure and other parameters of the reservoir will change. Four natural cores with different coring sites and different lithologies were selected for water-driven oil experiments to test the changes in porosity and permeability during the water-driven development stage (shown in table 1).

Table 1. Changes in permeability and porosity of Class II B core before and after water flooding

Reservoir Layer Number	Permeability ($\times10^{-3}\mu m^2$)			Porosity (%)		
	Before water drive	After water drive	Rate of change (%)	Before water drive	After water drive	Rate of change (%)
GII 1	207.43	195.8	-5.6	28.7	28.9	0.70
SI 7	452.32	501.52	10.9	27.4	27.5	0.36
GII 3	267.34	254.72	-4.7	28.6	28.8	0.70
PII 5	776.69	825.34	6.3	26.8	26.9	0.37

The experiments show that the porosity value of PII and SI formations is found to increase during long-term water injection, with an average change rate of roughly 0.53 %. Due to the high permeability of the core in this formation, the degree of cementation is relatively loose, the mud content is low, and the particles in the pore space are transported by scouring, but the pore channels are not blocked, thus leading to a larger porosity value.

At permeability lower than $260\times10^{-3}\mu m^2$, the permeability decreases by about 5 % after water drive. The main reason is the high content of clay minerals in this layer, and particle transportation plugs the pore channel, resulting in permeability decrease.

After the reservoir characterization of the Lamadian Class II-B formation, the geological model was developed using Eclipse. The model is a three-layer homogeneous model. The permeability is $200\times10^{-3}\mu m^2$, $400\times10^{-3}\mu m^2$, and $600\times10^{-3}\mu m^2$ from bottom to top.The specific model is shown in Figure 2 and table 2.

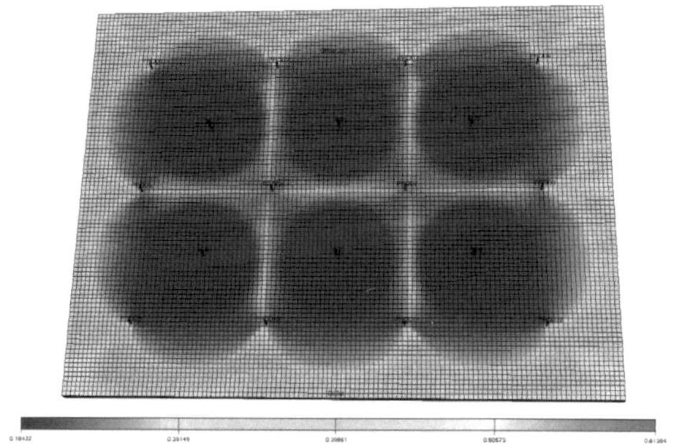

Figure 2. Model diagram

Table 2. Model parameter table

Parameters	Values
District-wide grid specifications	120×190×3
Grid size (m^3)	5×5×3
Average porosity	0.286
Average original oil saturation	0.54
Effective thickness (m)	3
Initialization pressure (MPa)	10

3. Characterization of polydynamic drive in Class IIB oil reservoirs

Numerical simulation analyses were performed with different injection pressures at the same polymer concentration, dosage, and molecular weight, using a five-point method well set. Under the actual well history conditions, wells were defined and production data were attached for three formations. Water was injected at injection wellhead pressures of 7MPa, 8MPa, 9MPa, 10MPa, and 11MPa, and transconcentrated at water contents of 95%, 96%, 97%, and 98%. History fitting of the well history was performed to obtain the relationship between recovery rate, water content, and the oil production contribution of each layer under different initial conditions with the injection volume, respectively, as shown in Figures 3-6 and Table 3.

https://doi.org/10.1051/e3sconf/202452801002

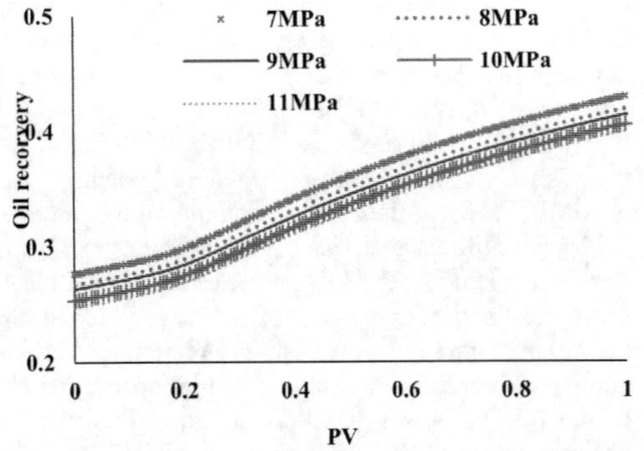

Figure 3. Variation of recovery rate with injection volume in polymer flooding driving under different injection pressures when the initial watercut is 95%

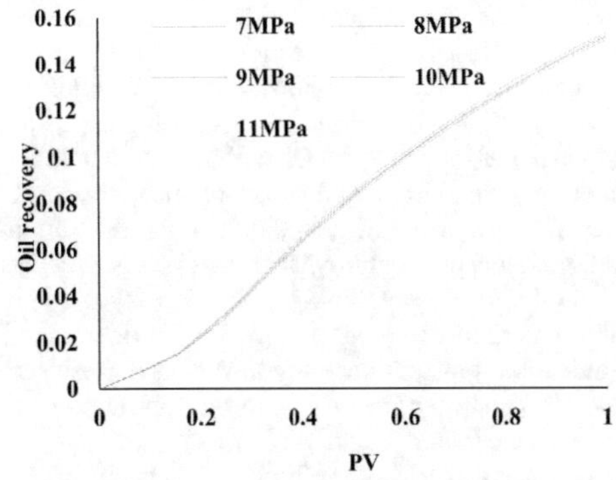

Figure 4. The variation of recovery with different injection pressures and injection volume at initial watercut of 95%

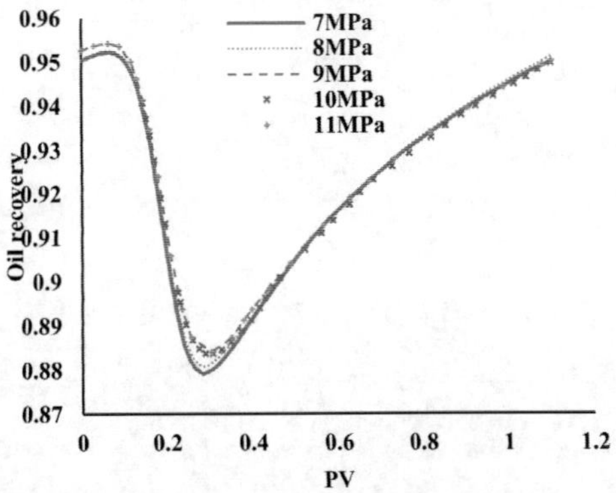

Figure 5. The variation of watercut with injection volume under different injection pressure when the initial watercut is 95%

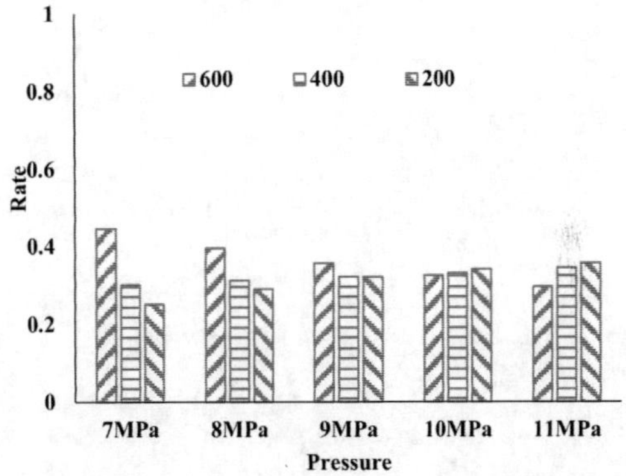

Figure 6. The effect of oil production in each layer under different injection pressure

Table 3. The changes of parameter under different injection pressure

different pressure (MPa)	Injection rate (PV/a)	Initial water content 95%(a)	Water content after polymerization 95%(a)	Productive response time (Month)	Minimum water content(%)
7	0.15	5.42	5.96	6.48	87.92
8	0.18	4.67	5.16	5.88	88.06
9	0.19	4.42	4.84	5.04	88.39
10	0.21	3.92	4.33	4.92	88.36
11	0.28	3.09	3.41	3.84	88.45

Under the initial water content of 95%, the injection pressure of 7MPa has the greatest impact on the overall recovery rate. For the polydynamic recovery, the highest polydynamic recovery was achieved at 7 MPa pressure.

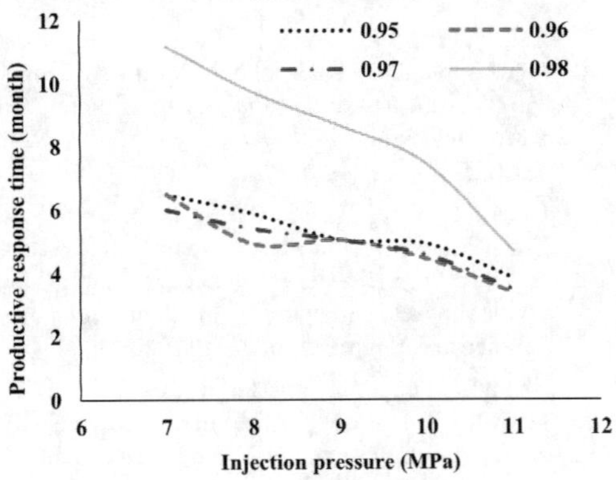

Figure 7. Permeability distribution in inresponse areas

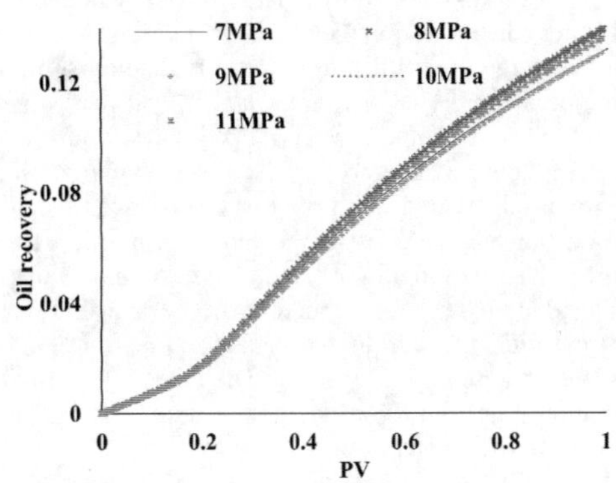

Figure 8. Variation of recovery with injection volume under different injection pressure when the initial watercut is 96%

However, the polydynamic effect time becomes shorter with the increase of injection pressure, and the decrease of water content decreases (shown in figure 7). When the injection pressure increases, judging from the cumulative oil production of each layer, the cumulative oil production of the polydrive in the high-permeability layer decreases obviously, and the cumulative oil production in the middle and low permeability layer increases gradually; the contribution rate of the high-permeability layer to the final oil production decreases, and the contribution rate of the middle and low permeability layer increases; the change of the contribution rate of the oil production in each layer decreases, as shown in Figure 8.

The overall recovery is the highest when the initial water content is 96% and 97% and the injection pressure condition is 7MPa, and the polydynamic recovery decreases sequentially under the injection pressure conditions of 8MPa, 10MPa and 11MPa. The high permeability layer is highly exploited at an injection pressure of 8MPa, and the middle and low permeability layers can also be mobilized, but the high permeability layer is still the main oil layer. When the injection pressure is 10MPa, 11MPa, the contribution rate of the middle and low permeability layer rises, and it can be well utilized, and then the middle and low permeability layer is the main oil layer. The lower the water content decline the higher the polydynamic recovery. The production as well as the contribution rate of the middle and high medium and low permeability layers still satisfy the law when the initial water content is 95%, see Figure 9.

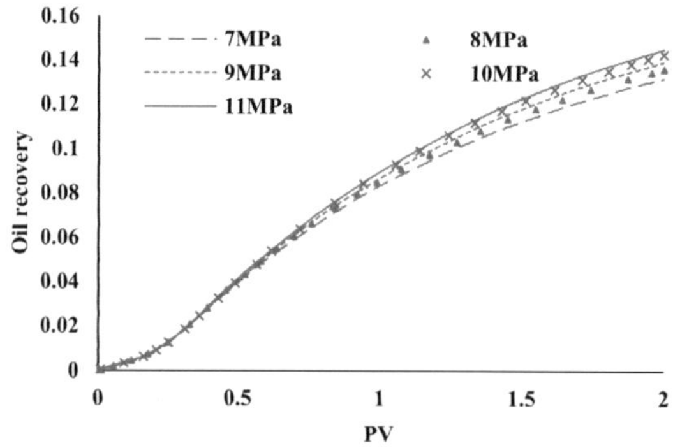

Figure 9. Variation of recovery rate with injection volume under different injection pressure when the initial watercut is 98%

When the initial water content is 98%, the recovery rate of polydynamic drive reaches the maximum value at 11MPa, because the production at this time mainly depends on the middle and low seepage layer. The polydynamic recovery time is characterized by accelerated recovery as the injection pressure increases, and at this time, the water content decrease increases with the increase of pressure.

4. Conclusion

(1) At the stage of injection polymerization when the initial water content is low, the low injection pressure can make the degree of mobilization of the remaining oil in the high-permeability layer in the Class IIB oil reservoir increase. At this time, the high permeability layer has the highest contribution rate of oil production and the highest overall oil production; the contribution rate of each layer's oil production decreases with the increase of injection pressure under the condition of the same initial water content and increases with the increase of the contribution rate of the high permeability layer and the contribution rate of the low permeability layer. The change value of contribution rate shows a decreasing trend with the increase of injection pressure.

(2) For Class II-B oil reservoirs, lower injection pressure should be adopted in the early stage to fully utilize the residual oil in the high-permeability layer, and with the increase of water content, the injection pressure should be elevated to exploit the low-permeability layer, so as to give full play to the development potential of the oil reservoir.

(3) Under the same initial water content condition, with the increase of injection pressure, the injection effect time will be shortened, and the decrease of water content will be lowered, and the value of the lowest water content will be smaller.

References

1. Lai H, Shi W, Wang J, Tang L, Lai N (2023) Development of Hydrophobic-Modified Nanosilica for Pressure Reduction and Injection Increase in an Ultra-Low-Permeability Reservoir. Applied Sciences, 13(9): 5248.
2. Cheng Z, Wang T, Bai Z, Wang L, Yuan C, Zhao Z, Song W (2023) A numerical simulation investigation on low permeability reservoirs air flooding: Oxidation reaction models and factors. Geoenergy Science and Engineering, 223: 211506.
3. Zhang Y, Feng Y, Li B, Han P (2019) Enhancing oil recovery from low-permeability reservoirs with a self-adaptive polymer: A proof-of-concept study. Fuel, 251: 136-146.

4. Liao H, Yu H, Xu G, Liu P, He Y, Zhou Y (2022) Polymer–surfactant flooding in low-permeability reservoirs: an experimental study. ACS omega, 7(5): 4595-4605.
5. Park, H, Han, J, & Sung, W (2015) Effect of polymer concentration on the polymer adsorption-induced permeability reduction in low permeability reservoirs. Energy, 84: 666-671.
6. Ghosh P, Mohanty K K (2020) Laboratory treatment of HPAM polymers for injection in low permeability carbonate reservoirs. Journal of Petroleum Science and Engineering, 185: 106574.

https://doi.org/10.1051/e3sconf/202452801003

Optimization design of oil well fracturing scheme in X block of Daqing Oilfield

Huizhuo Xu*

Development Program Room, Geological Research Institute, No.6 Oil Production Plant, Daqing Oilfield, Heilongjiang 163100, China

Abstract: With the continuous development of oil fields, the remaining oil in conventional old oil fields is gradually reduced after long-term water drive development. Fracturing can change the original water drive flow channel and increase the oil production of oil Wells. After long-term water flooding development, the oil production of oil Wells in Block X of Daqing oilfield has gradually decreased, and fracturing is needed to increase production. However, Block X has three target reservoirs, and the two formations are close to each other, so stratified fracturing cannot be realized by conventional stage fracturing technology. Therefore, flow limiting method is needed to control the fracture initiation of oil formations. And each layer can be better used. The overall fracturing effect is poor, and it is necessary to carry out a separate optimization design for the control technology of fracture height. The overall fracturing sequence is: first press the third stage, then press the first stage and the second fracturing stage. There is no interference between the 19th layer and the 20th layer, and the fracturing effect is good. For the 19th small layer, the fracture half-length is about 48m, and the 20th small layer half-length is about 55m, which achieves the expected fracturing effect.

1. INTRODUCTION

For the current oilfield development, fracturing is the most effective stimulation measure[1]. For the formation with multiple intervals, different fracturing processes are generally used for stratified fracturing, mainly including conventional flow-limiting fracturing[2], pumped bridge stage fracturing, packer ball fracturing and other stratified fracturing methods[3]. Among them, flow-limited fracturing is the earliest layered fracturing method. Because multiple layers are simultaneously fractured in the process of flow-limited fracturing, fractures and fracture time affect each other[4].

After long-term water flooding development, the oil production of oil Wells in Block X of Daqing oilfield has gradually decreased, and fracturing is needed to increase production. However, Block X has three target reservoirs, and the two formations are relatively close to each other. Therefore, stratified fracturing cannot be realized by conventional stage fracturing technology, so it is necessary to adopt flow limiting method to control the fracture initiation of oil formations. And each layer can be better used. This paper utilizes the stimplan fracturing software to simulate the fracturing of a typical well group in Block X of Daqing Oilfield by the flow-limiting method, and provides a construction plan for the

*Huizhuo Xu: E-mail:2387350331@qq.com; phone number: 13329502105

reservoir modification process. The calculation results can fully guide the fracturing construction process in the field, and the research method can effectively provide technical support for the production enhancement measures of the reservoir.

2. GEOLOGICAL SURVEY ANALYSIS OF BLOCK X

When the horizontal stress difference coefficient is 0-0.13, hydraulic fracturing can form complex fractures or network fractures. When the horizontal stress difference coefficient is 0.13-0.25, hydraulic fracturing can form more complex fractures at high net pressure. When the difference coefficient of horizontal stress is greater than 0.5, hydraulic fracturing can not form complex fractures. The average difference coefficient of horizontal stress in the block is 0.079, which can form complex cracks （shown in table 1）.

Table 1 Ground stress data table in adjacent areas

No.	Well Section(m)	Ground stress Azimuth(°)	Horizontal ground stress(MPa)	
			Horizontal	vertical
1	1490.0-1497.4	80	37.4	32.9
2	1498.0-1510.0	70	38.0	27.5
3	1510.0-1516.0	70	37.8	28.8
4	1517.0-1539.0	70	37.8	31.0
5	1544.4-1548.6	60	35.9	28.4
6	1550.6-1563.6	70	33.2	25.7
7	1564.6-1572.0	60	33.8	26.0
8	1572.0-1579.0	80	35.6	28.8
9	1580.4-1589.4	70	40.3	30.5

*Huizhuo Xu: E-mail:2387350331@qq.com; phone number: 13329502105

 https://doi.org/10.1051/e3sconf/202452801003

No.	Well Section(m)	Ground stress Azimuth(°)	Horizontal ground stress(MPa)	
			Horizontal	vertical
10	1590.0-1594.0	60	34.0	28.7
11	1594.6-1605.6	60	40.1	31.9
12	1606.4-1611.0	60	40.8	32.7
13	1613.0-1620.0	60	34.3	26.5
14	1621.0-1623.8	60	31.7	26.6
15	1625.0-1635.0	70	40.5	30.0
16	1636.4-1645.2	65	40.1	33.6
17	1645.2-1653.6	80	40.3	33.5
18	1654.0-1669.8	55	34.6	30.1
19	1671.6-1680.0	55	33.4	28.9
	average	66	36.8	29.6

In general, Young's modulus is 10-80GPa and Poisson's ratio is 0.2-0.4. It is believed that the larger Young's modulus is, the smaller Poisson's ratio is, the higher the fragility and complexity of fracturing fractures, and the lower the vice versa. It can be seen that the Young's modulus of the target interval is between 24-68GPa, and the Poisson's ratio is between 0.129-0.275.

According to the relationship between the horizontal principal stress and the vertical principal stress in the adjacent area, it can be judged that the vertical fracture structure at 1490m-1491.6m is basically vertical fracture structure, but it can not be judged at 1220m-1280m fracture structure. However, the stress change trend can be analyzed according to the stress gradient change law, and the change between the vertical and horizontal principal stress at the target well interval can be predicted. The average sandstone thickness of a single well drilled in this area is 57.95m, and the effective thickness is 23.93m. From the subdivision, the sandstone thickness of D2 layer is 22.31m, the effective thickness is 13.02m, and the drilling rate of channel sand is 30.98%. D1a small layer sandstone thickness 25.11m, effective thickness 8.78m; D1b small layer sandstone has a thickness of 9.46m, an effective thickness of 2.81m, and a drilling rate of 17.08%.

According to the logging curve, the formation stress, Young's modulus, Poisson's ratio and brittleness index corresponding to the depth of the reservoir are analyzed to find out the corresponding data at different positions of the reservoir, and the formation pressure distribution law is calculated according to the test production data and formation pressure analysis results.

The actual pressure gradient of the reservoir can be obtained as 0.0053MPa/m, while the pressure gradient of the overlying reservoir is 0.018MPa/m, the reservoir pressure is about 10.5MPa, and the overlying stress of the reservoir is 20.36MPa（shown in figure 1）.

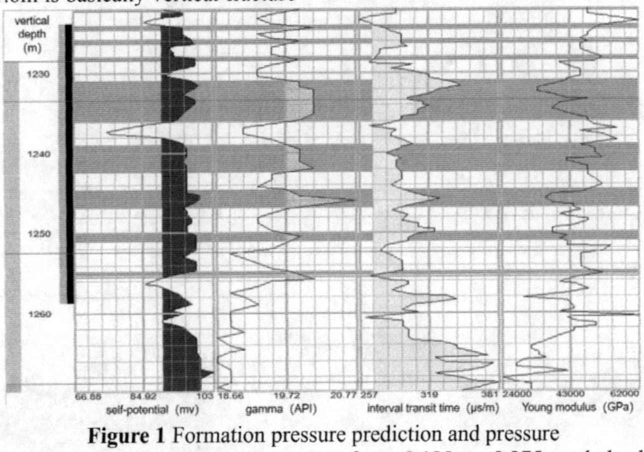

Figure 1 Formation pressure prediction and pressure gradient correction based on logging curves

Young's modulus and Poisson's ratio are the main rock mechanical parameters to characterize brittleness, and the characteristics of high Young's modulus, low Poisson's ratio and brittleness failure are the necessary conditions for reservoir compressibility[5]. Poisson's ratio can reflect the change of stress and elastic modulus can reflect the ability of maintaining crack propagation. Brittleness is a comprehensive property of materials, which is the ability to produce internal non-uniform stress under its own natural heterogeneity and external specific loading conditions, and lead to local failure, and then form a multi-dimensional fracture surface. To characterize brittleness, it is necessary to establish a specific brittleness index. It is generally believed that the higher the brittleness index is, the more likely it is to generate network cracks. The Young's modulus of the interval ranges from 24 to 68GPa, Poisson's ratio ranges from 0.129 to 0.275, and the brittleness index ranges from 24.24 to 46.96.

3. OPTIMIZATION OF FRACTURING SCHEME IN BLOCK X

Fracturing software is used to simulate whether pressure channelling will occur if comprehensive fracturing is performed under the formation conditions of the target well, and fracturing sequence and construction procedures are optimized according to the simulation results. The change law of total fracture width, effective fracture width, conductivity of sand-filled fracture and fracturing fluid loss in the first, second, third and fourth stages of the general fracturing process is analyzed[6]. The general post-fracturing simulation results are shown in figure 2:

 https://doi.org/10.1051/e3sconf/202452801003

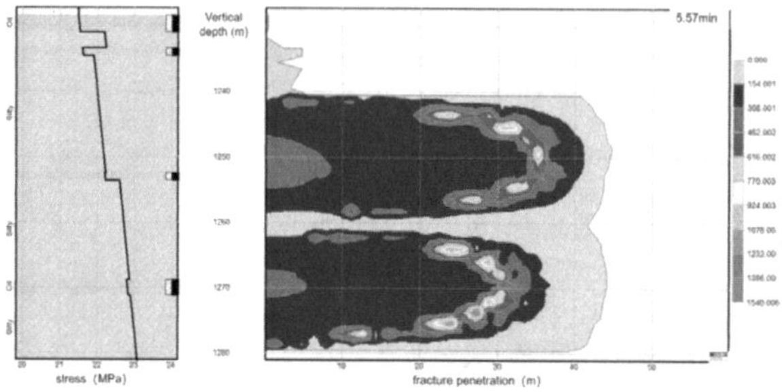

Figure 2 General fracturing simulation results

According to the simulation results, it can be seen that after general fracturing, there will inevitably be fracture interference between fractures, and the interference is serious. The first and second stages will be affected by the third and fourth stages in the fracturing process, and the last effective fracture half length is less than 10 meters, so it is necessary to perform staged fracturing simulation. In addition, as can be seen from Figure2, The fracture conductivity after fracturing is mainly affected by the proppant distribution, so the proppant distribution in the formation should be ensured to ensure the fracture has a certain conductivity during the fracturing simulation.

Stratified fracturing was carried out on the first and second stages, and the distribution of fractures after fracturing was observed. The specific simulation results are shown in Figure 3.

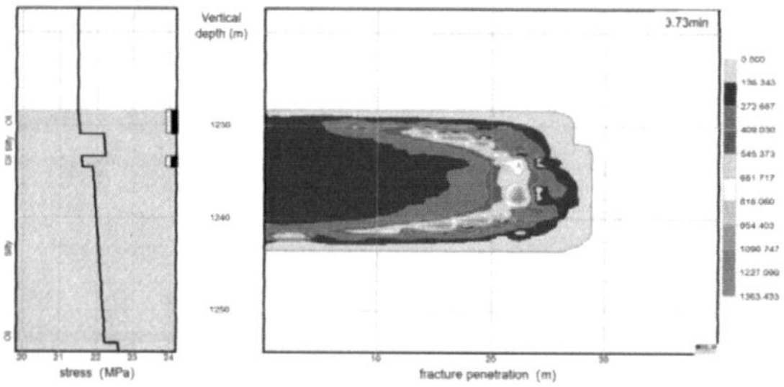

Figure 3 General fracturing of the first and second stages

According to the simulation results, it can be seen that the interference between fractures after fracturing is significantly reduced, and the overall fracturing effect is relatively good.

The numerical simulation study on the effect of construction displacement on the effective fracture width was carried out to simulate the change of the effective fracture width after sand removal and fracture closure. The simulation results are shown in figure 4.

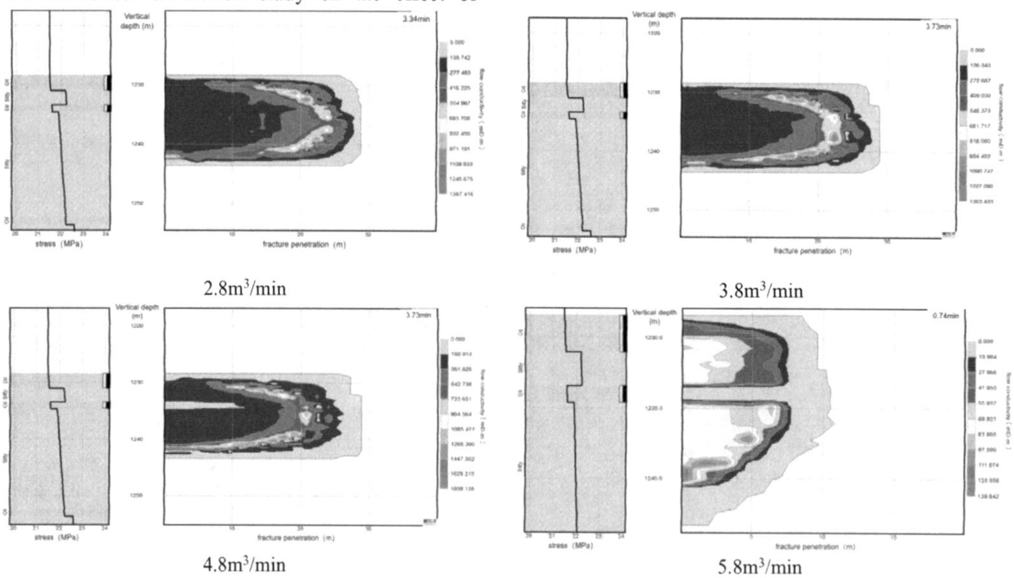

Figure 4 The change of seam width under different pumping speed

According to the simulation results, when the pumping speed is less than 2.8m^3/min, the fracturing fluid loses more, the proppant loses more during the flowback process, the effective fracture width is narrow, and the conductivity is reduced. When the pumping speed is greater than 3.8m^3/min, the fracture height is too high under unrestricted conditions, and the effective fracture half-length is less than 10m after fracturing, and the fracturing fails. Finally, according to the site construction situation, the construction displacement is determined to be 2.8m^3/min-3.8m^3/min.

According to the simulation results, the overall effect is poor in the fracturing process, and the focus is on the control of fracture height, so it is necessary to carry out a separate optimization design for the control technology of fracture height, and the overall fracturing order is: first pressure the third stage, then pressure the first stage and the second fracturing stage. According to the simulation results, the construction parameters are determined in table 2 and table 3:

Table 2 19th, 20th layers pumping procedures

No.	Stage Liquid volume	Stage type	Pumping rate	Fluid type	Proppant type	Proppant concentration
1	30	Lead fluid	3.0	Fracturing fluid	-	0
2	10	proppant	3.0	Fracturing fluid	silica sand	40
3	10	proppant	3.5	Fracturing fluid	silica sand	80
4	10	proppant	3.5	Fracturing fluid	silica sand	120
5	10	proppant	3.5	Fracturing fluid	silica sand	160
6	10	proppant	3.8	Fracturing fluid	silica sand	200
7	14	Displacing fluid	3.8	Flushing fluid	-	0

Table 3 21th layer Pumping procedures

No.	Stage Liquid volume	Stage type	Pumping rate	Fluid type	Proppant type	Proppant concentration
1	28	Lead fluid	3.0	Fracturing fluid	-	0
2	8	Lead fluid	3.0	Fracturing fluid	silica sand	40
3	8	Lead fluid	3.5	Fracturing fluid	silica sand	80
4	8	Lead fluid	3.5	Fracturing fluid	silica sand	120
5	8	Lead fluid	3.5	Fracturing fluid	silica sand	160
6	8	Lead fluid	3.8	Fracturing fluid	silica sand	200
7	14	Displacing fluid	3.8	Flushing fluid	-	0

4. CONCLUSION

(1) The formation brittleness index is higher, the Young's modulus is higher, the difference coefficient of ground stress is smaller, the bedding fractures are developed, and the reservoir compressibility is better.

(2) A separate optimization design is carried out for the control technology of fracture height, and the overall fracturing order is: first pressure the third stage, then pressure the first stage and the second fracturing stage.

(3) There is no interference between the 19th and 20th layers, and the fracturing effect is good. For the 19th small layer, the fracture half-length is about 48m, and the 20th small layer half-length is about 55m, which achieves the expected fracturing effect.

(4) The calculation results can comprehensively guide the field fracturing construction process, and the research method can effectively provide technical support for reservoir production enhancement measures.

REFERENCES

1. Meng Mianmo,Ge Hongkui,Shen Yinghao,Ji Wenming. Evaluation of the Pore Structure Variation During Hydraulic Fracturing in Marine Shale Reservoirs[J]. J. Energy Resour. Technol,2021,143(8).
2. Kalhori M ,Mehrabi H ,Sfidari E , et al. Target zone selection for hydraulic fracturing using sedimentological and rock mechanical studies with the support of the machine learning method of cluster analysis [J]. Geoenergy Science and Engineering, 2024, 237 212826-.
3. Kasala E E ,Wang J ,Lwazi M H , et al. The influence of hydraulic fracture and reservoir parameters on the storage of CO2 and enhancing CH4 recovery in Yanchang formation [J]. Energy, 2024, 296 131184.
4. Root E ,Guinn B ,Harris D , et al. 1125 Residential Proximity to Hydraulic Fracturing Wells Increased the Risk for Low Birth Weight [J]. American Journal of Obstetrics and Gynecology, 2024, 230 (1S): S590-S591.
5. Zhan J.M.,Yao X.H.,Zhang X.Q. Study on predicting the mechanical properties and fracturing behaviors of particle reinforced metal matrix composites by non-local approach[J]. Mechanics of Materials,2021,155.
6. Zhang Lifu,Hascakir Berna. A review of issues, characteristics, and management for wastewater due to hydraulic fracturing in the U.S.[J]. Journal of Petroleum Science and Engineering,2021,202.

https://doi.org/10.1051/e3sconf/202452801004

Improvement study of duct heater in VOCs adsorption concentration-catalytic oxidation process

Chongwei Li*, Zhongyang Liu

School of Electrical Engineering, Dalian University of Technology,Dalian,Liaoning, 116024, China

Abstract: Aiming at the catalytic oxidation process of VOCs in conjunction with fixed adsorption bed, the electric heating unit has the problem of large energy consumption, this paper understands the role of the electric heating unit in the process--providing energy for the process, and analyzes the energy consumption of the process, and the results show that the heat transfer efficiency of the electrically heated gas is a key factor affecting the energy consumption of the process. In order to improve the heat transfer efficiency, this paper explores the effect of flow rate on the heat transfer efficiency, and simulation verification is carried out by Fluent software, and the results show that increasing the flow rate can effectively improve the heat transfer efficiency and reduce the process energy consumption. Based on the above analysis, a pipeline heating scheme is proposed, which has the advantages of fast gas flow rate and high heat transfer efficiency in the pipeline, and effectively reduces the operational energy consumption of the process.

1. Introduction

The emission of VOCs (volatile organic compounds) is becoming more and more serious, and has now become a hot spot in the field of atmospheric environment treatment in China [1]. Catalytic oxidation method, as a highly efficient technology for treating VOCs, is widely used in the fields of painting and spraying because of a series of advantages such as high treatment efficiency and low temperature required for the reaction [2]. In the VOCs catalytic oxidation process, there are various heating methods to provide energy for the system, such as gas, electric heating, etc., in which the electric heating method has the advantages of cleanliness, easy to control, etc., so the process is usually used as a source of energy [3]. However, when the process is coupled with a fixed adsorption bed [3], the electric heating unit has the problem of high energy consumption, which needs to be further reduced. Currently, research on reducing process energy consumption focuses on two main areas: research on efficient catalysts to reduce operational energy consumption [4], and the addition of heat exchangers to recover heat from the exhaust gas. However, the effect of heating method on process energy consumption has not been reported in the literature, which restricts the application scope and effectiveness of catalytic oxidation technology. Therefore, this paper analyzes the energy consumption problem in the process and explores the influence of flow rate on fluid heat transfer, and proposes a pipeline heating scheme, which effectively reduces the energy consumption and is of great practical significance for the efficient removal of VOCs under the condition of lower energy consumption.

*Corresponding author's e-mail: 1595254632@qq.com

2. Analysis of the role and energy consumption of electric heating unit in adsorption-concentration-catalytic oxidation process

2.1. The role of electric heating units in the process

The process schematic is shown in Figure 1. When the catalytic oxidation process is coupled with a fixed adsorption bed, the adsorption bed desorption stage is divided into three phases: pre, mid and post [5]. In the pre-stage, the adsorption bed starts to warm up and the concentration of VOCs desorbed is low; in the mid-stage, the temperature of the adsorption bed is more stable and the concentration of VOCs desorbed is higher; in the post-stage, the temperature of the adsorption bed increases slightly and the concentration of VOCs desorbed gradually decreases.

The electric heating tube and catalyst bed are integrated in the catalytic oxidizer to provide energy for the process. In order to ensure the catalytic oxidation efficiency of VOCs, the gas flow rate in the catalytic oxidizer has to be maintained in a certain range, usually around 1 m/s.

The electric heating unit is the guarantee of the process effect and also determines the energy consumption of the process. Especially in the initial warming stage of desorption, the system needs a lot of energy to warm up, when the electric heating unit starts. In the pre-desorption stage, the adsorption bed is desorbed from the exhaust gas containing low concentration and low temperature VOCs, which firstly passes through the electric heating unit, which heats the low-temperature and low-concentration VOCs exhaust gas, and the temperature of the gas rises. Subsequently, the gas passes through the catalyst bed and undergoes catalytic oxidation reaction and releases a small amount of heat, and the gas temperature increases again. The gas enters the heat exchanger, the system discharges part of the exhaust gas and replenishes the new air at the same time, the new air and the exhaust gas exchange heat through the heat exchanger, and then the gas with a higher temperature enters the adsorption bed to warm it up. In the process of heating up the system, the electric heating unit is activated to provide heat for the system.

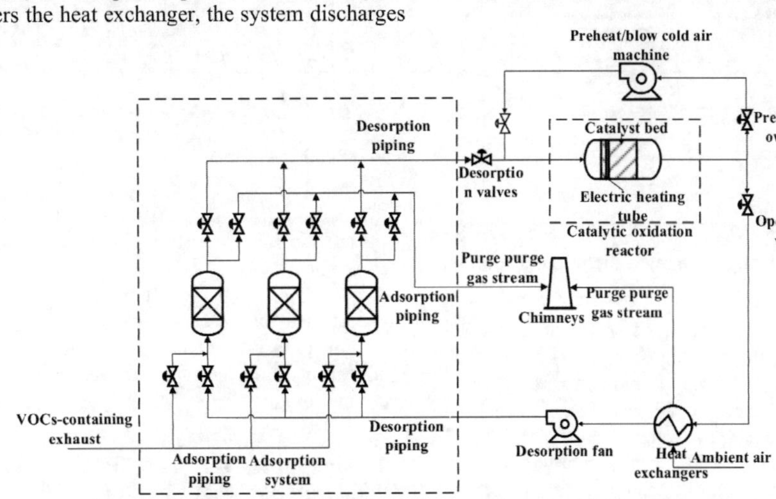

Figure 1. Schematic diagram of adsorption concentration-catalytic oxidation process

2.2. Process system energy consumption analysis

In the pre-stage of adsorption bed desorption, the required heat is completely provided by the hot air, which is provided by the electric heating unit and the VOCs reaction exothermic, in this stage the VOCs concentration is low and the reaction exothermic is less, so the heat is mainly provided by the electric heating unit. In order to ensure the desorption effect, the power of the electric heating unit has certain requirements.

The electric heating unit in the process needs to meet certain power requirements: when the desorption process is carried out, the adsorption bed, the gas in the system and the catalyst bed need to be warmed up, and the total heat required in the preliminary stage of the process is

$$Q = Q_1 + Q_2 + Q_3 \tag{1}$$

Where the heat required to warm the adsorption bed is

$$Q_1 = c_1 \cdot m_1 \cdot \Delta t_1 \tag{2}$$

Where c_1 is the specific heat capacity of the adsorption bed, m_1 is the mass of the adsorption bed, and Δt_1 is the temperature rise of the adsorption bed; the heat required for heating the gas in the system is

$$Q_2 = c_2 \cdot m_2 \cdot \Delta t_2, m_2 = \rho_2 \cdot V_2 \tag{3}$$

where c_2 is the specific heat capacity of the gas, Δt_2 is the temperature rise of the gas, ρ_2 is the density of the gas, and V_2 is the volume of the gas; the energy required to maintain the temperature of the catalyst bed is

$$Q_3 = c_3 \cdot m_3 \cdot \Delta t_3 \tag{4}$$

Where c_3 is the specific heat capacity of the catalyst bed, m_3 is the mass of the catalyst bed, and Δt_3 is the temperature rise of the catalyst bed. According to the requirements of desorption and purification, it can be seen that the system needs a certain time T to warm up, and the theoretical average distribution power is

$$P_1 = Q/T \tag{5}$$

The electric heating unit in the process is the electric heating tube, in the process of heat transfer of the electric heating gas, there is a heat transfer efficiency η_1, the energy that is not transferred to the air is the over-temperature loss, and the heat will be gathered in the interior of the electric heating tube, resulting in a rise in the surface temperature of the resistance wire. Hot air through the heat exchanger after the existence of energy loss (exhaust emissions), take the heat transfer efficiency of η_2. process system to the outside world of heat loss, take the efficiency of η_3, the effect of energy transfer diagram shown in Figure 2. Electrical energy is converted to the energy required for the system to warm up the conversion efficiency is

$$\eta = \eta_1 \cdot \eta_2 \cdot \eta_3 \tag{6}$$

The actual distribution power is

$$P_2 = P_1/\eta \tag{7}$$

In the process, the heat transfer efficiency η_2 and the heat loss to the outside world η_3 are usually fixed, i.e., η_2 and η_3 can be regarded as fixed values. At present, the heat transfer efficiency η_1 of the electrically heated air is low, which leads to a low conversion efficiency η

of the electrical energy into the energy required for the system to heat up the system, whereas the theoretical average distribution power P_1 is a fixed value in the process, which leads to the actual distribution power P_2 to be higher, resulting in a higher energy consumption.

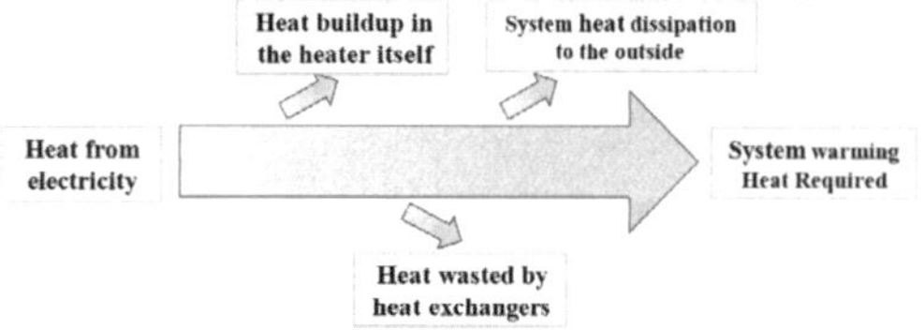

Figure 2. Energy transfer schematic

Therefore, the level of heat transfer efficiency η_1 of the electrically heated gas is the main factor determining the actual distribution power P_2, which also affects the energy consumption of the catalytic oxidation process. Therefore, improving the heat transfer efficiency of the electrically heated gas is the key to reducing the process energy consumption. In order to reduce the process energy consumption, it is necessary to explore the relevant factors affecting the heat transfer efficiency. According to the theory of heat transfer [6], the heat transfer mode of electrically heated air is thermal convection, and the key factors affecting the heat transfer efficiency are the heat transfer coefficient, temperature difference, and heat dissipation area, among which the heat transfer coefficient is related to the gas flow rate and perturbation. In the process, it is relatively easier to adjust the flow rate, therefore, the influence of flow rate on heat transfer efficiency is mainly explored next.

3. Effect of flow rate on gas heat transfer efficiency

In order to improve the gas heat transfer efficiency and solve the problems in the process, the effect of flow rate on the gas heat transfer efficiency will be explored next and analyzed using Fluent simulation software to provide theoretical support for the subsequent improvement program.

3.1. The role of electric heating units in the process

Heat transfer efficiency exists when a gas electric heater heats air. According to the principle of heat transfer [6], the heat transfer efficiency is positively correlated with the heat transfer coefficient, which indicates the ability of an object to transfer heat, and its high or low level responds to the high or low efficiency of heat transfer. The heat transfer coefficient is related to the gas flow rate, and their relationship is as follows

$$h = \frac{0.027 \cdot (\rho \cdot v)^{0.8} \cdot \mathrm{Pr}^{0.4} \cdot \lambda}{d^{0.2} \mu^{0.8}} \tag{8}$$

where v is the average flow rate; d is the inner diameter of the tube; μ is the kinematic viscosity coefficient; and ρ is the fluid density. When the heater size is fixed and the heating fluid is fixed, the nature of the fluid is determined, at which time ρ, μ, d, λ, and Pr are constants. By putting $C = \frac{0.027 \cdot \rho^{0.8} \cdot \mathrm{Pr}^{0.4} \cdot \lambda}{d^{0.2} \mu^{0.8}}$, C can be regarded as a constant to get the heat transfer coefficient and flow rate as a function of flow rate

$$h = C \cdot v^{0.8} \tag{9}$$

According to equation (9), it can be deduced that the heat transfer coefficient gradually increases with the increase of flow rate, which leads to the gradual increase of heat transfer efficiency. Therefore, in order to improve the heat transfer efficiency of the gas in the catalytic oxidation process, the flow rate should be increased as much as possible, so as to reduce the process energy consumption.

3.2. Simulation analysis of the effect of flow velocity on gas heat transfer

In order to investigate the numerical influence of different flow rates on the heat transfer efficiency under the same gas flow rate, simulation models of different sizes are established and simulated and analyzed with Fluent finite element software. In order to better approximate the engineering reality and simplify the calculation, a pipe heater model is constructed, which is composed of a cylindrical empty pipe and an internal electric heating pipe, and the model schematic is shown in Figure 3. The external cylindrical hollow tube is made of stainless steel and wrapped with an insulation layer to reduce heat dissipation, and the insulation layer is made of aluminum silicate cotton. The length of the electric heating tube is 20mm, the inner diameter is 3mm, the material is ferrochromium nickel alloy; the middle diameter is 5mm, the material is magnesium oxide; the most external diameter is 6mm, using stainless steel material. The schematic diagram of the electric heating tube model is shown in Figure 4.

In order to simplify the workload and improve the credibility of the results, three pipe heater models were established, the external cylindrical empty pipe size is different, but the internal electric heating pipe size is the same, and the power of the electric heating pipe is set to be the same, and the surface power density is 5W/cm^2. The size parameters of the external cylindrical hollow tube of each size pipe heater model are shown in Table 1. In the Fluent simulation analysis, the corresponding boundary conditions are set, and the specific contents are shown in Table 2. The simulation cloud diagram of the pipe heater at a flow rate of 20 m/s is shown in Figure 5.

https://doi.org/10.1051/e3sconf/202452801004

Table 1. Parameter list of model dimensions

	Inside diameter(mm)	External diameter(mm)	Lengths(mm)	Flow rates(m/s)
Model 1	116.2	120.2	100	1
Model 2	36.8	40.8	100	10
Model 3	26.0	30	100	20

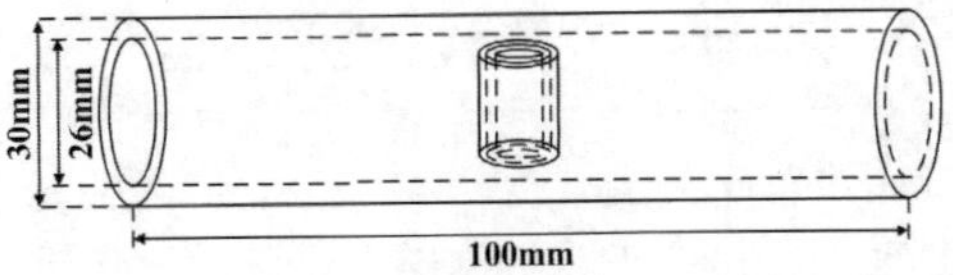

Figure 3. Pipe heater model for flow rate of 20 m/s

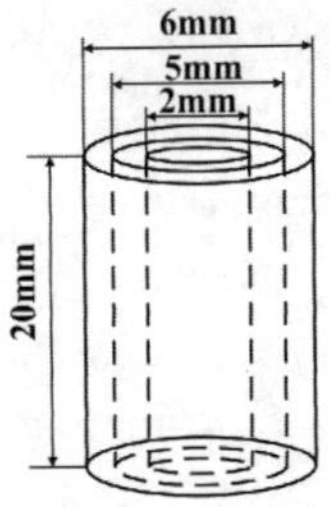

Figure 4. Schematic diagram of the electric heater model

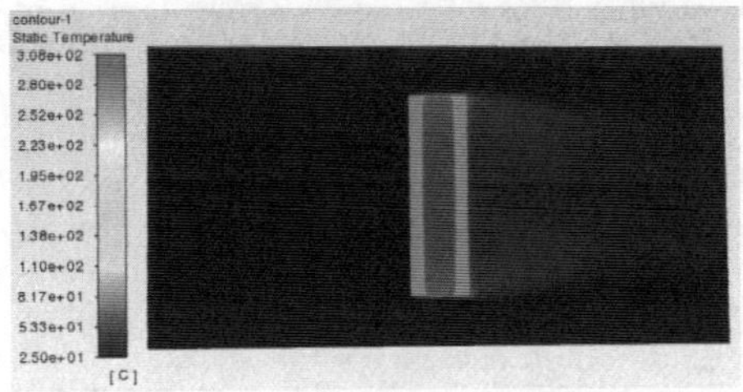

Figure 5. Simulation cloud of duct heater cross-section at 20 m/s flow rate

Table 2. Boundary condition settings

	Entrance Setting	Exit settings
Boundary Conditions:	Speed: 1m/s、10m/s、20m/s Direction: perpendicular to the entrance	Pressure outlet
Fluid properties:	25°C air	

Table 3. Parameters of simulation results

	Inner diameter of heater 116.3mm	Inner diameter of heater 36.8mm	Inner diameter of heater 26mm
Flow rates	1m/s	10m/s	20m/s
Electric heating tube resistance wire temperature	814°C	321°C	308°C
heat transfer coefficient	15.6W/(m^2·K)	81.6W/(m^2·K)	90.2W/(m^2·K)
Heating efficiency	70.8%	90.7%	93.0%

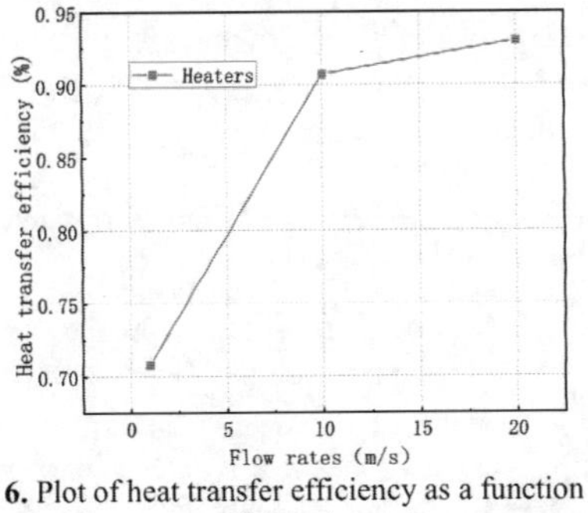

Figure 6. Plot of heat transfer efficiency as a function of inlet flow rate

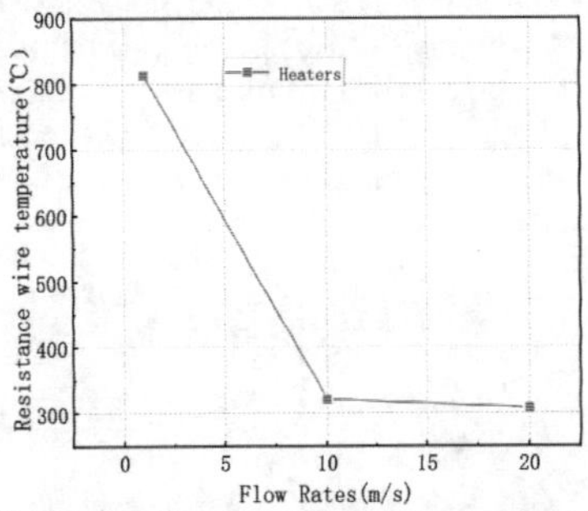

Figure 7. Plot of resistance wire temperature versus inlet flow rate

According to Table 3 and Figures 6 and 7, it can be seen that, in the case of the same gas flow rate and electric heating tube power, when the inner diameter of the pipe heater cross-section is reduced from 130 mm to 26 mm, the gas flow rate is increased from 1 m/s to 20 m/s, the electric heating tube resistance wire temperature is decreased from 814 ℃ to 308 ℃, and the heat transfer coefficient is increased from 15.6 $W/(m^2 \cdot K)$ to 90.2 $W/(m^2 \cdot K)$, the heating efficiency increased from 70.8% to 93.0%. This change is because when the gas flow rate increases, the Reynolds number increases, the electric heating tube on the gas perturbation enhancement, resulting in an increase in the heat transfer coefficient, which further leads to the gas from the surface of the electric heating tube to take away more heat, so that the temperature of the electric heating tube resistance filament decreases, which improves the efficiency of heat transfer of the gas.

From the previous 2.2 analysis, it can be seen that in the case of the same power distribution, the improvement of the gas heat transfer efficiency can shorten the adsorption bed heating time and reduce the energy consumption of the process, which provides a theoretical basis for the improvement of the process.

4. Duct heaters in process applications

Through the previous analysis, it can be learned that in the case of the same gas flow rate, increasing the flow rate is an important way to improve the heat transfer efficiency, but also a key means to reduce the energy consumption of the process. The current process gas flow rate is low, which leads to higher energy consumption. To cope with this problem, a pipeline heating scheme is proposed in this paper. In this program, the gas flow rate in the pipeline is faster, which can significantly improve the heat transfer efficiency and thus reduce the energy consumption of the process, thus effectively solving the problems in the process.

4.1. Components of duct heaters

The pipe heater consists of four parts: 1. Stainless steel pipe is used as the outer shell; 2. Electric heating pipe (used to generate heat) 3. Terminal (used to connect the electric heating pipe); 4. Flange (used to connect with the pipe). The schematic diagram of the pipe heater is shown in Figure 8, with the electric heating pipe built inside the stainless steel pipe. Inside the pipe heater is characterized by fast flow rate and wide range of applications, and is equipped with flanges at both ends to connect with pipes of different sizes.

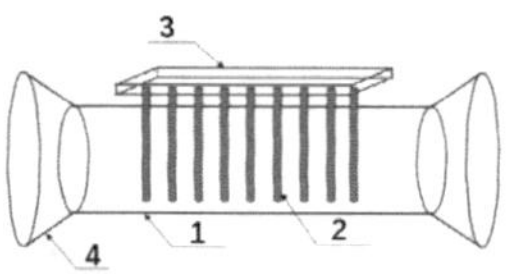

Figure 8. Schematic diagram of duct heater

4.2. Duct heaters in process improvement

The improved scheme is shown in Figure 9 for the catalytic oxidizer structure in the dashed box in Fig. 1, and the other parts remain unchanged. This scheme adds a duct heater in front of the inlet of the catalytic oxidizer to replace the electric heating tube in the catalytic oxidizer. In this scheme, the gas flow rate and the power of the electric heating tube in the duct heater were kept the same as the original scheme, but the gas flow rate in the duct heater was designed to be 20 m/s.

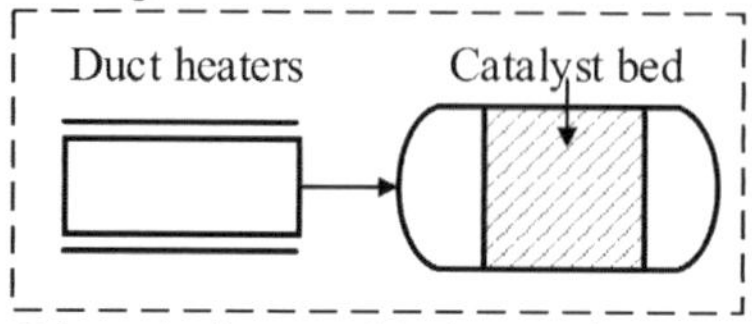

Figure 9. Schematic diagram of the improved program (partial)

With the duct heater method, the gas flow rate in the duct is faster but does not affect the flow rate in the catalytic oxidation reactor, thus not negatively affecting the catalytic conversion efficiency. However, the flow rate needs to be carefully controlled to avoid increasing the energy consumption of power devices such as fans. Therefore, according to the HG-T20570.6-1995 [7] standard, it can be seen that the gas flow rate is designed to be 20 m/s.

4.3. Advantages of the improved process

The pipeline heater improves the gas flow rate from 1m/s to 20m/s. From the previous simulation, it is known that the improvement of the gas flow rate, which in turn improves the heat transfer efficiency of the gas. The improvement of the gas heat transfer efficiency can shorten the time required for the adsorption bed to warm up under the same distribution power, thus reducing the energy consumption of the process. The following is the theoretical estimation: assuming that the heat transfer efficiency when the electric heater is integrated in the catalytic oxidation reactor is η_1, and the heat transfer efficiency of the improved pipeline heater is η_2, and the heat transfer efficiency is increased by $\eta_2 - \eta_1$. In the original process, the energy required for the process to warm up is Q, and the time for the warming up time is t. Assuming that the calculation is calculated by taking the heat transfer efficiency of η_1, the required power distribution power is P. When the flow rate is increased by using the pipeline heater. Assuming that the heat transfer efficiency is increased to η_2, and equipped with the same power, the warming time is $\frac{\eta_1}{\eta_2}t$. The warming time is reduced by $\frac{\eta_2 - \eta_1}{\eta_2}\%$.

At the same time, the increased flow rate reduces the wall temperature of the heater and extends the service life of the heater. In addition, the pipe heater, as an external component, can be flexibly added or removed during the process.

https://doi.org/10.1051/e3sconf/202452801004

5. Conclusion

To address the problem of high energy consumption of the process in the scenarios of intermittent operation or coupling with a fixed adsorption bed, this paper analyzes the influence of flow rate on fluid heat transfer based on the principle of catalytic oxidation process, and verifies the influence law of flow rate on heat transfer efficiency by Fluent finite element simulation software. A pipeline heating scheme is proposed to solve the problem of high energy consumption of the process in some application scenarios. The main conclusions drawn are as follows:

(1) In the adsorbent bed warming stage, the ducted heater improves the heat transfer efficiency of the air, reduces the energy consumption in the preheating stage, and under the condition of constant electric power input, thus shortening the preheating time and realizing faster preheating of the adsorbent bed.

(2) By adopting the pipeline heater, the gas flow rate is increased, the surface temperature of the electric heating tube is reduced, and the service life of the electric heating tube is effectively prolonged. This reduces the frequency of equipment maintenance and replacement, improves production efficiency and reduces production costs.

With the advancement of the "double reduction" goal, the use of pipeline heating method more effectively reduces the operational energy consumption and achieves energy saving effect, which is conducive to the further promotion of catalytic oxidation method for VOCs treatment. Future research could focus on the design optimization of the pipeline heater to further reduce the operating energy consumption.

References

1. Zhang Z, Jiang Z, Shangguan W. Low-temperature catalysis for VOCs removal in technology and application: A state-of-the-art review[J]. Catalysis Today, 2016, 264: 270-278.
2. Kim B R. VOC emissions from automotive painting and their control: A review[J]. Environmental engineering research, 2011, 16(1): 1-9.
3. LI Lei, Wang Xuehua, WANG Hao, et al. Adsorption-concentration-catalytic combustion process for treatment of low concentration and large air volume organic waste gas[J]. Journal of Environmental Engineering, 2015, 9(11):7.
4. Tian Z Y, Ngamou P H T, Vannier V, et al. Catalytic oxidation of VOCs over mixed Co–Mn oxides[J]. Applied Catalysis B: Environmental, 2012, 117: 125-134.
5. Liang Yulan. Study on the adsorption-desorption performance of honeycomb activated carbon on ethyl acetate[J]. Journal of Longyan College,2020,38(02):85-92.
6. Yang S-M, Tao W-Q. Heat Transfer (4th ed.) [Heat Transfer][M]. Higher Education Press, 2006.
7. Process system design technology center station of chemical ministry. Technical regulations for process system engineering design--pipe diameter selection: HG-T20570.6-1995 [S]. Beijing: Chemical Industry Press, 1995.

Heating performance analysis of solar energy-medium deep buried pipe

Xiaoxuan Zhao

School of Architecture and Civil Engineering, Xi'an University of Science and Technology, Xi'an 710054, China

Abstract. In order to promote the vigorous development of renewable energy. Regarding the thermal load of buildings in Yulin mining area. Establish a solar deep buried pipe system using TRNSYS software. The feasibility of the components is verified by experiments. After comparing three different operating systems, it is concluded that the solar-medium and deep buried pipe system has more advantages. By simulating the solar energy - deep buried pipe system, it is concluded that: (1) The water supply temperature is relatively stable and can be maintained at 48 °C. (2) The COP can be up to 6.2, and the heat supply can also meet the building demand. (3) The cost is also low after long - term operation. Therefore, it is feasible to use solar energy-deep buried pipe system in mining area.

1. Introduction

Solar energy and geothermal energy reserves are huge, and both belong to renewable clean energy. The combination of solar energy and geothermal energy is widely used. Zhang Zhenjun [1] solved the soil temperature imbalance caused by cold by using TRNSYS to establish a solar assisted ground source heat pump system (SGSHP). After 10 years of operation, the Coefficient of Performance (COP) of the solar auxiliary system increased by 28 %, and the soil temperature also eased. Cui Yunxiang and Liu Yanfang [2, 3] simulated different operation modes of SGSHP through TRNSYS, and display parallel mode as the optimal operating mode. Minwoo Lee and Hao Hong et al. [4, 5] pointed out that the minimum life cycle cost (LCC) of SGSHP is reduced by 7 %, and the maximum heat exchange efficiency of parallel system is 2.5 times that of series system. Oklahoma State University and Seyed Houman Razavi [6, 7] diverse scenarios simulating the combination of SGSHP. The results show that the SGSHP shows the greatest advantage in improving and maintaining the temperature of geothermal wells.

The state vigorously develops renewable energy. Renewable energy in mining areas includes waste heat, solar energy, ground heat exchanger and so on. Through literature review, it was found that the combination of solar energy and geothermal energy can meet the needs of system use and save energy. Therefore, the solar - medium - deep buried pipe heat exchange system (SMDBHE) is proposed for subsequent research to provide ideas for green energy applications.

2. Introduction of system model

2.1 Building load calculation

This paper mainly focuses on Yulin mining area, and selects a building with a total construction area of 6000 m^2. The domestic water hot water in the mine is a bathroom with a shower, with a value of 540 L per shower per hour. Calculate the required bath heat load. According to the following formula, the bath heat load Q_s can be obtained :

$$Q_s = \frac{cm(t_y - t_s)}{nT_s} \tag{1}$$

Where Q_s is the heat load of bathing, kW; c is the water's particular thermal capacity, kJ/(kg·°C); m is the amount of water used for bathing in a single day, kg; t_y is bath water temperature, °C; t_s is the starting bath temperature, °C; n is a single day shift; T_s is the length of a single heating time, h.

Add the mine shaft's antifreezing heat load after figuring out the building's heating and residential hot water needs. The inlet mixing temperature is required to be greater than 2 °C. Hot water for bathing has a temperature higher than 40 °C. The wellbore anti-freezing heat load is estimated by the following formula :

$$Q_j = a \times L \times \rho \times C_p \times (t_i - t_w) \tag{2}$$

Where Q_j is the wellbore anti-freezing heat load, kW; a is the correction coefficient, 1.1 is usually used ; L is the inlet air volume, m3 / s ; ρ is the density at 2 °C, using 1.284 kg / m^3 ; C_p is the constant pressure specific heat at 2 °C, using 1.01 kJ / (kg·°C) ; t_i is the design temperature of the inlet wellhead, 2 °C ; t_w is designed to calculate the temperature of outdoor air, °C.

Corresponding author's e-mail: 21204228096@stu.xust.edu.cn

This system is used in a mine in Yulin City, Shaanxi Province, with an average annual temperature of 9.58 °C. The temperature is introduced into the TRNSYS system to provide simulated environmental conditions. The TRNBuild in TRNSYS is used to construct the building model and calculate the annual heating and domestic heat load. The heating time is 11.1 ~ 4.1. The load results are shown in Figure 1. To satisfy the building's basic heat demand, a total load of up to 3902 kW is applied. The heat load for daily use is 126 kW.

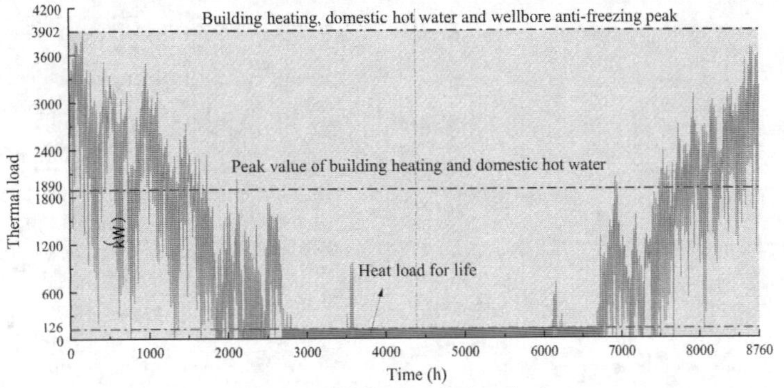

Figure 1. Building hourly heat load.

2.2 Model verification

The model is verified by experiments before the system is built. The experiment is located in Xi 'an, and is carried out by instruments such as temperature sensors, total radiation meters, handheld ultrasonic flow meters, and energy meters. Because of this system in the deep buried pipe heat transfer out into the heat pump unit. Therefore, the solar series water source heat pump model is verified. Figure 2 (a) is the simulation of the solar series water source heat pump model system. It is mainly composed of heat collecting plate, hot water storage tank, water source heat pump, water pump, temperature control and so on.

Figure 2 (b) is the system model verification result. The experiment was carried out on January 13, and 60 t of water with a temperature of 10 °C was heated. After the water tank temperature reached 50 °C, the system stopped running. The running time is 8 a.m.to 18 p.m. The temperature change of the water tank was recorded by testing. The comparison curve shows that, with an error of no more than 10%, the simulated water tank temperature is essentially comparable with the experimental measurement. The experimental value is slightly higher than the simulated value at 12 noon. Because the placement of solar panels in the experiment is better influenced by the sun. Will reach the set temperature before the analog value.

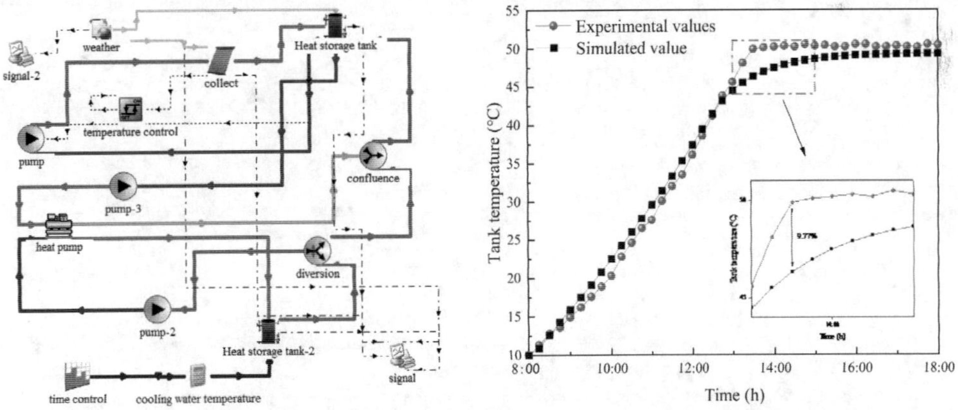

(a) (b)

Figure 2. Model and water tank temperature verification.

3. System mode ling

3.1 Middle and deep buried pipe parameters

The input module of buried pipe in TRNSYS is Type225, but because the depth of 225 buried pipes is up to 1860 m, the results after operation are not accurate for middle and deep layers. The system has been running for 20 years through OGS simulation [8]. Select data to perform static statistics on the temperature of medium to deep buried pipes replaced. And the calculated heat supply is also represented by a graph. Then it is input to the heat pump through Type 9e.

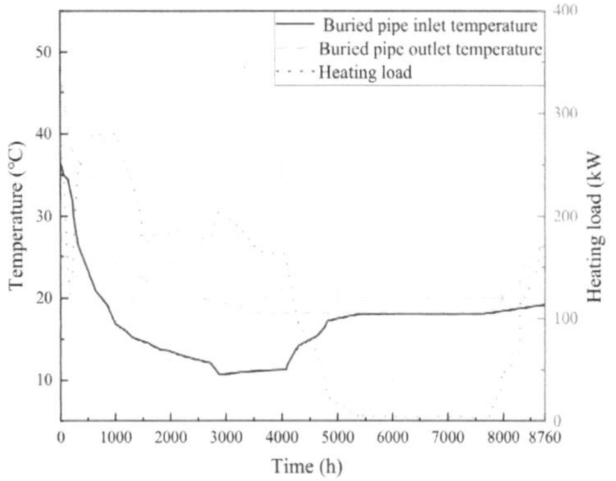

Figure 3. Heat transfer of buried pipe in middle and deep layer.

Figure 3 is the change curve of inlet and outlet temperature and heat supply of buried pipe in middle and deep buried pipe heat exchange. From the graph, it can be seen that the heat exchange of buried pipes is in the on state from 0 to 2184 hours (heating season), and in the off state from 2184 to 7320 hours (non heating season), without turning on the heating of buried pipes. Afterwards, the heating season resumed, providing heating for buildings.

3.2 Operation model

In the operation strategy of SMDBHE. Regarding the composition characteristics of solar energy and deep buried pipe systems, as well as the applicable conditions of each equipment. This system can achieve the operation of multiple subsystems according to different operating conditions. The operation process of different subsystems can be referred to as multiple operation modes of the system. The operating mode includes the following modes.

(1) Solar energy heating mode alone. Figure 4 (a) is the simulation model. In this mode, the amount of solar radiation collected. Heat is transferred to the heat storage water tank by a small amount of electric energy. Solar energy operates as a single heating source for a long term year-round operation. The heat storage water tank is used to provide heating for the buildings in the mining area. Obtain the annual water supply temperature variation curve after operation. (2) Medium-deep buried pipe heating mode. Figure 4 (b) is the simulation model. In this mode, the heat exchanged from the middle - deep buried pipes flows through the heat pump. After being lifted by a heat pump, it is supplied to buildings in the mining area. The heat exchange of underground pipe is used as a single heat source for a long time all year round. (3) Solar - medium deep buried pipe heating (SMDBHE) mode. Figure 4 (c) is the simulation model. In this mode, solar energy is used to assist the middle - deep buried pipe heating. Collect heat through solar panels. Buried pipes in the middle - deep layers also exchange the heat of soil and rock layers. The heat of both flows to the heat pump for heating. In order to protect the soil, the heat exchange of the middle and deep buried pipe is only carried out in the heating season, and it is stopped in the non-heating season for heat recovery.

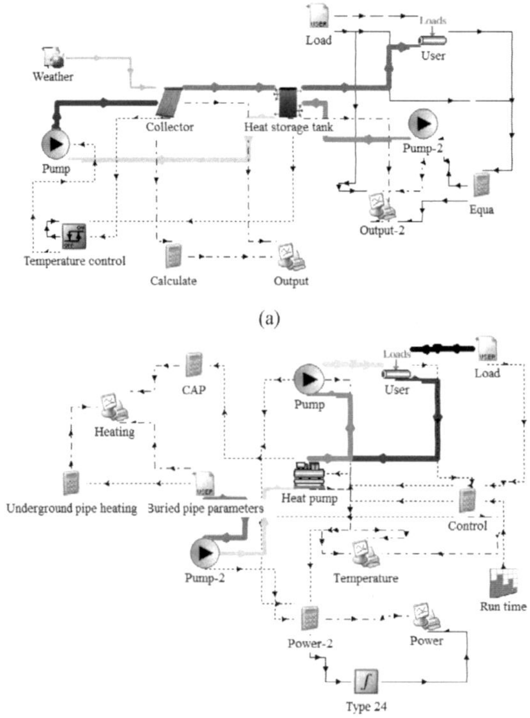

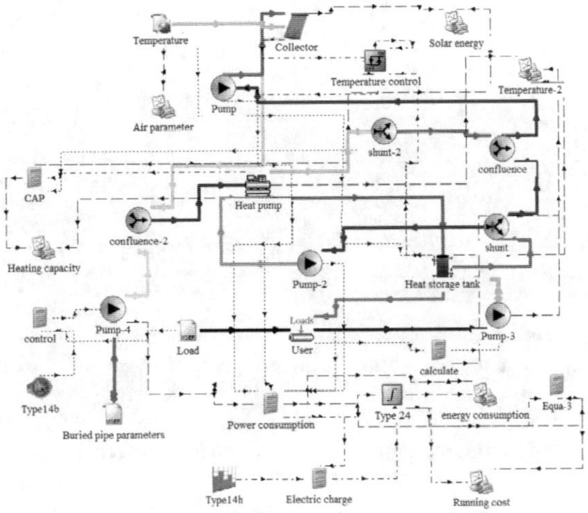

(c)

Figure 4. Three model simulation diagram.

4. Analysis of system simulation results

Through the simulation calculation of three operating modes, the comparison between water supply temperature and energy consumption is shown in the Figure 5.

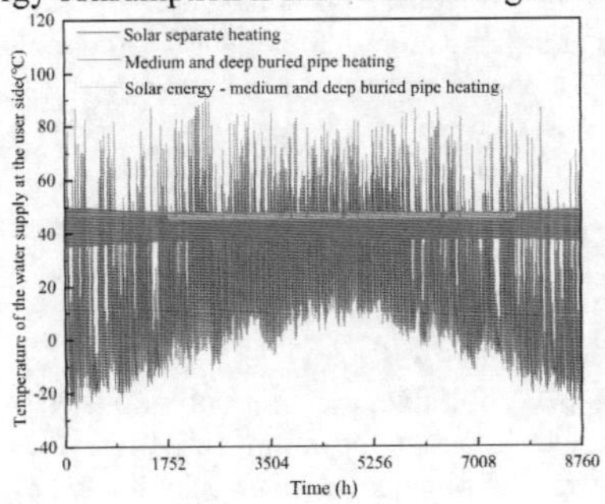

(a)

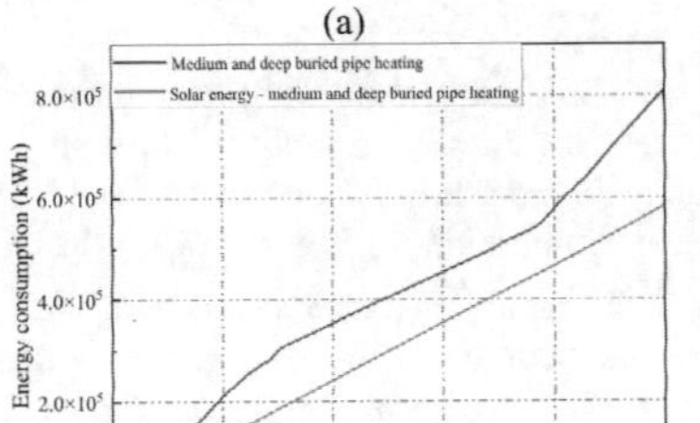

(b)

Figure 5. Comparison of three modes.

Compare the water supply temperature of the three modes. Calculate the temperature assurance rate. When the water supply temperature is greater than 42 °C. The temperature guarantee rate of long-term operation of solar energy alone is 26.28%. When the middle - deep buried pipes are used for long - term heating alone, it is 64.5%. Solar energy - deep buried pipe heating is 93%. The solar temperature guarantee rate is the lowest, and it is not suitable for single operation throughout the year. However, the temperature is high in the range of 3.15 ~ 10.15, which can be considered for use alone during this period.

When it comes to heating, especially for separate heat exchange of medium and deep buried pipes and SMDBHE, it can adequately address the primary heat needs of users. Then compare the system energy consumption as shown in Figure 5 (b). As can be seen from the figure, the energy consumption of the middle - deep buried pipe heating is 27.9% higher than that of the SMDBHE. Because the heat under the middle to deep buried pipes is extracted throughout the year, it is unfavorable for soil heat recovery. Therefore, SMDBHE are considered to be used for coordinated heating. The heating season is provided by the middle and deep buried pipe heat exchange and solar collector plate heat collection. In the unheated season, the deep buried pipe is discontinued. When the solar heat collection is sufficient, the solar energy is flowed to the hot water storage tank for heating ; when the heat collection capacity is insufficient, the solar energy is led to the heat pump for heating. Finally, it flows to the water tank for heating.

Then, the cost analysis and comparison of SMDBHE (Model 3) and medium to deep buried pipe heat exchange (Model 2) are carried out. The life cycle cost is used for calculation and comparison. The formula is as follows [9] :

$$LCC = \left[\frac{i(1+i)^m}{(1+i)^m - 1} \times C + Price \times E \right] \times m \quad (3)$$

Where LCC is the life cycle cost, Yuan ; C is the system initial investment, yuan ; i is the annual interest rate of bank loans, 4.9 % ; m is the system service time, year ; $Price$ is the local electricity price, using peak-valley electricity price, yuan ; E is all energy consumption of the system, kW.

The major upfront costs associated with this system are the equipment purchase price, the pipeline surcharge, and the installation cost. Typically, the pipeline surcharge and installation cost amount to 8% and 10% of the overall investment made in the system's primary equipment. The cost of electricity and maintenance management are included in the operation cost. Typically, the cost of maintenance management accounts for 2% of the system's entire expenditure in its primary equipment. The electricity cost is calculated according to the peak value of

Yulin 0.568 yuan / degree, the valley value 0.288 yuan / degree, the peak time is 8 : 00 ~ 22 : 00, and the valley value is 22 : 00 ~ 8 : 00. Detailed data are shown in Table 1.

Table 1. Life cycle cost statistics.

Initial investment			Model	LCC (million yuan)			
	Parameter	Price		1 a	5 a	10 a	15 a
Heat collector	19555 m^2	800 yuan/ m^2	2	785	1897	3499	5197
Well depth	2000 m	1040 yuan/m					
The volume of the tank	2785 m^3	500 yuan/m^3	3	824	1902	3478	5159
Heat pump	3920 kW	2600 yuan/kW					

According to the parameters and results, the comparison models 2 and 3 can be calculated. Although the initial investment model 2 is lower than 3, the long-term operation model 3 is more economical. The decrease of heat extraction capacity in long-term operation of model 2 leads to the increase of power consumption in the later period. The final operating cost is higher. Therefore, it is more appropriate to choose model 3 in terms of economy.

On this basis, the performance of SMDBHE is analyzed. In Figure 6, you can see the COP for the SMDBHE, along with the heat supply provided by the heat pump. From the COP point of view, the system can reach up to about 6, and the annual average can reach 5.16. According to the heat supply, the heating season is 2000 kW, and the non-heating season is 830 kW. When the designed hot water storage tank temperature reaches 50 °C during operation, the heat pump comes to a halt. After that, heat is supplied by the heat storage water tank. Then the heat supply of the heat pump may be slightly lower, but it can reach the use temperature of the building as a whole. From April to October, due to the cessation of heating, only domestic hot water needs to be met. The decrease of heat supply is due to solar energy running alone or combined with heat pump for heating. The system power also decreases, eventually leading to an increase in COP.

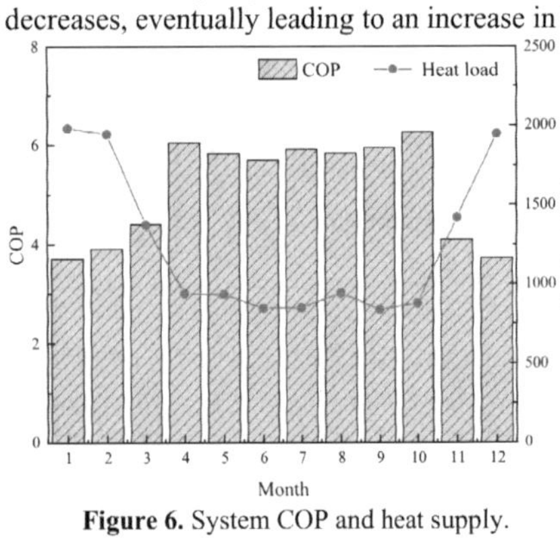

Figure 6. System COP and heat supply.

5. Conclusion

Through the discussion and analysis of the results, the following conclusions can be drawn :

(1) Under the three operating models, the water supply temperature of the SMDBHE is more stable, with a guarantee rate of 93 %. And the energy consumption is 5.89 × 105 kWh. For life cycle cost analysis, long-term operation is more cost-saving.

(2) On this basis, further analysis shows that the heat supply can reach the heat demand. The annual average COP is 5.16, and the performance is good. It is conducive to improving energy efficiency and laying a foundation for subsequent research.

References

1. Zhang Z J (2022) Performance analysis of solar assisted ground source heat pump heating device based on TRNSYS software. J. Energy saving., 41(12): 25-27. [CrossRef]
2. Cui Y X and Cai Y L(2019) Experimental study on combined operation of solar energy-ground source heat pump in winter in Shanghai. J. Fluid machinery., 47(02): 65-69. [CrossRef]
3. Liu Y F, Song M Y and Zhou Y, et al (2021) Research on the performance of the combined system of partitioned series-parallel solar-ground source heat pump cross-season heat storage. J. Journal of Solar Energy., 42(12): 71-79. [CrossRef]
4. Lee M, Ham S H and Lee S, et al (2023) Multi-objective optimization of solar-assisted ground-source heat pumps for minimizing life-cycle cost and climate performance in heating-dominated regions. J. Energy., 270. [CrossRef]
5. Hong H and Yang Y (2021) Study on solar-ground source heat pump and combined heating system in severe cold area. J. IOP Conference Series: Earth and Environmental Science., 825(1). [CrossRef]
6. Bose J E, Smith M and Spitler J D (2005) Ground source heat pump systems: A technology review. J.
7. Razavi S H, Ahmadi R and Zahedi A J A T E (2017) Modeling, Simulation and Dynamic Control of Solar Assisted Ground Source Heat Pump to Provide Heating Load and DHW. J. 129. [CrossRef]
8. Cai W L, Wang F H and Chen S, et al (2021) Analysis of heat extraction performance and long-term sustainability for multiple deep borehole heat exchanger array: A project-based study. J. Applied Energy., 289. [CrossRef]
9. Li C (2020) Study on the performance of domestic hot water supply system with solar energy and air source heat pump composite heat source. D. Tianjin University.

Numerical simulation of enhanced throughput of surfactant in Lamadian oilfield

Ruiqi Han

Dynamic room, Geological Research Institute, No.6 Oil Production Plant, Daqing Oilfield, Daqing Oilfield, Daqing Heilongjiang province, 163100, China

Abstract. Based on the actual data of Lamadian oilfield in terms of geological structure, reservoir fluid physical properties and production dynamics, a numerical simulation geological model of CMG STARS chemical flooding module is established, which is consistent with the static and dynamic characteristics of the oil reservoir. In order to study the effect of enhanced throughput of surfactants, based on the mechanism of enhanced throughput of surfactants, the numerical model was established by defining parameters such as oil-water interfacial tension at different concentrations, residual oil saturation, oil-water relative permeability curve, capillary pressure, isotonic point change and surfactant diffusion coefficient in the formation, etc. The numerical simulation of surfactant enhanced throughput was carried out to optimize the surfactant enhanced throughput parameters in Lamadian oilfield, and the sensitivity sequence of each injection and production parameter to the surfactant enhanced throughput and oil was determined. The final recommended injection and production parameters are surfactant injection amount (10000m^3), injection concentration (0.5%) and brazing time (2d). It is predicted that the cumulative oil increase can reach 1310m3, the initial daily fluid production is controlled at 80m3/d, the daily oil production can be increased by 10m3/d, and the water cut can be reduced by 10%.

1. Surfactant enhanced huff and puff mechanism

(1) Reduce the oil-water interfacial tension

In the determination of oil recovery efficiency, the conformance coefficient and washing efficiency of oil displacement agent are the decisive parameters[1]. The improvement of oil washing efficiency increases with the increase of capillary number.

(2) Reduce residual oil saturation and reduce injection pressure

The surfactant system has a strong emulsification effect on the crude oil[2]. Under the action of oil-water two-phase flow shear, the crude oil on the rock surface will be peeled and dispersed quickly, forming oil-in-water emulsion, improving the mobility ratio of the oil-water two phases and improving the sweep coefficient.

(3) aggregation and formation of oil belt mechanism

The oil is washed by surfactants from the formation surface to form oil droplets. The oil droplets collide with each other as they move forward, and the oil droplets converge and form oil bands. As the oil bands move, more oil beads converge and form larger oil bands, promoting the further movement of residual oil to the producing well[3].

(4) Reduce the thickness of boundary layer and improve the seepage capacity of oil-water phase

The adsorption of surfactant molecules on the surface of the boundary layer fluid in the low permeability reservoir reduces the spalling work of the boundary layer fluid, the thickness of the boundary layer fluid, the pore throat of the core flow becomes larger, the flow resistance of the fluid is reduced, and the seepage capacity of the oil-water phase is increased[4].

(5) Change the wettability of the rock surface

The wettability of rock closely affects the oil displacement efficiency. The oil displacement efficiency of hydrophilic surfactant is better, hydrophilic surfactants can increase the contact Angle of the interface between crude oil and rock, and further invert the lipophilicity of the rock surface, thus reducing the adhesion of oil droplets on the rock surface and promoting the stripping of crude oil[5].

Based on the mechanism of surfactant enhanced throughput, a numerical model was established to simulate surfactant enhanced throughput by defining parameters such as oil-water interfacial tension, adsorption, residual oil saturation, oil-water relative permeability curve, capillary pressure, isotonic point change, and surfactant diffusion coefficient in the formation at different concentrations[6].

Corresponding author's e-mail: hanruiqi1@petrochina.com.cn

2. Water drive history fitting and remaining oil distribution

Well 7P86-S522 with high water cut, low productivity and rich remaining oil near the fault is selected as the test object, and the model of well group 7P86-S522 is intercepted from the established geological model of Lamadian oilfield. The number of grids in the three directions is Nx = 19, Ny = 22 and Nz = 26 respectively, and the grid step in the x and y directions on the plane is 50m. In order to ensure the accuracy of the simulation results of increased throughput, the plane around well 7P86-S522 is encrypted within the range of 5×6 grids, and the single grid in the xy direction is encrypted into 5×5 grids. This well is located near the northeast fault of Lamadian Oilfield(shown in figure 1), the reservoir development is poor, the connectivity between oil and water is weak, and it is urgent to increase the development effect.

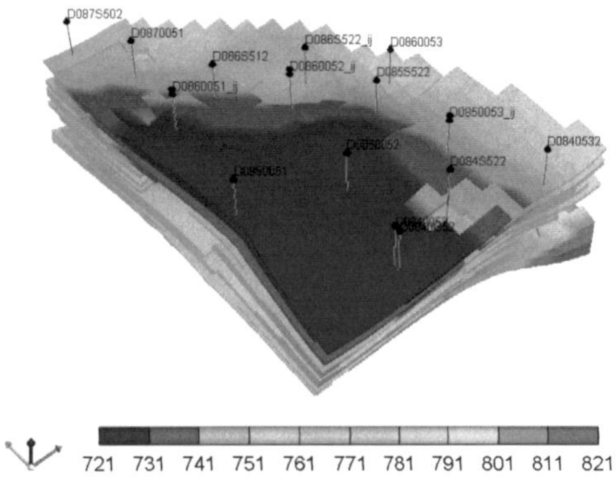

Fig 1 Geological model of well group 7P86-S522 in Lamadian Oilfield

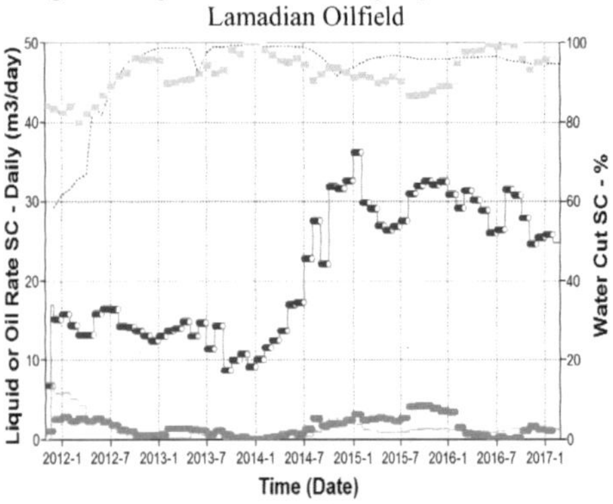

Fig 2 Historical fitting results

Historical fitting is an indispensable link in reservoir numerical simulation. Its essence is to use dynamic production data in the study area to test the accuracy and rationality of the established reservoir static model. When the predicted value of the static model is found to be inconsistent with the actual value during historical fitting, relevant parameters of the static model will be adjusted. The numerical simulation water drive parameters are gradually close to the actual values and meet the accuracy requirements.As can be seen from Figure 2, the numerical simulation fitting effect of well 7P86-S522 is good, which provides a basic guarantee for the subsequent research on remaining oil distribution, the optimization of throughput process parameters and the design of field test scheme.

3. Optimization of surfactant enhanced throughput injection and production parameters

3.1. Influence of surfactant injection amount on enhanced throughput effect

In order to investigate the effect of surfactant injection amount on the enhanced throughput, under the fixed conditions of surfactant injection concentration (0.4%), injection speed (200m^3/d), stew time (2d) and blowout speed (100m^3/d), Five injection rates of 4000m^3, 6000m^3, 8000m^3, 10000m^3 and 12000m^3 were designed to carry out surfactant enhanced throughput test on well 7P86-S522 in Lamadian oilfield. The oil increase effect in the forecast period (the period from the beginning of injection to the time when the daily oil production after the effect begins to be lower than the daily oil production before the measure) (the cumulative oil increase is the difference between the cumulative oil production of the program during the effective period and the cumulative oil production without the implementation of the measure).

Table 1 shows the numerical simulation and prediction results of oil increase effect of well 7P86-S522 under different surfactant injection amounts.

Table 1 Results of numerical simulation prediction

Argument / Program	injection (m^3)	injection concentration(%)	Injection speed(m^3/d)	stew time (d)	Discharge rate (m^3/d)	oil increase effect	
						Cumulative fuel increase(m^3)	tons of additive oil (m^3/t)
1-1	4000	0.4	200	2	100	222	13.88
1-2	6000					332	13.83
1-3	8000					508	15.88
1-4	10000					886	22.15
1-5	12000					1060	22.08

As can be seen from Table 1, with the increase of surfactant injection amount, the cumulative oil gain in well 7P86-S522 gradually increased, and the oil gain per ton agent showed a trend of first increasing and then decreasing. Considering the technical and economic effects, it is recommended that the injection volume of surfactant enhancement huff and puff in well 7P86-S522 is 10000m^3.

The relationship between daily oil production and water cut of well 7P86-S522 and time is shown in Figure 3 and Figure 4.

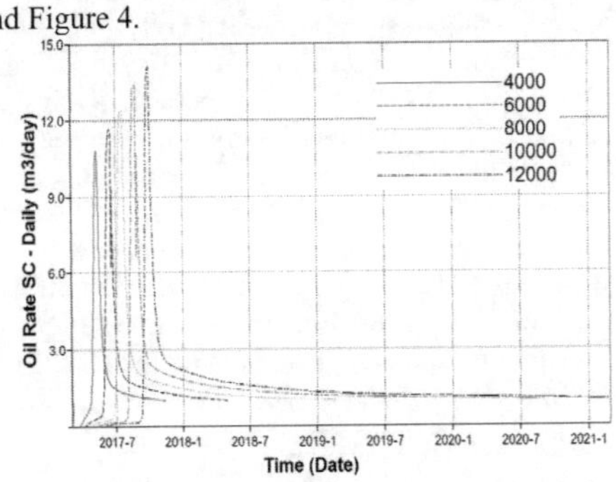

Fig. 3 Relation between daily oil production and time

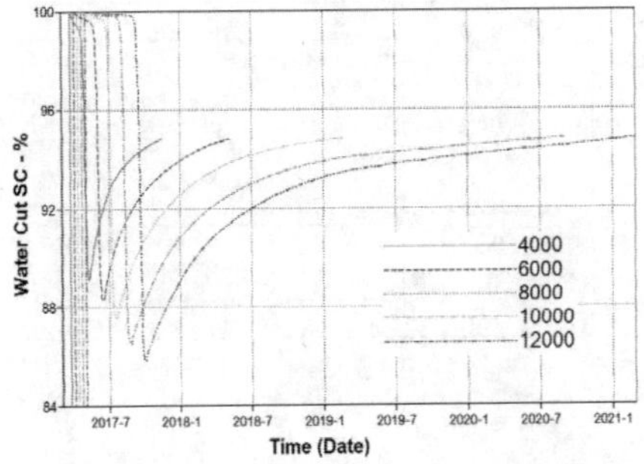

Fig. 4 Relation between water cut and time

As can be seen from Figures 3 and 4, with the increase of surfactant injection, the lower the water cut of well 7P86-S522, the higher the daily oil production, and the longer the oil increase period. When the injection amount is low, the surfactant system only spreads to the area near the well with low oil saturation, and the effect of oil increase and precipitation is poor. With the increase of injection amount, the surfactant system goes deep into the oil saturation rich area of the reservoir, and the effect of oil increase and precipitation is obvious. However, when the injection amount is too large, the crude oil will be displaced to the far well area, and can not be effectively produced, and the amplitude of oil increase precipitation will be weakened.

3.2. Influence of surfactant injection concentration on enhanced throughput

In order to investigate the effect of surfactant injection concentration on the enhanced throughput, five injection concentrations of 0.2%, 0.3%, 0.4%, 0.5% and 0.6% were designed under the fixed conditions of surfactant injection amount (10000m^3), injection speed (200m^3/d), stew time (2d) and blowout speed (100m^3/d). The surfactant enhanced throughput test was carried out on well 7P86-S522 in Lamadian Oilfield to predict the oil increase effect in the effective period.

Under the conditions of different surfactant injection concentrations, the numerical simulation and prediction results of oil increase effect of well 7P86-S522 are shown in Table 2.

Table 2 Results of numerical simulation prediction

Argument / Program	injection (m^3)	injection concentration(%)	Injection speed(m^3/d)	stew time (d)	Discharge rate (m^3/d)	oil increase effect	
						Cumulative fuel increase(m^3)	tons of additive oil (m^3/t)
2-1	10000	0.2	200	2	100	251	12.55
2-2		0.3				498	16.60
2-3		0.4				886	22.15
2-4		0.5				1284	25.68
2-5		0.6				1536	25.60

As can be seen from Table 2, with the increase of surfactant injection concentration, the cumulative oil gain in well 7P86-S522 gradually increased, and the oil gain per ton agent showed a trend of first increasing and then decreasing. Considering the technical and economic effects, it is recommended that the surfactant enhanced huff injection concentration of well 7P86-S522 be 0.5%.

The relationship between daily oil production and water cut of well 7P86-S522 and time is shown in Figure 5 and Figure 6.

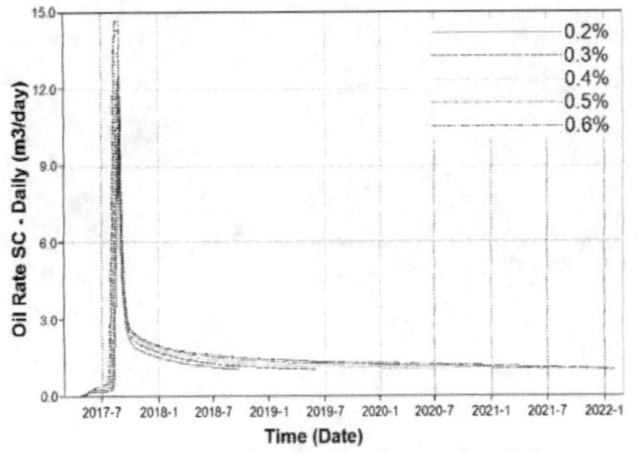

Fig. 5 Relation between water cut and time

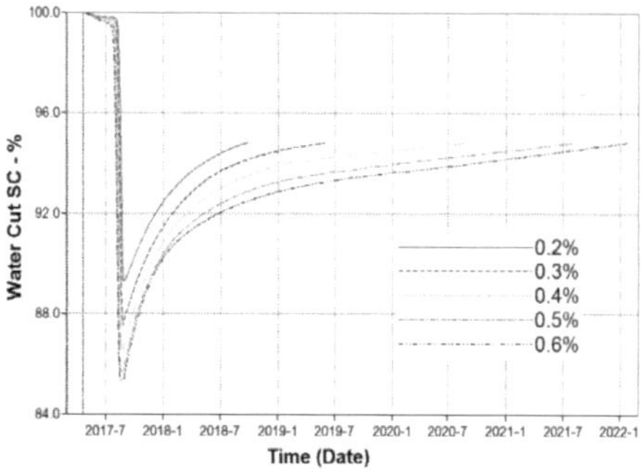

Fig. 6 Relation between water cut and time

As can be seen from Figures 5 and 6, with the increase of surfactant injection concentration, the lower the water cut of well 7P86-S522, the higher the daily oil production, and the longer the oil increase period. When the injection concentration is low, due to the large adsorption consumption of the injected surfactant, the effective concentration is too low, the oil-water interfacial tension is high, the oil washing effect is poor, and it is difficult to achieve a good oil increase precipitation effect. With the increase of injection concentration, the surfactant concentration in the formation is higher, the oil-water can reach ultra-low interfacial tension, the oil washing effect is better, and the oil increase precipitation effect is significant. However, when the concentration is too high, the decrease of oil-water interfacial tension is small, and the increase of oil precipitation is weakened.

3.3. Influence of brazing time on surfactant enhancement huff and puff effect

In order to investigate the effect of the brazing time on the surfactant enhanced throughput, five groups of brazing time of 1 day, 2 days, 3 days, 4 days and 5 days were designed under the condition that the surfactant injection amount (10000m^3), injection concentration (0.5%), injection speed (200m^3/d) and blowout speed (100m^3/d) were fixed. The surfactant enhanced throughput test was carried out on well 7P86-S522 in Lamadian Oilfield to predict the oil increase effect in the effective period.

The numerical simulation and prediction results of oil increase effect of well 7P86-S522 are shown in Table 3 under different conditions of well stew time.

Table 3 Results of numerical simulation prediction

Argument / Program	injection (m^3)	injection concentration(%)	Injection speed(m^3/d)	stew time (d)	Discharge rate (m^3/d)	oil increase effect: Cumulative fuel increase(m^3)	oil increase effect: tons of additive oil (m^3/t)
3-1				1		1262	25.24
3-2				2		1284	25.68
3-3	10000	0.5	100	3	100	1293	25.86
3-4				4		1302	26.04
3-5				5		1310	26.20

As can be seen from Table 3, the cumulative oil gain and ton agent oil gain of well 7P86-S522 gradually increase with the increase of the braising time, but the increase amplitude becomes smaller. Considering the technical and economic effects comprehensively, it is recommended that well 7P86-S522 surfactant enhanced huttling time be 2 days.

The relationship between daily oil production and water cut of well 7P86-S522 and time is shown in Figure 7 and Figure 8.

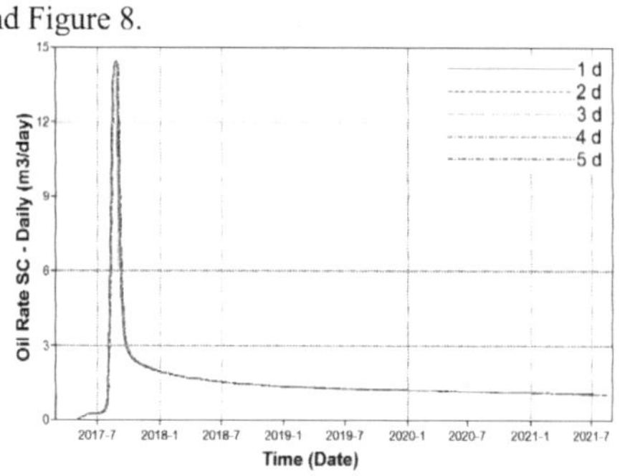

Fig. 7 Relation between water cut and time

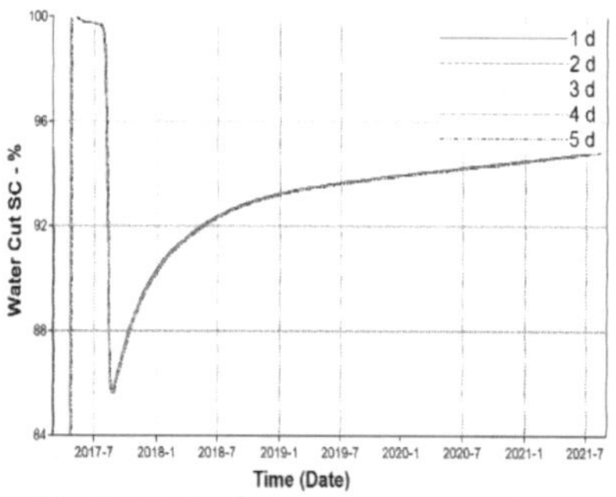

Fig. 8 Relation between water cut and time

It can be seen from FIG. 7 and FIG. 8 that the daily oil production and water cut of well 7P86-S522 change little under different braising time. However, with the increase of the stew time, the surfactant diffused and migrated to the high oil saturation area at the far end under high bottomhole pressure, which increased the effective range of the chemical agent, reduced the water content, increased the oil production, and had a certain oil increase effect.

3.4. Sensitivity analysis of injection-production parameters

In order to investigate the degree of influence of each injection and production parameter on the enhanced throughput of surfactant, the cumulative oil increase under different levels of each factor was selected as the test index, and the variance analysis of each factor was performed. The greater the variance, the greater the fluctuation of the test index at different levels of the factor, the more sensitive the influence of the factor on the test index. On the contrary, the smaller the fluctuation, the less sensitive. According to the data of the influence of various factors on the effect of increasing throughput and oil of surfactant, the variance analysis table 4 can be obtained.

Table 4 Analysis of variance

Factor / Parameter	injection rate (m^3)	injection concentration(%)	Stew time (d)
1	222	251	1262
2	332	498	1284
3	508	886	1293
4	886	1284	1302
5	1060	1536	1310
average	602	891	1290
Standard variance	321	476	17

It can be seen from the variance analysis in Table 4 that the sensitivity of each injection and production parameter to the enhanced throughput and oil increase effect of surfactant is different. The sensitivity order of each injection and production parameter is injection concentration > injection volume > time of brazing, in which injection concentration and injection volume are the most sensitive, while the sensitivity of time of brazing is weak. Therefore, the optimization of injection concentration and injection quantity should be emphasized in the optimization of injection and production parameters of surfactant enhanced throughput.

4. Conclusion

(1) The order of sensitivity of injection and production parameters to the effect of surfactant enhanced huttling and oil increase is as follows: injection concentration > injection volume > brazing time.

(2) The sensitivity of injection concentration and injection volume is the strongest, and the stew time is weak. The optimization of injection concentration and injection quantity should be emphasized in the optimization of injection and production parameters of surfactant enhanced throughput.

(3) The recommended injection and production parameters are surfactant injection amount (10000m3), injection concentration (0.5%), and brazing time (2d). Under the recommended scheme, it is predicted that the cumulative oil increase of well 7P86-S522 can reach 1310m^3, the initial daily fluid production can be controlled at 80m^3/d, the daily oil production can be increased by 10m^3/d, and the water cut can be reduced by 10%.

References

1. Anlun W ,Yinghe C ,Jianguang W , et al. Effect of surfactant on oil displacement efficiency of imbibition/huff and puff in low permeability reservoirs [J]. Energy Sources, Part A: Recovery, Utilization, and Environmental Effects, 2023, 45 (2): 6274-6289.
2. Guangsheng C ,Qingchao C ,Hongwei W , et al. Percolation Characteristics and Injection Limit of Surfactant Huff-n-Puff in a Tight Reservoir. [J]. ACS omega, 2022, 7 (34): 30389-30398.
3. Ya Y ,Yue Q ,Yue C , et al. Insights to surfactant huff-puff design in carbonate reservoirs based on machine learning modeling [J]. Chemical Engineering Journal, 2023, 451 (P1):
4. Ya Y ,Mingzhen W ,Baojun B . Descriptive statistical analysis of experimental data for wettability alteration with surfactants in carbonate reservoirs [J]. Fuel, 2022, 310 (PA):
5. Sharanya S ,J. J S ,Ziqi S . A Comparative Experimental Study of Huff-n-Puff Gas Injection and Surfactant Treatment in Shale Gas-Condensate Cores [J]. Energy & Fuels, 2018, 32 (9): 9121-9131.
6. Teklu W T ,Li X ,Zhou Z , et al. Low-salinity water and surfactants for hydraulic fracturing and EOR of shales [J]. Journal of Petroleum Science and Engineering, 2018, 162 367-377.

Study on the limit of ineffective cycle control in Lamadian Oilfield in the late period of ultra-high water cut

Hui Li

Development Program Room, Geological Research Institute, No.6 Oil Production Plant, Daqing Oilfield, Heilongjiang,163318, China

Abstract. The ineffective circulation in Lamadian oilfield is mainly banded ineffective circulation and dendritic ineffective circulation, and the ratio of ineffective circulation to the total well production is mostly concentrated below 40%. Therefore, this paper first divides the levels of ineffective circulation and simulates the proportion of liquid production in different levels of ineffective circulation to get the limits of ineffective circulation management. The results showed that the invalidation index $R < 0.2$ was a fourth-grade invalidation cycle, $0.2 \le R < 0.3$ was a third-grade invalidation cycle, $0.3 \le R < 0.4$ was a second-grade invalidation cycle, and $R \ge 0.4$ was a first-grade invalidation cycle. The reduction of water injection in the planar heterogeneous layer will lead to the decrease of cumulative oil production, and the discrimination index R is small, and the reduction range is almost unchanged. When the discriminant index R is above 0.4, the reduction of cumulative oil production increases significantly.

1. Classification of invalid cycles

1.1 Fuzzy comprehensive analysis method

According to the membership degree theory of fuzzy mathematics, the comprehensive identification method transforms qualitative evaluation into quantitative evaluation[1], which uses fuzzy mathematics to make an overall evaluation of things or objects restricted by many factors, and is suitable for solving various non-deterministic problems[2]. Based on the theory of fuzzy comprehensive identification, the geological and production factors affecting the formation of invalid circulation channels are selected by using static data and dynamic production data, and the fuzzy comprehensive identification index is calculated[3].

1.2 Determine influencing factor indicators

Permeability of oil and water Wells: the greater the permeability of the reservoir, the smaller the resistance of the fluid flow, and the greater the proportion of water absorption[4]. The easier the injected water is to rush along the high permeability interval, the stronger the scour transformation of the high permeability interval will be, and the easier it is to form the dominant seepage channel of water drive. Effective thickness of oil and water Wells: It is generally believed that the dominant seepage channel of water drive is mainly formed in the thicker interval[5]. The greater the thickness, the greater the proportion of water absorption, the stronger the scour effect of injected water on the bottom under the action of gravity differentiation, the easier it is to form the dominant seepage channel of water drive.Connectivity between oil and water Wells: Connectivity between oil and water Wells is described by the ratio of fluid production from Wells to water injection from connected Wells[6].Water injection amount: When the water injection pressure is unchanged, the water injection amount increases significantly, then the formation may appear large pores[7].

The value of the index R was between 0.12 and 0.78, which was refined into five intervals, namely, $R<0.2$, $0.2 \le R<0.3$, $0.3 \le R<0.4$, $0.4 \le R<0.5$, and $R \ge 0.5$, and the invalid cycle in different intervals was simulated.

2. Comparison of ineffective cycle governance schemes

According to the previous evaluation results of invalid cycle levels, aiming at different invalid cycle evaluation index intervals, the plane heterogeneous model was used to simulate invalid cycles in different intervals for governance, and the effects of different governance schemes were compared. In the planar heterogeneous model, the model parameters are extracted directly from the actual model, and are not added to other levels when controlling the water injection of ineffective circulation channels.

2.1 Invalid cycle comprehensive evaluation index R < 0.2

When the ineffective cycle comprehensive evaluation index $R < 0.2$, the amount of water injected in the control

Corresponding author's e-mail: lihui23@petrochina.com.cn

ineffective cycle layer changes to 80%, 60%, 40%, 20% and 0% of the original water injected, respectively, and the treatment effects under different schemes are compared. The corresponding water content under different treatment schemes is shown in the figure 1 and figure2.

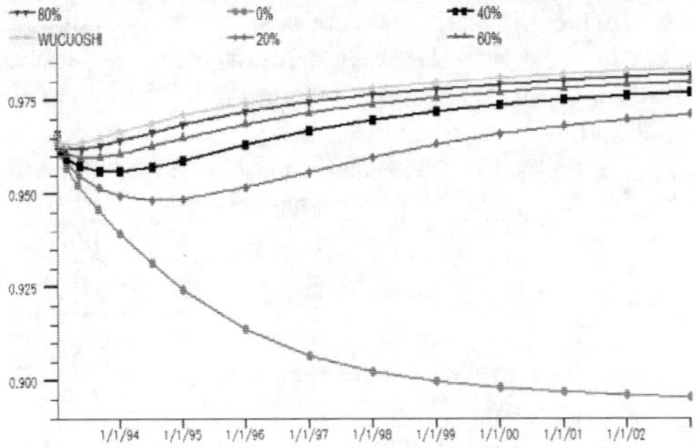

FIG. 1 Change of water content under different schemes

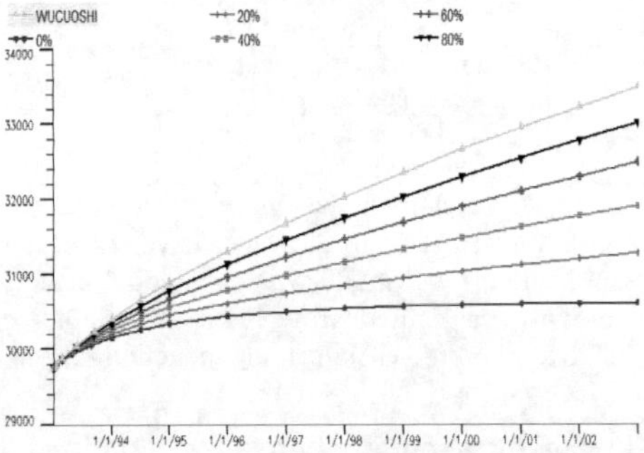

FIG. 2 Change of cumulative oil production under different schemes

It can be seen from the figure that with the decrease of water injection, the water cut also decreases. When the water injection volume changes to 80%, 60%, 40% of the original water injection volume, the water content is not much different; With the decrease of water injection, the decrease of water cut is gradually increased. When the water injection volume changed from 20% to 0%, the water cut decreased the most. In the planar heterogeneous model, reducing the water injection will reduce the cumulative oil production. Table 1 is a comparison of the cumulative oil production reduction under each scheme.

Table 1 Ineffective cycle comprehensive evaluation index R < 0.2 treatment effect

Scheme No.	Measure Scheme	Cumulative oil production reduction/m^3
1	Control the invalid circulation water injection to 80%	482.73
2	Control the invalid circulation water injection to 60%	1007.559
3	Control the invalid circulation water injection to 40%	1580.711
4	Control the invalid circulation water injection to 20%	2213.312
5	Control the invalid circulation water injection to 0%	2879.1

As can be seen from the table, reducing the water injection of Wells reduces the water injection of normal oil-well channels and ineffective circulation channels in well groups, resulting in a decrease in cumulative oil production, with almost no change in the reduction range.

2.2 Invalid cycle comprehensive evaluation index 0.2≤R < 0.3

When the comprehensive evaluation index of ineffective circulation is 0.2≤R < 0.3, the water injection volume of the controlled ineffective circulation layer is changed to 80%, 60%, 40%, 20% and 0% of the original water injection volume, respectively, and the reduction value of cumulative oil production under different schemes is compared. The corresponding water content under different treatment schemes is shown in figure 3 and figure 4.

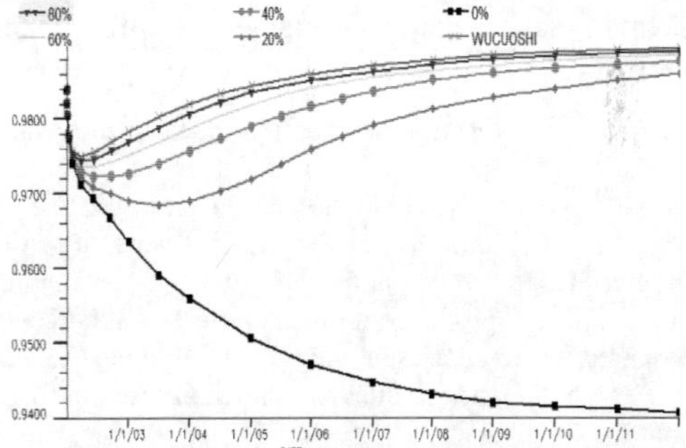

FIG. 3 Change of water content under different schemes

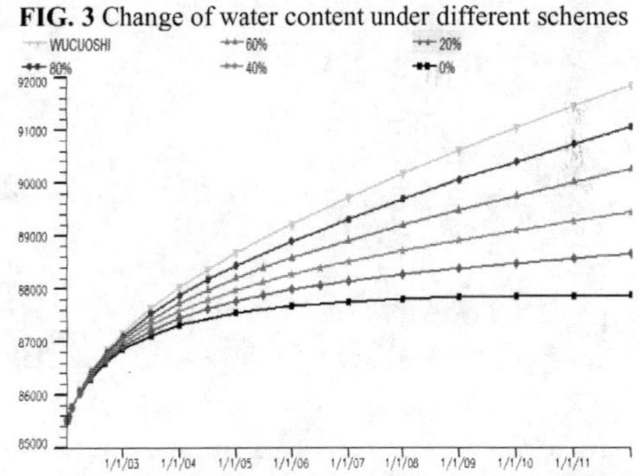

FIG. 4 Change of cumulative oil production under different schemes

It can be seen from the figure that with the decrease of water injection, the water cut also decreases. When the water injection volume becomes 80%, 60% and 40% of the original water injection volume, the water content has little difference and tends to be the same. With the decrease of water injection, the decrease of water cut is gradually increased. When the water injection volume changed from 20% to 0%, the water cut decreased the most.

In the planar heterogeneous model, reducing the water injection will reduce the cumulative oil production. Table 2 is a comparison of the cumulative oil production reduction under each scheme.

Table 2 Ineffective cycle comprehensive evaluation index 0.2≤R < 0.3 treatment effect

Scheme No.	Measure Scheme	Cumulative oil production reduction/m^3
1	Control the invalid circulation water injection to 80%	788.805
2	Control the invalid circulation water injection to 60%	1593.656
3	Control the invalid circulation water injection to 40%	2408.688
4	Control the invalid circulation water injection to 20%	3198.852
5	Control the invalid circulation water injection to 0%	3972.766

As can be seen from the figure, reducing the water injection of Wells reduces the water injection of normal oil-well channels and ineffective circulation channels in well groups, resulting in a decrease in cumulative oil production, with little change in the reduction range.

2.3 Invalid cycle comprehensive evaluation index 0.3≤R < 0.4

When the comprehensive evaluation index of ineffective circulation is 0.3≤R < 0.4, the water injection volume of the control ineffective circulation layer is changed to 80%, 60%, 40%, 20% and 0% of the original water injection volume, respectively, and the reduction value of cumulative oil production under different schemes is compared. The corresponding water content under different treatment schemes is shown in figure 5 and figure 6.

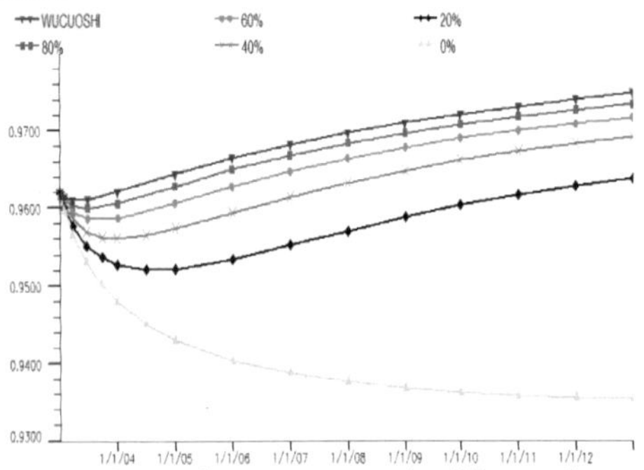

FIG. 5 Change of water content under different schemes

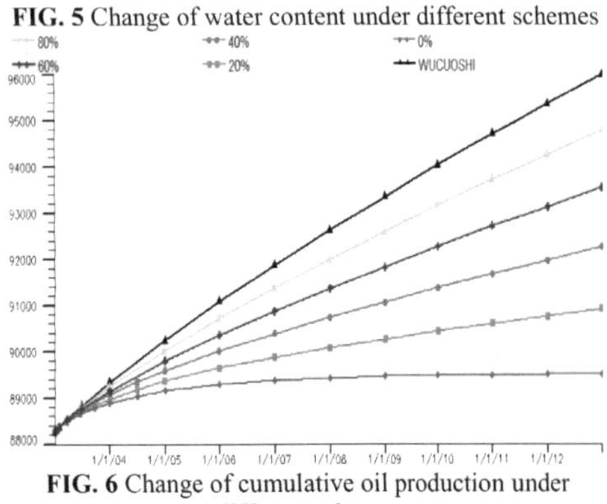

FIG. 6 Change of cumulative oil production under different schemes

It can be seen from the figure that with the decrease of water injection, the water cut also decreases. With the decrease of water injection, the decrease of water cut is gradually increased. When the water injection volume changed from 20% to 0%, the water cut decreased the most. In the planar heterogeneous model, reducing the water injection will reduce the cumulative oil production. Table 3 is a comparison of the cumulative oil production reduction under each scheme.

Table 3 Ineffective cycle comprehensive evaluation index 0.3≤R < 0.4 treatment effect

Scheme No.	Measure Scheme	Cumulative oil production reduction/m^3
1	Control the invalid circulation water injection to 80%	1198.39
2	Control the invalid circulation water injection to 60%	2438.039
3	Control the invalid circulation water injection to 40%	3720.882
4	Control the invalid circulation water injection to 20%	5070.617
5	Control the invalid circulation water injection to 0%	6474.437

As can be seen from the figure, reducing the water injection of Wells reduces the water injection of normal oil-well channels and ineffective circulation channels in well groups, resulting in a decrease in cumulative oil production, with little change in the reduction range.

2.4 Invalid cycle comprehensive evaluation index 0.4≤R < 0.5

When the comprehensive evaluation index of ineffective circulation is 0.4≤R < 0.5, the water injection volume of the control ineffective circulation layer is changed to 80%, 60%, 40%, 20% and 0% of the original water injection volume, respectively, and the reduction value of cumulative oil production under different schemes is compared. The following is the plane heterogeneous model when the comprehensive evaluation index of the invalid cycle is 0.4≤R < 0.5. The corresponding water content under different treatment schemes is shown in figure 7 and figure 8.

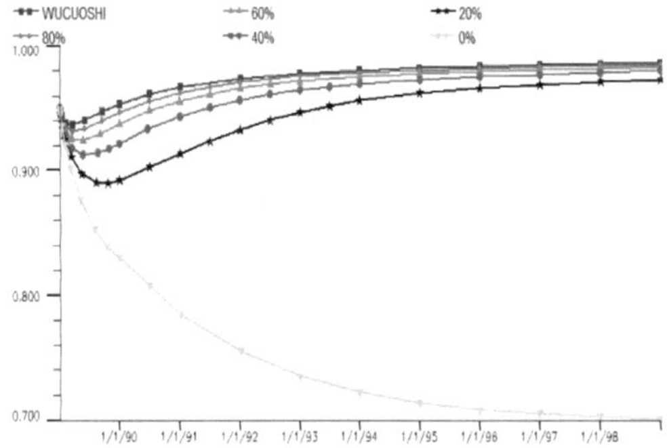

FIG. 7 Change of water content under different schemes

https://doi.org/10.1051/e3sconf/202452801007

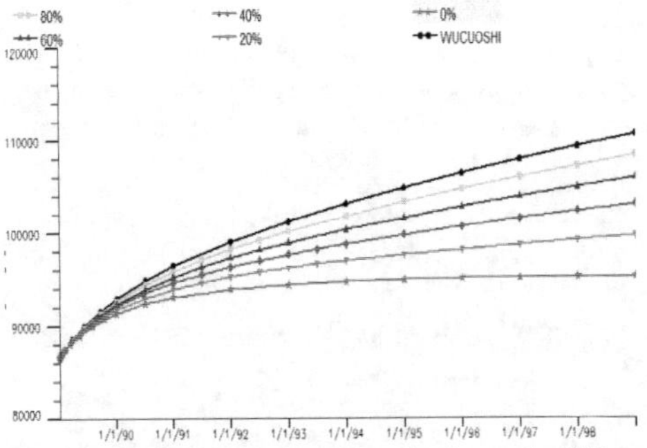

FIG. 8 Change of cumulative oil production under different schemes

It can be seen from the figure that with the decrease of water injection, the water cut also decreases. When the water injection volume becomes 80%, 60%, 40%, 20% of the original water injection volume, the water content tends to be the same; With the decrease of water injection, the decrease of water cut is gradually increased. When the water injection volume changed from 20% to 0%, the water cut decreased the most. In the planar heterogeneous model, reducing the water injection will reduce the cumulative oil production. Table 4 is a comparison of the cumulative oil production reduction under each scheme.

Table 4 Ineffective cycle comprehensive evaluation index 0.4≤R < 0.5 Treatment effect

Scheme No.	Measure Scheme	Cumulative oil production reduction/m³
1	Control the invalid circulation water injection to 80%	2223.76
2	Control the invalid circulation water injection to 60%	4719.54
3	Control the invalid circulation water injection to 40%	7548.23
4	Control the invalid circulation water injection to 20%	10925.496
5	Control the invalid circulation water injection to 0%	15362.567

As can be seen from the figure, reducing the water injection of Wells reduces the water injection of normal oil-well channels and ineffective circulation channels in well groups, resulting in a decrease in cumulative oil production. With the reduction of water injection, the reduction of accumulative oil production is increasing. When the water injection is controlled to 40%, the reduction of accumulative oil production is significantly increased.

2.5 Invalid cycle comprehensive evaluation index R≥0.5

When the comprehensive evaluation index of ineffective circulation R≥0.5, the water injection volume of the control ineffective circulation layer changes to 80%, 60%, 40%, 20% and 0% of the original water injection volume, respectively, and the reduction value of cumulative oil production under different schemes is compared. The corresponding water content under different treatment schemes is shown in figure 9 and figure 10.

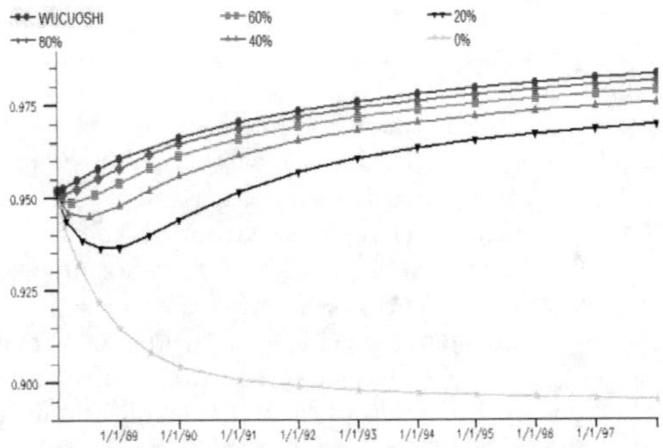

FIG. 9 Change of water content under different schemes

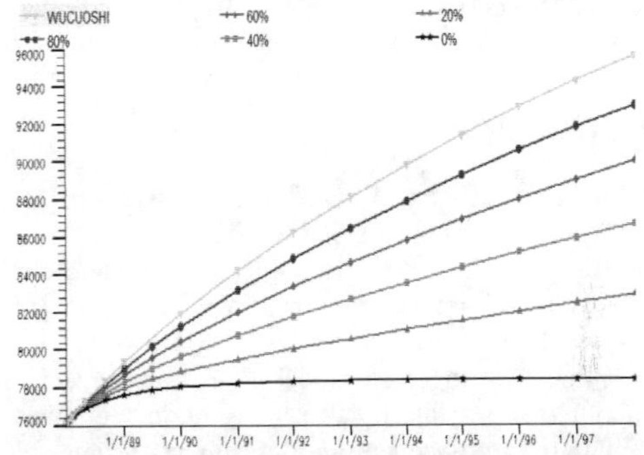

FIG. 10 Change of cumulative oil production under different schemes

It can be seen from the figure that with the decrease of water injection, the water cut also decreases. With the decrease of water injection, the decrease of water cut is gradually increased. When the water injection volume changed from 20% to 0%, the water cut decreased the most. In the planar heterogeneous model, reducing the water injection will reduce the cumulative oil production. Table 5 is a comparison of the cumulative oil production reduction under each scheme.

Table 5 Ineffective cycle comprehensive evaluation index R≥0.5 treatment effect

Scheme No.	Measure Scheme	Cumulative oil production reduction/m³
1	Control the invalid circulation water injection to 80%	2655.226
2	Control the invalid circulation water injection to 60%	5614.25
3	Control the invalid circulation water injection to 40%	8934.797
4	Control the invalid circulation water injection to 20%	12745.554
5	Control the invalid circulation water injection to 0%	17228.476

As can be seen from the figure, reducing the water injection of Wells reduces the water injection of normal oil-well channels and ineffective circulation channels in well groups, resulting in a decrease in cumulative oil production. With the reduction of water injection, the reduction of accumulative oil production is increasing. When the water injection is controlled to 40%, the reduction of accumulative oil production is significantly increased.

Through comprehensive analysis of the simulation results of the plane heterogeneous ideal model, it can be recognized that reducing the water injection of Wells reduces the water injection of normal oil-well channels and ineffective circulation channels in well groups, resulting in a decrease in the accumulative oil production. However, the accumulative oil production of well groups with ineffective circulation in different discriminant index ranges is different: When the discriminant index of the ineffective circulation channel $R < 0.4$, with the decrease of water injection, the reduction rate of cumulative oil production is almost unchanged. When the discriminant index R of invalid circulation channel is > 0.4, and the control water injection is 40%, the amount of water injection is continued to reduce, and the reduction of cumulative oil production is significantly increased.

3 Conclusion

(1) The classification of invalid cycle is divided. The invalid cycle discrimination index $R < 0.2$ is a fourth-level invalid cycle, $0.2 \leq R < 0.3$ is a third-level invalid cycle, $0.3 \leq R < 0.4$ is a second-level invalid cycle, and $R \geq 0.4$ is a first-level invalid cycle.

(2) Plane heterogeneity model simulation results in the reduction of water injection, resulting in the reduction of cumulative oil production, the discrimination index R is small, the reduction range is almost unchanged; When the discriminant index R is above 0.4, the reduction of cumulative oil production increases significantly.

References

1. Xianmin Z ,Yiming Z ,Haicheng L , et al. Dynamic Sweep Experiments on a Heterogeneous Phase Composite System Based on Branched-Preformed Particle Gel in High Water-Cut Reservoirs after Polymer Flooding. [J]. Gels (Basel, Switzerland), 2023, 9 (5):
2. Pawel Z ,Przemyslaw R ,Vitalii B , et al. Diagnostics of High Water Content Paper-Oil Transformer Insulation Based on the Temperature and Frequency Dependencies of the Loss Tangent [J]. Energies, 2022, 15 (8): 2813-2813.
3. Jan H ,Tomáš D ,Ján P . Drying Biomass with a High Water Content—The Influence of the Final Degree of Drying on the Sizing of Indirect Dryers [J]. Processes, 2022, 10 (4): 739-739.
4. Naweed K S ,Hongda H ,Huaizhu L , et al. Decision-Making Techniques for Water Shutoff Using Random Forests and Its Application in High Water Cut Reservoirs. [J]. ACS omega, 2021, 6 (50): 34327-34338.
5. Hui L H ,Qiao C C ,Na L , et al. Application of "oil-phase" microbes to enhance oil recovery in extra heavy oil reservoir with high water-cut: A proof-of-concept study [J]. Petroleum Science and Technology, 2023, 41 (6): 635-649.
6. Tong L ,Wenwu Y ,Dan L , et al. Synthesis and Plugging Mechanism of New Water-Swellable Rubber Particles for Fractured Pores in High Water-Cut Reservoirs [J]. Processes, 2022, 10 (12): 2469-2469.
7. Naweed K S ,Hongda H ,Huaizhu L , et al. Decision-Making Techniques for Water Shutoff Using Random Forests and Its Application in High Water Cut Reservoirs. [J]. ACS omega, 2021, 6 (50): 34327-34338.

https://doi.org/10.1051/e3sconf/202452801008

Study on the seepage characteristics and influencing factors of low permeability reservoir based on microscopic seepage experiments in real sandstones-taking Dongsheng gas field in Hangjinqi area as an example

Zhanyang Zhang[1], Tao Lei[1], Ting Hao[1], Jianbiao Wu[1], Hui Guo[1], Di Chen[1], Yushuang Zhu[2]

[1]Sinopec North China Oil &. Gas Branch,Zhengzhou,Henan 450006, China
[2]State Key Laboratory of Continental Dynamics/Department of Geology, Northwest University, Xi'an, 710069, China

Abstract: Taking the dense sandstone gas reservoirs of Shibang Group and Shanxi Group in Dongsheng area of Ordos Basin as an example, based on the characterization of the dense sandstone reservoirs in the study area in terms of petrological features, pore structure and diagenesis, we carried out real sandstone microscopic seepage experiments to analyze the gas-water seepage law and fluid storage state of different types of reservoirs, and systematically evaluated the efficiency of gas-water repulsion from the dense sandstone reservoirs and the microscopic influencing factors. The results show that the study area has strong non-homogeneity and different degrees of pore density, so a variety of fluid transportation channels are formed, and different transportation channels represent different types of evacuation. When the residual pore space and dense sandstone microporous space exist at the same time, gas and water are mainly transported along the residual pore space, and the microporous space of the dense particles almost does not enter; gas and water transport is very much affected by permeability, and the transport resistance will increase with the decrease of permeability; the pressure has a certain effect on the efficiency of the replacement, but it can only improve the efficiency of the replacement to a certain extent; the fluids in the process of rock transport, there will be varying degrees of bubbles stuck off In the process of fluid transport in the rock, there will be different degrees of bubble jamming phenomenon, resulting in the transport needs to overcome a large capillary resistance, which greatly increases the replacement pressure; for the low-permeability sandstones in Hangjinqi area, the good homogeneity of the pore structure of the conductor does not mean that the transport effect is good, in the area of the conductor pore space is strongly developed, the gas-water transport will form an advantageous channel, and its transport effect is better.

1 Introduction

Tight sandstone reservoir is a reservoir type with significant mine resource prospects. However, its heterogeneity is strong, which brings certain challenges to natural gas exploitation and reservoir evaluation [1,2]. The results of the study show that the efficiency of gas reservoir development is mainly affected by internal factors of the reservoir and external factors of the development system and thus is low due to the lack of energy supplementation; Visual seepage experiment of real sandstone is one of the main means to characterize the microscopic seepage characteristics of unconventional reservoirs, which has advantages in the study of microscopic seepage mechanism and fluid storage state because of its obvious intuition and authenticity [3]. Therefore, it is very important to study the influence of geological characteristics and development regime of tight sandstone gas reservoirs on production efficiency in order to use this technology.

At present, there are fewer researches to explore the microscopic seepage law and influencing factors for the low development efficiency of Dongsheng gas field in Hangjinqi area of Ordos Basin. In this study, Dongsheng gas field was taken as the research object, and representative samples were selected to carry out microscopic gas-driven water seepage experiments. The seepage law and fluid storage characteristics of gas drive water were explored by means of scanning electron microscopy, casting thin sections and high-pressure mercury injection, and the influencing factors were analyzed by combining the microscopic image of gas drive water seepage, in order to provide guidance for the development of similar reservoirs.

2 Geology

The Ordos Basin is the second largest sedimentary basin in China, with an area of $37\times10^4 km^2$, from Yinshan in

Corresponding author: Yushuang Zhu, Professor, Doctoral Supervisor, Northwest University, China.E-mail address: petroleum_gas@163.com.

the north to Qinling in the south, Lvliang Mountain in the east and Tengger Desert in the west (fig 1) [4]. The study area was in a long-term uplift state before the late Paleozoic, and began to be reaccepted for deposition in the Late Carboniferous. The Upper Paleozoic strata overlap on the mixed metamorphic rock basement from south to north. From bottom to top, the Carboniferous Taiyuan Formation, Permian Shanxi Formation, Lower Permian Lower Shihezi Formation, Upper Permian Upper Shihezi Formation and Shiqianfeng Formation were successively deposited. The Triassic, Jurassic and Cretaceous strata were successively deposited in the Mesozoic, and the Cenozoic strata were lost due to tectonic uplift. The layers of this study are the Shanxi Formation and the Lower Shihezi Formation.

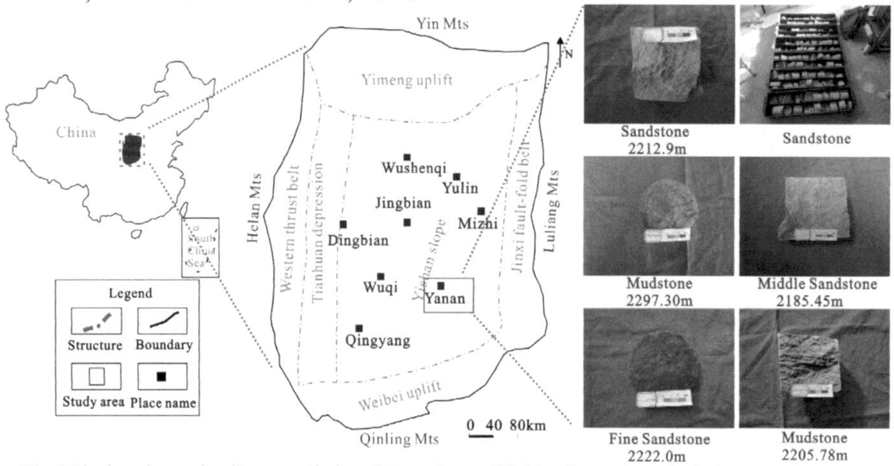

Fig.1 The location and sedimentary facies of Dongsheng oilfield region and synthetical stratum histogram

3 Basic reservoir characteristics

Dongsheng gas field is mainly located in the north of Ordos basin and the south of Yimeng uplift. The main producing formations in the study area are divided into Shanxi Formation and Shihezi Formation [5,6].

The Lower Shihezi Formation is developed in the study area. The lithology is mainly sandstone, with a small amount of argillaceous rock. The sandstone color is mainly light gray-green, gray-white and gray-yellow, and the grain size is mainly gravel-bearing coarse-medium sandstone and fine sandstone. The mudstone is mainly purple brown, brown and gray green. According to the sedimentary cycle, the Lower Shihezi Formation is divided into three sections, namely, He 1, He 2 and He 3 section. The content of lithic sandstone and quartz in the sandstone of the Lower Shihezi Formation is high, mainly developed lithic sandstone and feldspathic lithic sandstone, which are well sorted and rounded.

3.1 Physical characteristics

Based on the physical property test statistics of Dongsheng gas field reservoir samples and the relevant data collected on the physical properties of the target layer in the study area (fig 2), the porosity is distributed between 5 and 9%, with an average of 8.76%; the permeability is distributed between 0.19 and 1.4 mD, with an average of 0.92mD. The distribution of the porosity-permeability relationship shows that there is a good positive correlation between permeability and porosity. With the increase of porosity, permeability increases gradually.

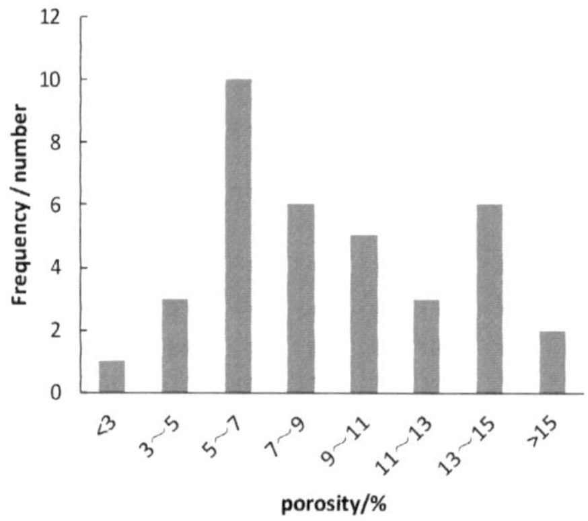

Fig.2 Characteristics of porosity and permeability in the study area

3.2 Detrital composition and interstitial material development characteristics

Through the analysis of the lithology data of the Lower Shihezi Formation in the collected samples, it is found that the lithology of different target intervals in the study area is quite different. The lithology of the Shan 2 Member and the Lower Shihezi Formation in the study area is similar, with higher lithic content, while the Shan 1 Member is significantly different, with lower lithic content, mainly quartz sandstone and sub-lithic sandstone [7]. Therefore, in this study of the target layer, the Shan 1 and Shan 2 members and the Lower Shihezi Formation were studied separately, and the differences in rock characteristics within the same target layer were analyzed .

https://doi.org/10.1051/e3sconf/202452801008

3.3 Distribution characteristics of pore structure

The high-pressure mercury injection testing was mainly used to study the reservoir pore structure (Table 1). Although the high-pressure mercury injection cannot separate the pores and throats, it can reflect the development of smaller pore and throats due to its large experimental pressure and is very suitable for low-permeability reservoirs. High-pressure mercury injection can characterize the pore throat size, heterogeneity and connectivity of the reservoir.

Table 1 Mercury injection characteristic parameters of different types of reservoirs in the study area

type	sample	por/%	sem/mD	median pressure /Mpa	median radius/μm	displacement pressure/Mpa	SHg/%	mercury withdrawal efficiency /%
I	D-35	13.623	2.198	2.446	0.300	0.280	83.563	29.778
II	D-9	11.218	1.003	15.835	0.046	1.164	76.253	32.920
	D-15	5.940	1.099	36.706	0.020	0.740	66.633	40.223
	D-18	9.241	1.409	2.538	0.290	0.271	71.759	29.974
III	D-5	5.253	0.418	116.268	0.006	1.804	57.498	35.255
	D-25	4.430	0.307	114.781	0.006	1.169	57.330	36.718

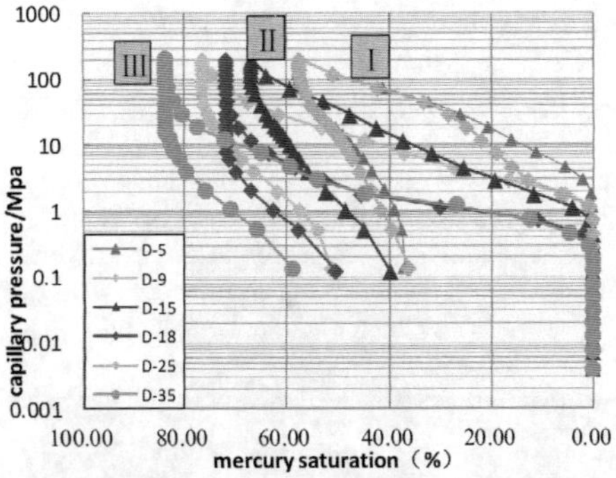

Fig.3 High pressure mercury intrusion curve

Based on the mercury injection data and capillary pressure curve (fig 3), combined with the relevant characteristics of the reservoir physical properties in the study area, the results show that the mercury injection curves can be divided into three categories (fig 1).

4 Reservoir seepage characteristics

4.1 Instruments and equipment and experimental steps

The microscopic simulation experiment system includes three parts: real sandstone microscopic model, microscopic observation system (mainly solid microscope, equipped with a camera and video system to monitor and observe the experimental process, and can take photos and videos at the same time) and pressurization system [3,8].

In the gas-water displacement experiment, the simulated gas used in the experiment is nitrogen, and the simulated water is distilled water.

4.2 Experimental result

4.2.1 Type- *I*

In the process of water flooding, when the inlet pressure is 79.4KPa, water fills the channel and enters from the middle first, and the speed is slow. After pressurizing, the water is rapidly fingered forward and reaches the outlet; if the displacement pressure is reduced to below the outlet pressure, the water stops flowing, but the water still retains the original channel. At this time, if the pressure is increased again, the water still advances along the original channel, and the outlet pressure does not change (fig 4).

Sample	Saturated water	Gas water driving	Microscopic characteristics
Type-I			
Type-II		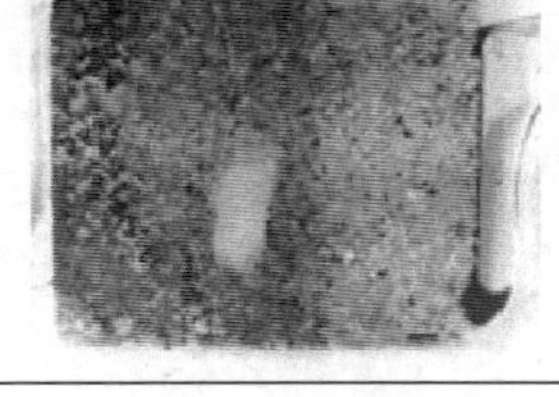	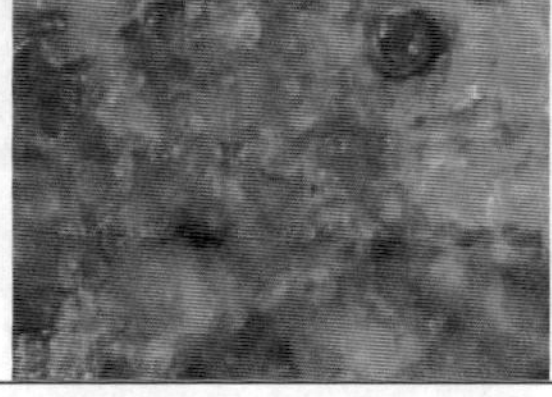

Type-III

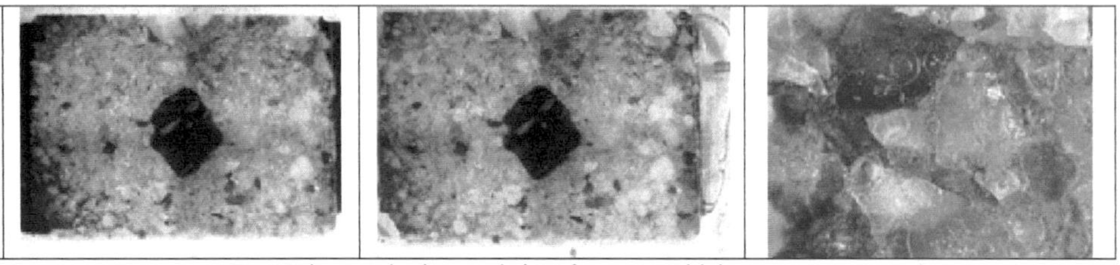

Fig.4 Microscopic characteristics of gas water driving

4.2.2 Type- II

In the process of water flooding (Fig.4), when the inlet pressure is 35.5KPa, the water filling groove enters from the bottom first, and the speed is slow. After pressurization, the water moves forward along the bottom and the range gradually expands. If the displacement pressure is reduced to below the outlet pressure, the water stops flowing, but the water still retains the original channel. With continuous pressurization, the water still advances along the original channel, and the water flooding range does not change.

4.2.3 Type-III

In the process of gas water driving (fig 4), when the inlet pressure is 45 KPa, the gas enters the matrix from the lower part after filling the groove at a slow speed. After continuous pressurization, the range of gas flooding has not been significantly expanded; If the displacement pressure is reduced to below the outlet pressure, the gas stops flowing, and the original channel is separated by the liquid to form a new resistance. At this time, the pressure is increased, and a new channel appears in the middle, and the outlet pressure increases.

In the process of saturated water entering the matrix, the saturated water enters very slowly. Finally, the saturated water enters the matrix more completely, but it takes a longer time.

4.3 Occurrence state of residual water

The original water saturation of Dongsheng ranges from 47.37% to 66.0%, with an average of 56.06%. The shapes of residual water observed after gas-driven water mainly include: Residual water at the corner, Membrane residual water, and Cluster residual water (fig 5).

Residual water at the corner	Membrane residual water	Cluster residual water

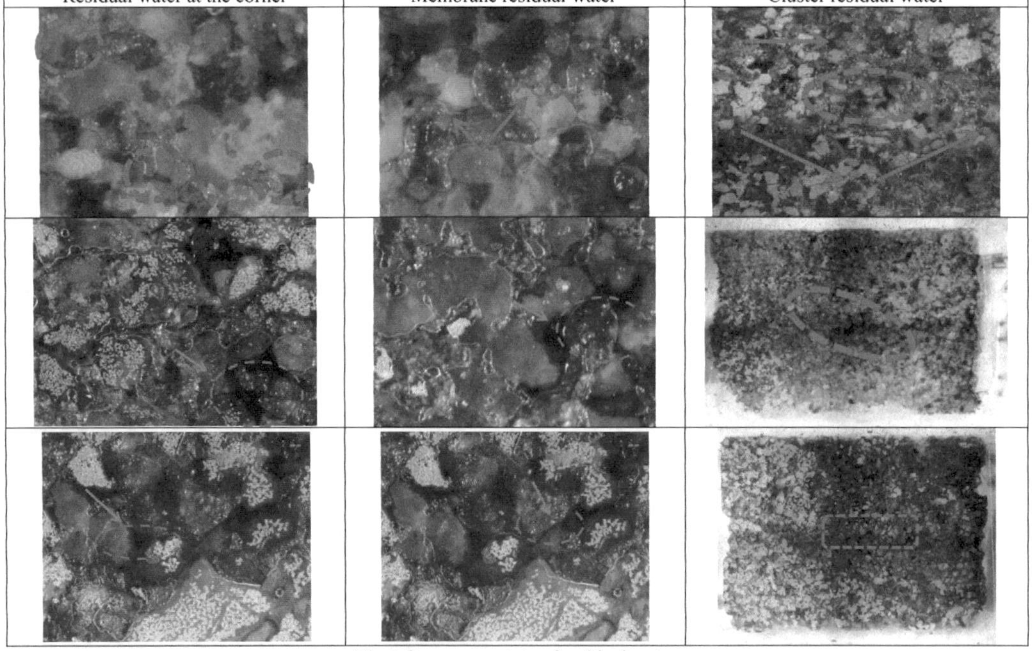

Fig.5 Occurrence state of residual water

Based on the above analysis, the occurrence state of formation water in the microscopic pore throat is divided into three types: free water, capillary water and bound water.

Bound water refers to groundwater that is bound to the surface of mineral particles or exists in the corner of pores and cannot flow.

Capillary water develops in the micro-pore throat of reservoirs with low porosity, low permeability and strong heterogeneity. It cannot move freely under the action of gravity, and the water output after fracturing is generally small.

 https://doi.org/10.1051/e3sconf/202452801008

5 Influencing factor

5.1 Permeability and gas migration resistance

As the permeability of the model decreases, the displacement pressure applied to the model increases significantly. Physical properties and irreducible water content control different types of formation water content and residual water saturation. (fig 6).

5.2 Pressure and displacement efficiency

The pressure can improve the displacement efficiency within a certain range, but after reaching a certain degree, that is, the injected water has formed a relatively stable seepage channel. Increasing the pressure can only increase the flow rate of water, but has little effect on the displacement efficiency [2,9].

With the increase of displacement pressure gradient, the residual water saturation decreases continuously (fig 7). The results show that the correlation coefficient on the intersection diagram of displacement pressure gradient and residual water saturation reaches 0.94, and the correlation coefficient between porosity and permeability and residual water saturation is low, only 0.23. Therefore, when the displacement pressure gradient is quite constant, the physical properties control the residual water saturation.

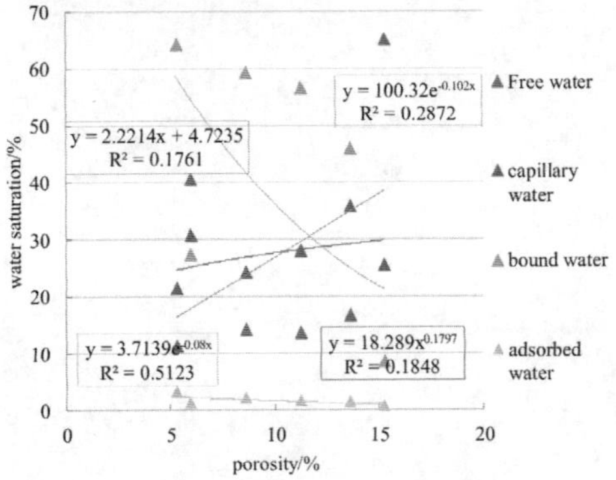

Fig.6 Relationship between porosity, permeability and water content

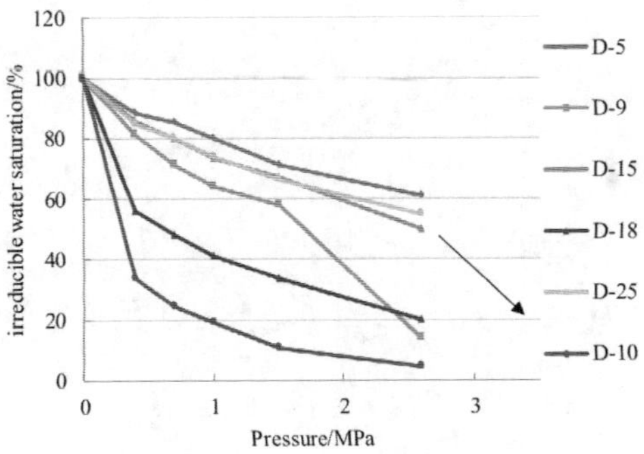

Fig.7 The relationship between displacement pressure gradient, different types of formation water, porosity, permeability and residual water saturation

5.3 Bubble breaking phenomenon in the migration process

During the displacement experiment, different degrees of bubble breakage were observed, that is, the advancing continuous gas phase was broken when passing through the pore throat. After the broken bubble moved forward, it was necessary to overcome the large capillary resistance, resulting in the abnormal outburst of the Jamin effect, thus increasing the displacement pressure.

When the pressure decreases, the continuous gas column will be disconnected from certain parts during the gas migration process, which destroys the continuity of the gas and also increases the capillary resistance of the gas migration.

6 Conclusion

1. Due to the differences in the size of the model particles and the developed cements, the pore density is different, thus forming a variety of fluid migration channels. When the residual pores and the micropores of tight sandstone exist at the same time, gas and water mainly migrate along the residual pores, and the micropores of tight particles almost do not intake air and water.

2. The migration of gas and water is greatly affected by permeability, and migration resistance will increase with the decrease of permeability; pressure has a certain impact on displacement efficiency.

3. During the migration of fluid in rocks, bubble droplets will get stuck, hich makes the migration need to overcome a large capillary resistance, thus greatly increasing the displacement pressure.

Acknowledgement

This work is financially supported by the scientific research project of Shaanxi Provincial Department of Science and Technology (No.2024JC-YBQN-0563); the Postdoctoral Research Project in Shaanxi Province, China (No.2023BSHYDZZ131).

Reference

1. Schmoker, J.W., 2002. Resource-assessment perspectives for unconventional gas systems. Aapg Bulletin, 86(11): 1993-1999.
2. Ruilan, L., Qun, L., Yongxin, H., Jinde, F. and Chunhong, M., 2010. A New Method for Forecasting Dynamic Reserve of Fractured Gas Well in Low-permeability and Tight Gas Reservoir, International Oil and Gas Conference and Exhibition in China, pp. SPE-130755-MS.
3. Guojian W, Qin Q, W. K R, et al. Experimental study on vertical light hydrocarbon microseepage mechanisms [J]. Petroleum Science and Technology, 2023, 41(23).
4. Jie H, Xiangdong G, Hongjun C, et al. Research on optimal production distribution of tight sandstone

gas reservoir [J]. Energy Reports, 2021, 7.

5. Mattar, L., Anderson, D. and Stotts, G., 2006. Dynamic material balance - oil- or gas-in-place without shut-ins. Journal of Canadian Petroleum Technology, 45(11): 7-10.
6. Muther, T., Syed, F.I., Dahaghi, A.K. and Negahban, S., 2022. Socio-Inspired Multi-Cohort Intelligence and Teaching-Learning-Based Optimization for Hydraulic Fracturing Parameters Design in Tight Formations. Journal of Energy Resources Technology-Transactions of the Asme, 144(7).
7. Yong T, Chengxi H, Youwei H, et al. Microscopic Flow Characteristics of Immiscible CO2 Flooding and CO2 Foam Flooding After Water Flooding in Fractured Porous Media: A Visual Investigation [J]. Transport in Porous Media, 2023, 149(1).
8. Zhang D, Hu H, Dong Y, et al. Seepage Simulation of Conglomerate Reservoir Based on Digital Core: A Case Study of the Baikouquan Formation in Mahu Sag, Junggar Basin [J]. Processes, 2023, 11(11).
9. Jie H, Xiangdong G, Hongjun C, et al. Productivity calculating method in tight gas reservoirs considering the stress sensitivity [J]. Arabian Journal of Geosciences, 2022, 15(2).

E3S Web of Conferences **528**, 01009 (2024) https://doi.org/10.1051/e3sconf/202452801009
NET-LC 2024

Research and practice of high-speed spiral composite drilling technology in complex and weak coal seams

Yubo Song

CCTEG Xi'an Research Institute (Group) Co., Ltd., Xi'an 710077, China

Abstract: In response to the difficult slag discharge and low hole formation rate encountered in the Wangpo Coal Mine, a study was conducted on high-speed spiral composite drilling technology. This study was integrated with the performance characteristics of the ZDY6000LR-type high-speed pit rig, resulting in the design of drill bits in three different specifications: Φ110/63.5 mm, Φ95/60.3 mm, and Φ88/50 mm. Additionally, the study optimized the three-wing drill pipe's feed speed, rotation speed, and airflow volume. The industrial field test, conducted on the 3206 working face of the Wangpo Coal Mine, resulted in the design of three drill bit combinations that exhibited high strength and stiffness. This design improvement led to enhanced hole integrity, elevated slag removal efficiency, and extended service life for both the drill pipe and bit. The high-speed spiral composite drilling technology, alongside the supporting equipment, effectively addresses the challenges of slag discharge and nozzle top drilling encountered during operations. It significantly improves the deep hole drilling rate, achieving a drilling depth of over 100 m with a success rate exceeding 70%. This technology provides vital support for the extraction of gas from coal seams within the mine. Notably, the Φ95/60.3 mm screw pipe and its corresponding drill bit demonstrated the optimal hole-forming effect, with an average hole depth of 111.2 m and a hole formation rate of 90.9%. These findings offer valuable insights for drilling operations under similar conditions of complex and weak coal seams.

1 Introduction

Mine gas poses a significant hazard, severely impacting the safety of mining operations [1-2]. As coal mining activities intensify, the emission of mine gas is increasing [3]. In recent years, the advancement of drilling technology and the subsequent development of drilling equipment for underground mine tunnels have become crucial in managing mine disasters, yielding positive results [4-5]. In China, high-gas mines frequently encounter complex and weak coal seams characterized by crushed coal layers, low permeability, high gas content, high gas pressure, and significant challenges in drilling [6]. Although various drilling equipment and techniques have been developed and successfully implemented [7-10], the variability in coal seam geological conditions continues to complicate drilling operations along the coal seam. Consequently, gas extraction from weak coal seams typically relies on short drilling holes for preliminary gas pumping, which suffers from limited coverage, high overall costs, and low equipment efficiency. Addressing the challenges associated with complex weak coal seams, this study explores the process and combination of spiral drilling to enhance the efficiency and success rate of hole formation in downstream drilling operations.

2 Overview of the mine

The Wangpo Coal Mine, situated in the Qinshui Basin, boasts a designed production capacity of 3 million tons per year. It is characterized by a high coal seam gas content, classifying it as a high-gas mine. The coal seams within the mine exhibit stability, with gentle inclinations and simple structures. The primary coal seam, identified as the No. 3 coal within the Shanxi Formation, has an average burial depth of approximately 500 m and an average thickness of 5.76 m. The coal seams are notably underdeveloped, resulting in a significantly reduced bearing capacity, which frequently complicates the drilling and construction process by making hole formation challenging. In response to these conditions, this study investigates high-speed spiral composite drilling techniques and designs various drilling combinations. An experimental study was conducted at the 3206 working face of the Wangpo Coal Mine to evaluate the efficacy of these techniques.

3 High-speed spiral composite drilling technology

The complex and weak coal seams encountered in high-gas mines are characterized by low mechanical strength, high gas content, and high gas pressure, which complicate

Corresponding author's email: songyubo@cctegxian.com

the drilling process, leading to the potential scrapping of drills and serious safety hazards. Addressing the issues of hole collapse, inadvertent hole deviation, and the inadequate slag removal capacity of the drill pipe, this study adopts a screw composite drilling technology. This technology primarily utilizes high-speed spiral slag removal, supplemented by pneumatic auxiliary powder discharge, to effectively address the challenges encountered during drilling operations in the Wangpo Coal Mine.

3.1 Spiral drilling

Spiral drilling is a drilling technique that transmits power from the drill machine to the bit at the bottom of the hole via a screw drill rod, facilitating rock-breaking drilling. The rock debris generated at the bottom and sides of the hole is transported upwards by the rotation of the screw drill rod's blade. The advantages of near-horizontal spiral drilling include: (1) The drill screw pipe makes contact only through the drill pipe blade and the wall of the drilling hole, minimizing disturbance to the wall and promoting the stability of the drilling hole wall in weak coal formations. (2) The high rotational speed of the drill screw enables timely discharge of coal powder produced within the hole, ensuring unobstructed flow and effectively reducing accidents within the hole.

3.2 Discharge analysis of high-speed screw drilling

During the drilling of near-horizontal holes in coal mines, the cinder accumulates on the lower side of the hole wall due to gravity and is transported towards the hole by the blade. In the process of high-speed spiral drilling, the cinder, constrained by the wall of the hole, experiences centrifugal force resulting from the high-speed rotation of the drill pipe. This force compacts the cinder under significant centrifugal action.

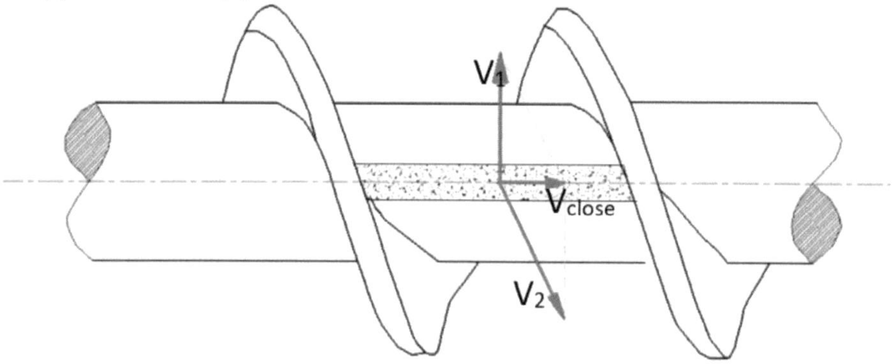

Figure 1. Schematic diagram of the high-speed spiral powder discharge movement

Within the coal cinder, there is no relative movement, and the motion of the coal cinder can be segmented into the traction speed V1 of the drill pipe's rotation and the relative sliding speed V2 between the coal cinder and the spiral groove. The resultant speed is directed axially along the drill pipe towards the hole, illustrating the slag discharge movement as depicted in Figure 1. The motor synthesis can be represented as follows:

$$V_{\text{close}} = \frac{n\pi R_0}{30}\tan\beta \qquad (1)$$

R0: The distance from the microcore to the center of the drill rod (m);

n: The rotational speed of the drill (r/min);

β: The helical ascent angle at point O (°).

From Equation 1, it is evident that the speed of cinder discharge is directly proportional to the rotational speed of the drill pipe.

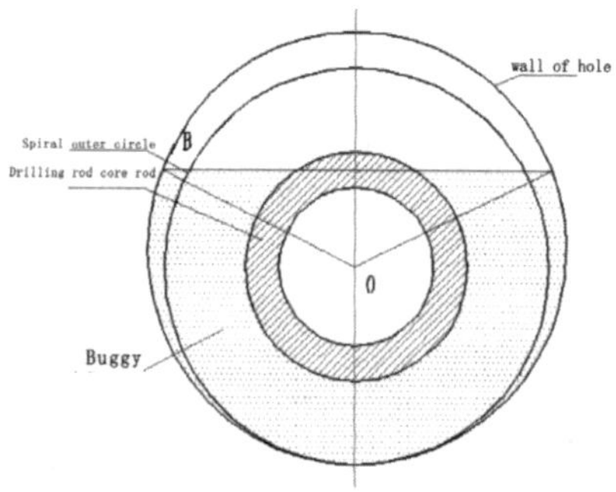

Figure 2. Cross-sectional schematic of spiral drill pipe discharging powder

As illustrated in Figure 2, the gravitational pull ensures that the drill pipe's weight remains closer to the lower side of the borehole wall. Consequently, during high-speed rotation, a portion of the material may escape from the spiral groove due to the lack of support. The disparity between the diameters of the drill bit and the drill pipe increases the annular space gap, thereby enlarging the space through which material can escape. In practical applications, the drill bit's diameter typically exceeds that of the drill pipe by merely 10 mm. This results in a minimal escape fraction, with the majority of the debris being expelled outward by the rotational force of the drill pipe.

3.3 Air drilling

Air circulation drilling is a technique that employs compressed air, conveyed through the drill pipe, to generate a high-velocity airflow within the borehole, which facilitates the removal of debris. Concurrently, this process can expedite the evacuation of gases from the borehole. In the context of drilling within complex and soft coal seams, pneumatic dust extraction offers several advantages: (1) The high-velocity airflow that is directed to the bottom of the hole via the drill pipe serves to cool the drill bit, mitigating the risk of toxic and hazardous gas emissions that may arise from coal combustion during prolonged drilling operations. (2) Such airflow is also effective in clearing pulverized coal that may be suspended in the borehole, preventing blockages at the drill bit that can occur due to an excess of fine coal particles. (3) Additionally, the high-velocity airflow exerts minimal impact on the borehole walls, reducing the likelihood of wall instability.

3.4 High-speed spiral composite drilling

High-speed spiral composite drilling[12] technology harnesses the high rotational speed of a drill to drive a screw drill, utilizing this primary power source for slag discharge through a high-speed rotational process and supplementing it with pressurized air to expel coal dust from the borehole. The main attributes of high-speed spiral composite drilling include: (1) It amalgamates the benefits of spiral drilling with air drilling, substantially enhancing the hole formation rate during drilling operations. (2) Through the precise management of the drill pipe's speed and airflow, the technology enables timely evacuation of rock powder from the borehole, minimizes in-hole accumulation, and significantly boosts slag removal efficiency. (3) The high-velocity airflow not only cools the drill bit but also rapidly removes toxic gases and vapors from the borehole, preventing gas build-up and subsequently reducing the incidence of drilling-related accidents.

4 Improvement and design of drilling tools

In alignment with the technical specifications of high-speed spiral drilling and the characteristics of the coal seam, three drilling assemblies have been engineered:

(1) A Φ110/63.5mm screw drill pipe combined with a Φ120mm three-wing taper polycrystalline diamond compact (PDC) bit, as illustrated in Figures 3 and 4. This setup yields a larger borehole diameter, which is advantageous for gas extraction. The drill rod is designed with a robust, break-resistant double-head screw structure, enhancing slag removal efficiency. The three-wing cone scraper PDC drill is durable, withstanding damage in the coal seam, thus extending its service life, facilitating smooth powder discharge, and contributing to efficient hole formation.

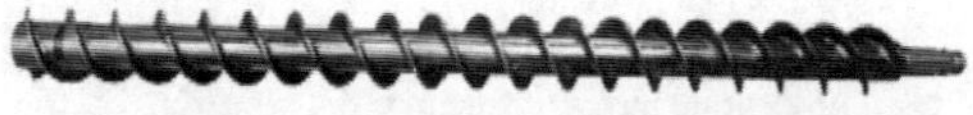

Figure 3. Φ110/63.5 mm screw drill pipe

Figure 4. Φ120 mm three-wing scraper bit

(2) A Φ95/60.3mm screw drill pipe paired with a Φ110mm three-wing taper PDC bit, as shown in Figure 5. The drill rod features a wide blade structure that lessens the impact on the borehole wall. The expansive spiral blade is capable of expelling large volumes of coal powder and chunks, ensuring clear passage within the borehole.

Figure 5. Φ95/60.3 mm screw drill pipe

(3) A Φ88/50mm screw drill pipe coupled with a Φ98mm three-wing taper PDC bit, presented in Figure 6. The smaller diameter of the hole is advantageous for hole formation. The drill pipe, with its double-headed screw structure, improves slag discharge efficiency.

Figure 6. Φ88/50 mm screw drill pipe

The research demonstrates that utilizing drilling tools equipped with wide blades and a single-head spiral configuration significantly enhances drilling efficiency. The wide blade contributes to the stability of the borehole wall, while the single-head spiral design facilitates the expulsion of large and granular coal powder. Nonetheless, the combined application of these drilling tools presents challenges, notably the wear and tear of "U" type pins and nuts. Furthermore, the conventional three-wing alloy bit exhibits limited rock-breaking capabilities and is prone to rapid wear, necessitating enhancements. To address the pronounced wear issues associated with drill pipe usage, both the design and materials have been optimized. From a design perspective, while maintaining the structural integrity of the drill pipe joint, the inner through-hole diameter is judiciously minimized, and the shoulder depth is increased. This adjustment ensures the nut is more effectively shielded by the screw groove. Material-wise, spring steel, characterized by its high tensile strength, high elastic limit, and superior fatigue resistance, has been chosen to supplant manganese steel. This selection is aimed at mitigating the swift wear experienced by the three-wing alloy bit. The introduction of the three-wing scraper PDC bit, known for its durability during drilling operations, aids in efficient hole formation. These enhancements to the drilling assembly not only minimize mechanical wear during construction but also significantly prolong the equipment's operational lifespan.

5 Control of drilling process parameters

5.1 Feed speed

Feed speed significantly impacts the efficacy of high-speed spiral drilling technology. Achieving an optimal feed speed is crucial for efficient drilling operations. The appropriate feed speed must adhere to the following equation (Equation 2):

$$\frac{\pi D^2}{4} v \le W \tag{2}$$

W: Powder discharge volume (m^3);

D: Diameter of the drill bit (m).

In practice, the feed speed should be adjusted to correspond with the coal rock layer's strength, preventing excessive coal dust production and accumulation within the borehole, especially when drilling through softer rock layers. An abrupt increase in drill torque is often indicative of a collapse hole burying the drill. Under such circumstances, drilling should be halted immediately, the drilling tool retracted repeatedly to facilitate the expulsion of accumulated coal dust, and drilling resumed only once the borehole is clear.

5.2 Drill pipe speed

The speed of the drill pipe significantly influences the efficiency of high-speed spiral drilling. Typically, the drill pipe's speed should be synchronized with the feed speed. If the discharge of powder within the borehole is not smooth, drilling should be immediately halted. The operator should then employ a back-and-forth sliding motion with the drill pipe's rotation to clear the coal powder before it is safe to resume drilling. In cases of coal instability, borehole collapse, the presence of large slag particles, or abnormal gas pressure — to prevent accidents — reversing the screw direction to agitate the coal within the hole, accompanied by rotational movements of the drill pipe, can effectively remove the pulverized coal and restore dynamic balance within the borehole. As the drilling progresses, the capability of the drill pipe to remove powder diminishes, leading to increased accumulation of coal debris and making the removal more challenging. To ensure effective drilling, a method involving a gradual increase in rotational speed, coupled with periodic reverse rotations and sweeping actions, is recommended to reduce borehole resistance.

5.3 Air volume

In air drilling operations, the volume of air serves as an auxiliary means for slag removal and gas expulsion from the borehole. The high-pressure airflow suspends the pulverized coal, which is then expelled from the hole when the airflow velocity exceeds the coal's free suspension speed. Empirical evidence suggests a direct correlation between the required air volume and drilling velocity. To ensure efficient removal of pulverized coal and maintain a clear path for the drill bit, airflow must adhere to non-clogging conditions. The optimal wind speed for air drilling is determined by the following equation:

$$v > v_b = 1.3 v_n m^{0.25} \tag{3}$$

v_b : Critical wind speed to prevent clogging (m/s);

m : Mass ratio of pulverized coal flow to air flow.

During high-speed spiral composite drilling[11], especially in soft and fragmented coal conditions that produce substantial amounts of pulverized coal, the rotational pressure is maintained between 5-18 MPa. Feed speed is adjusted to decrease as pressure increases, with rotation speeds regulated between 80-300 r/min. A stable drilling pressure necessitates a consistent speed, which should be reduced as borehole depth increases. An appropriate air volume facilitates the expulsion of pulverized material from the borehole. Insufficient air volume leads to inadequate removal of coal dust, causing accumulation, whereas excessive air volume can destabilize the borehole walls due to air turbulence. Therefore, the air volume should be maintained at 50-60% of that used in conventional air drilling.

6 Selection of drilling equipment

To address the challenges of drilling through complex and weak coal seams at Wangpo Coal Mine, specifically to enhance the drill rig's capability to navigate through zones of crushing and high-stress concentration, the ZDY6000LR drilling rig was selected for this study. This drilling rig boasts high-speed, high-power, and high-torque capabilities, with a maximum torque of 6000 Nm and a top speed of 400 r/min. It is equipped with a significant lifting pressure, which is particularly beneficial in instances of drilling and borehole collapse, as it enables the use of strong powder actuation to augment construction capabilities. The drilling rig features an amplitude variation mechanism that offers a wide range of adjustments, simplifying operations and providing strong adaptability to different drilling conditions. It is designed to automatically switch to a lifting state in response to resistance changes within the borehole, effectively preventing stuck and buried drill incidents and thereby enhancing construction efficiency. Operation of the drill rig is streamlined through a single-handle control, offering ease of use, a wide operational range, and a centralizer that minimizes friction between the drill pipe during intense drilling activities. The hydraulic system employs constant power and pressure-variable technology to mitigate system overheating. The main technical parameters of the drilling rig are summarized in Table 1.

Table 1 Main technical parameters of the drilling rig of ZDY6000LR

Classification	Appellation	Parameter
Overall unit	Appearance / m	4.46×1.25×2.12
	Drill mass / kg	8000
Gyroscope	Nominal torque / Nm	6000～700
	Rated rotation speed / r/min	80～400
Feed mechanism	Spindle inclination / °	-30～+30
	Maximum feed / pull force / kN	140/190
	Feed in / pull trip / mm	1300
Pumping station	Power rating / kW	90

7 Field industrial test

7.1 Overview of the test working face

The area selected for drilling construction is the 3206 working face of Wangpo Coal Mine. Located in the south of the four centralized lanes within the second mining area, the 3206 working face is characterized as an island working face. This face spans 2092 m in length and 147 m in width, with a cutting length of 153 m. The average coal seam thickness is 4.45 m, with geological reserves estimated at approximately 2.051 million tons. The working face encompasses three roadways: the 3206 transportation channel, the 3206 return air channel, and the 3206 high drainage channel, with the 3206 return air channel situated in the south wing of the second mining area.

7.2 Drilling design

The drilling site is positioned within the return air channel of the 3206 working face. Given the dimensions of the roadway and the ZDY6000LR drill, there is no necessity for a dedicated drilling site setup; drilling operations can be conducted directly within the roadway. Boreholes are planned to be drilled in parallel alignment along the roadway, extending vertically towards the 3206 working face. To account for the coal seam's stability and permeability, the drilling spacing is set at 4 meters, with an opening inclination ranging from 0 to -10 degrees relative to the coal seam dip. The design specifies hole depths of 132 m, with borehole diameters of Φ 98 mm, Φ110 mm, and Φ120 mm,Drill gap4m . The drilling design parameters are summarized in Table 2.The drilling plane layout is shown in Figure 7.

Table 2 Design parameters of coal seam bedding drilling

Drill class	Trompil dip (°)	Trompil position (°)	Design hole depth (m)	Drill gap (m)	Trompil altitude (m)
Along the layer drill	0°~10°	285.35	132	4	1.5

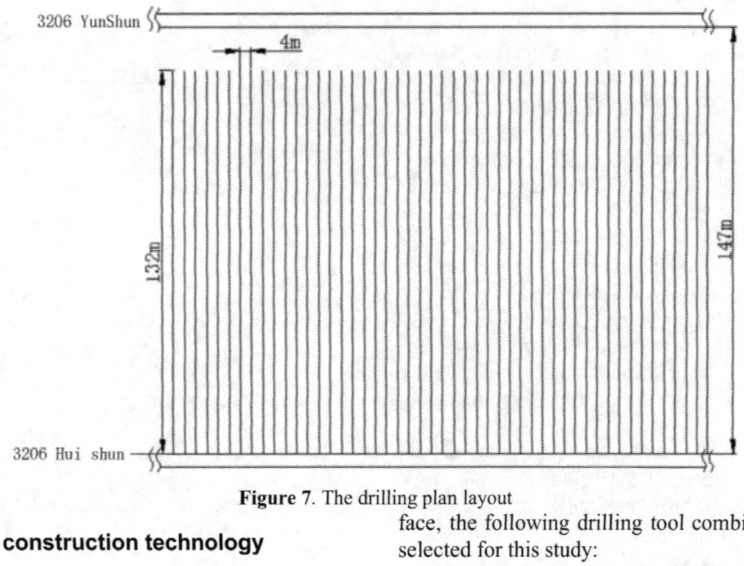

Figure 7. The drilling plan layout

7.3 Drilling and construction technology

7.3.1 Selection of drilling tool combination

Based on the coal seam conditions, drilling capabilities, and the designed drilling parameters for the 3206 working face, the following drilling tool combinations have been selected for this study:

For the hole section: After initiating the hole, proceed with a Φ 98 mm drill bit and active drill pipe. Following this step, employ the re-expansion bit in a step-by-step manner to widen the hole to its final diameter. If local conditions allow, utilize a Φ 300 mm drill bit coupled with a Φ 114/260 mm large screw drill rod for a single expansion to the final aperture.

For the bare section: Options include a Φ 98 mm triple drill bit with a Φ 88/50 mm screw drill rod, a Φ 110 mm triple drill bit with a Φ 95/60.3 mm screw drill rod, or a Φ 120 mm triple drill bit with a Φ 110/63.5 mm screw drill rod.

7.3.2 Drilling construction process

Despite the utilization of various drilling tool combinations for test drilling, the fundamental construction process remains consistent. This involves the sequence of initiating the hole, expanding it, conducting pipe grouting, installing the final hole apparatus, performing drilling operations, sealing the hole, and implementing continuous pumping.

Opening and expanding the hole: During drilling operations, strict adherence to the design parameters is maintained, ensuring the opening inclination and azimuth angle remain within ±2 degrees of the design specifications. The opening process should be characterized by stability, employing a gentle pressure and slow rotation principle. For expansion, a guided drill is used to maintain a consistent speed, ensuring a smooth borehole wall. Post-expansion, the hole is cleared to remove any accumulated rock powder. The selection of the opening point should strategically avoid areas near coal wall support anchor cables.

Installing the lower hole mouth tube: The primary objective here is to safeguard the stability of the hole section. This is achieved in conjunction with installing the hole apparatus for gas drainage and dust removal. A Φ159 mm × 1.5 m steel casing is selected for this purpose, ensuring a tight connection at the hole tube joint, which is then manually advanced into the borehole. The exposed section of the tube should measure 0.2 m, with the gap between the tube and borehole wall sealed using polyurethane.

Installing the orifice device: The device is secured using pipe hoops or flanges at the hole's four ports, which are connected to the gas drainage negative pressure pipe. The lower slag outlet links to the dust collector-steam separator, with the separator's upper end attached to the negative pressure pipe.

Normal drilling operations: During standard drilling, it is crucial to monitor slag return at the hole mouth and adhere to specified drilling parameters. In the event of any anomalies, drilling should be paused. Instead, employ the drilling tool to sweep the hole and evacuate slag, ensuring that pressure parameters are normalized before resuming drilling.

Completing the final hole and pumping: Upon reaching the designated depth, the drilling tool and orifice device are removed, the hole guard pipe is extracted, and the hole is then continuously sealed.

7.4 Test conditions

The field test encompassed 23 boreholes, achieving a total drilling length of 2022 m, with an average hole depth of 87.9 m. The deepest single borehole reached 132 m, and 73.9% of the boreholes exceeded 100 m in depth. The drilling construction outcomes are detailed in Table 3, while the comprehensive drilling plan is illustrated in Figure 8.

Table 3 Drilling construction in the 3206 coal face

Hole number	Inclination angle (°)	Orientation (°)	Depth (m)	Diameter (mm)	Completion reason
1#	2	285.35	23	98	Hole collapse
2#	3	285.35	107	98	Qualification met
3#	0	285.35	23	98	Hole collapse
4#	-2	285.35	103	98	Qualification met
5#	0	285.35	8	98	Hole collapse
6#	5	285.35	11	98	Hole collapse
7#	5	285.35	100	110	Qualification met
8#	3	285.35	102	110	Qualification met
9#	0	285.35	101	110	Qualification met
10#	-2	285.35	104	110	Qualification met
11#	-4	285.35	101	110	Qualification met
12#	-3	285.35	110	110	Qualification met
13#	4	285.35	100	120	Qualification met
14#	0	285.35	121	110	Qualification met
15#	-5	285.35	119	110	Qualification met
16#	-1	285.35	112	120	Qualification met
17#	-1	285.35	40	120	Hole collapse
18#	-2	285.35	115	120	Qualification met
19#	-2	285.35	101	120	Qualification met
20#	-2	285.35	122	110	Qualification met
21#	-5	285.35	132	110	Qualification met
22#	-4	285.35	67	110	Hole collapse
23#	-6	285.35	100	120	Qualification met

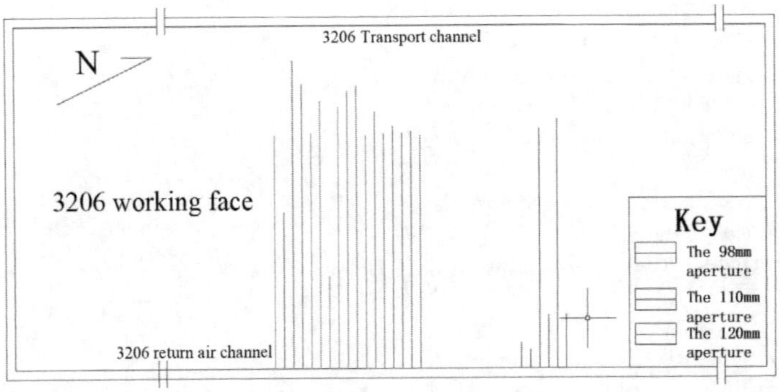

Figure 8. Drilling construction plan for the 3206 coal face

In this test, six boreholes with a diameter of Φ 98mm were drilled, achieving an average hole depth of 45.8 m and a hole formation rate of 33.3%. Additionally, eleven Φ 110 mm boreholes were completed, with an average depth of 111.2 m and a formation rate of 90.9%. Furthermore, six boreholes of Φ 120 mm were drilled, resulting in an average depth of 94.7 m and a formation rate of 83.3%. It has been determined that the rate of borehole formation is influenced not only by the conditions of the coal seam but also by the diameter of the borehole.

The substantial volume of slag generated within the coal seam boreholes highlights the limited slag discharge capacity of drills with smaller apertures, due to the narrow space between the drill rod and the borehole wall. This limitation hampers timely slag removal, especially when holes collapse or pulverized coal accumulates, potentially leading to buried drills and complicating hole formation. Drilling with larger apertures addresses this issue effectively; however, care must be taken to prevent the borehole diameter from becoming excessively large, which could destabilize the borehole wall. Based on the outcomes of these tests, selecting either a Φ 95/60.3 mm screw rod with a Φ 110 mm three-wing cone scraper PDC drill bit or a Φ 110/63.5 mm three-wing cone scraper PDC drill bit for subsequent construction activities is advisable. This choice aims to enhance the rate of hole formation and ensure the effectiveness of the construction process.

8 Conclusion

(1) Addressing the challenges of slag discharge and the low rate of hole formation in complex and weak coal seams, this study introduces high-speed spiral composite drilling technology. It explores critical parameters, including feed speed, drill pipe speed, and air volume, delineating the optimal range of construction parameters for the drilling process. This method leverages the high-speed rotation of the drill screw combined with pressurized air to expel the powder, enhancing the efficiency of powder discharge and reducing resistance within the borehole.

(2) To fulfill drilling requirements, three tool combinations have been developed: a Φ 98 mm three-wing drill with a Φ 88/50 mm screw drill, a Φ 110 mm three-wing cone scraper PDC drill with a Φ 95/60.3 mm drill pipe, and a Φ 120 mm three-wing cone scraper PDC drill with a Φ 110/63.5 mm screw drill. Improvements to the drill pipe material and the selection of the three-wing cone scraper PDC drill have notably extended the lifespan of drilling tools. The employment of the ZDY6000LR type drilling rig has proven to effectively meet the demands of deep-hole drilling.

(3) Field industrial testing at Wangpo Coal Mine demonstrates a hole formation rate exceeding 70%. Larger aperture holes have been shown to effectively address the issues of challenging slag removal and embedment during construction. For subsequent construction phases, it is recommended to use a combination of a Φ 110 mm three-wing cone scraper PDC drill with a Φ 95/60.3 mm screw drill or a Φ 120 mm three-wing cone scraper PDC drill with a Φ 110/63.5 mm screw drill, aiming to improve the hole formation rate.

Acknowledgment

This work was supported by the Shaanxi Province 2023 Key Research and Development Plan, entitled "Research on safe and efficient hole forming technology of coal mine underground roof" (Project No.: 2023-YBGY-319).

References

1. Li, X. (2024) Application of geophysical exploration method in coal mine geological disaster exploration [J]. Energy and Energy Conservation, (01):234-237.DOI:10.16643/j.cnki.14-1360/td. 2024.01. 011.
2. Mardor. (2023) Application analysis of comprehensive mechanized coal mining equipment and technology in coal mine [J]. Energy and Energy Conservation, (09):206-208+214.DOI:10.16643/j.cnki.14-1360/td. 2023.09. 064
3. Qi, C. (2024) Analysis of directional drilling technology in coal mine geological control water [J]. Energy and Energy Conservation, (02):133-135+140.DOI:10.16643/j.cnki.14-1360/td. 2024.02. 022.
4. Cai, B. (2023) Study on the characteristics and prevention measures of geological disasters in coal

mine [J]. Energy and Energy Conservation, (08):39-41.DOI:10.16643/j.cnki.14-1360/td. 2023.08. 048.

5. Li Q.X., Shi, Z.J., Shi, H.Q. (2024) Application of directional drilling technology in underground coal mine [J]. Coal field geology and exploration, (02): 85-88 + 92.
6. Zhao, J. (2023) Application of underground directional drilling technology in advanced water exploration and discharge drilling [J]. Engineering Technology Research,5 (24).
7. Zhang, J., Wang, Y., and Huang, H.J. (2020) Directional drilling technology and equipment of soft gas moving screw drilling tool [J]. Coal Geology and Exploration, 278 (02): 36-41.
8. Feng, C.S. (2024) Analysis of crack gas control technology in coal mining face [J]. Coal, 33 (01): 103-105.
9. Xue, Y.P. (2020) Improvement and application of large diameter drilling construction technology in the broken coal body of Wangpo Coal Mine [J]. Coal mine safety, 51 (12): 140-143.
10. Fang, J., Li, Q.X., Xu, C., et al. (2018) Construction technology and development trend of gas extraction drilling of soft outburst coal seam [J]. Coal science and technology. 46(5):130-137,176.
11. Liu Z ,Xiao W ,Cui J , et al. Improving compound fault diagnostic accuracy of drilling permanent magnet synchronous motor using information fusion[J]. Geoenergy Science and Engineering,2024,235.
12. Fengfei S ,Rui L ,Xi W , et al. Enhanced electrochemical performance by polyaniline (PANI) on covalent organic framework - Carbon nanotube (COF-CNT) as cathode catalyst for microbial fuel cells[J]. International Journal of Hydrogen Energy,2024,55.

E3S Web of Conferences **528**, 01010 (2024)
NET-LC 2024
https://doi.org/10.1051/e3sconf/202452801010

Synthesis of Nickel-Based Medical Alloys via Mechanochemical Processing and Powder Metallurgy Techniques

Yusi Cheng [1, *], Haoze Zhou [2]

[1] Xi'an Jiaotong University, Xi'an, Shaanxi, 710049, China
[2] North West Agriculture and Forestry University, Xianyang, Shaanxi, 712100, China

Abstract. In this study, medical nickel alloys suitable for dental applications were synthesised by a combination of force chemical synthesis and powder metallurgy. Titanium and base powders were used, processed in a hydrogen atmosphere to obtain sub-micron particle sizes and homogeneously mixed by mechanical alloying techniques to optimise sintering and compaction properties. This method ensures perfect consistency of particle composition and morphology, essential for the subsequent sintering process, which densifies the material to a relative density of 98.96% while maintaining the precise shape and weight specifications required. Manganese, boron and cerium are added in moderate amounts to enhance bond strength with porcelain veneers without sacrificing corrosion resistance or causing discolouration. Boron content is controlled below 0.1 weight percent to reduce brittleness, and silicon is adjusted to maintain mechanical strength. The alloy's coefficient of thermal expansion is optimised and determined by dilatometry techniques to ensure compatibility with porcelain and compliance with industry standards for nickel-based dental alloys. Tested mechanical properties including yield strength, tensile strength and Vickers hardness showed robustness and resistance that surpassed most nickel-based alloys and competed with cobalt-based alternatives. The study concluded that the advanced synthesis technology used to produce the nickel-containing medical alloy has improved mechanical, thermal and aesthetic properties, making it ideal for the manufacture of dental prostheses that require high precision.

1. Introduction

Advancements in dental biomaterials have been driven by mechanochemical synthesis and powder metallurgy to produce nickel-based medical alloys. These processes are critical for developing materials with required mechanical properties, biocompatibility, and aesthetic harmony with porcelain veneers for dental restorations [1]. Utilizing pure titanium and nickel powders, mechanical alloying ensures uniformity in particle size and morphology, crucial for sinterability and compressibility during densification. Correct elemental composition, including elements like manganese, boron, and cerium, is vital for enhancing the alloy-porcelain bond strength, thermal expansion compatibility, and aesthetic quality of dental prostheses [2]. The development of these robust and thermally optimized materials exemplifies the efficacy of these synthesis techniques in fulfilling the stringent requirements of dental alloy performance. This introduction foreshadows a comprehensive analysis of the production, characteristics, and clinical significance of these advanced nickel-based medical alloys.

2. Experimental materials

The uniform spherical shape and average particle size of 2-6 μm are characteristics of the nickel powder manufactured by Novamet by the carbonyl synthesis process. These particles have excellent sinterability because of their tiny particle size and chemical purity, which are produced using the carbonyl synthesis process. Chromium particles, with an average size of 150-200 μm, are obtained by grinding chromium blocks of 99.9% purity, and they exhibit an absolutely flat shape as a result of mechanical impact [3]. These particles are utilized as the initial constituents in a nickel-chromium mixture, making up the primary components of multi-component nickel-based dental alloys, with a nickel content of 20%. At 1200°C, molybdenum oxide ($MoO3$) undergoes hydrogen reduction to generate powder, which has an average particle size of 5 μm.

In the synthesis of nickel-based dental casting ingots, the chromium and molybdenum content accounts for 95.0%, with the remainder comprising silicon, manganese, and cerium. These are integrated through reagent-grade mechanical processing and cold pressing of a homogeneous powder mixture under 15.0 GPa to obtain particles of the desired shape [4]. The mixture is then

*Corresponding author: 17792231335@163.com

sintered under vacuum at a temperature of 1295°C for 30 minutes to yield high-density ingots. Standard rods suitable for mechanical testing are produced using the lost-wax centrifugal casting method. Mechanical tests on these rods include measurements of Vickers hardness, tensile strength, elongation limit, and toughness yield. Through dilatometric experiments, the nickel-based alloy's thermal expansion coefficient is found between 30 and 650°C.

The technical and cosmetic evaluation of dental alloys involves utilizing consistent metal-ceramic veneering methods and casting technologies to fabricate various dental structures. Such methodologies aim to fulfill both the functional and aesthetic criteria essential for patient satisfaction in dental applications.

In the preparation of equiatomic nickel-titanium alloys, high-purity reagents—98.5% titanium powder and sub-10 μm spherical nickel particles produced by hydrogen reduction of NiO at 600°C—are used. The titanium powder, predominantly interconnected particulate morphology with agglomerates exceeding 250 μm, is mechanically alloyed with nickel in a 1:1 atomic ratio [5]. This process occurs in a planetary ball mill under an argon atmosphere for 40 hours at 200 rpm, using stainless steel milling media to ensure a homogenous mixture and prevent oxidation.

Mechano-thermal synthesis is then employed, involving intensive mechanical treatment of the alloyed powders for 30 hours post-heating at 550°C in a hydrogen gas flow, followed by sintering at 800°C under hydrogen to achieve controlled porosity. This careful synthesis process yields nickel-titanium alloys with the desired microstructural attributes, suited for applications requiring both shape memory and superelastic characteristics [6].

3. Experimental results

Figure 1 presents an optimized micrograph of a Ni-20 wt.% Cr powder mixture after undergoing 6 hours of intense mechanical processing in a planetary ball mill. The image reveals spherical agglomerates with an average size of approximately 75-85 micrometers, composed of submicron particles. These agglomerates are indicative of the mechanical forces applied during the milling process, which result in the cold welding and fracturing of particles [7]. The micrograph illustrates the efficiency of the milling process in producing fine particles with a high degree of homogeneity and a sinterable size distribution, ideal for subsequent metallurgical processes.

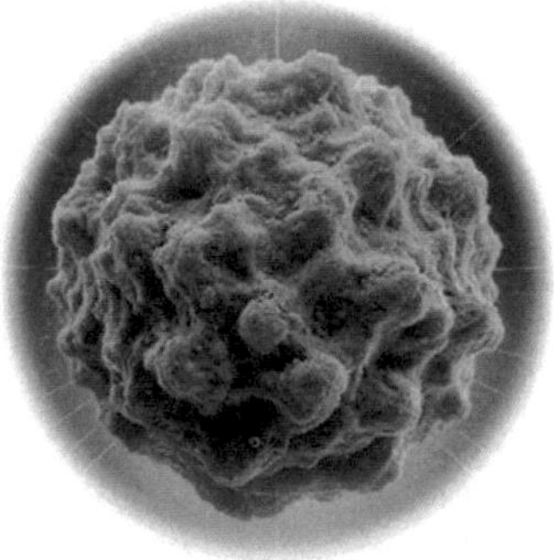

Figure 1. After four hours of mechanical treatment, an image of $NiCr$.

The provided image exemplifies a typical view of mechanically milled metallic particles, showcasing the morphological and structural alterations induced by mechanical impact. This visualization captures the homogenization of particles, which is a direct consequence of the collisional events and high-energy, oxygen-free surfaces that are characteristic of mechanical milling processes. Additionally, it displays an accumulation of various types of structural defects, a testament to the plastic deformation processes taking place at a microstructural level [8]. The development of such defects, along with the cold welding events, leads to the formation of agglomerates. These agglomerates are a fundamental aspect of the mechanical milling procedure, reflecting the balance between fracturing and welding mechanisms that define the final particle size and shape. The occurrence of these phenomena is one of the key factors that facilitates the sintering capability of the processed solids, making them suitable for advanced material synthesis and applications.

The image depicted in Figure 2 showcases the densification of dental alloy pellets synthesized through the sintering of mechanically alloyed $NiCr$ powder mixtures. The photograph illustrates a variety of geometrically diverse sintered bodies, indicative of the versatility of the sintering process to yield products suitable for a range of dental applications [9]. These densified bodies are the result of a carefully controlled sintering process, which not only preserves the desired shape but also ensures the metallurgical bonding of the powder particles to achieve a structure that is sufficiently robust for dental use. The successful sintering of these materials is critical, as it imparts the necessary mechanical properties to the dental alloys, such as hardness, wear resistance, and biocompatibility, all essential for their function in dental restorations.

Figure 2. Ni-based dental alloy pellets.

Figure 2 demonstrates the technical merits and versatility of the production method employed for dental alloy particles, showcasing the mechanical processing of parts, densification via powder metallurgy techniques, and the attainment of compact spheroids with 98.96% relative density, in the desired shapes and weights. The nickel-based alloy spheroids obtained in this manner exhibit impeccable castability, as evidenced by the casting of standard fine lattice models with consistent compositional makeup. Remarkably, the mechanical processing of the alloy components promotes compressibility and sinterability by producing perfect uniformity in the composition and shape of the metal particles. Because the cast ingots inherit the powder mixture's homogeneity, batch performance is guaranteed to remain stable. This holistic approach not only refines the mechanical characteristics of the alloy but also ensures the reproducibility of its properties, which is essential for high-precision applications in dental metallurgy [10].

Figure 3 illustrates the technical advantages and flexibility of the production methodology for nickel-based alloy particles. This method encompasses mechanical treatment of the components, densification through powder metallurgy techniques, and the creation of dense particulates with 99% relative density, tailored to the required shapes and weights. The nickel-based alloy particles, developed through this process, display excellent castability, which is validated by the casting of standard fine lattice models with a consistent compositional structure.

Significantly, the mechanical treatment of the alloy constituents results in ideal uniformity in the composition and morphology of the metal particles, thereby enhancing their compressibility and sinterability. This uniformity in the powder mixture is transferred to the steel ingots, ensuring consistency in the performance of batch-produced products. Figure 3 highlights the exceptional technical and aesthetic characteristics of the alloy, as well as the substantial potential for achieving highly complex porcelain-veneered dental structures.

The compositional balance of the nickel-chromium dental alloys is particularly noteworthy. The content of elements such as silicon, manganese, and cerium is calibrated to enhance certain properties of the alloy without compromising others. This balanced inclusion ensures that the performance enhancements are achieved without detracting from the alloy's overall functionality, affirming its suitability for high-precision dental applications.

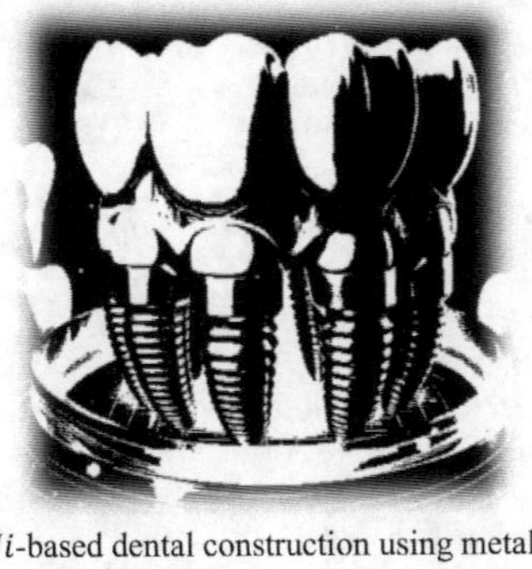

Figure 3. Ni-based dental construction using metal to ceramic.

The elemental composition of dental alloys is crucial for their mechanical integrity and aesthetic integration with porcelain veneers in prosthodontics. Precise quantities of Mn, B, and Ce are essential for optimal adhesion with ceramics, though excess manganese can reduce corrosion resistance and affect porcelain color. Additionally, boron levels above 0.1% can lead to brittleness, and silicon content over 6.23% may compromise mechanical strength, underscoring the importance of meticulous compositional control.

Thermal expansion compatibility is vital, with an ideal coefficient of thermal expansion (CTE) for nickel-based dental alloys around $13.9 \times 10^{-6} \times C^{-1}$ to ensure favorable compressive stresses.

Nickel-chromium alloys for dental applications exhibit high yield strength (~479 MPa), rivaling cobalt-based alloys, with a toughness yield of 4%, Vickers hardness (HV10) of 220, and tensile strength (Rm) of 668 MPa. These mechanical properties, resulting from a balanced composition, chemical purity, and ingot density, contribute to the superior performance of dental restorations.

4. Conclusion

The innovative approach of employing mechanochemical synthesis and powder metallurgy to produce medical-grade nickel-containing alloys has yielded materials with superior properties for dental applications. The process starts with high-purity base powders, which, through mechanical treatment, are transformed into homogenized and sinterable particles. This uniformity is crucial for the subsequent sintering process, which densifies the material to a relative density of 99%, preserving the precision in shape and mass necessary for dental prosthetics.

Careful attention to the alloy composition, particularly the addition of manganese, boron, and cerium, has been shown to enhance the bonding strength between the alloy and porcelain veneers, a key factor for the longevity and reliability of dental restorations. The meticulous calibration of these elements, alongside the control of boron and silicon content, demonstrates the ability to improve specific alloy properties without compromising overall performance.

Mechanical testing of these alloys has revealed excellent tensile strength, yield strength, and hardness, surpassing the performance of many existing nickel-based alloys and equating to that of cobalt-based alternatives known for their superior strength. The fine-tuning of the coefficient of thermal expansion further ensures compatibility with ceramic materials, a critical aspect for dental restorations that must withstand varying thermal conditions.

References

1. Erden, M. A., Uzun, F. G., Akgün, M., & Gökçe, H. (2023). Influence of Ti and Nb addition on the microstructure, mechanical, and machinability properties of 316L stainless steel fabricated by powder metallurgy. Materials Testing, (0). https://doi.org/10.1515/mt-2022-0358
2. Wang, F., Hu, J., Peng, Y., Wu, X., Xue, H., & Pang, H. (2023). Application and Modification of Nickel-based Metal-organic frameworks in Electrochemical Sensing. Advanced Sensor and Energy Materials, 100053. https://doi.org/10.1016/j.asems.2023.100053.
3. Lu, Y. B., Zhang, G. X., Yang, F. Y., Yao, M. Q., Liu, L. Y., & Pang, H. (2023). Synthesis methods and applications of high entropy nanoparticles. Rare Metals, 42, 3212-3245. 10.1007/s12598-023-02460-4
4. Afonnikova, S. D., Bauman, Y. I., Stoyanovskii, V. O., Volochaev, M. N., Mishakov, I. V., & Vedyagin, A. A. (2023). Effect of Cu on performance of self-dispersing Ni-catalyst in production of carbon nanofibers from ethylene. C, 9, 77. https://doi.org/10.3390/c9030077
5. Yang, X., Lu, X., Zhang, W., Guo, X., Ren, J., Xue, H., & Tang, F. (2023). Preparation and application of nano-Ni–Co alloy. Journal of Nanoparticle Research, 25, 152. 10.1007/s11051-023-05799-y
6. Fuku, X., Dyosiba, X., & Iftikhar, F. J. (2023). Green prepared nanomaterials from various biodegradable wastes and their application in energy. Nano-Structures & Nano-Objects, 35, 100997. https://doi.org/10.1016/j.nanoso.2023.100997.
7. Saleh, H. M., & Hassan, A. I. (2023). Synthesis and characterization of nanomaterials for application in cost-effective electrochemical devices. Sustainability, 15, 10891. https://doi.org/10.3390/su151410891
8. Gatou, M. A., Vagena, I. A., Lagopati, N., Pippa, N., Gazouli, M., & Pavlatou, E. A. (2023). Functional MOF-based materials for environmental and biomedical applications: a critical review. Nanomaterials, 13, 2224. https://doi.org/10.3390/nano13152224
9. Al-Senani, G. M., Al-Fawzan, F. F., Alshabanat, M., Abd-Elkader, O. H., Nasrallah, M., & Nasrallah, M. (2023). One Pot Synthesis, Surface, and Magnetic Properties of Ni–NiO@ C Nanocomposites. Crystals, 13, 1497. https://doi.org/10.3390/cryst13101497
10. Kalhoro, K. A., Khalid, M., Naz, K., Akhtar, M. S., Nadeem, M., & Somaily, H. H. (2023). Sol–gel synthesis and physical properties of Ni–Cu based spinel ferrites nanoparticles. Journal of Materials Research, 38, 3764-3775. 10.1557/s43578-023-01098-3

https://doi.org/10.1051/e3sconf/202452801011

Waste to Energy-Effects of Complex Components of Organic Wastewater on Coal Water Slurry Stabilizers

Dedi Li[1,2,*] , Biao Feng[2], Yuanlin Luo[2], Yuechao Wu[2], Junyang Xu[2], Jun Zhao[1]

[1]School of Mechanical Engineering, Tianjin University, Tianjin, 300072, China
[2]Power China Huadong Engineering Corporation Limited, Hangzhou, 311122, China

Abstract: Coal water slurry is an advanced and efficient coal utilization technology, prepare coal water slurry by gasification wastewater can realize the recycling of wastewater and improve the combustion or gasification efficiency of the slurry. Stabilizer is an important component to ensure fluidity and stability in industrial applications of coal water slurry, and the complex substances in wastewater for coal water slurry preparation have a great influence on the stabilizer. Therefore, in this paper, the influence of the major substances in gasification wastewater on the stabilizer is studied. Results show that (a) Ammonia nitrogen substances have adverse effects on stabilizers, because the ionized NH^{4+} in the solution will bond with the polar groups on the stabilizer, which will affect the stabilizer to lose its effectiveness and reduce the stability of the slurry. (b) Small organic molecules have little effect on stabilizers, this is because the binding strength and quantity of stabilizer to water molecules is much larger than that of organic matters such as phenolic substances. (c) High-valent metal ions have a greater impact on stabilizers, the metal ions easily form gels or precipitates with the stabilizer, causing it to lose its effectiveness, and also have an electrostatic effect with the carboxyl group on the side chain of the stabilizer, which makes the stabilizer easy to aggregate together and lose the steric hindrance effect, finally cause the stability of the slurry reduced. The conclusion of the influence of substances in wastewater on stabilizer can provide a technical reference for the selection of suitable wastewater and stabilizer to prepare coal water slurry, then promoting the efficient use of energy.

1. Introduction

The composition of organic wastewater is complex, and it comes from many sources and routes such as chemical industry[1], paper mill industry[2], pharmaceutical industry[3], textile industry[4], aquaculture[5] and other industries. In addition to containing a large amount of organic matter, it also contains metal cations, suspend solids, ammonia nitrogen, etc. If it is discharged without treatment, it will cause water pollution and do harm to environment. The conventional treatment of organic wastewater is mostly costly, the process is complicated, and the wastewater cannot be uniformly treated. Therefore, it is necessary to develop a new technology for resource utilization of organic wastewater with low cost and uniform treatment.

Coal water slurry(CWS) is a new type of coal-based fuel. It has been widely used in the field of power boilers in China. It has the characteristics of high combustion efficiency and low pollutant emission[6]. Therefore, for China, where coal is the main energy source, the CWS technology is an environmentally friendly and energy-saving technology suitable for utilization of coal. Developing CWS technology, using coal to produce clean fuel, and replacing oil with coal is a strategic and realistic choice for China's long-term and stable development of energy. Studies have shown that CWS can be prepared from wastewater for combustion and gasification, the application of CWS technology to treat organic wastewater can transform and utilize organic matter from wastewater, and achieve the purpose of organic wastewater's resource utilization[7].

CWS is a coarse dispersion system of solid-liquid two-phase. Under the action of gravity or other external forces, the particles are easily aggregated together to cause precipitation. In order to prevent the slurry from hard precipitation, a certain amount of stabilizer needs to be added to the slurry. The stabilizer can effectively prevent the coal particles from agglomerating when the slurry is at rest, and the viscosity of the slurry can decrease rapidly when it is sheared[8].

In this paper, the influence of the main components of organic wastewater on CWS stabilizers was explored, and data support was provided for selecting suitable stabilizers according to the characteristics of wastewater, so as to prepare CWS energy with better performance.

2. Experimental

The coal sample used in this paper is Shenhua coal. After the raw coal is exposed to sunlight and dried for 3-4 hours, it is put into a steel ball coal mill for 6 hours. A 100-mesh sieve was used for screening to obtain a experimental

*Corresponding author's e-mail: li_dd3@hdec.com

sample, whose proximate analyses and ultimate analyses data are shown in Table 1. A laser particle size analyzer (BT-9300ST, Baite, China) was used for particle size determination. The particle size distribution of the coal sample was unimodal as shown in Figure 2, and the average particle size of the coal sample was about 31.19 μm.

Table 1. Proximate and ultimate analyses of Shenhua coal

sample	Proximate analyses (%)				Q_{ad} (MJ/kg)	Ultimate analyses (%)				
	M_{ad}	A_{ad}	V_{ad}	FC_{ad}		C_{ad}	H_{ad}	N_{ad}	$S_{t,ad}$	O_{ad}
Shenhua coal	11.04	5.09	29.67	54.2	27.64	68.82	3.86	0.79	0.44	9.96

The gasification wastewater used in this paper, include gaswashing wastewater, sulfur wastewater, carbonation wastewater and industrial wastewater, comes from CWS gasification co-production ammonia process in ZJFD Co., Ltd.

The main water quality indicators of the four types of wastewater are shown in Table 2, it can be seen that organic matter, ammonia nitrogen and metal ions are the main components contained in the wastewater.

Table 2. The main water quality indicators of the four types of wastewater

Sample	COD ($mg \cdot L^{-1}$)	BOD_5 ($mg \cdot L^{-1}$)	Potassium ($mg \cdot L^{-1}$)	Sodium ($mg \cdot L^{-1}$)	Phenol ($mg \cdot L^{-1}$)	Ammonia nitrogen ($mg \cdot L^{-1}$)
Gas washing wastewater	2.12×10^3	777	1.02×10^3	7.66×10^3	17.1	285
Carbonization wastewater	496	174	2.80×10^3	35.9	0.01	2.91×10^4
Sulfur wastewater	5.36×10^4	1.8×10^4	0.50	1.83×10^5	0.46	187
Industrial wastewater	2.87×10^5	3.1×10^4	3.25	4.87×10^3	0.17	8.88×10^3

3. Results and discussion

3.1. Influence of ammonia nitrogen in wastewater on stabilizer

In order to study the influence of ammonia nitrogen on stabilizers, amines were used to simulate wastewater to prepare CWS in this paper. Ammonium chloride (NH_4Cl), Ammonium sulfate ($(NH_4)_2SO_4$), Hexamethylenetetramine ($C_6H_{12}N_4$), ammonium acetate (CH_3COONH_4), and ammonia ($NH_3 \cdot H_2O$) were selected as experimental materials. The stabilizer adopts xanthan gum which commonly used in CWS preparation, and the addition ratio is 0.1%. According to the calculation, add the corresponding mass of amines to ensure that the content of NH^{4+} is about 10^4ppm. Figure 1 is the rheological characteristic curve and WSR of different CWS prepared with ammonia-nitrogen.

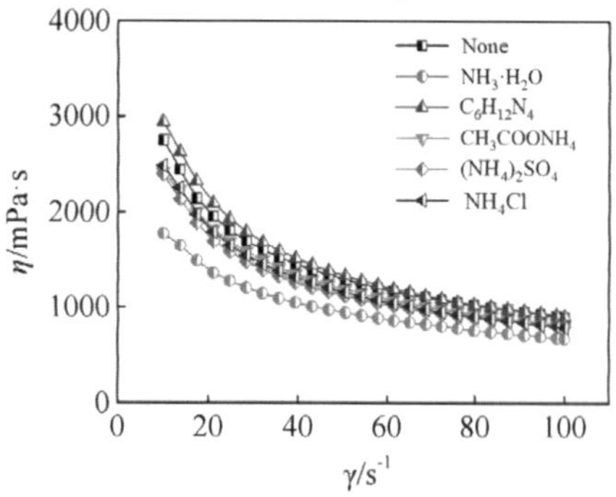

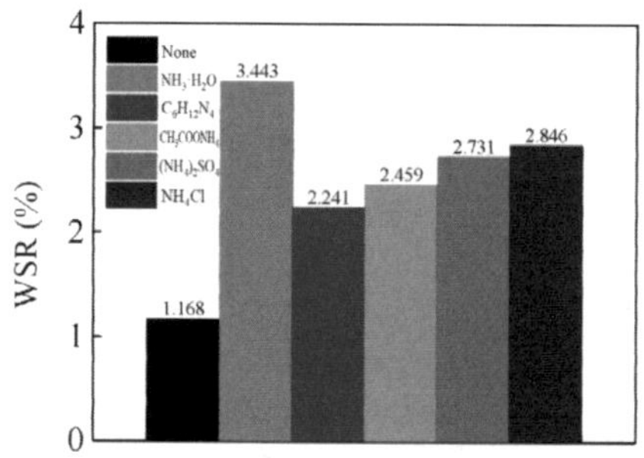

Figure 1. Rheological characteristic curve and WSR of different CWS prepared with ammonia-nitrogen

From Figure 1, compared with the apparent viscosity of ordinary CWS, the apparent viscosity of CWS with ammonia nitrogen substances is reduced to some extent. And WSR of CWS with ammonia nitrogen substances is higher than that of ordinary CWS. This indicates that ammonia nitrogen substances can cause the apparent viscosity of CWS to decrease and the WSR to increase. Among several different ammonia nitrogen substances, ammonia water has the greatest influence on the properties of CWS, indicating that the addition of ammonia nitrogen has an adverse effect on the stabilizer, as a result, the stability of the slurry is reduced, and the pseudoplasticity is also weakened.

The reasons for the analysis are as follows: There are many polar groups in the structure of xanthan gum. It is through the action of these groups that xanthan gum can form a stable network structure through bonding to enhance stability of slurry. Through the hydrogen bonds formed between hydroxyl and carboxyl groups and water molecules, the stabilizer can coat the water molecules in the structure and form a hydration film on the surface of the particles[9], when the slurry is sheared, water molecules are released and the fluidity of the slurry is enhanced. However, with the addition of ammonia nitrogen, NH^{4+} enters the solution after ionization, and the hydrogen atoms in NH^{4+} can form hydrogen bonds with the oxygen on the hydroxyl and carboxyl groups, which affects the interaction between xanthan gum and water molecules. The number of water molecules that can be bound or coated by the stabilizer is reduced, so the apparent viscosity of the slurry will be reduced and the stability will be deteriorated.Influence of organic matter in wastewater on stabilizers

In order to study the effect of organic matter on stabilizer, this paper adopts some typical organic matter in wastewater to simulate wastewater to prepare CWS. Phenol, o-cresol, m-cresol, pyridine and ethanol were selected as experimental materials. The stabilizer adopts xanthan gum, the addition ratio is 0.1%, and the organic matter of the corresponding quality is added according to the calculation. Figure 2 is the rheological characteristic curve and WSR of the CWS prepared with different organic matters.

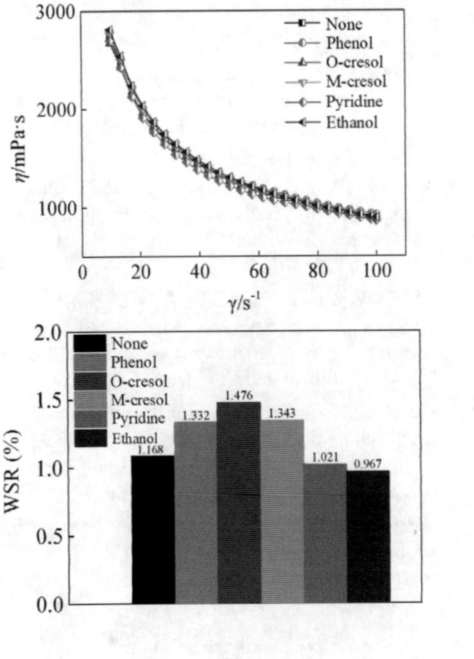

Figure 2. Rheological characteristic curve and WSR of the CWS prepared with different organic matters

From Figure 2, the rheological curve of the CWS prepared with organic matters is almost coincident with that of the original CWS, indicating that the addition of organic matter has little effect on the apparent viscosity of the slurry. Studies have shown that the addition of small molecular organic compounds such as phenolic substances will increase the apparent viscosity of the slurry, mainly because the phenolic substances can be adsorbed to the surface of coal particles, making the hydration film thicker and free water reduced[10]. The reason for the phenomenon in the figure may be due to the formation of the network structure of xanthan gum between coal particles and water, which encapsulates a large number of water molecules through bonding energy, the binding of small molecular organics to water molecules on the surface of coal is far less strong than that of stabilizers. That is to say, the amount of the binding and coating of water molecules by xanthan gum is far greater than that of small molecular organic matter, so the addition of organic matter has little effect on the apparent viscosity. Figure 1 also shows the effect of different organics on the WSR. It is found that the WSR of the slurry added with the organic matter is not much different from that of the original slurry. From the above analysis, it can be concluded that the incorporation of small molecular organics such as phenol into the CWS will not have much influence on the stabilizer.

3.2. Influence of metal ions in wastewater on stabilizers

In order to study the effect of metal ions on stabilizers, this paper uses some typical metal ions in wastewater to simulate wastewater to prepare CWS. NaCl, KCl, $MgCl_2$, $CaCl_2$, $CuCl_2$, $FeCl_3$ was selected as experimental materials. The stabilizer adopts xanthan gum, and the addition ratio is 0.1%, and the corresponding mass of metal ion solution is added according to the calculation. Figure 3 is the rheological characteristic curve and WSR of the CWS prepared with different metal ions solution.

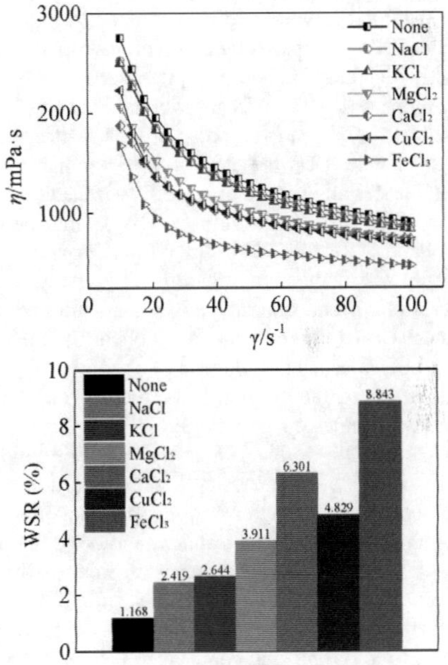

Figure 3. Rheological characteristic curve and WSR of the CWS prepared with different metal ions solution

From Figure 3, the apparent viscosity of CWS prepared from metal ion solution is lower than that of ordinary CWS, and the higher the valence state of metal ions, the greater the degree of apparent viscosity reduction. The WSR of the CWS prepared by metal ions solution has increased, and roughly shows the law that the higher the valence state of the metal ion, the higher the WSR. The reason for this phenomenon is that the incorporation of metal ions will cause the stabilizer to lose its effect, studies have shown that polyvalent metal salts can form gels or precipitates with xanthan gum, especially trivalent metal salts, which can form gels or precipitates at lower pH values[11], this will render the stabilizer ineffective; In addition, xanthan gum molecules contain strong polar groups such as -COO- and –OH, in saline solution, metal cations can have electrostatic effects with carboxylic acid groups on the side chains of xanthan gum and neutralizes the charge, thereby reducing the repulsion of the side chain, making it easier for the side chain to entangle with the main chain and easier for xanthan molecules to aggregate[12]. Finally, the steric hindrance of xanthan gum molecules in the slurry is weakened

4. Conclusion

As a clean and efficient coal-based fuel, CWS can prepared from gasification wastewater to realize resource recycling and sustainable development. Stabilizer is an important component to ensure fluidity and stability in industrial applications of CWS. The influence of the main components of gasification wastewater on the stabilizer of CWS is studied. The conclusion can provide a technical reference for the selection of suitable wastewater and stabilizer to prepare high concentration coal water slurry, then promoting the efficient use of energy.

(1) Compared with ordinary CWS, the apparent viscosity of CWS prepared by ammonia nitrogen solution is slightly decreased, and the water separation rate is slightly increased. This is because the ionized NH^{4+} in the solution will bond with the polar groups in the xanthan gum, which will affect the effect of the xanthan gum on water molecules and reduce the stability of the CWS.

(2) Compared with ordinary CWS, the apparent viscosity and WSR of CWS prepared with organic matter solution do not change significantly. This is because the binding strength and quantity of xanthan gum to water molecules is much larger than that of phenolic substances to water molecules, so the addition of phenolic substances has little effect on the stabilizer, and the properties of the slurry does not change significantly.

(3) Compared with ordinary CWS, the apparent viscosity of CWS prepared with metal ion solution is obviously decreased, and the WSR is obviously increased, and the higher the ion valence state, the more obvious the difference is. This is because the metal ions easily form gels or precipitates with the stabilizer, causing it to lose its effect, and also have an electrostatic effect with the carboxyl group on the side chain of the stabilizer, which makes the xanthan gum molecules easy to aggregate together and lose the steric hindrance effect, so the stability of the slurry is reduced.

Acknowledgements

This work was supported by the National Key Research and Development Program of China (Grant No. 2016YFB0600505). The support is gratefully acknowledged.

References

1. Ma H, Pedersen C, Zhao Q, et al. (2022) Utilizing 3D DOSY NMR in the characterization of organic compounds in coal chemical wastewater. Magnetic Resonance Letters 2.2:69-79.
2. Srikantha H, Guruswamy A, Mahesh S, et al. (2022) Batch Electrochemical Treatment of Raw Pulp and Paper Mill Wastewater Using Electrode Combinations. ECS Transaction 107.
3. Samal K, Mahapatra S, Hibzur A. (2022) Pharmaceutical wastewater as Emerging Contaminants (EC): Treatment technologies, impact on environment and human health Energy Nexus 6.
4. Zhou L, Lai Y, Zeng R. (2022) Core carbon fixation pathways associated with cake layer development in an anoxic-oxic biofilm-membrane bioreactor treating textile wastewater. Science of the Total Environment 835:155483-155483.
5. Singh A, Nenavathu B, Dasauni K, et al. (2022) ROS-Mediated Aquaculture Wastewater Remediation Using TeO2-Deposited ZnO Nanotubes Water, Air, & Soil Pollution 233.6.
6. Romanov D, Vershinina K, Dorokhove V, et al. (2022) Rheology, ignition, and combustion performance of coal-water slurries: Influence of sequence and methods of mixing. Fuel 322.
7. Wang X, Gu G, Liu M, et al. (2022) Progress in preparation of coal water slurry with industrial wastewater. Shandong Chemical Industry (China) 51.6:130-132.
8. Zhang Y, Hu S, Yang X, et al. (2021) Performance and mechanism of polyacrylamide stabilizers in coal water slurry Colloids and Surfaces A: Physicochemical and Engineering Aspects 630.
9. Ge L, Li J. (2019) "Study on stability of coal water slurry under different coal water slurry dispersants by using stability analyzer" IOP Conference Series: Materials Science and Engineering 631.2.
10. Wang J, Liu J, Wang S, et al. (2018) "Slurrying property and mechanism of coal-coal gasification wastewater-slurry[J]. Energy & Fuels, 32(4):4833-4840.
11. Chen L, Zhang L, Lei F, et al. (2022) "Preparation of pumpkin seed oil O/W emulsion by whey protein isolate-xanthan gum compound emulsifier and its stability" China Oils and Fats:1-13.
12. Wang P. (2019) "Preparation and characterization of Xanthan-Metal salt composite flocculant and its application in red tied mitigation" YanTai University of China.

Present situation of comprehensive utilization of coal-based solid waste

Haichao Li

National Energy Group Xinjiang Energy limited liability company, Urumqi, Xinjiang, 830000, China

Abstract: At present, the annual discharge of coal-based solid waste in China has exceeded 800 million tons, but the mainstream treatment methods are still simple treatment methods such as goaf backfilling and open-pit stacking. The massive stacking of coal-based solid waste is not only a serious waste of resources itself, but also brings a series of problems such as environmental pollution.
Under the "Carbon Peak" and "Carbon Neutrality" strategy in China, in order to achieve green and rapid development of the coal industry, it is necessary to solve the problem of coal-based solid waste treatment and achieve green and comprehensive utilization of coal-based solid waste. Taking coal gangue, gasification slag and fly ash as representative coal-based solid waste, the present situation of coal-based solid waste production in China and the basic properties of coal-based solid waste are introduced. The utilization status and application characteristics of coal-based solid waste in cement concrete, roadbed materials, soil improvement and so on are focused on, and the resource utilization potential of coal-based solid waste is analyzed and summarized. Then the application prospect of coal-based solid waste in high-value utilization in the future is analyzed and prospected.

1. Introduction

At present, coal is still the main primary energy in China, and the situation is unlikely to undergo significant alterations in the near future [1]. The extensive use of coal will result in the generation of a large amount of coal-based solid wastes, which will cause serious environmental problems and even endanger personal safety. Coal-based solid waste is mainly a by-product in the process of coal mining, processing, combustion and conversion, including coal gangue, fly ash, coal gasification slag, etc. [2].

Coal gangue is mainly waste formed during coal mining, with annual emissions reaching nearly 800 million tons [3]. It has lower carbon content and higher ash content. At present, the main treatment methods of coal gangue are goaf backfilling and open-air stacking. Many coal gangue mountains will spontaneous combustion in the process of open-air stacking, releasing a large amount of sulfur dioxide and nitrogen oxides, seriously polluting the local air. At the same time, coal gangue contains heavy metal elements such as Mn, Cr, and As [4]. Long-term open-air stacking will cause harmful elements in coal gangue to spread to the surrounding environment, destroying the local ecological equilibrium and affecting the lives of local residents [5].

Coal gasification slag primarily comprises the by-product generated during the coal gasification process. According to statistics, the cumulative yearly discharge of coal gasification slag has surged to 60 million tons in 2021[6]. The current main treatment method in our country is still open-air stacking. The open-air stacking of large amounts of coal gasification slag will not only cause dust and pollute the atmospheric environment, but also contain heavy metal ions such as Mo, Sr, and Sb that are relatively easy to leach and flow in the gasified slag [7]. It will seriously pollute nearby soil and groundwater resources.

Fly ash is mainly a waste produced during the combustion of coal. In some thermal power plants, pulverized coal is burned as a source of power, and a large amount of fly ash is produced after combustion. As of the end of 2021, the national fly ash emissions in 2021 exceeded 650 million tons [8]. The simple accumulation of large amounts of fly ash will not only waste land resources, but harmful substances such as heavy metal ions will gradually penetrate into the groundwater, contaminating the groundwater and destroying the local ecological environment [9]. Moreover, the fine particles in fly ash will be suspended in the air to form fine particulate matter (PM2.5 and PM10), which will have a negative impact on air quality and seriously endanger the health of local residents.

Under the "Carbon Peak" and "Carbon Neutrality" strategy in China, in order to achieve green and rapid development of the coal industry, addressing the issue of coal-based solid waste management is imperative to attain environmentally sustainable and all-encompassing utilization practices for coal-based solid waste.

Based on this, taking coal gangue, coal gasification slag and fly ash as representative solid wastes, on the basis of summarizing the classification and characteristics of

Corresponding author's e-mail: 20028672@ceic.com

https://doi.org/10.1051/e3sconf/202452801012

coal-based solid waste, the ways of comprehensive utilization of coal-based solid wastes are focused on. The prospective development trajectory of coal-based solid waste's comprehensive utilization is forecasted to offer insights for further research in this field.

2. Output of coal-based solid waste

2.1 Output of coal gangue

Coal gangue is a kind of black rock with low carbon content, low heat utilization value and hard texture [10]. Coal gangue output represents 10% to 15% of the total raw coal production, originating from three primary sources: gangue during open-pit mining stripping and roadway construction, gangue extracted from the roof, floor, and interlayers during mining operations, and gangue separated during coal washing processes. The respective proportions of these sources are 45%, 35%, and 20%.In recent years, China's coal output has increased year by year, and coal gangue production has also increased. According to the data calculated in the "2021-2022 China Bulk Industrial Solid Waste Comprehensive Utilization Industry Development Report", the coal gangue production in 2021 will be approximately 743 million tons, an increase of 5.84%, a significant increase [11].

The production of coal gangue in coal mines varies greatly in different parts of China. The production of coal gangue in Taiyuan, Tangshan and other places is about 30%, while that in the newly built mines in Ordos and other places is about 10%. At the same time, most of the coal gangue production in China is concentrated in Shanxi Province, Shaanxi Province, Inner Mongolia Autonomous region and Xinjiang Uygur Autonomous region, where coal gangue production accounts for more than half of the national coal gangue production.

2.2 Output of coal gasification slag

The application of coal gasification technology is becoming more and more widespread under the domestic energy structure and actual demand, and the amount of coal gasification slag is also increasing [12]. Coal gasification is a technology in the coal chemical industry that converts incomplete combustion of coal into CO and H_2 in an oxygen environment while producing tar at the same time [13]. Gasification slag is a solid waste formed by reacted inorganic matter and residual carbon particles in coal. According to statistics, the annual output of waste residue from the million-ton coal gasification process can reach more than 900,000 tons. Coal gasification slag as the main component of waste residue, it can account for as much as 95% [14]. According to statistics, the annual emission of coal gasification slag in China is more than 33 million tons [15].

2.3 Output of fly ash

China's current main method of producing electricity is thermal power generation, and the main fuel is still coal. 250~300Kg fly ash is produced for every ton of coal burned [16]. The energy policy of "power transmission from west to east" implemented in China has increased the production of fly ash in Northwest China, and a large amount of fly ash has been stranded in Northwest China due to the limitation of transportation cost. According to the 2019 China fly Ash Industry Development report, the production of fly ash in China in 2019 is about 655 million tons [17], and it is expected that subsequent production will continue to grow.

A large number of coal-based solid wastes in China have been treated by simple methods such as open-air stacking and landfill for a long time, which has seriously affected the environment. Hence, there is an urgent need to advance novel technologies to enhance the efficiency of coal-based solid waste utilization.

3 Properties of coal-based solid waste

3.1 Properties of coal gangue

The common coal gangue is grayish black or dark brown. Its shape will be obviously different according to the density of coal gangue and the way of treatment. The density of coal gangue is usually low, which depends on the content of carbon, the proportion of rock and mineral composition, and the water content. The size of particles in coal gangue ranges from powder to bulk. The carbon content of coal gangue is low, generally 20% to 30%. Coal gangue has multi-phase structure such as kaolinite, illite and Muscovite, and its chemical composition is complex and diverse. The main components are SiO_2, Al_2O_3 and C, accounting for more than 80% [18], as shown in Table 1. There are also some substances such as Fe_2O_3, CaO, MgO and trace heavy metal elements.

It is worth noting that due to the differences in the generation age and mining methods of coal mines in different mining areas, the composition and properties of coal gangue are also quite different. Du Yanzhe [19] analyzed the physical and chemical characteristics of coal gangue in specific regions of Shanxi Province, and found that the contents of Fe, Ca and K in coal gangue are quite different. The main mineral composition is kaolin and quartz, and some samples also contain hematite, calcite and dolomite. Huang Cheng [20] and other researchers found that after long-term storage or spontaneous combustion of coal gangue, the toxic and harmful elements in coal gangue are more easily leached, and the properties of coal gangue may change.

Table 1. main composition and content of coal gangue [18].

Composition	SiO_2	Al_2O_3	Fe_2O_3	CaO	MgO	K_2O	C	Na_2O
wt/%	30~60	15~40	2~10	1~4	1~3	1~2	20~30	1~2

3.2 Properties of coal gasification slags

Gasification slag is a byproduct generated during coal gasification, typically classified into coarse slag and fine slag fractions. The gasification slag formed after condensing along the furnace wall in the molten state becomes coarse slag, the general particle size is 1.19mm~4.75mm, and the residual carbon content is about 0 ~ 10% [21], which is mainly concentrated in the slag discharge outlet of the gasifier, accounting for 60% and 80% of the total gasification slag. The slag brought out by the gas becomes fine slag, the particle size is generally less than 0.075mm, and the residual carbon content is about 10%-30% [22], which is mainly concentrated in the gas dedusting device, accounting for 20%-40% of the total gasification slag. Because of its formation mode, coal gasification slag is mainly composed of minerals and unburned carbon. SiO_2, Al_2O_3, Fe_2O_3 and C are the main components of coal gasification slag.

The physical and chemical characteristics of coal gasification slag are influenced by various factors such as coal rank, gasification process, and operating temperature. Li Yu [23] found that the preliminary quality classification of coal gasification slag can be realized only by screening and classification, which can be divided into three types according to its carbon content. The characteristics of high, medium, and low carbon slag exhibit substantial variations. High carbon slag primarily consists of porous, irregular, flocculent particles characterized by a high specific surface area and well-developed pore structure. Medium carbon slag is mainly composed of dense and smooth spherical, irregular massive and flaky particles, as well as some irregular flocculent particles. Low carbon slag almost has no crystal phase and is mainly composed of dense and smooth irregular particles with small specific surface area. Ge Xiaodong [24] found that the surface of coal gasification fine slag is severely oxidized, has many voids, and contains a large amount of water. However, its surface is mainly composed of hydrophobic functional groups, so it has a certain degree of hydrophobicity, and carbon can be recovered through flotation.

3.3 Properties of fly ash

Fly ash is a solid waste produced in the process of coal combustion, and its color is brown to black affected by residual carbon particles. The properties of fly ash are mainly affected by many factors, including the coal type, combustion mode, combustion temperature, boiler type and so on. The clay minerals in coal are in the melting state at high temperature environment in the pulverized coal furnace. After condensation, spherical fly ash with smooth surface is formed due to surface tension [25]. The crystalline minerals are mainly quartz and mullite, and the amorphous minerals are mainly amorphous glass phases composed of SiO_2 and Al_2O_3. However, the combustion temperature of the circulating fluidized bed is low, so most of the fly ash produced is irregular flocs [26]. The crystalline minerals are mainly quartz and the amorphous minerals are mainly clay minerals such as metakaolin.

Fly ash is mainly composed of inorganic substances in coal, mainly SiO_2, Al_2O_3, Fe_2O_3, CaO and a small amount of MgO, K_2O, Na_2O and other metal oxides, as well as a small amount of unburned carbon particles[27], in which there is less unburned carbon in pulverized coal furnace and more unburned carbon in circulating fluidized bed fly ash. Due to the addition of calcareous desulfurizer in fluidized bed fly ash, the content of calcium oxide in fly ash is higher [28]. The fly ash of pulverized coal furnace is generally small and hollow because of the high temperature, and the average particle size is less than 10 μm [29]. Because the surface atoms of fly ash are in an unsaturated state, the surface has certain activity. Fly ash has a porous structure inside, so it has a large specific surface area and is adsorbable.

Analysis of the composition and characteristics of coal gangue, coal gasification slag, and fly ash reveals significant variability in the properties of coal-derived solid waste, influenced by factors such as coal rank, mining location, and processing techniques. This variability poses a major challenge in the comprehensive utilization of coal-based solid waste. To achieve truly efficient utilization, different methods must be adopted based on the specific composition and properties of the waste. Presently, China faces a major obstacle in the lack of classified utilization of coal-based solid waste, hindering progress towards comprehensive utilization in the future.

4. Utilization mode of coal-based solid waste

Coal-based solid waste is not only one of the bulk industrial solid waste, but also a valuable resource disposed of. For a long time, the utilization of coal-based solid waste in China mainly adopts the way of direct treatment, including open-pit accumulation, goaf backfilling and incineration treatment. The storage of coal-based solid waste has the advantages of low cost and simple treatment process. However, long-term storage requires a lot of land resources, and there is still a risk of pollution.

Using coal-based solid waste to backfill mined-out areas can effectively reduce waste volume, but poses the risk of soil and groundwater contamination. Direct incineration of coal-based solid waste diminishes the overall waste volume, yet releases hazardous gases like sulfur oxides, leading to environmental pollution. Thus, secondary processing and utilization following classification is recommended. This approach aligns with the green development principles of the modern era. According to statistics, the utilization rate in China of coal gangue, coal gasification slag and fly ash is 20%, 16% and 65% respectively [30]. Due to their unique properties and diverse valuable components, coal-based solid waste finds extensive applications in resource extraction, construction materials, agriculture, environmental protection, and high-value-added sectors. Notably, the cement industry stands as a prominent sector for comprehensive utilization of coal-based solid waste in China, with coal gangue and fly ash commonly utilized in the production of building materials like bricks, mortar, and concrete blocks. Coal-based solid waste is also widely used as roadbed stabilizer and pavement material in highway, railway and other road engineering. Coal-based solid waste accounts for a relatively small but also of great significance in soil improvement. Therefore, three utilization methods of coal-based solid waste including as cement and concrete, roadbed material, and soil improver are discussed in the present paper.

4.1 Used as cement and concrete

Coal-based solid waste can be used to make cement, since its the main component is SiO_2 and Al_2O_3, which is similar to clay. Sun Daosheng [31] studied the thermal technology and reasonable formula for preparing cement clinker by using coal gangue as silica-alumina raw material, mixed with limestone, iron tailings, and quartz sand. Controlling lime saturation coefficient in 0.89 ~ 0.95, aluminum ratio in 1.2 ~ 1.6, silicon ratio in 2.1 ~ 2.6, the prepared raw meal calcined at 1400 °C, the safety of the cement clinker is qualified. When the clinker is calcined at 1450 °C according to the proportion of lime saturation coefficient 0.95, silicon ratio 2.55 and aluminum ratio 1.55, the 3d static compressive strength of the cement produced is 45.3MPa, and the 28d slurry compressive strength is 95.3 MPa.

Su Wenjun [32] found that adding phosphorus slag and fluorite compound mineralizer to coal gangue raw meal can improve the burnability of raw meal. And phosphorus slag contains CaO, which can save limestone resources. The composite mineralizer can accelerate the solid state reaction when the calcination temperature is in the range of 1300 °C ~ 1400 °C.

Bai Guoliang [33] studied the physical and chemical properties of coal gangue, such as calorific value and carbon content, and the mechanical properties of coal gangue concrete. The study found that the surface of coal gangue coarse aggregate is rough and there are many internal pores. Therefore, after the gangue concrete is stressed, the gangue coarse aggregate cannot effectively control the crack extension. In the compression test, the gangue concrete cubes were mostly broken by the gangue coarse aggregate, and the interface bond between gangue and cement exhibited significant strength, indicating a robust connection. Carbon content emerged as the primary factor influencing the mechanical characteristics of coal gangue concrete. Regression analysis was conducted to formulate a predictive equation for the mechanical strength of coal gangue concrete based on the carbon content of the gangue.

Liu Chaoqun [34] studied the effect of activated coal gangue mixed with cement on cement. The study revealed that the fluidity of cement blended with activated coal gangue surpasses that of standard concrete, with an optimal water-to-binder ratio of 0.36. The enhanced mechanical properties of concrete when incorporating activated coal gangue can be attributed to its robust secondary hydration reaction activity, leading to the effective filling of cement pores with secondary hydration reaction products.

Although the main components of coal gasification slag are SiO_2 and Al_2O_3, according to the national standard of "Fly ash used in cement and concrete" (GB/T 1596Mel 2005), the loss on ignition of fly ash that can be used in cement and concrete should not be more than 15%. However, the loss on ignition of gasification slag is about 20%, so it can't be directly used to prepare cement and concrete. In view of the precedent of fly ash being used as cement raw material ingredients, Dang Liwen [35] mixed a small amount of coal gasification slag into cement raw material and calcined it. The compressive and flexural strength of the obtained gasification slag cement is higher than that of cement without adding gasification slag.

Hang Meiyan [36] et al. studied the effect of activators on coal gasification slag powder. The loss on ignition of coal gasification slag was controlled below 6% by grinding and screening. The coal gasification slag powder was treated with an activator and then added with cement. Scanning analysis was carried out through SEM and SED. It was found that the compressive strength of the coal gasification slag cement after activation was greatly improved.

Fly ash is mainly composed of SiO_2 and Al_2O_3, and its particle size is fine and its surface is rough, so it is suitable to be used as cement and concrete. Liu Quan [37] and others studied the physical and chemical properties of fly ash in China, and found that with SiO_2+Al_2O_3+Fe_2O_3 content and loss on ignition as constraint conditions, fly ash in accordance with mixing mortar and concrete accounted for 53.2%, and fly ash in accordance with cement active mixture accounted for 73.1%. This shows that most of the fly ash in China is suitable for making cement and concrete.

In a study by Bai Jiajia [38], the impact of fly ash content on carbonated cement paste was investigated, revealing that the addition of fly ash can accelerate the carbonation process of cement paste, enhance its pore structure, and promote crack healing. Optimal reduction in crack width occurred at a 30% fly ash content.

Chen Hongfei [39] explored the effects of fly ash substitution for Portland cement clinker on cement's physical properties, noting that fly ash inclusion prolongs setting times and significantly alters the mechanical properties of cement mortar. While compressive strength decreased notably, flexural strength exhibited an initial rise before declining, with peak flexural strength achieved at a 10% fly ash content.

Peng Yufa [40] studied the influence of fly ash on concrete performance, it was established that fly ash content significantly influences concrete's mechanical properties, with compressive strength initially increasing and then decreasing with rising fly ash content. The maximum compressive strength was achieved at a 30% fly ash content. Overall, variations in the source, coal quality, and processing methods of coal-based solid waste can impact the ultimate performance of cement and concrete. Notably, the calcination process of coal-based solid waste may lead to the conversion of elements like S and N into pollutants such as SO2 and NOx, posing environmental concerns. Therefore, it is not only necessary to further ensure the stability of the indicators of products in different regions and explore the mechanical properties of coal-based solid waste cement and concrete, but also to explore the performance differences caused by differences in product raw materials, as well as potential risks such as environmental pollution risks during utilization.

4.2 Used as subgrade material

Coal gangue physically has a certain compressive strength and stability, can withstand the role of vehicles and traffic load, and coal gasification slag can also be used to build road base or subbase. Coal gangue as a by-product, does not need additional mining costs, so coal gangue as a roadbed material can save construction costs. Although coal gangue may be slightly fragile compared with traditional roadbed materials such as crushed stone, it is easy to break up in the process of practical application, resulting in changes in gradation. Therefore, it is very important to use graded coal gangue as roadbed material.

Zhang Zongtang [41] studied the influence of grading on the compaction and characteristics of coal gangue roadbed filling. It has been found that the compaction characteristics of coal gangue roadbed filling have a significant impact on the grading. When the value limit of good gradation is obtained through the model equation as $1.8867 \leq D \leq 2.6309$, the gradation is uniform. As the fine particles increase, the compacted dry density of the coal gangue roadbed first increases and then decreases. The optimal grading range is $2.04 \leq D \leq 2.55$.

Zhang Rui [42] studied the road performance of coal gangue. The physical and chemical properties of coal gangue was analyzed through tests on its chemical composition, loss on ignition, liquid plastic limit and plasticity index, water immersion expansion, resistance to disintegration, and crushing value. The road performance of coal gangue was studied by combining screening tests, compaction tests, and California bearing ratio tests.

Hao Liping [43] studied the settlement of coal gangue roadbed in practical engineering applications using practical engineering as an example. After being subjected to static pressure for 1 time, vibration pressure for 6 times, and static pressure for 1 time, the cumulative settlement of the roadbed tends to stabilize. The final compaction settlement difference is 0.6mm<2mm, which meets the construction technical requirements. And the compaction amount of each experimental point is above 95%. This provides a scientific compaction method for the practical engineering application of coal gangue as roadbed.

Guo Kaichuan [44] studied the performance of coal gangue road and the construction technology of coal gangue roadbed filling using highway engineering as an example. It is found that the natural grading of coal gangue cannot meet the construction requirements. But liquid plastic limit, compaction characteristic test, bearing capacity test, etc. all indicate that coal gangue meets the requirements of road performance. Therefore, it is necessary to discard or crush the large particles in coal gangue to improve its grading. And in order to prevent rainwater from seeping into the coal gangue roadbed and affecting its performance, it is necessary to add a construction process of edge soil to the roadbed to ensure construction quality.

From the point of view of composition and microstructure, the main components of coal gasification slag and fly ash are SiO_2 and Al_2O_3, and the surface is rich in pore structure and good permeability, so it can replace the fine aggregate of concrete for road engineering. Rational use of gasification slag and fly ash can effectively improve road durability and freeze-thaw resistance.

Sheng Yanping [45] studied the properties of gasification slag cement stabilized macadam base material, and found that gasification slag can replace part of the mineral composition in cement to be used in roadbed material. When gas slag cement is used as roadbed material, the compressive strength and splitting strength are slightly lower than that of PC32.5 cement base material, but it still meets the requirements of roadbed material, and the crack resistance of gas slag cement is better than that of PC32.5 cement base material, which can improve the crack resistance of roadbed.

Gao Peng [46] and others have studied the use of gasification slag as road base material. Through an examination of the impact of cement dosage, aggregate gradation, and other variables on the mechanical properties of coal gasification slag pavement base material, it was observed that the compressive strength of cement-stabilized gas slag material is marginally lower than that of cement-stabilized macadam at equivalent dosages. Nevertheless, by employing a rational gradation approach, the mechanical properties of cement-stabilized gasification slag can be significantly enhanced. Zhang Bo [47] conducted research on the use of fly ash in highway engineering. It was found that fly ash has the characteristics of good permeability, high structural strength, and light self-weight. Therefore, it is very suitable for road engineering. Gao Funing [48] studied the road performance of CFB fly ash. It was found that CFB fly ash has excellent road performance, high shear strength, and is not easy to settle. The compaction method has a relatively small impact on the compaction degree of the roadbed. When compacted 4 times and subjected to strong vibration 2 times, the strength requirements can be met. The optimal moisture content of CFB fly ash is between 37.1% and 39.8%. During the construction process, water should be replenished in a timely manner.

Wang Lei [49] studied fly ash cement stabilized macadam and found that fly ash cement stabilized macadam has better crack resistance than cement stabilized macadam, and the crack resistance is the best when the content of fly ash reaches 6%.

Generally speaking, coal-based solid waste is widely used in engineering, but there is still little basic theoretical research. The interaction mechanism between coal-based solid waste and other roadbed materials should be studied in depth, and the long-term road performance of roadbed materials should be studied in depth. At the same time, trace harmful elements in coal-based solid waste should be analyzed to avoid affecting the surrounding environment.

4.3 Used as a soil improver

In some areas of China, the content of organic matter in

soil is less, and the porosity is small, which is not conducive to plant growth. Coal-based solid waste is rich in various trace elements needed for plant growth, and the content of organic matter is between 15% and 20%. Therefore, coal-based solid waste can be used as soil improver, improve soil porosity, increase soil trace elements, and make soil suitable for plant growth.

Zhang Yuhang [50] investigated the enhancement impact of coal gangue on the physical and chemical characteristics of saline-alkali soil. Coal gangue was incorporated into the saline-alkali soil based on varying dosages and particle sizes, with improvement assessed through alfalfa pot culture for analysis of the soil's physical and chemical properties. When the amount of coal gangue was 20%, the effect of small particle size and mixed particle size on saline-alkali soil quality and plant growth was the best.

Fan Qiuyun [51] and others studied the effect of adding fly ash to coal gangue matrix on plant growth. After mixing coal gangue and soil at 1:1, different amounts of fly ash were added respectively. It is found that adding fly ash to coal gangue matrix can greatly improve the properties of coal gangue matrix and promote plant growth.

Li Yilin [52] studied the effect of coal-based solid waste combined with AM fungi on improving degraded soil in mining area. Weathered coal, coal slime and coal gangue were quantitatively mixed with cow dung and AM fungi respectively. It was found that coal-based solid waste combined with AM fungi could promote plant growth, improve soil nutrients and improve soil microbial community structure.

The utilization of coal gasification slag as a soil conditioning agent following appropriate treatment is a viable approach for incorporating coal gasification slag in ecological agriculture. Zhu Dandan[53] investigated the utilization of coal gasification slag for soil improvement and observed that gas slag exhibits favorable outcomes in enhancing desertified soil. By adding coal gasification slag, the soil water retention rate increased significantly, pH decreased, carbon content increased, and available silicon content increased significantly. And gas slag can be used as humic acid slow-release agent for organic fertilizer.

Wang Shaoli [54] studied the effect of coal gasification slag conditioner on the growth and quality of ryegrass. It was found that gas slag conditioner could effectively increase the yield of ryegrass and the content of crude protein in ryegrass. When the application rate was 2.25kg/m2, the yield and quality of ryegrass were the best. Moreover, gasification slag conditioner can improve soil nutrients, nitrogen and potassium content increased.

Ai Guo [55] studied the effect of coal gasification slag combined with bacterial bran on soil improvement. With the increase of coal gasification slag dosage, the root activity of alfalfa seedlings increased at first and then decreased. When the gasification slag is 30% and bacterial bran is 15%, the effect of combined application is the best.

Fly ash has rough surface, high porosity and large specific surface area, which can effectively improve soil porosity and water retention. Fly ash contains a variety of trace elements, such as Ca, Mg, K, etc., which can effectively improve soil fertility. Fly ash is weakly alkaline which can neutralize acidic soil, making it more suitable for plant growth.

Wu Lin [56] studied the nutrient index and risk of fly ash as soil improver, and found that fly ash is rich in available phosphorus, available potassium and organic matter, and most of the heavy metals such as Co, Ni, Cr, Zn, Pb and Cu are stable residue state, which exhibits limited migratory and transformative properties, resulting in a low risk of heavy metal pollution.

Zheng Peng[57] investigated the impact of fly ash and soft rock on enhancing soil water characteristics in mining areas, revealing that the incorporation of fly ash elevated the fine soil particle content. Optimal soil water retention was achieved at a fly ash to soft rock ratio of 3:1.

Ou Yanjun[58] explored the mechanisms underlying yield enhancement through fly ash-based soil conditioners, determining their ability to immobilize heavy metal contaminants, enhance soil water retention, and boost corn yield by up to 52.3%.Generally speaking, the current studies are mainly focused on the properties of soil, but there are few studies on the plant properties of reclaimed soil. Although the leaching amount of heavy metal elements is lower than the national standard in the short term, the increase of long-term leaching amount will inevitably affect plant growth. Long-term monitoring of reclaimed soil should be carried out, and the physiological properties of growing plants in reclaimed soil should be further studied.

5. Conclusions

To sum up, coal-based solid waste is a huge bulk industrial solid waste in China. If resources cannot be properly utilized and disposed of randomly, it will not only have a huge impact on the environment, but also waste the utilization value of coal-based solid waste. At present, coal-based solid waste is used as cement and concrete, preparation of roadbed materials, soil improvement and remediation, etc., which has the advantages of abundant raw materials and simple treatment, as well as typical successful cases and economic and environmental protection value. Although the products need to be regularly tested to eliminate related risks, it still provides a green application model for the development of coal-based solid waste in China. Nevertheless, there remains a significant disparity compared to developed nations. Further advancements in new technologies are essential to enhance the overall utilization rate of coal-based solid waste.

In addition to the aforementioned utilization approaches, numerous scholars have conducted extensive research on extracting useful components from coal-based solid waste and exploring high-value utilization strategies. Despite these efforts, a large-scale and mature application technology for coal-based solid waste resources has yet to be established. Therefore, the comprehensive utilization of coal-based solid waste requires further research and development of high-value utilization methods on the basis of optimizing existing methods, reducing resource waste and accelerating green development.

Reference

1. Hua Yang, Dong fang Li, Shun ge Shen. Analysis on the Green Development of Coal Chemical Industry in China[J]. Inner Mongolia Science Technology & Economy, 2018(22): 10-12.
2. Cheng Gu, Yu Li. Study on progress in comprehensive utilization of coal based solid waste[J], Coal and Chemical Industry, 2020,43(09): 98-101.
3. Xiao yun Wang, Yan xia Niu. Review of research on coal gangue with its classification, hazards and comprehensive utilization[J], Industrial Minerals & Processing, 2023: 1-9.
4. Jablonska, B., et al., The structural and surface properties of natural and modified coal gangue. JOURNAL OF ENVIRONMENTAL MANAGEMENT, 2017. 190: p. 80-90..
5. Peng hui Li, Hui Ren. Study on Fine Utilization of Coal Gangue[J], Guangzhou Chemical Industry, 2021,49(10): 27-29.
6. Ying Li, Hui Li, Feng Wu, et al. Preparation of Non Autoclaved Aerated Concrete with High Content of Coal Gasification Slag[J], Clean Coal Technology, 2023: 1-10.
7. Wen yu Fu, Wen qiang Sun, Lian yong Wang. Advances in Resource Utilization Technologies for Coal Gasification Slag[J], Environmental Engineering, 2023: 1-11.
8. Peng Yuan. Present situation and Development trend of Comprehensive Utilization of fly Ash in China[J], Fujian Building Materials, 2022(07): 116-118.
9. Li ping Wang, Chao Li. Research Progress on Resource Utilization of Aluminum and Silicon in Fly Ash[J], Multipurpose Utilization of Mineral Resources, 2023: 1-7.
10. Huan Li, Chun ming Yang. Research Progress in Comprehensive Utilization of Bulk Industrial Solid Waste Coal Gangue, Journal of Natural Science of Hunan Normal University, 2023: 1-11.
11. Ji wen Chang, Gen jie Du, Jian lei Du, et al. Current Situation of the Comprehensive Utilization of Coal Gangue in China and the Related Problems and Recommendations[J], China Environmental Protection Industry 2022(08): 13-17.
12. Fang Li. Analysis and treatment on waste in coal chemical production process[J], Coal and Chemical Industry, 2018,41(05): 155-157.
13. Jiang shan Qu, Jian bo Zhang, Zhi gang Sun, et al. Research progress on comprehensive utilization of coal gasification slag[J], Clean Coal Technology 2020,26(01): 184-193.
14. Wei dong Ai. The study on preparation and properties of coal gasification slag/polymer composites[D], Jilin University, 2020.
15. Ju fen Zhu, Jian Li, Long Yan, et al. Research progress and application prospect of coal gasification slag resource utilization[J], Clean Coal Technology, 2021,27(06): 11-21.
16. Long Jiang. Comprehensive utilization situation of fly ash in coal-fired power plants and its development suggestions[J], Clean Coal Technology, 2020,26(04): 31-39.
17. Comprehensive Utilization of Fly Ash is Developing in an All-round Way in China industry Development Report of Fly Ash Materials Branch in 2020[J], China Concrete, 2021(10): 28-29.
18. Qi hui Li. Research progress on properties and comprehensive utilization of coal gangue[J], Applied Chemical Industry, 2023,52(05): 1576-1581.
19. Yun zhe Du, Yi ling Liu, Wei Shen, et al. The basic physical and chemical properties of coal gangue in some typical areas of Shanxi Province[J], Coal Processing & Comprehensive Utilization, 2023(04): 80-83.
20. Cheng Huang, Jie Zhang, Min Li, et al. Evaluation of Properties and Comprehensive Utilization Suitability of Coal Gangue in Main Coal-producing Areas[J], Energy and Energy Conservation, 2023(10): 10-13.
21. Xiang gang Feng, Fen fei Ge, Yin min Zhang, et al. Research progress and application prospect of high-value utilization of coal gasification slag[J], Clean Coal Technology, 2023: 1-11.
22. Jinyi Zhang. Preparation of coal gasification slag-based composite conductive filler and its application in antistatic polymer materials[D], Jilin University, 2022.
23. Yu Li, Li hua Hou, Yun fei Wang, et al. Study on Classification and Properties of Coal Gasification Coarse slag[J], Multipurpose Utilization of Mineral Resources: 1-9.
24. Xiao dong Ge. Surface properties analysis of coal gasification coal cinder and flotation extraction research[J], China Coal, 2019,45(01): 107-112.
25. Zhi juan Gao. Comparison of physical and Chemical Properties between pulverized Coal fired Boiler fly Ash and circulating Fluidized bed fly Ash[J], Environmental Protection and Circular Economy, 2018,38(09): 68-73.
26. Cui bin Li, Da gen Su. Formation and Property of Fly Ash Produced from Recycle Fluid Bed[J], Cement Technology, 2010(03): 29-30.
27. Kumar, K., et al., Characteristics Of Fly Ash In Relation Of Soil Amendment. MATERIALS TODAY-PROCEEDINGS, 2017. 4(2): p. 527-532.
28. Jie Wang. Study on Physicochemical Properties of Modified Ash from Different Sources[D], Taiyuan University of Technology, 2020.
29. Jing Xu, Ting Li, Xiao ling Liang. Review on the Research Progress of Utilization Technology of fly Ash[J], Journal of Chifeng University(Natural Science Edition), 2019,35(06): 108-110.
30. Qiao yan Tian, Fu ren Kang, Kai yu Zhang, et al. Research Progress on Ecological Utilization of Coal-based Solid Waste[J], Journal of Yulin University, 2021,31(06): 57-62.
31. Dao sheng Sun, Wen jun Su, Ai guo Wang, et al.

Experimental Research on Utilization of Coal Gangue as Silicon-Aluminum Raw Material in the Preparation of Cement Clinker[J], Materials Reports, 2016,30(16): 130-134.

32. Wenjun Su, Experimental research on utilization of coal gangue as silicon- aluminum material in the preparation of cement clinker[D], Anhui Jianzhu University, 2016.
33. Guo liang Bai, Han qing Liu, Hui Liu, et al. Study on physicochemical properties of coal gangue and mechanical properties of coal gangue concrete[J], Journal of Building Structures, 2023,44(10): 243-254.
34. Chao qun Liu, Ze wen Zhu, You hua Zhang, et al. Hydration Mechanism and Properties of Activated Coal Gangue Cement[J], Bulletin of the Chinese Ceramic Society, 2023,42(10): 3660-3670.
35. Li wen Dang, Wen bin Jia. Application of Gasification Slag in cement Raw meal proportioning[J], Cement, 2017(07): 26-27.
36. Mei yan Hang, Xue tao Lu, Yan mei Guo, et al. Experimental Study on Activation Effect of Micropowder of Coal Gasification slag[J], Bulletin of the Chinese Ceramic Society, 2019,38(03): 878-883.
37. Quan Liu, Zhi min Bai, Dong Wang, et al. Chemical Composition and Physicochemical Properties of Fly Ash and Its Application[J], China Non-metallic Minerals Industry, 2021(01): 1-9.
38. Jia jia Bai, Nan Lin, Jian min Wang, et al. Effect of fly ash dosage on microstructure of carbonized cement paste[J], Journal of Water Resources and Water Engineering, 2022,33(05): 154-159.
39. Hong fei Chen. Effect of Fly Ash Replacing Silicate Clinker on Physical Properties of Cement[J], Jiangxi Building Materials, 2023(02): 37-39.
40. Yu fa Peng, Yun hua Zhu, Xun Li, et al. Experimental Study on Effect of Fly Ash on Concrete Properties[J], Chongqing Architecture, 2022,21(09): 70-72.
41. Zong tang Zhang, Wen hua Gao, Chang ping Liu, et al. Experimental Study of Gradation Effect on Compaction and Strength Characteristics of Coal Gangue Subgrade Filler[J], Journal of Engineering Geology, 2023,31(05): 1774-1781.
42. Rui Zhang. Study on Bearing Characteristics of Coal Gangue Modification and Filling Roadbed[D], Shandong Agricultural University, 2023.
43. Li ping Hao. Application of Coal gangue material in filling Construction of Expressway subgrade[J], Communications Science and Technology Heilongjiang, 2019,42(12): 15-16.
44. Kai chuan Guo. Road performance of Coal gangue and its Application in subgrade filling Construction[J], TranspoWorld, 2023(09): 111-113.
45. Yan ping Sheng, Xin Ji, Gang Xu, et al. Study on the performance of coal gasification slag cement stabilized macadam base[J], Applied Chemical Industry, 2020,49(06): 1407-1412.
46. Peng Gao, Qing hong Li, Jian ping Tian, et al. Research and Application of Road Base Material Prepared by Coal Gasification Slag[J], Journal of Wuhan University of Technology(Transportation Science & Engineering), 2021,45(01): 155-160.
47. Bo Zhang. Application of fly Ash in Expressway Construction[J], Transport Construction & Management, 2015(Z2): 97-99.
48. Fu ning Gao, Xin Zhou, Zhi xian Zhao, et al. Research on Design and Construction Technology of CFB Fly Ash as Subgrade Materia[J], Highway, 2021,66(09): 145-149.
49. Lei Wang. Study on Engineering Application of cement fly Ash stabilized Macadam Base[J], TranspoWorld, 2023(20): 111-113.
50. Yu hang Zhang, Zi ling Song, Tao Kong, et al. Amelioration Effect of Coal Gangue on Physical and Chemical Properties of Saline-alkaline Soil[J], Ecology and Environmental Sciences, 2021,30(01): 195-204.
51. Qiu yun Fan, Chao ying Zhang, Yu qing Geng, et al. Effects of fly ash application on the properties of coal gangue matrix and plant growth[J], Science of Soil and Water Conservation, 2022,20(05): 85-92.
52. Yilin Li. The Reclamation of coal-based solid Waste combine AM Fungus on degraded soils in mining areas[D], Inner Mongolia University of Science & Technology, 2022.
53. Dandan Zhu. Study on the utilization of coal gasification fine slag in soilimprovement and water pollution control[D], Jilin University, 2021.
54. Shaoli Wang. Effects of coal gasification slag conditioner on growth, quality, and soil nutrients of ryegrass[D], Yulin University, 2023.
55. Guo Ai, Sheng juan Hui, Zhong xun Deng, et al. Effects of Coal Gasification Slag Combined with Pleurotus ostreatus Medium Residue on Root Growth of Alfalfa Seedlings in Soil of Coal Mine Areas[J], Hunan Agricultural Sciences, 2022(11): 40-44.
56. Lin Wu, Yong hong Zheng, Zhi guo Zhang, et al. Assessment of Nutrient and Pollution Risk of Fly Ash as a Soil Amendment[J], Environmental Science & Technology, 2020,43(09): 219-227.
57. Peng Zheng, Ting hui Dang, Jiang Xue. Experimental Study on the Improvement of Soil Moisture Characteristics of Coal Mine Dump by Fly Ash and Arsenic Sandstone[J], Acta Pedologica Sinica, 2023,60(02): 399-408.
58. Yanjun Ou. Mechanism on increasing crop yield and improving its quality of fly ash-based new multifunctional soil conditioner[D], University of Chinese Academy of Sciences (Institute of process Engineering, Chinese Academy of Sciences), 2021.

https://doi.org/10.1051/e3sconf/202452801013

Preparation of phosphorylated conjugated microporous organic polymers and extraction of low-concentration rare earth elements

Zhiqian Liu[1,2], Yace Mi[1,a], Qing Liu[2], Meng Rong[2,b], Liangrong Yang[2,c]

[1] College of Chemical Engineering, Inner Mongolia University of Technology, Hohhot, 010051, China
[2] CAS Key Laboratory of Green Process and Engineering, Institute of Process Engineering, Chinese Academy of Sciences, Beijing 100190, China

Abstract. Aiming at the extraction and recovery of rare earth elements (REEs) in low-concentration rare earth leaching tailings, a kind of phosphorylated conjugated microporous organic polymer (P-CMP) was prepared and applied to adsorb rare earth ions (Nd^{3+}, Dy^{3+}, Y^{3+}). The structure of P-CMP was characterized by FT-IR. SEM and EDS were utilized to illustrate the surface morphology and element distribution of P-CMP, respectively.The adsorption mechanism of P-CMP was determined by XPS. The effects of pH, ion concentration and adsorption time on adsorption capacity were investigated. The isotherm of the rare earth adsorption process was fitted by Langmuir and Freundlich models, and the kinetics was fitted by Pseudo-first-order and Pseudo-second-order. The results show that P-CMP is an amorphous microporous material, which has strong interaction with Nd^{3+} (30.18 mg/g), Dy^{3+} (37.84 mg/g) and Y^{3+} (22.01 mg/g), and the synergistic effect of amino and phosphate groups provides good selectivity (S.F.(Nd^{3+}/Me^{n+}) = $5*10^3$~$5*10^4$). The P-CMP selectively adsorbs rare earth ions in the actual system and cycles 8 times.

1. Introduction

Rare earth has always been known as an industrial vitamin, because of its special magnetic, optical, electrical properties, which is widely used in aerospace, information manufacturing, green energy, biochemistry and other fields [1]. The need of rare earth elements has elevated due to the industry developing rapidly, and the waste of rare earth leaching tailings is serious [2]. The content of rare earth in leaching tailings is low along with high-concentration impurity ions (Al^{3+}, Ca^{2+}, Mg^{2+}, Mn^{2+}, K^+, Na^+), thereby difficult to make secondary utilization [3].

At present, the separation methods of rare earth ions mainly include precipitation, extraction, membrane separation and adsorption, among which the adsorption stands out of economical utilization and stable cycle performance [4], and is more suitable for rare earth tail liquid with low concentration and high interference ions. The types of adsorbents encompass porous organic polymers, inorganic materials, and cost-effective biomaterials [5]. There are many kinds of porous organic polymers, with high adjustable degree, stable adsorption performance and low cost, but the traditional porous polymers have disadvantages of low adsorption capacity, slow kinetics and poor selectivity [6]. Conjugated microporous polymers prepared by coupling reaction have low monomer requirements and high crosslinking degree, providing a method for introducing functional groups [7].

Most adsorbents introduce a high density of single functional groups. Although the adsorption capacity of the adsorbents is pretty high, it is unsatisfying for their indistinguishable adsorption and sluggish kinetics. In this work, the synergistic action of phosphoric acid and amino group was introduced to adsorb rare earth ions, leading to enhancing the kinetics and selectivity to interfering ions [8-9].

2. Preparation and characterization

2.1 Preparation of materials

Firstly, by Knoevenagel condensation reaction, the conjugated microporous polymer with double bond was formed by pyroquinol and p-phenyldiformaldehyde under the condition of catalyzing by dimethylamine. Secondly, the cyano-rich polymer was subsequently subjected to modification in phosphorous acid at a temperature of 80 °C. The initial cyanide group was replaced by a secondary amino group and two phosphate groups, following by washing and drying, and then a powder adsorbent with dark red was formed for use. As shown in Fig 1.

Corresponding author's e-mail: miyacetx@163.com[a], mrong@ipe.ac.cn[b], lryang@ipe.ac.cn[c]

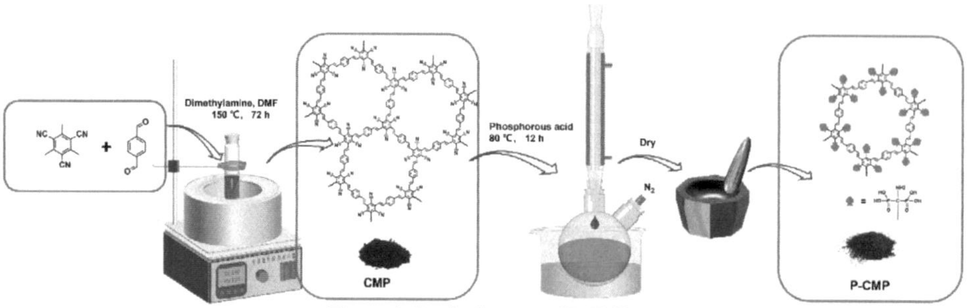

Figure 1. Synthesis route of P-CMP.

2.2 Characterization

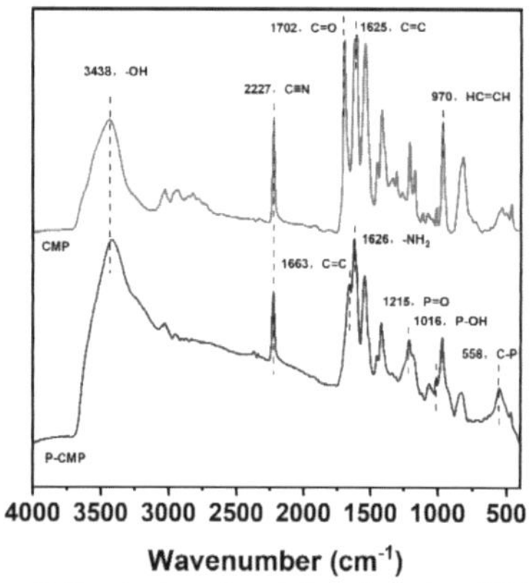

Figure 2. Fourier transform infrared spectra of P-CMP

The materials before and after modification were characterized by Fourier infrared spectroscopy (Fig 2). Because of the transformation from C≡N to $-NH_2$, the vibration peak of C≡N corresponding to 2227 cm^{-1} had a significant decrease, and the characteristic peak of secondary amino group appeared at 1626 cm^{-1}. A new tensile vibration peak observed at 558 cm^{-1} was speculated to C-P bind, signifying the successfully grafting phosphoric acid groups onto the conjugated microporous skeleton surface. The stretching vibration peaks of P=O and P-OH at 1215 cm^{-1} and 1016 cm^{-1} were also observed from FTIR spectra.

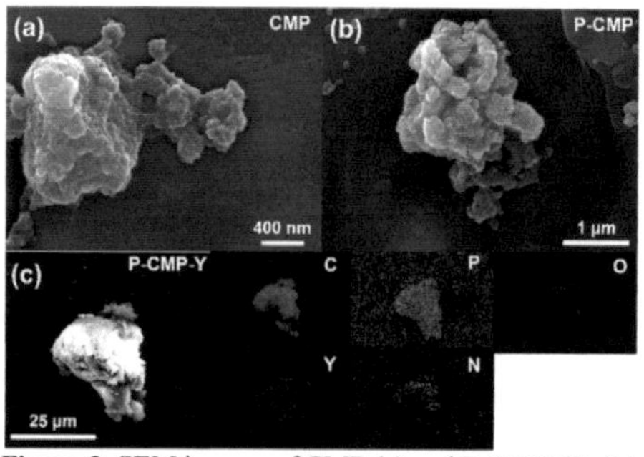

Figure 3. SEM images of CMP (a) and P-CMP (b); (c) Elemental mappings of P-CMP after adsorption loading Y^{3+}

The surface morphology was characterized by scanning electron microscopy along with energy dispersion spectrometer for element distribution of the adsorbent. As shown in Fig 3, the surface of the adsorbent was displayed as compact agglomeration of irregular spheroids with diameters changing between 40 to 86 nm, The morphology of raw materials did not change after phosphorylation. which indicated that the material belonged to amorphous polymer. According to Fig 3c, the adsorbent was composed of C, O, P and N elements, and loaded with a large amount of Y^{3+} proving that P-CMP was successfully prepared and had good an adsorption effect on rare earth ions.

3. Adsorption experiment

3.1 pH effects

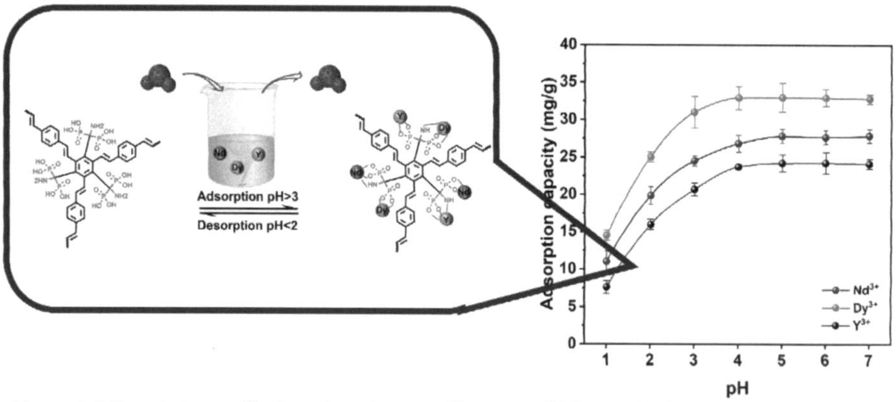

Figure 4. Effect of pH on Nd^{3+} adsorption of P-CMP ($C_{0(REE^{3+})}$=200 ppm, T=298 K, S/L=1 g/L, 24 h)

From Fig. 4, it can be concluded that the application of P-CMP had a wide pH range: the adsorption of rare earth ions was stable in the range of 23-35 mg/g at pH>2. All adsorption sites in this range can electrostatically interact with rare earth ions. When the pH continued to decrease (pH<2), H^+ in the solution will compete with rare earth ions for adsorption sites, leading to protonation of the adsorption site, which carried a positive charge to repel rare earth ions. Such investigation also proved the probability of rare earth ions desorption after adsorbing.

3.2 Isotherms and dynamics

Table 1. Fitting data of the adsorption isotherm of P-CMP.

REEs	Langmuir			Freundlich		
	Q_m(mg/g)	K_L(L/mg)	R^2	n	K_F (mg^{1-n}L^n/g)	R^2
Nd	30.18	8.163	0.991	19.89	22.04	0.909
Dy	37.84	6.127	0.994	14.43	23.57	0.990
Y	22.01	4.041	0.995	14.74	13.84	0.701

Table 2. Kinetic fitting parameters.

REEs	Pseudo-first-order			Pseudo-second-order		
	Q_e(mg/g)	K_1(min^{-1})	R^2	Q_e(mg/g)	K_2(g/mg^{-1}·min^{-1})	R^2
Nd	29.79	0.445	0.996	30.76	0.0322	0.998
Dy	38.11	0.523	0.995	39.14	0.0334	0.999
Y	21.27	0.247	0.988	63.17	0.0027	0.991

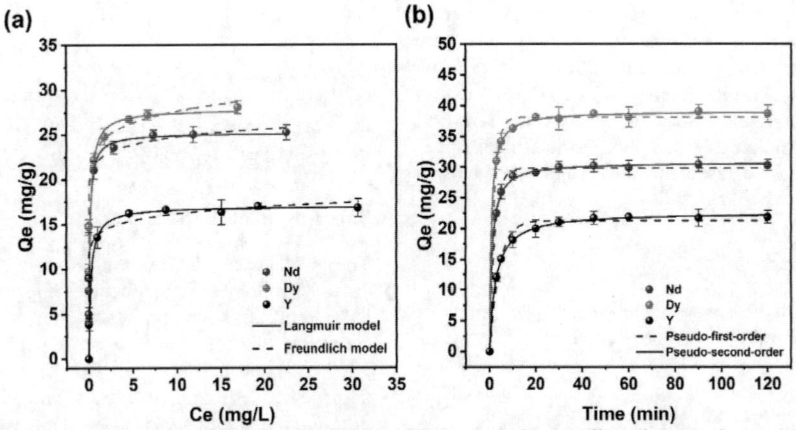

Figure 5. (a) Adsorption isotherms of P-CMP (pH = 4, T=298 K, S/L=1 g/L, 24 h); (b) Effect of contact time on REE^{3+} adsorption capacity of PO-PPR-1 ($\mathbf{C_{0\,(REE^{3+})}}$=200 ppm, T=298 K, pH=4, S/L=1 g/L).

3.2.1 Isotherm model

P-CMP was used to perform isotherm tests for Nd^{3+}, Dy^{3+} and Y^{3+} respectively, and the rare earth solution concentration was 5-50 ppm. The Langmuir model and Freundlich model were explored to fit the data and the results were shown in Fig. 5a. The adsorption capacity augmented with the increase of the initial concentration and finally stabilized to 20-30 mg/g. The adsorption capacity of P-CMP for three rare earth ions followed the order of $Nd^{3+} < Dy^{3+} < Y^{3+}$ (mole), which is consistent with the general rule that heavy rare earth elements have a higher affinity for phosphate ester adsorbents. Details of the fitting results were shown in Table 1. It can be clearly seen that the adsorption process is more consistent with the langmuir adsorption model, which signifies that the adsorption process is a single layer adsorption, and the adsorption sites are uniformly distributed on the surface of the adsorbent without interference.

3.2.2 Dynamic model

The pseudo-first-order kinetic equation and pseudo-second-order kinetic equation were carried out kinetic fitting of the P-CMP adsorption rate of rare earth ions, and the fitting results were shown in Figure 5b. The adsorption capacity increased with the increase of adsorption time, and did not change at 20min, meaning that P-CMP had a fast adsorption rate for rare earth ions, and reached adsorption equilibrium within 20 min. Referring to Table 2 for details of the fit: the value of Qe obtained from the pseudo-second-order dynamics coms near Qm, and R^2 is higher than the other's model, indicating that the adsorption process belongs to chemisorption.

https://doi.org/10.1051/e3sconf/202452801013

3.3 Selectivity

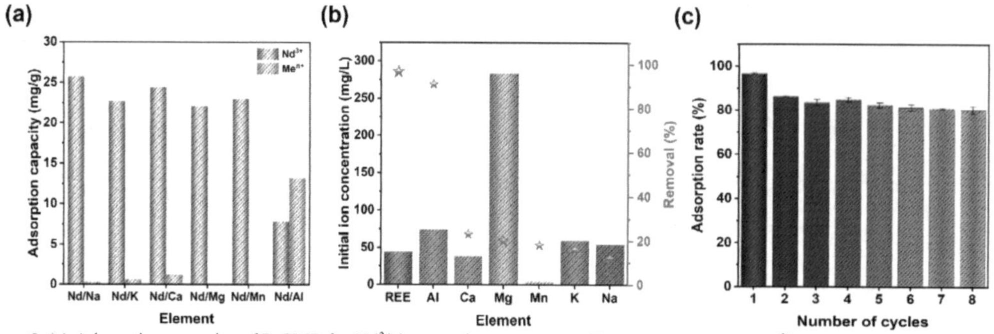

Figure 6. (a) Adsorption capacity of P-CMP for Nd^{3+} in two-element system ($C_{0\,(REE^{3+})}$=50 ppm, Nd^{3+}: Me^{n+}=1:5 (mole), S/L=2 g/L, T=298 K, 24 h, pH=4); (b) Selective adsorption of P-CMP in real tail liquid (S/L=10g/L, T=298K, 24 h, pH=4); (c) Recyclability of P-CMP in real tail liquid (S/L=10g/L, T=298K, 24 h, pH=4).

The design of a two-element simulation system incorporating Nd^{3+} was undertaken for the purpose of conducting selective adsorption experiments, wherein the interfering ions comprised Al^{3+}, Ca^{2+}, Mg^{2+}, Mn^{2+}, K^+ and Na^+. The result was shown in Figure 6a. The presence of Me^+ and Me^{2+} was no interference to the adsorption of rare earth, and the removal rate of rare earth was above 86%, while less than 3% for impurity ions. Al^{3+} still had an inhibitory effect on the adsorption of rare earth, removed by 56%. The structure of aminophosphoric acid enhances the selectivity of the adsorbent to Me^+ and Me^{2+}, but the separation of Al^{3+} still needs further investigation.

3.4 Real system

Real rare earth leach tailings were obtained from Changting, Fujian Province. The ion composition of the tailing liquid was shown in Figure 6b. When using 100 mg of adsorbent, the adsorption of rare earth ions in solution was 96.6% while 90% for Al ions, but the adsorption rate of other impurity ions were less than 22%. Al^{3+} and REE^{3+} have similar physical and chemical properties, so they are usually co-adsorbed [10].

3.5 Recyclability

The recyclability of P-CMP was carried out in real system, and 1 M HCl (30ml) was used to desorb the P-CMP at the end of each adsorption. The result after 8 cycles was shown in Figure 6c. After 8 cycles, the adsorption rate of rare earth was still above 80%, and 83% of the original adsorption properties was retained. The results showed that P-CMP had a good cycle stability and cost effectivity.

3.6 Adsorption mechanism

To elucidate the adsorption mechanism, the P-CMP was characterized by X-ray photoelectron spectroscopy (XPS) before and after adsorption. As shown in figure 7 (a), a characteristic peak of Nd^{3+} appeared at binding energy of 1000 eV indicated that the REE^{3+} has been successfully adsorbed by the constructed P-CMP. According to Fig. 7(b), the P fine spectrum is divided into two deconvolution peaks ($2P_{1/2}$ and $2P_{3/2}$). Compared with the original adsorbents, the binding energy of $2P_{1/2}$ decreased from 133.77 to 133.62 eV after adsorption. In addition, according to the O 1s spectrum, the binding energy of P=O and P-OH increased to 531.44 and 532.88 eV, respectively, implying that the phosphate group is strongly coordinated with the Nd^{3+}. Moreover, from the N 1s plot, the binding energy of the primary amine group decreased to 401.58 eV after adsorption, further indicating the synergistic adsorption of REE^{3+} by amino and phosphate groups.

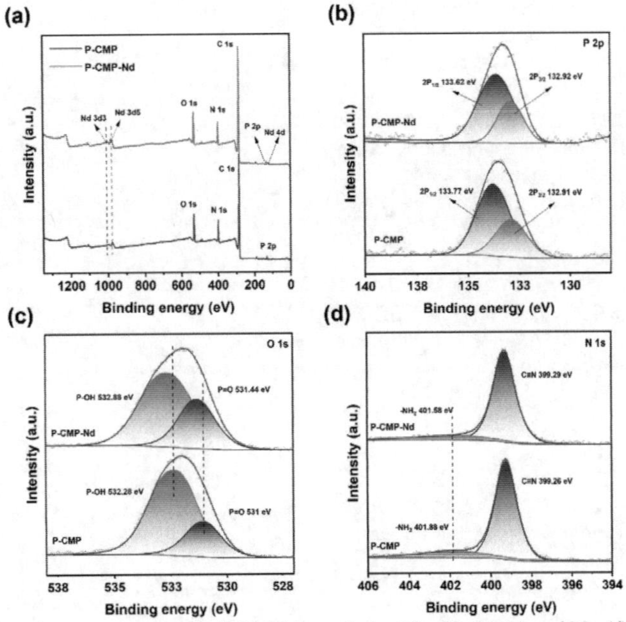

Figure 7. (a) XPS survey spectra of P-CMP; High-resolution N1s (b), O1s(c), and P2p (d) of P-CMP.

4 Conclusion

The present study proposed the utilization of phosphorylated conjugated organic polymer for rare earth enrichment in low-concentration solutions. Furthermore, an investigation was conducted to explore the potential of selective adsorption of rare earth using an aminophosphoric acid structure. Aminophosphoric acid can effectively improve the selectivity of adsorbent to Me^{2+} and Me^{+}. As mentioned above, when the removal rate of rare earth ions reached more than 85%, the removal rate of $Mg^{2+}$and Mn^{2+} could be ignored. The results showed that the optimal adsorption range of pH was 3-7, and the adsorption of P-CMP was monolayer chemisorption. The maximum adsorption capacities of Nd^{3+}, Dy^{3+} and Y^{3+} were 29.79, 38.11 and 27 mg/g, respectively, accompanied by reaching the equilibrium within 20 minutes. P-CMP was used for selective adsorption of rare earth ions in real system and 8 times of cyclic adsorption experiments were carried out, which showed that P-CMP had good selectivity and cyclic stability.

Acknowledgments

The authors gratefully acknowledge financial support from the Jungar Banner applied technology research and development project (2023YY-03) and basic research expenses of universities directly under the Inner Mongolia Autonomous Region (RZ200002361).

References

1. Cheisson, T., and E. J. Schelter (2019) Rare earth elements: Mendeleev's bane, modern marvels. Science, 363: 489-93.
2. Moldoveanu, G. A., and V. G. Papangelakis (2018) An overview of rare-earth recovery by ion-exchange leaching from ion-adsorption clays of various origins. Mineral. Mag., 80: 63-76.
3. Moira Feil, Lukas Rüttinger, Adelphi, Berlin (2011) Sustainable Prevention of Resource Conflicts: New Risks Raw Materials for the Future? Case Study and Scenarios for China and Rare Earths. Germany Federal Environment Agency, 7: 15-160.
4. Brewer, A., J. Florek, and F. Kleitz (2022) A perspective on developing solid-phase extraction technologies for industrial-scale critical materials recovery. Green Chem., 24: 2752-65.
5. Asadollahzadeh, Mehdi, Rezvan Torkaman, and Meisam Torab-Mostaedi (2020) Extraction and Separation of Rare Earth Elements by Adsorption Approaches: Current Status and Future Trends. Sep. Purif. Reviews, 50: 417-44.
6. Wu, Dongbei, Yawei Gao, Wenjun Li, Xiangning Zheng, YongGui Chen, and Qigang Wang (2016) Selective Adsorption of La^{3+} Using a Tough Alginate-Clay-Poly(n-isopropylacrylamide) Hydrogel with Hierarchical Pores and Reversible Re-Deswelling/Swelling Cycles. ACS Sustainable Chem. Eng., 4: 6732-43.
7. Bi, Shuai, Zhi-An Lan, Silvia Paasch, Wenbei Zhang, Yafei He, Chao Zhang, Feng Liu, Dongqing Wu, Xiaodong Zhuang, Eike Brunner, Xinchen Wang, and

Fan Zhang (2017) Substantial Cyano-Substituted Fully sp^2-Carbon-Linked Framework: Metal-Free Approach and Visible-Light-Driven Hydrogen Evolution. Adv. Funct. Mater., 27.

8. Wu, L., M. Yang, L. Yao, Z. He, J. X. Yu, W. Yin, and R. A. Chi (2022) Polyaminophosphoric Acid-Modified Ion-Imprinted Chitosan Aerogel with Enhanced Antimicrobial Activity for Selective La(III) Recovery and Oil/Water Separation. ACS Appl. Mater. Interfaces, 14: 53947-59.
9. Zhang, Gang, Wenxia Ma, Yuqin Feng, Ming Yang, Zheng Fang, Yongsheng Yang, Weiyan Yin, and Zehao Li (2022) Recyclable Amino-Phosphorylated Polyacrylonitrile Fiber for Gd (III) Adsorption in Aqueous Solution. ACS Appl. Polym. Mater., 4: 8396-406.
10. He, Q., J. F. Chen, L. M. Gan, M. L. Gao, M. M. Zan, and Y. F. Xiao (2023) Insight into leaching of rare earth and aluminum from ion adsorption type rare earth ore: Adsorption and desorption. J. Rare Earths, 41: 1398-407.

https://doi.org/10.1051/e3sconf/202452801014

Research on the key technology of Turkish gas storage shaft

Yaohui Wang*

Zhong Wei Energy Service Co., Ltd, Sinopec, China

Abstract: The construction procedure of the gas storage well is complicated, and each technical index request is extremely high, does not allow any mistake. The well deviation is difficult to control. Because of the large dip angle of formation, it is difficult to ensure the wellbore quality by conventional anti-deviation measures, and the change of well deviation and azimuth can not be monitored at any time . So it's almost impossible for conventional anti-incline measures to satisfy the design. In the upper large hole section, the tower pendulum bha is used. In the section below the second open hole, the MWD steering system is used to prevent deviation and reduce deviation, and in the second open large hole section, the slim hole is used first, after reaming, the difficulty of deviation control is reduced, the ROP is increased, and the contradiction between wellbore quality and ROP can be solved by using MWD steering system to prevent deviation and straighten, because the casing and the surface casing and the production casing belong to the large diameter thick wall casing, the rigidity is very big, and the casing is also the abnormal oil drilling pipe, this well summarizes the experience, after strengthening the well opening measures, the safe and smooth running.

1 Introduction

Wellbore structure design of salt cavern gas storage in Turkey:a Φ914.4 mm wellbore with φ762 mm casing at a depth of 45m. The Φ660.4 mm hole is designed at the first open hole, and the φ508 mm casing depth is 600 m to 1050 m. The Φ444.5 mm wellbore and the Φ339.7 mm gas-sealed casing are designed with the second open hole, and the depth of the casing is 1130m. The Φ215.9 mm hole to 1484m and the hole reamed to Φ444.5 mm after completion of drilling. This kind of large borehole construction is also rare in the world. Secondly, because the dip angle of salt dome structure is more than 40 °after drilling into salt bed, it is difficult to control well deviation by conventional anti-deviation measures, and well deviation is required to be less than 1°in drilling design, which is a major difficulty in the whole construction[1-6].

2 drilling construction difficulties

The UGS-1 well was drilled in Cihanbeyli, Mezgit and Tuz. For details, see Table1 and Figure 1.

Table 1 Reference table of UGS-1 well formation

Condition	Well section(m)	Thickness(m)	lithology
Cihanbeyli Kochisar	0-693	693	The upper earth yellow marl is mainly, and the middle and lower parts are mainly gray marl and limestone
Anhydrite	693-706.5	13.5	Anhydrous hard-gypsum is mainly used
Tuz	706.5-1500	793.5	salt rock

*Corresponding author email: slbhwwh@163.com

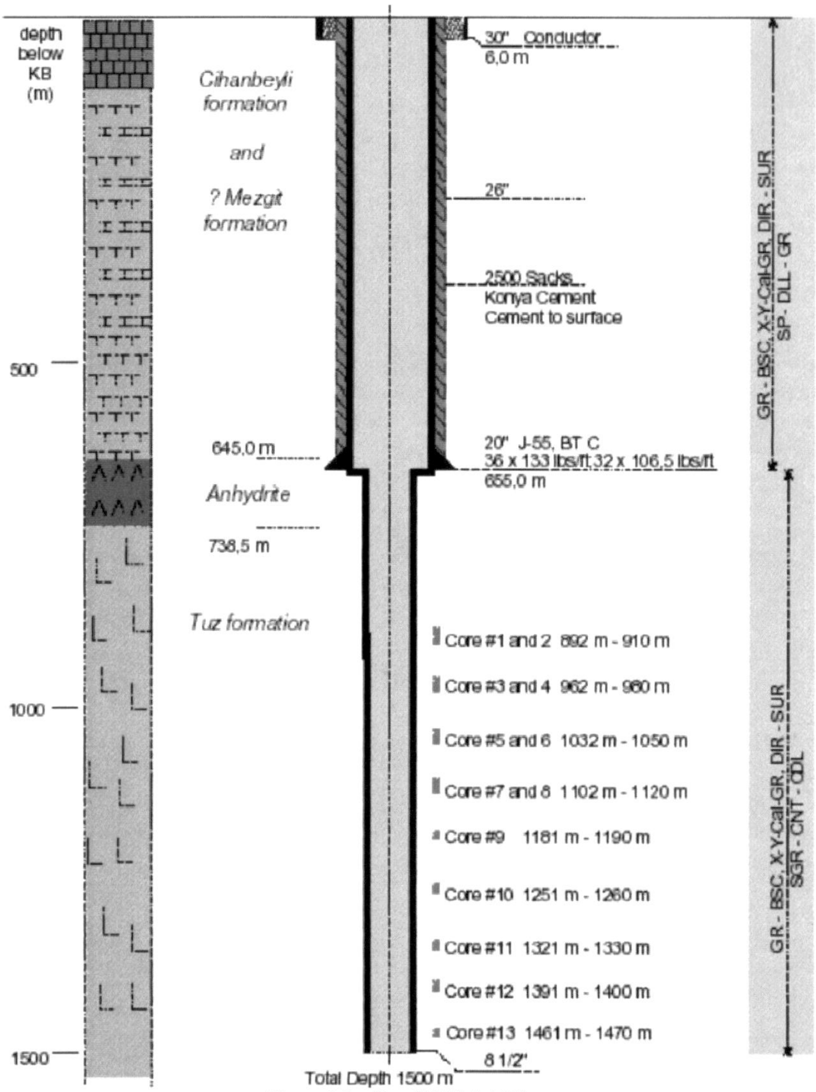

Figure 1 Profile of UGS-1 drilling

The construction procedure of the gas storage well is complicated, and each technical index request is extremely high, does not allow any mistake. The well deviation is difficult to control. Because of the large dip angle of formation, it is difficult to ensure the wellbore quality by conventional anti-deviation measures, and the change of well deviation and azimuth can not be monitored at any time[6-9] . Therefore, it is almost impossible to satisfy the design requirements of conventional deviation control measures to make the deviation less than 1°, and the design depth quality requirements are shown in table 2.

Table 2 Allowable range of hole deviation and level shift

Well depth(m)	Well deviation(°)	Full angle rate of change(°/25m)	Horizontal displacement(m)	Diameter enlargement rate(%)
0-635	≤3°	≤1.0	≤20	<10
635-1459	≤1°	≤1.0	≤20	<10

3 Selection of anti-incline measures

At present, the technology of anti-incline straightening at home and abroad can be divided into passive anti-incline straightening technology and active anti-incline straightening technology.

3.1 passive anti-slant straightening technique

1. The pendulum BHA straightening technique can achieve good anti-slant and straightening effect for the formation with various dip angles. However, the

https://doi.org/10.1051/e3sconf/202452801014

technique is only effective at low bit weight, which means that the drilling speed must be sacrificed in exchange for wellbore quality.

2. The technology of BHA straightening is to improve the rigidity of BHA and reduce the clearance between BHA and wellbore, increase the bending resistance of bottom hole assembly to effectively reduce the deviation caused by BHA bending. But this method is only suitable for the strata with small dip angle.

3.2 active anti-deviation and straightening technology

1. Vertical drilling system is a set of fully automatic rotary steering vertical drilling system, which is a continuous deviation control during rotary drilling using reliable rotary steering technology, the drilling parameters are completely liberated to ensure that the wellbore is vertical and maximum ROP is increased. The system is especially suitable for the stratum with strong natural tilting force (the stratum dip angle is more than 40°).

2. MWD steering system is a kind of wireless measure-while-drilling tool, which can monitor the deviation azimuth at any time with the single bending power drilling tool. The steering drilling tool is a downhole power drilling tool assembly with a single bend and a stabilizer. In the case of small deviation angle, composite drilling produces similar pre-bending dynamic deviation prevention effect, which produces stronger cutting on the down-hole wall and is beneficial to deviation prevention. The passive anti-deviation measures have some disadvantages, such as low penetration rate and limited adaptability to formation. Although vertical drilling system can ensure that the well is vertical and maximize drilling speed, but its expensive use costs are often prohibitive. In contrast, MWD steering system is cheap and can effectively control the deviation, so in the construction of gas storage wells in the selection of MWD steering system anti-deviation straight .

4 Construction measures of each well section

The tower pendulum BHA is used in the first open interval (0-650m) and the second in order to increase the rigidity and size of the lower bha as much as possible. In drilling, small bit weight is combined with drilling and lifting. When drilling fast, the bit weight is controlled without following pressure blindly. Every 50-100m, after finding the inclination increasing trend, lift in time and strengthen the interval of inclination measurement, and control the deviation by using pendulum drill. BHA: Φ660.4 mm bit + Φ279.4 mm drill collar × 3 + shock absorber + φ654 mm centralizer + Φ228.6 mm drill collar × 4 + Shock Absorber + Φ139.7 mm drill pipe in the second open interval (650-1130m) , when the BHA is about 10T and the rotary table speed is 60R/min, the deviation is measured and recorded after each single drill, and the deviation trend is analyzed. When the deviation rises to about 0.7 ° , directional deviation correction drilling is carried out. BHA: Φ444.5 mm bit + φ286 mm single bend power drill (0.75 °) + Φ228.6 mm non-magnetic drill collar + Φ203.2 mm drill collar × 3 pieces + Φ139.7 mm drill pipe. There are some disadvantages in the structure of the BHA. Because of the rigidity of the BHA, the eyeful centralizer and the friction of the brine slurry, it is very difficult to slide drilling and the ROP of the composite drilling machine is very slow. And its deviation correction ability is limited, in the formation with strong deviation-making ability can not be good deviation correction, to ensure wellbore quality. In order to solve this problem, the 215.9 mm bit and φ172 mm power drilling tool are used to drill to the target depth first, and then the 215.9 mm hole is reamed to Φ444.5 mm hole. This method not only ensures that the deviation will not exceed the standard during drilling, but also greatly improves the ROP. The total drilling and reaming time is also less than the original construction plan.

In 3 open-hole Section (1130-1459m) , 3 open-hole assembly using MWD + φ172mm power drilling tool (0.75 °) , combined drilling pressure is about 4 t, rotary table drilling speed is 60R/min, a single deviation is measured after drilling and the deviation data is recorded well, directional deviation rectification in time when well deviation increases. BHA: Φ215.9 mm bit + φ172 mm single-bend power drill (0.75 °) + φ210 mm centralizer + Φ177.8 mm non-magnetic drill collar + 158.8 mm drill collar × 9 pieces + Φ139.7 mm drill pipe. The difficulty of inclination control in three-open hole drilling is that three coring operations should be carried out in three-open hole drilling, each coring length is 18m. This means that at least 18m of the well deviation is out of control with each coring. The coring barrel has low rigidity and poor bending resistance. When the formation with strong deflecting ability is drilled, the deviation will be increased to more than 1 ° in the 18m section of coring drilling. In order to solve this problem, the well deviation should be adjusted to 0 ° as far as possible before drilling to the hole depth for coring, so that the deviation after coring is not too large. Secondly, control well drilling parameters to avoid artificial deviation in coring drilling.

5 Construction effect analysis at present

10 wells have been drilled in the project of salt cavern gas storage in Turkey, and various anti-deviation and straightening techniques have been continuously practiced in the construction, and the concrete construction scheme has been perfected, all the well quality indexes meet the design requirements, as shown in table 3.

Table 3 Statistical table of bore hole of completion well

num	Section	Maximum well deviation	Bottom hole displacement
UGS-4	656-1459	0.71	5.47
UGS-5	685-1459	0.86	1.01
UGS-6	760-1459	0.83	3.2
UGS-7	693-1459	0.95	7.81

UGS-8	777-1459	0.98	4.46
UGS-9	816-1459	0.94	2.90
UGS-10	882-1539	0.71	2.39
UGS-11	890-1484	0.95	5.62

Table 3 shows that the MWD steering system can effectively control the deviation to meet the design requirements.

6 Large diameter gas-sealed casing run-in and cementing operation

6.1 Wellbore preparation before running casing

Because the casing and the surface casing and the production casing are both large diameter thick-wall casing, the rigidity is very big, and the casing is also an abnormal oil drilling pipe, the 30-inch pipe and 20-inch surface 13-3/8-inch production casing in this well can not go down if the sidewall is slightly irregular and the wellbore quality is slightly bad. Because the deviation of the well dropped from 0.5 ° to 0.1 ° at the depth of 25m to 40m, the catheter could not go down, and because there was no 36-inch centralizer, the welder teeth of the amplifier were slapped. After the hole was enlarged in this section, the catheter went down smoothly. Considering that the surface casing and production casing are rigid, and the length of the Well Run is long, in order to prevent the difficulty of running casing in the lower section, the method of gradually increasing the rigidity of the through hole drilling tool is adopted, that is, using the single centralizer drilling tool structure to open the well first, after the well is cleared, the double centralizer full hole drilling tool structure is used to open the well, strengthen the drilling in case of obstruction, strengthen the circulation in the end, after the well is sealed, the short start-up and dry-out method is adopted to ensure the well is clear, and the large size rigid casing is lowered smoothly[10-11] .

6.2 Preparation for cementing

Well depth confirmation: drilling and logging engineers confirm that the actual well depth is 786.00 m.equipment personnel: all casing and cementing accessories, equipment must be adjusted in place to ensure normal operation. The casing runner receives an open-top cementing education and understands the entire process and operating standards. Wellbore preparation: after Reaming to 786 m cycle, short up and down 6 columns, after recirculation, using 0.5 T plastic ball lubricant to add thick mud 15 square well, short up and down 5 columns dry down to bottom normal after up to a depth of 120 m, in order to verify the bearing capacity of the upper formation and ensure the cement slurry to return to the surface, a pressure test is carried out with a 40-square small displacement circulation of weighted drilling fluid containing 10% plugging agent density of 1.45 g/cm3, seal the upper section with 30 squares of drilling fluid containing 1.5 tons of plastic ball lubricant.

6.3 Notes for setting casing

Casing body: under the casing root diameter, check the male and female thread buckle and bending, remove the unqualified casing. Clean the buckles on the floor and apply silk sealant on the floor. The rig is operated smoothly by crane with wire guards. Bushing string structure: floating shoe + 1 bushing + floating hoop + 1 bushing + floating hoop + bushing string with inner socket, the inner socket is coated with butter, and the drill floor has 1 m seat with 2 lifting clamps. The back clamp is struck below the middle of the female buckle collar. Keep the casing centered when bucking. It is qualified to buckle up to 4 mm below the bottom edge of the triangle and up to 7 mm above the bottom edge of the triangle. The floating hoop and the above 5 bushings of the floating shoe are all coated with thread lock fixing glue. Next casing: the first five full of water, electrical measurement to do the scale. No falling objects during casing running. Lower the drilling fluid at a uniform speed, and make sure that no more than 5 wells are hollowed out, and observe the return of drilling fluid from the wellhead at any time. According to the requirements of the centralizer, the first 20 under each two casing centralizer, a pipe in a centralizer, the middle of 3-5 casing string plus a centralizer can be considered, not only the center of the casing, but also to take into account the location of leakage layer. Butter the last two bushings up and down 5m[12-17].

6.4 Low temperature high salt resistance micro expansion gas channeling prevention cement slurry system technology

At low temperature, it is difficult to make cement slurry with salt resistance, low water loss, micro-expansion, high strength, anti-channeling, impact resistance and good bonding interface. Based on the characteristics of salt bed cementing and gas storage, a new type of cement slurry system-low temperature, high salt resistance, micro-expansion and gas channeling prevention cement slurry system is developed. This cement slurry system is a low density cement slurry technology, which is developed on the basis of the theory of compact packing and the particle size distribution of materials, taking full advantage of the activity of the constituent materials and the physical chemistry between them, the active minerals with a wide range of particle sizes are packed tightly together, i.e. the cement slurry contains more solids per unit volume, the formed low density cement slurry has the characteristics of high strength, good stability, gas channeling prevention, corrosion prevention, free fluid low and so on . The batching is shown in table 4 and the cementing quality figure is shown in Figure2.

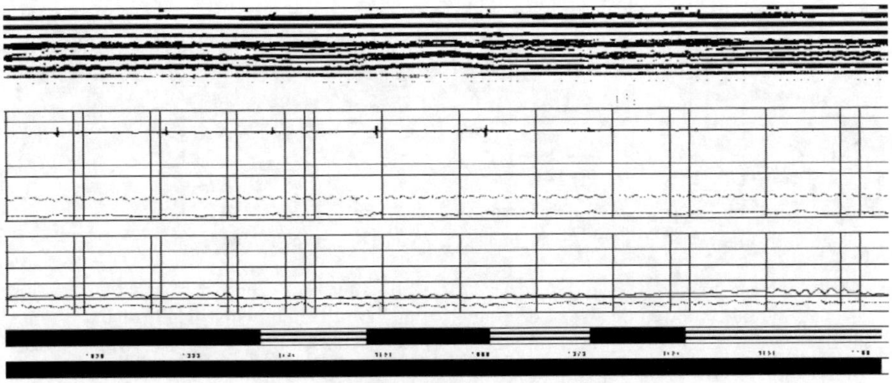

Figure2 cementing quality curve of UGS-9

Table 4 main materials

Material Organization	name	No.	Function	amount added
Admixture	Salt-resistant low-temperature activity enhancer	KCMO50	Filling material, improve the strength of cement stone	5%-20%
	Salt-resistant multi-functional anti-gas channeling agent	KCMO51	Anti-channeling, improve the corrosion resistance of cement stone	5%-15%
	Salt-resistant anti-gas channeling agent	KCMO62	Filling material can reduce the density of cement slurry	Adjust for density
	High anti-depressant	KCMO63	effectively reduce the density of cement slurry	1%-4%
	Salt-resistant expansion agent	KCMO25	Cement Stone to form micro-expansion, sealing the annulus	0.37-3.021
Other auxiliary additives	Salt-resistant anti-channeling fluid loss agent	KCMO75	Effectively control the water loss of salt-bearing cement slurry	0.1%-1.5%
	Salt-resistant fluid improver	KCMO13	Improve the stability of cement slurry	Adjust for density
	Salt-resistant dispersant	KCMO67	Improve the slurry mixing ability, reduce the viscosity	0.4-1.91
	Salt-resistant retarder	KCMO68	Effective control of cement slurry thickening time	0.4-0.8 l
	Antifoaming agent	KCMO03	Salt-containing cement slurry effectively inhibit foam	0.2-0.41
	Defoamer	KCMO15	Effectively eliminate bubbles	0.2-0.41

7. The principle and calculation of cement slurry natural balance method for plugging

The principle diagram of balanced cement plugging is shown in Figure 3. Because all lost circulation is of the lost circulation type and a certain liquid column height can be established, the cement slurry natural balance method is used for plugging. That is, after injecting cement slurry, the liquid column pressure in the wellbore is greater than the pressure of the lost circulation layer, causing 1/2 or 2/3 of the cement slurry to enter the lost circulation layer. Subsequently, the liquid column pressure in the wellbore is balanced with the pressure of the lost circulation layer. In this equilibrium state, the cement slurry is set to achieve the purpose of plugging.

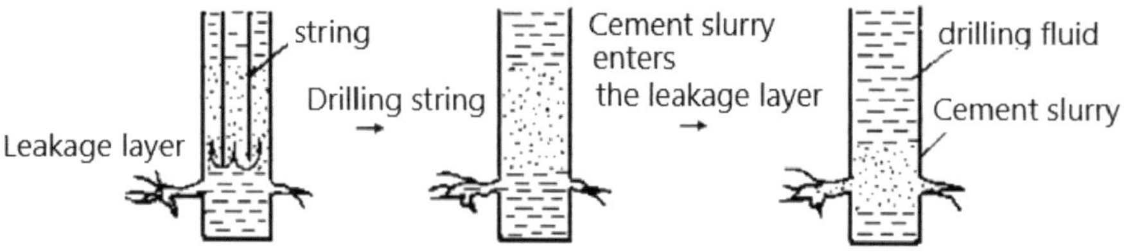

Figure 3 Schematic diagram of balance method of cement leaking stoppage

Firstly, the leakage layer pressure is calculated based on the measured static liquid level well depth and drilling fluid density inside the well.

Pl = 0.01 m (H-B1).....................(1)

In the equation, Pl - leakage layer pressure is MPa; ρ M is drilling fluid density of g/cm3; H is depth of lost circulation well of m; B1- static level well depth before injecting mud is m.

After injecting cement slurry, there are two liquid columns, forming the following equilibrium relationship.

Pl = 0.01 ch1 + 0.01 mh2...............(2)

In the equation, ρ C is the cement slurry density of g/cm3; H1 is the Design cement plug height of m; H2 is the length of the drilling fluid column in the well after plugging construction of m.

By obtaining from Equations (1) and (2)

h2 = (H-B1) - h1ρc/m........... ...(3)

To plug the cement slurry and meet the design requirements, the length of the drilling fluid column in the well after construction must meet the requirements of Equation (3). Therefore, the initial depth of the drilling column is

H1 = B1 + h2 = H - h1 ρc/ρm...... ......(4)

From Equation (4), it can be seen that the depth of the drilling string in the plugging construction is related to the density of the drilling fluid in the well, the density of the cement slurry and the design height of the cement plug. That is, the greater the difference is between the density of the water slurry and the drilling fluid, the higher the design height of the cement plug is and the lower the depth of the drilling tool is.

After the plugging construction, if the density of the cement slurry is not consistent with the density of the drilling fluid, the depth of the static liquid level will change from B1 to B2.

B2 = B1 + (ρc/ρm-1) h_l......... ......(5)

Equation (4) is an expression for the depth of the drill string in equilibrium. During plugging operations, the static level of the drilling fluid will change to a dynamic level, meaning that during the injection and replacement of cement slurry into the well, the static level will rise. As the liquid level rises, the pressure of the liquid column continues to increase. Under the increasing pressure, the leakage rate also continues to increase. When the leakage rate is equal to the construction displacement, the drilling fluid level in the well will maintain a dynamic equilibrium state, which is the dynamic liquid level.

Due to the presence of dynamic liquid levels, to ensure that the length of the drilling fluid column in the well after construction meets the requirements of Equation (3), the actual depth of the drilling column is calculated according to the following equation.

$H_l' = H_l - \Delta h = H - (\rho_c/\rho_m)\, h_l - \Delta h$............(6)

$\Delta h = B_l - b$

Among them, b is the depth of the dynamic liquid level well (m). B1 and b can be measured and obtained before plugging construction.

8. Optimize well-depth structure and plugging plan

In response to design defects, due to the lack of professional manufacturers producing standard conduits for the petroleum industry, the strength of the 30-inch conduit is greatly reduced. According to load calculations, the maximum depth is 100 meters. As the last position of the leakage layer is usually before 120 meters, even if the design is modified, it cannot completely seal the leakage layer which at least clears the obstacles for the later judgment of the leakage layer. To solve the problem of plugging in the region, it is still necessary to find better plugging methods.

Based on the characteristics of multiple leakage layers, large cracks and connectivity with surface water, we have improved the cement slurry near the equation equilibrium plugging method. Firstly, we use a three-high (high water loss, high solid phase, and high viscosity) plugging fluid with a certain setting strength to act as a partition wall around the leakage layer, blocking the erosion of formation water and isolating the formation water from the cement slurry, ensuring that the cement slurry is not corroded by formation water before solidification. After the cement slurry is used, the well sealing device is closed to apply pressure to the well, forcing the three-high plugging fluid and cement slurry to enter the formation at the same time which allows the cement slurry not to be corroded by water to travel further in the formation. After hardening, a solid cement wall is formed to achieve the goal of complete sealing.

9. On-site implementation effect

The Ugs-8 well is the first well of underground gas storage in Turkey. It is decided to use a polymer-saturated brine anti-sloughing drilling fluid system to solve the wellbore stability problem of the formation. The system is composed of 500 ml water + 0.3% soda + 5% sodium earth + 0.2% naoh + 3% SPNH + 0.3% DSP-2 or PAC-LV + 1% WNP-1 or CMC-LV +

0.2% pam + NaCl + other processors. Other processors include lubricant, defoamer, viscosity reducer and weight agent.

Basic formulation properties of the system(shown as table 5):

Table 5 Basic formulation properties

Density (g/cm³)	1.20	Ø600	43
Viscosity (S)	60	Ø300	26
APIFilter loss (ml)	3.6	Ø200	20
Mud thick (mm)	0.5	Ø100	12
Static shear force (Pa)	1/19	Ø6	2
Plastic viscosity (mpa.s)	17	Ø3	2
Dynamic shear force (pa)	4.5	Total solid phasor	13%
PH	9.5	CL-(mg/L)	185857
HTHP Filter loss120°C (ml)	13	Viscosity	0.1853

10 Conclusion

(1)Tower pendulum BHA is used in the upper large hole section, which can completely meet the quality requirements of upper wellbore, and is easy and economical to operate.

(2)Using MWD guide system anti-deviation and straightening technology is beneficial to anti-deviation and anti-deviation, and also can ensure well quality. At the same time, the small hole drilling and reaming are adopted in the second large hole section, which reduces the difficulty of deviation control and improves the ROP and safety.

(3)Using MWD steering system to prevent deviation and straighten can solve the contradiction between wellbore quality and ROP, which is beneficial to increase rop, shorten drilling cycle and increase economic benefit.

(4)It is necessary to strengthen the research on the influence of drilling parameters and the structure of long-bore coring tools in salt beds on the deviation of Wells in order to meet the strict requirements of wellbore quality.

(5)In general, using the MWD steering system is more economical and economical than using the vertical drilling system, with a smaller investment in exchange for a larger profit.

(6)The drilling fluid system used in the UGS block plays a very good role in the wellbore stability and in the construction process to ensure the effective content of various anti-sloughing agents in the drilling fluid as far as possible to control the loss of drilling fluid filtration. One-open API filter loss control below 6 ml, two-open and three-open API filter loss control below 4 ml, thus effectively ensuring the wellbore stability, wellbore unimpeded and down-and-out drilling smoothly. The success rate of electrical measurement is 100%. Under the condition of a long drilling cycle, there is no blocking and sticking phenomenon.

Reference

1. Ding Guosheng, Wei Huan. Review and prospect of underground gas storage construction in China for 20 years [J]. Oil and gas storage and transportation, 2020, 39 (01): 25 - 31.
2. Zhang Guanghua. Present construction situation and development suggestion of Sinopec underground gas storage [J]. Natural gas industry, 2018, 38 (08): 112 - 118.
3. Optimization and effectiveness of plugging technology for old wells in oil and gas reservoir underground gas storage. Chemical Engineering and Equipment Journal 2018.
4. Wei Guoqi, Zheng Yali, Qiu Xiaosong, etc. Geological theory and application of underground gas storage in China [J]. Journal of Petroleum, 2019, 40 (12): 1519 - 1530.
5. Qi Xingtao, JIA Bo. Research and Application of anti-gas channeling toughening and high-strength plugging agent for gas storage [J]. Technical Supervision of Petroleum Industry, 2018, 34 (05): 23 - 26 33.
6. Zhao Jinzhou. Wen 23 key engineering technology of underground gas storage [J]. Oil drilling technology, 2019, 47 (03): 18 - 24.
7. Kuru, E. The Buckling Behavior of Pipes and Its Influence on The Axial Force Transfer in Directional Wells[J]. SPE52840
8. Mitchell, R. New Buckling Solution for Extended Reach Wells[J]. SPE74566
9. WU Xiongjun, LIN Yongxue, SONG Bitao, et al. Oil base drilling fluid technology for drilling broken ordovicianformation in Shunbei block[J]. Drilling Fluid & Completion Fluid, 2020, 37(6)：701-708.
10. NING Fangxing. The main control factors of shale oil enrichment in Jiyang depression[J]. Acta Petrolei Sinica, 2015, 36(8)：905-914.
11. WANG Min, MA Rui, LI Jinbu, et al. Occurrence mechanism of lacustrine shale oil in the Paleogene Shahejie formation of Jiyang Depression[J]. Petroleum Exploration and Development, 2019, 46(4)：789-802.
12. Du Xulin, Su Yanchun, Fang Maojun, et al. A semi-analytical model coupled gas-water two-phase flow-geomechanics for dynamic closure analysis of fractures in tight sandstone. Natural Gas Geoscience, 2024, 62(02):1672-1690.
13. Feng Xinyu, Xiao Hui, Li Quanhui, et al. Experimental study on hydraulic fracture characteristics of tight sandstone. Unconventional Oil & Gas, 2023, 10(4):132-138.
14. Zhou Changjing, Sun Zhonghua, Xiao Yuanxiang, et al. Study on Brittleness Characteristics and Fracturing Crack Propagation Law of Deep

Thin-Layer Tight Sandstone in Longdong, Changqing. Processes, 2023, 11, 2636.

15. Ma Dongdong, Wu Yu, Yin Jiadi, et al. Effect of initial pore pressure on the hydraulic fracturing of tight sandstone: an experimental study. Geomech. Geophys. Geo-energ. Geo-resour. 2023, 9-15.
16. Wu Xiaolong, Guo Yintong, Chang Xin, et al. Experimental Study on Cyclic Hydraulic Fracturing of Tight Sandstone under In-Situ Stress. Processes 2023, 11, 875.
17. Zhang Qi, Zhang Xiaoping, Sun Wei. Hydraulic fracturing in transversely isotropic tight sandstone reservoirs: A numerical study based on bonded-particle model approach. 2020,136:104068.

E3S Web of Conferences 528, 01015 (2024) https://doi.org/10.1051/e3sconf/202452801015
NET-LC 2024

Research on the technology of casing inspection through tubing in the Saertu Development Zone of Daqing Oilfield

Zhonghe Zhao

No.1 Oil Production Company of Daqing Oilfield Company Ltd., Daqing Heilongjiang 163001, China

Abstract: With the further development of Daqing Oilfield, the occurrence of widespread casing damage in the Saertu Development Area has been on the rise annually.In order to promptly prevent, detect and address casing damage, the management departments of the oil extraction plants must schedule regular casing inspection operations. This approach not only raises production costs but also prolongs the well shutdown period, further impacting crude oil output.This article focuses on the technical needs of oilfield development and has developed tubing and casing detection technology, the radial magnetic scanning system is integrated with the sixteen arm wellbore diameter, the synchronous tubing and casing inspection instrument represents the latest in concurrent inspection technology, developed specifically for the technical demands of oilfield development Without displacing the tubing column, it simultaneously captures data on tubing inner diameter, tubing continuity, the extent of casing damage via magnetic scanning, and continuous 3D imaging of the tubing at standard inspection speeds, the workload of checking and matching tasks were significantly reduced. Between 2022 and 2023, the through-tubing casing inspection technology was utilized in 525 wells, with 356 wells verified by mobilizing the tubing column, achieving a predictive accuracy of 87.6% and generating economic benefits of 69.6 million yuan.

1. Introduction

The conventional inspection process for casing damage is currently the most commonly used traditional method, which occupies limited resources of the operation team, costs around 39000 yuan, and requires shutting down the well for 3 days, directly resulting in an increase in production costs and a shortened production cycle for oil and water wells. This article has conducted research on tubing and casing inspection technology[1]. Firstly, it reduces the workload of casing inspection, lowers production costs, and reduces the duration of well shut-in; Secondly, before moving the tubing, perform logging to predict casing damage, optimize the well repair plan, and timely scrap the original tubing, which can provide technical means to reduce difficult major repairs and terminate the well.

2. Principle of Through-Tubing Casing Inspection Technology

Through-tubing casing inspection technology refers to the use of inspection instruments that take the oil tubing as the inspection channel, utilizing ultra-small diameter wellbore measurement technology combined with radial magnetic scanning technology. Not only can the oil casing synchronous inspection instruments use ultra-small diameter 16-arm or 40-arm calipers to detect changes in the inner diameter of the tubing, but they also employ the effect of a low-frequency, high-energy magnetic field that penetrates the tubing to scan the casing wall. When the casing is deformed, worn, or corroded due to formation stress, the magnetic scanning[2] through-tubing simultaneous inspection technology can clearly and accurately reflect the condition of the casing damage[3]. It accomplishes this by going down the tubing once to capture data on changes in the tubing's inner diameter, casing deformation, and corrosion leak points.

2.1 Combination of radial magnetic field scanning and ultra-small diameter sixteen-arm caliper

When casing wear occurs without the tubing being compressed or deformed, the magnetic scanning data curve can reflect different curve changes based on the casing's shape, clearly identifying features such as casing couplings, tubing couplings, expansion joint reinforcing sections, and downhole tool segments. At the same time, the ultra-small diameter sixteen-arm within the tubing can precisely measure the data of the tubing's bore, and combining these two data sets allows for the assessment of the condition of both the oil and casing. In cases of severe casing wear, disconnection, or when the tubing is severely compressed, bent, or deformed, leading to the operational tubing column becoming stuck, the magnetic field detection curves for both tubing and casing will show prominent curve changes. Traditional

zhaozhonghe0128@126.com

https://doi.org/10.1051/e3sconf/202452801015

electromagnetic flaw detection techniques[4] may not be able to distinguish between oil and casing damage due to the confusion of the two magnetic field curves. By integrating the radial magnetic field scanning curves with the data from the ultra-small diameter sixteen-arm caliper within the tubing, this technology overcomes the limitation of being unable to distinguish the extent of damage to the oil casing using magnetic scanning curves alone. It can effectively differentiate the real data changes in casing and tubing, and by combining the data from the sixteen-arm caliper within the tubing, it is possible to accurately detect the true shape changes of the tubing and estimate the degree of casing deformation, the composition of the instrument is show in Figure 1. The synergy between radial magnetic scanning technology and the ultra-small diameter sixteen-arm imaging technology within the tubing complements each other's advantages, surpassing the limitations of conventional electromagnetic inspection techniques. The comprehensive data collection of the entire system accurately represents the full wellbore profile of both tubing and casing, achieving precise determinative grading of casing damage and accurate quantitative detection of tubing damage.

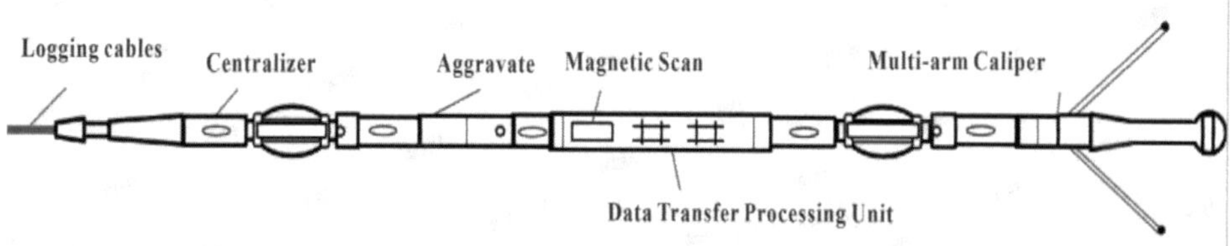

Figure 1: Through-Tubing Casing Inspection Tool Structure Diagram

2.2 MMS360 Magnetic Scanning Technology

The MMS360 magnetic scanning technology shares the same basic principle with electromagnetic flaw detection, using the principle of electromagnetic induction to detect material defects. When the alternating magnetic field generated by the detection instrument passes through the material, any internal defects (corrosion, perforation, fracture) will disturb the magnetic field, thereby generating anomaly signals. The detector then determines the presence of defects by analyzing these signals.When like-pole emitting coils are at specific distances, the magnetic field lines from the emitting coils form a flattened radial cross-sectional magnetic field. Utilizing the effect of low-frequency, high-energy magnetic fields, radial magnetic scanning lines are formed within the oil casing. The magnetic field cross-section can penetrate the tubing and casing, forming a magnetic circuit in the formation and scanning both the casing and tubing simultaneously. The distribution of magnetic field is shown in Figure 2. By using a like-pole magnetic field layout, the intensity of the emitted magnetic fields from above and below creates an overlap, increasing the scanning depth radius of the magnetic field. This avoids most of the magnetic field's return flow along the tubing column, truly achieving a planar scanning effect. The increased scanning radius provides a more sensitive magnetic response to the presence, absence, and proximity of tubing and casing, breaking through the detection bottleneck of conventional electromagnetic flaw detection technology.

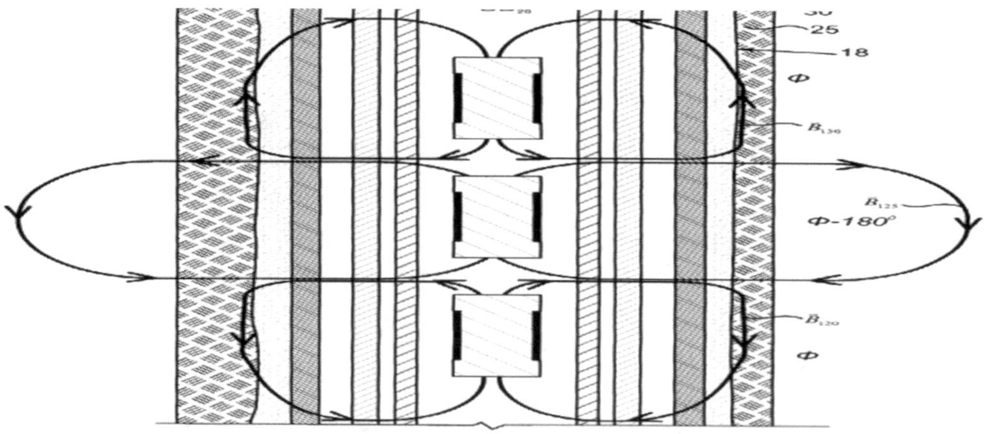

Figure 2: Magnetic Field Distribution Diagram for MMS360 Magnetic Scanning Technology

https://doi.org/10.1051/e3sconf/202452801015

2.3 The data processing for through-tubing casing inspection technology

The through-tubing casing inspection instrument integrates radial magnetic cross-sectional scanning technology with an ultra-small diameter sixteen-arm borehole imaging system[5]. It detects changes in the tubing's inner diameter and performs 3D imaging. Simultaneously, it collects continuous data on casing deformations and breaks through radial magnetic cross-sectional scanning. The data processing system[6] presents various data curves and 3D images, including curves of tubing inner diameter changes, full wellbore tubing 3D imaging diagrams, oil casing couplings, casing reinforcement segment morphology curves, and casing damage morphology curves. The result curve is shown in Figure 3.

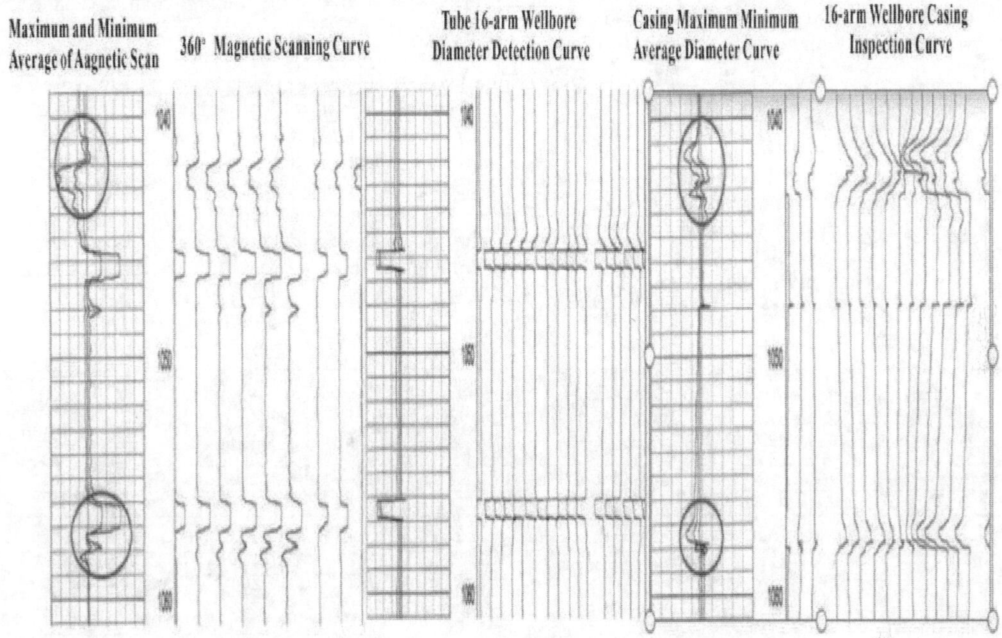

Figure 3: Test Curve Graph of Through-Tubing Casing Inspection Technology

3. Fields of Application for Through-Tubing Casing Inspection Technology

In recent field trials, through-tubing casing inspection technology has fully demonstrated its advantages, including a fast and simple inspection process, strong system throughput, and complete data reproducibility.This technology is expected to gradually reduce the workload of conventional operations teams in future routine casing inspections, allowing them to focus more time and energy on measures to increase oilfield production.Its application before major overhauls and during well repair processes provides more accurate technical data for devising well repair strategies, preventing issues such as tubing dropout, casing disconnection, and lost tools in the wellbore due to blind tubing extraction.The adoption of through-tubing casing inspection techniques in future oilfield operations will save significant costs associated with casing inspection tasks for oil extraction plants, reduce the consumption of downhole tools, and provide good economic benefits to oil production.It will play a positive role in reducing downtime and shutdown time for production units and management departments, saving on casing inspection costs, and reducing the expenses of major overhauls.

This entry outlines the applicability of the through-tubing casing inspection magnetic scanner in various oilfield production well scenarios, emphasizing its wide range of uses in the industry:

(1)The through-tubing casing inspection process, replacing the conventional casing inspection operations, enables simultaneous inspection of tubing and casing in a single operation without moving the tubing column.It provides comprehensive data including full-well tubing shape curve diagrams, full-well tubing 3D imaging diagrams, casing magnetic scanning curve diagrams, and accurate measurements of casing wear depth and length.

(2)Casing Wear Prediction and Forecasting: The through-tubing casing inspection process, by regularly cycling through observations of the well network, establishes a casing wear monitoring database for the block. Combined with big data analysis techniques, it provides more timely and accurate data for predicting, governing, and finely managing casing wear areas in oilfields.

(3)In cases of problematic wells, where tubing gets stuck during well repair or operational processes, the use of a through-tubing casing inspection device can accurately identify the reasons for the tubing getting stuck, promptly detect casing wear conditions, and prevent the aggravation of casing wear due to misjudgment.

(4)Optimizing well repair plans, during well repair or operational processes, detecting casing deformation parameters due to existing casing wear before well repair can benefit the optimization of well repair techniques.When conventional wellbore instruments cannot be used, the through-tubing casing inspection instrument can fully utilize its advantages of being precise, small, short, and capable of combining multiple parameters.It can provide accurate real data on casing wear, full well oil/tubing 3D imaging diagrams, casing disconnection depth, casing disconnection length, tubing dropout, and downhole object determination, among other data.

(5)Well Repair Quality Inspection. The through-tubing casing inspection instrument[7] can provide more comprehensive continuous data on downhole tubing changes than traditional inspection instruments. It enhances the quality of well repairs and ensures a higher success rate of well repairs. It provides well repair technicians with more intuitive and accurate downhole data.

4. Field Application and Benefit Analysis

Between 2022 and 2023, the through-tubing casing inspection technology was applied to 525 wells, with 356 verified by moving the tubing column, achieving a predictive accuracy of 87.6%. The completion is shown in Table 1. The application of through-tubing casing inspection technology reduced the workload for casing inspection operations and decreased the number of problematic wells awaiting repair, achieving a total economic benefit of 69.6 million yuan. Of this, casing inspection operations accounted for an economic benefit of 63.3 million yuan, and the optimization of repair strategies based on predictive diagnostics before moving the tubing column saved 6.3 million yuan in operational costs.

Table 1: Completion Status of Through-Tubing Casing Inspections

Years	Total Wells Inspected with Through-Tubing Casing Inspection Technology	Wells Inspected During Well Repair Processes	Wells Inspected for Casing Inspection in Water Wells	Wells Verified by Moving the Tubing Column	Accuracy Rate of Verification by Moving the Tubing Column (%)
2022	79	60	19	79	86.1
2023	446	205	241	277	88.1
Total	525	265	260	356	87.6

5. Conclusion

(1)Using the through-tubing casing inspection instrument in water well casing inspection operations can effectively shorten the construction period, reduce the impact on water yield, and decrease the downtime of wells.

(2)Utilizing the through-tubing casing inspection instrument during well repair processes allows for the precise formulation of subsequent repair measures, enhances the success rate of techniques, and reduces the addition of problematic wells requiring repairs.

(3)With the widespread application of through-tubing casing inspection technology, the accumulation of the number of wells inspected and monitoring data contributes to the establishment and continual enhancement of a casing wear survey database.By summarizing curve characteristics and data to achieve a technical level of numerical simulation calculation methods, it gradually refines the qualitative indicators of casing wearl severity, laying the foundation for the next step of achieving quantitative interpretation of casing wear severity through tubing.

References

1. Yu R M Y H B .Magnetic detection technique for double-layer steel tubing and casing damage[J].INSIGHT,2017,59(2):81-84.
2. Wang J ,Li Z ,Zhao K , et al.Isolated scan unit and scanning tunneling microscope for stable imaging in ultra-high magnetic fields[J].Ultramicroscopy,2024,261113960-.
3. Yan Zhengguo, Fan Yizhou. Analysis of the Application of VideoLog Visualization Logging Technology in Casing Damage Detection in Oil and Gas Fields [J]. Yunnan Chemical, 2019.
4. Yotsuji J ,Ikushima K ,Yamada H .Flaw Detection for Thin Sheet Using Acoustic Stimulated Electromagnetic Wave Technique[J].ISIJ International,2020,60(5):948-953.
5. Wang Jiabing. Development of VideoLog Downhole TV Video Processing Software [D]. Xi'an Shiyou University, 2020.
6. Wang C ,Wang Y ,Zou X , et al.Study of a borehole panoramic stereopair imaging system[J].International Journal of Rock Mechanics and Mining Sciences,2018,104174-181.
7. Geophysics; Findings from E.M. Viggen and Co-Authors Reveals New Information on Geophysics (Analysis of outer-casing echoes in simulations of ultrasonic pulse-echo through-tubing logging)[J].Journal of Technology Science,2017,160-.

A Life Cycle Assessment Of Energy-saving And Carbon Reduction Potential ——Empirical research On A Road Reconstruction Project In China

Ke Wang*

1CCCC HIGHWAY CONSULTANTS CO.,LTD., Beijing, China

Abstract: A life cycle standardized assessment module of energy-saving and carbon reduction is established based on the current road cost management software. The assessment module combines the domestic road material consumption quota and the main LCA database CLCD in China. Quantitative energy-saving and carbon reduction benefits of material and technical alternative measures could be calculated autocratically. Empirical project data shows that reducing concrete, reducing mixing temperature and reusing material of the old road are essential measures. It is possible to reduce carbon by 22-28% and save energy by 45-50%.

1. Introduction

With the requirement of energy conservation and carbon reduction, road reconstruction design become important such as the use of new materials, energy and masonry reduction design measures[1]. In 1999, Piantanakulchai took the lead in assessment of carbon emissions for a highway[2].Yue Huang measuring the carbon footprint[3].The life-cycle carbon of asphalt paving proves reducing waste is important [4,5].It is also found that context up to 34% of GHG emissions can be avoided based on current technology[6].

An empirical research is conducted on a road reconstruction project in Hainan province, China. The assessment module is set based on the cost management software in which the material consumption is calculated using a unified social average labor quota in China. Thus the assessment method could be copied and popularized easily.

2. Energy consumption and carbon emissions assessment

2.1 assessment scope and method

The assessment module of energy consumption and carbon emission is the basis of optimal design[7]. Generally, the ISO14040 life cycle evaluation standard is adopted from cradle to grave evaluation perspective, including raw materials, construction, maintenance, and abandonment stages[8].

Firstly, the bill of three stages and 18 materials are selected to represent the material consumption which covers 99.2% of the road reconstruction project as shown in Table 1. Considering the error rate of in environmental is up to 25% from different databases[9], the application of recognized local databases CLCD is the key method to ensuring the applicability and comparability of assessment results.

Table 1 Carbon and Energy Coefficients of Road Reconstruction

stages	Materials	Coefficients		Data sources
		f_i (carbon)	p_i (Energy)	
Raw material production	Steel reinforcement	2.64	39.05	CLCD
	Asphalt	0.86	56.00	
	Steel pipe	2.84	42.27	
	Coal	0.21	25.98	
	Electric	0.97	12.79	
	Steel plate	2.91	43.72	
	Heavy oil	0.66	40.82	
	Aluminium alloy	23.86	281.42	
	Clay	0.003	0.03	

*339232370@qq.com

	Gasoline	0.93	50.61	
	Cement	0.74	4.50	
	Water	0.0002	0.003	
	Crushed stone	0.002	0.03	
	Diesel oil	0.80	47.34	
	Iron piece	2.36	29.08	
	Sand	0.003	0.04	
	Natural gas	0.28	15.49	
Construction and operation	Coal	1.9	20.91	Domestic specifications
	Heavy oil	3.17	41.82	
	Gasoline	2.93	43.07	
	Diesel oil	3.1	42.65	
	Natural gas	2.16	38.93	
Abandonment	Construction waste of old road	0.01	0.18	ELCD

Note:data corms from China Life Cycle Core Database of Sichuan University,General Principles for the Calculation of Comprehensive Energy Consumption (GB/T 2589-2020) and the Guidelines for the Preparation of Provincial Greenhouse Gas Inventory (FGBG [2011] No. 1041).

Secondly, the bill is imported into the Energy consumption and carbon emissions assessment module based on the cost management software. Then the coefficients of carbon or primary energy consumption set in the software module could be used to account the GWP and PED according to formula (1) and (2).

$$\sum_{i=1}^{n} GWP = Q_i \times f_i \tag{1}$$

$$\sum_{i=1}^{n} PED = Q_i \times p_i \tag{2}$$

GWP is the carbon emissions, with the unit kg CO2 eq/kg. PED is the primary energy consumption, with the unit MJ/kg. Q_i is the inventory quantity of i materials. f_i is the carbon coefficients of i materials. p_i is the primary energy consumption coefficients of i materials.

2.2 Assessment result

According to the assessment result, the total carbon emission of the project is 1.8e6kgCO2eq/km, which is mainly concentrated in the raw material production stage, assessment for 83%, followed by 10% in the construction stage and 5% in the original road abandonment stage, and 2% in the maintenance stage. The primary energy consumption is 2.58E+07MJ/km. The main energy consumption and carbon emission is the construction stage.

Reducing concrete is the key to carbon reduction, as shown in Fig. 1. Cement raw materials produce a large amount of carbon dioxide in the production process of calcination. The cement consumption is mainly reflected in the pavement base by 19%, bridge and culvert structures by 36% and concrete ditches by 12% that do have alternative technical measures and materials. From the perspective of primary energy consumption in Fig. 2, asphalt is the highest by 40% in Fig. 3, mainly because asphalt mixing uses a hot mixing process, which consumes much energy.

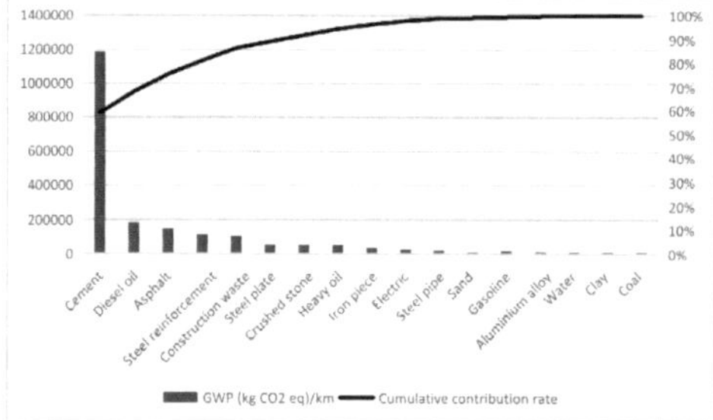

Fig. 1 Cumulative contribution of carbon (left)

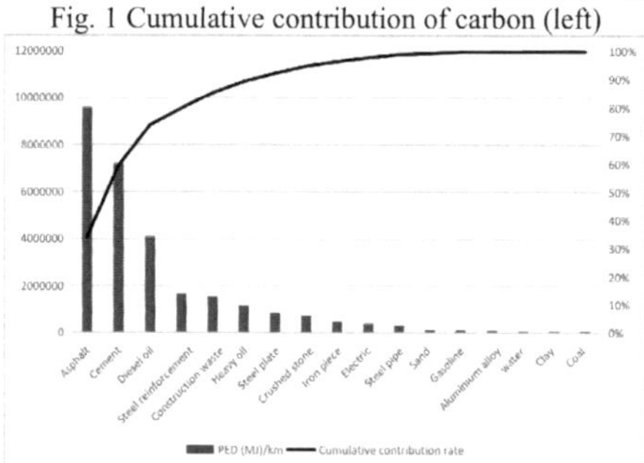

Fig. 2 Cumulative contribution of primary energy(right)

3. Application of optimized design

The goal of the assessment is to guide the optimization of road design to lower energy consumption and carbon emission.The comparison of energy conservation and carbon reduction measures show that warm mix asphalt is the most effective measure for energy conservation. At the same time, concrete replacement is the effective measure for carbon reduction, as shown in Table 2. Other measures include such as replacing oil with gas in the mixing station, reusing old road materials, and arranging material transportation which contribution less compared with the main materials as pavement and slope protection measures.

https://doi.org/10.1051/e3sconf/202452801016

Table 2 energy conservation and carbon reduction contribution

measures	GWP	PED	Implied in the provincial road	Introduction
Foam warm mix asphalt	3%	28%	0	Future possible
Rubble concrete replacing foam light soil protection	17%	9%	0	Future possible
Mortar rubble protection replacing rubble concrete	11%	6%	0	possible
25% Plant mixed hot regeneration dosage	0	9%	0	possible
Rubble concrete ditch replaced by rubble concrete ditch	3%	1%	0	possible
Changing oil to gas in mixing station	1%	3%	100%	
Reuse of old road as sub-base	1%	1%	30%	
Granular material reuse to abutment backfill	1%	1%	100%	
Reuse of mortar rubble	1%	1%	100%	
Reuse of waste excavation in backfill	1%	1%	100%	
Optimize transportation organization	1%	1%	100%	

(1) Comparison and selection of pavement measures

The carbon emission of the concrete measure are much higher than that of the asphalt measure because the carbon of the cement product itself is amplified, and the cement structure has poor reprehensibility, which will be abandoned after the design life,shown in Table 3.

Table 3 Energy saving and carbon reduction of pavement measures

	Asphalt	Cement concrete	25%Asphalt plant mixing hot regeneration	Foam asphalt cold regeneration
GWP (kg CO2 eq/m2)	30	91	30	24
PED (MJ/m2)	1212	736	981	485
Cost per square meter	134	145	94	137
GWP contribution	16%	50%	16%	13%
PED contribution	46%	28%	37%	18%
	Implementation		Conservative optimization	

(2) Comparison and selection of slope protection measures

Rubble structure is superior to concrete structure from the carbon reduction aspect in Table 4.Considering the construction inefficiency of rubble structure, increasing the labour subsidy of mortar rubble can increase the use of rubble structure.

Table 4 Energy saving and carbon reduction of the protection measure

	Rubble concrete+8% foam light soil	Rubble	Rubble concrete	foam light soil
GWP (kg CO2 eq/m3)	218	80	211	309
PED (MJ/m3)	1542	551	1503	1994
Cost per cubic meter	746	400	750	700
GWP contribution	21%	8%	21%	30%
PED contribution	11%	4%	10%	13%
	Implementation	Conservative optimization		

According to different implementation scenarios including "no energy conservation and carbon reduction measures"、 "Implied in the provincial road "、 "conservative optimization" and "future optimization" as shown in Fig. 3. Based on the existing engineering technology, the "conservative optimization" scenario is expected to further reduce carbon by 17-22% and energy consumption by 13-16%. It is expected to reduce carbon by 22-28% and save energy by 45-50% in future.

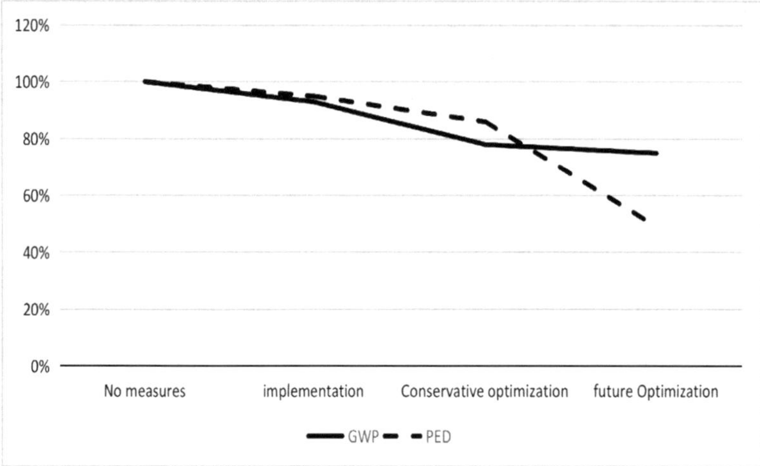

Fig. 3 Prediction of Energy Conservation and Carbon Reduction Scenarios

4. Conclusion

A life cycle standardized assessment module of energy-saving and carbon reduction is established based on the current road cost management software. The assessment module combines the domestic road material consumption quota and the main LCA database in China. Quantitative energy-saving and carbon reduction benefits of material and technical alternative measures could be calculated autocratically.

Reducing concrete, reducing mixing temperature and reusing material are essential measures that reduces energy consumption and carbon emissions throughout the life cycle of the road reconstruction . In addition, the optimized technical measures include using new energy, reusing old road waste, and arranging material transportation, are also essential measures to reduce energy consumption and carbon.It is possible to reduce carbon by 22-28% and save energy by 45-50%.

References

1. ZHENGYi.The Embodiment of Low-carbon Concept in Highway Design[J].Foreign science and technology journal database (abstract edition) Engineering technology, 2022.
2. Piantanakulchai M, Inamura H, Takeyama Y. A life cycle inventory analysis of carbon dioxide for a highway construction project using input-output scheme a case study of the Tohoku expressway construction works[J]. Doboku Gakkai Ronbunshuu D, 2010, 16.
3. Yue Huang, Bachar Hakim, Susanna Zammataro. Measuring the carbon footprint of road construction using CHANGER[J]. International Journal of Pavement Engineering, 2013, 14(6):1-11.
4. Van Rompu J , Olard F , Pouget S ,et al.Mechanical and Environmental Performances of Warm Mix Asphalt with Reclaimed Asphalt and Plant-Based Binder[J].Journal of Testing and Evaluation: A Multidisciplinary Forum for Applied Sciences and Engineering, 2023(4):51..
5. Castigliego J R , Pollack A , Cleveland C J ,et al.Evaluating emissions reductions from zero waste strategies under dynamic conditions: A case study from Boston[J].Waste Management, 2021(126-):126.
6. Fernández-Sánchez G, Álvaro Berzosa, Barandica J M, et al. Opportunities for GHG carbons reduction in road projects: a comparative evaluation of carbons scenarios using CO2NSTRUCT[J]. Journal of Cleaner Production, 2015, 104:156-167.
7. White P, Golden J S, Biligiri K P, et al. Modeling climate change impacts of pavement production and construction[J]. Resources Conservation & Recycling, 2010, 54(11):776-782.
8. Geng J , Zhou D , Bai J ,et al.Quantification of the carbon emission of urban residential buildings: The case of the Greater Bay Area cities in China[J].Environmental impact assessment review, 2022(Jul.):95.
9. Azarijafari H, Yahia A, Amor M B. Life cycle assessment of pavements: reviewing research challenges and opportunities[J].Journal of Cleaner Production 2016, 112:2187-2197.

Study on Wax Deposition Law in Daqing Gulong Shale Oil

Xuefeng Zhao[1,2], Yunchen Wang[3], Zhongshan Zhao[2,4], Dongmei Leng[2,4], Qin Liu[2,4], Jionghao Li[3]*

[1]Daqing Oil Field Co, Daqing, Heilongjiang, 163002, China
[2]National Key Laboratory of Continental Shale Oil,Daqing, Heilongjiang, 163002, China
[3]China University of Petroleum,Beijing, Changping, Beijing, 102249, China
[4]Daqing Oilfield Design Institute Co.,Ltd, Daqing, Heilongjiang, 163002, China

Abstract. Daqing Gulong shale oil exhibits elevated wax content, a high wax precipitation point, and a high freezing point. However, its oil density and viscosity are relatively low, making it susceptible to wax crystal precipitation. The risk of pipeline clogging is heightened in lower ambient temperatures, posing a hazard to the safe production of shale oil. This study focuses on testing the fundamental physical properties of crude oil. The wax deposition flow loop was employed to replicate wax deposition in extracted fluids under actual high-flow gas-liquid conditions within a pipeline. The deposition patterns were determined based on the experimental findings. The results indicate that, irrespective of whether in single-phase, gas-liquid two-phase, or oil-gas-water three-phase conditions, the deposition mass rises with decreasing oil temperature. Furthermore, the deposition mass rises with an increased temperature difference between the oil and the pipeline wall. Conversely, the deposition mass declines with an increased gas-liquid ratio, and the deposition thickness declines with elevated water content. The observed deposition patterns highlight the significant influence of molecular diffusion and the impact of gas phase shear. These experimental findings offer valuable references and insights for the transportation of Daqing shale oil through pipelines.

1. Introduction

Pipeline transportation plays a crucial role in the conveyance of oil and natural gas, serving as an economically and environmentally friendly method with substantial conveying capacity [1]. However, challenges arise in the pipeline transportation of crude oil containing wax. When the temperature of the pipe wall falls below both the oil temperature and the wax precipitation point, wax molecules within the oil stream gradually precipitate. This leads to the formation of deposits on the pipe wall, causing an increase in friction coefficient, pressure drop, and a reduction in circulation area, among other issues [2]. Ultimately, these challenges contribute to the occurrence of pipe clogging, presenting significant risks to the safe production of oil. Addressing this issue has become a prominent research focus in the field of flow assurance.

Numerous scholars have delved into the exploration of diverse factors influencing wax deposition, contributing to a more comprehensive understanding of the characteristics of wax deposition under various conditions. Fan and Ehsani used a cold finger device for wax deposition experiments to study the wax deposition phenomena[3-4]. M. Lashkarbolooki conducted single-phase wax deposition experiments on crude oil from the Kershman field using a loop channel. This research experimentally investigated the impacts of oil temperature, wall temperature, and the temperature difference between the oil and the wall on wax deposition [5]. Rainer and Lene explored the influence of flow rate on wax deposition through experiments using a loop channel with condensate from the North Sea [6]. Singh scrutinized the effect of deposition time on wax deposition through microscopic observation and gas chromatography [7]. At the University of Calgary, Kasumu conducted experiments on wax deposition in a water-in-oil (W/O) emulsion loop channel, varying water contents, flow rates, and temperatures using wax-containing simulated oil and tap water [8]. Similarly, Ekarit, from the University of Tulsa, performed experiments on wax deposition from a water-in-oil (W/O) type emulsion annulus under different conditions of water content, flow rate, and deposition time, utilizing South Pelto crude oil [9]. Matzain and Nilufer individually investigated wax deposition under gas-liquid two-phase flow conditions using crude oils with varying wax contents [10-11].

While the aforementioned experimental research results offer valuable insights for actual pipeline transportation, their applicability to cases involving highly waxy oil may be limited. Crude oils in China typically exhibit high wax content, often exceeding 20%. Consequently, the wax deposition characteristics of high wax content oils in China may differ significantly due to the diverse properties of different oils. Daqing Gulong shale oil, being a representative crude oil with both high wax content and a high wax precipitation point, experiences extremely easy wax precipitation.

*Corresponding author's e-mail: 1019394325@qq.com

Given these unique characteristics, it becomes imperative to conduct specific research on the wax deposition patterns and control methods for Daqing Gulong shale oil. This research aims to elucidate the wax deposition behavior of Gulong shale oil and formulate an economically and logically sound wax deposition control program accordingly.

2. Experimental section

2.1. Experimental shale oil properties

The physical properties of the shale oil used in the experiment are shown in Table 1, with a Pour point of 30 °C, a wax appearance temperature of 50.86 °C, and a wax content of 23.29 wt%.

Table 1. Physical properties of the shale oil sample.

Density at 25°C(kg/m^3)	Pour point(°C)	WAT(°C)	wax (wt.%)
812.4	30	50.86	23.29

2.2. Wax Deposition Flow Loop

The schematic representation of the wax deposition test is illustrated in Figure 1, depicting the wax deposition flow loop. The experimental setup predominantly includes an oil storage tank, a transfer pump, a water supply system, a gas supply system, a hot bath temperature control system, a cold bath temperature control system, a measuring section, a measurement system, and a data acquisition system.

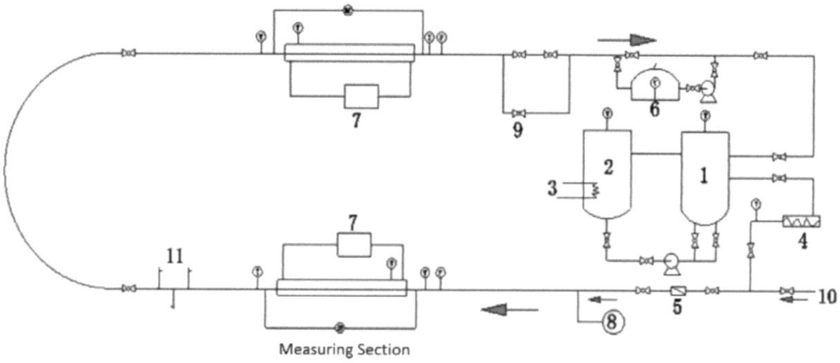

1-oil tank ; 2-circulating water tank ; 3 - water tank heating coil ; 4 - screw pump ; 5-liquid mass flowmeter ; 6 - gas-liquid separator ; 7 - temperature controlled water bath ; 8-gas injection section ; 9-fast closing pipe section ; 10 - Air purge ; 11 - Local Sampler

Figure 1. Schematic diagram of the flow loop apparatus for wax deposition.

The entire loop spans a total length of 25 meters and is subject to temperature control achieved by circulating pure water through a water bath temperature control system, ensuring compliance with experimental temperature requirements. Within this setup, the test section, responsible for wax deposition, features a diameter of 25.4mm and extends over a length of 2 meters. It is temperature-controlled through a dedicated thermostatic water bath. The oil storage tank, acting as the source for the oil supply system, boasts a volume of 120 liters. Surrounding this tank is a heat bath responsible for elevating the oil temperature to the desired level. To ensure uniform heating of the crude oil, a stirring paddle is employed within the tank. A screw pump facilitates the conveyance of the oil sample from the heated tank to the annulus and back to the tank. The flow rate is adjustable within a specified range, regulated by the pump, and monitored by a mass flow meter. The gas supply system has the capacity to provide gas for the entire loop, facilitating multi-phase mixed transport experiments. The measurement system and data acquisition system enable the transmission and recording of sensor data, including pressure, temperature, and flow rate, to a computer.

The wax deposition flow loop is designed to conduct wax deposition experiments under both single-phase and multiphase conditions. It is equipped to investigate the influence of key experimental factors, including fluid temperature, wall temperature, flow rate, deposition time, and gas-oil ratio, on the wax deposition characteristics.

2.3. Experimental Process

The experimental procedure involved initial preheating of shale oil to 70 °C for a duration of 6 hours, followed by injection into the annulus tank. To maintain the crude oil at the desired temperature, both the tank and the pipe were heated, with the test pipe section specifically held at the experimental temperature. Once the system temperature stabilized, the experiment commenced. The oil flowed inside the pipe, and the flow was halted when the designated deposition time was achieved. Subsequently, the air blowdown process was initiated, and after completion, the test section was disassembled. The thickness of the wax deposition layer was determined using the volumetric method. To ensure accuracy, the experiment was repeated several times, and the average value of the deposition thickness was calculated. The investigation into the wax deposition law under different experimental conditions involved modifying air volume, liquid volume, and water content while repeating the aforementioned experimental steps.

Table 2. Experimental scheme.

No.	Oil Temp. (°C)	Coolant Temp. (°C)	Temp. difference (°C)	Gas-liquid ratio (%)	Water content
1	40	35	5	0	0
2	35	30	5	0	0
3	35	25	10	0	0
4	35	30	5	4:1	0
5	35	30	5	13:1	0
6	35	30	5	0	20
7	35	30	5	4:1	20
8	35	30	5	13:1	20
9	35	30	5	0	40
10	35	30	5	0	60

The experimental scheme is shown in Table 2. Throughout the experiment, a constant flow rate of 0.9 m/s was maintained, while the deposition time varied at intervals of 1hour, 2hour, 3hour, 4hour, and 5hour. The study involved a comparative analysis to investigate the impacts of experimental time, temperature, temperature difference, gas-liquid ratio, water content, and other influencing factors on the thickness of wax deposition.

3. Result and discussion

Figure 2 demonstrates the effect of time, temperature interval, and temperature gradient on the deposited thickness under single-phase conditions. It can be observed that the trend of the increase of the deposited thickness gradually slows down with the increase of time. This is due to the thermal resistance effect resulting from the formation of the wax deposition layer, which acts as an inhibitor to the diffusion of wax molecules. Fixing the temperature difference between the oil walls, there is a tendency to exacerbate the deposition in the low temperature range. At the beginning of wax deposition formation, an initial gel layer is first formed on the cold wall. Obviously, in the low-temperature interval, the further the dissolved wax molecules crystallize away from the wall, the further the highest temperature isotherm for gel formation is from the cold wall, resulting in a thicker initial gel layer. Therefore, the gel layer gradually grows into a thicker wax deposition layer in the low temperature interval. Fixing the oil temperature, there is a tendency to increase the temperature gradient to intensify the deposition. This is due to the enhancement of the deposition driving force due to the temperature gradient. All of the above behaviors are wax deposition effects triggered by molecular diffusion [12].

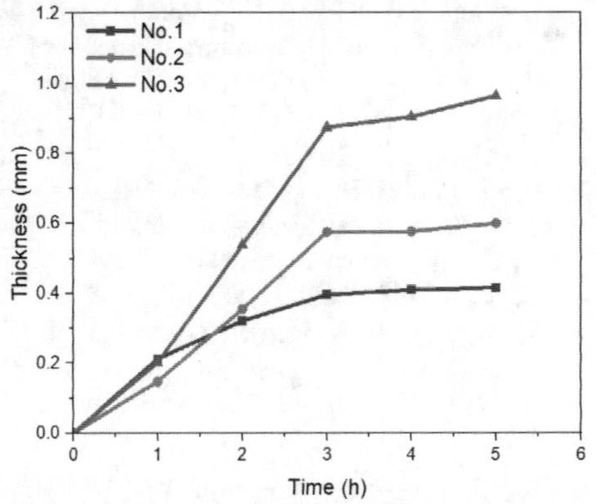

Figure 2. The effect of time and temperature difference on the thickness of wax deposition.

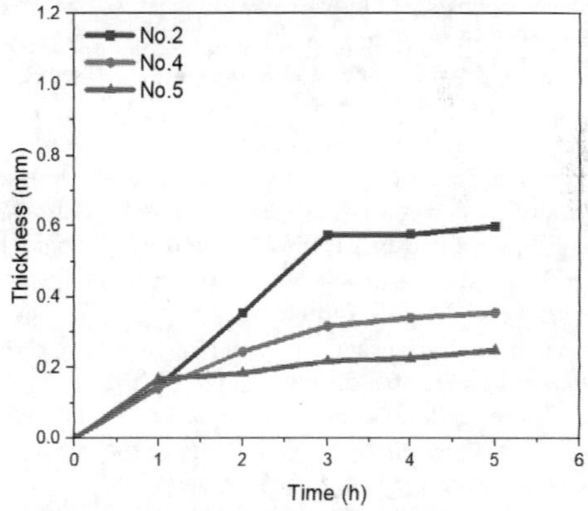

Figure 3. Effect of gas-liquid ratio and time on deposition thickness.

Figure 3 illustrates the influence of gas-liquid ratio and time on the deposition thickness in gas-oil two-phase conditions. Notably, the trend of deposition thickness growth exhibits a gradual slowdown with increasing time, mirroring observations in single-phase scenarios. Conversely, as the gas-liquid ratio rises, resulting in an augmented gas-phase velocity, the deposited layer undergoes the impact of gas shear, leading to a substantial reduction in the thickness of the deposited layer.

In Figure 4, the impact of gas-liquid ratio and time on deposition thickness is explored under oil-gas-water three-phase conditions with a constant 20% water content. Notably, the trend reveals a gradual deceleration in the increase of deposition thickness with extended time. Concurrently, a rise in the gas-liquid ratio corresponds to a reduction in deposition thickness, aligning with the observed behavior in gas-oil two-phase systems.

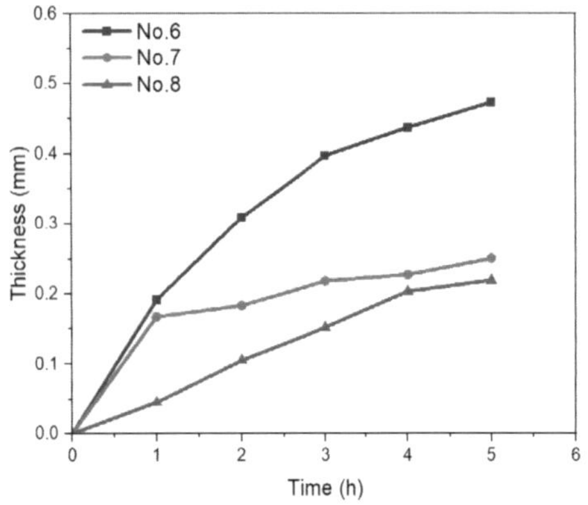

Figure 4. Effect of gas-liquid ratio and time on deposition thickness at 20% water content.

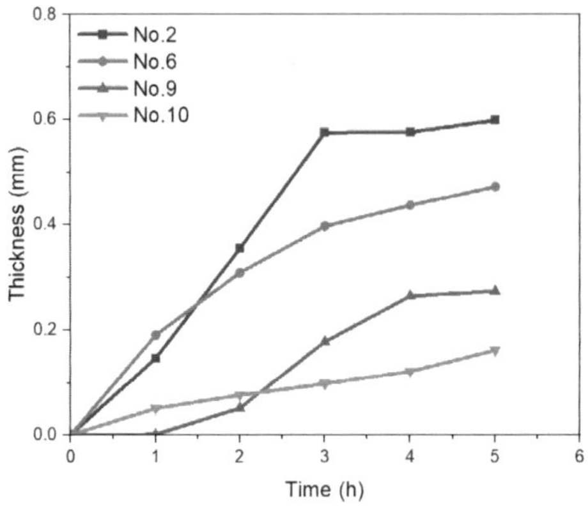

Figure 5. Effect of varying water content and time on sediment thickness.

In Figure 5, the influence of varying water content and time on sediment thickness in both oil and water phases is showcased. Notably, the trend indicates a gradual deceleration in the increase of deposited thickness with prolonged time. Furthermore, as the water content rises, there is an observed decrease in deposition thickness. This phenomenon is attributed to the potential inhibitory effect of water, which hinders the diffusion of wax molecules and serves as a barrier to deposition.

4. Conclusion

The experimental investigation of oil deposition in Daqing Gulong shale, conducted using a flow loop device, highlights the substantial impact of various factors on wax deposition thickness. Time, temperature difference, gas-liquid ratio, and water content emerge as key influencers in this study. Notably, as time progresses, there is a significant increase in wax deposition thickness, a phenomenon attributed to the diffusion of wax molecules. Furthermore, under consistent water content conditions, elevating both the gas-phase flow rate and water content correlates with a reduction in wax deposition thickness. This observed trend is linked to the inhibitory effects of gas-phase shear and water molecule diffusion, which collectively impede the diffusion of wax to the pipe surface. These findings contribute valuable insights and establish a solid foundation for optimizing the pipeline transportation of Daqing Gulong shale oil.

References

1. Diwen C , Ying X , Fangrui M ,et al.Research progress in wax deposition in oil-water two-phase flow[J].Petrochemical Technology, 2018.
2. Li, Bingfan, et al. A comprehensive review of wax deposition in crude oil systems: Mechanisms, influencing factors, prediction and inhibition techniques[J].Fuel,2024,357:129676.
3. Fan K , Li S , Li R .Development of wax molecular diffusivity correlation suitable for crude oil in wax deposition: Experiments with a cold-finger apparatus[J].Journal of Petroleum Science and Engineering, 2021, 205(7):108851.DOI:10.1016/j.petrol.2021.108851.
4. Ehsani S , Haj-Shafiei S , Mehrotra A K .Experiments and modeling for investigating the effect of suspended wax crystals on deposition from 'waxy' mixtures under cold flow conditions[J].Fuel, 2019, 243(MAY 1):610-621.DOI:10.1016/j.fuel.2019.01.089.
5. M. Lashkarbolooki, A. Seyfaee, F. Esmaeilzadeh, and D. Mowla. Experimental Investigation of Wax Deposition in Kermanshah Crude Oil through a Monitored Flow Loop Apparatus. Energy Fuels 2010, 24, 1234–1241
6. Mason S L , Meyer J C , Needham D J .The Development of a Wax Layer on the Interior Wall of a Circular Pipe Transporting Heated Oil -- The Effects of Temperature Dependent Wax Conductivity[J]. 2021.DOI:10.1007/s10665-021-10171-x.
7. Fan K F, Li S, Huang Q Y, et al. Research progress on radial properties of wax deposits in crude oil pipelines[J]. Chemical Industry and Engineering Progress, 2020, 2020: 1-19.
8. KASUMU A S, MEHROTRA A K. Solids Deposition from Wax–Solvent–Water "Waxy" Mixtures Using a Cold Finger Apparatus[J]. Energy & Fuels, 2015,29(2):501-511.
9. Ekarit Panacharoensawad. Wax deposition under two-phase oil-water flowing conditions [D]. Tulsa, Oklahoma, University of Tulsa, 2012.
10. Liu H S , Duan J M , Li J ,et al.Wax deposition modeling in oil-water stratified pipe flow[J]. Petroleum Science, 2023, 20(1):526-539.
11. Nilufer.K.Multiphase paraffin deposition behavior of a Garden Banks Condensate:(Doctoral Dissertaiton).Tusla:University of Tusla.2003.
12. Aiyejina A , Chakrabarti D P , Pilgrim A ,et al.Wax formation in oil pipelines: A critical review[J].International Journal of Multiphase Flow, 2011, 37(7):671-694. DOI:10.1016/j.ijmultiphaseflow.2011.02.007.

Study on Characteristics of Six Polysaccharides and Cellulose Blends

Jing Liu[1], Tao Zhang[1], Meiwei Zhao[2], Huifan Li[3], Zengjin Zhu[4], Fan Xie[4], Lei Yang[1], Mengfei Jiang[1], Song Miao[1], Guiyuan Zhou[1,*]

[1]Research and Development Center, China Tobacco Yunnan Industrial Co., Ltd., Yunnan Kunming, 650201, China
[2]Kunming College, Yunnan Kunming, 650201, China
[3]China Tobacco Schweitzer (Yunnan) Reconstituted Tobacco Co., Ltd, Yuxi, Yunnan Province, 653100 China
[4]Institute of health sciences and engineering, University of Shanghai for Science and Technology, Shanghai 200093, China

Abstract. In order to study the internal cross-linking mechanism between cellulose and polysaccharides as wet-end additives in papermaking reconstituted tobacco, crystalline cellulose (MCC) was used as the target in this paper, sodium alginate (SA) , pullulan (Pul) , κ-carrageenan (κ-C), carboxymethyl chitosan (CMCS), hydroxypropyl chitosan (HPCS) and tamarind polysaccharide (TSP) were selected, the structure and composition of the monomer polysaccharides were determined by dynamic light scattering and ion chromatography, at the same time, the characteristics of the above-mentioned polysaccharides and MCC blend were analyzed, the results showed that SA contained a lot of negative charge, which resulted in weak adhesion and poor adsorption ability in cellulose blend; Pul was the most compact chain conformation, effectively improve the adhesion of cellulose, can be quickly adsorbed on cellulose; κ-C increased the viscosity of MCC blend and could adhere the 30μm size fragments of cellulose CMCS had strong adhesion to cellulose edge; due to the introduction of carboxymethyl group, the hydrophilicity of HPCS and the apparent viscosity of MCC blend increased greatly; the complex branched-chain structure of TSP results in high storage modulus of TSP/cellulose blend, and the amount of TSP adsorbed on cellulose is large and lasts long. Pul, TSP, κ-c, CMCS and TSP could adsorb cellulose well.

1. Introduction

Polysaccharides are hydrophilic, high-molecular-weight polymers formed by polymerizing monosaccharide units such as glucose, fructose, galactose, and arabinose in certain proportions. Chemically, the structural diversity of polysaccharide polymers, arising from variations in sugar unit types, polymerization degrees, inter-sugar bonding and arrangement, degree of group substitution, and multi-level structures, gives rise to both common and specific properties such as viscosity, fluidity, gelation, film formation, among others.

As one of the crucial components in cigarette manufacturing, the paper-making process for reconstituted tobacco leaves has garnered increasing attention. This process transforms waste materials (such as tobacco dust, scraps, trimmings, ash rods, and low-grade tobacco leaves) into valuable resources and supports functions like reducing tar and harm, lowering consumption, and highlighting style in cigarette production. In the tobacco industry, food colloids, mainly wet-end retention aids, fiber dispersants, and adhesives, are applied to produce reconstituted tobacco leaves. In order to further explore the application of food colloids[1-2] in the reconstituted tobacco leaf industry, this study selected six typical food colloids, including tamarind polysaccharide, pullulan, sodium alginate, carboxymethyl chitosan, hydroxypropyl chitosan, and κ-carrageenan as additives for reconstituted tobacco leaves. These were blended with microcrystalline cellulose to investigate their interaction with cellulose and explain the internal crosslinking mechanism behind the macroscopic effects, providing a basis for applying food colloids in reconstituted tobacco leaves.

2. Materials and Methods

2.1 Materials and Instruments

Sodium alginate (SA) from MACKLIN; pullulan (Pul) from Usolf; κ-carrageenan (κ-C) from Usolf; carboxymethyl chitosan (CMCS) from Usolf; hydroxypropyl chitosan (HPCS) from Usolf; tamarind polysaccharide (TSP) with purity ≥97%, from Yunnan Maodouli Group Food Co., Ltd.; microcrystalline cellulose (MCC) with a length ≥30 μm.

Discovery HR-3 rotational rheometer from TA Instruments, USA; BC-R202 rotary evaporator from Shanghai Beikai Biological Chemical Equipment Co., Ltd.; HPSEC-MALLS multi-angle laser light scattering

*Corresponding author's e-mail: 13769374852@163.com

instrument from Wyatt, USA; ICS-5000+ ion chromatography system from Thermo, USA; JSM 7610 FPLUS thermal field emission scanning electron microscope from JEOL, Japan; electronic analytical balance from METTLER TOLEDO, USA; BPZ-6933LC vacuum drying oven from Shanghai Yiheng Scientific Instrument Co., Ltd.

2.2 Experimental Methods

2.2.1 Analysis of Food Colloid Monosaccharide Composition.

Clean chromatography bottles were used to weigh appropriate amounts of polysaccharide samples and add 1 mL of 2M TFA acid solution. The mixture was then heated at 121°C for 2 hours. After nitrogen flushing and drying, the sample was washed with 99.99% methanol and dried again. This methanol washing process was repeated 2-3 times. The sample was then dissolved in sterile water and transferred to a chromatography bottle for analysis[3].

The chromatographic system employed was the Thermo ICS 5000+ ion chromatography system (ICS 5000+, Thermo Fisher Scientific, USA), utilizing an electrochemical detector to analyze and detect monosaccharide components.

Dionex™ CarboPac™ PA20 (150*3.0 mm, 10 μm) liquid chromatography column was used with an injection volume of 5 μL. The mobile phases consisted of A (H2O), B (0.1M NaOH), and C (0.1 M NaOH, 0.2M NaAc), with a flow rate of 0.5 mL/min. The column temperature was maintained at 30°C. The elution gradient was as follows: 0 min A/B/C (95:5:0, V/V), 26 min A/B/C (85:5:10, V/V), 42 min A/B/C (85:5:10, V/V), 42.1 min A/B/C (60:0:40, V/V), 52 min A/B/C (60:40:0, V/V), 52.1 min A/B/C (95:5:0, V/V), and 60 min A/B/C (95:5:0, V/V)[4].

2.2.2 Analysis of Food Colloid Molecular Weight and Conformation.

The molecular weight and conformation of the six polysaccharides were determined using a high-performance liquid chromatography system comprising a multi-angle laser light scattering detector (MALLS, DAWN HELEOS-II), refractive index (Optilab T-Rex, RI) detector, and differential pressure viscometer (ViscoStar III, DP) (Wyatt Technology, Santa Barbara, CA, USA). Following slight modifications to the experimental method outlined by Strydom et al. [5], 5 mg of the sample was dissolved in 5 mL of mobile phase (0.1 mol/L NaNO3) and hydrated at room temperature with continuous stirring at 500 rpm for 24 hours. Molecular weight analysis of prepared samples was conducted using high-performance liquid chromatography-multi-angle laser light scattering detector-refractive index detector-viscometer analysis. The analysis column comprised a guard column (OHpak SB-G), Shodex OHpak 803 HQ, and 805 HQ, with a flow rate of 0.6 mL/min and a column temperature of 40°C. Data analysis was performed using ASTRA7.1.3 software.

2.2.3 Preparation of Food Colloid-Cellulose Blend Solutions.

2.0 g of cellulose was diluted with 10 mL of water, stirred thoroughly until fully swollen, and mixed with 10 mL of polysaccharide solution (0.1%, w/v).

2.2.4 Adsorption Kinetics.

Adsorption kinetics of cellulose were determined by adding 10 mg/mL cellulose and 1 mg/mL polysaccharide to deionized water under 25°C and 150 rpm shaking conditions. Samples were collected at 3, 6, 9, and 12 minutes, followed by centrifugation at 10000 rpm for 10 minutes. The supernatant was analyzed using the phenol-sulfuric acid to determine the total sugar content. The change in total sugar content before and after adsorption onto cellulose was used to calculate the amount of polysaccharide adsorbed onto the cellulose sample.

Total sugar content was determined using the phenol-sulfuric acid method. Specifically, 12.5 g of anhydrous glucose was dissolved in distilled water and made up to 50 mL, resulting in a 0.25 mg/mL glucose standard solution. Varying volumes (0, 0.1, 0.2, 0.3, 0.4, and 0.5 mL) of the glucose standard solution were mixed with deionized water to 1.0 mL in test tubes, followed by the addition of 0.5 mL of 3% phenol solution and 5 mL of concentrated sulfuric acid (added slowly). After mixing, the tubes were left to react for 0.5 hours at room temperature, and absorbance at 490 nm was measured using a spectrophotometer. A standard curve was plotted using the mass concentration of the standard solution as the abscissa and A490 as the ordinate. The experiment was performed in triplicate[6].

Accurately weigh 5.0 mg of the sample and dissolve it in 10 mL of deionized water to obtain a solution with a concentration of 0.5 mg/mL. Take 0.1 mL of the solution and dilute it to 1.0 mL with water. Measure the absorbance at 490 nm according to the procedure above. Calculate the total sugar content based on the standard curve. Perform triplicate measurements.

2.2.5 Rheological Property Measurement.

The steady flow characteristics and dynamic viscoelasticity of the six polysaccharide blends were determined using a rotational rheometer. Steady-state flow behavior was analyzed at shear rates of 0.1-100 s-1, while dynamic viscoelasticity, including storage modulus (G') and loss modulus (G"), was measured by frequency scanning in the range of 0.1 to 100 rad/s. The rheometer employed a plate-plate fixture (40mm) with a gap set at 1.0 mm. All tests were performed within the linear viscoelastic region, and strain scans (1Hz,0.01% ~20%) were conducted to obtain all measurements, with a strain value of 1%.

2.2.6 Viscosity Measurement.

1.36 g of cellulose was dissolved in 100 mL of DMSO, followed by the addition of 13.6 mg of polysaccharide. After thorough mixing and swelling, the viscosity of the resulting solution was measured using an Ubbelohde viscometer (constant temperature of 20°C with an error of ± 0.01°C). The solution's flow time (t) at each concentration was determined as the average of four parallel measurements.

In a polymer/solvent system, the relationship between the reduced viscosity of the polymer and its concentration follows the classical Huggins equation. By calculating ηsp=(t-t0)/t0, the specific viscosity ηsp/c of the cellulose and polysaccharide solution and the specific viscosity ηspm/c of the blend can be obtained. Plotting ηsp/c and ηspm/c against c allows for determining characteristic viscosity [η]. A series of data on the relative viscosity and specific viscosity of the blend were used to derive the viscosity average molecular weight (μ). The calculation formula is as follows:

$$b_m = {w_2}^2 b_2 + {w_3}^2 b_3 + 2w_2 w_3 b_{23} \quad (1)$$

$$b_2 = {[\eta]_2}^2 k_2 \, , b_3 = {[\eta]_3}^2 k_3 \quad (2)$$

$$[\eta]_m = [\eta]_2 w_2 + [\eta]_3 w_3 \quad (3)$$

$$VB = (b_m - b)/2(2w_2 w_3) \quad (4)$$

$$\mu = \frac{VB}{([\eta]_3 - [\eta]_2)^2} = \frac{\left\{\frac{b_m - b_2}{[\eta]_m - [\eta]_2} - \frac{b_3 - b_2}{[\eta]_3 - [\eta]_2}\right\}}{2([\eta]_3 - [\eta]_m)} \quad (5)$$

Among them, wi = Ci/C, i = 2,3, where C is the total polymer concentration; [η]2 and [η]3 are the characteristic viscosities of each polymer. The compatibility of the blend was studied using the μ criterion. μ≥0 indicates miscibility, while μ< 0 indicates phase separation [7].

2.2.7 Microstructural Observation.

The polysaccharide blend solution was placed in a vacuum-drying oven at 55°C for 48 hours to remove moisture. The dried blend samples were adhered to conductive adhesive and sputter-coated. The dried blend samples' surface microstructure was analyzed using a field emission scanning electron microscope.

2.2.8 Thermal Gravimetric Analysis (DTG).

Thermogravimetric analysis was performed using a DTG-60H thermogravimetric analyzer with a sample loading of approximately 7 mg and nitrogen gas as the carrier with a 50 mL/min flow rate. The heating program was set to ramp at 20°C/min from 30°C to 800°C. The weight loss of the sample was calculated according to the following formula:

$$w(\%) = \frac{m_0 - m_t}{m_0} \times 100 \quad (6)$$

Where: w denotes the weight loss of the sample, % (mass fraction); m0 denotes the initial mass of the sample, mg; and mt denotes the mass at temperature t, mg.

3. Results and Analysis

3.1 Monosaccharide Composition Analysis

Monosaccharide composition analysis is a necessary step in studying polysaccharides' structural characteristics and structure-function relationships. Monosaccharide composition analysis was conducted on the six polysaccharides mentioned above, as detailed in Table 1.

Table 1. Monosaccharide composition of six polysaccharides.

Monosaccharide / Sample	Ara	Gal	Glc	Xyl	Man	Gal-UA	Gul-UA
SA	0.35%	0.00%	96.72%	0.00%	0.00%	1.09%	1.84%
Pul	0.00%	1.04%	96.84%	0.00%	2.12%	0.00%	0.00%
κ-C	0.00%	75.41%	22.65%	0.00%	0.00%	1.94%	0.00%
CMCS	0.00%	4.72%	91.24%	0.00%	0.00%	0.00%	0.00%
HPCS	0.00%	0.46%	84.35%	0.00%	0.00%	0.00%	0.00%
TSP	1.78%	21.34%	39.59%	35.17%	0.00%	2.13%	0.00%

It can be observed that Sodium Alginate (SA), Pullulan (Pul), Carboxymethyl Chitosan (CMCS), and Hydroxypropyl Chitosan (HPCS) are primarily composed of glucose. κ-Carrageenan (κ-C) predominantly comprises galactose and glucose, while Tamarind Seed Polysaccharide (TSP) consists mainly of glucose, xylose, and galactose.

According to literature reports [8-14], Sodium Alginate (SA) is a copolymer composed of β-D-mannuronic acid (M units) and α-L-guluronic acid (G units) connected by β-1,4-glycosidic bonds, forming different proportions of GM, MM, and GG segments. Sodium Alginate (SA) typically carries a negative charge over a wide pH range. Pullulan (Pul) is an α-glucan composed of repeating maltotriose units linked by α-1,4-glycosidic bonds and interconnected by α-1,6-bonds, resulting in short branch lengths, minimal steric hindrance, and high rigidity. Chitosan is a linear polysaccharide; in Carboxymethyl Chitosan (CMCS), the incorporation of carboxymethyl groups enhances the water solubility of chitosan, alters the secondary structure of chitosan molecules and reduces crystallinity, while Hydroxypropyl Chitosan (HPCS), a derivative of chitosan, exhibits improved water solubility due to the

https://doi.org/10.1051/e3sconf/202452801018

introduction of hydroxypropyl groups. κ-Carrageenan (κ-C) is a linear polymer chain composed of galactose and glucose units, characterized by a large molecular weight and long chain length. Tamarind Seed Polysaccharide (TSP) comprises galactose, glucose, and xylose, featuring a relatively hydrophobic main chain. In TSP, d-glucopyranose (β-d-Glc p) is substituted at O-75 by pyranose xylose residues (α-d-Xyl p) and further connected to pyranose galactose units (β-d-Gal p), resulting in a more complex branched structure compared to other polysaccharides.

3.2 Light Scattering Analysis

In polymer conformation, the structure parameter ρ (Rg/Rh) can be used to characterize the conformational features of polymer chains in solution:

- When ρ=0.77, it indicates a spherical conformation.
- As ρ increases, the polymer chains become looser, more expanded, and more flexible.
- When ρ=1.5-1.8, polymer chains are in a linear and flexible state.
- When ρ is around 2, the dispersion system becomes irregular aggregates or highly branched structures.
- When ρ is much larger than 2, the dispersion system turns into a rigid rod-like structure[15].

The Mark-Houwink-Kuhn-Sakurada equation ([η] = kMαη) with α (M-H) values of 0.22, 0.5-0.8, and 1-2, respectively, reflects the conformation of spheres, flexible chains, and rigid rods[16].

Table 2. Results of light scattering.

Name	Mw/×105Da	Rg/nm	Rh/nm	α（M-H）	ρ
SA	5.4	83.8	50.4	0.779	1.66
Pul	3.3	17.7	16.7	0.659	1.06
κ-C	7.4	83.0	52.4	0.747	1.58
CMCS	6.6	53.6	33.2	0.668	1.62
HPCS	7.4	61.3	42.1	0.696	1.45
TSP	8.1	63.1	43.3	0.687	1.46

Table 2 presents the results of light scattering for the six polysaccharides. From Table 2, it can be observed that the order of conformational looseness to compactness is as follows: Sodium Alginate (SA) > Carboxymethyl Chitosan (CMCS) > κ-Carrageenan (κ-C) > Tamarind Seed Polysaccharide (TSP) > Hydroxypropyl Chitosan (HPCS) > Pullulan (Pul). The polydispersity indices (Mw/Mn) of the six polysaccharides range from 1.0 to 2.0, indicating the broad distribution of molecular weights. Pullulan (Pul) exhibits high molecular weight, branched structure, and low viscosity, with a ρ value of 1.06, indicating a semi-rigid chain [17]. The extension of the molecular chains of Sodium Alginate (SA) is due to the ionization of its carboxyl groups into negative ions in solution, leading to electrostatic repulsion between chain segments. The molecular weights of these six polysaccharides are relatively large, with α values ranging from 0.5 to 0.8, indicating that the conformation of the molecular chains is flexible and irregular.

3.3 Adsorption Kinetics Analysis

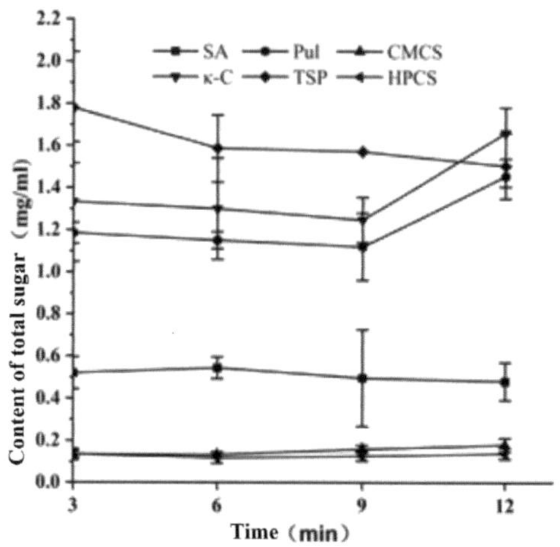

Figure 1. Content of total sugar of supernatant time.

The adsorption of polymers onto cellulose is one mechanism by which polymers prevent pulp coagulation [18]. Therefore, analyzing the adsorption of food colloids on cellulose can elucidate their anti-coagulation effects in reconstituted tobacco pulp. Changes in the total sugar content of the supernatant can be used to evaluate the adsorption of food colloids on cellulose, as shown in Figure 1. From Figure 1, it can be observed that the adsorption concentration of Hydroxypropyl Chitosan (HPCS) and Carboxymethyl Chitosan (CMCS) on MCC surfaces showed minimal variations, indicating their stable adsorption on the MCC surface. Over the 3-12 minute period, the total sugar concentration of Tamarind Seed Polysaccharide (TSP) continuously decreased, indicating ongoing adsorption. The side-chain structure of TSP facilitates its cross-linking and entanglement with cellulose, although the adsorption process is relatively slow. Due to its high molecular weight, κ-Carrageenan (κ-C) becomes entwined with cellulose upon adsorption, significantly reducing the equilibrium concentration of κ-C. As the mixed system oscillates, the entangled κ-C molecules disentangle, leading to a subsequent increase in concentration.

3.4 Rheological Properties Analysis

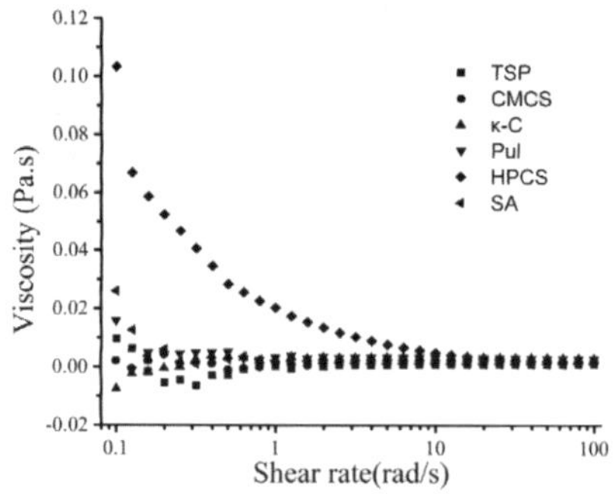

Figure 2. The viscosity change of different groups of blend liquid.

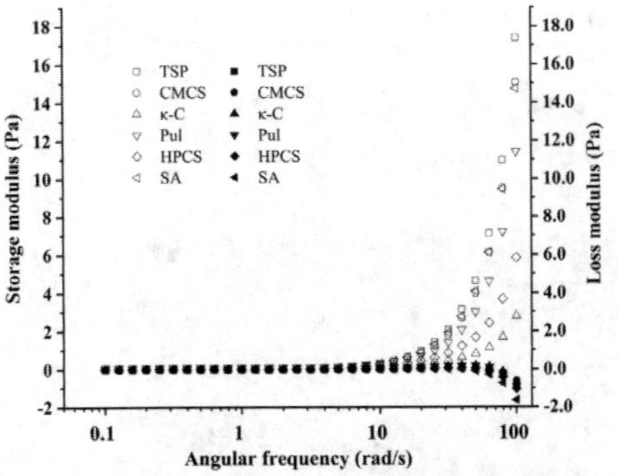

Figure 3. The modulus of different groups of blend liquid.
Note: Hollow symbols represent storage modulusG′, while solid symbols represent loss modulusG″

Figures 2 and 3 depict the viscosity and modulus curves of MCC blends with added polysaccharides. Polymer solution viscosity analysis revealed that shear thinning occurred in all five blend solutions except for the κ-C group at low shear rates. The shear thinning behavior aligns with the trends observed in the frequency scan, indicating the loss of ordered structures. At low frequencies, all six blend systems' storage modulus (G′) exceeded the loss modulus (G″). Both storage and loss moduli increased with increasing frequency, with the storage modulus increasing more significantly than the loss modulus. At high frequencies, the storage moduli of all six polymer solutions exceeded the loss moduli. This is typical behavior of semi-dilute entangled polymer solutions, indicating weak associations among polymer chains that are not strong enough at this concentration to exhibit gel-like behavior across the entire frequency range [19]. Tamarind Seed Polysaccharide (TSP) can form weak gels with cellulose through hydrogen bonding arrangements and associations, explaining the modulus levels. With its high molecular weight and numerous branching structures, TSP exhibits stronger interactions among polymer chains [20], resulting in a higher storage modulus than the other five groups.

3.5 Viscosity Analysis

Tables 3 and 4 present polysaccharides' viscosity and blend compatibility test results with crystalline cellulose MCC. It can be observed that the addition of Tamarind Seed Polysaccharide (TSP) and Hydroxypropyl Chitosan (HPCS) groups resulted in a slight decrease in viscosity. In contrast, the addition of Pullulan (Pul), Sodium Alginate (SA), and Carboxymethyl Chitosan (CMCS) groups led to a slight increase in viscosity. The addition of the κ-Carrageenan (κ-C) group resulted in a significant increase in viscosity. All groups exhibited phase separation, forming microphase-separated blocks, indicating poor compatibility [21]. Due to the presence of numerous side chains in TSP and the influence of its molecular chain conformation by bonding modes, branching structures, intramolecular hydrogen bonds, and electrostatic repulsion, it formed larger aggregates after adhering to cellulose crystalline monomers, accelerating the settling of cellulose in the aqueous solution, consistent with the cellulose adsorption results.

Carboxymethyl Chitosan (CMCS) possesses viscosity and weakly increases viscosity through interaction with cellulose [22]. κ-Carrageenan (κ-C) solution exhibits high viscosity at low concentrations and significantly increases viscosity after blending with cellulose, forming a gel readily. Pullulan is a linear macromolecule with low solution viscosity. Its hydrogen bonds with cellulose are weak, resulting in poor compatibility. However, blending with cellulose can increase viscosity.

Table 3. Viscosity of blend liquid.

	MCC/SA	MCC/Pul	MCC/κ-C	MCC/CMCS	MCC/HPCS	MCC/TSP	MCC
μ（mPa·s）	2.25±0.04	2.3±0.01	4.68±0.07	2.27±0.07	1.86±0.01	1.95±0.01	2.12±0.03

Table 4. Compatibility of blend liquid.

Number	System	W2[a]	[η] (g/L)	μ
1	MCC	-	0.011	0.0
2	SA	-	0.013	0.0
3	Pul	-	0.199	0.0
4	κ-C	-	1.990	0.0
5	CMCS	-	0.011	0.0
6	HPCS	-	0.103	0.0
7	TSP	-	0.591	0.0
8	MCC/SA	0.01	0.015	-0.196
9	MCC/Pul	0.01	0.016	-0.203
10	MCC/κ-C	0.01	0.081	-0.306
11	MCC/CMCS	0.01	0.016	-0.196
12	MCC/HPCS	0.01	0.006	-0.199
13	MCC/TSP	0.01	0.029	-0.220

[a] Polysaccharide content by a mass fraction in the blend.

3.6 SEM Analysis

Figure 4 illustrates the scanning electron microscopy (SEM) images of blends of six polysaccharides with crystalline cellulose. At a magnification of 500 times, Carboxymethyl Chitosan (CMCS) and Hydroxypropyl Chitosan (HPCS) groups exhibit similar structures. Cellulose forms a compact cluster as the main body, with the layered edges of cellulose folding and adhering to each other through the swelling caused by water absorption and the action of CMCS and HPCS in the blend state. Each cellulose fragment forms spherical shapes during the blending drying process. In the Tamarind Seed Polysaccharide (TSP) group, cellulose fragmentation is evident, but there is no significant change in size, with noticeable internal pores and no smooth outer surface. In the Sodium Alginate (SA) group, numerous small particles smaller than 100 μm appear, indicating that the adhesive ability of SA to

https://doi.org/10.1051/e3sconf/202452801018

connect cellulose fragments is the weakest among the six groups, especially at a magnification of 500 times, where cellulose layers are peeled off individually. In the Pullulan (Pul) group, there is also a weak ability to bond individual large cellulose skeletons, but the internal interaction among cellulose is stronger, resulting in a higher proportion of filamentous entities than in other groups[23]. In the κ-Carrageenan (κ-C) group images, fewer cellulose fragments smaller than 30 μm are present, indicating that its electrostatic interaction can adsorb individual fragments effectively.

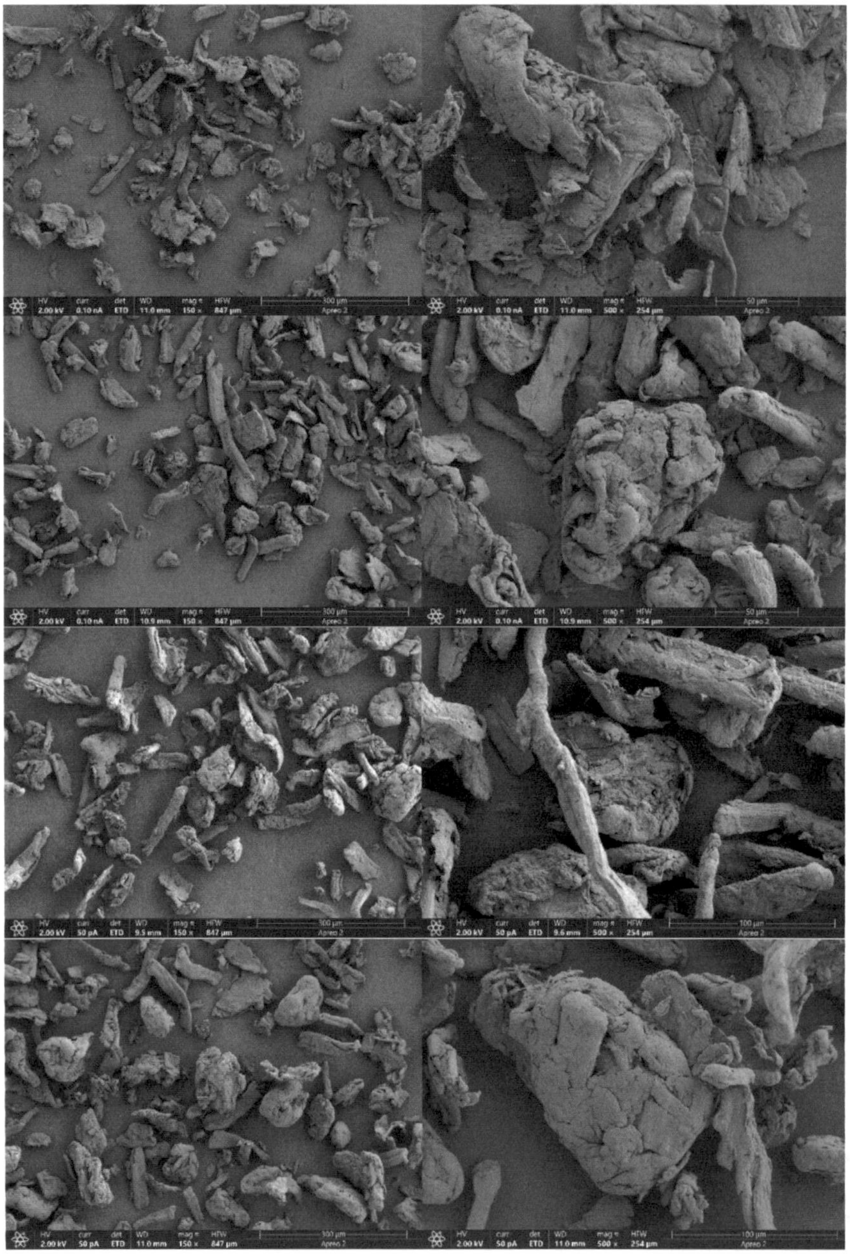

Figure 4. SEM images of six blends of food colloid and MCC(From top to bottom, they are: MCC/SA, MCC/Pul, MCC/k-C, MCC/CMCS, MCC/HPCS, and MCC/TSP).

3.7 Research on the application of six polysaccharides in the papermaking practice of reconstituted tobacco sheet

In the practical application process, six kinds of polysaccharides, including sodium alginate (SA), pullulan (Pul), κ-carrageenan (κ-C), carboxymethyl chitosan (CMCS), hydroxypropyl chitosan (HPCS), and tamarind polysaccharide (TSP), were added to the reconstituted tobacco pulp as wet-end additives in papermaking reconstituted tobacco base sheet. The effects of polysaccharides on the retention rate of the reconstituted tobacco pulp and the physical properties of the leaf base are as follows:

Table 5. Effects of polysaccharides on the retention rate of the pulp and the physical properties of the base sheet

Polysaccharide samples	Static retention rate (%)	Tensile strength of base sheet (kN/m)	Bulk of base sheet (cm3/g)	Air permeablility of base sheet（K CU）
Blank control	85.13%	0.060	5.664	15.336
TSP 0.5%	87.18%	0.133	5.147	8.233
Pul 0.5%	86.86%	0.161	4.690	5.278
SA 0.5%	86.55%	0.179	5.163	5.240
κ-carrageenan 0.5%	87.90%	0.165	4.658	5.966
CMCS 0.5%	86.86%	0.212	4.481	4.209
HPCS 0.5%	86.16%	0.201	4.336	4.315

Note: Adding polysaccharide is 0.5% of the absolute dry weight of reconstituted tobacco sheet pulp.

Adding polysaccharide as a wet-end additive in the papermaking process, the six kinds of polysaccharides all contribute to the improvement of reconstituted tobacco sheet pulp retention rate, among which κ-carrageenan and TSP have better effects; the impact on the physical properties of reconstituted tobacco sheet base is as follows: the addition of polysaccharides will enhance the tensile strength of base sheet, which has a positive impact on improving the addictive properties of the product, but it will also reduce the bulk of sheet base to a certain extent, especially the permeability of the base will be significantly reduced. Therefore, the dosage of polysaccharides is worth further study.

4. Discussion

The discussion s are obtained as below:

(1) The total monosaccharide content of CMCS and HPCS is less than 100%, which may be related to the impurity of the samples.

(2) Generally, due to the structural similarity between polysaccharides and cellulose molecules, there is a high affinity between polysaccharides and the cellulose surface, leading polysaccharides to adsorb on the cellulose surface preferentially[24]. However, the negative charge of polymers hinders their adsorption to cellulose under conditions without electrolytes[25].

(3) CMCS itself has viscosity and weakly enhances the viscosity through interaction with cellulose.

κ-Carrageenan exhibits high viscosity at low concentrations in aqueous solution. After blending with cellulose, the viscosity increases significantly, and gel formation is easily induced[26]. As a linear macromolecule, Pullulan does not have high viscosity in aqueous solution. Its hydrogen bonding with cellulose is weak, leading to poor compatibility. However, blending with MCC can increase viscosity[27].

5. Conclusion

This study draws the following conclusions through analysis of monosaccharide composition, light scattering, adsorption kinetics, rheological properties, viscosity, and scanning electron microscopy:

(1) The main monosaccharide components of SA, Pul, CMCS, and HPCS are glucose, while κ-Carrageenan consists mainly of galactose and glucose, and TSP consists mainly of glucose, xylose and galactose. The degree of compactness from loose to compact is as follows: SA > CMCS > κ-C > TSP > HPCS > Pul.

(2) When polysaccharides are blended with MCC, SA exhibits the weakest adhesion ability. Pul has a weak ability to bond individual large cellulose skeletons, but the internal interaction among cellulose is stronger. The adsorption of HPCS and CMCS on the MCC surface is stable. In the blend state, the layered edges of cellulose first swell by water absorption and then adhere to each other through the action of CMCS and HPCS, resulting in each cellulose fragment forming into spherical shapes during drying. The side chain structure of TSP is conducive to its cross-linking and entanglement with cellulose, but the adsorption process is relatively slow. The cellulose fragmentation of TSP is significant, with noticeable internal pores and no smooth outer surface. κ-Carrageenan adsorbs fragments through electrostatic interactions.

(3) The storage modulus of TSP is higher than that of other polysaccharides, possibly due to its higher molecular weight and more abundant side chain structure, resulting in stronger interactions between polymer chains.

(4) The viscosity of TSP and HPCS blended with MCC decreases slightly, while that of Pul, SA, and HPCS blended with MCC increases slightly. κ-Carrageenan blended with MCC significantly increases viscosity. All groups exhibit phase separation, forming microphase-separated blocks, indicating poor compatibility.

Acknowledgments

This work was financially supported by the China Tobacco Yunnan Industrial Corporation Youth Science and Technology Project "Application of Natural Plant Polysaccharides in Reconstituted Tobacco" and the China Tobacco Yunnan Industrial Corporation Project "Application Technology Research of Natural Plant Polysaccharides in Reconstituted Tobacco" (2022YL04).

References

1. Yang R, Zang Y, Wu P, et al (2023) Research progress on functional properties of polysaccharide food colloids and their applications in food processing. https://kns.cnki.net/kcms2/detail/11.2206.TS.20230530.0958.028.html
2. Fang Y, Zhao Y, Sun C, et al (2022) Research progress and future trends in food colloids: Component interaction, future food structure design, and colloidal nutrition perspective. Food Science, 43(15): 1-20.
3. Ye J (2022) Research on the fine structure, solution conformation of peanut polysaccharides, and its application in stabilizing water-in-oil emulsions. Wuxi: Jiangnan University.
4. Salvador L D, Suganuma T, Kitahara K, et al (2000) Monosaccharide composition of sweet potato fiber and cell wall polysaccharides from sweetpotato, cassava, and potato analyzed by the high-performance anion exchange chromatography with pulsed amperometric detection method. Journal of Agricultural & Food Chemistry, 48(8):3448-54.
5. Strydom D J (1994) Chromatographic separation of 1-phenyl-3-methyl-5-pyrazolonederivatized neutral, acidic and basic aldoses. Journal of Chromatography A, 678(1): 17-23.
6. Zhang J (2021) Isolation, purification, structural identification, and hypoglycemic activity of nettle polysaccharides. Xi'an: Shaanxi University of Science and Technology.
7. Song W, Li X, Fu Q, et al. Determination of compatibility of polymer blends by dilute solution viscosity method[J]. Chinese Journal of Tissue Engineering Research and Clinical Rehabilitation, 2008, 12(6):1161-1164.
8. Łabowska M B, Michalak I, Detyna J, et al (2019) Methods of extraction, physicochemical properties of alginates and their applications in biomedical field–a review, Open Chemistry, 17 (1):738-762.
9. K.I. Draget Alginates, G.O. Phillips, P.A. Williams (2000) (Eds.), *Handbook of hydrocolloids*, CRC Press, Boca Raton, pp. 379-395.
10. Wani S M, Masoodi F A, Mir S A, et al (2023) Pullulan production by Aureobasidium pullulans MTCC 1991 from apple pomace and its characterization, Food Bioscience, 51: 102254.
11. Chen L (2022) Preparation of modified chitosan materials and their adsorption properties for heavy metal ions in wastewater. Nanchong: China West Normal University.
12. Gao Y, Gong X, Yu S, et al (2022) Immune enhancement of N-2-Hydroxypropyl trimethyl ammonium chloride chitosan/carboxymethyl chitosan nanoparticles vaccine,International Journal of Biological Macromolecules, 220: 183-192.

https://doi.org/10.1051/e3sconf/202452801018

13. Yan J, Wang Y, Zhang Z, et al (2023) Study the improvement of curcumin's physical and chemical stability and release properties based on κ-carrageenan/Argopecten irradians gel beads. *Chinese Institute of Food Science and Technology, Proceedings of the 20th Annual Meeting of the Chinese Institute of Food Science and Technology*.pp. 123-124.
14. Zhao T, Nie C, Ai L, et al (2021) Research progress on structure, function, and modification of loquat polysaccharide. Food and Fermentation Technology, 57(06):67-76.
15. Zhang H, Nie S, Ai L, et al (2020) Research progress on the structure and characterization methods of Ganoderma lucidum polysaccharides. Journal of Chinese Institute of Food Science and Technology, 20(01): 290-301.
16. Xu X, Chen A, Ge X, et al (2020) Chain conformation and physicochemical properties of polysaccharide (glucuronoxylomannan) from Fruit Bodies of Tremella fuciformis. Carbohydrate Polymers, 245: 116354.
17. Xu X, Chen P, Wang Y, et al (2009) Chain conformation and rheological behavior of an extracellular heteropolysaccharide Erwinia gum in aqueous solution. Carbohydrate Research, 344(1): 113-119.
18. Beghello L (1998) Some factors that influence fiber flocculation. Nordic Pulp & Paper Research Journal, 13(4):274-279.
19. Rochefort W E, Middleman S (1987) Rheology of xanthan gum: Salt, temperature, and strain effects in oscillatory and steady shear experiments. Journal of Rheology, 31 (4): 337-369.
20. Zhang H, Zhao T, Wang J, et al (2020) The interactions in ternary system made of xanthan gum, Carbopol 940 and anionic/nonionic surfactant. International Journal of Biological Macromolecules,149:1189-1197.
21. Song W, Li X, Fu Q, et al (2008) Determination of compatibility of polymer blends by dilute solution viscosity method. Chinese Journal of Tissue Engineering Research and Clinical Rehabilitation, 12(6): 1161-1164.
22. Chang H, Shao Z, Wang X, et al (2008) Effect of carboxymethyl chitosan on the properties of hydroxypropyl methylcellulose solution. Cellulose Science and Technology,16(04):24-28.
23. Maja Milanović, Dejan Ćirin, Veljko Krstonošić, The interactions in ternary system made of xanthan gum, Carbopol 940 and anionic/nonionic surfactant[J], Journal of Molecular Liquids, 2021, 344:117696.
24. Myllytie P, Salmi J, Laine J (2009) The influence of pH on the adsorption and interaction of chitosan with cellulose.BioResources,4(4):1647-1662.
25. Junka K, Filpponen I, Johansson L, et al (2014) A method for the heterogeneous modification of nanofibrillar cellulose in aqueous media,Carbohydrate Polymers, 100:107-115.
26. Guo R, Ding E (2007) Research on gelation of κ-carrageenan with nano-microcrystalline cellulose.Forest Chemicals and Industry, 27(02):120-122.
27. Xiao Q, Tong Q (2009) Research on the properties of edible packaging film prepared by pullulan/sodium alginate/carboxymethyl cellulose sodium composite. Food Industry Science and Technology,30(05):286-288.

Research on gas drainage technology and equipment of high-level directional drilling in coal mine

Yubo Song

CCTEG Xi'an Research Institute(Group)Co., Ltd.，Xi'an, 710077, China

Abstract: The gas overrun at the upper corners of the goaf and return air lanes has been restricting the development of China's coal mining industry. To solve the problem of over-limit gas concentration at the upper corners, the fundamental reason is to reduce the gas concentration in the "fracture zone" of the coal roof . At present, the most commonly used methods such as high extraction roadway, buried pipe extraction, and high-level drilling in the "fracture zone" gas control of the goaf have the disadvantages of excessive cost and heavy workload, and the drilling trajectory of conventional drilling rigs cannot be accurately positioned problem. Therefore,it is proposed to use high-position roof directional long drilling technology to replace traditional high-drainage roadway for gas control, and analyze the influence of directional drilling diameter, position, depth and other parameters on gas drainage efficiency. Based on the analysis of four drilling cases of the directional drilling rig ZYWL-13000DS of China Coal Science and Industry Group Chongqing Research Institute Co., Ltd. at 33 (4) 13 working face of Shuangliu Coal Mine, from the perspective of construction effect,the directional drilling is a more reasonable construction level 5-8 times the mining height above the roof is appropriate; the construction diameter should be as large as possible when conditions permit;when the drilling depth is 20m deep into the fissure zone, the pumping capacity will increase greatly, and with the construction of the mining, pumping Capabilities tend to stabilize. Compared with the traditional high-drainage road control effect, the use of directional long drilling technology for gas drainage can reduce the cost by more than 70% and shorten the effective construction period by more than 75%. It is finally verified that the use of high-directional directional long drilling technology can replace the high-drainage roadway for "fracture zone" gas drainage.

1 Introduction

Coal mine is currently one of the main energy sources in China, because coal is a porous medium, the coal seam is usually rich in gas, and about 90% of the coal mines in China are high gas mines, gas concentration is high will seriously threaten the life safety of underground workers [1-2], restricting the development of coal mine industry.

In recent years, China's coal mining capacity has been continuously strengthened, the coal mining depth has been continuously deepened, the geological conditions are becoming more and more complex, and the difficulty of gas control has been significantly increased, leading to an increasing number of high gas mines. Under the influence of mining, the overlying rock of the coal seam roof in the goaf produces a large number of cracks, which increases the air permeability of the rock layer and significantly enhances the gas passage rate. The "gas channel" formed by the overlying rock fissure makes the gas move along the fissure and form the gas enrichment zone. Restricted by the shape of the channel, a secondary vortex will inevitably form in the upper corner. With the generation of secondary vortices, the energy of gas flowing through the upper corner is consumed, the flow rate is slowed down, and the gas in the vortex cannot flow out in time, resulting in aggregation, causing the gas concentration in the upper corner and the return air lane, affecting the construction safety of coal mining face construction. Therefore, how to carry out the fissure band gas extraction efficiently and quickly is the key to solve the gas concentration overlimit in the upper corner and the return air lane.

Generally speaking, before coal mining, the ground rig should be used to pump gas in the coal seam. However, in the actual construction process, affected by the terrain, cost and other factors, ground extraction is rarely used, and the commonly used control scheme is underground gas control. In China, the gas control in underground goaf in coal mines usually uses the methods of high extraction roadway, buried pipe extraction, high drilling and other [3], etc., among which the most important one is the use of high extraction roadway extraction, that is, a special gas extraction channel is dug out in the crack zone formed by the coal seam mining at the top of the mining layer. This kind of practice extraction efficiency is high, but the engineering quantity is large, the construction cycle is long, and the early cost input is too high. Relatively speaking, it is more

Corresponding author's e-mail:songyubo@cctegxian.com

economical to use high drilling for gas extraction, but the traditional high drilling adopts conventional drilling rig, and the drilling track is uncontrollable and cannot accurately locate the extraction crack band. With the deeper and deeper tunneling surface, the geological conditions are more and more complex, and the limitations of conventional drilling RIGS are increasingly prominent. Therefore, it is urgent for a high drainage roadway extraction technology and equipment to realize the efficient extraction of the underground roof crack zone gas.

2 Kong for lane technique

In the process of coal seam mining, affected by the formation pressure, the coal wall support area in mining direction, from the layer and compaction area [4], in the longitudinal collapse on the top of the bending subsidence, the bottom formation collapse, the middle formation is affected by both growth fissure to make the fissure concentration extended to full thickness, form the crack zone. The gas in the coal seam is divided into free gas and attached gas. Affected by the mining of coal mine, the gas originally attached to the coal seam rises up along the crack, and the concentration of gas rises from bottom to up. In the subsidence zone, because the rock did not grow more cracks, internal attached gas less, so most of the gas will gather in the junction of the fracture zone, but affected by the underlying stress here, lead to extraction negative pressure extraction efficiency is low, and the falling zone to the fracture zone this position due to crack development, uniform gas distribution, become the core area of gas extraction.

With the mature application of long-hole directional drilling technology [5] in coal mine, it has gradually become one of the important methods of conventional drilling because of its simple construction method and controllable construction track for underground gas extraction. And with the innovation of drilling equipment technology, the diameter of hole is getting bigger and bigger, the efficiency of hole is getting higher and higher, and the influence radius of drilling gas is getting bigger and bigger. The process of "substituting the hole for the hole" has been greatly developed. Compared with the traditional extraction way, roof large diameter high long drilling trajectory can be accurately controlled, large hole aperture, high efficiency, drainage range, long cycle, at the same time, in the mining area of roadway construction, can effectively avoid the current situation of the overlap between extraction and tunneling, the construction cost is lower, shorter cycle.

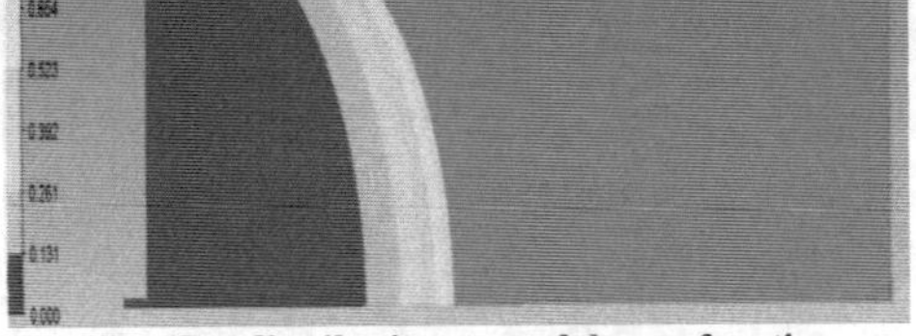

(1) Gas distribution map of the goaf section

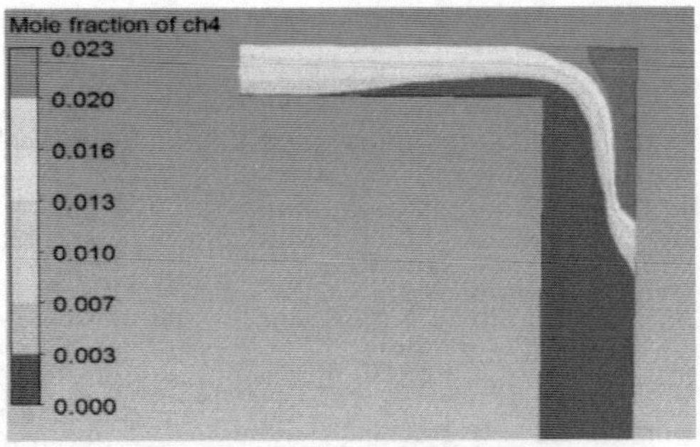

(2) Gas distribution diagram in the upper corner

Figure 1. gas distribution without drainage.

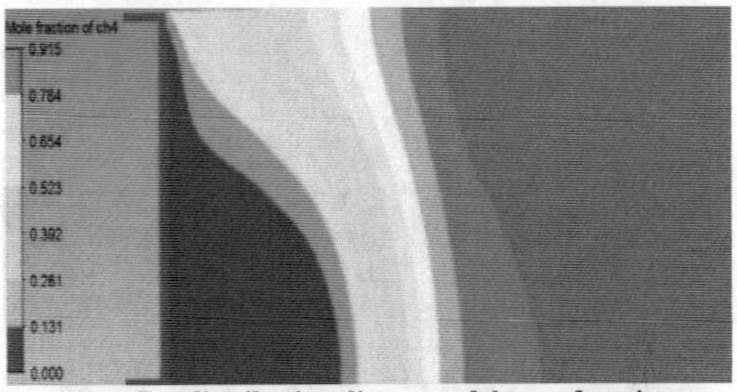

(1) Gas distribution diagram of the goaf section

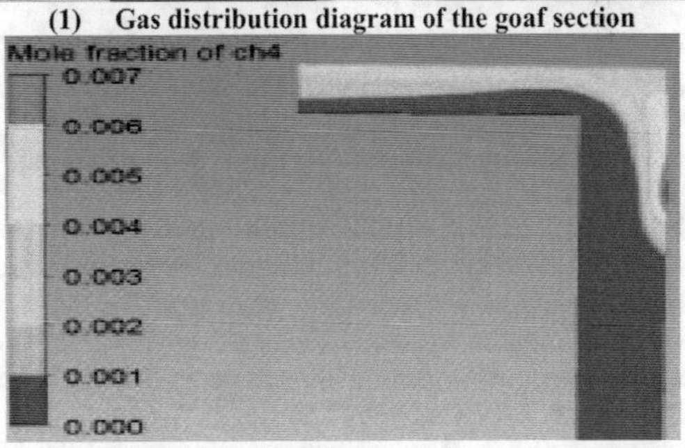

(2) Gas distribution diagram in the upper corner

Figure 2. gas distribution after high-level drilling.

As shown in figure 1, figure 2 is the method of numerical simulation for gas extraction analysis results[6], by comparison, can obviously see that with the high drilling gas extraction, along the working direction gas concentration distribution, reduce the gas content, the gas transition more smooth, low concentration gas area to the goaf, the upper corner gas content is significantly reduced, so using high directional drilling has a very positive influence on safe coal mining. At present, the three technical problems of high directional drilling gas extraction are mainly: drilling layer position, drilling diameter and drilling depth.

2.1 Selection of drilling location

At present, the division methods of "three belts" of goaf attached rock are mainly divided into empirical formula method [7], simulation calculation method [8-9] and drilling measurement method [10].

At present, the most accurate coal seam division method is the drilling measurement method. Distribution of the "three belts" of covered rock can be accurately and intuitively determined by geological drilling with [11], so as to guide drilling. However, this method has high economy and time cost, and it is difficult to implement in the drilling process.

In recent decades, it has been found that the position of rock cover on the mining surface affected by coal mining is exponential to the height of coal mining, but it is less affected by the length. Therefore, the empirical formula of the gas conduction zone (falling zone and fissure zone) is proposed:

$$H_1 = \frac{100M}{aM + b} + c \quad (1)$$

Where H1 represents the thickness of the guide gas belt, M represents the thickness of the coal mining, and the values of the constants a, b and c are affected by the hardness of the rock layer.

Due to falling zone upper fracture is given priority to with layer fissure development, the lower fissure is more broken fracture[12], and falling zone is more likely to appear by the top fracture belt compaction gas extraction efficiency is low, so the lower layer of the fracture zone for gas extraction of gold area, to determine the thickness of the fracture zone, can introduce falling band thickness empirical formula:

$$H_2 = \frac{M}{(K-1)\cos\alpha} \quad (2)$$

Where H2 is the maximum thickness of the falling band, M is the thickness of the coal seam, K is the coefficient of layer breaking, and α is the inclination of the coal seam.Formula 1 and formula 2 can find out the position of the crack band in the coal seam.

The position of the crack band obtained by using the empirical formula is relatively rough, When the coal seam thickness is greater than 3 m, The fracture band calculated using the empirical formula are inaccurate[13], 11 Occasionally, ineffective drilling, To address this situation, At present, the optimal solution is to use a small amount of drilling data combined with a numerical simulation method to simulate the recovery face, Wei Yousheng[14], Lu Huiqing[15], Through ANSYS and FLCE3D, The trend of the overrock simulation is similar to the actual rock formation, It is believed that the simulation results can guide the actual production.

2.2 Selection of drilling diameter

The diameter of the borehole plays a decisive role in the efficiency of gas extraction. Studies have shown that the flow of gas in the crack is mainly in two forms of penetration and diffusion, which is affected by the size of the crack. Secondary stress generated during drilling will lead to deformation of the rock mass around the drilling hole, affecting the permeability coefficient of the rock layer. The pressure relief radius R around the drilling hole can be expressed by formula (3) [16]:

$$R = \left[1 + \frac{1}{\sqrt{2} f_r K_p} \ln\left(\frac{\sigma_0}{\sigma_c}\right)\right] r_0 \quad (3)$$

$$K_p = (1 + \sin\varphi)/(1 - sin\varphi) \quad (4)$$

In formula: σ_0 :Rock mass stress before drilling, MPa;σ_cUniaxial compressive strength, MPa;f_rFriction factor between moving rock mass and moving rock mass;r_0Drilling radius;φ:Internal friction angle of the coal.

It can be seen that the pressure relief radius is inversely proportional to the strength of the rock mass and is proportional to the diameter of the borehole. The larger the pressure relief radius, the smoother the gas flow, and the better the extraction effect.

Wu Aijun[17] et al. conducted numerical simulation on the second _2 coal seam in Xuehu Coal Mine, and analyzed the drainage hole with a diameter of 80mm / 100mm / 120mm respectively under the fixed gas pressure, extraction negative pressure and the constant permeability coefficient of the coal seam. When the aperture of 80mm, 0.35 m3/min, 0.4 m3/min at 100mm and 0.45 m3/min at 120mm.

It can be seen that with the increase of the diameter of directional drilling, the gas extraction capacity shows a trend of enhancement, and the larger gas drainage hole can provide a larger hole wall for gas circulation. Therefore, in order to achieve a good extraction effect, large aperture and multiple channels are mostly used for gas extraction operations. However, combined with rock mechanics, the larger the hole diameter, the more likely it is to cause the hole collapse, but cause losses. Qi Qingjie[18] et al. simulated a working face of flat coal and found that the drilling diameter had little influence on the gas pressure attenuation in the early stage of extraction. According to incomplete statistics, at present, only Chengzhuang Mine of Jincoal Group adopts high directional drilling with diameter ofφ193, while other mines such as Sihe Mine of Jinmei Group, Fenxi Group, Hecoal Group, China Coal Xinji and other groups all adopt high directional drilling with diameter ofφ153. The construction process is after the directional drilling hole toφ120, the hole expansion toφ153 /φ193 drilling process.

2.3 Selection of drilling depth

Because the falling zone is a porous medium, the natural movement speed of the gas distribution is slow, so the depth of the borehole is required to increase the efficiency of the gas contacting the hole wall. The drilling depth will affect the distance of the invalid drilling section. In order to draw out more gas, after the drilling line, at least the boundary between the center of the fissure zone or the zone and the compaction zone should be extended. The minimum drilling length calculated according to the dividing line is:

$$L = \frac{(1.5h + a + h_1)(w + l - h_1\cot\beta + h_1\tan\alpha)}{(w - h_1\cot\beta + h_1\tan\alpha)\sin\theta} \quad (5)$$

$$\theta = \arctan\frac{1.5h + a + h_1}{w + (1.5h + a)\cot\beta + h_1\tan\alpha} \quad (6)$$

Where: h is the mining height of coal seam (m), h1 is the height of the coal seam roof (m),Width of the w coal pillar along the inclined plane (m),The l is the period step (m),βis fall (°),αis coal seam inclination (°),θIs the angle (°) of drilling hole and coal seam tendency.

According to the analysis of the gas extraction situation of ZYWL-6000DS directional drilling rig in [19] high mazhuang Mine, the gas extraction effect is significantly improved after 16-20 meters. With the excavation of the mining face, the distance between the borehole deep and the cracks gradually increases, and the effect of gas extraction continues to increase. With the increase of time, the content of the gas enrichment area in the fissure zone decreases, and the gas extraction in the drilling hole reaches the peak and gradually decreases. In the later stage, with the advance of the mining surface, the formation of new cracks will permeate the new free gas, making the high borehole flow out into a balance state. However, when the hole collapse phenomenon occurs in the borehole, the gas content extracted in the hole will decrease sharply. Therefore, the construction of deep holes requires more on the geological conditions of the coal mine, the skills of the boreand the performance of the drilling equipment. At present, the group mine uses high directional drilling for gas control in the upper corner, the depth is mostly 450-500m.

3 Construction cases of high-level and large-diameter directional drilling

In order to solve the problem of high large diameter directional hole, Chongqing Research Institute of China Coal Science and Industry Group Co., Ltd. launched three types of large diameter directional drilling rigs, namely ZYWL-6000DS[20], ZYWL-13000DS and ZYWL23000DS[21] drilling rigs in various coal mines and achieved good results. The technical parameters of the three drilling RIGS are shown in Table 1. Unlike traditional conventional rig, directional rig will form a complete set with drilling measurement system, make inclined bits, track display computer for real-time drilling trajectory control, with the drilling measurement system is in the process of the drilling of the orientation, inclination Angle, tool for Angle and other technical parameters of equipment[22], the mainstream use wired with drilling measurement system and wireless mud pulse with drilling measurement system.

Table 1. Comparison of three drilling rig parameters.

parameter	ZYWL-6000DS	ZYWL-13000DS	ZYWL23000DS
Rated output torque/(N.m)	6000~1500	13000～3000	23000~4000
Rated output speed/(r/min)	50~210	45～150	35～110
Advance / pull out force/kN	170/220	280/280	400/400
Advance / pull trip/mm	850	1200	1200
Final hole diameter/mm	93/133	120/153	153/193/300
Main motor power/kW	55	132	200

3.1 Drill-accompanying measurement system

The wired drilling measurement system is to connect the measurement system to a random computer through a special cable drill bar to transmit data. Traditional drilling measurement device adopts the built-in large capacity rechargeable nickel metal hydride battery [23] , each charge need to take the measurement system from the hole, increase the working hours, China coal Chongqing research institute developed using the cable drill rod signal transmission line for power supply new type with the drilling measurement system, effectively overcome this problem.

Mud pulse technology is gradually mature new technology in recent years, the equipment abandon the traditional directional drill pipe cable, using drill medium can transfer pulse signal to know the drill azimuth, tool Angle information, the system for battery power, to save electricity, in normal drilling operations for sleep. When need for azimuth determination, stop drilling, awaken the measurement system through a pulse signal, then open the mud pump to detect the system orientation information, finally close the mud pump, the system will seal some water in the drill pipe, through the part of the water signal to the sensor on the braid, decoding after display location information on a random computer. At present, Chongqing Research Institute and Xi'an Research Institute of China Coal Science and Industry Group have mature solutions[24-25].

3.2 Construction cases of Shuangliu Coal Mine

Shuan gliu Coal Mine is affiliated to Shanxi Coking Coal Fenxi Mining Group, located in Mengmen Town, Liulin County. The mine is rectangular and covers an area of about 29.6km2. Since the construction of the mine in 1998, the current output is 3 million tons / year.

The drilling rig serving the mine is ZYWL-13000DS directional drilling rig, which is standard with BC480 pump truck, mud pump flow reaches 480 L/min, ZSZ1500 drilling measurement system, directional drilling φ120mm, reaching the final hole φ153mm through one hole expansion, the drilling efficiency is higher than that of ZYWL-6000DS.

The construction site is 33 (4) 13 working face in the third mining area of Shuangliu Coal Mine, the working face length is 1673m, the tendency length is 201m, the gas emission is 14-26 m3/min, and the average thickness of coal seam is 3.56m. The coal seam is directly topped by gray-black mudstone, horizontal bedding, homogeneous, semi-smooth, about 3m thick. The old top is gray black medium grain sandstone, semi-hard, horizontal bedding, thickness of about 3.23m. Above the old top is mostly gray medium grain sandstone, mainly quartz, feldspar, hard texture.

According to the operation requirements, design 4 large diameter directional boreholes with horizontal spacing 1m, inclination 9, diameter 153mm and depth 300m. The 1 # hole is 26m from the coal seam roof, bending length 168 and horizontal length 132. 2 # hole is 28m from the coal seam roof, the bending length is 162m and the horizontal length is 138m. The 3 # hole is 30m

 https://doi.org/10.1051/e3sconf/202452801019

away from the coal seam roof, the bending length is 156m, and the horizontal length is 144m. The 4 # hole is 30m away from the coal seam roof, the bending length is 150m, and the horizontal length is 150m. The drill hole trajectory map is shown in Figure 3.

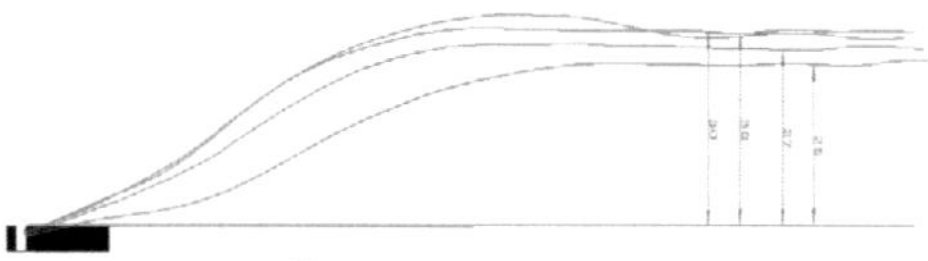

(1)Drilling hole drift up and down

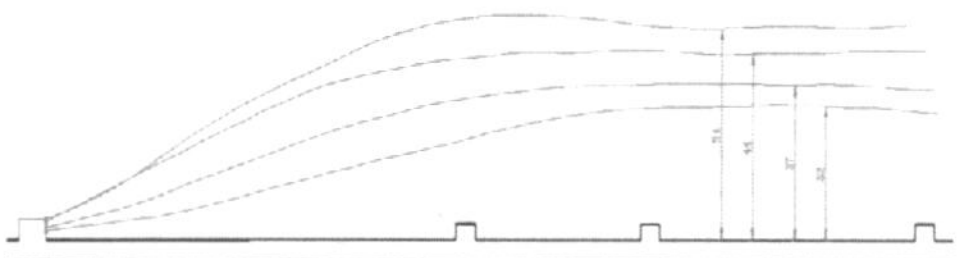

(2)Track diagram of left and right offset of borehole

Figure 3. Drilling track of Shuangliu mine.

All boreholes were compared by high and low negative pressure extraction system and observed for 40 days. The purity performance of gas in each hole is shown in Figure 4:

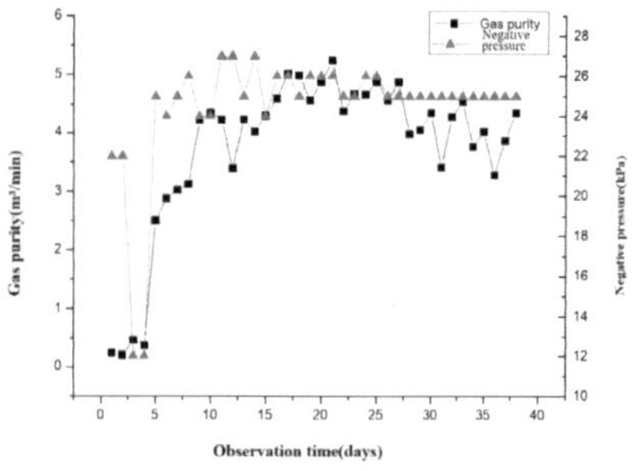

A: 1 # hole

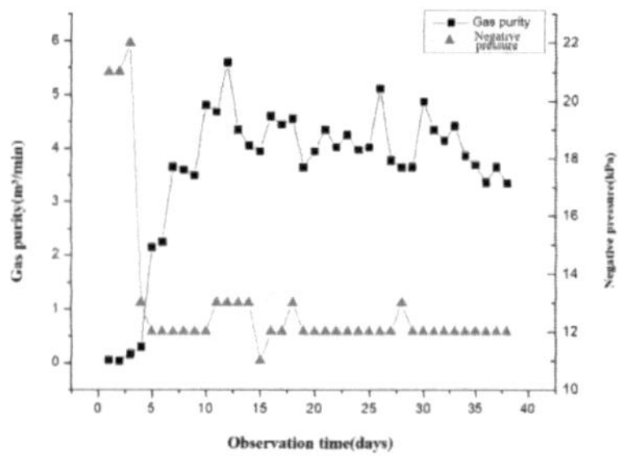

B: 2 # hole

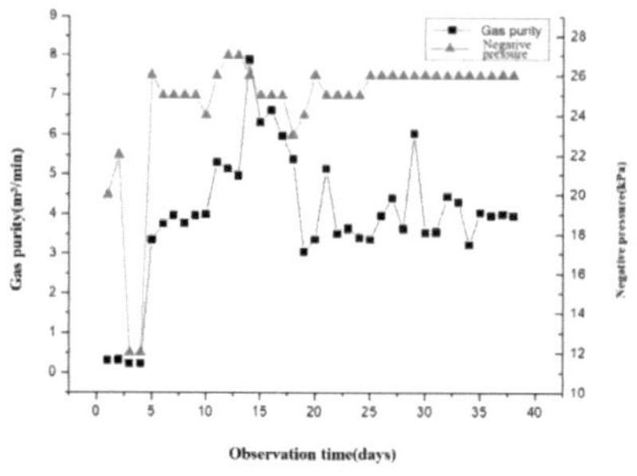

C: 3 # hole

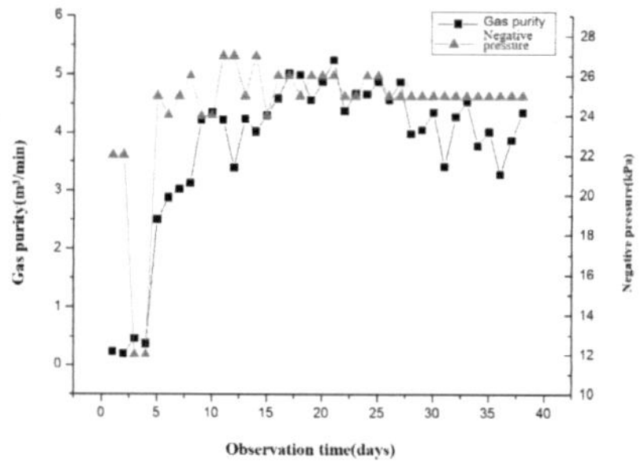

D: 4 # hole

Figure 4. Comparison of suction negative pressure and gas purity of each hole.

It can be seen from Figure 5 that the purity Observation time (days)volume of 1 # hole gas has been maintained below 0.5%, so it can be judged that the hole does not play the role of gas extraction. Figure. 5 shows the total purity volume of large diameter drilling on the high roof compared with that of high extraction roadway:

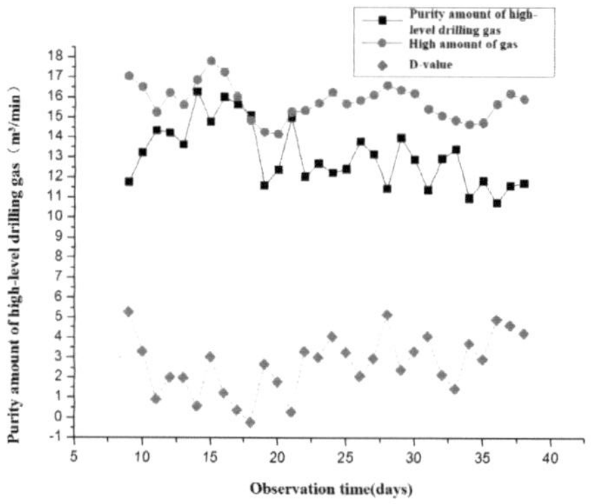

Figure 5. Comparison of gas drainage effect.

As can be seen from Figure 5, in the case of a failure of the high roof directional drilling, the difference between the total amount of gas extraction and the high roadway is also maintained at 2 m3/min, which proves the feasibility of the large diameter directional drilling rig of the high roof replacing the high roadway.See Table 2 for the economic comparison of high-rise directional long hole, through hole and high mining roadway.

Table 2. Economic comparison of high-level directional long hole, common hole and high extraction roadway.

High-level directional long drilling hole	Ordinary drilling	High lane
Coal lane drill yard 5×3.4×9m	Coal lane drill yard 4.5×3.4×4m	rock gangway 2.7*2.6*300m
Φ153mm 4 boreholes with a total of 1200m	Φ 75mm drilled 50 holes totaling 5500m	300m gun digging rock roadway
4 People / class	3 People / class	About 10 people / class
360 Thousand	three hundred and ninety-seven thousand	one million, five hundred and thirty-four thousand
30 Days	50 Days	140 Days

4 Conclusion

Introduced the hole generation lane gap zone gas control theory, analyzes the influence of extraction drilling parameters, and introduces the roof mine case large diameter directional drilling construction technology, the conclusion proves that using roof high large diameter directional long drilling process instead of high roadway gas control is feasible, can effectively reduce the cost of gas mining, shorten the gas control cycle. The gas control results meet the requirements of "Coal Mine Safety Regulations", and it has a good application prospect for high drilling with large diameter directional rig.

Combined with the construction case, it can be seen that the reasonable construction level of directional drilling is 5-8 times the height of the roof. The drilling diameter should be constructed in the light of the actual capacity of the drilling equipment. The drilling depth is mainly affected by the current construction technology. In the case of no hole collapse, the gas extraction efficiency will be significantly improved after the drilling goes deep into the crack zone of about 20m. Combined with the quality of operators and drilling equipment, 500m-750m is a more reasonable construction depth for high directional drilling.

Each high drilling field generally has 4-5 holes arranged close to the return air side. The hole spacing is 10 m. It can achieve the problem of exceeding the gas concentration in the upper corner. The gas concentration in the upper corner after the extraction of high directional drilling is similar to that of high drainage lane and ordinary high penetrating layer drilling, but the cost is low, the time is short, and the benefit is remarkable.

References

1. ZHANG Bingxuan. Study on Gas Control Measures in High Gas Mine[J]. ENERGY AND ENERGY CONSERVATION, 2019, No.167(8): 19-20.
2. NING Yu. Progress on Technology of Coal Bed Methane Development andUtilization in China Coal Mining Area [J]. Coal Science and Technology, 2013, 41 (1): 12-15, 73.
3. Li Yanming. Upper corner gas control based on high level directional long borehole [J]. Coal Science and Technology, 2018, 46 (1): 215-218.
4. HU Binqiang. Research on distribution law of dominant gas extraction area of high-level borehole [J]. Industrial and mining automation, 2019, 45 (7): 102-108.
5. Yan Baoyong. Study on drilling technique of high level directional long borehole [J]. Coal Science and Technology, 2016, 44 (4): 55-58.
6. Baoyong YAN;Xiaoze ZHANG;Yibin LI. Study on the Optimum Position of Gas Extraction in Goaf of High Position Borehole Based on CFD[C]. Conference Proceedings of the 7th International Symposium on Project Management(ISPM2019),2019:811-818.
7. Liu Yang. Study on Gas Drainage Technology with Large Diameter High Level Sector Boreholes of the Goaf in High Gas Mine [D]. Taiyuan University of technology, 2013.
8. Zhang Xiaoming, Hou Zhongjie. Numerical simulation of overburden three zones for mining operation with thick overburden and less depth [J]. Coal Science and Technology, 2007, no.387 (2): 93-96.
9. Huang Zhian, Tong Haifang, Zhang Yinghua, et al. Definition criteria and simulation determination of "three zones" of overburden in goaf [J]. Journal of University of Science and Technology Beijing, 2006, 609-612.
10. Chen Quanqiu. Study on gas control in high mining face [J]. Journal of Shanxi Institute of Energy, 2018, 31 (4): 36-38, 41.
11. Li Shugang, Wang Linhua, Lin Haifei, et al. Similar Simulation Experiment and Analysis on Evolution Characteristics of"Three Zones"in Overburden Strata of Stope [J]. Mining safety and environmental protection, 2013, 40 (3): 17-20.
12. Wang Fuhou. Determination Method of Caving& Fracture Zone Height for GasDrainage by High Level Borehole [J]. Coal technology, 2008, No.173 (8): 75-76.
13. Cao Wentao, Zhao Zhongyi, Wang Yanhong. Study on the technology of gas extraction in goaf by "hole instead of lane" [J]. Coal mine modernization, 2015, No.129 (6): 37-39.
14. Wei Yousheng. Numerical Simulation Study on Distribution Characteristics of "Three Zone" Height in Goaf of Fully-mechanized Caving Face [J]. Coal Science and Technology, 2019, 40 (1): 32-35.
15. Lv Huiqing, Xu Yongcheng, Liu Chunjiang, et al. Measurement and simulation analysis of "Three

Zones" in goaf of fully-mechanized mining face in Hongqinghe Coal Mine [J]. China Energy and Environmental Protection, 2019,41 (12): 43-48.

16. Yang Ningbo, Wang Zhaofeng. Research on Change Law of Permeability in Coalseam around Boreholes [J]. Coal technology, 2008, No.177 (12): 67-69.
17. Wu Aijun, Wang Qiaoli, Numerical simulation and analysis on gas extraction with bedding drillings in the 22 coal seam of Xuehu Mine [J]. Journal of Safety Science and Technology, 2014, 10 (6): 90-95.
18. Qi Qingjie, Qi Yun, Zhang Jianguo, et al. Study on optimum drilling parameters for gas extraction in fully mechanized caving face [J]. Journal of Safety Science and Technology, 2018, 14 (6): 76-83.
19. Lv Zhaohui, GUI Hongwei. Application of High Level Long Directional Drilling Technology in Yanmazhuang Coal Mine [J]. Shandong coal technology, 2018, no.217 (9): 73-74, 77.
20. Ruijun. Evaluation of gas drainage effect of high-level directional long borehole and roof strike borehole [J]. China chemical trade, 2017,9 (36): 243245.
21. Yan Baoyong. Several key technologies about ZYWL-23000DS high-power directional drill rig [J]. Mining machinery, 2020, Volume 48 (3): 12-15.
22. Shi Zhijun, Dong shuning, Yao Ningping, et al. Technology and Equipment of Horizontal Measuring Directional Drilling in Underground Coal Mine [J]. Coal Science and Technology, 2013, 41 (3): 1-6.
23. Qin Yi. Research on Measurement Device for Mine While Drilling Suitable for Screening and Lowering Process, 2018, 39 (10): 25-28.
24. Shi Zhijun, Yao Ke, Tian Hongliang, et al. Present situation and prospect of directional drilling technology and equipment while drilling measurement in underground coal mine [J]. Coal Science and Technology, 2019, 47 (5): 22-28.
25. Hu Zhenyang, Zhang Jiadi, Hao Shijun, et al. Mud pulse measuring while drilling system used in high position directional drilling [J]. Exploration Engineering(Rock & Soil Drilling and Tunneling), 2019, 46 (3): 20-24.

E3S Web of Conferences **528**, 01020 (2024) https://doi.org/10.1051/e3sconf/202452801020
NET-LC 2024

The Analysis of Gas-liquid Separation Technology Research Progress

Yupeng Yuan

College of Electromechanical Engineering, Wuhan University of Technology; Wuhan China.

Abstract—Gas-liquid separators are commonly utilized in the energy and chemical industries. This paper briefly introduces the separation principles and research progress of the primary gas-liquid separation equipment domestically and internationally. The paper briefly summarizes the working principles of gravity separation, filtration separation, and centrifugal separation equipment. Next, attention is focused on optimizing the separation performance of centrifugal separation equipment, which offers advantages of small size and high separation efficiency. Finally, a comparative analysis of these gas-liquid separators is provided.

1. Introduction

Gas-liquid separation is a technology used to separate the liquid phase from the gas phase through physical or chemical means. It finds wide applications in the oil and gas industry, chemical industry, energy production, environmental protection, etc. It primarily aims at separating and removing undesirable substances while enhancing the recycling efficiency of valuable ones [1]. For instance, in the oilfield extraction process, gas-liquid separation enhances oil extraction efficiency, facilitating the extraction of pure crude oil for subsequent processing and transportation [2]. Similarly, in chemical processes, gas-liquid separation facilitates the obtainment of pure chemicals and organic solvents. Moreover, in refrigeration systems, gas-liquid separation technology ensures the normal circulation and operation of the refrigerant, preventing the occurrence of liquid strike phenomena [3].This paper introduces the current state of the main gas-liquid separators, focusing on their advantages, disadvantages, and research progress. The principles of commonly used gas-liquid separation, namely gravity separation, filtration separation, and centrifugal separation, are discussed. Furthermore, existing issues are analyzed. Emphasis is placed on optimizing the separation performance of centrifugal separation equipment, which boasts small equipment size and high separation efficiency. Finally, a comparative analysis of these gas-liquid separation devices is conducted, aiming to provide insights into future research directions and the prospects of gas-liquid separation technology.

2. Gravity Gas-Liquid Separator

2.1 Separation principle

The gravity gas-liquid separator is the most common gas-liquid separation device. Its separation principle relies on the density difference between the gas and liquid phases. When the gas-liquid two-phase mixture passes through the equipment, the heavier liquid phase precipitates to the bottom under the influence of gravity, forming a liquid layer, while the lighter gas phase rises to the top, forming a gas layer. This separation results in distinct layers of gas and liquid phases, achieving the separation of the gas-liquid mixture. Gravity gas-liquid separators are primarily categorized into vertical and horizontal forms [4], as illustrated in Fig. 1.

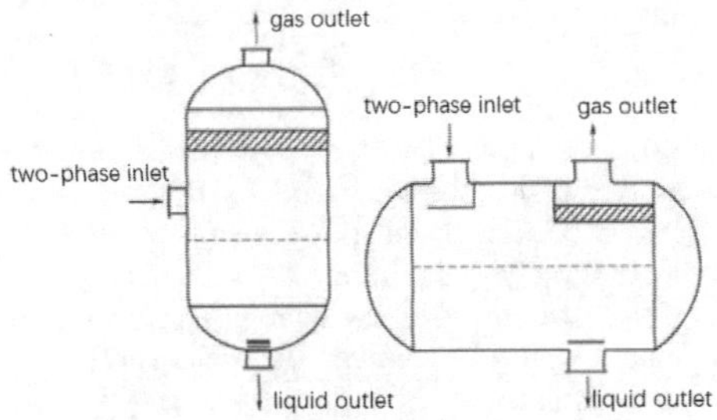

Figure 1. Gravity gas-liquid separator .left: Vertical ,right:Horizontal

2.2 Research status

Gravity gas-liquid separators are characterized by simple structures, easy fabrication, and low resistance. However, they necessitate long residence times, resulting in large,

3436198291@qq.com.

https://doi.org/10.1051/e3sconf/202452801020

bulky separators with poor separation efficiency. They are typically capable of separating only larger droplets, with a typical limit of 100μm [5]. Optimization of performance has always been at the core of research endeavors. Researchers have sought to enhance separation efficiency, minimize pressure drops, and cater to diverse operational scenarios by refining the design and operational parameters of the separator. Yushan Zhang [6] identified the key factors influencing the processing capacity of gravity gas-liquid separators through a study of their structure and operational principles. Meanwhile, Xindong Lu [7] assessed the offshore adaptability of both vertical and horizontal separators through numerical simulations of their internal flow fields under transverse rocking conditions. Secondly, the utilization of hydrodynamic simulation technology is progressively emerging as a pivotal research approach. Numerical simulations offer comprehensive insights into the flow dynamics of gas and liquid phases within the separator, thereby furnishing robust support for optimized design strategies. Weidong Ye [8] conducted an analysis of the working principle of gravity gas-liquid separators through numerical simulations, quantifying the impact of structural parameters on separation efficiency. Meanwhile, Yang Qin [9] investigated the internal flow field using the particle model of the Eulerian-Eulerian method within a multiphase flow framework. The simulation results were in good agreement with experimental findings, demonstrating the feasibility of the turbulence model and algorithm.

3. Filtered gas-liquid separator

3.1 Separation principle

The filtered gas-liquid separator separates the liquid phase medium from the gas phase medium using a filtering medium. Its core component is the filter element, typically made of metal wire mesh or glass fiber [10]. Fig. 2 illustrates the schematic diagram of the wire mesh. The wire mesh gas-liquid separator is a typical filtration separation equipment. As the gas-liquid two-phase mixture enters the separator, larger liquid particles are trapped within the internal tiny gaps of the wire mesh structure. These trapped liquid particles gradually aggregate to form droplets, which, under the influence of gravity, eventually collect at the bottom of the separator in the liquid collection chamber. Meanwhile, relatively pure gas passes through the wire mesh structure at the top or side of the separator, thus achieving the separation of the gas-liquid two-phase mixture. The relatively pure gas passing through the screen structure is released at the top or side of the separator, thereby achieving the separation of gas and liquid.

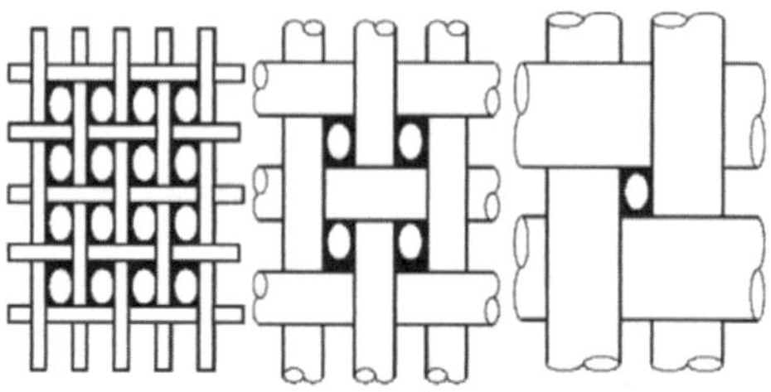

Figure 2. Wire Mesh

3.2 Research status

The screen type gas-liquid separator offers advantages of a simple structure, small volume, high separation efficiency, and minimal pressure drop loss. For droplets with diameters greater than 5μm, the separation efficiency can reach 98% to 99.5%, with a pressure drop of only 250 to 300Pa [11]. Experimental research by Hisham [12] indicates that, for achieving the same separation effect, the silk screen separator exhibits the lowest total pressure loss compared to cyclone separators, fiber silk beds, and vane inertial separators. Yonghong Shi [13] has conducted a detailed analysis of the separation mechanism of the silk screen gas-liquid separator, providing quantitative insights into the different separation mechanisms affecting gas-liquid separation. There has been extensive research on factors that primarily influence the separation efficiency of screen-type gas-liquid separators, including the density of the gas-liquid two-phase, the diameter of liquid droplets, gas flow rate, etc., leading to the development of relatively mature methods and theories. Guocheng Guo [14] investigated the effects of various factors, such as the two-phase flow rate, screen material, diameter, density of the screen grid, droplet diameter, and gas stream distribution density, on the separation efficiency. However, the wire mesh gas-liquid separator is not suitable for two-phase separation with large gas-liquid two-phase flow rates and high carryover liquid content, as this may lead to wire mesh clogging. Additionally, cleaning the metal wire mesh is challenging, resulting in higher operational costs. Furthermore, there is a lack of research on the variation in separation efficiency and the model prediction of liquid droplet separation efficiency after secondary carryover.

4. Centrifugal gas-liquid separator

4.1 Separation principle

The centrifugal gas-liquid separator primarily utilizes centrifugal force to separate the gas and liquid phases. Due to the centrifugal force being dozens of times or even more powerful than gravity, centrifugal gas-liquid separation exhibits higher efficiency compared to gravity separation. Moreover, it features a shorter gas-liquid retention time, smaller equipment volume and footprint, easy installation, flexible operation, stable and continuous operation, as well as easy maintenance, making it the most researched method of gas-liquid separation [10]. The cyclone separator is a typical centrifugal separation equipment, commonly featuring a tangential reversal inlet structure, as depicted in Fig. 3. Upon entry into the separator in a straight line, the gas flow containing liquid droplets is constrained by the cylindrical wall, transitioning into a downward spiral motion along the wall, forming an "outer vortex flow". During this process, droplets with higher density are propelled towards the wall due to the action of centrifugal force. Upon contact with the wall, these droplets descend along it under the combined effect of centrifugal force and gravity, eventually exiting through the discharge port at the bottom. Upon reaching the cone part, the contraction of the wall surface leads to a dramatic increase in velocity. As the diameter decreases, the pressure on the outer wall increases, creating a low-pressure zone at the center of the cone. Influenced by the high-pressure area, the airflow moves towards the low-pressure area at the center, spiraling upward from the bottom until discharged through the exhaust pipe, forming an "inner vortex flow". This process achieves high-efficiency separation of the gas and liquid phases [15].

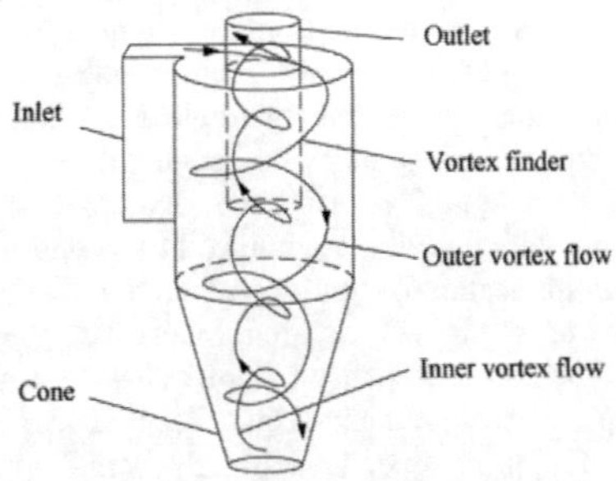

Figure 3. Schematic structure of tangential reversing cyclone separator

4.2 Research status

Separation efficiency and pressure loss are two primary indicators of the performance of gas-liquid cyclone separators, and they are often positively correlated [16]. Researchers primarily investigate and analyze the characteristics of the flow field inside the separator through both experimental observation and numerical simulation to provide a theoretical basis for enhancing separation performance. For experimental studies, LDA (laser Doppler anemometer) and PIV (particle image velocimetry) are commonly employed to measure the internal flow field [17-20]. For numerical research, many researchers at home and abroad utilize CFD (computational fluid dynamics) techniques to investigate the internal flow characteristics of the separator, providing theoretical support for enhancing separation efficiency. The author summarizes various numerical simulation methods in the literature, including turbulence models (RSM, Reynolds Stress Model), pressure-velocity coupling methods (SIMPLE, Semi-Implicit Method for Pressure Linked Equations; SIMPLEC, SIMPLE Consistent), and discrete formats (QUICK, Quadratic Upstream Interpolation for Convective Kinetics; PRESTO, Pressure Staggering Option), as presented in Table 1.

Table 1 Numerical simulation methods

Author	NUMERICAL SIMULATION METHODS		
	Turbulence model	**Pressure-velocity coupling**	**Discrete format**
Kun Fang[21]	RSM	SIMPLE	First Order Upwind
Baoyu Cui[22]	RSM	SIMPLE	QUICK
K.Elsayed[23]	RSM	SIMPLEC	QUICK/PRESTO
M.Wasilewski[24]	RSM	SIMPLEC	PRESTO/Second Order Upwind
Lei Li[25]	RSM	SIMPLEC	QUICK
S.B.Kuang[26]	RSM	SIMPLE	Second Order Upwind
L.Huang[27]	RSM	Coupled	QUICK
M.Azadi[28]	RSM	SIMPLEC	Second Order Upwind
Wenjing Li[29]	RSM	SIMPLE	QUICK
Liedong Li[30]	RSM	SIMPLE	QUICK

Juan Li [31] compared the tangential and axial velocities in the internal flow field predicted by different turbulence models, discrete formats, and pressure-velocity coupling methods with experimental data [32]. The study concluded that: the RSM turbulence model provides more accurate predictions of the cyclone's flow field; the prediction results of the discrete formats QUICK and Second Order Upwind are relatively similar; the SIMPLE algorithm is more suitable for pressure-velocity coupling approaches in calculating the flow field.

The internal flow field of the separator constitutes a complex turbulent flow field. External rotation plays a positive role in droplet trapping, while internal rotation consumes energy [33]. Consequently, a preliminary mechanism to enhance separation performance can be derived: to bolster the rotational strength of external cyclones while suppressing the development of internal cyclones. Research on optimizing separation performance primarily revolves around four aspects: geometry optimization, geometry parameter optimization, operational parameter optimization, and internal component optimization.

1) Geometry Optimization: Geometric structure optimization modifies the original design concept to optimize separation performance. By appropriately altering or optimizing the inlet structure, this improves the stability of the inlet structure and strengthens the stability of the internal flow field, thereby enhancing separation efficiency. Zhao B[34] employed the RSM turbulence model to compare the separation performance of tangential and helical dual inlet structures. The results indicate that the latter structure enhances flow symmetry and increases separation efficiency. M. Wasilewski [24] investigated the impact of inlet angle and other factors on separation performance, which showed a significant effect on pressure drop changes.

2) Geometric parameter optimization: Optimization of geometric parameters solely alters the separator's geometry. The separator has the following eight crucial dimensions: inlet height (a), inlet width (b), exhaust pipe diameter (Dx), exhaust pipe insertion length (S), column height (h), total height (Ht), and cone tip diameter (Bc), as depicted in Fig. 4.

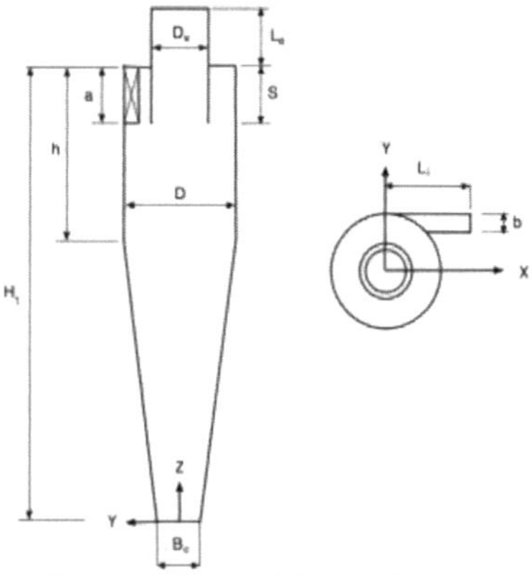

Figure 4. Tangential reversal geometry

Adjustment of part size can enhance separation performance. The influence of five distinct inlet sizes on flow pattern and performance was examined utilizing the Reynolds Stress Turbulence Model (RSM). Findings revealed that with increasing inlet size, the maximum tangential velocity decreases, along with a reduction in pressure drop, albeit resulting in lower overall separation efficiency [23].

3) Optimization of operating parameters: Optimization of operational parameters can enhance separation performance by adjusting the velocity or concentration of the gas-liquid two-phase flow. Gao.Z [35] analyzed the internal flow under different inlet velocity conditions, revealing that flow stability was superior at 12 m/s compared to 15 m/s and 18 m/s, consequently leading to improved separation performance.

Internal component optimization: Optimization of internal components entails adding additional components to the original basic framework to enhance separation performance. In the event of equipment failure, instead of replacing the entire equipment, only the faulty components need replacement, thereby reducing maintenance costs and labor. Currently, internal components are primarily categorized into 8 types: center body structure, drag reducing rod, guide plate, drag reducing frame, guide vane, centerpiece, and vortex stabilizer [36]. In conclusion, despite the cyclone separator's simple structure, its flow characteristics are complex, necessitating further research for accurate simulation of its internal flow field. Additionally, during the separation process, droplets may undergo phenomena such as liquid film formation, secondary droplet formation, and crushing, yet there is relatively scant research in this domain. Finally, structural optimization is frequently multifaceted, necessitating exploration of the influence of each factor, which remains a research hotspot.

5 Conclusions

In summary, gas-liquid separation is required in many instances, with a variety of methods and equipment available. Table 2 summarizes the advantages and disadvantages of each.

Table 2 Advantages and disadvantages of gas-liquid separators

Types	advantages	drawbacks
gravitational	Simple structure, suitable for large flow, low flow resistance	Low separation efficiency,long residence time, large space occupation
filtered	High separation efficiency, suitable for small particle sizes	separation effect is poor when the gas-liquid flow is large, the screen is easy to be clogged, and needs to be leaned regularly
centrifugal	High separation efficiency, small size, less maintenance	Complex internal flow and high pressure loss

Through the introduction and comparison of various gas-liquid separation methods and equipment, it becomes apparent that each type of equipment has certain limitations, and their scope of application is relatively narrow and not universal. Therefore, the development of high-efficiency and low-resistance gas-liquid separation technology with universal applicability, along with the combined application of multiple separation technologies, will be the focus of research. Moreover, secondary carryover of liquid droplets, a significant factor affecting separation efficiency, has received limited attention in terms of studying its mechanism, influencing factors, and strategies to minimize it.

References

1. Kunteng H, Jianyong C, and Ying C, Research status of gas-liquid separation technology [J]. Journal of Chemical Engineering, 2021, 72 (S1): 30-41.
2. Yazhou L, Xiaowei L, and Feng L, Research progress of oil and gas separation devices [J]. China Special Equipment Safety, 2020, 36 (07): 22-25+30.
3. Li S, Simulation study on separation effect of gas-liquid separator [J]. Refrigeration and Air Conditioning, 2017, 17 (02): 22-26.
4. A.Bahadori, Chapter 14 - Basic Engineering Design for Natural Gas Processing Projects [M].2014:633-721.
5. Chengzhao L, Development and field testing of separators for skid-mounted gas gathering units [J]. Natural Gas and Oil, 1994, (01): 36-40.
6. Yushan Z, Weiqing Z, and De Y, Analysis of processing capacity of gravity gas-liquid separator [J]. Petroleum Mining Machinery, 2017, 46 (01): 25-28.
7. Xindong L, Gravity gas-liquid separator offshore adaptability study [D], 2015.

8. Weidong Y, Tingting C, Structural optimization and numerical simulation of separation performance of gravity gas-liquid separator [J]. Mechanical Manufacturing and Automation, 2018, 47 (02): 133-136
9. Qin Y, Jin F, and Haibing L, CFD simulation of internal flow field in a gravity gas-liquid separator [J]. Mechanical Research and Application, 2007, (02): 72-74
10. Xiangjun R, Zhenbo W, and Youhai J, Advances in gas-liquid separation technology and equipment [J]. Filtration and Separation, 2008, (03): 43-47
11. Y.N. Lebedev, I.A. Zilberberg, and Y.P. Lozhkin, High-Efficiency Mist Eliminators [J]. Chemistry and Technology of Fuels and Oils, 2002, 38 (1): 42-45.
12. H.T. El-Dessouky, I.M. Alatiqi, and H.M.Ettouney, Performance of wire mesh mist eliminator [J]. Chemical Engineering and Processing: Process Intensification, 2000, 39 (2): 129-139.
13. Yonghong Shi, Analysis of separation performance and pressure loss of screen gas-liquid separator [J]. Petrochemical Equipment, 2006, (03): 35-37.
14. Guocheng G, Mansheng W, Analytical study on separation mechanism of silk screen gas-liquid separator [J]. Food and Machinery, 2010, 26 (04): 98-101+108.
15. Gao Z, Wei Y, and Liu Z, Internal components optimization in cyclone separators: systematic classification and meta-analysis [J]. Separation & Purification Reviews, 2021, 50 (4): 400-416.
16. Qiuping L, Yongjie X, and Lihong D, Numerical simulation study of JLX-type cyclone separator with guide vanes [J]. Chemical Engineering, 2015, 43 (01): 37-41.
17. G. Solero, A. Coghe, Experimental fluid dynamic characterization of a cyclone chamber [J]. Experimental Thermal and Fluid Science, 2002, 27 (1): 87-96.
18. Liu Z, Zheng Y, Jia L, Stereoscopic PIV studies on the swirling flow structure in a gas cyclone[J]. Chemical Engineering Science, 2006, 61 (13): 4252-4261.
19. Liu Z, Zheng Y, Jia L, An experimental method of examining three-dimensional swirling flows in gas cyclones by 2D-PIV [J]. Chemical Engineering Journal, 2007, 133 (1): 247-256.
20. A.J. Hoekstra, Gas flow field and collection efficiency of cyclone separators [J]. TU Delft, PhD Thesis, Delft University of Technology, 2000.
21. Kun F, Jingbo W, and Xina W, Numerical simulation of flow field in cyclone [J]. Coal Technology, 2008, (12): 5-7.
22. Baoyu C, Dezhou W, and Qingxiang Z, Numerical study of flow field inside hydrocyclone[J]. Journal of Northeastern University (Natural Science Edition), 2014, 35 (06): 894-897.
23. K. Elsayed, C. Lacor, The effect of cyclone inlet dimensions on the flow pattern and performance [J].Applied Mathematical Modelling, 2011, 35 (4): 1952-1968.
24. M. Wasilewski, L.S. Brar, Effect of the inlet duct angle on the performance of cyclone separators [J].Separation and Purification Technology, 2019, 213: 19-33.
25. Lei L, JunQing J, and Kun C, Structural optimization of cyclone separator based on FLUENT software analysis [J]. Natural Gas and Oil, 2014, 32 (03): 71-75+13.
26. S.B. Kuang., K.W.Chu , and A.B. Yu, Numerical study of liquid-gas-solid flow in classifying hydrocyclones: effect of feed solids concentration [J]. Minerals Engineering, 2012, 31: 17-31.
27. Huang L, Deng S, and Guan J, Development of a novel high-efficiency dynamic hydrocyclone for oil-water separation [J]. Chemical Engineering Research and Design, 2018, 130: 266-273.
28. M. Azadi, A. Mohebbi, A CFD study of the effect of cyclone size on its performance parameters [J]. Journal of Hazardous Materials, 2010, 182 (1): 835-841.
29. Wenjing L, Numerical simulation of natural gas cyclone gas-liquid separator [D], 2009.
30. Liedong,M, Numerical simulation of structural optimization of gas-liquid cyclone separator [D], 2014.
31. Juan L, Xiaokang Y, and Xiaoheng L, Research on numerical simulation method of flow field inside cyclone [J]. Journal of Coal, 2019, 44 (10): 3250-3257.
32. K. Hsieh., K. Rajamani, Phenomenological model of the hydrocyclone: Model development and verification for single-phase flow [J]. International Journal of Mineral Processing, 1988, 22 (1-4): 223-237.
33. Aiqin Z, Jing Z,and Wei G, Numerical simulation study of an efficient low-resistance cyclone separator [J]. Science and Technology Information, 2015, 13 (36): 146+149.
34. Zhao B, Su Y, and Zhang J, Simulation of Gas Flow Pattern and Separation Efficiency in Cyclone with Conventional Single and Spiral Double Inlet Configuration [J]. Chemical Engineering Research and Design, 2006, 84 (12): 1158-1165.
35. Gao Z, Wang J, and Wang J, , Analysis of the effect of vortex on the flow field of a cylindrical cyclone separator [J]. Separation and Purification Technology, 2019, 211: 438-447.
36. Gaozuo W, Juan W, Jiangyun W,. Advances in the application of internal component technology within cyclone separators [J]. Journal of Petroleum (Petroleum Processing), 2019, 35 (02): 393-402.

https://doi.org/10.1051/e3sconf/202452801021

Research and Patterns of Rock Acid Pressure Corrosion Tests

Jiangchun Hu , Xuewei Huang* , Luyang An, Chenxu Wang, Mengyuan li, Bing Liu and Miao Liu

School of Civil Engineering and Architecture, Zhongyuan University of Technology, Zhengzhou 450007, China

Abstract: In China, tight sandstone gas reservoirs are widely distributed, yet their development is hindered by issues such as low extraction efficiency and a monotonous extraction approach. To enhance the efficiency of tight sandstone gas extraction, this study investigates the internal fracture network characteristics of sandstone and its correlation with chemical solutions. Employing the sandstone acid fracturing technique, an experimental setup under gravitational conditions is established. Through this apparatus, fracturing experiments are primarily conducted on rocks utilizing the acid fracturing technique. The experimental findings suggest that the acid-fracturing coupling approach effectively induces rock fracturing. Uniaxial compression tests also indicate a noticeable alteration in the elastic-plastic behavior of rock samples after chemical solution corrosion, becoming more pronounced with increasing duration. The research demonstrates that this experimental setup significantly enhances the fracturing of sandstone and consequently improves the extraction of tight sandstone gas, holding crucial reference value.

1 Introduction

In recent years, with the continuous discovery of tight sandstone gas in basins such as Ordos, Tarim, and Sichuan, tight sandstone gas refers to a type of gas reservoir where natural gas is stored in sandstone with low porosity and permeability. The characteristics of this type of reservoir include the microscopic pore structure of sandstone rock, resulting in the difficulty of gas flow. The low porosity and permeability of the rock restrict the movement of natural gas in the rock matrix, making extraction relatively challenging. The development of tight sandstone gas requires the use of unconventional techniques, such as hydraulic fracturing and horizontal wells, to fracture the rock, increase permeability, and maximize gas release through multilayer development. Due to the vast potential reserves of this type of reservoir, the development of tight sandstone gas is strategically important for meeting energy needs, and acid pressure is one of the important methods for effectively developing such sandstone oil and gas reservoirs[1]. Acid pressure is a common reservoir modification technology, especially suitable for sandstone reservoirs. The core objective of this technology is to improve the physical properties of the reservoir by injecting acidic solutions, thereby increasing the oil and gas recovery rate. The main mechanism of acid pressure on sandstone reservoir development is through the injection of acidic solution, which reacts with minerals such as calcium carbonate and silicates in the reservoir rock, producing water-soluble products such as salts and carbon dioxide, thereby dissolving the reservoir rock, effectively increasing porosity and permeability. Additionally, the cleaning action of the acidic solution enables effective cleaning of reservoir pores, dissolving or flushing impurities such as clay particles and mud grains, enlarging effective pores, and further increasing permeability. Furthermore, the injection of acidic solution can generate pressure, inducing crack formation, which helps to increase the permeability of the reservoir, making it easier for oil and gas to flow through the reservoir and be extracted. Finally, the acidic solution has a dissolving effect on deposits, cements, or shales in the reservoir, preventing pore blockage and effectively enhancing the fluidity of oil and gas. Therefore, acid pressure technology comprehensively enhances the physical properties of sandstone reservoirs through multiple mechanisms, creating favorable conditions for increasing recovery rates.

Gong Wei optimized the acid solution formula and process to complete the technology for jet acid pressure transformation[2]. Guo Yanjie studied the acid pressure method for reservoirs to achieve difficult production transformation[3]. He Chunming et al. researched acid pressure technology suitable for deep-seated low-permeability fractured carbonate reservoirs[4]. Zhang Long et al. found through triaxial compression tests of acid-base chemical solutions that both acidic and alkaline solutions have a negative effect on sandstone[5]. Hou Bing et al. used the phase field method to simulate the phenomenon of hole chaining and found that acid pressure cracking is more effective in multiple hole chains to increase flood discharge area[6]. Guo Jianchun used stereoscopic acid pressure technology to improve the development of deep-seated rock oil and gas[7]. Youliang Chen and Peng Xiao studied the damage constitutive model of rocks under freeze-thaw-confining pressure and acid erosion and found the damage evolution of rock

*Corresponding author e-mail:1769341005@qq.com

under acid erosion and freezing-thawing and confining pressure, and derived the model parameters to reveal the reliability of the model[8]. Fanhui Zeng provided a method of acidifying rocks to reduce fracturing pressure, and the research results showed that acidification can reduce rock strength and critical stress intensity factor of cracks, thereby reducing crack pressure combination[9]. Jianchun Guo found through quantitative prediction models that the greater the acid concentration, the greater the rock damage variables and shear strength intensity[10]. Ahmadi, M.A., & Shadizadeh, S.R experimental investigation on the adsorption behavior of natural surfactants extracted from Zyziphus Spina Christi leaves onto shale sandstone reservoir rocks under static and dynamic conditions, with implications for chemical EOR applications[11]. Greibrokk conducted a study on the application of multi-stage supercritical fluid extraction in samples from Kimmeridge Clay and Posidonia shale formations, aiming to assess its potential as a substitute for liquid extraction of petroleum source rocks[12]. These studies indicate that most scholars use separate methods of chemical solution and gravity pressure to fracture rocks, and there are few models to simultaneously handle stress and chemical solution for rock fracturing.

This paper starts from the characteristics of the internal crack network of sandstone and the relationship between seepage flow to improve the production of tight gas. It establishes a model of a gravity-pressure-coupled test system for rock acid pressure under the gravity pressure, which integrates a corrosion simulation device, a gravity-pressure device, and an electrochemical detection device. Based on this model, the influencing factors such as solution dosage ratio and load value during the test process are analyzed, and then the optimized supporting devices are realized, providing theoretical support for the effective extraction of sandstone gas.

2 Overview of the Experimental Model

2.1 Overview of the Experimental Model

The experimental setup primarily consists of a gravity pressure apparatus and a corrosion simulation device, complemented by a working platform and pillars, creating a coupled rock acid-fracturing testing system. In essence, this apparatus simulates the subsurface environment by employing the gravity pressure apparatus and corrosion simulation device installed on the working platform. The setup includes sandstone specimens placed within the test chamber and a gravity pressure apparatus containing chemical solutions. The upper part of the rock specimen is connected to the pressurized container, enabling simultaneous application of chemical solutions and gravitational pressure on the rock sample. This system achieves the coupling of gravity pressure and chemical corrosion. Furthermore, an electrochemical workstation is connected to the rock specimen via conductive wires and linked to a computer through data cables, forming a closed circuit to monitor the electrochemical changes in the rock sample.

This experimental design facilitates real-time monitoring of the microstructural changes in shale under the combined influence of gravity pressure and chemical solution pressure. It is suitable for indoor investigations into the microscopic structural variations of rocks subjected to the simultaneous effects of pressure and acid corrosion, achieved through laboratory experiments and continuous monitoring. The equipment is designed for simplicity, featuring a clear operational procedure and user-friendly interface, with the advantage of straightforward instrumentation.

2.2 Design of the Experimental System Model

This apparatus primarily comprises a corrosion simulation device, a gravity pressure apparatus, and an electrochemical detection device to facilitate the experimental process. In the corrosion simulation device, the core operation involves the interaction between the rock sample, chemical solution within the container, and the transmitted load, occurring through a flexible board during pressurization. The gravity pressure apparatus applies pressure to the experimental setup by increasing the weight of the weights, allowing for the acquisition of electrochemical data as the load is continuously applied. The electrochemical detection device operates by inducing reactions through an electrochemical workstation, and the reaction data are then transferred to a computer via conductive wires. Consequently, the computer accurately receives and records data reflecting the internal changes in the rock specimen.

The specific system models are illustrated in Fig 1 and Fig 2.

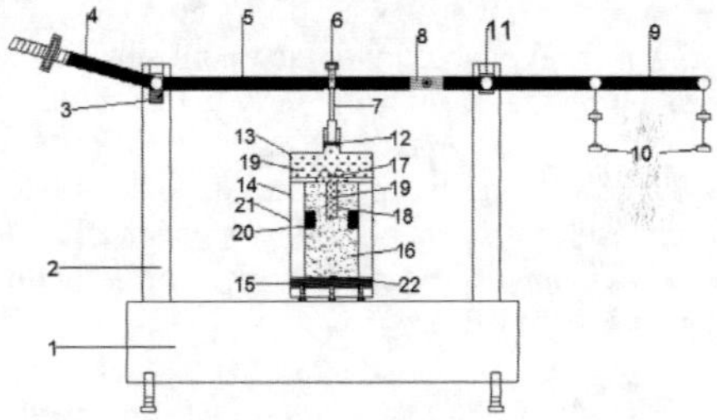

Fig. 1 Three-dimensional Schematic Diagram of the Experimental Structure

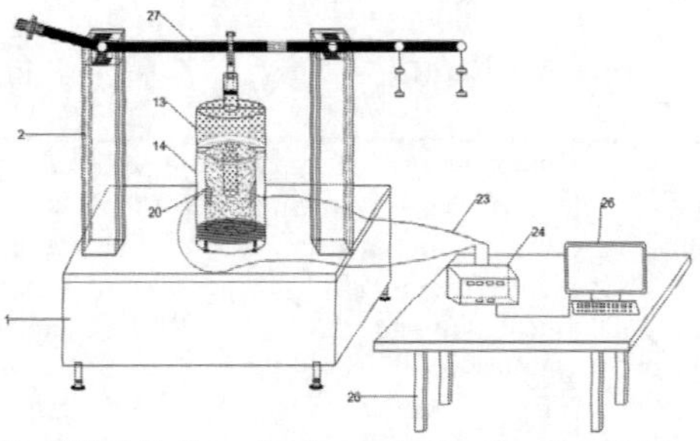

Fig. 2 Model's Corrosion-Pressurization Coupling Device Layout

E3S Web of Conferences **528**, 01021 (2024)
NET-LC 2024
https://doi.org/10.1051/e3sconf/202452801021

2.3 Operation of the Experimental System Model

1)Device Installation: Initially, boreholes are drilled at the upper part of the rock sample, and the subsequently drilled rock sample is placed in a chemical solution for corrosion. The bottom of the test chamber accommodates the required rock specimens, adjusting the test chamber to an appropriate height. Subsequently, the gravity pressure container and baffle are connected through the test chamber, with the container holding the chemical solution connected to the pillar using a straight rod, thereby completing the installation of the gravity pressure apparatus.

2)System Calibration: The vertical rod is adjusted to a horizontal position using a horizontal adjustment balance weight. The adjustable pressure rod is then mounted on the vertical rod, and pressure loading is achieved by adding weights to the tray. Finally, the signals from the collected electrodes are transmitted to the electrochemical workstation via conductive wires.

3)Experimental Testing: Gradual pressure loading is applied by incrementally adding weights to the tray at different positions, assessing the compressive strength of the rock sample. The experimental data recorded by the electrochemical workstation are then analyzed to reveal the characteristics of crack development and the dynamic evolution process.

2.4 Acid-Fracturing Experiment on Sandstone

To investigate the impact of the sandstone acid-fracturing test apparatus on the reaction characteristics of sandstone, a simplified experimental procedure was devised based on the setup of the test apparatus. Two key factors, namely the primary acid concentration and corrosion time, were defined during the experiment to assess their influence on the reaction characteristics of sandstone,As shown in Table 1.The aim was to explore the temporal variations in sandstone under different pH values of chemical solutions, thereby validating the feasibility of the experimental apparatus design.

Table 1 Solution-Water Ratio Schemes for Chemical Solutions

Solution	Composition	Solution	pH
1	HCL	0.2	3
2	HCL	0.2	6
3	Distilled	---	7
4	NaOH	0.2	9
5	NaOH	0.2	12

The sandstone specimens, with a diameter of 50mm and a height of 150mm, were subjected to the chemical solution corrosion as outlined in Table 1.1 The specimens were divided into five groups based on different pH values of the chemical solutions. Subsequently, the corroded sandstone samples underwent boiling in distilled water to remove surface chemical residues. Afterward, the sandstone samples were dried in an oven and allowed to stand. Following this, the statically placed sandstone samples were introduced into the corrosion and pressure apparatus with chemical solutions of varying pH values. A constant load, provided by a 2kg weight, was applied for 30 days, 60 days, and 90 days to obtain sandstone samples at different corrosion time points, along with the corresponding chemical solutions post-treatment. The specimens, having undergone chemical interaction, were then subjected to uniaxial compression tests. The changes in stress-strain curves of the sandstone samples after acid-fracturing and the variations in the pH values of the chemical solutions were observed.

By establishing the experimental apparatus described above, the process of rock acid pressure corrosion can be observed clearly, allowing for a more detailed examination of the patterns of change in rocks under acid pressure corrosion conditions. This provides a novel approach to more efficiently extracting tight sandstone gas, thereby achieving increased production of tight sandstone gas.

3 Results Analysis

3.1 Analysis of Acid-Fracturing Test Results

3.1.1 Variation Patterns in Rock Mass Over Time

Acid-Fracturing experiments induce chemical erosion on sandstone samples, resulting in a loss of mass. The Mass Loss Rate serves as a quantitative indicator to measure the extent of this loss. Calculating the difference between the initial and final masses allows determining the numerical value of the mass loss rate, reflecting the sandstone samples' erosion resistance in acidic solutions. Thus, to better illustrate the changes in sandstone mass in chemical solutions, the Mass Loss Rate (Q) is defined as follows:

$$Q=\frac{S_t-S_0}{S_0}\times 100\% \tag{1}$$

Where: S_0 is the initial mass of the rock sample; S_t is the mass of the rock sample after t time of interaction with the chemical solution

The Mass Loss Rate curve effectively depicts the extent of the rock's interaction with the chemical and aqueous solutions. By incorporating the initial mass of the rock sample and the mass obtained after corrosion with the chemical solution into Equation (1), followed by data processing and graphical representation using software, as shown in Fig 3. the trend can be analyzed.

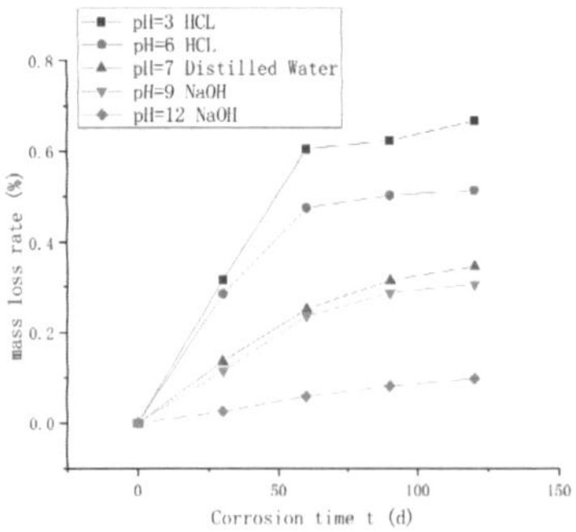

Fig. 3 Sandstone Mass Loss Dynamics Across Varied pH Over Time

(1) The mass loss rate of sandstone gradually increases with the extension of corrosion time. Overall, the trend indicates that in the early stages of corrosion, the degree of mass change in sandstone is higher, while in the later stages, the mass change becomes more gradual.

(2) Under various pH conditions and corrosion time levels in chemical solutions, sandstone exhibits different degrees of mass loss. In the presence of chemical solutions, the mass loss rate of the rock increases gradually with the extension of corrosion time, reflecting a continuous deterioration of rock mass. During the initial corrosion in chemical solutions, sandstone experiences the fastest corrosion in HCl solution with pH=3 and NaOH solution with pH=6, showing higher degrees of mass loss rate change, approximately 0.38% and 0.26%, respectively. As the corrosion progresses into the middle and later stages, the mass change of sandstone in HCl solution with pH=3 and NaOH solution with pH=6 stabilizes, and the mass loss rate gradually tends to stabilize, maintaining at 0.67% and 0.57%, respectively. The results demonstrate that the higher the concentration of hydrogen ions in the chemical solution, the faster the corrosion of sandstone, leading to a greater mass loss rate. Conversely, as the concentration of hydrogen ions decreases, the mass loss rate gradually stabilizes. For distilled water with pH=7, sandstone exhibits minimal mass change, and the mass loss rate steadily approaches a slow, stable state, remaining below 0.1%.

Based on the observed patterns, it can be seen that sandstone is more effectively corroded under chemical solutions with pH values of 3 and 12, leading to more significant mass loss. Therefore, using high-concentration chemical acid solutions can better corrode sandstone, thereby achieving increased production of tight sandstone gas.

3.1.2 Effect of Acid Pressure Corrosion Time on Different pH Values of Chemical Solutions

In the acid pressure corrosion experiment, sandstone samples were used as experimental materials. During the corrosion process of sandstone samples with chemical solutions, certain substances inherent in the sandstone could interact with the dissolved substances in the chemical solution, leading to changes in the pH value of the chemical solution over time. As shown in Figure 4. Therefore, the degree of pH reaction detected can explicitly indicate the effectiveness of the interaction between sandstone samples and chemical solutions.

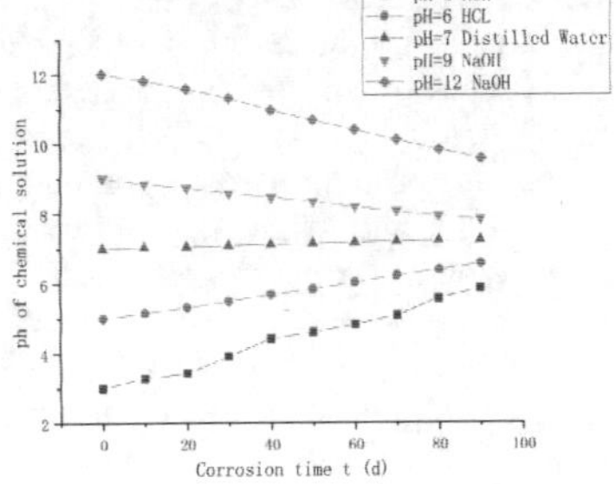

Fig. 4 pH Variation During Acid Pressure Corrosion

Figure 4 shows the variation of different pH values of solutions with acid pressure corrosion time：

(1) As the corrosion time progresses, the pH values of the solutions slowly approach neutral values.

(2) In the initial stages of corrosion, the pH values of chemical solutions at different initial pH values exhibit both upward and downward trends, repeating these fluctuations with prolonged corrosion time. Significant variations are observed in the chemical solutions with initial pH values of 3 and 12. The chemical solution with an initial pH of 7 shows smaller pH variations and eventually exhibits weak alkalinity.

(3) In the acidic solutions with pH values of 3 and 6, as the corrosion time increases, the pH of the chemical solution gradually shifts towards neutral. The former undergoes faster changes, stabilizing around 4.7 in the later stages of the experiment, while the latter stabilizes around 6.8.

(4) In the alkaline solutions with pH values of 9 and 12, the pH of the chemical solution gradually shifts towards weak acidity with increasing corrosion time. The effect is more pronounced in the early stages, with a more significant decrease in pH. In the later stages, both solutions maintain nearly the same rate of decrease, stabilizing around 7.6 for pH=9 and 9.3 for pH=12.

(5) Analyzing the results of the corrosion of sandstone samples by chemical solutions at different pH values, the pH of the chemical solution gradually tends towards neutralization with increasing corrosion time. The solution with an initial pH of 3 tends to neutralize the fastest, followed by pH=12, pH=6, pH=9, and pH=7. These observations suggest that acidic solutions have a more pronounced impact on the pH system of chemical solutions during the sandstone pressure corrosion experiment compared to alkaline and neutral solutions. Additionally, the analysis indicates that minerals present in the sandstone samples interact with the ions in the chemical solution during corrosion, causing the partial decomposition of minerals and resulting in the chemical solution trending towards a neutral state.

Based on the observed patterns, it is found that during the process of acid-pressure corrosion over time, the intense reaction of sandstone gradually diminishes into a slower process. It can be analyzed that with the prolongation of acid-pressure corrosion time, sandstone is continuously corroded, leading to a continuous change in the chemical solution. Therefore, prolonging the corrosion time can more effectively corrode the rock, thereby reducing the difficulty of extracting tight sandstone gas and achieving increased production of tight gas.

3.2 Sandstone Uniaxial Compression Test Result Analysis

During the acid pressure corrosion experiment, sandstone samples were extracted for uniaxial compression testing in cycles of 30 days. The TestExpert software corresponding to the rock creep apparatus was employed for parameter control and setup. The experiment data obtained were used to create stress-strain curve graphs using the software. Finally, stress-strain curve features for samples

https://doi.org/10.1051/e3sconf/202452801021

with the same pH values extracted at 30, 60, and 90 days were analyzed. As shown in Fig 5.Among them, Fig (a), (b), (c), (d), and (e) represent stress-strain curves at different pH values

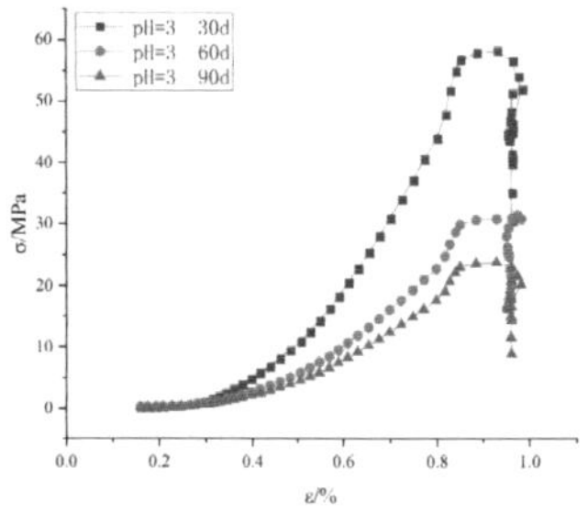

(a) ph=3

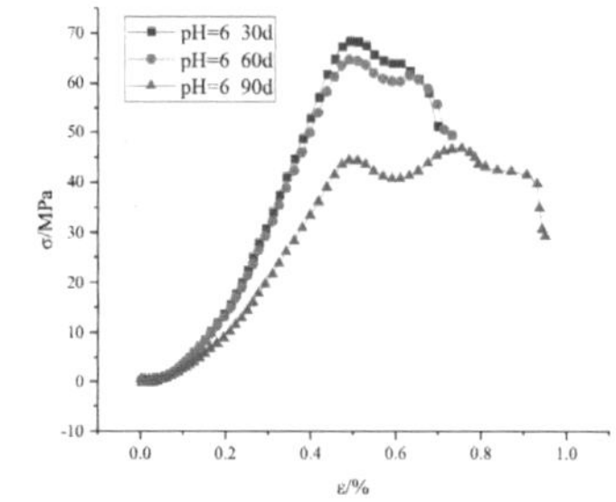

(b) ph=6

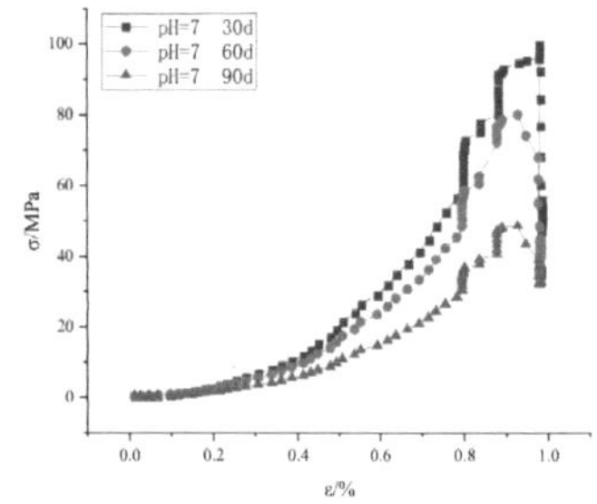

(c) ph=7

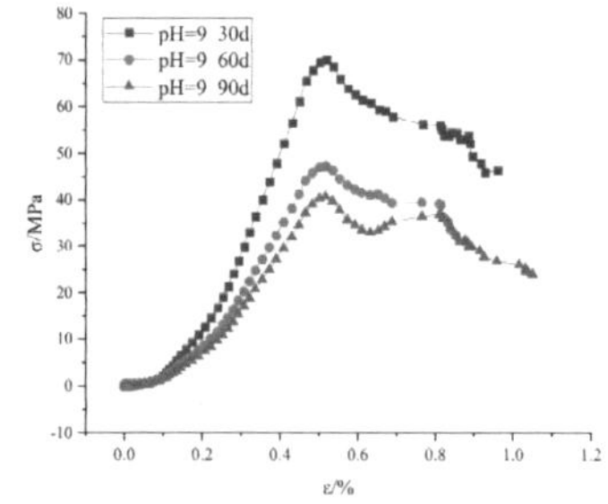

(d) ph=9

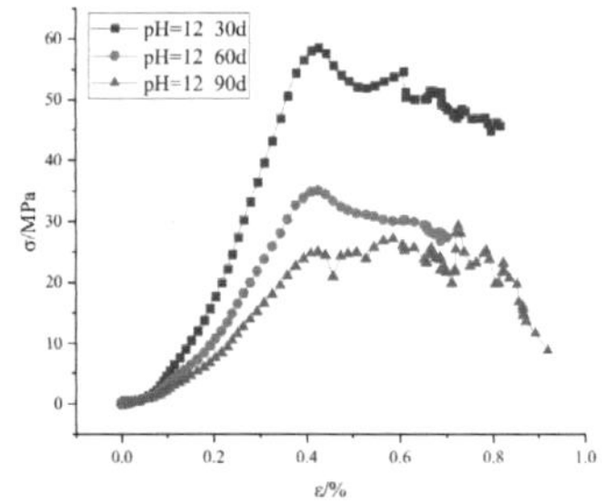

(e) ph=12

Fig. 5 Stress-Strain Curves at Different pH Values Corresponding to Days

Based on Fig 5, the stress-strain curves under corrosion conditions exhibit four main stages, with the following patterns:

(1)Initial Uniaxial Fracture Stage: In this stage, the sandstone sample, subjected to axial stress, initially experiences the closure of the original tensile cracks and micro-fissure structures. This leads to a compaction process, resulting in the early nonlinear deformation shape, represented by an upward concave curve on the stress-strain curve graph (Fig 5). Notably, the higher acidity levels demonstrate a more pronounced concave curve, indicating a more significant impact of chemical solution corrosion. As time progresses, the softening of the sample increases, and the strain compaction point becomes significantly higher. Analysis suggests that, under axial stress, the chemical solution induces greater changes in the internal porosity of the sandstone sample, leading to an extended compaction time. This reflects the more apparent development of porosity in the sandstone sample due to the corrosion by the chemical solution.。

(2)Initial Uniaxial Fracture Stage: In this stage, the sandstone sample, subjected to axial stress, initially experiences the closure of the original tensile cracks and micro-fissure structures. This leads to a compaction process, resulting in the early nonlinear deformation shape, represented by an upward concave curve on the stress-strain curve graph (Fig 5). Notably, the higher acidity levels demonstrate a more pronounced concave curve, indicating a more significant impact of chemical solution corrosion. As time progresses, the softening of the sample increases, and the strain compaction point becomes significantly higher. Analysis suggests that, under axial stress, the chemical solution induces greater changes in the internal porosity of the sandstone sample, leading to an extended compaction time. This reflects the

https://doi.org/10.1051/e3sconf/202452801021

more apparent development of porosity in the sandstone sample due to the corrosion by the chemical solution.

(3)With the continuous application of axial stress, new cracks develop in the sandstone sample. The stress-strain curve in Figure 5 shows a fluctuating trend due to the repeated rupture of new cracks. Notably, the plastic curve for the solution with a pH of 7 indicates a direct descending trend without the observed fluctuation. In contrast, solutions of different concentrations show varying degrees of fluctuation. The peak value on the stress-strain curve represents the deformation when the sandstone reaches failure. A higher peak value indicates a more significant corrosion effect on the sandstone sample, with a more pronounced decrease in the value, suggesting a more apparent softening effect as the corrosion time extends. The sandstone sample gradually exhibits flexible characteristics during this stage.

(4)Ultimate Failure Stage: After reaching the peak axial stress, internal damage occurs in the sandstone sample. In this stage, cracks propagate rapidly, and the load-bearing capacity of the sample decreases rapidly but does not reach zero, indicating that the sample still retains some load-bearing capacity. Figure 5 shows a reverse growth trend in the curve during this stage, with an overall descending trend.

In summary, under the corrosion of different chemical solutions, there are similarities in the stress-strain curves. Figure 5 illustrates that, under the corrosion of chemical solutions with higher pH values, the curve characteristics of the sandstone sample become more pronounced. With an increase in the duration of chemical solution exposure, both the elastic and plastic deformation stages exhibit significant changes. The curves in the plastic stage are more apparent than those in the elastic stage, with more significant peak values. As the corrosion time lengthens, the curves show a descending trend, indicating the gradual manifestation of flexible characteristics in the sandstone sample. During the uniaxial compression process, it is also evident that under conditions of higher chemical solution concentration, the stress variation of sandstone is more pronounced. Furthermore, with the prolongation of corrosion time, the internal stress variation of sandstone becomes more noticeable. Therefore, under conditions of higher concentration and longer corrosion time, it is more suitable for extracting tight gas from sandstone, thus effectively achieving increased production of tight gas.

4 Conclusions

This study employed a self-developed acid pressure corrosion test apparatus to assess the impact of chemical solutions with varying pH values on sandstone samples. The evaluation encompassed measurements of rock mass loss, variations in solution pH post-reaction, and subsequent uniaxial compression experiments following acid pressure corrosion. The feasibility of the acid pressure corrosion apparatus was verified through these experiments, yielding the following conclusions:

(1)Verification of Rock Acid Pressure Corrosion Design:Conducting indoor experiments validated the design for rock acid pressure corrosion. Relevant indicators were monitored during the sandstone's reaction with chemical solutions. The results indicated a gradual increase in rock mass loss rate during the acid pressure corrosion process. Notably, the mass loss rate escalated more rapidly with higher concentrations of chemical solution pH.

(2)Evolution of Solution pH and Rock Behavior:After a certain duration of corrosion, the pH of the solution within the container gradually approached weakly acidic or weakly alkaline conditions, trending towards a neutral value. In acidic and alkaline chemical solutions, the chemical reaction intensity of the rock in acidic solutions was stronger than in alkaline solutions. It can be inferred that ions in alkaline solutions react with ions in the sandstone sample, leading to dissolution or precipitation, resulting in a reduction in sandstone mass and a tendency for the solution's pH to approach neutrality.

(3)Uniaxial Compression Testing and Stress-Strain Curve Evolution:Uniaxial compression tests were conducted on rock samples subjected to chemical solution and gravity pressure. The evolution of stress-strain curves was observed. Over time, the strain at the peak point gradually increased on the curve, indicating a slow reduction in peak intensity and elastic modulus. In contrast, curves representing peak intensity and elastic modulus exhibited a gradual decline.

In summary, the experiments provided valuable insights into the behavior of sandstone under acid pressure corrosion conditions. The study successfully verified the feasibility of the acid pressure corrosion apparatus and highlighted the influence of chemical solution pH on rock mass loss and mechanical properties.

(1)In general, this study conducted experimental verification using a self-developed acid pressure corrosion test apparatus. It involved subjecting rock samples to corrosion experiments with chemical solutions of varying pH values, observing the resulting mass loss of the rocks, and subsequently conducting uniaxial compression tests after acid pressure corrosion. The experimental findings are as follows:

(2)During the acid pressure corrosion process, the mass loss rate of the rock gradually increases with the duration of chemical solution corrosion, with a faster increase observed with higher concentrations of solution pH.

(3)After a certain period of acid pressure corrosion, the pH value of the solution gradually approaches neutrality. In acidic and alkaline solutions, the chemical reactions within the rocks are more pronounced under acidic conditions, leading to a reduction in rock mass, while the solution pH tends towards neutrality.

Uniaxial compression tests indicate that with prolonged chemical solution corrosion time, the peak strain of the rock gradually increases, while the peak strength and elastic modulus exhibit a declining trend.

These patterns suggest that the properties of rocks under acid pressure corrosion conditions are influenced significantly by both the pH value of the chemical solution and the duration of corrosion, thereby affecting both the mass loss and mechanical performance of the rocks. Moreover, under conditions of high concentration chemical solutions and prolonged corrosion time, the enhancement of tight sandstone gas production can be more pronounced.

References

1. Li C, Zhao L, Zhu Q, Li Y, et al. Current Status and Development Trends of Acid Fracturing Technology[J]. Oil & Gas Geology and Recovery Efficiency, 1-12.
2. Gong W. Hydraulic Jet Acid Fracturing Technology for Horizontal Wells in Deep Fractured Carbonate Reservoirs[J]. Fault-Block Oil & Gas Field, 2020, 27(6), 808-811.
3. Guo Y. Research on Acid Fracturing Technology in Sandstone Reservoirs[D]. Northeast Petroleum University, 2016.
4. He C, Cai B, Tang B. Z, et al. Fine Study on Volume Acid Fracturing Technology in Fractured Carbonate Reservoirs[J]. Petroleum and Natural Gas Chemical Industry, 2014, 43(05), 534-538.
5. Zhang L, Lu Y, Wu A, et al. Experimental Study on Mechanical Properties of Sandstone Under Triaxial Compression under Chemical Corrosion[J]. Journal of Hubei University of Engineering, 2023, 43(06), 117-122.
6. Hou B, Fan M, et al. Numerical Simulation of Acid Fracturing Connectivity Holes Based on the Phase Field Method[J]. Acta Petrolei Sinica, 2022, 43(06), 849-859.
7. Guo J, Gou B, Qin N. et al. Innovation of Deep Carbonate Reservoir Transformation Concept: Three-Dimensional Acid Fracturing Technology[J]. Natural Gas Industry, 2020, 40(02), 61-74.
8. Chen Y & Xiao P. Study on the Damage Constitutive Model of Rock under Freeze-Thaw-Confining Pressure-Acid Erosion[J]. Geophysics and Geological Engineering, 2021, 11(20), 123-145.
9. Zeng F & Guo J. A Finite Element Model to Predict Wellbore Fracture Pressure with Acid Damage[J]. Journal Title, 2015, 44(10), 256-274.
10. Guo J & Zeng F. A Model for Predicting Reservoir Fracturing Pressure of Perforated Wells after Acid Damage[J]. Journal Title, 2011, 38(2), 345-364.
11. Ahmadi M A, Shadizadeh S R. Experimental investigation of a natural surfactant adsorption on shale-sandstone reservoir rocks: Static and dynamic conditions[J]. Fuel, 2015, 159: 15-26.
12. Greibrokk T, Radke M, Skurdal M, et al. Multistage supercritical fluid extraction of petroleum source rocks: application to samples from Kimmeridge clay and Posidonia Shale formations[J]. Organic geochemistry, 1992, 18(4): 447-455.

The methodology for calculating the apparent water absorption index from test injection data

Peng Li

Daqing oilfield Co., LTD. No. 6 oil production plant geological research institute planning and planning office, Daqing Heilongjiang province, China

Abstract: After the oilfield enters the medium-high water content period, the treatment problem of produced water becomes more and more prominent, and the same-well injection technology can realize the oil-water separation of the produced fluid at the well. The separated water is injected back into the oil reservoir by the injection pump, while the oil-water mixture with higher oil content is lifted to the surface by the extraction pump. However, once the tampering operation is carried out on the production layer, the fluid production capacity of the layer cannot be judged, and it is impossible to determine the reasonable production allocation of the injection and recovery layer. Moreover, in the normal production process, the production tubing can not be taken out normally, so it is difficult to determine the production capacity after blocking by logging method. Therefore, it is necessary to carry out theoretical research on the production capacity of the same injection and recovery wells after tampering according to the method of reservoir engineering, so as to provide a theoretical basis for the reasonable production allocation in the field.

1. INTRODUCTION

In the process of carrying out a certain operation in the same-well injection and production wells, due to the higher pressure of the injected layer section, there is a certain pressure difference between the injected layer and the wellbore, so part of the injected water of the injected layer section flows backwards into the wellbore, while the formation pressure of the extracted layer section is lower than the wellbore pressure, and there is also a certain pressure difference between the wellbore and the extracted layer section, therefore, the produced water layer seeps into a part of the wellbore water[1-3]. This leads to the fact that the amount of water injected into the well after the operation is often larger than the actual amount of water injected to make up for the pressure deficit in the formation, but since this process occurs simultaneously in the wellbore, it is not possible to monitor the actual amount of water coming in and out of the formation, so the actual amount of water injected after the operation is difficult to predict, and it is not possible to judge the impact of this operation on the production capacity of the formation and the ability of the formation to absorb the water[4-6]. Therefore, there is an urgent need to find a practical and accurate method to calculate the actual water injection rate after the operation to determine whether the operation will cause damage to the formation[7]. After calculating by various methods, it was finally determined that the unstable well test method was used to evaluate whether the injection of the same well caused any harm to the reservoir. The water absorption index and visual water absorption index were calculated by test injection data, load data and formation coefficient parameters respectively, so as to measure the water absorption of the reservoir and provide a new way for rational production allocation in the field.

2. THEORETICAL DEDUCTION

After the operation, test injection with a certain pressure P_1, at this time can know a flow rate Q_1, and then change a pressure P_2 and then test injection, by can get a flow rate Q_2, through the two sets of test injection data, and then combined with the method of unstable test wells can be found out in the injection of water after the pressure of the P_{wf1} and stratigraphic coefficients K_h.

Based on P_1 and P_2, the bottomhole pressures P_{wf1} and P_{wf2} can be found according to equation (1).

$$P_{wf} = P + \rho g h \tag{1}$$

Where P_{wf} is the bottomhole pressure before well shut-in, MPa; P is the wellhead injection pressure, MPa; ρ is the density of the wellbore fluid, g/m³; g is the acceleration of gravity, m/s²; h is the wellbore fluid column pressure, m.

The variation of wellbore flow pressure and injection flow rate with time is shown in Figure 1.

Peng Li: E-mail:2185126860@qq.com, phone number: 15776541860

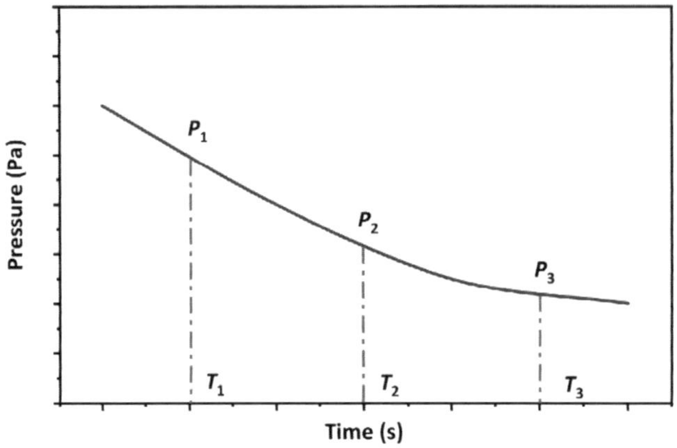

Figure. 1 Variation of wellbore flow pressure and injection volume with time

After Δt time of injection with injection volume Q, the bottomhole pressure rises to:

$$\Delta P = \frac{Q\mu}{4\pi Kh}\ln\frac{2.25\eta\Delta t}{r_w^2} \tag{2}$$

$$\Delta P = P_{wf(i)} - P_{wf(i-1)} \tag{3}$$

Where ΔP is the wellbore flow pressure difference at time t; Q is the displacement at the time of trial injection, *t/d*; μ is the formation fluid viscosity, Pa·s; K is the formation permeability, m^2; h is the thickness of the oil layer, m; P_{wf} is the bottomhole pressure before well shut-in, Pa; r_w is the diameter of the wellbore, m; η is the coefficient of pressure conductivity, which indicates the area of pressure propagation per unit time.

$$\eta = \frac{K}{\phi\mu C_t} \tag{4}$$

3. EXAMPLE CALCULATIONS

According to the method of unstable well test combined with well test injection data to calculate the bottomhole pressure recovery law after shutting off the well for a period of time, the test injection data are shown in the table:

Table 1 Test injection data

Well name	Stratigraphic coefficient of the reinjection layer(μm²·m)	First test injection			Second test injection		
		Test injection pressure(MPa)	Flow rate(m3/d)	Test water injection time (min)	Test injection pressure(MPa)	Flow rate(m3/d)	Test water injection time (min)
G158-463	0.861	9	167	30	10	208	15
G161-483	2.239	11	103.8	10	9.0	103.6	10

Taking μ=0.5×10^{-3}Pa·s; η=0.607; r_w=0.1m combined with the data in Table 1 from Eq. (3) and Eq. (4) the system of equations can be linked to have:

$$\Delta P_1 = \frac{Q_1\mu}{4\pi Kh}\ln\frac{2.25\eta\Delta t_1}{r_w^2} = P_{wf1} - P_{wf0}$$

$$\Delta P_2 = \frac{Q_2\mu}{4\pi Kh}\ln\frac{2.25\eta\Delta t_2}{r_w^2} + \frac{Q_1\mu}{4\pi Kh}\ln\frac{2.25\eta\Delta t_1}{r_w^2} = P_{wf2} - P_{wf0}$$

Bring in the data to get:

$$17.8\text{-}P_{wf0} = \frac{0.00144\times 0.5}{4\pi Kh}\ln\frac{2.25\times 0.607\times 10\times 60}{0.1^2}$$

$$19.8\text{-}P_{wf0}=\frac{0.00138\times0.5}{4\pi Kh}\ln\frac{2.25\times0.607\times10\times60}{0.1^2}$$

The calculated results were collated to obtain the apparent water absorption index as shown in Table 2:

Table 2 Calculation results of water absorption index

Well name	Actual stratigraphic factor	Calculation of stratigraphic coefficients using test injection data	Calculation of water absorption index using test injection data	Water absorption index based on load	Calculation of water absorption index using stratigraphic coefficients
G158-463	0.861	2.971	6.54	5.61	2.02
G161-483	2.239	1.697	3.74	10.52	5.26

In order to facilitate the work construction of the field personnel, it is necessary to convert the water absorption index into the apparent water absorption index, for this purpose, we compared the calculated water absorption index, and the results are shown in Table 3.

Table 3 Calculation results of visual water absorption index

Well name	Calculation of water absorption index using test injection data	Calculation of apparent water absorption index using test injection data	Calculation of water absorption index using load	Calculation of apparent water absorption index using load	Calculation of water absorption index using stratigraphic coefficients	Calculation of apparent water absorption index using stratigraphic coefficients
G158-463	6.54	6.69	4.6	4.62	2.02	2.02
G161-483	3.74	3.78	6.15	6.22	5.26	5.31

$$J_{awai}=\frac{Q_{inj}}{P_{inj(well\ head)}} \quad (5)$$

$$J_{wai}=\frac{Q_{inj}}{P_{inj(well\ head)}+\rho gh-P_e} \quad (6)$$

Where the depth of the middle of the oil reservoir is 1000m, the formation pressure is 10MPa, due to the depth of the middle of the oil reservoir static liquid column pressure and formation pressure are similar to about 10MPa, so the formation pressure and the static liquid column pressure can be about off, that the absorption index is the same as the apparent absorption index, you can get the approximate relationship between the apparent absorption index and the parameters of the test injection.

In order to facilitate the field operation, need to design the corresponding test injection pressure plate according to the calculation results, the first and second test injection pressure difference of 2MPa, two groups of test injection time are 10 minutes, then the second test injection displacement and the visual water absorption index can be obtained as follows between the relationship between the relationship between the curve, as shown in the figure for the operation of the different injection displacement prior to the second test injection displacement and the rule of law between the visual water absorption index.

From the Figure 2, it can be seen that the water absorption index increases with the increase of the test injection displacement and varies in an approximate linear relationship. The categories of the curves indicate the changes under different first injection displacements, and the water absorption index shows a decreasing trend in the first injection displacement uniformly increasing and decreasing with the increase of the first injection displacement. The highest decrease in water absorption index was observed in the range of 30-40 t/d. In the range of 60 to 70 t/d, the decrease in water absorption index slows down.

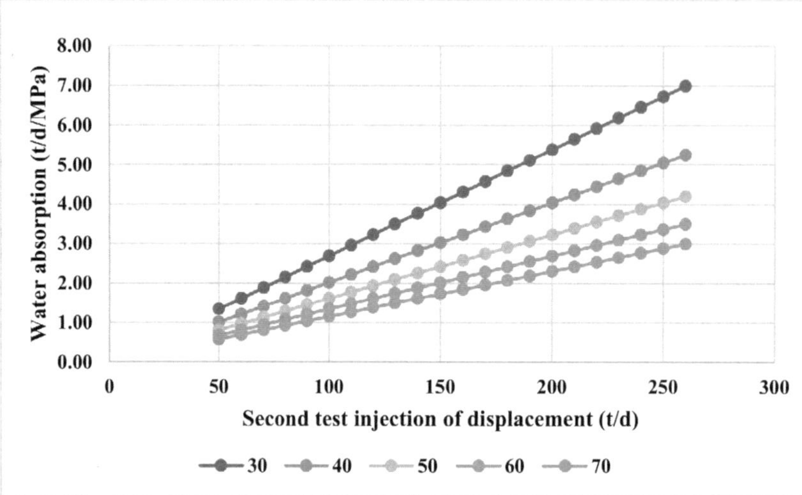

Figure. 2 Variation between test injection displacement and apparent water absorption index for the second time with different injection displacements before operation

4. SUMMARIZE

(1) The actual amount of water inflow and outflow from the formation cannot be monitored in the wellbore, so it is difficult to predict the actual amount of water that will be injected after the operation, and thus it is impossible to judge how the operation will affect the production capacity and water absorption capacity of the formation.

(2) Calculate the bottomhole pressure recovery law after shutting down the well for a period of time based on the method of unstable well test combined with well test injection data.

(3) According to the calculation results to design the corresponding test injection pressure plate, the first and second test injection pressure difference of 2 MPa, two groups of test injection time are 10 minutes, then the second test injection displacement and visual water absorption index can be obtained between the test injection displacement and visual water absorption index change rule.

REFERENCES

1. Ling Zongfa, Hu Yongle, Li Baozhu, Wang Lijuan. Optimization of horizontal InjectionProduction patterns [J](in Chinese). Journal of Petroleum Exploration and Development 01 (2007).
2. Trimonova, M., Baryshnikov, N., Zenchenko, E., Zenchenko, P.. Turuntaev, S. The Study of the Unstable Fracure Propagation in the Injection Well: Numerical and Laboratory Modeling. In SPE Russian Petroleum Technology Conference. OnePetro (2017, October).
3. Wang, B., Sun, J., Shen, F., Li, W., Zhang, W., Mechanism of wellbore instability in continental shale gas horizontal sections and its water-based drilling fluid countermeasures. Natural Gas Industry B, 7(6), 680-688 (2020).
4. Kerunwa, A., Ekwueme, S. T., An Approach to Curbing Wellbore Instability in Shales through Nanoparticles-Augmented Water-Based Drilling Muds. Petroleum & Coal, 62(4) (2020).
5. Choi, S., Jo, J. W., Kim, J., Song, S., Kim, J., Park, S. K., Kim, Y. H. Static and dynamic water motion-induced instability in oxide thin-film transistors and its suppression by using low-k fluoropolymer passivation. ACS Applied Materials & Interfaces, 9(31), 26161-26168 (2017).
6. Li, Y., Jiang, G., Li, X., Yang, L., Quantitative investigation of water sensitivity and water locking damages on a low-permeability reservoir using the core flooding experiment and NMR test. ACS omega, 7(5), 4444-4456 (2022).
7. Abdullah, A. H., Ridha, S., Mohshim, D. F., Yusuf, M., Kamyab, H., Krishna, S., Maoinser, M. A., A comprehensive review of nanoparticles in water-based drilling fluids on wellbore stability. Chemosphere, 136274 (2022).

Research on methods and scale of deeply exploitable hydropower resources

Fei Wang*, Yusheng Liu , Qilong Li, Da Xu, Jialiang Yu

PowerChina Huadong Engineering Corporation Limited, Hangzhou 311122, China

Abstract. Deep development of hydropower is an important measure to optimize the structure of clean energy, meet the peak shaving and backup needs of the power system, and achieve the national dual carbon goals. This article conducts a systematic analysis of the development potential of developed and undeveloped hydraulic resources, including unplanned river sections, undeveloped planned power stations, unit renovation and capacity expansion, and the expansion of existing/ongoing power stations. We focused on studying the factors that affect the deep development of hydropower, including the reasonable operation mode of hydropower with different regulation performance in the power grid, the energy structure characteristics of the power system, and the comprehensive water use requirements. A calculation method for the reasonable installation and utilization hours and deep development resources of hydropower stations with different regulation performance after deep development has been proposed. According to calculations, Zhejiang Province can increase its scale by 3331MW, with great potential for deep development.

1. Introduction

According to the 2023 Global Climate Report released by the World Meteorological Organization at the end of 2023, multiple climate records were broken in 2023, with greenhouse gas levels reaching historic highs, sea level rise reaching historic highs, and sea ice extent in Antarctica dropping to historic lows [1]. China has also proposed voluntary emission reduction targets, achieving carbon peak before 2030 and carbon neutrality before 2060 [2-3].

Building a clean energy structure system consisting of wind power, solar energy, tidal energy, hydropower, biomass energy, and other clean energy sources is an important measure to promote economic development and achieve CCER goals [4]. Hydropower is widely recognized worldwide due to its mature technology, renewable energy, zero emissions, high energy efficiency, and strong comprehensive utilization functions [5]. At the same time, hydropower has the advantages of fast startup and shutdown speed, strong adjustable output, and can effectively suppress the random fluctuations of hydropower and solar power output. It plays a huge role in stabilizing the peak valley difference and load fluctuations of the power system, and improving the power generation efficiency of other power stations. Meanwhile, hydropower is also a key factor in building a "hydropower wind power photovoltaic energy storage" base.

The hydropower resources in the eastern and central regions of China have been basically developed. Many hydropower stations were built earlier and have encountered a series of problems that affect the safe operation of power plants, such as turbine wear, cavitation, decreased average efficiency of turbines, outdated electrical equipment, and low level of automation [6]. Meanwhile, some hydropower stations, especially those with better regulation capabilities, have higher utilization hours and smaller installed capacity, which suppresses the peak shaving advantage of hydropower in the power grid. Therefore, it is very necessary and urgent to carry out deep development of hydropower.

2. Main types of deep development

Hydraulic resources can be divided into undeveloped, under development, and developed according to their development status. Among them, according to whether the river has completed the hydropower planning, the undeveloped situation can be divided into two types: the undeveloped river section and the planned but not constructed hydropower station [7-8]. The key contents of deep development of hydropower mainly include: combining the current status of hydropower production and operation, further reviewing and adjusting the planning of rivers with development potential, expanding the existing hydropower stations with conditions, transforming and increasing the capacity of some old hydropower units, and constructing new hydropower stations in large water storage areas. There are four main methods for deep hydropower development:

(1) Unplanned development of river sections. Unplanned river sections refer to river sections that have not been developed or utilized, and have not undergone hydropower planning. For the utilization of such river

*Corresponding author's e-mail: wang_f6@hdec.com

sections, the method of hydropower resource survey can be used to evaluate the amount of resources for deep development based on the water head, average flow rate, water utilization rate, etc. of the river section.

(2) Undeveloped planned power plants. Undeveloped planned power stations refer to hydropower stations that have undergone hydropower planning but have been delayed in construction due to difficulties in development and other reasons. For the tapping of potential in the absence of power stations, the main approach is to analyse the reasonable working positions of hydropower stations in the power system based on the development of power demand, and to analyse the installed capacity/guaranteed output ratio of hydropower stations from the perspective of reasonable capacity utilization; At the same time, it is necessary to analyse the amount of electricity that can be generated by increasing installed capacity, estimate the cost of electricity prices, and compare it with the marginal cost of new energy in the power system. Taking into account the analysis results of rational utilization of capacity and reasonable increase in power generation, combined with the specific situation of the power system and relevant engineering cases, this paper analyses the reasonable annual utilization hours of installed capacity for hydropower stations with different regulation performance in the power system after deep development, and then estimates the potential resources that can be tapped by the uncompleted power stations.

(3) Unit capacity expansion and renovation. Unit renovation is aimed at old hydroelectric power plants that have been built and operated for more than 20 years. After the unit renovation, the efficiency of the unit is improved, which can increase the output of the unit nameplate and achieve the effect of increasing the installed capacity. The amount of resources required for unit renovation and capacity increase can be estimated based on the improvement in efficiency after the unit renovation.

(4) Expansion of existing/ongoing power stations. Considering that with the development of the economy, the power supply structure has taken on diverse forms, and the external conditions during the design of the original power station have undergone significant changes. For example, in the 1980s and 1990s, the requirements of the power grid for hydropower stations were mainly electricity, so the installed annual utilization hours of hydropower stations were often higher. However, with the development of the economy, the power supply structure has undergone significant changes, and the demand for peak shaving power sources in the power grid is increasing day by day, the annual utilization hours of installed capacity of hydropower stations can be appropriately reduced. Therefore, it is necessary to conduct in-depth analysis of the installed capacity of existing/ongoing hydropower stations, and hydropower stations with conditions for expansion can be expanded.

3. Key factors for deep development

3.1. Reservoir regulation performance

With the increasing electricity load year by year, the requirements for electricity quality and reliability are constantly improving [9]. The proportion of renewable energy sources such as wind and solar power, as well as power plants with poor controllability and rapid response ability such as nuclear power, in the power grid is constantly increasing. The power system has put forward new requirements for the utilization of conventional hydropower stations [10]. In the power system, the role of conventional hydropower is gradually shifting from being primarily focused on generating electricity, to balancing power generation with auxiliary services such as peak shaving and backup.

For hydropower stations without regulating capacity, the output size depends on the natural inflow, so it is not suitable to bear variable loads. In order to reduce water loss, it is generally operated at base load. During periods of low electricity consumption, there may be water abandonment due to the minimum technical output of thermal power.

For hydropower stations with daily regulation capacity or above, the natural incoming water within a day can be regulated by the reservoir to adapt to changes in load. Therefore, in the power system, it is mainly responsible for peak or waist load work. It can fully leverage the operational flexibility and economy of hydropower stations, maximize the replacement of thermal power plant capacity, make more reasonable use of hydraulic resources, and reduce water waste; At the same time, it can improve the power generation efficiency of hydropower stations with poor regulation ability and reduce water waste during low electricity consumption periods.

For hydropower stations with annual or multi-year regulation, the regulation and storage of natural runoff flowing through reservoirs within or even between years can be redistributed according to load changes.

According to the requirements of ecological and environmental protection, there should be no dehydrated river section downstream of the hydropower station, and a certain amount of ecological base flow should be discharged to meet the ecological water use requirements in the river. Downstream power stations with good regulation capabilities usually have reverse regulation reservoirs or cascade hydropower stations with basic water level connections. Otherwise, it is also necessary to consider releasing ecological flow.

3.2. Peak shaving capacity demand in power systems

Take 2035 as the planning level year. On the basis of planning level year power load forecasting and clear power source planning, considering hydropower stations with regulating capacity as a whole, analyzing the suitable working positions and tasks of the hydropower station group in the power grid, in order to determine the appropriate installed capacity/guaranteed output ratio (referred to as "installation guarantee ratio") of the hydropower station group.

For hydropower stations with regulating capacity, their winter output is usually small, and the working capacity for peak shaving in the system is relatively large. Therefore, based on the above principles, the winter power consumption of the system is balanced, and hydropower stations with regulating capacity are analyzed as a whole. After appropriately simplifying the above model functions, the following methods and steps can be used for calculation and analysis.

1) Determine the typical daily winter load curve L1 of the power system for the planning level year 2035.

2) Priority should be given to absorbing the electricity generated by wind and solar power plants, and the daily process of wind and solar power generation should be formulated. Deductions should be made on the typical daily load curve L1 to form a new daily load curve L2.

3) Using load curve L2, analyze the reasonable working positions of pumped storage power stations, gas turbines, nuclear power stations, unregulated hydropower stations and other power stations that are currently under development and have been clearly planned and developed in the power grid, and form residual load curve L3.

4) The residual load curve L3 is the combined load curve of a hydroelectric power plant group with regulating capacity and coal-fired power. The maximum peak shaving capacity of the hydroelectric power plant group is calculated, that is, the maximum working capacity of the hydroelectric power plant with regulating capacity is calculated under the daily guaranteed electricity quantity.

5) If the hydropower station is not considered to bear the rotating reserve, the maximum working capacity of the hydropower station group is its installed capacity. If it is considered to bear a certain amount of reserve capacity, its installed capacity can be increased by a certain amount in addition to the maximum working capacity. In this study, the hydropower reserve capacity is temporarily considered to not exceed 30% of the working capacity. For power stations with significant impact on the power grid, separate analysis should be conducted. For example, the reserve capacity of Xin'anjiang Hydropower Station should consider the allocation in the East China power grid of Zhejiang, Jiangsu, Shanghai, Anhui and other regions in 2035, as well as the total reserve capacity demand.

6) Based on the above analysis, the working position of a hydroelectric power plant group with regulating capacity on a typical daily load curve can be determined, and the installed capacity of the hydroelectric power plant group can be calculated according to step 5). The installed capacity to guaranteed output ratio of the entire hydroelectric power plant group can be calculated.

3.3. Other influencing factors

Based on the analysis of the installation guarantee ratio of hydropower groups in the power system, combined with the analysis results of the increased power generation of hydropower stations with different regulation capacity types, further analyse the reasonable installation and utilization hours of each regulation capacity type of power station, and use this as the discrimination basis and reference index for the deep development of power stations.

If the proportion of water and electricity in the power system is small, the installation and maintenance ratio calculated by the above analysis method is large, and some power stations bear the role of peak shaving. The daily electricity consumption in the control month is basically completed within 1 hour to 1.5 hours. In this case, it is necessary to reasonably consider adding other types of peak shaving energy, such as increasing the peak shaving capacity of the power system through air compression energy storage and solar thermal energy storage in the future. Therefore, the installation and maintenance ratio should not be too large, and the annual utilization hours should not be too low. The reasonable installed capacity of a specific power station should be determined through an economic rationality analysis of the expansion scale.

For low head hydropower stations, as the installed capacity increases, the power generation flow increases, the head loss increases, and the downstream water level rises, resulting in a decrease in the actual utilization head. Therefore, the rationality of the installed capacity utilization hours and the economic operation mode of the reservoir should be considered based on the specific situation of the project. If the comprehensive utilization requirements of the power station, such as irrigation, water supply, navigation, ecological water use, etc., are met, the rationality, feasibility, and economy of increasing installed capacity should be reasonably analysed on the basis of meeting the engineering comprehensive utilization requirements.

4. Method for estimating the amount of resources for deep hydropower development

4.1. Unplanned development of river sections

For unused river sections with development potential, the amount of hydraulic resources can be preliminarily estimated for power generation based on data such as utilization head and flow rate. Preliminary calculation of installed capacity is based on the reasonable water utilization rate of daily regulation power stations in the same basin, taking into account the daily regulation performance, and the reasonable installation and utilization hours of daily regulation power stations in the planned level year. The calculation formula for annual power generation $\boldsymbol{E}$ and installed capacity $\boldsymbol{P}$ is as follows:

$$E = \eta AQHT_y \quad (1)$$

Where: $\boldsymbol{E}$ represents the average annual power generation over many years; $\boldsymbol{\eta}$ for water utilization efficiency; $\boldsymbol{A}$ is the comprehensive output coefficient; $\boldsymbol{Q}$ is the average annual flow rate at the dam site; $\boldsymbol{T_y}$ is the number of hours in a year, which is 8760h.

$$P = E/T_1 \quad (2)$$

Where: ***P*** represents the installed capacity; $\boldsymbol{T_1}$ is the reasonable installation and utilization hours of the daily regulating power station in the power system.

4.2. Undeveloped planned power stations

Due to the fact that most undeveloped planned power stations were planned or designed earlier, and the current power system and expected construction and operation time have undergone significant changes, the electricity load demand and the role of hydropower in the planning level year have undergone significant changes. Based on the reasonable utilization hours of hydropower deep development in the power system, adjust the installed capacity utilization hours of uncompleted power stations and make reasonable adjustments to the installed capacity. The change in electricity quantity caused by the adjustment of installed capacity is determined by comparing and analysing the water energy parameters of similar hydropower stations of different types with the change in installed capacity. Due to differences in storage capacity regulation capacity, runoff characteristics, and other factors, the relationship curve between installed capacity and utilization hours also varies. Different watersheds should be analysed separately.

The original planned installed capacity of the undeveloped power station was $\boldsymbol{P_0}$, with an annual power generation of $\boldsymbol{E_0}$ and an installed utilization hour of $\boldsymbol{T_0}$. Based on the analysis of the power system demand and the regulation capacity of the power station, the planned level year requires an increase in the installed capacity of the power station, resulting in a decimal value of $\boldsymbol{T_1}$ for the installed utilization of the power station. The installed capacity $\boldsymbol{P_1}$ and annual power generation $\boldsymbol{E_1}$ of the power station in the planning level year satisfy the following equations:

$$E_1 = E_0 + \Delta E = P_0T_0 + (P_1 - P_0)\Delta T = P_1T_1 \quad (3)$$

Where, $\Delta\boldsymbol{E}$ and $\Delta\boldsymbol{T}$ represent the annual power generation and utilization hours of supplementary installed capacity $\boldsymbol{\Delta P = P_1 - P_0}$.

According to equation (3), the installed capacity $\boldsymbol{P_1}$ and annual power generation $\boldsymbol{E_1}$ of the power station in the planning level year can be calculated:

$$P_1 = P_0(T_0 - \Delta T)/(T_1 - \Delta T) \quad (4)$$

$$E_1 = P_1T_1 \quad (5)$$

4.3. Unit capacity expansion and renovation

The capacity expansion and renovation of the unit is only applicable to hydropower stations that have been put into operation for 20 years or more. The capacity expansion and renovation of other existing power stations will not be considered temporarily. Specific analysis will be conducted when the conditions are suitable in the later stage, combined with the needs of expansion.

Due to the utilization of the original water transmission line and volute position, the increase in installed capacity of the unit is relatively limited. Therefore, it is preliminarily considered that the installed capacity will increase by 15% after the unit capacity expansion and renovation.

After the renovation and capacity increase of the unit, due to the improvement of installed capacity and unit efficiency, as well as the increase in water utilization rate, the power generation of the power station increased after the renovation and capacity increase of the unit. In terms of the impact of improving unit efficiency on electricity consumption, the efficiency increase of different types of power plants is basically the same; In terms of the impact of improving water utilization efficiency on electricity consumption, for power plants with good regulation ability, there is less increase in electricity consumption, while for power plants with poor regulation ability, there is a greater increase in electricity consumption.

Based on the research results and implementation experience of similar power plant unit capacity expansion and renovation projects that have been carried out, the principles for unit capacity expansion and renovation of each power plant are proposed as follows:

① After the unit capacity expansion and renovation, the increase rate of installed capacity ρ Considering factors such as the actual operating efficiency of the power plant unit, the size of the water transmission pipeline and volute, this stage can temporarily take around 10% to 20%;

$$P_1 = P_0 \times (1 + \rho) = P_0 \times (1 + 10\%\sim20\%) \quad (6)$$

② After the unit capacity expansion and renovation, the proportion of annual average power generation increase of the power station is determined based on its regulation performance ξ, Specific proportion values ξ As follows: 5% is taken for years of regulating power stations; 7.5% for annual/quarterly regulation of power stations; Take 10% for daily/non regulated power plants.

$$E_1 = E_0 \times (1 + \xi) \quad (7)$$

4.4. Expansion of Existing Power Stations

After analysing the capacity expansion and renovation of existing hydropower stations, compare the installation and utilization hours of different types of hydropower stations with the reasonable utilization hours of deep hydropower development in the power system in the planning level year, and analyse their unit expansion space. When the utilization hours of the installed capacity of the power station are less than or close to the reasonable utilization hours of this type of power station in the planning level year, no expansion will be carried out. For power stations with installed capacity utilization hours higher than the reasonable installed capacity utilization hours of this type of power station in the planned level year power grid, by expanding the original power station and increasing the installed capacity, the installed capacity utilization hours of the power station will be close to the reasonable utilization hours of deep development of this type of hydropower station in the province. For low head through flow units with a power generation head of 20m or less, expansion should not be considered temporarily, as the increase in full head discharge may lead to an increase in downstream water level and a decrease in head, which may not have significant electricity benefits. If the expansion capacity of the hydropower station is less than half of the unit capacity, it is considered to be reserved for future unit renovation and expansion, and not included in the expansion capacity.

The installed capacity of the original hydropower station is P_0, the annual power generation is E_0, and the utilization hours of the installed capacity are T_0. Based on the analysis of the power system demand and the regulation capacity of the power station, the planning level year requires an increase in the installed capacity of the power station, making the decimal of the installed capacity of the power station T1. Based on the analysis results of hydropower stations of the same type in similar watersheds, the relationship between the additional installed capacity utilization hours dT and the installed capacity utilization hours $dT=dE/dP=f(T)$ is used to determine the average additional installed capacity utilization hours of the power station from T_0 expansion to T_1 installation utilization hours ΔT. According to the planning level year, the reasonable utilization hours of installed capacity T1 and the decimal utilization of supplementary installed capacity are determined ΔT. Calculate the installed capacity P_1 and annual power generation E_1 of the power station in the planning level year using the following formula:

$$P_1 = P_0(T_0 - \Delta T)/(T_1 - \Delta T) \quad (8)$$

$$E_1 = P_1 T_1 \quad (9)$$

4.5. Analysis of the scale of hydropower deep development in Zhejiang Province

Although there are many rivers in Zhejiang Province, the degree of development of hydropower resources is not uniform. Some easily developed sections and sites have been overdeveloped, while some remote or technically difficult sections and sites have been underdeveloped. The Changfeng~Misai section of Qiantangjiang having a head of 26 meters, the Qingshandian~Huaguangtan-II section of Fenshuijiang having a head of 123 meters, and the Dajun~Yongku section of Xiaoxi having a head of 46 meters that have not been utilized. However, due to limited information available, it is not yet possible to determine the limiting factors of this section of the river. Therefore, the utilization of these river sections will not be considered in this study.

There are four undeveloped hydropower stations (with an installed capacity of 50MW or above) in Zhejiang Province, namely Dajun (with an installed capacity of 250MW, the same below), Nanxijiang (64MW), Guihu (80MW), and Yangxi (80MW). The total installed capacity of the four power stations is 474MW. After calculation, by tapping the potential of the installed capacity of the undeveloped power stations, the installed capacity can be increased to 762MW. As shown in Table 1.

Table 1. Deep Development Scale of Undeveloped Planned Power Stations in the Zhejiang.

Undeveloped planned power stations	Regulating ability	Planed Scale		Adjuested Scale	
		Installed capacity (MW)	Annual power generation (GWh)	Installed capacity (MW)	Annual power generation (GWh)
Dajun	Annual	250	433	308	462
Nanxijiang	Multi-annual	64	128	144	144
Guihu	Seasonal	80	211.1	130	261.1
Yangxi	Annual	80	220.4	180	270.4
Total		474	992.5	762	1137.5

There are a total of 6 hydropower stations (with an installed capacity of 50MW and above) operating in Zhejiang Province for more than 25 years, namely Fuchunjiang, Hunanzhen, Huangtankou, Xin'anjiang, Jinshuitan, and Shitang. Among them, Fuchunjiang and Xin'anjiang underwent unit capacity expansion and renovation in 2003 and 2004, respectively. Therefore, only Hunanzhen, Huangtankou, Jinshuitan, and Shitang power stations will be considered for unit capacity expansion and renovation in this project. After the capacity expansion and renovation of the four power plant units mentioned above, the installed hydropower capacity can be increased by about 104MW, and the average annual power generation can be increased by about 101GWh. As shown in Table 2.

Table 2. Deep Development Scale of Capacity Expansion and Renovation Stations in the Zhejiang.

Hydropower stations	Regulating ability	Present Scale		Adjuested Scale	
		Installed capacity (MW)	Annual power generation (GWh)	Installed capacity (MW)	Annual power generation (GWh)
Hunanzhen	Multi-annual	270	540	300	567
Huangtankou	Daily	82	176	96	194
Jinshuitan	Annual	300	490	348	527
Shitang	Daily	78	189	90	208
Total		730	1395	834	1496
Added scale				104	101

Nine hydropower stations (with an installed capacity of 50MW or above) have been built in Zhejiang Province. After analysis, the main hydropower stations that can be considered for expansion include Huaguangtan First Class, Tankeng, Hunanzhen, Huangtankou, and Xin'anjiang. The total installed capacity before expansion is 2130MW, with a total annual average power generation of 4060.8GWh; After expansion, the total installed capacity is 4475MW, with a total annual average power generation of 4542GkWh, an increase of 2465MW compared to the total installed capacity before expansion, and a total annual average power generation increase of 576GkWh. As shown in Table 3.

Table 3. Deep Development Scale of Expansion of Existing Hydropower Stations in the Zhejiang.

Hydropower st ations	Regulating abilit y	Present Scale		Adjuested Scale	
		Installed capacity (MW)	Annual power ge neration (GWh)	Installed capacity (MW)	Annual power ge neration (GWh)
Huaguangtan I	Annual	60	127	100	147
Tankeng	Multi-annual	600	1021	1125	1126
Shanxi	Multi-annual	200	355	390	393
Hunanzhen	Multi-annual	300	567	630	633
Xinanjiang	Multi-annual	850	1896	2230	2243
Total		2010	3966	4475	4542
Added scale				2465	576

According to the above research, the hydropower resources in Zhejiang Province can be further developed through four aspects: utilization of undeveloped river sections, tapping into the installed capacity of undeveloped power stations, increasing the capacity of existing power station units, and expanding and renovating existing/under construction power stations. This can increase the installed capacity of hydropower by 3331MW and the annual power generation by 1813GWh. There is great potential for deep development of hydropower. As shown in Table 4.

Table 4. Estimation of Hydropower Deep Development Scale in the Zhejiang Province.

Types of deep development	Installed capacity (MW)	Annual power generation (GWh)	Capacity ratio (%)
Unplanned development of river sections [a]	/	/	/
Undeveloped planned power stations	762	1137	22.88
Unit capacity expansion and renovation	104	100	3.12
Expansion of Existing Power Stations	2465	576	74
Total	3331	1813	100

[a] The utilization of unplanned river sections is considered because of development conditions.

5. Conclusion

Based on the analysis presented above, the conclusions are obtained as below:

(1) The deep development of hydropower has good capacity benefits. It plays a role in peak shaving, frequency regulation, and backup in the system, improving the safety, stability, and economy of power grid operation. It can also increase the inclusion rate of renewable energy such as wind power and solar energy, and promote the achievement of carbon reduction goals.

(2) There are four main methods for deep hydropower development, including: utilization of unplanned river sections, potential installed capacity of undeveloped planned power stations, capacity expansion and renovation of existing power station units, and expansion of existing and ongoing power stations.

(3) The scale of deep hydropower development should mainly consider the reservoir regulation performance, peak shaving demand of the power system, comprehensive water use requirements, and technical economy. After studying the hydropower development situation in Zhejiang Province, it is found that the scale of deep development is 3331MW, which can increase the annual power generation by 1813GWh and has great potential for deep development.

To further leverage the role of hydropower resources in energy transformation and sustainable development, it is recommended to establish a sound policy system and market mechanism for deep hydropower development, clarify the responsibilities and rights of each department, and coordinate the interests of all parties. At the same time, we will increase policy support and financial investment in the deep development of hydropower, encourage enterprises and social capital to participate in the deep development of hydropower, and promote the healthy development of the hydropower industry.

https://doi.org/10.1051/e3sconf/202452802002

Acknowledgments

This work was financially supported by the PowerChina Huadong Engineering Corporation Research Fund(KY2022-GH-02-03).

References

1. LUO Xiaolin, HUANG Li, WU Sainan, Approaches to Expanding Flexibility of Hydropower to Achieve the Carbon Peaking and Carbon Neutrality Target and Their Assessment Framework [J]. Hydropower and Pumped Storage 2023(1): 1-8.
2. HUANG Qiang, GUO Yi, JIANG Jianhua et al. Development Pathway of China's Clean Electricity Under Carbon Peaking and Carbon Neutrality Goals [J]. Journal of Shanghai Jiaotong University, 2021(12): 1499-1509.
3. YANG Bowen, The Systematic Thinking, Economic Logic and Scientific Path for the Realization of Carbon Summit and Carbon Neutral Target Strategy under Xi Jinping's New Development Concept [J]. Economist, 2021(9): 5-12.
4. DUAN Jingdong, LING Shihe, XIAO Lingjuan. Based on the "double carbon" target economic research of large hydropower for sustainable high-quality development[J]. Water Resources and Hydropower Engineering, 2022(S02): 344-348.
5. Dhaubanjar, Sanita, et al. A systematic framework for the assessment of sustainable hydropower potential in a river basin - The case of the upper Indus[J]. The Science of the total environment 786(2021):147142.
6. CHEN Guiya, CHEN Xiao, CAI Shubing, et al. Study on improving water energy resources utilization efficiency of Changjiang River Basin in new stage [J]. Yangtze River, 2023(10) :1-6.
7. YIN Jing. Feasibility Analysis of Hydropower Station Expansion and Reconstruction Project Abroad [J], Water Conservancy Science and Technology and Economy, 2020(4):72-75
8. THUY T. B. Nguyen, NINH H. Bui. Proposing Adjustments to the Dry Season Inter-Reservoir Operation Rules of the Red River System to Improve Water Use Efficiency of the Reservoirs[J]. Journal of geoscience and environment protection,2023,11(5):316-334.
9. BARROS M T L, TSAI F T C, YANG S L, et al. Optimization of large-scale hydropower system operations[J]. Journal of Water Resources Planning and Management, ASCE, 2003, 129(3):178-188.
10. PENG Caide, HAN Lingli, ZONG Wanbo, Discussion on Capacity Upgrading of Generating Units of Large and Medium-sized Conventional Hydropower Stations [J] . Water Power,2022(4) :1-3,36.

The energy-saving design of prefabricated farm house in the middle of Anhui Province

Lan Wei

College Of Geology And Construction Engineering (Jewelry College),Anhui Technical College Of Industry And Economy,Anhui 230051, China

Abstract. Masonry building has low construction requirements and it is widely used in agricultural houses in the middle of Anhui Province. However, the old farm houses have problems such as aging structure, high building energy consumption, irrational floor plan layout and poor living comfort. Nowadays, the structure of many farm houses has undergone major changes, among which the prefabricated farm houses have the advantages of high performance, short construction period and flexible construction, and the materials used and construction process conform to the principle of sustainable development. It is of great significance to carry out research on the construction of light steel prefabricated farm houses and its energy-saving optimization. Firstly, through field investigation and case analysis, the present situation of traditional farm houses in central Anhui was summarized. Secondly, the prefabricated energy-saving design of farm houses was optimized.

1. Introduction

For a long time, most of the traditional farm houses in Anhui are still masonry structures, which have serious problems such as poor earthquake resistance, unreasonable house layout, and waste of resources. However, the prefabricated farm houses have the advantages of low construction cost, fast progress, environmental friendliness, and guaranteed construction quality[1-3]. At the same time, they are constructed in a modular factory and assembled only on the construction site. The process of masonry and plastering is eliminated, which ensures the construction speed, and also ensures the reduction of construction waste on the construction site, which is conducive to environmental protection.

2. Survey of traditional farm houses in the middle of Anhui Province

The survey was carried out in the middle of Anhui Province, which mainly adopted questionnaire survey, field mapping, exchange and interview. Through questionnaires, rural residents were asked about their family size, housing needs and housing conditions. Field observation and measurement are carried out on the outside and inside of the rural farm house to understand the actual use area and space utilization of the farm house, and the natural energy utilization of the house.

2.1. Forms of farm houses

In the process of research on the construction period of rural buildings, most of them were built in the 1990s, 2000 and later, relatively few were built before the 1980s, and houses in the 1960s were even rarer. The structure is divided into traditional farm houses and new farm houses, most of the traditional farm houses are masonry structures while the new farm houses are concrete structures (Figure 1). Through the field calculation of the farm house, it is concluded that the insulation of the farm house achieved by the masonry structure solely relying on the maintenance structure of the brick wall cannot meet the requirements of energy saving. However, as the Windows of the newly built farm house generally adopt wood-framed Windows, the performance of wood-framed Windows deteriorates with the passage of time, which is very unfavorable to the insulation of the house.

Corresponding author's e-mail:476677401@qq.com

https://doi.org/10.1051/e3sconf/202452802003

Figure 1. Aerial photo of farm house in the middle of Anhui Province

2.2. Layout

Through the investigation of rural farm houses in the middle Anhui area, it is found that the buildings in the middle Anhui area are represented by Jianghuai buildings. Because of the flat and open terrain, the buildings occupy a larger area than the buildings in the south of Anhui, forming a form of large surface and large room. Most of the traditional farm houses adopt the layout form of independent courtyards, while the middle Anhui area integrates the culture of southern Anhui and northern Anhui, and the buildings contain both courtyards in the south and courtyards in the north. Jianghuai type courtyard farm house takes the courtyard as the center of the building sequence, forming the patio - hall axis, the layout is compact and exquisite. With the courtyard as the center, different functional rooms are arranged around it, and the two floors of the building form a sense of staggered height. The whole building sequence is connected by the courtyard, forming the central axis order of hall - patio - courtyard - patio - hall.

2.3. Analysis of farm house problems in the middle of Anhui Province

In the process of investigating the problems of the villagers' houses, the construction quality of the farm houses is not good, the moisture resistance measures are insufficient and the heat preservation measures are insufficient, and the ventilation and lighting are poor, the function is missing, and the space design is unreasonable[4].As shown in Table 1.

Table 1. Summary of farm house problems in the middle of Anhui Province

Problem type		Concrete problem
Use of function	Plane layout	The space layout is chaotic and the courtyard lacks reasonable design
		The area of the homestead is too large, and the land resources are wasted
	Lighting and ventilation	The room depth is large, the window to ground ratio does not meet the standard, lighting and ventilation effect is not good
	Building vacancy	The phenomenon of building vacancy is serious, a large number of farm houses are uninhabited, and some houses have security risks
Facade appearance	Facade style	The architectural styles are diverse, not uniform, and the colors are not coordinated
	Facade shape	The shape blindly follows the trend, abandons the local traditional farm house form, and is not coordinated with the natural ecological environment
Structural construction	Structural design	Most of them are self-built by residents, with unreasonable design and structure
	Construction quality	The technical level of construction personnel is low, and the quality is not guaranteed
Low-carbon and energy saving	Enclosure structure	Wall, roof, doors and Windows and other envelope structure thermal insulation and air tightness is poor
	Weak awareness of energy conservation	The popularization of many energy-saving measures is not high, and the villagers' awareness of energy-saving is poor
	Low utilization of renewable resources	Solar vacuum tube water heater installation disorder; Solar photovoltaic panel application is less; Low utilization of biogas and other renewable energy sources such as soil geothermal energy

3. Promotion strategy of prefabricated energy-saving construction of farm houses in the middle of Anhui Province

3.1. Application advantages of prefabricated energy-saving farm houses

(1) Ensure the quality of rural housing construction

The prefabricated components of the prefabricated farm houses are produced in standardized factories, which strictly ensures the accuracy and quality of the construction, reduces the size error, and ensures the accuracy of the on-site installation. The use of prefabricated construction technology can effectively reduce the construction quality problems caused by site construction, so as to ensure the quality of prefabricated farm houses.

(2) Improving energy conservation

At present, the residents in the rural housing area of Anhui Province have no awareness of energy saving, and the energy utilization rate of rural housing construction is low. More than 80% of the new prefabricated farm house building system, including the main structure, can be recycled and reused, and the construction is prefabricated in the factory, which can reduce the dust, sewage and construction waste emissions from the construction site from the source.

(3) Improve the living experience

The assembly building is implemented on the basis of strict architectural, structural, energy-saving and other detailed design, to ensure the use of function and size requirements, to ensure lighting and ventilation, thermal insulation, waterproof and moisture-proof and other functional requirements. At the same time, wall, roof, floor and other factory production (Figure 2), quick assembly on site, the quality of its waterproof and thermal insulation layer is easy to be guaranteed, to achieve good performance.

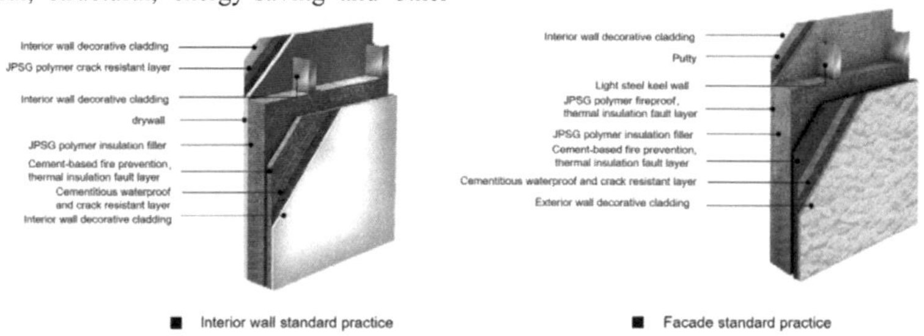

Figure 2. Structural practice diagram

3.2. Design concept of prefabricated energy-saving farm house

With the development of society, the rural environment is changing with each passing day. For the residential farm house design in Anhui, it is necessary to comprehensively analyze the local culture, technical level and energy saving, and find out the suitable design strategy of prefabricated energy-saving farm house according to local conditions. Rural revitalization is promoting the development of rural farm houses to resource-saving and environment-friendly direction, so there are several aspects to be considered in the design of prefabricated energy-saving farm houses[5].

(1) Extract local architectural features

The appearance of the building should match the local architectural style and culture. Huizhou culture is a typical representative of Anhui. After the continuous improvement and development of Song Dynasty and Ming Dynasty, Huizhou culture has reached its peak in Qing Dynasty and has a strong cultural heritage until today[6]. Taking the architecture of central Anhui as an example, because it is located at the border between north and south, the central Anhui area has absorbed the architectural characteristics of the south and the north, where there are both southern patios and northern courtyards. On the basis of satisfying the living function, the architectural characteristics of rural rural houses are optimized to achieve the design purpose of "energy saving, livable and characteristic" (Figure 3).

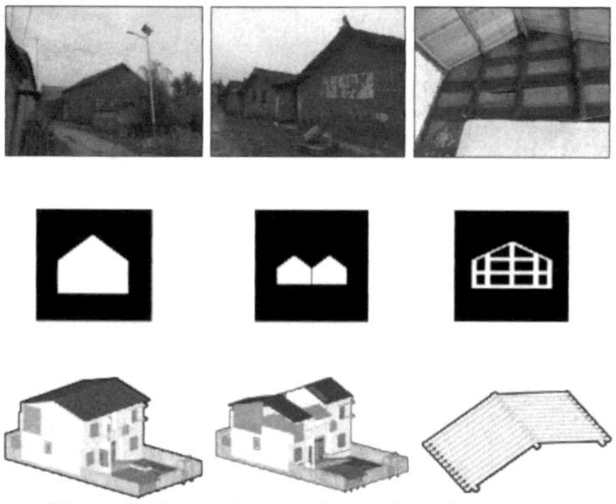

Figure 3. Elements extraction of farm house style

(2) Diversification of energy-saving technologies

Building energy saving technology is diverse, so building energy saving needs to consider the actual situation of the local use of local conditions, but also to introduce the latest technology, and constantly promote the update of energy saving technology. For example, through the design and use of building materials, the building can maximize the use of solar energy to achieve indoor comfort, reduce energy consumption and reduce dependence on traditional energy sources, without the use of active mechanical or electrical equipment. Through the redesign of the farm house plane, the farm house is divided into different areas, such as residential area, kitchen, storage room, etc., in order to realize the functional requirements of different Spaces and avoid energy waste. Public areas such as living rooms and dining rooms should be as close as possible to south-facing Windows to maximize the use of solar energy.

(3) Principle of economic applicability

The principle of economical application should be established when the agricultural house assembly energy saving transformation is carried out, and the analysis should be made according to the family economic level of most residents in the rural area of central Anhui province, combined with the cost in the transformation process. Finally, to achieve a balanced state, within the acceptable transformation cost, strive to achieve the maximum energy saving transformation effect, meet the principle of economic applicability. Building components can be in accordance with the standard modules of building parts in southern Anhui, the factory prefabricated three kinds of module plates 300mm,600mm,900mm[7-8]. The parts are processed and designed in the factory, and only assembled on the construction site, which has little impact on the environment during the whole cycle of building construction. Villagers can choose their own modular construction meter to meet the needs of different family populations (Figure 4).

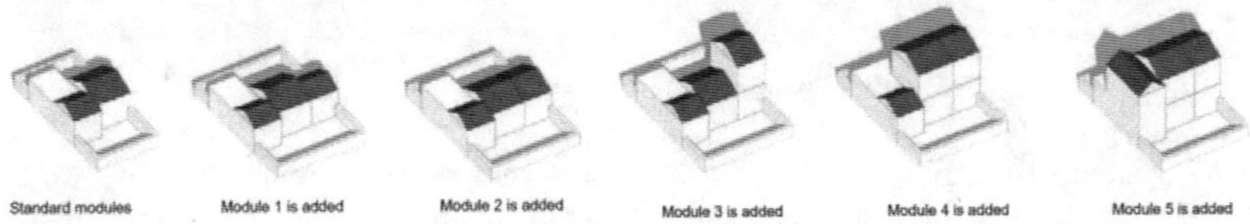

Figure 4. Modular design

3.3. Prefabricated energy saving design of farm houses in the middle of Anhui Province

(1) Solar photovoltaic technology

Active solar building technology refers to the use of solar energy as the main source of energy, using a series of technical means to control the collection, conversion and utilization of solar energy, in order to achieve energy saving, environmental protection, economy and comfort and other multiple goals of the building[9]. According to the current technology and trends, the future new rural building power supply and distribution system will be based on solar photovoltaic power generation technology.

(2) Rainwater collection system

Rainwater collection system includes five parts: collection system, transportation system, purification system, water storage system and transfer system. The collection system is generally set on the roof, and the transportation system sends the collected rainwater to the reservoir, and then through the filter system, filter the impurities in the rainwater, collect the filtered rainwater through the water storage system, and finally through the transmission system, the water is sent to the place where it is needed for irrigation and reuse. The roof of a building is usually used as a catchment surface to collect rainwater, but gullies, gutters and depressions can also be used to collect water (Figure 5).

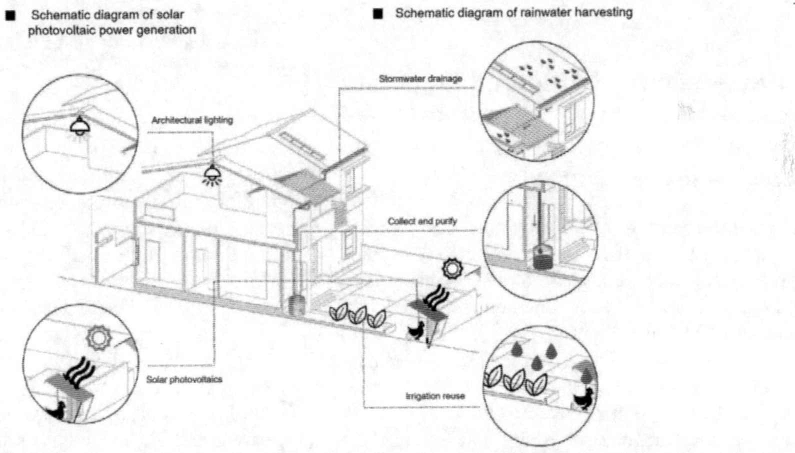

Figure 5. Schematic illustration of solar photovoltaic power generation and rainwater harvesting

(3) Wood grille energy saving

Taking advantage of the characteristics of rural green space, the positive role of local plants in building energy conservation is rationally utilized, and the role of garden planting is sun protection and heat insulation. It can also optimize the design of the courtyard of the farm house, increase the greening, and reduce the consumption of energy while improving the quality of living (Figure 6).

 https://doi.org/10.1051/e3sconf/202452802003

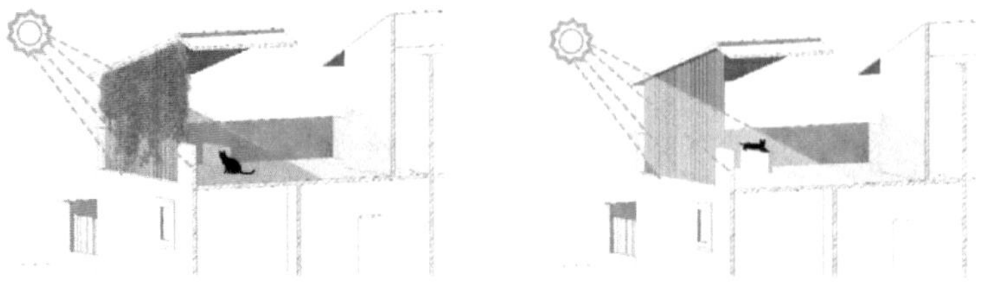

Figure 6. Energy saving diagram of wooden grille

(4) Utilization of geothermal energy

Geothermal energy is a green, low-carbon, recyclable renewable energy, with large reserves, wide distribution, clean environmental protection, stable and reliable. In summer, the closed cycle system will exchange heat between the higher indoor temperature and the lower underground temperature, thereby reducing the indoor temperature. In winter, the closed circulation system will exchange heat between the indoor temperature and the underground temperature, thereby increasing the indoor temperature (Figure 7).

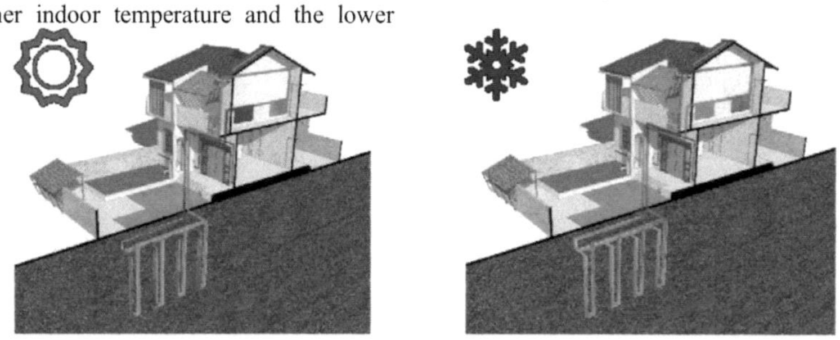

Figure 7. Diagram of geothermal energy utilization

4. Suggestions for promotion and application of prefabricated energy-saving farm houses in the middle of Anhui Province

Limited by the region and the economic development level of some areas in the middle of Anhui Province, although the light steel prefabricated building technology has been relatively mature, the promotion and application of prefabricated farm houses in the new rural construction in central Anhui is still in the development stage. With the gradual progress of new rural construction, Anhui Province needs to build high level, high quality prefabricated housing, and to achieve reasonable layout, in order to meet the needs of rural comfort[10-11].

(1) First of all, it is necessary to combine the culture of middle Anhui to shape the rural landscape. Influenced by Huizhou area, the element of horsehead wall often appeared in the residential buildings in central Anhui area. Most of the buildings in central Anhui area adopted cyan tiles instead of ash treatment, thus forming cyan hanging mountain roofs or hard mountain roofs. The cultural elements of middle Anhui are integrated into the prefabricated farm house, the rural layout is planned uniformly, the national characteristic tourism is developed, and the rural economy is promoted.

(2)Secondly, testing and promotion at selected points, and strengthening policy guidance is an important part. Due to the influence of traditional concepts, most rural residents are in a state that they have never heard of or seen before, while residents in central Anhui are more accepting of traditional brick-concrete structures. Therefore, to vigorously promote the prefabricated farm houses, the government needs to lead, carry out pilot promotion, and actively introduce the prefabricated construction enterprises to arrange the prefabricated base in the rural area of central Anhui, so that the people truly realize the advantages of the prefabricated farm houses.

(3) Finally, it is very important to improve the degree of standardization, improve supporting facilities and technology, and establish after-sales service system. The research and development, production, transportation, construction and after-sales of prefabricated farm houses are integrated into a complete industrial chain, and the cost of prefabricated farm houses is controlled while improving the quality. From the economic point of view, the cost of prefabricated farm house is high, which is one of the difficulties restricting the promotion of the farm house, and improving the degree of standardization is one of the effective measures to control the cost. And provide regular maintenance services for farmers to ensure the performance and service life of farmers' houses. Improving the supporting technology can also increase farmers' trust in prefabricated farm houses and

promote the application of prefabricated farm houses in new rural construction.

5. Conclusion

At present, rural revitalization is being carried out all over China, and rural housing construction is an important part of rural revitalization. This paper probes into the energy saving design strategy of prefabricated farm houses in central Anhui, hoping to provide reference and ideas for the future rural unified construction of farm houses. It is believed that more experts will devote their attention to this field in the future, and the prospects of rural construction in China will become better and better.

Acknowledgments

This work was financially supported by the 2021 Provincial Key Project of Humanities and Social Sciences 《Exploration of New Spatial Forms of Vegetable Markets in the Post-epidemic Era Based on UWB (Indoor Positioning) and Ventilation and Lighting Simulation》.(numbering: Sk2021A1121).

References

1. Huang L (2019) Research on the application of assembled light steel structure residence in western rural area. Sichuan architecture, 39(2): 28-29.
2. Mu X L (2021) The application and development of prefabricated building in Shanxi's rural housing construction. Shanxi architecture,47(8): 28-30.
3. Xiao S, Hao S Y and Ren X (2018) Study on the countermeasures of rural prefabricated building development in our country. Journal of Engineering Management, 32(01): 7-11.
4. Li X(2018) The problem of light steel structure house in new countryside and characteristic small town popularizing. Building Technology, 17(2): 43-44.
5. Jiang C Z, Xia W and Zhai Y (2005) Some thoughts on the authenticity protection of ancient vernacular dwellings in southern Anhui. Journal of Anhui University of Architecture, 01: 1-4.
6. Maureen M (2018) Interior Design Visual Presentation: A Guide to Graphics, Models and Presentation Methods.John Wiley,Hoboken.
7. Eriksson H, Sandberg M and Jansson G (2021) Exploring Product Modularity in Residential Building Areas. Buildings (Basel), 11(7): 281.
8. Vural N, Vural S, Engin N, et al (2007) Eastern Black Sea Region-A sample of modular design in the vernacular architecture.Building and environment,42(7): 2746-2761.
9. Wu R (2018) Study on the application of green energy-saving building with assembled steel structure. Green building materials, 08: 53-56.
10. Zhang W (2021) Study on the extension and application of fabricated light steel structure rural housing in Yangzhou. Sichuan building materials, 47(10): 47+67.
11. Yang C H, Liu X J, Jiang J J and Tao Y (2023) The popularization and application of assembled farm house in Guangxi new rural construction.Construction, 55(03): 63-68.

https://doi.org/10.1051/e3sconf/202452802004

Quantitative Study of Post-Pandemic Electricity Productivity Fluctuations Based on LMDI

Shanshan Wu*[1], Guoqiang Ji[1], Shujun Zhang[2], Jiaxi Chen[3], Haiqing Lu[2]

[1]State Grid Energy Research Institute Co.Ltd., Beijing, China
[2]State Grid Zhejiang Electric Power Co., Ltd, Hangzhou,China
[3]State Grid Zhejiang Electric Power Co., Ltd. Economic and Technological Research Institute, Hangzhou,China

Abstract. This study focuses on China's electricity productivity post-pandemic, employing the Logarithmic Mean Divisia Index (LMDI) method and the Kaya identity. Despite the clarity of the electricity productivity indicator, factors like industrial structure, energy efficiency, and electrification rates contribute to fluctuations. Quantitative analysis reveals a decisive role of the terminal electrification rate in declining electricity productivity. The proposed analytical method provides a more accurate assessment of electricity's contribution to economic growth, aiding informed policy decisions for sustainable energy development in China.

1. Introduction

China's economic development efficiency is a crucial issue of widespread concern in various sectors of society. Due to the advantages of timeliness and authenticity, electricity data is often utilized to calculate electricity productivity, which involves directly dividing the GDP of a particular region by its electricity consumption. Electricity, as a clean, efficient, and convenient secondary energy source, is increasingly prevalent in economic production. Many studies focus on the relationship between electricity consumption and economic growth in China[1]. China's proportion of electricity in final energy consumption and the share of electricity in primary energy consumption both exceed the levels seen in developed countries at similar stages of development. Furthermore, there is considerable room for increased electrification, indicating a growing role for electricity in the economy in the future[2]. In recent years, the global outbreak of the COVID-19 pandemic has significantly impacted China's electricity usage and economic situation, attracting widespread attention in society. Studying the post-pandemic trends in China's electricity productivity and conducting a thorough analysis of the reasons for fluctuations is essential for objectively understanding the relationship between the economy and electricity and guiding societal expectations appropriately.

The advantage of the electricity productivity indicator lies in its clear logic, ease of understanding, and simple calculation. It enables a quick assessment of the impact of electricity factors on economic growth and allows for comparative analysis between regions and over time. However, ongoing adjustments in industrial structure, energy transition, and the rapid promotion of green and low-carbon initiatives will influence the trajectory of electricity productivity from various perspectives. In terms of energy transition, there is a relationship of mutual substitution between electricity and other energy types, and changes in the electrification rate are a crucial factor affecting electricity productivity. Regarding economic structural adjustments, China continues to advance industrial restructuring, and the core content and direct results of initiatives such as "three reductions, one compensation," "supply-side structural reform," and "high-quality development" lead to changes in economic structure[3], resulting in objective differences in electricity consumption density among industries. In terms of new technologies and energy strategies, digitization and intelligence will significantly increase the dependence of various industries on electricity. Simultaneously, China has proposed an energy strategy prioritizing conservation, which will to some extent reduce the electricity demand. These changes ultimately manifest in industry energy utilization efficiency.

In summary, changes in electricity productivity do not directly reflect the efficiency of economic and social development. This paper will establish a quantitative decomposition model for the factors influencing electricity productivity, conducting an in-depth quantitative analysis of the post-pandemic trends in China's electricity productivity. This approach aims to clarify the reasons for the indicator's fluctuations, providing a more comprehensive and objective understanding of electricity productivity.

*Corresponding author's e-mail: wushanshan@sgeri.sgcc.com.cn

2. Analysis of Changes in Electricity Productivity Post-Pandemic

The productivity of electricity in China exhibits an "M"-shaped trend within the year, influenced significantly by heating and cooling electricity demand during the winter and summer seasons. The first and third quarters show relatively smaller values of electricity productivity. It is noteworthy that China's electricity productivity has shown a continuous decline over the past five years (shown as Figure 1), dropping from 12.3 yuan/kilowatt-hour in 2019 to 11.0 yuan/kilowatt-hour in 2023. This decline may be attributed to various factors, including but not limited to technological advancements, energy structure adjustments, and improvements in energy utilization efficiency. This not only reflects the transformation and development of China's power system but may also be correlated with the national policy orientation towards clean energy and efforts to promote the sustainable development of the power industry.

To delve into the reasons behind the trend in electricity productivity, the next section of this paper establishes a decomposition model for electricity productivity. This model aims to comprehensively and systematically analyze the various factors influencing the changes in electricity productivity. The model will assist in gaining a profound understanding of the composition structure of electricity productivity and reveal the driving mechanisms behind it.

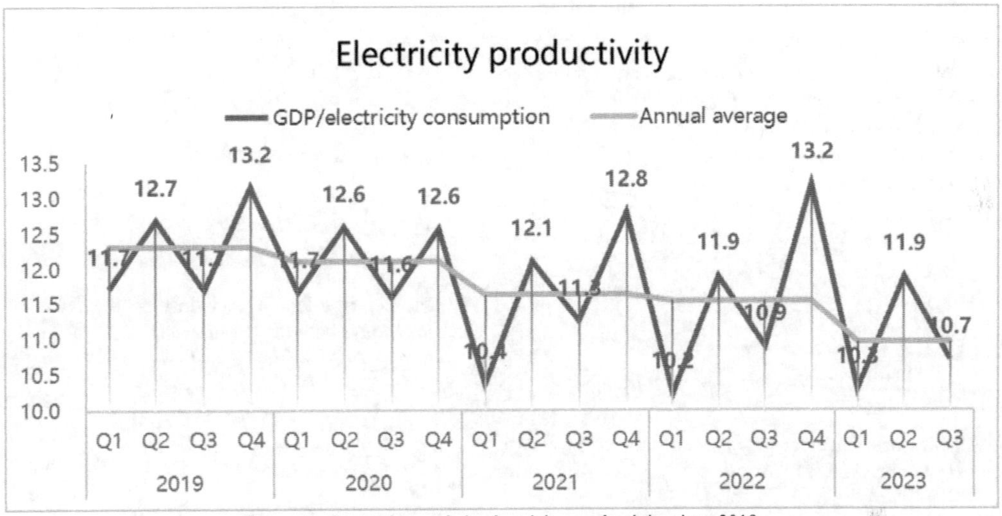

Figure 1. Quarterly trends in electricity productivity since 2019.

3. Establishment of the Decomposition Model for Electricity Productivity

The Logarithmic Mean Divisia Index (LMDI) decomposition method is a technique based on decomposing the target variable. Its advantage lies in the absence of unexplained residuals after the decomposition, and it allows for relatively straightforward expressions using both additive and multiplicative decomposition. This method is widely applied in analyzing trends in factors affecting energy or environmental impacts to gain in-depth insights into the contributions of individual factors to overall changes[4]. In the field of electricity, the LMDI method is employed to decompose electricity flows data or electricity consumption for conducting analyses of the development of the electric power sector[5], [6]. Therefore, in this paper, to analyze the reasons for changes in electricity productivity, the LMDI method is employed to decompose the electricity productivity. The foundation of the LMDI model is the Kaya Identity, a commonly used international method for examining the sources of carbon dioxide emissions, breaking down the target function of carbon dioxide emissions into factors such as energy use, economic output, and population[7]. Drawing inspiration from the Kaya Identity, this paper extends the concept to decompose the factors influencing electricity productivity into four elements: industry structure, industry energy utilization efficiency, energy structure, and end-use electrification rate. Logarithmic averaging is then utilized to analyze these influencing factors. While the decomposition model delves into the trends of electricity productivity, it comprehensively enhances the understanding of the main factors affecting electricity productivity and reveals their interrelationships.

Firstly, industry structure, as a crucial factor in electricity productivity, reflects the demand characteristics for electricity across different industries. Through decomposing industry structure, we can understand the extent of electricity consumption in different sectors, thereby clarifying the contributions of different industries to electricity productivity. Secondly, the role of industry energy utilization efficiency in electricity productivity is paramount. This factor considers the efficiency of energy use in different industries, subsequently influencing the fluctuations in electricity productivity. By decomposing industry energy utilization efficiency, a deeper understanding of the energy utilization status in different sectors can be obtained, providing targeted recommendations for improving electricity production efficiency. Energy structure is another indispensable factor directly related to the types of energy used in electricity production. Decomposing energy structure enables the assessment of the contributions of different energy sources to electricity productivity, thereby revealing the proportion of clean and traditional energy in power output. Lastly, the end-use electrification rate considers the proportion of electricity ultimately converted for end-user consumption. Decomposing the end-use electrification rate reflects the losses and benefits in the conversion process from electricity generation to end-user terminals, offering crucial information for enhancing power utilization efficiency.

Establish the Kaya Identity, whose expression is given by Equation (1).

$$\frac{GDP}{EC}=\frac{GDP}{GDP_{\text{idustry}}}\times\frac{GDP_{\text{idustry}}}{EN_{\text{idustry}}}\times\frac{EN_{\text{idustry}}}{EN}\times\frac{EN}{EC}$$
$$=(\textit{industry structure})^{-1}\times(\textit{industry energy use efficiency})^{-1}\times\textit{energy consumption structur}\times(\textit{terminal electrification rate})^{-1} \quad (1)$$

Where EC, EN, denote electricity and energy consumption respectively, and the subscript of industry denotes an industry. The inverse of the industry structure, the inverse of the industry energy use efficiency, the energy consumption structure, and the inverse of the terminal electrification rate are denoted as $Str=\frac{GDP}{GDP_{\text{idustry}}}$, $Eff=\frac{GDP_{\text{idustry}}}{EN_{\text{idustry}}}$, $EStr=\frac{EN_{\text{idustry}}}{EN}$, $ER=\frac{EN}{EC}$, and the expression (2) is obtained:

$$\frac{GDP}{EC}=Str\times Eff\times EStr\times ER \quad (2)$$

The change in value electricity productivity from year t to year t+j can be expressed as follows:

$$\Delta\frac{GDP}{EC}=\Delta Str+\Delta Eff+\Delta EStr+\Delta ER \quad (3)$$

Where, $\Delta\frac{GDP}{EC}$ denotes the overall change in electricity productivity from year t to t+j degrees, this change can be decomposed into the change in the inverse of the industry structure $\Delta\frac{GDP}{EC}_{Str}$, the change in the inverse of the industry's energy use efficiency $\Delta\frac{GDP}{EC}_{Eff}$, the change in the structure of energy consumption $\Delta\frac{GDP}{EC}_{EStr}$, and the change in the inverse of the terminal electrification rate $\Delta\frac{GDP}{EC}_{ER}$. The specific calculations are as follows:

$$\Delta\frac{GDP}{EC}_{Str}=\frac{\frac{GDP}{EC}(t+j)-\frac{GDP}{EC}(t)}{In\frac{GDP}{EC}(t+j)-In\frac{GDP}{EC}(t)}*In\frac{Str(t+j)}{Str(t)} \quad (4)$$

$$\Delta\frac{GDP}{EC}_{Eff}=\frac{\frac{GDP}{EC}(t+j)-\frac{GDP}{EC}(t)}{In\frac{GDP}{EC}(t+j)-In\frac{GDP}{EC}(t)}*In\frac{Eff(t+j)}{Eff(t)} \quad (5)$$

$$\Delta\frac{GDP}{EC}_{EStr}=\frac{\frac{GDP}{EC}(t+j)-\frac{GDP}{EC}(t)}{In\frac{GDP}{EC}(t+j)-In\frac{GDP}{EC}(t)}*In\frac{EStr(t+j)}{EStr(t)} \quad (6)$$

$$\Delta\frac{GDP}{EC}_{ER}=\frac{\frac{GDP}{EC}(t+j)-\frac{GDP}{EC}(t)}{In\frac{GDP}{EC}(t+j)-In\frac{GDP}{EC}(t)}*In\frac{ER(t+j)}{ER(t)} \quad (7)$$

The contribution of each influencing factor to the electricity productivity from year t to year t+j is $\Delta\frac{GDP}{EC}_{Str}/\Delta\frac{GDP}{EC}$, $\Delta\frac{GDP}{EC}_{Eff}/\Delta\frac{GDP}{EC}$, $\Delta\frac{GDP}{EC}_{EStr}/\Delta\frac{GDP}{EC}$, $\Delta\frac{GDP}{EC}_{ER}/\Delta\frac{GDP}{EC}$.

4. Analysis of the Reasons for the Change in the Electricity Productivity

Considering the accessibility of data, the calculation time interval was selected from 2019 to 2021, taking the industrial sector as an example. During this period, the proportion of industrial value added to GDP rose from 30.8% to 31.4%, contributing 51.3% to the decrease in electricity productivity. The energy efficiency of the industrial sector improved, with energy consumption reducing from 0.75 tons of standard coal per 10,000 yuan to 0.71 tons of standard coal per 10,000 yuan, which, paradoxically, contributed negatively at -130.1% to the decrease in electricity productivity. The proportion of industrial energy consumption to total energy consumption increased from 51.0% to 51.3%, contributing -16.2% to the decrease in electricity productivity. Finally, the rate of electrification in end-use increased from 7.8% to 8.5%, with a positive contribution rate of 195% to the decrease in electricity productivity.

In the premise of data accessibility, the temporal scope for the analysis was set from the year 2019 to 2021, focusing on the industrial sector. Throughout this designated interval, the share of industrial value added to the Gross Domestic Product (GDP) exhibited an upward trend from 30.8% to 31.4%. Decomposing the impact of the changes in electricity productivity reveals that the industrial sector plays a pivotal role, contributing 51.3% to the decline in this metric.

Initially, there was a shift in the energy efficiency utilized by the industry, decreasing from 0.75 tons of standard coal per 10,000 yuan to 0.71 tons of standard coal per 10,000 yuan. The slight decrease in energy efficiency in industrial production reflected a negative contribution rate of -130.1% when calculating the changes in electricity productivity, indicating an adverse effect of declining energy efficiency on the economic productivity of electricity within the industrial processes. Subsequently, the proportion of industrial energy consumption within the total energy consumption rose from 51.0% to 51.3%. This change manifested a contribution rate of -16.2% to the decline in electricity productivity. This might be associated with the higher demand for energy within the industrial sector and a relatively lower energy utilization efficiency, resulting in a reduction in the economic productivity of electricity. Lastly, the electrification rate in end-use applications experienced an increase from 7.8% to 8.5%. This elevation yielded a contribution rate of 195% to the decrease in electricity productivity, signifying that the heightened rate of electrification in the industrial sector has exercised a positive impact on the value productivity of electricity, possibly reflecting the adoption of advanced technologies in the utilization of power.

5. Conclusion

Based on a comprehensive evaluation of multiple influencing factors, this study provides a more practical and comprehensive analysis method for electricity productivity, assisting policymakers in better understanding the relationship between power and the economy. In our research, we decomposed the electricity productivity using the Logarithmic Mean Divisia Index (LMDI) method and established a decomposition model based on the Kaya identity, conducting an in-depth analysis of fluctuations in electricity productivity from aspects such as industrial structure, energy utilization efficiency, energy mix, and the rate of terminal electrification. The enhanced practicality and operability of our electricity productivity analysis method is a significant advancement. The findings indicate that increases in the rate of terminal electrification and changes in industrial structure are causes for the decline in electricity productivity, while improvements in industry-specific energy efficiency and shifts in the energy consumption structure are reasons for an increase in electricity productivity. The rise in terminal electrification played a decisive role, ultimately leading to a decrease in electricity productivity.

Looking ahead, as the rate of terminal electrification continues to climb, electricity productivity may further decline. Thus, it is necessary to have a more comprehensive and objective understanding; changes in electricity productivity do not directly reflect the efficiency of economic and social development, and a comprehensive judgment based on multiple influencing factors is required. Our study enhances the accuracy and scientific rigor of electricity productivity analysis, dissecting the indicator with greater precision through technical means. From an economic impact perspective, our research more accurately assesses the actual contribution of electricity to economic growth, offering robust support for the scientific formulation of economic policies and power planning. The comprehensive analysis of industrial structure and energy utilization efficiency provides targeted energy optimization recommendations for different regions, further promoting the sustainable development of the energy industry.

Acknowledgments

This work was financially supported supported by Science and Technology Project of State Grid Corporation of China（Research on Multi temporal and Spatial Scale Power Supply and Demand Forecasting and Early Warning Technology Considering Complex and Multiple Factors，No.5700-202257453A-2-0-ZN）.

References

1. Zhang, C. et al. (2017) 'On electricity consumption and economic growth in China', Renewable and Sustainable Energy Reviews, 76, pp. 353–368. Available at: https://doi.org/10.1016/j.rser.2017.03.071.
2. Wang, X. and Tan, X. and Jin. Y. (2020) Ideas and methods for measuring the economic value of electricity, China Power Enterprise Management, (1), pp. 56–57.
3. Guo, K. (2019) Trends in China's Industrial Structure Adjustment and Upgrading and Policy Ideas for the 14th Five-Year Plan Period, China Industrial Economics, (7), pp. 24–41. Available at: https://doi.org/10.19581/j.cnki.ciejournal.2019.07.002.
4. Ang, B.W. (2015) LMDI decomposition approach: A guide for implementation, Energy Policy, 86, pp. 233–238. Available at: https://doi.org/10.1016/j.enpol.2015.07.007.
5. Zhang, C. et al. (2018) 'Analysis of electricity consumption in China (1990–2016) using index decomposition and decoupling approach', Journal of Cleaner Production, 209. Available at: https://doi.org/10.1016/j.jclepro.2018.10.246.
6. Koval, V. et al. (2023) 'Fundamental Shifts in the EU's Electric Power Sector Development: LMDI Decomposition Analysis', Energies, 16(14), p. 5478. Available at: https://doi.org/10.3390/en16145478.
7. Peng, D. and Liu, H. (2022) Measurement and Driving Factors of Carbon Emissions from Coal Consumption in China Based on the Kaya-LMDI Model, Energies, 16(1), p. 439. Available at: https://doi.org/10.3390/en16010439.

https://doi.org/10.1051/e3sconf/202452802005

Study on energy saving of non-uniform air supply in office environment

Han Wu*, Guiqiang Wang

School of Municipal and Environmental Engineering, Shenyang Jianzhu University, Shenyang, 110168, China

Abstract. There is a situation that the personnel in the office building are not concentrated and scattered, and some personnel will not always work at the workstation, and the creation of local comfort at the unmanned workstation will only cause great waste, so it is necessary to create an environment with uneven distribution of indoor parameters. In view of this situation, we simulate two working conditions: uniform and inconsistent airflow provision within the office space. analyze the comfort of these two working conditions in the office room, discuss the unevenness and energy efficiency of the ventilation openings above and near personnel in the office room. The simulation results show that creating non-uniform indoor air supply can greatly save energy.

1.Introduction

Currently, the ventilation system approach of most public buildings is ceiling air supply, which creates a uniform state of indoor environment with a single parameter [1]. According to research, most public buildings have some spaces that are not used in daily lives [2]. Unused areas and spaces no longer need to maintain the comfortable environment required by the human body. In this case, opening the entire floor of air vents to maintain the indoor human comfort environment will only cause a waste of resources. Personalized ventilation (PV) represents a solution by installing a personalized terminal near office personnel to achieve independent control of the local environment around each individual [3]. E.Biyik designed a personalized air supply system in the early stage in the guise of a table-mounted units, called the Tabletop Environment Conditioning (TAC) system [4], this concept is gradually being applied to other indoor conditions. Current research on vent layout is concentrated on air quality management and energy efficiency saving [5]. The traditional end control method is to adjust the end damper opening based on the temperature of the return air duct [6]. Yang et al. [7] proposed the use of ceiling-type personalized terminals to provide individualized air jets, which avoid the installation of air ducts in areas of human activity, but this increases the distance between the occupants and the terminals, thus creating an overlapping effect in adjacent terminal wind areas [8]. According to the current research status, based on satisfying the basic comfort of human and not changing the original air-conditioning system of the building, the indoor environment with uneven distribution of indoor environmental factor is created to achieve the effect of energy saving.

*Corresponding author's e-mail: 1033687056@qq.com

2. Model and research techniques

2.1 Physical model

One of the offices in an office building is taken as the main object of study, the office model was identified as a standard layer, the thermal exchange between the ceiling and the floor was negligible, and the office was surrounded by internal walls. The model parameters are shown in Table 1, the office floor plan is shown in Figure 1, and the geometric model is shown in Figure 2.

Table 1. Model parameters.

Name	Size
Architectural entities	14.5 m×13.5 m×3.0 m
Air outlets	0.24 m×0.24 m
Return air outlets	0.77 m×0.19 m
Table	1.2 m×0.6 m×1.1 m
Computer	0.6 m×0.1 m×0.3 m

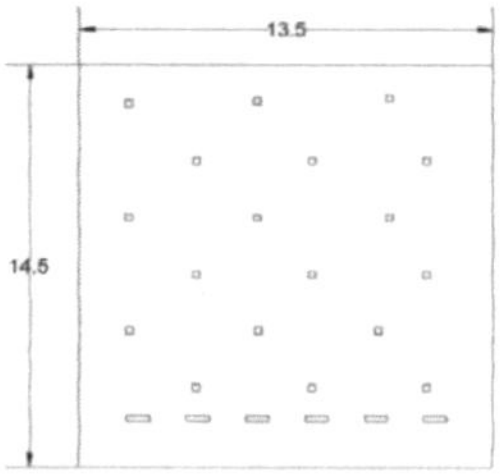

Figure 1. Office Floor Plan.

Figure 2. Geometric diagram of office space.

2.2 Mathematical model

For the actual situation and specific simulation analysis, our turbulence model is solved by the RNG k-ε model. The fundamental equations encompass: continuity equation, momentum equation, the energy conservation equation, and k and ε equations. The specific equation is as follows: the continuity equation is denoted by (1)、the momentum equation is represented by (2)、the energy conservation equation is represented by (3)、the RNG k-ε equation model is derived from the transport equations (4) and (5):

$$\frac{\partial \rho}{\partial t} + \nabla \cdot (\rho v) = 0 \tag{1}$$

where: ρ denotes the fluid's mass density., kg/m³; v denotes fluid velocity, m/s; t denotes time, s.

$$\frac{\partial u_i}{\partial t} + u_j \frac{\partial u_i}{\partial x_j} = -\frac{1}{\rho}\frac{\partial p}{\partial x_i} + \frac{\mu}{\rho}\left[\left(\frac{\partial u_i}{\partial x_j} + \frac{\partial u_j}{\partial x_i}\right) - \frac{2}{3}\frac{\partial u_k}{\partial x_k}\right] + g_i \tag{2}$$

where: u_i denotes the velocity of the fluid in the x_i direction, m/s； p denotes stress； μ denotes dynamic viscosity，N·s/m; g_i denotes the same gravitational force as in the x_i direction，9.81N/kg.

$$\frac{\partial}{\partial t}(\rho h) + \nabla \cdot (\rho h \omega) = \nabla \cdot \left[(k + k_t)\nabla T\right] + S_h \tag{3}$$

where: h denotes the total energy per unit mass of the fluid，J/kg; k denotes the molecular thermal conductivity，W/(㎡·K)； k_t denotes the thermal conductivity due to turbulent transport，W/(㎡·K)； S_h indicates that all user-defined volumetric heat source terms are included；

$$\frac{\partial}{\partial t}(\rho k) + \frac{\partial}{\partial x_i}(\rho k u_i) = \frac{\partial}{\partial x_j}\left(\alpha_k \mu_{eff} \frac{\partial k}{\partial x_j}\right) + G_K + G_b - \beta\varepsilon - Y_M + S_K \tag{4}$$

$$\frac{\partial}{\partial t}(\rho \varepsilon) + \frac{\partial}{\partial x_i}(\rho \varepsilon u_i) = \frac{\partial}{\partial x_j}\left(\alpha_k \mu_{eff} \frac{\partial \varepsilon}{\partial x_j}\right) + C_{1\varepsilon}\frac{\varepsilon}{k}(G_K + C_{3\varepsilon}G_b) - C_{2\varepsilon}\rho\frac{\varepsilon^2}{k} - R_\varepsilon + S_\varepsilon \tag{5}$$

where: k a denotes turbulent energy，J; ε a denotes the turbulent dissipation rate，%; u_i denotes the velocity component, m/s; u_t denotes the turbulent viscosity coefficient; x indicates the coordinate location; G_k denotes the turbulent energy due to the laminar velocity gradient，J; G_b denotes turbulent energy due to buoyancy, J; Y_m denotes the fluctuations due to the diffusion of the transition in compressible turbulence; $C_{1\varepsilon}$、$C_{2\varepsilon}$、$C_{3\varepsilon}$ are 1.44, 1.92, and 0.09; α_k and α_ε are expressed as the turbulent reaction Plante numbers of the k and ε are 1.0 and 1.3; S_k and S_ε are user-defined data.

2.3 Research Techniques

In order to further explore the non-uniformity of the indoor environment, we introduce the non-uniformity coefficient k_y, the unevenness in the office is judged by calculating the unevenness coefficient of the office [9], its expression is (6):

$$k_y = \frac{\sigma_y}{\bar{y}} \tag{6}$$

where：y denotes temperature or velocity；σ_y denotes the root mean square difference；$\bar{y}$ denotes the arithmetic average；k_y indicates that the smaller the value, the more uniform the distribution, and the opposite is not uniform [10]. The energy-saving cooling capacity is calculated as formula (7):

$$Q = c * \rho * \upsilon * \Delta t \tag{7}$$

where: c denotes the specific heat of air，J/（kg·k）; ρ denotes the fluid's mass density，kg/m³; v indicates the air volume，m³/h; Δt indicates the temperature difference between the inlet and outlet air，℃.

3.Numerical solving

3.1 Meshing

For most employees in the office, they tend to work in a sedentary state, to satisfy meet the requirements of human comfort, the air emitted by the air outlet located in the ceiling needs to reach at least the height of the human chest in order to effectively adjust the surrounding environment of the office personnel [11]. To save calculation time while ensuring simulation accuracy, supply vent, return vent, computer, personnel, tables and chairs are partially encrypted, and the generated grid elements are 3781919, the minimum orthogonal mass of the grid is 0.22, and the grid quality is good, which meets the calculation requirements, and the office grid is shown in Figure 3.

https://doi.org/10.1051/e3sconf/202452802005

Figure 3. Grid diagram of the office.

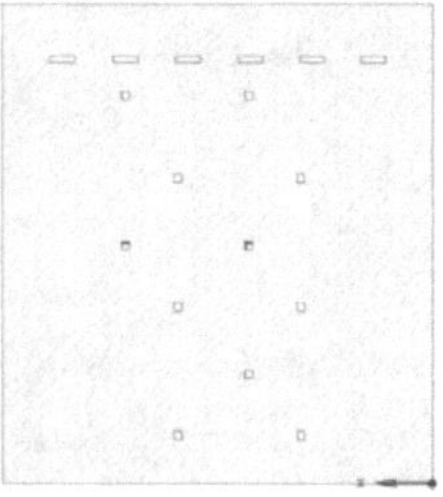

Figure 4. Tuyere opening and closing diagram

3.2 Setting of simulation conditions

The office model is a closed model, the temperature of the supplied air is established at 20 °C, the supply air velocity is set to 2.8 m/s, and the return air outlet is set to the pressure outlet. To facilitate the computation process, the floor slab and the ground can be regarded as adiabatic boundary conditions without temperature difference, and the surrounding walls are all internal walls, and the heat transfer is assumed to be a constant heat flow boundary condition, and The coefficient of heat transfer is 3.2 W/(㎡· K); The free flowing wall surface temperature is 28 °C, the human and computer are used as indoor heat sources, with a heat generation of the human is 55 W/㎡, and the heat generation of the computer is 100 W/㎡ [12]. The opening and closing of the tuyere is controlled by user-defined functions, and the opening of the tuyere when creating a non-uniform air supply environment is shown in Figure 4.

4. Simulation results and analysis

According to the above conditions, the model is simulated under two working conditions, and the findings from the simulation are presented below:

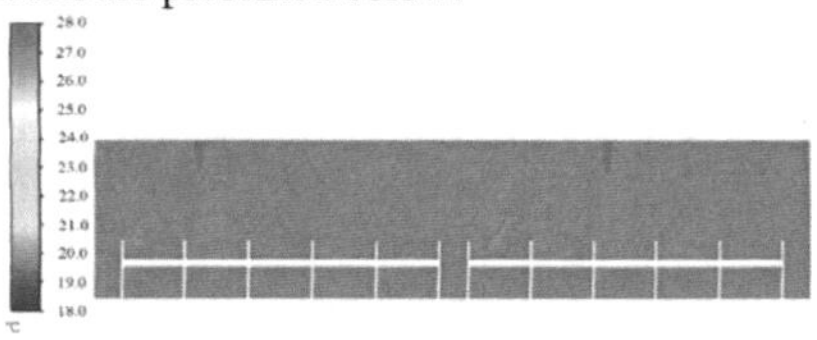

Figure 5. Temperature distribution map on the side section of the office room.

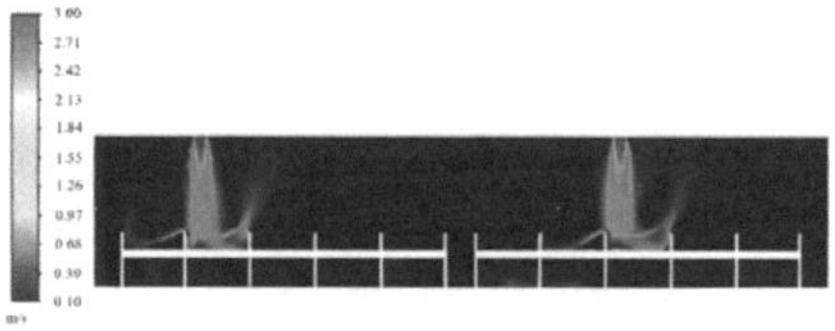

Figure 6. Map of wind speed distribution on the side section of target area

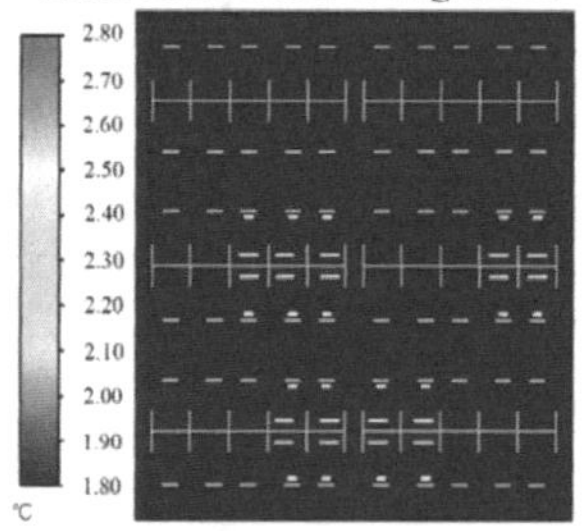

Figure 7. Temperature distribution cloud map of the target area.

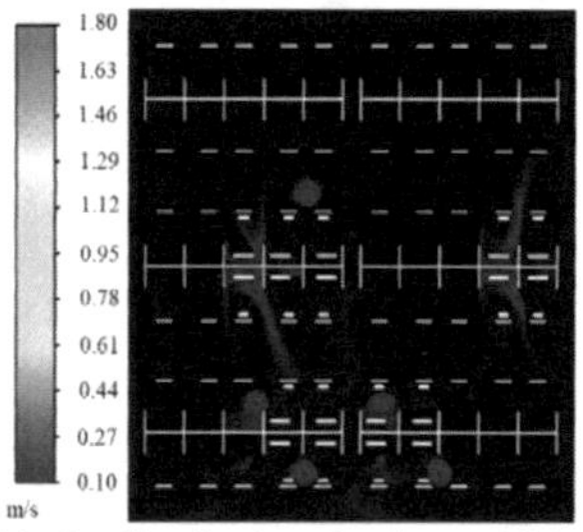

Figure 8. Velocity distribution cloud map of the target area.

When all the indoor air outlets are opened, it can be concluded from Figure 5~8 that the overall ambient temperature profile within the room is about 22 °C, and the overall indoor temperature distribution is 0.1 m/s. The internal airflow velocity and temperature in the room can satisfy the criteria for comfort of the human.

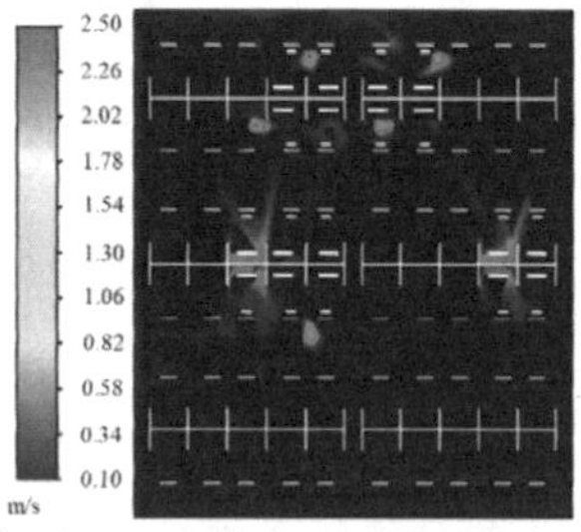

Figure 9. Cloud map of wind speed distribution in the target area.

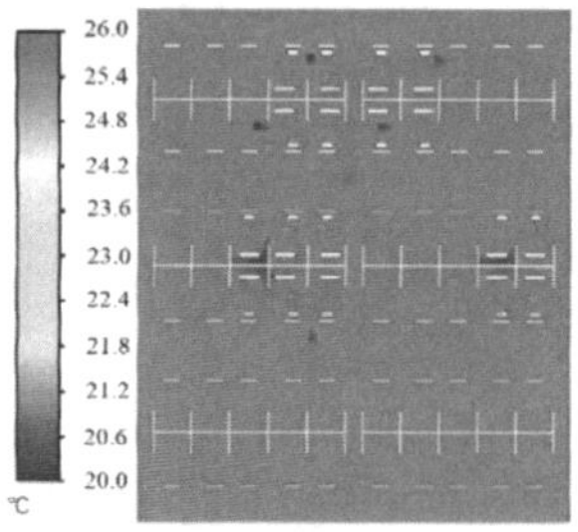

Figure 10. Temperature distribution cloud map of the target area.

Figure 11. Temperature distribution map.

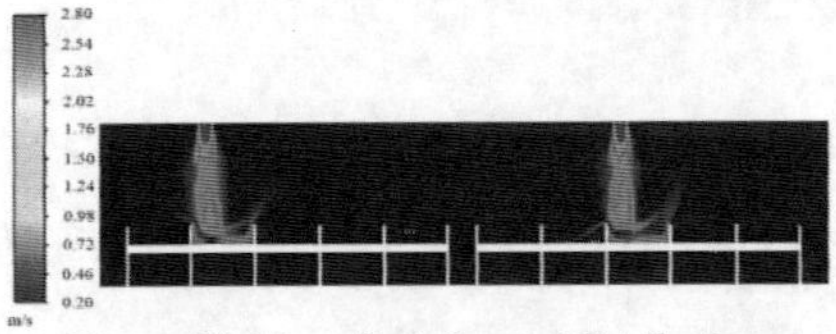

Figure 12. Map of wind speed distribution.

When the air outlet above or near the person is opened, it can be concluded from Figure 9~12 that the indoor temperature is roughly 21.8 °C, and the overall internal airflow velocity. is about 0.1 m/s. The internal airflow velocity and temperature can still meet the comfort requirements of the human body. According to equation (6), the non-uniformity coefficient of temperature and wind speed is are 0.4 and 2.18. From this, it can be concluded that the distribution of the environment in the office is uneven, creating an uneven indoor environment. The energy-saving situation is mainly considered in the following aspects:

(1) Energy-saving cooling capacity:

According to equation (7), the cooling capacity savings can be calculated:

$$Q_{uniform}=1.003\ \text{kJ/（kg·k）} \times 1.29\ \text{kg/m}^3 \times 2.8 \times 0.24 \times 0.24 \times 3600\ \text{m}^3/\text{h} \times (273.15+10)\ \text{k} = 59.086\ \text{kW} \quad Q_{non\text{-}uniform}=1.003\ \text{kJ/（kg·k）} \times 1.29\ \text{kg/m}^3 \times 2.8 \times 0.24 \times 0.24 \times 3600\ \text{m}^3/\text{h} \times (273.15+6)\ \text{k} = 58.252\ \text{kW} \quad Q_{save}= Q_{uniform} - Q_{non\text{-}uniform}= 59.086\text{kW} - 58.252\ \text{kW} = 0.834\ \text{kW}$$

(2) Energy-saving electricity:

Fans and water pumps are important components of the central air conditioning system, fans and water pumps account for a significant portion of electricity consumption during operation [13]. The air conditioning on workdays is calculated based on 11 hours.

Save electricity= Cooling capacity*Time =0.834 kW×11 h= 9.174 kWh

5. Conclusion

In this paper, the indoor non-uniform environment of traditional tuyeres is simulated. Through the comparative analysis of the outcomes of the modeling analysis, it can be concluded that the indoor non-uniform air supply can save about 0.834 kW of cooling capacity per day, and about 9.174 kWh of electricity can be saved per day. Since the actual office situation cannot be realized in the setting process, the precision of the simulation outcome data needs to be further demonstrated and tested by actual experimental data.

References

1. Yang B, Melikov AK, Kabanshi A, et al. A review of advanced air distribution methods-theory, practice, limitations and solutions[J]. Energy and Buildings, 2019, DOI: 10.1016/j.enbuild.2019.109359.
2. Energy Consumption Statistics Committee of China Building Energy Conservation Association. 2018 China Building Energy Consumption Research Report [J]. Architecture, 2019 (2): 26-31
3. Shao X, Li X, Ma X, et al. Optimising the supply parameters oriented to multiple individual requirements in one common space[J]. Indoor and Built Environment, 2014, 23(6): 828-838.
4. Bin Huang. Energy saving control of variable air volume (VAV) air conditioning systems [J]. Urban Construction Theory Research: Electronic Edition, 2012001 (002): 1-4
5. Davardoost F, Kahforoushan D. Evaluation and investigation of the effects of ventilation layout, rate, and room temperature on pollution dispersion across a laboratory indoor environment[J]. Environmental Science and Pollution Research, 2019, 26(6): 5410-5421.
6. Jafarinejad T, Erfani A, Fathi A, et al. Bi-level energy-efficient occupancy profile optimization integrated with demand-driven control strategy: University building energy saving[J]. Sustainable Cities and Society, 2019,48:101539
7. Parameshwaran R, Harikrishnan S, Kalaiselvam S. Energy efficient PCM-based variable air volume air conditioning system for modern buildings[J]. Energy and Buildings, 2010, 42(8): 1353-1360.
8. Shao X, Li X. Evaluating the potential of airflow patterns to maintain a non-uniform indoor environment[J]. Renewable energy, 2015, 73: 99-108.
9. Wang Y, Quan M, Zhou Y. Effect of velocity non-uniformity of supply air on the mixing characteristics of push-pull ventilation systems[J]. Energy, 2019, DOI: 10.1016/j.energy.2019.115962.
10. Zhai Z J, Metzger I D. Insights on critical parameters and conditions for personalized ventilation[J]. Sustainable Cities and Society, 2019, 48: 101584.
11. Hu J, Shan C, Wu J, et al. A cooperative ceiling air supply method to satisfy personal thermal preferences in a discretionary indoor position[J]. Journal of Building Engineering, 2020, DOI: 10.1016/ j. jobe.2020.101367.
12. Melikov A K. Advanced air distribution: improving health and comfort while reducing energy use[J]. Indoor Air, 2016, 26(1): 112-124.
13. Haoqiang Huang. Energy saving measures analysis of central air conditioning supply systems in high-rise buildings [J]. Science and Technology Outlook,2016,26(20):45.

Dynamic prediction of steam consumption in beer production process based on Attention Mechanism CNN-BiLSTM

Kun Lu[1*], Jiong Liu[2] and Yan Shi[2]

[1] Institute of Electric Power, Inner Mongolia University of Technology, Hohhot, Inner Mongolia, 010051, China
[2] Inner Mongolia Lingyi High-tech Group, Hohhot, Inner Mongolia, 010010, China

Abstract. During the normal production process of the brewery, the steam consumption can be accurately predicted, and the boiler steam output can be planned to achieve a balance between steam production and use. In order to improve the accuracy of steam consumption prediction, this paper proposes a dynamic prediction method of steam consumption based on attention mechanism-convolution-bidirectional long short-term memory neural network (CNN-BiLSTM). In this paper, the real-time monitoring data of the brewery energy management system is selected as the experimental data for analysis, and the CNN-BiLSTM network model based on attention mechanism is compared with the results of CNN net-work, LSTM network, BiLSTM network and CNN-BiLSTM network prediction. The experimental results show that the mean absolute error, root mean square error and R-Square of the model are 0.0329, 0.0449 and 97.5%, which are better than the other four models, and can predict the steam consumption of brewery more accurately.

1. Introduction

In the beer production process, breweries use a large amount of steam as a heat source [1]. Most breweries use steam produced by their own boilers, and the actual amount of steam used varies according to the beer production schedule. In the process of beer production, the goal of energy saving and consumption reduction can be achieved by accurately predicting the amount of steam consumption, adjusting the production capacity of the boiler in real time, and achieving a balance between steam production and use.

In recent years, many researchers have made energy consumption predictions. Mughees uses a deep bidirectional long short-term memory based sequence to sequence (Bi-LSTM S2S) regression method to predict the day-ahead peak power demand, and has achieved initial success [2]. Shahid uses the strength of recurrent neural networks with long short-term memory (LSTM) embedded in wavelet kernels to propose a wind power prediction model [3]. Lu combined time series decomposition with CNN neural network, and used convolutional neural network to predict the energy consumption data of the workshop [4]. Liu used fuzzy C-means clustering and LSTM network to predict the output of CSP power station [5]. Li introduced a CNN-BiGRU model based on artificial bee colony optimization into the Hadoop framework to predict the main steam flow rate of thermal power plant boilers [6].

In beer production, the number of equipment using steam is large and the structure is complex, and conventional forecasting cannot solve the problem of steam consumption prediction well. In deep learning, CNN can retain the connection between adjacent data and capture local data features, BiLSTM can obtain more available information through bidirectional learning of data, and the attention mechanism can effectively capture the relationship between global and local data. Therefore, this paper combines the three models to construct a dynamic prediction model of steam consumption in breweries based on CNN-BiLSTM based on attention mechanism to improve the accuracy of steam consumption prediction.

2. Model building

The model is composed of three parts: CNN network, BiLSTM network and attention mechanism module, because the time series prediction results of a single model have low accuracy and cannot meet the prediction requirements, so this paper uses the functions of each module to construct a CNN-BiLSTM prediction model based on attention mechanism, and its structural model is shown in Figure 1.

*Corresponding author's e-mail: LJ-LY@163.com

https://doi.org/10.1051/e3sconf/202452802006

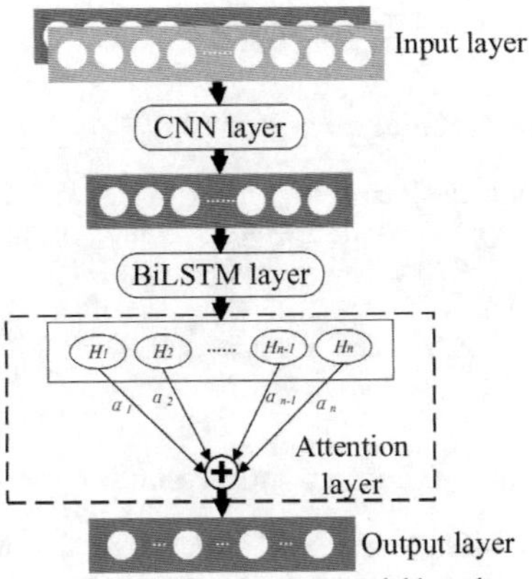

Figure 1. CNN-BiLSTM structure model based on attention mechanism.

2.1. CNN

Convolutional Neural Network (CNN), as an important achievement of deep learning, can extract features on one-dimensional time series. CNN is composed of multiple filters superimposed and have the ability to mine latent features in the data, and the convolutional neural network structure is shown in Figure 2.

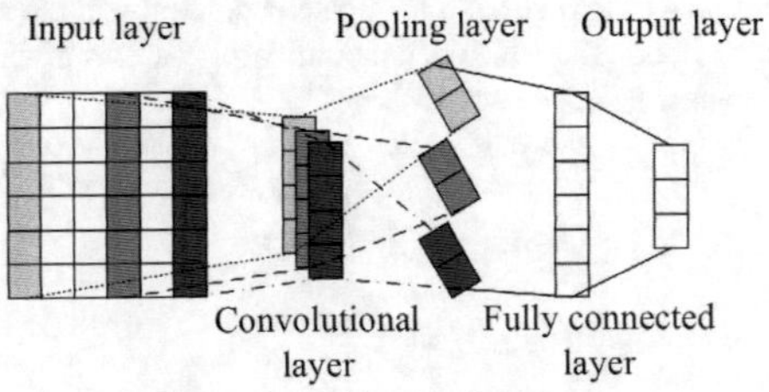

Figure 2. CNN network structure.

CNN consists of five parts: input layer, convolutional layer, pooling layer, fully connected layer, and output layer [7]. The convolutional layer extracts and maps the features of time series data, and the feature information of various scales can be obtained through convolutional kernels of different sizes, reducing the number of parameters.

A pooled layer is an output sample of a convolutional layer that retains only important information and prevents overfitting. There are two common pooling methods: max pooling and mean pooling, and the average pooling method is used to calculate the local optimal solution as the input of BiLSTM.

In CNN, parallel training can be achieved while greatly reducing the network weight parameters through weight sharing. Finally, the weight of the network is adjusted according to the error backpropagation algorithm to minimize the error.

2.2. BiLSTM

As an optimization model for RNN, long short-term memory (LSTM) can better obtain long-term dependencies in time series sequences and effectively solve the gradient vanishing problem of traditional RNN models [8]. LSTM introduces memory cells into each hidden layer neuron, and uses the forgetting gate, input gate and output gate to control the memory unit, which solves the problem that RNNs rely on input information for a long time, so that LSTM can learn the input information in a long-term sequence. The internal structure of the LSTM neural network is shown in Figure 3.

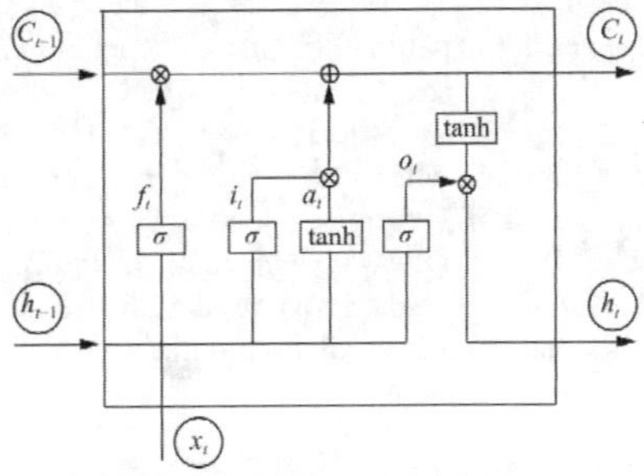

Figure 3. LSTM network structure.

In Figure 3, the current time is t, C_{t-1} and C_t are the memory cell states at time (t-1) and time t, respectively, h_{t-1} and h_t are the outputs at time (t-1) and time t, x_t is the input vector at time at current time, and σ and tanh represent the feedforward network layer with activation functions of sigmoid and tanh. The calculation process is shown in Equation 1-6, W is the weight and b is the bias.

$$f_t=\sigma\,(W_f\,[h_{t-1}\ x_t])+b_f \tag{1}$$

$$i_t=\sigma\,(W_i\,[h_{t-1}\ x_t])+b_i \tag{2}$$

$$\widetilde{C}_t=tanh\,(W_C\,[h_{t-1}\ x_t])+b_C \tag{3}$$

$$C_t=f_t{\cdot}C_{t-1}+i_t{\cdot}\widetilde{C}_t \tag{4}$$

$$o_t=\sigma\,(W_o\,[h_{t-1}\ x_t])+b_o \tag{5}$$

$$h_t=o_t{\cdot}tanh\ C_t \tag{6}$$

Bidirectional Long Short-Term Memory Neural Network (BiLSTM) is different from LSTM in that BiLSTM can perform forward and reverse bidirectional observation of sequence data through bidirectional structure. Compared with LSTM, BiLSTM can capture more information in the sequence, which is conducive to improving the model's understanding and prediction ability of the sequence [9].

2.3. Attention mechanisms

The attention mechanism is to simulate the attention distribution mechanism of the human brain, focusing on key information and screening more important information.

The attention mechanism allocates and accumulates the implicit state of the BiLSTM output, and different weights represent the degree of attention to different time points. Let the input sequence be k eivectors h_i, i=1,2,... k, the attention weight a_i is normalized by the softmax function, and the calculation process is shown in equations 7-8：

$$a_i=softmax(s_i)=\frac{e^{s_i}}{\sum e^{s_i}} \tag{7}$$

$$s_i=tanh(w_i^{\,T}h_i+b_i) \tag{8}$$

The final environment vector c is obtained by multiplying and accumulating h_i and a_i.

$$c=\sum_{i=1}^{k} a_i h_i \tag{9}$$

3. Experimental data and preprocessing

3.1. Experimental data acquisition

In this paper, the data collected from the energy system of a domestic brewery from August 1, 2023 to August 31, 2023 were taken as the research object, and 13 variables were selected for experiments, such as the amount of malt added by the pulverizer, the amount of water added to the pulverizer, the amount of water added to the boiling pot, and the filling volume of the packaging line, with a sampling period of 5 minutes, a total of 9000 sets of data, the first 8550 sets of data were taken as the training data, and the last 450 sets of data were used as the test data, and the steam consumption prediction model was established on this basis.

3.2. Experimental data preprocessing

The box plot discriminant method was used to screen the outliers of the experimental data, and the lower and upper quartiles V_1 and V_3 of the data were first calculated by calculation, and the I_{QR} of the interquartile range was obtained according to Equation 10:

$$I_{QR}=V_3\text{-}V_1 \tag{10}$$

According to the I_{QR}, the normal value range $[F_1, F_u]$ of the sample data is determined, and the data beyond the $[F_1, F_u]$ interval is the outlier, and the calculation process is shown in Equation 11:

$$[F_1,F_u]=\left[V_1\text{-}\frac{3}{2}I_{QR},V_3+\frac{3}{2}I_{QR}\right] \tag{11}$$

For missing values and cleared outliers, the data are supplemented by interpolation, as shown in Equation 12:

$$x_i=\frac{x_{i\text{-}1}+x_{i+1}}{2} \tag{12}$$

The dimensions of different evaluation indexes in the experimental data are different, and direct use will affect the experimental analysis results. To eliminate dimensional effects between indicators, the data need to be normalized. In this paper, the MaxMin method is used for normalization, and the calculation process is shown in Equation 13:

$$x_{std}=\frac{x\text{-}x_{min}}{x_{max}\text{-}x_{min}} \tag{13}$$

4. Design of Experiments

4.1. Experimental environment

The experimental environment is Windows 11 64-bit operating system, and the programming is based on MATLAB R2022b environment. The hardware configuration is CPU In-tel(R) Core(TM) i5-12500H 2.50 GHz, 16 GB RAM, and graphics card NVIDIA GeForce RTX3050Ti.

4.2. Parameter setting and evaluation indicators

The CNN-BiLSTM network model based on attention mechanism includes input layer, 2 convolutional layers, 2 pooling layers, 1 BiLSTM layer, 1 attention layer, and 1 fully connected layer. The number of convolution kernels of the two convolutional layers is 32 and 64, respectively, and the size is 3×1×1, the activation function is ReLU, stride=1, the number of neurons in the BiLSTM layer is 128, and the Dropout is set to 0.2. The training times were 1000 times, the learning rate was 0.01, and the Adam gradient descent algorithm was used.

In order to better evaluate the prediction ability of the model and ensure the effectiveness and scientificity of the model, MAE, RMSE, and R^2 are selected as the evaluation indexes, and the evaluation indicators are shown in equations 14-16.

$$MAE=\frac{1}{n}\sum_{i=1}^{n}\left|y_i\text{-}\hat{y}_i\right| \tag{14}$$

$$RMSE=\sqrt{\frac{1}{n}\sum_{i=1}^{n}(y_i\text{-}\hat{y}_i)^2} \tag{15}$$

$$R^2=1\text{-}\frac{\sum_{i=1}^{n}(y_i\text{-}\hat{y}_i)^2}{\sum_{i=1}^{n}(y_i\text{-}\bar{y})^2} \tag{16}$$

4.3. Comparative experiments

In order to better evaluate the performance of the proposed model and prove the advantages of the network model in prediction, this paper compares the CNN, LSTM, BiLSTM, CNN-BiLSTM and CNN-BiLSTM-Attention models, and the prediction results of the models are shown in (a)-(e) in Figure 4.

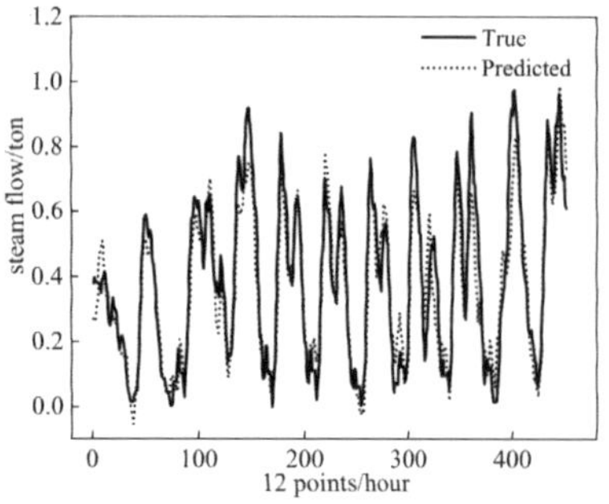

(a)

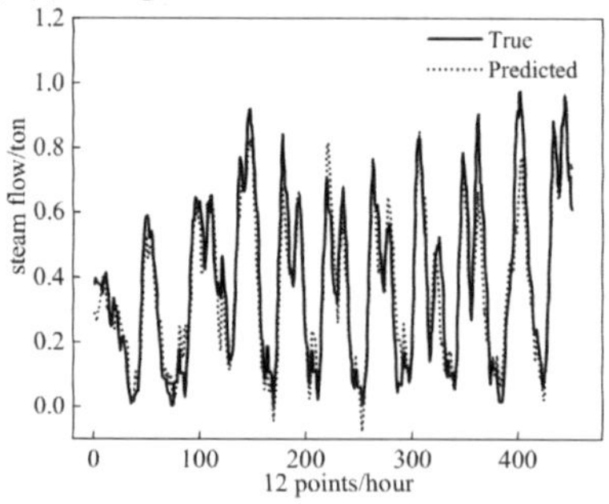

(b)

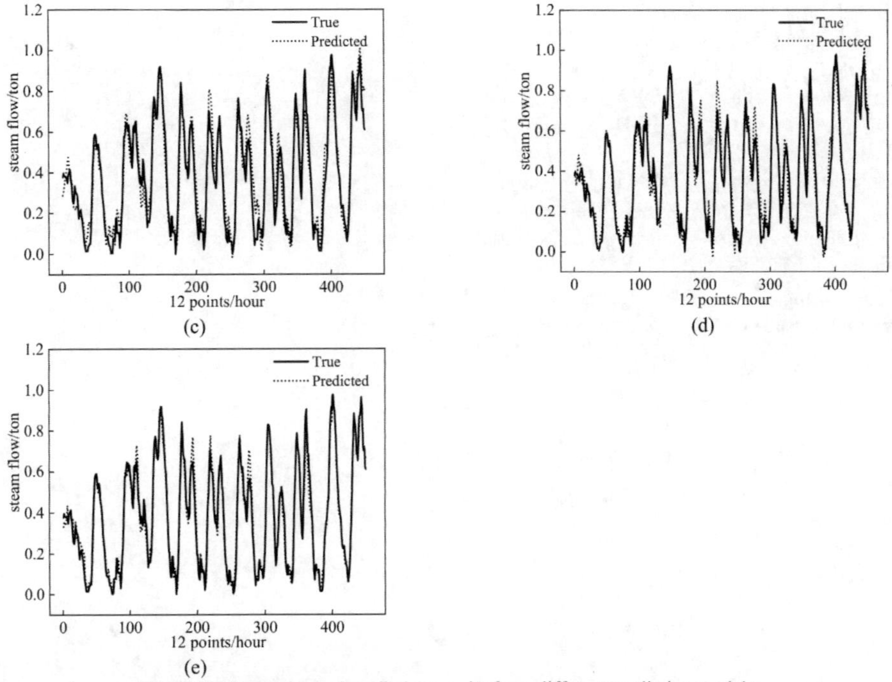

Figure 4. Comparison of prediction results from different prediction models.

As can be seen from the five graphs in Figure 4, the overall trend of the prediction results of the five models is basically consistent with the real value, and the prediction effect of the CNN-BiLSTM model is better than that of the single model. After the attention mechanism is added to the CNN-BiLSTM model, the prediction effect of the CNN-BiLSTM-Attention model is further improved.

In order to further reflect the accuracy of the prediction effect of the CNN-BiLSTM-Attention model, Table 1 lists the evaluation indicators of the five models.

Table 1. Comparison of evaluation indexes of different prediction models.

Model	MAE	RMSE	R^2
CNN	0.0708	0.0914	87.2%
LSTM	0.0670	0.0885	87.9%
BiLSTM	0.0601	0.0761	91.1%
CNN-BiLSTM	0.0451	0.0580	94.7%
CNN-BiLSTM-Attention	0.0329	0.0449	97.5%

As can be seen from Table 1, compared with the CNN, LSTM, BiLSTM and CNN-BiLSTM models, the MAE and RMSE of the CNN-BiLSTM-Attention model are reduced by 53.5%, 50.9%, 45.3% and 27.1%, respectively, and the RMSE is reduced by 50.8%, 49.3%, 41.0% and 22.6%, which further indicates that the prediction results of the model are more accurate and the performance is more superior.

5. Conclusion

In order to improve the steam utilization rate, this paper proposes a CNN-BiLSTM steam consumption prediction model based on attention mechanism, and compares the prediction results of the prediction model with the CNN, LSTM, BiLSTM and CNN-BiLSTM models..

Acknowledgments

This work was financially supported by the Inner Mongolia Autonomous Region Scientific and Technological Achievements Transformation Guidance Project . (2019CG040).

References

1. Huang W L, Yang M and He D K (2023) Research on Reducing Packaging Steam Consumption in Breweries. J. Sino-foreign Liquor Industry, 03: 18-24.
2. Mughees N, Mohsin S. A, Mughees A and Mughees A (2021) Deep sequence to sequence Bi-LSTM neural networks for day-ahead peak load forecasting. J. Expert Systems with Applications, 175.
3. Shahid F, Zameer A, Mehmood A and Raja M. A. Z (2020) A novel wavenets long short term memory paradigm for wind power prediction. J. Applied Energy, 269.
4. Lu Y K, Sheng B Y, Zhang Z H and Zhang Y Q (2021) Workshop Energy Consumption Prediction Method Based on Time Series Decomposition and CNN. J. Computer Applications & Software, 38(06): 339-344.
5. Liu Z L, Guo J H, Li W, Jia H T and Chen Z (2024) Short-term prediction of CSP output based on FCM-LSTM. J. Chinese Journal of Engineering, 46(01): 178-186.
6. Li M, Zong X J, Lian L, He K and Yang Z J (2023) Data Analysis and Application of Thermal Power

Plant Based on CNN-ABC-BiGRU. J. Electronic Design Engineering, 31(22): 11-15, 20.

7. Mohammadpour L L, Ling T C, Liew C S and Aryanfar A (2022) A Survey of CNN-Based Network Intrusion Detection. J. APPLIED SCI-ENCES-BASEL, 12(16): 8162-8162.
8. Liang, B S, Wang S Y, Huang Y Q, Liu Y L and Ma L P (2023) F-LSTM: FPGA-Based Hetero-geneous Computing Framework for Deploying LSTM-Based Algorithms. J. ELECTRON-ICS, 12(5): 1139-113.
9. Li Y and Wang L P (2022) Human Activity Recognition Based on Residual Network and BiLSTM. J. SENSOR-S, 22(2): 635.

A New Calculation Method of Well Leakage in Gas Well with Annular Pressure

Shaoping Yuan, Heng Luo*, and Haijun Wu

Safety and Environmental Protection Branch, CNOOC Energy Technology & Services, Haikou, Hainan, China

Abstract. The gradual promotion of the concept of gas well integrity has made major oil and gas fields pay more and more attention to the integrity management and evaluation of gas wells with annular pressure. Annular leakage rate is one of the key parameters for gas well integrity evaluation. However, the leakage rate measurement of high-temperature, high-pressure, and acid-containing gas wells has high risks and costs, which are often difficult to obtain in field operations. In order to obtain the bottom hole leakage rate parameters of the gas well to assist the integrity evaluation, this article is based on the safety valve leakage model and the small hole leakage model, combined with the characteristics of the gas well with annulus pressure, the idea of applying the annulus leakage rate calculation from the perspective of model mechanism is analysed, and the model parameters are clarified as well. The application of the two leak models is compared by case wells calculation, which verifies their rationality and feasibility. The research of this paper provides a theoretical basis for the leakage evaluation of gas well with annular pressure, and promotes development of gas well integrity management research.

1 Introduction

With the development of oil and gas industry, the phenomenon of abnormal annulus pressure in oil and gas wells is gradually increasing, which has become an important problem affecting the safe production of gas wells[1-3]. One of the important reasons for causing annular pressure is the failure of pipe integrity, and the rate of annular leakage is one of the most core parameters for determining whether the integrity of gas wells has failed[4-6]. Due to the problems of high temperature, high pressure and sour gas, how to safely and effectively determine the leakage area and leakage rate of gas wells in the annulus is currently facing technical challenges[7, 8].

Alves proposed a method for evaluating the probability of uncontrolled external leaks in subsea wells during the production phase[9]. Hanh conducted simulation research on the fate and impact of oil spills from abandoned oil wells[10]. Zhang established a continuous annular pressure model caused by multiple oil pipe leakage points[11]. Some researchers also use special methods to determine the leakage of wells[12].

In order to enhance the accuracy of calculating the leakage rate under pressure in the annulus and improve the reliability of judging the integrity of gas wells through the leakage rate in the annulus, this paper will use the safety valve leakage model and the small hole leakage model to clarify the idea of applying the two methods to the calculation of the leakage rate under pressure in the annulus from the perspective of the fundamental mechanism of the model. The feasibility of the method will be verified through the calculation results of example wells.

2 Safety valve leakage model

2.1 Calculation Model and Formula

The calculation methods of downhole safety valves in GB/T 22342-2008 and API RP 14B-2015 are unified and applied to gas wells, and the leakage rate is calculated by the corresponding deviation factor and differential pressure:

$$q = 2.84\times10^{-3}\left(\frac{p_f}{Z_f}-\frac{p_i}{Z_i}\right)\left(\frac{1}{t}\right)\left(\frac{V}{T}\right) \quad (1)$$

where, q is leakage rate, m^3/s; p_i is initial pressure at the beginning of testing, Pa; p_f is the pressure after a specific time interval during testing, Pa; Z_i is initial compression coefficient, dimensionless; Z_f is specific interval compression coefficient, dimensionless; t is cumulative time to target pressure, s; V is downhole safety volume of gas in the tubing column above the valve, m^3; T is the absolute temperature of the downhole safety valve, K.

2.2 Analysis of model applicability

The rate of pressure increase in the upper oil pipe space of the safety valve corresponds to the leakage rate of the safety valve, as shown in Figure 1.

* Corresponding author: luoheng2@cnooc.com.cn

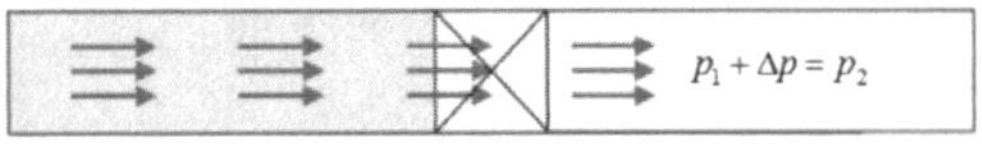

Fig. 1. Leakage model of safety valve.

During the actual mining process, the annulus pressure is regularly monitored, and if the annulus pressure exceeds 6.9×10^5 Pa, a pressure relief/pressure recovery test is performed on the annulus. When the pressure relief reaches pressure equilibrium, all pressurized gas accumulated from the bottomhole leakage is released, and the gas that continues to be released is the real-time leakage volume from the bottomhole leakage point. When the relief line is closed at this point, the annulus above the surface is equivalent to a closed cavity, and using this time point as a starting point, the pressure increment in the tubing column space is all contributed by the leakage volume within a specific time interval, as shown in Figure 2.

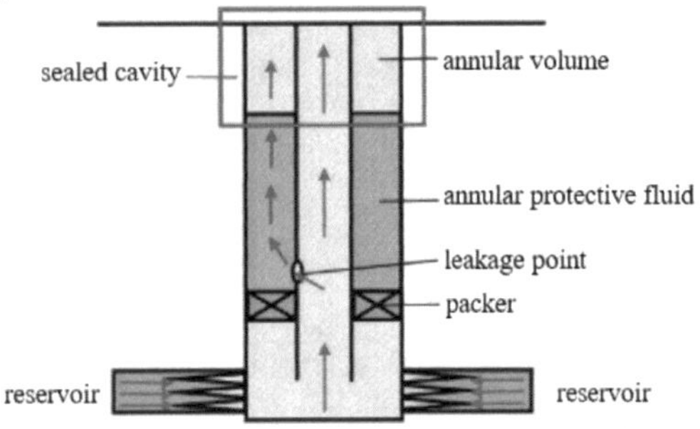

Fig. 2. Leakage modeling of annulus pressure wells.

The application of the safety valve leakage model applied to the leakage calculation of annulus pressure wells the parameter value should be taken as: (1)the pressure relief equilibrium pressure of the annulus as the initial pressure p_i; (2)After the pressure relief equilibrium, close the pressure relief line, after a period of time of the ring After a period of time, the annular pressure is taken as the final pressure p_f; (3)The time required to increase the pressure to the final pressure after closing the relief line is the cumulative time t.

It should be noted that, when using the safety valve method, the temperature, artificial factors, pressure change process are ignored.

3. Small hole leakage model

3.1 Calculation models and formulas

When the fluid passes through the nozzle, needle valve, downhole safety valve, gas lift valve air holes, the flow state can be summarized as nozzle flow, and the small hole leakage leakage model to calculate the leakage, as shown in Figures 3 and 4.

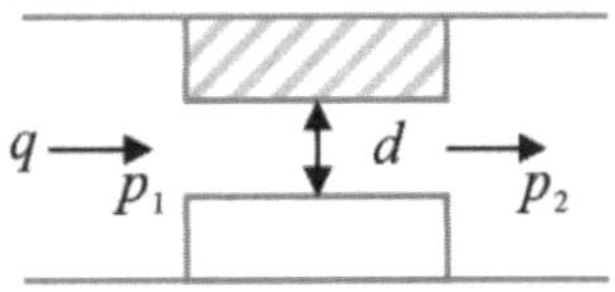

Fig. 3. Schematic diagram of nozzle flow.

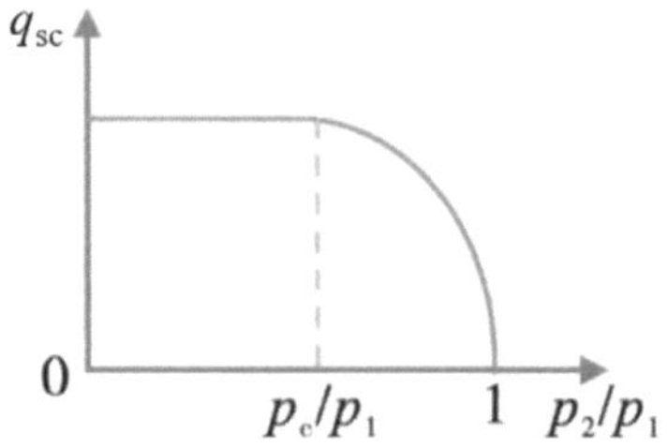

Fig. 4. Nozzle flow characteristics.

Angle pressure regulator valve nozzle critical flow refers to the fluid through the nozzle orifice high speed flow, the speed reaches the pressure wave propagation speed in the fluid medium, that is, the speed of sound when the flow state.

In order to predict the nozzle flow dynamics and the relationship between the gas flow rate and the throttle pressure drop, it is necessary to determine whether the critical flow state is reached. According to thermodynamic principles, the critical pressure ratio is:

$$R_{cp} = \left(\frac{2}{k+1}\right)^{\frac{k}{k-1}} \tag{2}$$

where, k is natural gas adiabatic index, dimensionless. The natural gas adiabatic index is 1.25-1.30, R_{cp} = 0.546-0.555.

When $R_{cp} \geq p2/p1$, it is a critical flow; otherwise it is subcritical. According to the isentropic principle of gas nozzle flow, for the subcritical flow state, the leakage rate model can be expressed as follows:

$$q_{sc} = \frac{4.72\times10^{-2} p_1 d^2}{\sqrt{r_g T_1 Z_1}} \sqrt{\frac{k}{k-1}[\left(\frac{p_2}{p_1}\right)^{\frac{2}{k}} - \left(\frac{p_2}{p_1}\right)^{\frac{k+1}{k}}]} \tag{3}$$

where, q_{sc} is non-critical flow rate at the nozzle outlet, m^3/s; d is pressure relief pipeline diameter, m; r_g is relative density of natural gas, dimensionless; T_1 is gas temperature, K; Z_1 is gas deviation coefficient, dimensionless; p_1 is pressure value at the inlet of the nozzle, Pa; p_2 is pressure value at the outlet of the nozzle, Pa.

For the critical flow, the maximum flow rate of nozzle flow is independent of p_2, its flow rate model can be expressed as:

$$q_{max} = \frac{4.72\times10^{-2} p_1 d^2}{\sqrt{r_g T_1 Z_1}} \times \sqrt{\frac{k}{k-1}[\left(\frac{2}{k+1}\right)^{\frac{2}{k-1}} - \left(\frac{2}{k+1}\right)^{\frac{k+1}{k-1}}]} \tag{4}$$

3.2 Model applicability analysis

In order to comply with the original physical meaning and modeling assumptions of the nozzle flow method, the pressure relief state should be brought to equilibrium, which is unfolded using the isentropic principle.

At this time, the pressure released by the pressure relief line is comparable to the pressure replenished at the annulus leakage point, and the annulus leakage rate is considered to be the same as the wellhead leakage rate. Therefore, the leakage rate of the annulus can be determined by taking the leakage rate of the pressure relief line at the time of pressure relief equilibrium.

Considering the entire annulus as a "pool", all the leaks at the bottom of the well are equated to the inlet of the pool, and the pressure relief line is considered as the outlet of the pool. Under this model, no matter how many leaks there are at the bottom of the well, the sum of the leaks is only one outlet, the pressure relief line.

Treating the pressure relief pipeline as a small orifice flow nozzle, the total leakage rate of all leakage points at the bottom of the well can be calculated by using the small orifice leakage model for the "annulus-pressure relief pipeline". This allows for the determination of the leakage level at the bottom of the well, as shown in Figure 5.

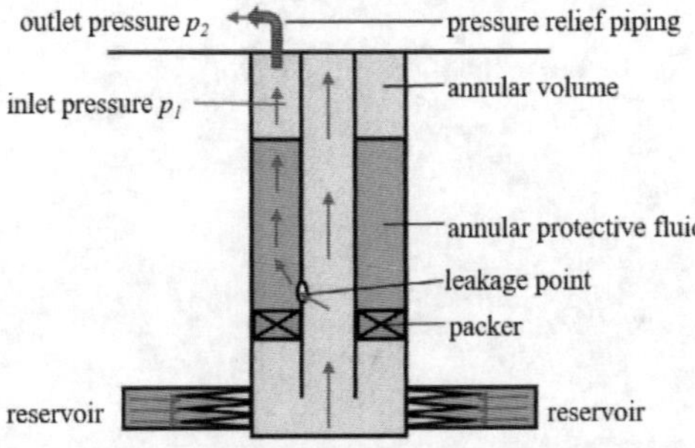

Fig. 5. Leakage pool modeling for annulus pressure wells.

When using the small hole leakage model, it must be ensured that the relationship between the equivalent diameter of the small hole d and the equivalent diameter of the flow channel D is: $d/D \le 0.2$, that is, the nozzle diameter should be less than 20% of the diameter of the pipe.

Currently, the straight diameter of gas well tubing is 88.9 mm or 101.6 mm, then the equivalent diameter d of all leakage points at the bottom of the well should be less than 17.78 mm or 20.32 mm.

The diameter of the pressure relief pipeline currently used is generally 9.525 mm or 6.35 mm, the outer diameter of the A annulus is generally 279.4 mm, and the diameter of the oil pipe is 88.9 mm or 101.6 mm. Based on the same area rule, the equivalent diameter of the annulus can be obtained:

$$S_{\text{eff1}} = \pi\left[\left(\frac{279.4}{2}\right)^2 - \left(\frac{88.9}{2}\right)^2\right] = \pi r_{\text{eff1}}^2$$
$$S_{\text{eff2}} = \pi\left[\left(\frac{279.4}{2}\right)^2 - \left(\frac{101.6}{2}\right)^2\right] = \pi r_{\text{eff2}}^2 \quad (5)$$

So the equivalent diameter of the annulus can be obtained as:

$$r_{\text{eff1}} = 132.4\text{mm} \quad r_{\text{eff2}} = 130\text{mm} \quad (6)$$

$$6.35mm / r_{\text{eff1}} = 4.8\%$$
$$9.525mm / r_{eff2} = 7.3\% \quad (7)$$

From equation (7), it can be seen that when using the pool model, the relationship requirement of the small hole leakage model $d/D \le 0.2$ is fully met.

When using the small-hole leakage model for the "annulus-pressure relief line", the inlet of the small-hole is the annulus pressure, and the outlet of the small-hole is the relief outlet. It should be noted that the calculation of the pressure relief line is based on the state of pressure relief equilibrium, so the pressure of the small hole leakage model should be taken as follows: (1)Inlet pressure p_1 is taken as the annulus pressure in pressure relief equilibrium; (2)Outlet pressure p_2 is taken as the pressure relief outlet pressure.

4. Calculation example

A pressure relief/recovery test was conducted on a pressurized well in an annular space of a certain block in Weizhou using a pipeline with a diameter of 9.525 mm. The pressure changes in the A, B, and C annular spaces over time were recorded, and the results are shown in Figure 6. From Figure 6, it can be seen that there is a significant change in the pressure in the A annulus. The well with pressure in this annulus is mainly due to the presence of pressure in the A annulus between the oil pipe and the production casing, while there is no annulus pressure in the B and C annulus.

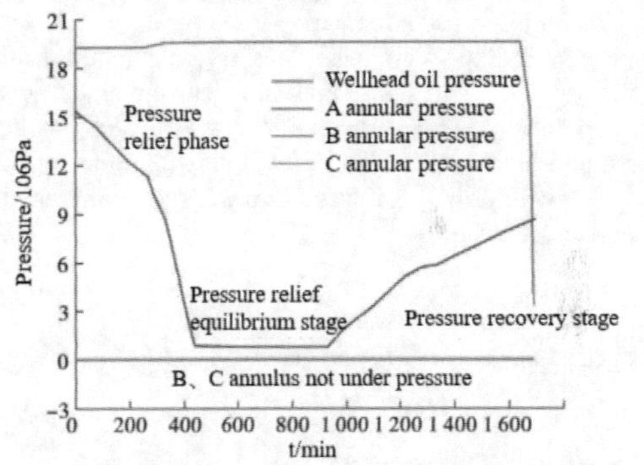

Fig. 6. Results of pressure relief/recovery testing for pressure wells in Weizhou Environmental Control Zone.

In order to verify the calculation method of bottomhole leakage in annulus pressure gas wells feasibility, the above mentioned annulus pressure wells are used for calculation.

Based on logging data and natural gas basic physical properties parameters, after on-site testing, the initial parameters required for the two calculation methods in Table 1 and Table 2 were obtained, and the model was validated using these data.

Table 1. Statistics of initial parameters for safety valve method.

Parameters	Difficulty	Value
Inlet pressure (p_1/ Pa)	medium	0.8×10^6
outlet pressure (p_2/ Pa)	medium	8.6×10^6
Gas temperature (T/ K)	low	324.6
annular volume (V/ m^3)	high	/

Recovery testing time (t/ s)	medium	4.482×10^4
Gas deviation factor (Z)	low	0.93

Table 2. Initial parameter statistics for the small-hole leakage modeling method.

Parameters	Difficulty	Value
Gas temperature (T/ K)	low	324.6
Pressure relief pipeline specifications (d/ m)	low	9.525×10^{-3}
inlet pressure (p_1/ Pa)	medium	0.103×10^6
outlet pressure (p_2/ Pa)	low	0.8×10^6
Gas deviation factor (Z)	low	0.93
Gas relative density (r)	low	0.58-0.62
Gas adiabatic index (k)	low	1.25-1.30

Two different methods were used to calculate pressurized wells under the same conditions, and the comparison results are shown in Figures 7-9.

Due to the fact that the injection time of annular protective fluid is generally relatively long, there is no accurate measurement and record of the amount injected and the leakage of the bottom annular protective fluid after several injections, which makes it difficult to determine the liquid level height in the wellbore. Therefore, when using the safety valve method for calculation, the liquid level height H is used as a variable.

Figure 7 shows a comparison of the calculation results using a safety valve and a small hole leakage model, with the top downhole packer as the lowest reference position for liquid level and the liquid level height of the annular liquid level from the lowest reference position as the calculation variable.

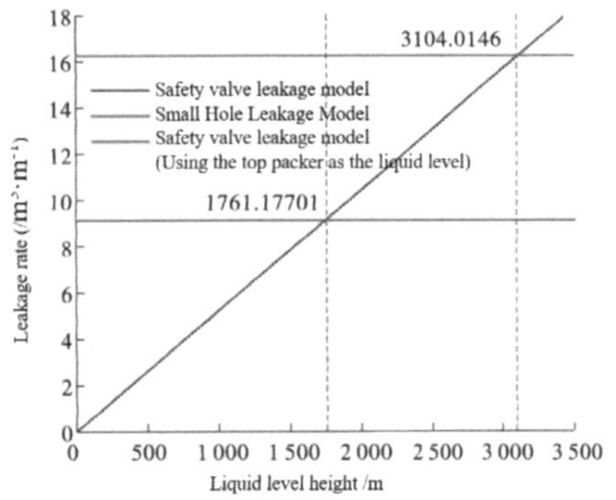

Fig. 7. Comparison of calculation results for a certain well in Weizhou.

From the comparison between the small hole leakage method and the safety valve method in the case of variable liquid level, it can be seen that in the range of 1550-1920 m, the deviation of the two methods is within 10%, and it can be assumed that the liquid level in the annulus is within this range. The depth of the topmost packer is about 3100 m, but the actual liquid level must be higher than the depth of the topmost packer.

The liquid level height calculated from the back calculation of the small hole leakage model is about 1760 m, which is higher than the depth of the uppermost packer, indicating that the small hole leakage model is feasible in the calculation of leakage in annulus pressure wells.

Figure 8 and Figure 9 compare the results of the safety valve method and the small hole leakage method by considering both the liquid level and the temperature change. From Figures 8 and 9, it can be seen that there is a big difference between the two methods when the temperature is low, and the deviation of the results of the two methods decreases as the temperature rises, and the corresponding liquid level rises at high temperatures, i.e., the safety valve method gets closer and closer to the results of the nozzle flow method with the liquid level as the independent variable as the temperature rises.

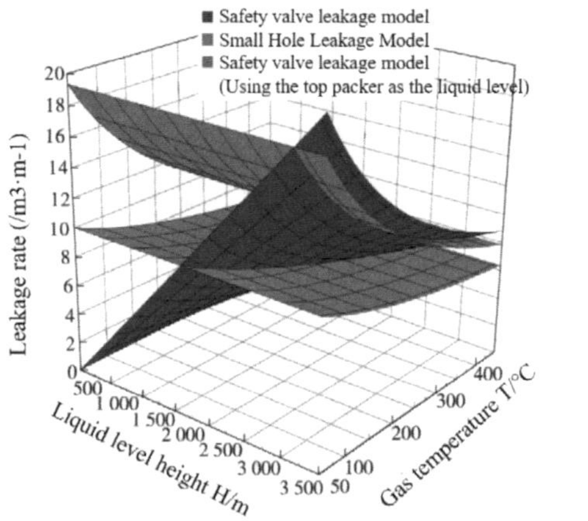

Fig. 8. The influence of liquid level height and temperature on calculation results.

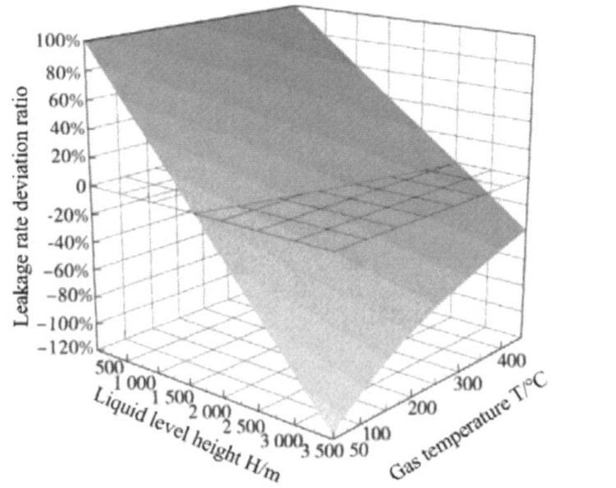

Fig. 9. The influence of liquid level height and temperature on the deviation ratio of leakage rate.

The main standards for evaluating the safety consequences of leakage are: UK-HSE, Det Norske Veritas, Lloyd's Register and domestic in-service evaluation standards. Domestic in-service evaluation standards are shown in Table 3. The evaluation standards of companies and enterprises in different countries are slightly different, but since the standards are more detailed for the leakage rate grading evaluation, the domestic in-service evaluation standards are used as the basis for the evaluation of leakage rate in this paper.

Table 3. Domestic in-service assessment standards for leakage volume

Safety consequence level	Description of safety consequences
Minor(1)	Highly flammable medium < 0.01 kg/s
Average(2)	Highly flammable medium 0.01-0.1 kg/s
Moderate(3)	Highly flammable medium 0.1-1 kg/s
Major(4)	Highly flammable medium 1-10 kg/s
Disaster(5)	Highly flammable medium > 10 kg/s

Currently, the leakage rate is roughly evaluated by using 10% of the production rate as the leakage amount. Combining the results of the example wells of the two methods of leakage rate calculation with the domestic in-service evaluation standard, the leakage rate of this Weizhou annulus pressure well is 0.115-0.205 kg/s, which belongs to the medium leakage. The results of the two methods are the same, and they are more rigorous than the production estimation method, which can more accurately evaluate the leakage rate of the annulus pressure well.

5. Conclusion

(1) The model mechanism of the safety valve leakage model was analyzed, and it was concluded that the use of the safety valve leakage model to calculate the bottomhole leakage rate of annulus pressure wells should correspond to the recovery stage of pressure relief/pressure recovery.

(2) Analyzing the model mechanism of the small hole leakage model, it is believed that when calculating the bottom hole leakage rate of an annular pressurized well using the small hole leakage model, the pressure relief stage should be considered in response to the pressure relief/pressure recovery, and the pressure relief pipeline should be considered as a small hole and calculated under the pressure relief equilibrium state.

(3) By calculating examples and comparing the calculation deviations of the two methods, the feasibility of the two methods was verified by combining the calculation results and leakage assessment standards.

References

1. Al-Sarkhi, Z.A., Ghendour, N., Azzi, A., 2023. Two-phase flow in a vertical concentric annulus: Pressure drop and flow configuration. Proceedings of the Institution of Mechanical Engineers, Part C: Journal of Mechanical Engineering Science. 237, 4454-4467.
2. Mohamed, M., Panchal, Y., Mounir, N., Woolsey, G., Abou-Sayed, O.A., Abou-Sayed, A. S., 2021. Evaluation of annulus pressure buildup during injection operations. Journal of Energy Resources Technology. 143, 073002.
3. Liu, Z., Liu, C., Zhang, W., Wang, Y., Cao, J., 2022. Modeling fluctuating pressure in the eccentric annulus with a four-parameter rheological method. Energy Reports. 8, 1405-1417.
4. Doble, R., Mallants, D., Huddlestone-Holmes, C., Peeters, L.J.M., Kear, J., Turnadge, C., Wu, B., S., Noorduijin, E.A., 2023. A multi-stage screening approach to evaluate risks from inter-aquifer leakage associated with gas well and water bore integrity failure. Journal of Hydrology. 618, 129244.
5. Chu, S., Viswanathan, H., Moodie, N., 2023. Legacy well leakage risk analysis at the Farnsworth unit site. Energies. 16, 6437.
6. Das, A.K., Hassanzadeh, H., 2023. Analytical solutions for brine invasion across caprock and leakage through an abandoned well in CO_2 storage sites. Journal of Hydrology. 625, 130046.
7. Askar, A.H., Illangasekare, T.H., 2023. Controlling potential far-field brine leakage from CO_2 storage formations using deep extraction wells: Numerical and experimental testing. Greenhouse Gases: Science and Technology. 13, 3-30.
8. Shen, X., Liu, H., Zhang, Y., You, L., Guo, M., Ma, L., 2022. An integrated model for carbon geosequestration considering gas leakage. Journal of Petroleum Science and Engineering. 217, 110899.
9. Alves, A.L.R., Netto, T.A., 2021. Probability of uncontrolled external leakage during the production phase of a subsea well. Journal of Offshore Mechanics and Arctic Engineering. 143, 011702.
10. Pham, H.H.T., Lohne, H. P., Arild, Ø., Schlenk, D., Pampanin, D. M., 2021. Modeling of environmental fate and effects of oil leakages from abandoned subsea wells using an environmental impact factor tool. Integrated Environmental Assessment and Management. 17, 626-638.
11. Zhang, B., Xu, Z., Lu, N., Liu, H., Liu, J., Hu, Z., Xie, J., Wang, K., 2021. Characteristics of sustained annular pressure and fluid distribution in high pressure and high temperature gas wells considering multiple leakage of tubing string(Article). Journal of Petroleum Science and Engineering. 196, 108083.
12. Moeinikia, F., Ford, E.P., Lohne, H.P., Arild, O., 2018. Leakage calculator for plugged-and-abandoned wells(Article). SPE Production and Operations. 33, 790-801.

 https://doi.org/10.1051/e3sconf/202452802008

Research on Intelligent Management of New Integrated Distribution Network: An Example of New Energy Engineering Management System

Zhaokai Liang[1], Yaodong Lao[2], Yihao Luo[3,*]

[1]School of Materials and Energy, Guangdong University of Technology, Guangzhou,Guangdong,China
[2]Guodian Investment (Jiangmen) Energy Development Co., Ltd.,Kaiping,Guangdong, China
[3]School of Materials and Energy, Guangdong University of Technology, Guangzhou,Guangdong,China

Abstract. The rapid development of new energy makes the distribution network project face higher management requirements, the traditional distribution network project mainly focuses on transmission and transformer unidirectional circulation, to today's optical storage and charging, source network and load storage, micro-grid, virtual power plant and other "multi-terminal active module" together to form a multi-directional concurrency of the new distribution network. If the distribution stage is not well coordinated, the existing distribution network may be impacted. In order to improve the management pain points, this paper first discusses the status quo and pain points of traditional distribution network project management, and then takes a self-developed new energy project management system as a typical case, which is expected to explore a management system paradigm applicable to the new integrated distribution network project, and how to integrate the old and new energy sources from the perspective of project management, which will indirectly have a positive effect on the growth of the investment benefits.

1. Introduction

New energy sources including sun, wind, biomass, geothermal, ocean, and so on are referred to as new energy. Encouraging energy transformation and green development is a calculated decision. China's installed capacity for power generation from non-fossil energy sources accounted for 49.6% in 2022[1-2]. New energy modules have grown to be a significant component of distribution networks, which are evolving from the conventional "passive network" to the "active multi-terminal network." 77% of China's new installed capacity in 2023 was grid-connected new energy installed capacity[3], demonstrating the strong collaboration between distribution network engineering and new energy engineering.

To be able to accomplish the efficient grid connection, transmission, distribution, and consumption of new energy, the development of new energy necessitates the connection of the corresponding new integrated distribution network engineering[4]. Its attributes include a flexible and dynamic operation mode, a diverse and elastic user base, and a variety of distributed and variable power supply sources[5]. This increases the demands and difficulties for distribution network engineering planning, design, construction, operation, and maintenance.

In the past, traditional distribution network engineering primarily relied on manual methods to manually follow up on the data, materials, progress, changes, personnel, and settlement aspects of the engineering[6]. These methods included dispersed paper materials, electronic documents and tables, instant communication, and other methods. In the past, there have been significant drawbacks to this type of management approach, especially difficulty in adjusting to the engineering of the distribution network[7] after the addition of new energy modules. This has resulted in a major bottleneck for the undertaking capacity, project turnover speed, and subsequent income of the construction unit.

2. Method

This research aims to address these issues by focusing on a photovoltaic storage and charging new energy integration investment project in a large industrial zone. As shown in Table 1,two sections that are similar in scale but independent of one another are chosen as typical cases for analysis and contrast experiments. The goal is to realize the smart implementation of new integrated distribution network engineering and improve the management level of new energy distribution network engineering.

*Corresponding author's e-mail: luoyihao@gdut.edu.cn

Table 1. The table of implementation strategies and countermeasures

Factor	Countermeasure
A variety of sectors, areas, tools, and data are used in new energy engineering, and this data could be mined and evaluated for valuable insights[8].	bcentralized multi-field data mining, presentation, and governance
To include novel equipment like optical storage and "soft equipment" like virtual power plant algorithms[9]. Materials and equipment complexity vastly beyond traditional distribution network engineering complexity.	Label-based mixed-material hardware and software management
The attributes of "stage investment" "agile iteration" "step-by-step benefit"[10] and other terms related to new energy engineering establish the necessity of coordinating a large number of dispersed generation, supply, and use objects in real time and achieving amicable cooperation between different source, grid, load, and storage elements.	coordination of several engineering divisions and subprojects Synchronous or asynchronous changes and automatic association
The new energy engineering may have synchronous and asynchronous changes, that is, the same change may have different impacts on different parts at different time points, which requires effective management of the relevance and timeliness of the changes	Personnel safety and qualification, real-time allocation and status monitoring, and stopping unauthorized construction
A shortage of unified management and evaluation among new energy engineering personnel results in uneven personnel quality. Additionally, since multiple engineering units and asynchronous construction types are frequently involved in new energy engineering, personnel are numerous and dispersed.	Personnel safety and qualification, real-time allocation and status monitoring, and stopping unauthorized construction
The capacity of modern energy engineering to produce electricity has a significant impact on the distribution network's installation due to the benefits' distribution and the assessment and forecast[11] of power generating output.	Complete process accessible cost calculation and resolution

3. Discussion

To finish the first part of the project and acquire the blank control group data, we first employ the old management mode, which entails using the old system in conjunction with manual methods; the new system is then used for management in the second portion. As described in Tables 2, 3, 4, 5, 6, and 7,we gather use data, do statistical analysis, and make comparisons. Thirty days after the first and second portions go into operation, respectively, some data (such the cost and settlement module) are collected and compared.

3.1 Data Governance Module

Table 2. Comparison before and after using the new system

Data indicator	Before improvement	After improvement	Change	Improvement point
Instant messaging (cell phone, SMS, chat program, etc.; number/week)	100+	20-	-80%	There's no need to utilize instant messaging apps for working interactions.
Unification ratio of data label format	50%	100%	0.5	Label and group by category all information on materials, projects, employees, adjustments, costs, and settlements.
Cross-role data transmission time (average duration/hour)	3	0.5	-500%	All data is accessible at all times as long as the network is stable.

3.2 Material management module

Table 3. Comparison before and after using the new system

Data indicator	Before improvement	After improvement	Change	Improvement point
Material arriving and departing warehouse audit time (average time/h)	5	0.5	-90%	One click is all it takes for construction workers to apply for materials, and for warehouse employees to enter and exit the warehouse.
Inventory of materials and mistake rate (%) for arriving and departing warehouses	2%	0.50%	-75%	Reduce human mistake by choosing pre-programmed material labels on the system.
Rate of material waste and loss (%)	1%	0.50%	-50%	Each material is traceable, and important materials are location-specific.

Types of manageable materials	70%	95%	25%	Assist with the system label administration of new energy "soft materials"

3.3 Project Coordination Module

Table 4. Comparison before and after using the new system

Data indicator	Before improvement	After improvement	Change	Improvement point
Cost of project follow-up time (average man-hour/h)	24	8	-66.70%	All subprojects and subengineering could be added, deleted, modified, and checked in one place.
Rate of timely project execution (%)	80%	95%	0.15	increased productivity despite having fewer managerial staff members

3.4 Asynchronous Change Association Module

Table 5. Comparison before and after using the new system

Data indicator	Before improvement	After improvement	Change	Improvement point
Time spent on the project modification process (average man-hour/h)	72	8	-88.80%	Implement multi-unit and multi-role responsibilities with a comprehensive prepared change procedure.
Project alteration post-calculation deviance percentage (%)	23%	5%	-78.30%	Real-time estimation of the amount of time following the modification, with data updated for reference in real-time

3.5 Personnel Allocation and Supervision Module

Table 6. Comparison before and after using the new system

Data indicator	Before improvement	After improvement	Change	Improvement point
Conflict rate (%) of monthly figures for multi-unit construction schedules	12%	0%	-12%	Reduce disputes by flexibly scheduling the building period.
Reach rate (%) for the expiry reminder of the personnel safety regulation	20%	100%	4	After an expiration, the platform limits and suspends work.
Rate of personnel violations (%)	5%	0%	-100%	When a violation reaches a certain threshold, the platform limits and suspends work.
Construction personnel safety regulation score on-site (average score, total score of 12 points)	10	11.5	0.15	The safety regulation late rate is significantly decreased following management.

3.6 Cost and Settlement Module

Table 7. Comparison before and after using the new system

Data indicator	Before improvement	After improvement	Change	Improvement point
Expected/actual investment return for new energy engineering (%)	72%	90%	18.00%	modified in accordance with the dynamic absorption capacity assessment of the distribution network
Cost of design for new energy engineering/total cost after modification (%)	90%	96.50%	0.065	dynamically managed the expense resulting from the modification
Delivery time for completion data (average time/hour)	4	0.5	0.875	All pertinent information is finished on the platform.
Time spent throughout the settlement process score (1–5)	3.5	4.5	0.428	The platform allows all parties' settlement details to be transparently and automatically determined.

4. Conclusion

This research highlights the flaws and inadequacies of the conventional distribution network engineering management method while analyzing the traits and difficulties of modern energy engineering[12]. These issues restrict the effectiveness of new energy engineering in terms of construction, grid connection rate, consumption impact, and benefit. They may even have an influence on the current distribution network.

We also build a native integrated distribution network management system that supports new energy engineering and illustrates the particular application and benefits of the new management mode. It does this by drawing on the findings of earlier research and using two sections of a large-scale science and technology park's photovoltaic storage and charging distribution network engineering and energy investment project as typical cases.

Additionally, there are some limitations in this research[13]. For example, there is still room for improvement and optimization in system design and implementation . More work needs to be done to promote the new type of integrated distribution network engineering implementation management mode's universality, and more consideration of factors like artificial intelligence[14],policies, regulations, and the market environment needs to be brought to the standard new energy engineering management process.

We continuously improve the consumption level and investment output ratio of the new type of integrated distribution network engineering by putting these particular plans into practice. This increases the practical value of new energy engineering from the fundamental perspective of the distribution network and its engineering management. There is no doubt that implementing these steps will hasten the arrival and utilization of new energy, mitigate adverse effects like grid impact, and steer the sector toward greater prosperity in the future.

Acknowledgments

This research was supported by the School of Materials and Energy, Guangdong University of Technology.We would like to express our gratitude to Professor Zhipeng Sun and Lecturer Yihao Luo from the School for their guidance.

References

1. Liu G, Cai B, Li Q, et al. China's pathways of CO2 capture, utilization and storage under carbon neutrality vision 2060[J]. Carbon Management, 2022, 13(1): 435-449.
2. Clark A, Zhang W. Estimating the employment and fiscal consequences of thermal coal phase-out in china[J]. Energies, 2022, 15(3): 800.
3. Yu S, Zhou S, Dai Y, et al. Impacts of RPS and FIT on inter-regional power transmission line layout in China: Considerations of high renewable energy penetration[J]. Energy Policy, 2023, 178: 113615.
4. Eriksson E L V, Gray E M A. Optimization and integration of hybrid renewable energy hydrogen fuel cell energy systems–A critical review[J]. Applied energy, 2017, 202: 348-364.
5. Sperstad I B, Degefa M Z, Kjølle G. The impact of flexible resources in distribution systems on the security of electricity supply: A literature review[J]. Electric Power Systems Research, 2020, 188: 106532.
6. Pilo F, Celli G, Ghiani E, et al. New electricity distribution network planning approaches for integrating renewable[J]. Wiley Interdisciplinary Reviews: Energy and Environment, 2013, 2(2): 140-157.
7. Zeng B, Zhang J, Yang X, et al. Integrated planning for transition to low-carbon distribution system with renewable energy generation and demand response[J]. IEEE Transactions on Power Systems, 2013, 29(3): 1153-1165.
8. Zhou K, Fu C, Yang S. Big data driven smart energy management: From big data to big insights[J]. Renewable and sustainable energy reviews, 2016, 56: 215-225.
9. Kasaei M J. Energy and operational management of virtual power plant using imperialist competitive algorithm[J]. International Transactions on Electrical Energy Systems, 2018, 28(11): e2617.
10. Ullah S, Haidar A M A, Hoole P, et al. The current state of Distributed Renewable Generation, challenges of interconnection and opportunities for energy conversion based DC microgrids[J]. Journal of Cleaner Production, 2020, 273: 122777.
11. Dunbar A, Tagliaferri F, Viola I M, et al. The impact of electricity price forecast accuracy on the optimality of storage revenue[C]//3rd Renewable power generation conference (RPG 2014). IET, 2014: 1-6.
12. Georgilakis P S, Hatziargyriou N D. A review of power distribution planning in the modern power systems era: Models, methods and future research[J]. Electric Power Systems Research, 2015, 121: 89-100.
13. Nissen U, Harfst N. Shortcomings of the traditional "levelized cost of energy"[LCOE] for the determination of grid parity[J]. Energy, 2019, 171: 1009-1016.
14. Yang S, Zhang Z, Cao R, et al. Implementation for a cloud battery management system based on the CHAIN framework[J]. Energy and AI, 2021, 5: 100088.

Research on the law of hydrogen production by supercritical water gasification of oily sludge

Yuanwang Duan, Shuzhong Wang *, Yanhui Li, Yulong Wang

Key laboratory of Thermo-Fluid Science and Engineering of MOE, School of Energy and Power Engineering, Xi'an Jiaotong University, No. 28 Xianning West Road, Xi'an 710049, China

Abstract. As China's oil consumption continues to increase, the amount of oily sludge generated increases yearly. Aiming at the limitations of the traditional oil-containing sludge treatment technology, supercritical water is used to dissolve and convert organic components into hydrogen for hydrogen energy recovery to realize the reduction, harmlessness, and resourcefulness of the oily sludge treatment. The influences of reaction temperature, pressure, time, material ratio, and other parameters on hydrogen production from supercritical water gasification of oily sludge were experimentally examined. Results indicated that hydrogen yield increases with increased response temperature, pressure, and time, and hydrogen yield decreases with increased response mass ratio.

1. Introduction

Currently, a considerable amount of oily sludge is generated in China in the process of crude petroleum extraction, gathering and transportation, and oil refining, and the production of oily sludge is increasing yearly. The composition of oily sludge is very complex, and the on-site sludge often contains crude oil, emulsion breakers, viscosity reducers, inorganic salts, and a small number of heavy metal compounds. It also contains benzene, phenolics, anthracene, and other toxic substances with a lousy odor[1]. When oily sludge is not handled correctly, it will pollute the surrounding environment and deteriorate the ecological environment[2].

The generation of oily sludge is unavoidable, and oily sludge contains a high number of hydrocarbons, metals, inorganic salts, etc., which contains a large amount of hydrocarbons is also a vital resource if the direct reduction treatment not only causes a resource waste but also lead to CO_2 pollution. Traditional oily sludge treatment technologies include fluidization, conditioning, centrifugation, solvent extraction, electrochemical treatment, ultrasonic treatment, membrane separation technology, etc. However, each sludge treatment technology has certain defects, such as the shortcomings of fluidization-conditioning-centrifugal technology treatment, which has a low sludge treatment capacity, and the process being more complex. In solvent extraction technology, many chemicals are added, and the excess chemicals must be reused later, which increases the operating cost of oily sludge. In treating oily sludge, Membrane separation technology can only block the solid phase components of large particles. It cannot upgrade the quality of oil products, so the subsequent waste membrane must still be treated. The disadvantage of electrochemical treatment technology is that too much sludge treatment will lead to an unstable electric field and poor electric dehydration quality when recovering sludge. The disadvantage of ultrasonic treatment technology is that the requirements for ultrasonic sound intensity processing time and other parameters are relatively strict. Otherwise, it is easy to cause re-emulsification [3].

Supercritical water gasification hydrogen production technology places organic matter in supercritical water, and supercritical water not only provides a good reaction environment for organic solvents but participates in hydrogen production reaction as a reactant[4]. The advantage of this technology is that the reactants are water and organic matter, and new drugs do not need to be added to the reactants with high water content, avoiding the shortcomings of the traditional treatment technology that drugs are not easy to separate and the cost of dehydration is high[5].

It is considered to use supercritical water to treat oily sludge, dissolve organic matter in oily sludge, and convert it into hydrogen. Firstly, the cost of dewatering traditional oily sludge can be avoided; secondly, oily sludge can produce a large amount of combustible gas such as hydrogen to achieve energy recovery; finally, organic components in solid phase residue after supercritical water treatment can be rapidly reduced. The oily sludge's reduction, harmlessness, and resorization are completely realized [6]. However, because oily sludge contains a significant amount of crude petroleum and other organics, and different organic waste gasification laws have different impacts on hydrogen production, this paper examines the feasibility of hydrogen production from oily sludge gasification, optimizes the reaction parameters of hydrogen generation from oily sludge gasification, and

*Corresponding author's e-mail: SZWang@aliyun.com

explores the experimental mechanism of hydrogen production from gasification. It offers theoretical and experimental evidence for applying supercritical water gasification of oily sludge for hydrogen generation.

2. Experimental stage

The crude oil is obtained from the field. As shown in **Figure 1**. To ensure the repeatability and uniformity of the extracted crude oil, the crude oil is pre-treated before the experiment. The specific method is: the crude oil of the Xinjiang oilfield is placed in a grinding bottle with a grinding bottle stopper, and the grinding bottle is placed in the oven. Slowly heat to 80°C, constant temperature for 2 hours, remove and put into vacuum drying box to cool to chamber temperature. Then, please take out the grinding bottle and put it in a place with little difference in ambient temperature for 48 hours to complete the crude oil pretreatment[6].

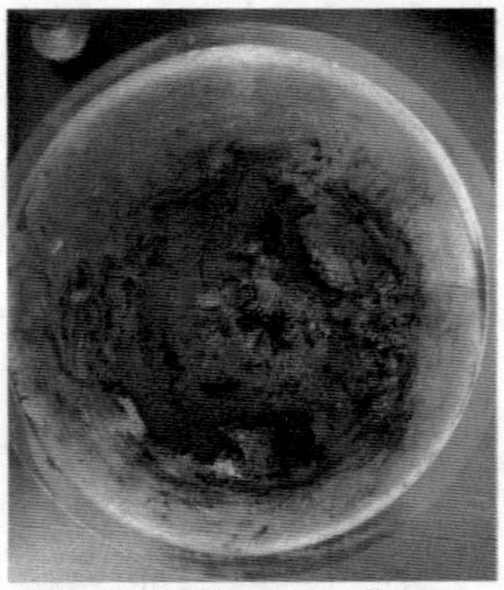

Figure 1. Oily sludge diagram.

After adding the quantitative oily sludge and distilled water into the reactor, tighten the kettle cover bolts, connect the argon cylinder, adjust the pressure of the argon cylinder to 1MPa, and slowly open the exhaust valve to purge the inside of the reactor for 5 min. Then, close the vent valve and pressurize the kettle at the original pressure of the experimental design scheme. After adding to the experimental pressure, close the air inlet valve and gas cylinder, carry out heating according to the experimental design program, keep the heating rate each time at 400°C/h, and calculate the heating time according to the required reaction temperature. The heating time inside the reaction kettle is usually 20-30 min longer than the heating time set by the heating furnace. When the temperature at the top of the reaction kettle reaches the set temperature, start timing, stop heating after reaching the target time, remove the reaction kettle and rapidly cool it down to room temperature, then collect the gases from the exhaust port by applying the drainage collection method, and then introduce it into the gas chromatograph for analysis, and ultimately get the percentage of the various gas-phase components in the product.

2.1. Influence of temperature on the hydrogen constant

The variation of experimental results of hydrogen production from the supercritical water gasification of oily sludge at 400°C~ 544°C was investigated under initial reactor pressure of 1.8 MPa, time of 60 min, and constant concentration of 10%. From **Figure 2**, It was observed that the hydrogen production increased from 0.128 mmol/g to 3.179 mmol/g with the increase in reaction temperature. Methane yield increased from 0.21 mmol/g to 6.62 mmol/g, carbon dioxide yield increased from 0.704 mmol/g to 2.691 mmol/g, and carbon monoxide yield decreased to 0.002 mmol/g at 490 °C. The results showed that hydrogen, methane, and carbon dioxide yields increased significantly with increasing temperature, and carbon monoxide yield increased and then decreased[7-8].

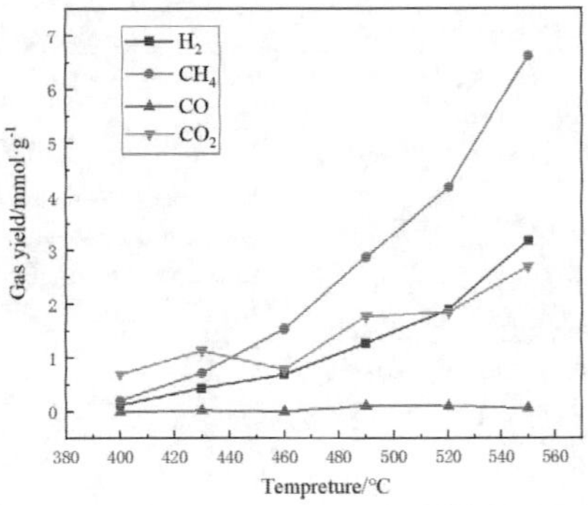

Figure 2. Temperature on gas yield of different species.

2.2. Influence of pressure on the hydrogen constant

From the results in **Figure 3**, it can be shown that the hydrogen production increases from 3.89 mmol/g to 4.46 mmol/g with increasing reaction pressure. methane production decreases from 6.61 mmol/g to 6.34 mmol/g and then rises to 6.78 mmol/g with pressure production. CO_2 production is strongly affected by the pressure production rate, increasing from 7.59 mmol/g to 2.84 mmol/g and then to 2.95 mmol/g. The pressure had the least effect on the CO yield in the range of 0.186 mmol/g ~ 0.009 mmol/g. As the reaction pressure increased, the hydrogen yield gradually increased, the methane and carbon dioxide yields decreased and then increased, and the carbon monoxide yield decreased.

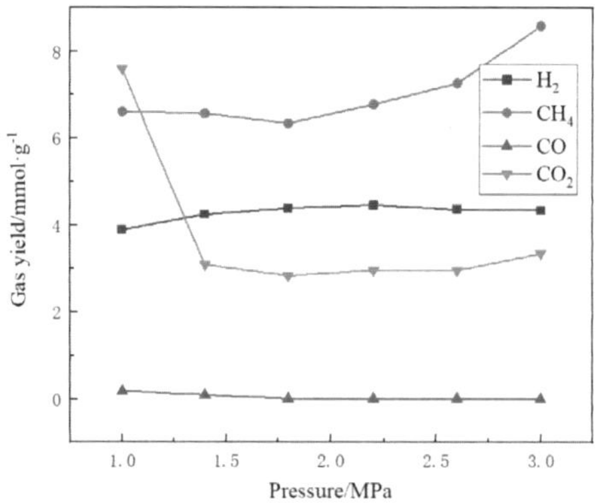

Figure 3. Effect of reaction pressure on gas production

2.3. Influence of reaction time on the hydrogen constant

As shown in **Figure 4**, the reaction time varied from 15 to 150 min, and the hydrogen production per unit of sludge increased from 2.38 mmol/g to 5.92 mmol/g as the reaction time increased. Methane production also increased dramatically from 4.90 mmol/g to 8.32 mmol/g. CO production increased from 0.072 mmol/g at 20 min to CO production increased from 0.072 mmol/g at 20 min to 0.0025 mmol/g at 150 min. 150 min co2 yield increased from 1.90 mmol/g to 4.05 mmol/g. With the increase in reaction time, the hydrogen and methane yields significantly increased, the carbon dioxide yield increased, and the carbon monoxide yield decreased.

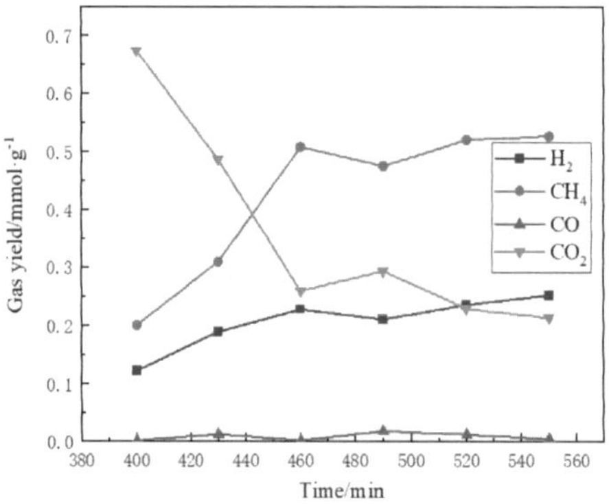

Figure 4. Effect of reaction time on gas production.

2.4. Influence of Material Ratio on the Hydrogen Constant

As seen in **Figure 5**, along with the increase in the material ratio from 10% to 100%, the hydrogen production per unit of sludge decreased from 5.86 mmol/g to 1.65 mmol/g. The methane production increased from 8.29 to 9.71 mmol/g. The carbon monoxide production increased from 0.0012 to 0.44 mmol/g and then decreased to 0.25 mmol/g. The carbon dioxide production increased from 13.67 to 2.46 mmol/g. The carbon dioxide production decreased from 0.0012 to 0.44 mmol/g. Carbon dioxide production decreased from 13.67 mmol/g to 2.46 mmol/g finally.

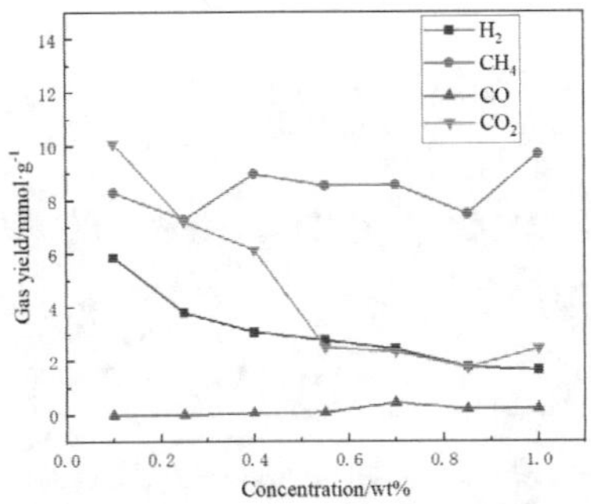

Figure 5. Effect of material ratio on gas production.

3. Conclusion

In this thesis, the influence of each reaction factor on the supercritical water gasification hydrogen production experiment of oily sludge was investigated by using a one-way experimental design, and the specific conclusions are as follows:

(1) The reaction temperature plays a role in promoting the hydrogen yield experiment of oily sludge. Accompanied by the increase of reaction temperature from 400°C to 544°C, hydrogen yield increased from 0.128 mmol/g to 3.179 mmol/g.

(2) The reaction pressure played a facilitating role in the hydrogen yield experiments with oily sludge. Accompanied by the increase of reaction pressure from 1.0 MPa to 2.2 MPa, hydrogen yield increased from 3.89 mmol/g to 4.46 mmol/g.

(3) The reaction time plays a facilitating role in the oily sludge hydrogen production experiment. Accompanied by the increase in reaction time from 15 min to 150 min, hydrogen yield increased from 2.38 mmol/g to 5.92 mmol/g.

(4) The reaction material ratio inhibited the hydrogen yield experiment of oily sludge. Along with the increase of reaction material ratio from 10% to 100%, the hydrogen yield of 5.86 mmol/g

(5) This paper mainly explores the feasibility of hydrogen production from supercritical water gasification of oily sludge and finds that oily sludge can still produce a large amount of hydrogen in supercritical water. On this basis, the hydrogen production law under different reaction parameters is explored, which provides a theoretical basis for realizing hydrogen production from supercritical water gasification of oily sludge. The following research direction should be to solve the energy consumption problem of heating water and oily sludge to a supercritical state, which can be combined with supercritical hydrothermal combustion technology to produce energy after oily sludge combustion to realize the whole process of treating and utilizing oily sludge.

Acknowledgments

This work was financially supported by the National Natural Science Foundation of China (22008190; 52176162; 22378322) and the China Postdoctoral Science Foundation (2022M722526).

References

1. Hoang S A , Bolan N , Madhubashani A M P ,et al.Treatment processes to eliminate potential environmental hazards and restore agronomic value of sewage sludge: A review[J].Environmental Pollution, 2022(Jan.):293
2. Li Y , Feng H , Wang J ,et al.Current status of the technology for utilizing difficult-to-treat dust and sludge produced from the steel industry[J].Journal of cleaner production, 2022.
3. Chang X , Li X , Ge S .Insight into the oil removal mechanism of quaternary ammonium ionic liquid microemulsions for oily sludge treatment[J].Sustainable Energy & Fuels, 2023.
4. Liang X, Zhao Q, Dong Y, et al. Experimental Investigation on Supercritical Water Gasification of Organic-Rich Shale with Low Maturity for Syngas Production[J], Energy & Fuels, 2021, (9):35.
5. Lu Yj, et al. "Hydrogen production by biomass gasification in supercritical water over Ni/γAl_2O_3 and Ni/CeO_2-γAl_2O_3 catalysts[J], " International Journal of Hydrogen Energy (2010).
6. Wang F , Zhu S , Gong X .Gasification of oily sludge in supercritical water[J].Oxidation Communications, 2015, 38(3):1391-1400.
7. Ibtissem H , Nawel O , Hassen M A ,et al.Supercritical water gasification of glycerol for Hydrogen production using response surface methodology[J]. 2019.
8. Zaker, Azadeh , et al. "Chemicals from heavy oils by ZSM-5 catalysis in supercritical water: Model compound and reaction engineering." AIChE Journal 66.7(2020).

Study on synthesis and properties of environmental protection water treatment agent

Zhan Liu*, Meifang Yan, Yuhua Gao, Na Li, Haihua Li, Zhenfa Liu

Institute of Energy Sources, Hebei Academy of Science, Shijiazhuang Hebei 050081, China

Abstract. The copolymers were obtained by a water-based polymerization reaction using maleic anhydride and acrylic acid as the main raw materials. And the scale inhibition performance of the resulting scale inhibitor was investigated by static scale inhibition experiments. The experimental results proved that the scale inhibitor has good scale inhibition performance for calcium carbonate and calcium sulfate. When the dosage is 100mg/L, the scale inhibition rate of the agent reaches 100%. This avoids the hidden danger of affecting normal production due to system scaling.

1. Introduction

Fouling is one of the common problems in the circulating water system[1], the ionic oil bath in the circulating water is concentrated and enriched after fouling, it will cause system blockage problems, affecting normal production[2-4]. At present, the scale inhibitors used in industry are generally based on phosphorus-containing agents, phosphorus with the wastewater discharged into the water body is easy to cause eutrophication of the water body, is not conducive to the sustainable development of the environment[5-7].

In this paper, maleic anhydride, acrylic acid as the main raw material, in the role of the initiator synthesized an excellent performance of the scale inhibitor, its calcium carbonate, calcium sulfate have good scale inhibition, and does not contain phosphorus, belongs to a kind of environmentally friendly green water treatment agent.

The use of this agent avoids the hidden danger of affecting normal production due to system scaling, and does not cause water eutrophication and other water environment problems caused by the large amount of phosphorus enrichment. The agent can be widely used in chemical, pharmaceutical, petroleum and other industries, the sustainable development of the environment has great significance.

2. Experimental part

2.1 Apparatus and drugs

Maleic anhydride (MA), acrylic acid (AA), ammonium persulfate, isopropanol, sodium bicarbonate, anhydrous calcium chloride, sodium tetraborate, sodium sulfate, etc., the above agents are analytically pure.

Electronic balance, constant temperature magnetic stirrer, electric thermostatic water bath, syringe pump, etc.

2.2 Synthesis of copolymers

After dissolving a certain amount of MA and distilled water in a four-necked flask, the temperature was raised to 75°C and then AA was added, the chain transfer agent isopropanol was added, while N_2 was passed through the four-necked flask, and after a few minutes, an aqueous solution of ammonium persulfate was injected into the four-necked flask using a syringe pump. Holding and stirring at a certain temperature for 6 hours, a yellow liquid was obtained as the final product. After washing with anhydrous ethanol and drying, a solid product was obtained as the purified copolymer.

2.3 Scale inhibition properties of polymers on calcium carbonate

Referring to GB/T16632-2008, add the prepared water containing certain chemicals into the container. The container was then held at 80°C for 10 hours. After removal and cooling, the calcium ion concentration in the upper layer of the clear liquid in the container was titrated, and the scale inhibition rate η_1 was calculated.

$$\eta_1 = \frac{V_1 - V_0}{V_2 - V_0} \times 100\% \tag{1}$$

Where: V_0 (mg/L) - Calcium ion concentration in the blank solution after the experiment.

V_1 (mg/L) - Calcium ion concentration after the experiment during dosing.

V_2 (mg/L) - Ion concentration of calcium in raw water samples at room temperature.

*Corresponding author's e-mail address:651882920@qq.com

https://doi.org/10.1051/e3sconf/202452802010

2.4 Determination of scale inhibition performance of calcium sulfate

The scale inhibition performance of calcium sulfate is determined by referring to the petroleum standard SY/T5673-93, adding a certain amount of synthesized agent to 500 mL of prepared water (Ca^{2+}=5000 mg/L , HCO_3^-=7500 mg/L (in terms of $CaSO_4$)), the resulting solution was thermostated in a water bath at 70°C for 25 h. The concentration of calcium ions in the cooled solution was titrated with a certain concentration of EDTA solution, and the scale inhibition rate of the copolymer on calcium sulfate was calculated as η_2.

$$\eta_2 = \frac{X_1 - X_0}{X_2 - X_0} \times 100\% \quad (2)$$

Where: X_0 (mg/L) - calcium ion concentration after heating without adding chemicals

X_1 (mg/L) - Calcium ion concentration after heating with additives

X_2 (mg/L) -Concentration of calcium ions in the original water sample at the room temperature

2.5 Evaluation of iron oxide dispersion performance

Preparation water containing polymers (150 mg/L $CaCO_3$ for calcium and 10 mg/L for iron ions) is added to the vessel with strong stirring and then held in a 50°C water bath for 5 hours. After cooling the transmittance of the upper layer of the solution is measured (3 cm cuvette, wavelength 450 nm).

2.6 Evaluation of biodegradation performance

The biodegradability of the agent was determined by using the biological shaking table experimental method, 500ml of inoculum and aqueous solution containing a certain amount of agent were added to the conical flask used for the experiment, and then placed in the constant temperature oscillating shaking table at 25°C, and the chemical oxygen demand of the solution, i.e., COD was determined at 1, 7, 14, 21 and 28 days, respectively. The calculation can be made from the following formula to get the biodegradation rate of the agent.

$$\text{Biodegradation rate} = (1 - \frac{C_t - C_{bt}}{C_0 - C_{b0}}) \times 100\% \quad (3)$$

Where: C_{bt} (mg/L) - COD in the inoculated reaction solution without agent at time t.

C_t (mg/L) - COD in the inoculation reaction solution containing the agent.

C_{b0} (mg/L) - Initial value of COD in the inoculation reaction solution without pharmaceutical agent.

C (mg/L)- COD in the inoculation reaction solution with agent.

3. Results and discussion

3.1 Study on the scale inhibition performance of agents on calcium carbonate

The agent will be tested for calcium carbonate inhibition performance, to determine the scale inhibition performance of its different concentrations, the results are shown below. As can be seen from Figure 1, the scale inhibition rate of the agent on calcium carbonate increases with the increase of dosage, and the scale inhibition rate reaches 100% when the dosage of the agent is 10mg/L. This indicates that the synthesized agent has a good scale inhibition performance on calcium carbonate. This indicates that the synthesized polymers have high scale inhibition properties against calcium carbonate.

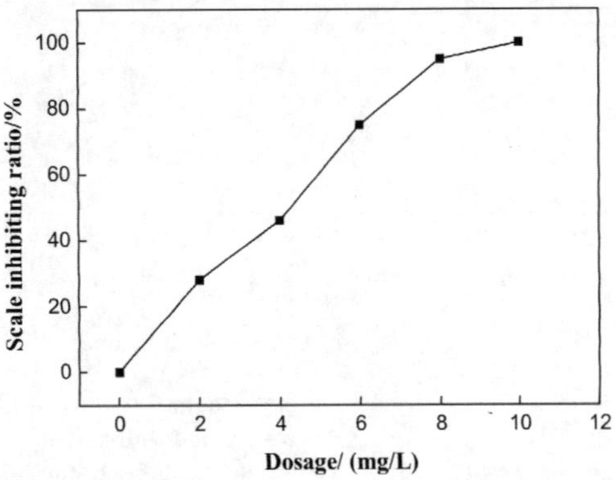

Figure 1. Scale inhibition of calcium carbonate by the polymer.

3.2 Research on scale inhibition of calcium sulfate by pharmaceutical agents

The scale inhibition performance of calcium sulfate is tested as shown in Figure 2 that the scale inhibition rate of calcium sulfate is excellent, and the scale inhibition rate of calcium sulfate can reach 100% when the dosage of the agent is 8mg/L.

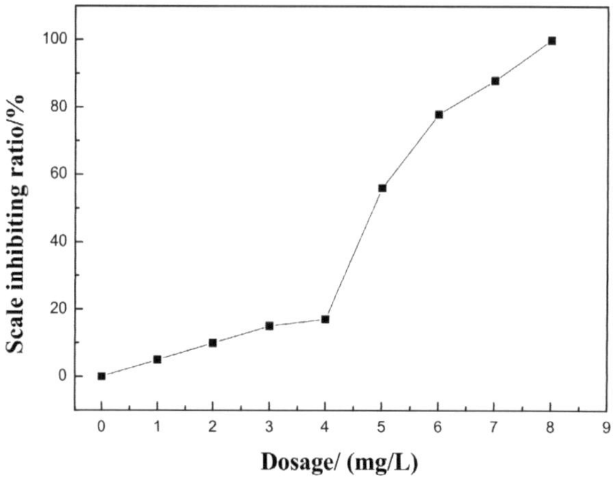

Figure 2. Scale inhibition of calcium sulfate by the polymer.

3.3 Study on the dispersing properties of agents on iron oxide

The dispersion performance of the agent was investigated, and the product showed good dispersion performance after the two monomers were copolymerized. As can be seen in Figure 3, when the dosage was increased to 100 mg/L, the dispersion effect of the agent reached the best, and the transmittance of the solution reached 47%. At this time, the groups contained in the agent and the iron ions in the solution chelate, so that the iron ions will not precipitate in the form of precipitation, to achieve the role of dispersion.

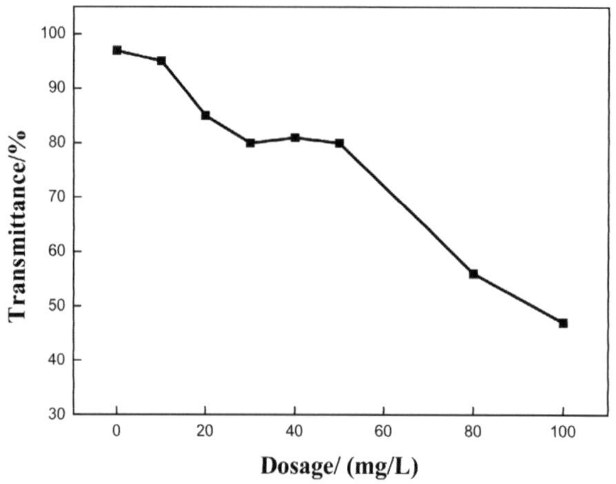

Figure 3. Study on the dispersing performance of the pharmaceutical agent

3.4 Degradability study of the agent

The results of the biodegradability tests are shown in Figure 4. It can be seen that the biodegradation rate of the agent increases with the prolongation of the number of days, in the 28th day, its degradation rate can reach 60%, which indicates that the agent belongs to the easy biodegradability of the substance, can avoid the long-term use of the water body caused by the problem of eutrophication and pollution of the water environment and so on.

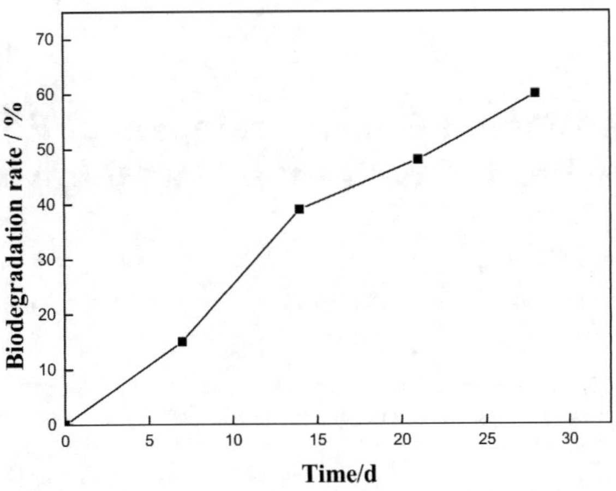

Figure 4. Biodegradability of the agent.

4. Conclusion

(1) A green water treatment polymer agent that is very friendly to the environment -MA/AA copolymer was synthesized using MA and AA as raw materials.

(2) The static experiment was used to determine a number of properties of the agent, the scale inhibition rate of calcium carbonate and calcium sulfate can reach 100%, and the biodegradation rate of the agent was 60% in 28 days, indicating that MA/AA copolymer has good scale inhibition and dispersion properties and biodegradation properties, and is an effective water treatment agent.

(3) The agent does not contain phosphorus is an environmentally friendly green water treatment agent, the use of the agent, to avoid the discharge of phosphorus-containing wastewater, environmentally friendly, to avoid the long-term use of the agent caused by the eutrophication of the water body and the water environment pollution and other issues, has a very good prospect for development.

Acknowledgments

This article was supported by the Science and Technology Program of Hebei Academy of Sciences (No. 24702).

References

1. Liu Zhan, Wang Chongpu, Li Na et al. Synthesis and research of a phosphorus and nitrogen-free circulating cooling water treatment chemicals. Applied Chemical Engineering, 2023, 52(07): 2068-2071+2079.
2. Wang Meiqiang, Li Yanqi, Lin Yuanping et al. Preparation and mechanism of MA/AA/DETA terpolymer corrosion and scale inhibitor. Fine Petrochemicals, 2024, 41(01): 10-14.
3. Guo Shunliang. Synthesis and research on phosphine-containing copolymer scale and corrosion inhibitors. Lanzhou Jiaotong University, 2024.
4. Jia Yu, Ping Shuangrong, Xu Xiangdong et al. Research and application of new phosphorus-free furnace water treatment agent for industrial boilers. Petrochemical Technology, 2023, 30(10):4-6.
5. Zhang Qing, Wang Lulu. Research on environmentally friendly low phosphorus recycled water treatment technology and its application. Cleaning World, 2023, 39(11): 47-49.
6. BEAULIEU J J, DELSONTRO T, DOWNING J A. Eutrophication will increase methane emissions from lakes and impoundments during the 21st century. Nature Communications, 2019, 10(1): 1375.
7. MARTINOD A, NEVILLE A, EUVRAD M, et al. Electrodeposition of a calcareous layer: Effects of green inhibitors. Chemical Engineering Science, 2009, 64(10): 2413 2421.

E3S Web of Conferences **528**, 02011 (2024)
NET-LC 2024
https://doi.org/10.1051/e3sconf/202452802011

Non-Cooperative Bidding Game Strategies of Power Generation Companies in the Electricity-Carbon Market Environment

Qingkai Sun*, Xiaoxuan Zhang, Zheng Zhao, Menghua Fan, Chen Lv

State Grid Energy Research Institute Co., Ltd., Beijing 102209, China

Abstract. This research scrutinizes the non-cooperative bidding strategies of power generation firms within the integrated electricity and carbon market framework. It aims to furnish pivotal insights for market strategy formulation and underpin policy development toward a greener power sector. Key innovations include a non-cooperative game model elucidating the interplay and competitive conduct among power providers in dual markets, alongside an empirical analysis showcasing how strategic bidding influences economic and environmental outcomes. This study significantly contributes to understanding market dynamics in promoting sustainable and environmentally sound power systems.

1. Introduction

In the context of escalating global climate change and environmental conservation challenges, the research and development, as well as the application of sustainable energy, have emerged as focal points of international attention [1]. Within this framework, power generation companies, serving as principal energy suppliers, play a pivotal role in enhancing energy efficiency and diminishing environmental impacts [2]. These entities are tasked with the dual objective of maintaining economic viability amidst fierce market competition while actively adopting measures to reduce carbon emissions, thereby facilitating the development of a low-carbon economy and addressing the challenges posed by climate change.

Currently, research on the market competition strategies of power generation companies, both domestically and internationally, primarily focuses on the singular perspective of the electricity market, with insufficient attention to the interaction between the electricity and carbon markets. For instance, the study in Reference [3] analyzes competition strategies in the electricity market without addressing the potential impact of carbon emissions trading. Conversely, research on carbon emissions trading mechanisms, such as the work presented in References [4] and [5], often overlook the complexities of the electricity market. While these studies provide insights into understanding market dynamics, they exhibit limitations in the context of the intricate environment combining both electricity and carbon markets. In recent years, with the gradual advancement of power systems towards intelligence and increased attention to carbon emissions issues, new research has emerged. For example, Reference [6] investigates the market equilibrium issues of the electricity market under the consideration of carbon market factors. Meanwhile, Reference [7] analyzes the scenario where the electricity and carbon markets operate in coordination. Although these studies offer some understanding of the interaction between the electricity and carbon markets, there is still a need for more in-depth exploration, particularly in terms of the competitive strategies of power generation companies, especially those based on non-cooperative game theory.

This article focuses on the non-cooperative bidding game strategies of power generation companies within the electricity-carbon market environment, aiming to provide critical references for market decision-making by power generation companies and theoretical support for the formulation of energy policies, further promoting the development of the power system towards a more sustainable and environmentally friendly direction.

2. A Model of Electricity-Carbon Coordinated Trading Strategies for Power Generation Companies Based on Non-Cooperative Game Theory

This section aims to delve into the competitive behaviors and bidding strategies of various power generation companies within the electricity and carbon markets through the construction of a model framework based on non-cooperative game theory. The model focuses on exploring the complex interactions between market mechanisms and corporate strategies.

2.1. Theoretical Framework for Electricity-Carbon Market Coordinated Trading Mechanisms

In the model constructed in this section, the coupling operation mechanism between the electricity market and

*Corresponding author's e-mail: sunqingkai_123@163.com

the carbon market is elaborately explained and analyzed. The core of the electricity market's operating mechanism lies in how the relationship between electricity supply and demand influences the market price of electricity.

For power generation companies participating in both markets, the interaction between the electricity and carbon markets introduces a series of complex operational challenges. On one hand, the price formation mechanism in the electricity market directly affects the company's revenue from electricity sales; on the other hand, the trading cost of carbon emission rights in the carbon market directly impacts the company's operational expenses. Against the backdrop of this dual coupled market mechanism, power generation companies must seek an equilibrium between economic benefits and environmental responsibility in their electricity output and carbon emissions. Therefore, when formulating electricity-carbon coordinated trading strategies, power generation companies must comprehensively consider the supply and demand dynamics of the electricity market and the fluctuations in the carbon emission rights market, aiming to optimize their overall economic benefits and environmental impact. The relationship between them is shown in Figure 1 below.

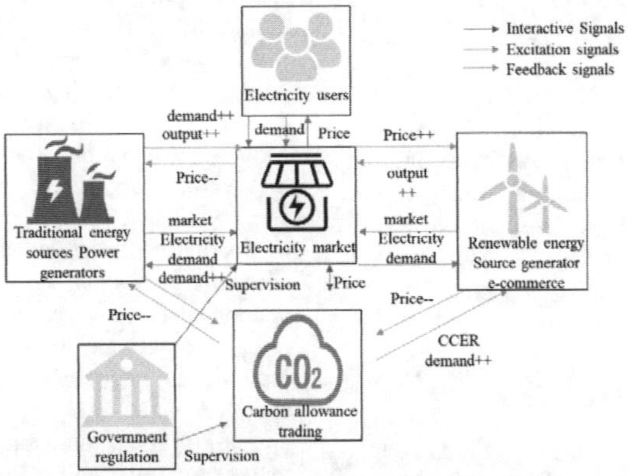

Figure 1. Mechanism framework of power generation enterprises participating in electricity carbon market collaborative trading.

2.2. Non-Cooperative Game Model for Electricity-Carbon Coordinated Trading among Power Generation Companies

In this study, we constructed a model for analyzing the non-cooperative bidding behaviors of power generation companies A and B in both the electricity and carbon markets. Considering the complexity of the actual electricity carbon synergy market environment, this study adopts a simplified form with the aim of analyzing practical problems. The model is based on the following objective function and constraints, which can be specifically expressed as:

(1) The objective function for maximizing the profit of power generation company A:

$$\Pi_A = \sum_{t=1}^{T}(P_t \cdot Q_{A,t} - C_A \cdot Q_{A,t} - E_{A,t} \cdot \alpha_t) \tag{1}$$

Where, Π_A represents the total profit of company A; $Q_{A,t}$ denotes the electricity generation of company A in period t; C_A is the cost per unit of electricity; $E_{A,t}$ represents the carbon emissions of company A in period t; P_t is the selling price of electricity in the power market during period t; α_t is the selling price of carbon emission rights in the carbon market during period t; T is the total number of periods.

(2) The objective function for maximizing the profit of power generation company B:

$$\Pi_B = \sum_{t-1}^{T}(P_t \cdot Q_{B,t} - C_B \cdot Q_{B,t} - E_{B,t} \cdot \alpha_t) \tag{2}$$

Where, Π_B represents the total profit of company B; $Q_{B,t}$ denotes the electricity generation of company B in period t; C_B is the cost per unit of electricity; $E_{B,t}$ represents the carbon emissions of company B in period t.

(3) Constraints

1) Generation Capacity Constraint

The electricity generation of companies A and B must not exceed their maximum generation capacities. This can be specifically expressed as:

$$\begin{cases} 0 \le E_{A,t} \le E_{max,A}, \forall t \\ 0 \le E_{B,t} \le E_{max,B}, \forall t \end{cases} \tag{3}$$

Where, $Q_{max,A}$ represents the maximum electricity generation capacity of power generation company A; $Q_{max,B}$ represents the maximum electricity generation capacity of power generation company B.

2) Carbon Emission Constraint

The carbon emissions from power generation companies A and B must comply with the environmental regulations' limitations. This can be specifically expressed as:

$$\begin{cases} 0 \le E_{A,t} \le E_{max,A}, \forall t \\ 0 \le E_{B,t} \le E_{max,B}, \forall t \end{cases} \tag{4}$$

Where, $E_{max,A}$ represents the maximum carbon emission limit for power generation company A; $E_{max,B}$ represents the maximum carbon emission limit for power generation company B.

3) Market Equilibrium Condition

The supply and demand in both the electricity market and the carbon market should satisfy the market equilibrium. This can be specifically expressed as:

$$\begin{cases} \sum_{i=A}^{B} Q_{i,t} = D_t, \forall t \\ \sum_{i=A}^{B} E_{i,t} = E_t, \forall t \end{cases} \tag{5}$$

Where, D_t represents the total electricity demand in period t; E_t represents the total carbon emission limit in period t.

3. Case Study Analysis

In this study, we simulate the trading behaviors of power generation companies A and B in both the electricity and carbon markets to analyze their game strategies and the impact on market prices and corporate profits. The specific data comes from the actual transaction data of a provincial power trading center in China. The following is a case study analysis based on simulation results:

3.1 Analysis of the Non-cooperative Game Bidding Process at a Typical Moment

The game process of power generation and carbon emissions between Company A and Company B in a typical moment is shown in Figure 2. At the beginning of the game process in hour 14, we observed that the electricity generation and carbon emissions of companies A and B exhibited significant fluctuations. These fluctuations were mainly due to the companies trying different strategies at the early stage of the game to adapt to the uncertainty of market prices and demand. For example, company A's electricity generation started at 250 megawatts and, after several fluctuations, eventually stabilized around 500 megawatts, showing an increase of about 100%. Similarly, company B's electricity generation started from 200 megawatts and, after initial fluctuations, finally stabilized at over 550 megawatts, with a growth rate exceeding 175%. The trend in carbon emissions was similar to that of electricity generation, where company B's carbon emissions grew by more than 150%, while company A's growth was about 100%.

It is noteworthy that as the number of games increased, especially around the 30th game, the electricity generation and carbon emissions of both companies tended to stabilize. This trend indicates that with the accumulation of market information and strategy adjustments, the companies gradually found the optimal response strategies, thereby reducing their sensitivity to market price fluctuations. This phenomenon also reflects the learning and adaptation process of market participants in repeated games, i.e., through continuous trial and error and adjustment, companies can more accurately predict market dynamics and optimize their electricity generation and carbon emission strategies accordingly.

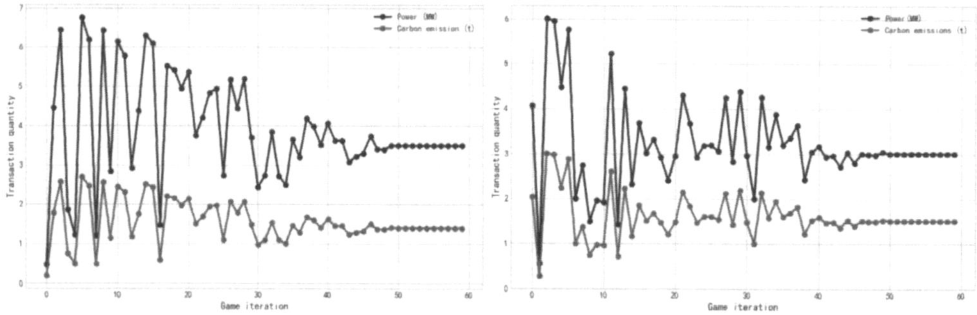

Figure.2 Game Process of Power Generation and Carbon Emissions between Company A and Company B at Typical Moments

3.2 Comparative Analysis of 24h Electricity-Carbon Trading Volume between Power Generation Companies A and B

In the analysis of the 24-hour trading volume, companies A and B exhibited different market participation patterns, especially in terms of electricity and carbon emission trading volumes.

The 24-hour electricity carbon trading volume between Company A and Company B is shown in Figure 3. The analysis of company A's trading volume indicates that it maintained lower electricity generation and carbon emissions during low demand periods, reflecting its robust strategy in controlling operational costs and reducing carbon emissions. However, during high demand periods, such as the 14th and 20th hours, company A significantly increased its electricity generation to nearly 80% of its maximum capacity, with a corresponding increase in carbon emissions. This change suggests that although company A is relatively conservative in its daily operations, it still actively adjusts its strategy to capture higher profit opportunities during market demand peaks. In contrast, company B showed a more aggressive market participation attitude throughout the day. In most periods, company B's electricity generation was close to or reached its maximum capacity, especially during peak electricity demand periods, where its generation even exceeded that of company A, reaching more than 90% of its maximum capacity. Correspondingly, company B's carbon emissions also remained at a higher level throughout the day, particularly during peak electricity demand periods, significantly exceeding those of company A. This reflects its strategy of pursuing market share and profit maximization, even if it means higher carbon emissions and potential environmental costs.

Overall, company A's strategy is more focused on balancing cost control and environmental impact, while company B is more inclined to maximize revenue by seizing market opportunities, even if it entails higher carbon emissions and potential environmental costs. The difference in strategies reflects the diverse trade-offs between market participation, environmental responsibility, and profit pursuit among different companies. Analyzing these trading volumes can provide deeper insights into the behavioral patterns of different companies in response to market dynamics and environmental policies, as well as their underlying strategic considerations.

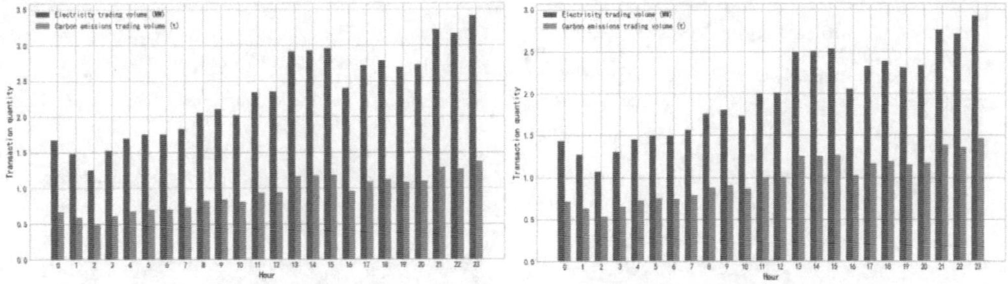

Figure.3 24h electricity carbon trading volume between Company A and Company B

4. Conclusion

This paper focuses on the critical issue of sustainable energy development, delving into the competitive strategies of power generation companies in the electricity and carbon markets, with a particular emphasis on the significant role these strategies play in enhancing energy efficiency and reducing environmental impact. By constructing and analyzing a non-cooperative game model of power generation companies' participation in the electricity-carbon market, this study not only fills a gap in the existing literature regarding the interaction between the electricity and carbon markets but also provides a solid theoretical foundation for power generation companies to formulate strategies in the face of an increasingly complex market environment.

Acknowledgments

This work was financially supported by the SGCC Science and Technology Project "Electricity-Carbon Market Synergy and Green Electricity under the Dual Carbon Target Research on Key Technologies for Certification"(SGBJDK00KJJS2200522)

References

1. Gabriela Ileana Iacobuţă, Clara Brandi, Adis Dzebo, Sofia Donaji Elizalde Duron. Aligning climate and sustainable development finance through an SDG lens. The role of development assistance in implementing the Paris Agreement [J]. Global Environmental Change, May 2022, 102509.
2. Hongjie Li, Yitao Han, Xiuli Wang, et al. Optimal Strategies of Power Generation Companies and Electricity Customers Participating in Electricity Retailing Trading[J]. IEEE Access, 2023, 129660 – 129670.
3. Ana R. Silva, H.M.I. Pousinho, Ana Estanqueiro, et al. A multistage stochastic approach for the optimal bidding of variable renewable energy in the day-ahead, intraday and balancing markets [J]. Energy, Volume 258, 1 November 2022, 124856.
4. Xiaopeng Guo, Xinyue Zhang, Xingping Zhang. Incentive-oriented power-carbon emissions trading-tradable green certificate integrated market mechanisms using multi-agent deep reinforcement learning [J]. Applied Energy, Volume 357, 1 March 2024, 122458.
5. Xinyue Zhang, Xiaopeng Guo, Xingping Zhang. Assessing the policy synergy among power, carbon emissions trading and tradable green certificate market mechanisms on strategic GENCOs in China [J]. Energy, Volume 278, Part B, 1 September 2023, 127833.
6. Christos N. Dimitriadis, Evangelos G. Tsimopoulos, Michael C. Georgiadis. Optimal bidding strategy of a gas-fired power plant in interdependent low-carbon electricity and natural gas markets[J]. Energy, Volume 277, 15 August 2023, 127710.
7. Yuan Yan, Zihan Lin, Fushuan Wen. Optimal Operation Strategy for an Integrated Low-Carbon Energy System in Coupled Electricity and Natural Gas Markets[J]. 2019 IEEE Innovative Smart Grid Technologies - Asia (ISGT Asia), May 2019, 21-24 .

 https://doi.org/10.1051/e3sconf/202452802012

Sensitivity analysis of underwater launch based on Morris method

Jian Xie*[1,2], Dong Cheng[1,2], Shuli Chang[1,2], Pengtao Li[1,2]

[1]Henan key laboratory of underwater intelligent equipment, Zhengzhou, Henan 450000, China
[2]Zhengzhou Mechanical and Electrical Engineering Research Institute, Zhengzhou, Henan 450000, China

Abstract. The discharge velocity, pressure difference and acceleration of underwater launch performance are affected by many factors, including depth, sailing speed, burning rate, water temperature, density and other operating parameters. The existing underwater launch sensitivity analysis is a local sensitivity qualitative analysis based on single-step perturbation analysis, which can only investigate the influence of a single uncertain factor on the model results in a small range. In this paper, based on the underwater launch model under multi-factor coupling, 10 input parameters were selected to calculate the multi-factor sensitivity of the underwater launch model based on the three objectives of velocity, pressure difference and acceleration at the time of discharge, and the Morris global sensitivity analysis method was used to calculate the sensitivity of the underwater launch model, and the sensitivity of the 10 input parameters to the model results was sorted. The results show that the five input parameters of burning rate, depth, propellant temperature, density and water temperature have the highest sensitivity to velocity, pressure difference and acceleration, and the sensitivity order is slightly different. The influence of water temperature on the three targets is almost evenly distributed in the range of values, and the other parameters have different sensitive ranges for different targets.

1. Introduction

Underwater load launching technology is a launching method that accelerates the load underwater to a certain speed to get out of the launcher barrel through the combustion and expansion of the working medium. Now it has been developed to a more advanced variable depth underwater launch technology, adding an energy adjustment control mechanism on the original basis.The useful energy reaching the tail of the load is regulated by controlling the amount of cooling water injected to achieve emission at different depths. Due to the huge economic cost and labor cost of underwater launch test, it becomes an important means to study underwater launch technology to construct mathematical model of underwater launch simulation to simulate the underwater launch process and obtain the underwater launch state parameters[1-4]. The mathematical simulation model of underwater launch needs to be diagnosed and predicted by sensitivity analysis to ensure the accuracy and reliability of the model.

Sensitivity analysis is an analytical calculation method based on mathematical model, which draws relevant conclusions by calculating the influence degree of input parameters on model output results. It is used to evaluate the influence of each input parameter of the model on model results qualitatively or quantitatively, including the influence of model output and model performance[5]. Sensitivity analysis can be divided into local sensitivity analysis and global sensitivity analysis[6]. Local sensitivity analysis only examines the impact of small changes in each input parameter of the model within a local range on the model results. Only when the model input and output are linear or approximately linear dependent, can local sensitivity analysis better grasp the impact of each input parameter of the model on the overall performance of the model. Global sensitivity analysis studies the global impact of each input parameter of the model on the model results within its allowable variation range, and it does not require the model to be a linear system. Therefore, global sensitivity analysis also includes examining the impact of the interaction between model parameters on the model results[7]. Global sensitivity analysis can be divided into qualitative analysis and quantitative analysis. Qualitative analysis is to study the relative size of response of different parameters to output variables, such as Morris method, Delta Test method, etc. Quantitative analysis can obtain the contribution rate of different parameters to the output variables, such as Sobol method and EFAST method[8-11].

In the calculation of previous underwater launch sensitivity studies, researchers only used local sensitivity analysis, and only calculated the influence of small changes in the value range of input parameters on the output results, without considering the interaction between input parameters. In this paper, the input parameters of the underwater launch model were analyzed based on the Morris global sensitivity analysis method, and the significant order of the influence of the input parameters (depth Ht, speed Vt, water temperature Ts, burning rate Vr, propellant

*Corresponding author's e-mail: sherjam@hrbeu.edu.cn

temperature Ty, density ρ, atmospheric pressure P, air initial temperature Tk, friction coefficient μ and structural inner diameter D) on the exit velocity, pressure difference and acceleration was determined, which provided a basis for the subsequent quantitative analysis of underwater launch sensitivity.

2. Underwater launch model

2.1. Basic Assumptions

Underwater launch is a complex process of energy exchange and energy conversion, and there are many factors that affect the change of the whole working system. The following assumptions are made when establishing the model of underwater launch:

(1)Without considering the change of gas parameters along the pipeline, the flow of gas along the pipeline is regarded as the transmission process of working medium energy from the power source to the launching cylinder.

(2)The working medium gas, water vapor and the air in the launching cylinder can be uniformly mixed every instant to exchange energy and form a mixture of uniform state parameters at each point.

(3)For the energy loss such as heat transfer, the energy coefficient is taken into account, and its value is assumed to be a constant in the whole working process.

(4)For the macroscopic kinetic energy contained in the working medium gas, kinetic energy coefficient is introduced, and its value is assumed to be a constant in the whole working process.

2.2. Underwater launch model

2.2.1. Equation of flow.

In the launching process, the flow per second (G_g) and working medium quality (m_g) are

$$G_g = \frac{1}{C^*} A \mu_f P_c \quad (1)$$

$$m_g = \int_0^t G_g dt + m_{g0} \quad (2)$$

In the equation(1), C^* is the characteristic velocity, which represents the influence of the combustion process of working medium on the mass flow rate, and its dimension is the same as that of the velocity; A is the cross section area of the conveying pipeline; μ_f is the flow coefficient; P_c is the pressure in the combustion chamber.

2.2.2. Equation of energy.

The mixed gas formed by the working medium gas, water vapor and air builds up the pressure in the closed space, overcomes the resistance of the load movement, and ejects the load out of the launching cylinder. The energy equation is as follows.

$$U_{g1} + U_{l1} + U_{a1} + Q = U_{g2} + U_{l2} + U_{a2} + \int_1^2 PdV \quad (3)$$

Cooling water is added in the working process, and the cooling water absorbs the latent heat of evaporation into water vapor. In this process, a certain saturation pressure will correspond to a certain saturation temperature. The Clausius-Claperon equation is used to calculate the change of saturation pressure with temperature.

$$\frac{dP_{sat}}{P_{sat}} = \frac{h_{fg}}{R} \frac{dT_{sat}}{T_{sat}^2} \quad (4)$$

2.2.3. Equation of motion.

The actual gas state equation of 55 Martin-Hou is used to describe the state of the working medium gas in the emission cylinder.

$$P = \frac{RT}{V-b} + \frac{A_2 + B_2T + C_2e^{-5.475Tr}}{(V-b)^2} + \frac{A_3 + B_2T + C_2e^{-5.475Tr}}{(V-b)^3} + \frac{A_4}{(V-b)^4} + \frac{B_5T}{(V-b)^5} \quad (5)$$

2.2.4. Equation of state.

The kinetic energy of the emitted load is calculated by subtracting the work done by the resistance during the motion. The resistance includes gravity, friction, atmospheric pressure, hydrostatic pressure, lateral load force and additional inertial force.

$$\int_1^2 PdV = \frac{1}{2} MV^2 + FL \quad (6)$$

$$F = Mg + F_z + P_0S_t + \rho g S_t L + F_h + \psi Ma \quad (7)$$

3. Underwater launch sensitivity analysis

This paper takes the input parameters of the underwater launch model as the research object, and uses Morris method to analyze the global sensitivity of the parameters. The effects of depth Ht, speed Vt, water temperature Ts, burning rate Vr, propellant temperature Ty, density ρ, atmospheric pressure P, air initial temperature Tk, friction coefficient μ and structural inner diameter D on the velocity, pressure difference and acceleration of underwater ejector cylinder are studied. The specific steps are as follows.

3.1. Morris sample design

Assuming that the number of sampling points of a single parameter is P, the parameter values are discretized to obtain discrete points $\left\{0, \frac{1}{\text{p-1}}, \frac{2}{\text{p-1}}, \cdots, 1-\Delta\right\}$, $\Delta = \frac{1}{p-1}$ .In this paper, P=6 is taken, and the value range of input parameters is evenly divided into (P-1) isometric intervals.

3.2. Generate a randomized matrix

The randomization matrix of the sampling matrix $B^*_{(l,n)}$ can be expressed as:

$$B^*_{(l,n)} = (M_{(l,n)} \times Z^*_{(l,n)} + \frac{\Delta}{2} \times J^*_{(l,n)}) \times P^*_{(n,n)} \quad (8)$$

In equation (8), $l = n+1$, n is the number of parameters; $B^*_{(l,n)}$ is a $l \times n$ dimensional matrix, which is characterized by the difference between two adjacent rows and only one parameter value, the other parameter values are the same; $M_{(l,n)}$ is a l dimensional column vector in which all elements are 1; $Z^*_{(l,n)}$ is the n dimensional row vector, $Z^*_{(l,n)} = \left[x_1, \cdots, x_i, \cdots, x_n\right]$. The parameter x_i is randomly taken from $\left\{0, \frac{1}{\text{p-1}}, \frac{2}{\text{p-1}}, \cdots, 1-\Delta\right\}$. $P^*_{(n,n)}$ is a $n \times n$ dimensional random permutation matrix, each row and column of the matrix has only one value of 1, and the rest values are all 0. $J^*_{(l,n)}$ is a $l \times n$ dimensional matrix, and the matrix $J^*_{(l,n)}$ calculation equation is

$$J^*_{(l,n)} = (2B_{(l,n)} - J_{(l,n)}) \times D^*_{(n,n)} + J_{(l,n)} \quad (9)$$

In equation (9), the matrix $B_{(l,n)}$ is a strictly lower triangular matrix with all element 1; $J_{(l,n)}$ is a matrix where all the elements are 1; $D^*_{(n,n)}$ is a n dimensional diagonal matrix, and each diagonal element can only be randomly valued ± 1 with equal probability.

In this paper, the number of input parameters of underwater launch model is 10. According to equation (8) and (9), a random sample matrix $B^*_{(11,10)}$ can be obtained by programming with Matlab. In this paper, 6 random sample matrices are obtained after 6 times of sampling. The random sample matrix $B^*_{(11,10)}(1)$ is listed below, and other generated random sample matrices will not be described again.

$$B^*_{(11,10)}(1) = \begin{bmatrix} Ht_2 & Vt_1 & Ts_3 & Vr_4 & Ty_6 & \rho_5 & P_6 & Tk_4 & \mu_6 & D_6 \\ Ht_2 & Vt_2 & Ts_3 & Vr_4 & Ty_6 & \rho_5 & P_6 & Tk_4 & \mu_6 & D_6 \\ Ht_2 & Vt_2 & Ts_3 & Vr_4 & Ty_6 & \rho_5 & P_6 & Tk_4 & \mu_5 & D_6 \\ Ht_2 & Vt_2 & Ts_3 & Vr_4 & Ty_5 & \rho_5 & P_6 & Tk_4 & \mu_5 & D_6 \\ Ht_2 & Vt_2 & Ts_4 & Vr_4 & Ty_5 & \rho_5 & P_6 & Tk_4 & \mu_5 & D_6 \\ Ht_2 & Vt_2 & Ts_4 & Vr_4 & Ty_5 & \rho_6 & P_6 & Tk_4 & \mu_5 & D_6 \\ Ht_1 & Vt_2 & Ts_4 & Vr_4 & Ty_5 & \rho_6 & P_6 & Tk_4 & \mu_5 & D_6 \\ Ht_1 & Vt_2 & Ts_4 & Vr_3 & Ty_5 & \rho_6 & P_6 & Tk_4 & \mu_5 & D_6 \\ Ht_1 & Vt_2 & Ts_4 & Vr_3 & Ty_5 & \rho_6 & P_6 & Tk_4 & \mu_5 & D_5 \\ Ht_1 & Vt_2 & Ts_4 & Vr_3 & Ty_5 & \rho_6 & P_6 & Tk_5 & \mu_5 & D_5 \\ Ht_1 & Vt_2 & Ts_4 & Vr_3 & Ty_5 & \rho_6 & P_5 & Tk_5 & \mu_5 & D_5 \end{bmatrix}$$

3.3. Calculation of basic sensitivity factors

Assuming that the value of the i-th parameter changes in two adjacent rows of the randomized matrix $B^*_{(l,n)}$, the basic influence factor calculation equation of the i-th parameter is as follows:

$$SN_i = \frac{y(z_1, \cdots, z_{i+1}, \cdots, z_n) - y(z_1, \cdots, z_i, \cdots, z_n)}{\Delta} \quad (10)$$

In equation(10), y(x) is the objective function. All l group of adjacent row elements in the randomized matrix $B^*_{(l,n)}$ are taken as input parameters of the model, and the basic factors of all n parameters can be obtained according to equation(10), a basic influence factor can be obtained for each sample with this parameter using a randomized matrix $B^*_{(l,n)}$. The expressions of the absolute mean E_i and standard deviation σ_i of the basic impact factor for the response of the i-th input parameter are:

$$E_i = \sum_{j=1}^{r} \left|SN_{ij}\right| (r)^{-1} \quad (11)$$

$$\sigma_i = \left[\sum_{j=1}^{r} \left(SN_{ij} - \sum_{j=1}^{r} SN_{ij} / r\right)^2 (n)^{-1}\right]^{0.5} \quad (12)$$

In formula (12), The absolute value of SN_{ij} is taken in order to avoid the positive and negative balance at $SN_{ij} < 0$.

4. Conclusion

For parameter sensitivity, the absolute mean E_i mainly describes the importance of the input variable x_i. The greater the value of E_i, the greater the influence of parameter x_i on the model output results. Standard deviation σ_i describes the importance of a variable from the other hand. If variable x_i has a higher standard deviation, it indicates that parameter x_i affects the output value of the model and the interaction with other parameters is greater, indicating that the impact of this parameter on the output of the model is non-linear.

4.1. Sensitivity ranking

Morris global sensitivity analysis method and underwater launch model were used to conduct sensitivity analysis on the input parameters of 10 underwater launch models, and the sensitivity factors of input parameters to the three targets of exit cylinder velocity, pressure difference and acceleration as well as the absolute mean and standard deviation of input parameters were calculated. The results are shown in the following table.

Table 1. Calculation table for velocity sensitivity of underwater launch.

Input parameter	SN_i						Absolute mean	Standard deviation	Rank
	1	2	3	4	5	6	E_i	σ_i	
Vr	-2.271	-2.149	-2.628	-3.855	-3.623	4.590	3.186	3.139	1
Ht	2.093	-2.102	2.148	2.311	2.151	-2.116	2.153	2.214	2
Ty	-0.954	0.2475	-0.481	0.2585	-0.708	-0.252	0.4832	0.4978	3
ρ	-0.191	0.3575	0.261	0.2385	-0.224	-0.212	0.2472	0.2739	4
Ts	0.2020	-0.204	-0.207	0.2095	-0.183	-0.194	0.1998	0.2081	5
D	-0.105	-0.112	0.1115	0.113	0.11	0.1095	0.1101	0.1133	6
Vt	-0.077	0.0755	0.085	-0.083	0.0825	0.0775	0.0801	0.0828	7
μ	0.064	-0.069	-0.062	0.0745	-0.058	-0.063	0.0651	0.0685	8
P	0.0405	0.037	-0.043	-0.042	-0.043	-0.041	0.0409	0.0417	9
Tk	0.013	-0.012	-0.013	-0.02	-0.021	-0.016	0.0157	0.0124	10

Table 2. Calculation table of pressure difference sensitivity of underwater launch.

Input parameter	SN_i						Absolute mean	Standard deviation	Rank
	1	2	3	4	5	6	E_i	σ_i	
Ht	0.09182	-0.0909	0.0901	0.09205	0.09323	-0.9316	0.0919	0.0949	1
Vr	-0.0059	-0.0010	-0.0179	-0.0308	-0.0314	0.02044	0.0179	0.0199	2
Ty	-0.0286	-0.0139	-0.0082	-0.0079	-0.0060	0.0148	0.0133	0.0140	3
ρ	-0.0117	0.0095	0.0155	0.0135	-0.0128	-0.0126	0.0126	0.0139	4
Ts	0.00704	-0.0069	-0.0076	0.0075	-0.0081	-0.0079	0.0075	0.0077	5
D	-0.0036	-0.0036	0.00369	0.00459	0.00368	0.00359	0.00379	0.00389	6
P	1.81E-3	1.84E-3	1.82E-3	1.84E-3	1.82E-3	1.81E-3	1.82E-3	0.00188	7
Tk	-4.3E-4	2.2E-4	4.6E-4	3.7E-4	3.7E-4	4.9E-4	3.89E-4	3.93E-4	8
μ	1.1E-4	-7.5E-5	-8.5E-5	-2.5E-5	0.5E-5	1E-5	5.17E-5	7.09E-5	9
Vt	1E-5	1E-5	1E-5	1E-5	-1.5E-5	1E-5	1.08E-5	1.11E-5	10

Table 3. Calculation table of acceleration sensitivity of underwater launch.

Input parameter	SN_i						Absolute mean	Standard deviation	Rank
	1	2	3	4	5	6	E_i	σ_i	
Vr	-6.107	-6.782	-7.047	-6.903	-2.558	11.4	6.799	7.254	1
Ht	2.364	-4.031	1.973	4.6655	1.0335	-1.831	2.649	3.131	2
Ty	-1.148	1.363	2.519	1.076	-1.075	-1.265	1.407	1.617	3
ρ	-0.252	4.388	0.133	0.236	-0.221	-0.227	0.9095	1.8302	4
Ts	0.1635	-0.323	-0.206	0.177	0.1865	-0.219	0.2125	0.2364	5
μ	0.085	-0.174	-0.083	0.427	-0.092	-0.116	0.1628	0.2228	6
D	-0.123	-0.202	0.1195	0.117	0.118	0.132	0.1351	0.1488	7
Tk	0.028	-0.509	0.079	-0.044	-0.039	-0.026	0.1208	0.2129	8
P	0.0345	-0.037	-0.031	0.026	-0.034	-0.033	0.0325	0.0332	9
Vt	0	-5E-4	-5E-4	5E-4	0	0	0.00025	0.00038	10

https://doi.org/10.1051/e3sconf/202452802012

4.2. Result analysis

Based on the results and discussions presented above, the conclusions are obtained as below:

(1) In the calculation of underwater ejector velocity sensitivity (Table 1), it can be seen that the influences of burning rate (Vr), depth (Ht), propellant temperature (Ty), density (ρ) and water temperature (Ts) on the ejector velocity are in order from high to low, and the influence of water temperature on the ejector velocity is approximately evenly distributed within the value range. The burning rate, depth, propellant temperature and density are sensitive in the value range.

(2) In the calculation of the pressure difference sensitivity of the underwater launch cylinder (Table 2), it can be seen that the influences of depth (Ht), burning rate (Vr), propellant temperature (Ty), density (ρ) and water temperature (Ts) on the pressure difference of the outlet cylinder are in order from high to low, and the influences of depth, density and water temperature on the pressure difference of the outlet cylinder are approximately evenly distributed within the value range. The burning rate and propellant temperature are sensitive to each other.

(3) In the calculation of pressure difference sensitivity of underwater launch cylinder (Table 3), it can be seen that the influences of burning rate (Vr), depth (Ht), propellant temperature (Ty), density (ρ) and water temperature (Ts) on the discharge cylinder velocity are in order from high to low, and the influences of propellant temperature and water temperature on the discharge cylinder pressure difference are approximately evenly distributed within the value range. The burning rate, depth, propellant temperature and density are sensitive in the value range.

References

1. Zhou H,Wu M L and Wang X E (2019) Mathematical Model and Simulation Analysis of a New Underwater Launching System.In: Materials Science and Engineering.WuHan.pp.012015.
2. Numerical investigation of a muzzle multiphase flow field using two underwater launch methods
3. Zahid M Z,Nadeem M and Ismail M (2020) Numerical Study of Submarine Launched Underwater Vehicle.In:IBCAST.Natl Ctr Phys.pp.472-476.
4. Li W N,Zhang Z Y and Lu J X (2023) Investigations on the flow characteristics and the structural response of the launch tube during the underwater launching process.J.Ocean Engineering. Commun.,279:114603-114616.
5. MORRIS M D (1991) Factorial sampling plans for preliminary computational experiments.J. Technometrics. Commun., 33(2): 161-174.
6. Nabi S,Ahanger M A and Dar A Q (2021) Investigating the potential of Morris algorithm for improving the computational constraints of global sensitivity analysis.J.Environmental Science and Pollution Research. Commun., 28(43):60900-60912.
7. Feng K X,Lu Z Z and Yang C Q (2019) Enhanced Morris method for global sensitivity analysis: good proxy of Sobol' index.J.Structural and Multidisciplinary Optimization. Commun., 59(2):373-387.
8. Wang C Y, Peng M J and Xia G L(2020) Sensitivity analysis based on Morris method of passive system performance under ocean conditions.J.Annals of Nuclear Energy. Commun.,137:107067-107075.
9. Li D L,Ju Q and Jiang P (2023) Sensitivity analysis of hydrological model parameters based on improved Morris method with the double-Latin hypercube sampling.J.Hydrology Research. Commun., 54(2):220-232.
10. Wu J C,Wang S P and Wang Y H (2020) Sensitivity Analysis of Design Parameters in Transverse Flux Induction Heating Device.J.IEEE Transactions on Applied Superconductivity. Commun., 30(4):1629-1651.
11. Pang M,Xu R C and Hu Z B (2021) Uncertainty and Sensitivity Analysis of Input Conditions in a Large Shallow Lake Based on the Latin Hypercube Sampling and Morris Methods.J.Water. Commun., 13(13):1861-1872.

Simulation of the effect of horizontal airflow channel on natural ventilation in apartments

Xichen Yu[1] ,Mingyuan Cheng[2]

Shenyangjianzhu University,Liaoning,Shenyang,110000, China

Abstract: Natural ventilation has an important impact on people's work and life. Reasonable use of natural ventilation can effectively improve indoor air quality and thermal comfort, reduce the use of air conditioning and reduce energy consumption. With the continuous development of urbanization, many apartments cannot effectively use natural ventilation. This paper uses CFD simulation to analyze the improvement effect of indoor ventilation after the installation of airflow channel in residential public transportation. The results show that the horizontal airflow channel can improve the indoor air environment and effectively reduce the dissatisfaction rate of indoor thermal comfort and indoor air age.

1. Introduction

Natural ventilation is widely used as a passive cooling method, and a reasonable natural ventilation method can not only effectively improve indoor air quality, improve human thermal comfort, but also reduce the use of air conditioning system and save energy consumption[2122]. Natural ventilation is often difficult to control and is affected by a variety of factors, including wind speed, air temperature, relative humidity, window opening behavior, building shape, and urban density[23]. With the continuous development of global urbanization, the penetration rate of multi-unit housing and urban density will increase in the future[24], which will seriously affect the effective use of natural ventilation. In particular, the multi-family apartment can only carry out unilateral natural ventilation, and can not carry out cross ventilation, the ventilation effect is poor, the indoor environment is hot, and the pollutants are difficult to be effectively discharged, which seriously affects the air quality and thermal comfort. In response to this problem, Zhang et al[25]. proposed a new method to enhance natural ventilation, adding horizontal airflow channels in apartment buildings can change the ventilation mode of apartments and effectively enhance the potential of natural ventilation. Based on the research, this paper further analyzes the effect of airflow channels on the improvement of indoor air environment. It provides a reference for the practical application of airflow channels.

2. Models and Methods

2.1 Turbulence model

The turbulence model mainly selects two-equation models. The two-equation model in fluent includes k-ε model, RNG k-ε model and Realizable k-ε model. The RNG model improves the calculation accuracy of fast shear flow, which can be applied to the calculation of Reynolds number in each interval[2627]. Therefore, the RNG k-ε model is used for the turbulence model.

The RNG k-ε equation:

$$\frac{\partial(\rho k)}{\partial t}=\frac{\partial}{\partial x_i}\left[\left(\alpha_k\mu_{\varepsilon ff}\right)\frac{\partial k}{\partial x_i}\right]+\mu_i\left(\frac{\partial v_i}{\partial x_k}+\frac{\partial x_k}{x_i}\right)-\rho C_d\varepsilon \quad (1)$$

$$\frac{\partial(\rho\varepsilon)}{\partial t}=\frac{\partial}{\partial x_i}\left[\left(\alpha_\varepsilon\mu_{\varepsilon ff}\right)\frac{\partial \varepsilon}{\partial x_i}\right]+\frac{\varepsilon}{k}\left(C_{1\varepsilon}G_k-C_{2\varepsilon}\rho\varepsilon\right)-R_\varepsilon \quad (2)$$

2.2 Physical model

The study model is the middle house of a ladder of three households. There are five areas inside the house, namely bedroom, living room, kitchen, and bathroom. There is one person in the bedroom and in the living room. Due to apartment graphic design lead to the family only unilateral can open the window, in order to solve the apartment in the natural ventilation condition of different wind can have good ventilation environment, in the residential apartment public transport into the horizontal airflow channel, the greater the horizontal airflow channel opening area, the better the ventilation effect, the research of horizontal airflow channel area 0.2m×1m size duct. The specific physical model is shown in the **Figure 1**.

[1]Corresponding author's e-mail: 954090381@qq.com
[2]Corresponding author's e-mail: 56210653@qq.com

https://doi.org/10.1051/e3sconf/202452802013

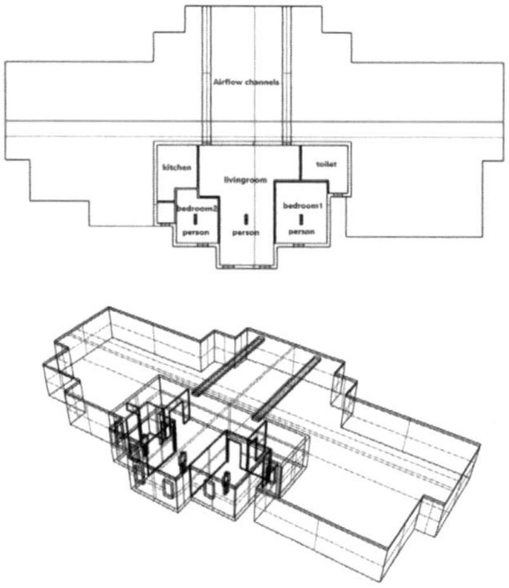

Figure 1. Arrangement of the horizontal airflow channel

2.3 Boundary condition

The site selected for this study was Shenyang, and the seasons were summer. The calculated temperature of outdoor ventilation in Shenyang in summer is 28°C, and the average outdoor wind speed is 2.6 m/s. The heat transfer coefficient of the exterior wall of the building is 0.65w /㎡·k, and the interior wall is the insulation wall surface. There are only personnel as internal heat source, the heat source power is 60W / ㎡, and the influence of other heat sources is ignored. The opening area of the window is 0.6m×1m. Indoor and outdoor coupling method is used for the simulation, and the gradient wind setting is used for the outdoor wind speed inlet. The expression is as follows:

$$U(z) = U_r \left(\frac{H}{H_r}\right)^{\alpha} \tag{3}$$

3 Evaluation criteria

3.1 aPMV model

Based on the engineering theory and various factors of human heat sensation, Yao Running [28]proposed the adaptability coefficient λ to correct the PMV equation, and the model considers the difference between the PMV model and the adaptability model, and establishes a connection. The expression is as follows:

$$aPMV = \frac{PMV}{1-\lambda * PMV} \tag{4}$$

Where λ is greater than 0 in the warm condition and less than 0 in the cold condition.According to the relevant evaluation criteria. **Table 1** shows the value of λ.**Table 2** and **Table 3** are the evaluation indicators of aPMV and air age

Table 1. λ value specification

Building climate zone	PMV	Residential,store,offices
cold areas	PMV≥0	0.24
	PMV<0	-0.50

Table 2. The aPMV sensory scaling

thermal inductance	cold	cool	Slightly cool	moderate	Slightly warm	warm	hot
PMV	-3	-2	-1	0	1	2	3

Table 3. Evaluation criteria for air age

Air Age T (s)	Air freshness standards
T<225	Fresh air
225≤T<400	The air is generally fresh
T≥400	The air is not fresh

Take the adaptive coefficient, and write the aPMV formula as an udf file, and model the model using the udf file written by fluent read.

4 Results and analysis

4.1 Improvement of air age under different wind direction incidence angles

Figure 2. For the comparison of the average air age in different areas of the apartment without ventilation and no airflow passage. It can be seen from the figure below that the average indoor air age at the incidence Angle of each wind direction is significantly improved. When the incidence Angle was 0, the average air age of bedroom 1 decreased from 750s to 447s, where the air age decreased significantly, but the air was in the stale range. The average air age of bedroom 2 decreased from 560s to 293s. Bedroom 2 improvement effect is obvious, from the air age value can see the indoor air from the fresh improvement to relatively fresh. The average air age of bedroom 1 after adding the horizontal airflow channel at 90°and 225°. and the wind Angle at 315°, the average air age of bedroom 2 has an upward trend, but all are within the range of fresh air. Among them, the living room is the best improvement effect, not ventilated to the incidence Angle under the indoor average air age is reduced.

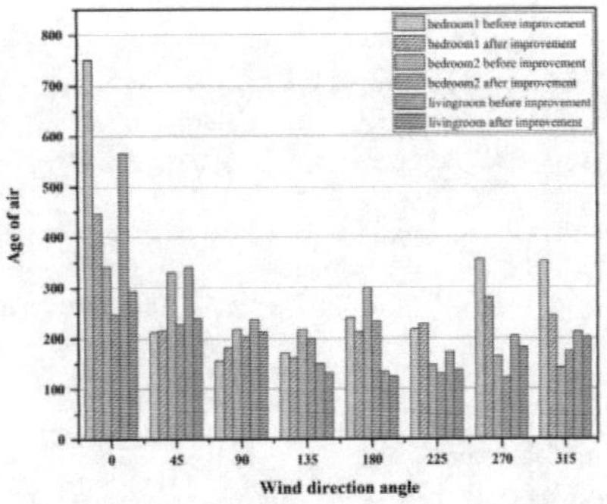

Figure 2. Bar chart of air age under different incidence angles of wind direction before and after improvement

4.2 Improvement of PMV-PPD at different wind angles

Figure 3.shows the comparative analysis of indoor PMV-PPD with and without horizontal airflow channels. As can be seen from the figure, the summer ventilation indoor hot comfort in Shenyang area is within the level II standard range, and the whole indoor area is between slightly warm and warm. When the wind direction Angle is at 0°and 180°, the improvement effect is obvious. Average PMV of bedroom 1 decreased from 1.36 to 1.28, mean air age of bedroom 2 decreased from 1.28 to 1.22, and dissatisfaction rates of both bedroom 1 and bedroom 2 decreased by 4%. The wind direction is in 180°bedroom 2 improvement effect is better, dissatisfaction rate decreased by 3%. Among them, the wind Angle is not improved at 90°, and the indoor dissatisfaction rate of the apartment increases. The reason for this phenomenon may be that the airflow channel is not added, which causes the formation of the pressure difference between the room outside the window and then the window, and then the wind prevents the wind from the window after adding the airflow channel. It can be seen from the chart that after adding the horizontal airflow channel, the indoor thermal comfort can be improved in most of the incident wind Angle, which will reduce the dissatisfaction rate of personnel.

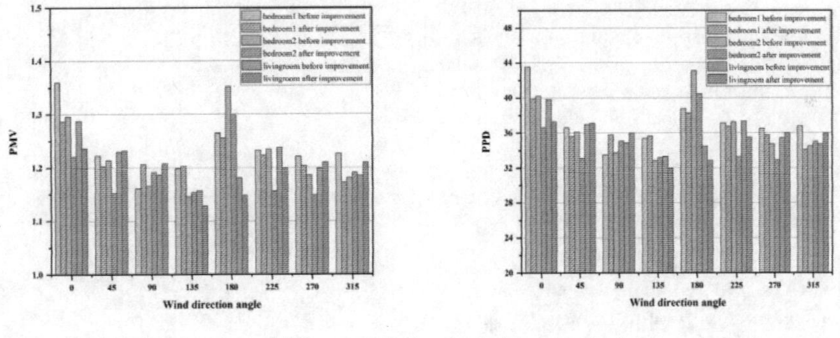

Figure 3. PMV-PPD bar chart at different wind direction incidence angles before and after improvement

4.3 Leeward indoor environment improvement effect

It can be seen in the above analysis that the improvement effect is obvious when the wind direction Angle of adding the horizontal airflow channel is 0°. **Figure 4.** is the CFD cloud map of the improvement effect when the wind direction Angle is 0°. It can be seen from the cloud chart that when the horizontal airflow channel is not added, the indoor air age in bedroom 1, living room and kitchen is large and poor ventilation. Although the air at the window is relatively fresh, the wind is difficult to enter the room, and the indoor heat is also difficult to discharge. After adding the airflow channel, the air age of each room is improved significantly, and the air quality of the location of the personnel is better. It can be seen from the PMV cloud diagram that when the horizontal airflow channel is not added, the heat of the personnel is difficult to be effectively discharged. when the airflow channel is added, the heat can be effectively taken away by the wind, and the overall heat and comfort of the room is evenly distributed. When the bedroom joins the airflow channel, the dissatisfaction rate of the indoor room has decreased evenly, not only limited to the location of the personnel. Due to the addition of the horizontal airflow channel, there are obvious wind speed changes at the door of the room, and the dissatisfaction rate is the lowest here. However, after adding the horizontal airflow channel, the ventilation effect in the toilet is inhibited, and the installation position of the airflow channel should be considered.

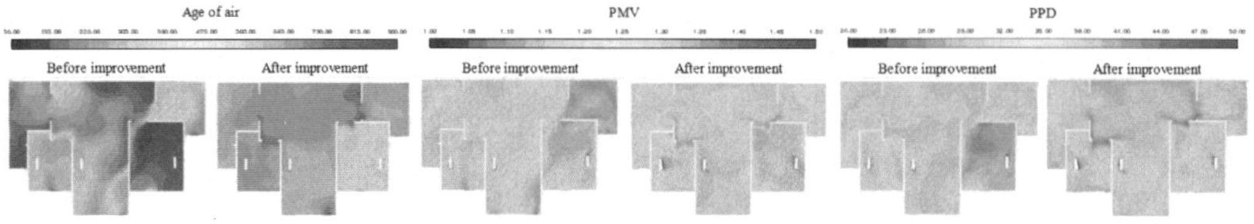

Figure 4. CFD cloud map of the leeward side improvement effect

5 Conclusion

In this study, aiming at the factors of poor natural ventilation effect in unilateral ventilation apartments, the improvement effect of natural ventilation was analyzed by increasing the airflow channel. Through numerical simulation, the following conclusions are drawn.

1. Horizontal airflow channels can improve the indoor air environment and thermal comfort under different wind direction incidence angles. When the wind angle is 0, the indoor improvement effect is the best. The average air age of bedroom 1 dropped from 750s to 447s and bedroom 2 decreased from 560s to 293s, which can meet the freshness standard and reduce the thermal comfort dissatisfaction rate by 4%. The addition of airflow channels has the best effect on improving the natural ventilation of the living room. The location of the airflow channel will affect the ventilation of individual rooms, but the overall ventilation effect is better.

2. Under the trend of urbanization and increasing urban density, the installation of horizontal airflow channels in unilateral ventilated apartments can change the way apartments are ventilated. The use of airflow channels can increase the cross-ventilation effect in the apartment, improve the utilization rate of natural ventilation, improve indoor thermal comfort and air quality, reduce the use of mechanical ventilation, and save energy consumption.

Reference

1. Iskandar L, Bay-Sahin E, Martinez-Molina A, et al. Evaluation of passive cooling through natural ventilation strategies in historic residential buildings using CFD simulations[J]. Energy and Buildings, 2024: 114005.
2. Bay E, Martinez-Molina A, Dupont W A. Assessment of natural ventilation strategies in historical buildings in a hot and humid climate using energy and CFD simulations[J]. Journal of Building Engineering, 2022, 51: 104287.
3. Chen J, Augenbroe G, Zeng Z, et al. Regional difference and related cooling electricity savings of air pollutant affected natural ventilation in commercial buildings across the US[J]. Building and Environment, 2020, 172: 106700.
4. Bao X, Zhang T, Dewancker B J, et al. Exploring the Unit Spatial Layout Preference for Urban Multi-Unit Residential Buildings: A Survey in Beijing, China[J]. Sustainability, 2023, 15(15): 12013.
5. Zhiyi Z, Wei Y, Tianwen W, et al. Potential of cross-ventilation channels in an ideal typical apartment building predicted by CFD and multi-zone airflow model[J]. Journal of Building Engineering, 2021, 44: 103408.
6. J. Franke, A. Hellsten, H. Schlunzen, B. Carrissimo, Best practice guideline for the CFD simulation of flows in the urban environment, COST office (2007)
7. B. Blocken, W.D. Janssen, T. van Hooff, CFD simulation for pedestrian wind comfort and wind safety in urban areas: general decision framework and case study for the Eindhoven University campus, Environ. Model. Software 30 (2012) 15–34
8. Yao R, Li B, Liu J. A theoretical adaptive model of thermal comfort–Adaptive Predicted Mean Vote (aPMV)[J]. Building and environment, 2009, 44(10): 2089-2096.

E3S Web of Conferences **528**, 02014 (2024) https://doi.org/10.1051/e3sconf/202452802014
NET-LC 2024

Design of a Testing Tool Based on Fault Injection for Functional Safety

Hang Yan*, Shanshan Li, Jianmei Lei, Tao Yan, Chunpeng Tang

China Automotive Engineering Research Institute Co.,Ltd. Chongqing 401122,China

Abstract. ISO 26262 provides testing requirements for functional safety development and testing to mitigate unacceptable risks arising from system functional failures. Fault injection plays a vital role in assessing system robustness and validating the efficacy of safety mechanisms. This paper explores the fault types and implementation process of fault injection in the context of functional safety confirmation and validation. A fault injection board specifically designed for electrical faults is developed and integrated with Controller Area Network (CAN) messages to verify the effectiveness and correctness of safety mechanisms. The proposed approach offers the advantages of flexible switching and ease of control, making it a valuable tool in ensuring the functional safety of automotive systems.

1. Introduction

The rapid development of automotive electronic and electrical technology has brought convenience, but at the same time, product safety issues have become increasingly prominent. To standardize the safety behaviors involved in product design, research and development, testing, and post-processing of automotive electronics and electrical systems, the International Organization for Standardization released ISO 26262 "Road vehicles -- Functional safety" in 2011[1], aiming to reduce the uncontrollable risks caused by failures in automotive E/E systems. The standard covers the life cycle of automotive systems and provides a basis for product development testing based on the V-model, encompassing the concept phase and system design (software and hardware levels)[2]. Fault injection testing (FIT)[3,4], as a recommended testing method in various stages, injects potential faults into the system under test to detect whether there are failure risks under fault conditions, thereby verifying the effectiveness of the system's safety mechanisms. However, in the actual vehicle validation and verification testing process, there are issues such as complex operations and cumbersome procedures. To reasonably and effectively implement fault injection testing, this paper designs a set of fault injection boards that provide diverse fault types. Compared with manual injection or software simulation, it can achieve precise and controllable fault injection. Furthermore, by integrating CAN (Controller Area Network) message instructions, it simplifies the operation in the actual process.

2. Fault injection

2.1. Types of Fault Injection

The purpose of fault injection testing is to verify the effectiveness of the safety mechanisms or safety measures proposed during the system design and software/hardware design processes, including attributes such as response time and whether the system enters a safe state[5]. The specific implementation methods for fault injection can be broadly categorized into two types: hardware-based faults and software-based fault injection[6,7]. For hardware-based fault injection, the main categories are as follows:

(1) Signal interference: Generating magnetic fields or radio frequency energy using electromagnetic interference (EMI)[8] equipment to affect circuit operation, simulating real-world electromagnetic compatibility issues.

(2) Power faults: Altering the supply voltage, including overvoltage, undervoltage, transient voltage dips or surges, to test the system's adaptability and protection mechanisms against power fluctuations.

(3) Temperature/humidity variations: Subjecting the system to extreme or unstable humidity environments, inducing potential increases in component failure rates.

(4) Structural stress: Applying vibration, shock, or mechanical pressure, which may lead to loose connections, cracked solder joints, or other physical damage.

(5) Short circuits/open circuits: Temporarily short-circuiting or disconnecting a component, pin, or signal line on the circuit board, simulating component failure or electrical connection breakage.

(6) Signal injection: Injecting erroneous signals into circuit nodes, such as changing logic levels or introducing clock signal disturbances.

*yanhang@caeri.com.cn

https://doi.org/10.1051/e3sconf/202452802014

2.2. Process of Fault Injection

In Part 4 of ISO 26262[9](Table 1), fault injection is recommended at both the software/hardware integration and system levels. Fault injection serves the following purposes: (1) Demonstrating the diagnostic coverage of safety mechanisms; (2) Proving the correct implementation of safety requirements.

Table 1. Correct implementation of functional safety and technical safety requirements at the system level

	Methods	ASIL			
		A	B	C	D
1a	Requirement-based test[a]	++	++	++	++
1b	Fault injection test[b]	+	+	++	++
1c	Back-to-back test[c]	o	+	+	++

a A requirements-based test denotes a test against functional and non-functional requirements.

b A fault injection test uses special means to introduce faults into the system. This can be done within the system via a special test interface or specially prepared elements or communication devices. The method is often used to improve the test coverage of the safety requirements, because during normal operation safety mechanisms are not invoked.

c A back-to-back test compares the responses of the test object with the responses of a simulation model to the same stimuli, to detect differences between the behaviour of the model and its implementation.

Within the same system, there is traceability between different Automotive Safety Integrity Levels (ASILs), and safety analysis can identify critical elements and hazard propagation paths in the architecture.

The fault injection process[10] typically involves the following steps: (1) Defining the fault model; (2) Fault activation; (3) Analysis of experimental results and evaluation of safety measures. Fault models can be extracted from potential failure causes, critical components, or safety analyses of critical paths[11]. The fault injection strategy is based on verifying the identified failure models against the implemented safety mechanisms, such as the failure modes in a Failure Mode and Effects Analysis (FMEA)[12]. To meet the technical requirements for functional safety verification testing, a system analysis of the object under test is necessary, including its functional logic and implementation methods. Then, considering the ASIL levels required for the system and its components, an analysis of fault points that directly violate safety objectives is conducted, enabling targeted fault injection testing (Figure 1).

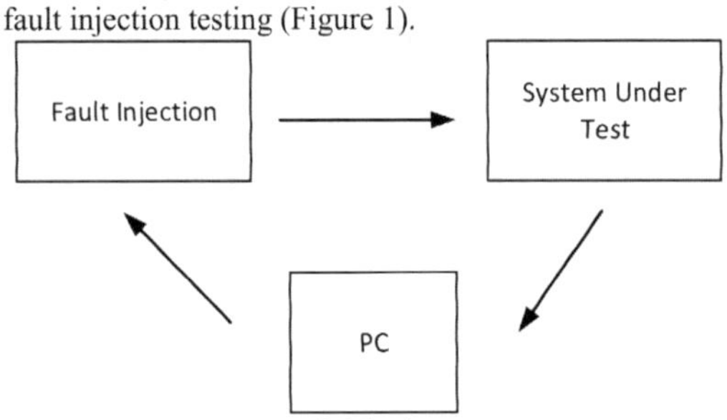

Figure 1. Fault injection test process

2.3. Theory and Implementation of Fault Injection

Combining the actual testing requirements and the ASIL level requirements specified in ISO 26262, this paper focuses on developing designs for several types of electrical faults, including power short circuits, ground short circuits, open circuits, and short circuits to other pins. The main principle is illustrated in Figure 2. It primarily utilizes cascaded control of multiple I/O (Input/Output) lines, combined with an MPU (Micro Processor Unit) to enable control of the I/O chain by sending a single command. This allows for the selection of different fault injection methods and provides the advantage of flexible switching between multiple integrated fault types.

Since the CAN bus is widely used in automotive communications, to minimize operational steps during actual testing, the fault injection commands are integrated with CAN messages. Specific CAN messages are sent to trigger fault injection. The fault injection command index and values are shown in Table 2. The physical picture is shown in Figure 3 and 4.

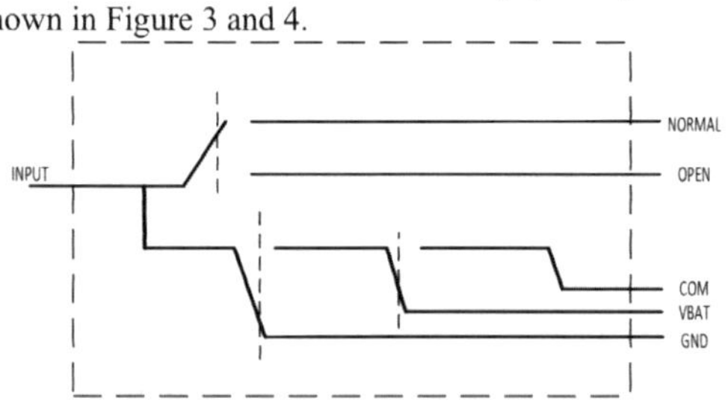

Figure 2. Theory of fault injection

Table 2. Control protocol

Index	Value	Remarks
0	02	Starting bit
1	01	Control mode
2	1~10	Channels
3	1~5	Fault types
4	03	End Bit

Where fault types, 1-normal; 2-open circuit; 3-short circuit to power; 4-short circuit to ground; 5-COM terminal. Reset message：02 01 00 00 03.

Figure 3. Fault Injection Board

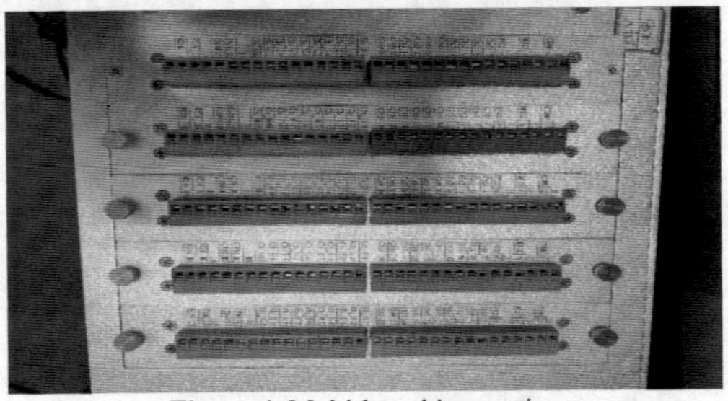

Figure 4. Multi-board integration

3. Solution Validation and Comparison

In a certain project test, targeting the functional safety requirement (Table 3) "The headlight controller should avoid erroneous control causing unintended extinguishing of the low beam headlights, leading to loss of forward illumination for the vehicle." During fault injection testing at the full vehicle level, a short circuit test needed to be performed on the control signal for the front left low beam headlight. According to the pin definition interface list for the headlight controller, the signal wire for the front left illumination lamp was connected to the fault injection board. Utilizing the self-defined protocol and with the assistance of the CANoe software, a message with the content "02 01 01 04 03" was transmitted (Fig. 5), and the change in the fault signal was observed in the Graphic window.

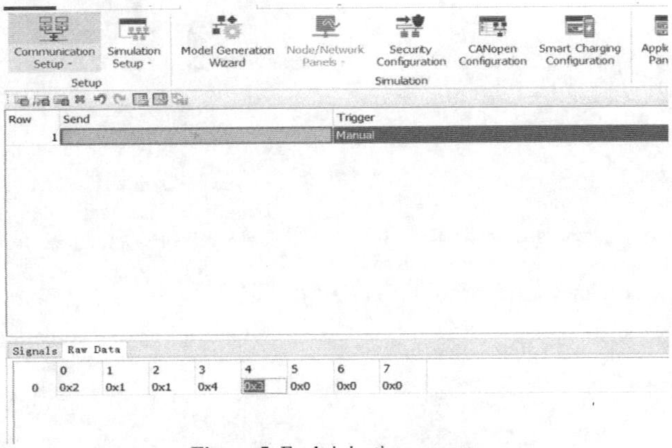

Figure 5. Fault injection message

Table 3. Functional safety requirement

Requirement Description	Testing Category	Test Description
The headlight controller should avoid erroneous control causing unintended extinguishing of the low beam headlights, leading to loss of forward illumination for the vehicle	Fault Injection	1.vehicle powered on and the low beam headlights remaining illuminated; 2. a short-circuit to ground was applied to the LDM_FR signal. LDM_FR transmitted LCU_xxxFlt as 0x2: Fault

As demonstrated in Figure 6, after sending custom fault messages, the observed value of the "LCU_xxxFlt" signal jumps, changing from "no fault" to "fault". In actual testing processes, there are situations where various faults are combined for testing. If signal and harness changes are made one by one, it would result in a significant waste of time and resources. To realize the combination of multi-signal and multi-fault, a host computer was developed. As shown in Figure 7, the host computer can implement different types and quantities of test combinations, and it supports the import and export of test results. This not only simplifies the operation process but also enhances testing efficiency.

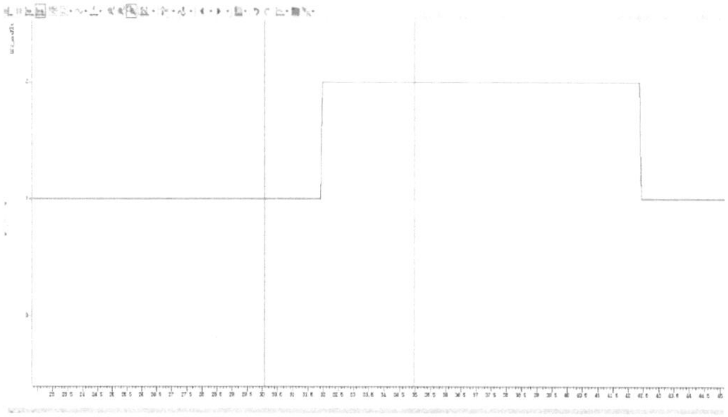

Figure 6. Fault message response

Figure 7. Multi-channels fault combination injection

4. Conclusion

In this paper, an analysis of the fault injection methodology was conducted in accordance with the ISO 26262 "Road Vehicles – Functional Safety" standard. Building upon this analysis, a fault injection board was designed to meet the requirements for electrical fault injection in the context of functional safety. This board enables fault injection through the selection of fault types via CAN messages, a solution that was subsequently validated. The integration with CAN messages streamlines the operation process. The versatility of multiple fault injection boards allows their application across various testing levels, such as hardware-in-the-loop (HIL) testing and vehicle validation testing.

References

1. ISO 26262:2011 Road Vehicles—Functional safety.
2. Luo, Y., Saberi, A. K., & den Brand, M. V. (2019). Safety-driven development and ISO 26262. Automotive Systems and Software Engineering: State of the Art and Future Trends, 225-254.
3. Da Silva, F. A., Bagbaba, A. C., Hamdioui, S., & Sauer, C. (2019, December). Combining fault analysis technologies for ISO26262 functional safety verification. In 2019 IEEE 28th Asian Test Symposium (ATS) (pp. 129-1295). IEEE.
4. Kongjian, Q., Tong, Z., Kuiyuan, G., Hongwei, Z., Yu, W., & Haoxin, C. (2020, September). The Method of Functional Safety Validation Test of AEBS Based on Fault Injection. In 2020 IEEE 3rd International Conference on Information Systems and Computer Aided Education (ICISCAE) (pp. 377-381). IEEE.
5. Breier, Jakub, and Xiaolu Hou. "How practical are fault injection attacks, really?" IEEE Access 10 (2022): 113122-113130.
6. Juez, G., Amparan, E., Lattarulo, R., Rastelli, J. P., Ruiz, A., & Espinoza, H. (2017, June). Safety assessment of automated vehicle functions by simulation-based fault injection. In 2017 IEEE International Conference on Vehicular Electronics and Safety (ICVES) (pp. 214-219). IEEE.
7. Given-Wilson, T., Jafri, N., & Legay, A. (2020). Combined software and hardware fault injection vulnerability detection. Innovations in Systems and Software Engineering, 16(2), 101-120.
8. Dumont, M., Lisart, M., & Maurine, P. (2020). Modeling and simulating electromagnetic fault injection. IEEE Transactions on Computer-Aided Design of Integrated Circuits and Systems, 40(4), 680-693.
9. ISO 26262-4:2011 Road vehicles--Functional safety—Part 4: Product development at the system level.
10. Ziade, H., Ayoubi, R. A., & Velazco, R. (2004). A survey on fault injection techniques. Int. Arab J. Inf. Technol., 1(2), 171-186.
11. Schmid, T., Schraufstetter, S., Wagner, S., & Hellhake, D. (2019, November). A safety argumentation for fail-operational automotive systems in compliance with iso 26262. In 2019 4th International Conference on System Reliability and Safety (ICSRS) (pp. 484-493). IEEE.
12. Sulaman, S. M., Beer, A., Felderer, M., & Höst, M. (2019). Comparison of the FMEA and STPA safety analysis methods–a case study. Software quality journal, 27, 349-387.

https://doi.org/10.1051/e3sconf/202452802015

Influence of decontamination process parameters on ultrasonic decontamination efficiency of pipeline

Chengwei Liu [1], Xinming Huang [2], Zhimeng Yao [3], Ke Li [4], Zhiyuan Chen [2], Shaochong Wei [1], Yang Zou [1], Jingcan Shi [1], Zhuang Lu [1], Guoxing Chen [1*]

[1]Suzhou Nuclear Power Research Institute Co., Ltd., Suzhou, 215004,China
[2]Daya Bay Nuclear Power Operations and Management Co., Ltd, Shenzhen, 518124,China
[3]Yangjiang Nuclear Power Co., Ltd, Yangjiang, 529500,China
[4]Taishan Nuclear Power Joint Venture Co., Ltd; Taishan, 529228,China

Abstract: Many radioactive pollutants will be deposited inside the nuclear power plant pipelines after long-term service, which will form radioactive hot spots and increase the radiation dose rate level of operators. The ultrasonic decontamination efficiency was studied through the establishment of ultrasonic decontamination dynamic test bench, under different decontamination process parameters. It has also been applied in the engineering filed. The results show that: Ultrasonic decontamination efficiency increases with the increase of decontamination power, and the maximum decontamination efficiency of pollutants is up to 62.69%. The main reason is that the mechanical effect and cavitation effect of ultrasonic are significant with the increase of power, which is conducive to the removal of pollutants in the pipeline. When the decontamination time is 60min, the ultrasonic decontamination efficiency is more than 90%. Meanwhile, after ultrasonic decontamination of the pipeline hot spots, the contact dose rate at the pipeline hot spots is reduced by 75.20% at the highest, and the environmental dose rate is reduced by more than 90%, which reduces the collective dose level of operators working near the pipeline hot spot greatly, and ensures the health of the operators and environmental safety.

1. Introduction

Due to the advantages of pipeline transportation with high efficiency, environmental protection, low cost and high safety, it is currently the most widely used transportation mode of fluid medium transmission currently, which is used in petroleum, chemical industry and electric power industry [1-2]. When the liquid medium flows inside the pipeline, it is affected by the flow rate, tube structure, temperature difference changes and flow resistance. Many pollutants are deposited inside the pipeline after long-term service, which will reduce the normal transmission flow and utilization efficiency of the pipeline and affect the normal use of the pipeline, it is a common and long existing problem that is difficult to solve in the industrial field. It will not only affect the stability of the pipeline, but also cause serious economic losses [3-4]. Especially in the field of nuclear power, the deposited pollutants contain a large amount of radioactivity. The continuous deposition of radioactive pollutants causes the radiation dose level around them to increase continuously. Which will form radioactive radiation hot spots after deposition in the pipeline, increase the collective dose level of maintenance personnel, threaten personal safety, and pollute the environment [5]. Therefore, developing an efficient, clean, low-cost and convenient pipeline decontamination technology has become an urgent problem to be solved.

At present, pipeline decontamination technology mainly includes two methods: physical decontamination and chemical decontamination. Physical decontamination is peeling the pollutants adhered to the inside of the pipeline off by applying external forces, and then completing the removal of pollutants from the pipeline. Common methods include mechanical cleaning, high-pressure water flushing, etc. [6-7]. Chemical decontamination is the process of reacting chemical reagents with pollutants inside pipelines to achieve the purpose of decontamination. Common methods include acid washing, alkaline washing, and organic solvents [8-9]. The current decontamination technology with many problems, which spends a large amount of manpower, water resources, and produces secondary waste that pollutes the environment and corrodes pipe walls. Ultrasonic decontamination technology, as a new type of physical cleaning method, has the advantages of being green, clean, low-cost and highly efficient compared to traditional decontamination methods. Especially, service pipeline decontamination can be achieved without damage in the pipeline structure [10].

Ultrasonic decontamination is a decontamination technology with good decontamination effect, high decontamination efficiency and the most promising

*Corresponding author's e-mail: meguoxing@foxmail.com

https://doi.org/10.1051/e3sconf/202452802015

application prospects. It is mainly a technology that uses tiny bubbles generated by high-frequency ultrasonic waves to explode in the medium and make pollutants be removed, which has attracted wide attention from many scholars in the decontamination application field. Wang et al. [11] studied the influence of fluid temperature on effect of ultrasonic decontamination, and the research results show that decontamination effect increased with the increase of temperature, but at 90°C, the decontamination efficiency basically reached saturation, and the decontamination efficiency was about 85%. Worapol et al. [12] studied the influence of ultrasonic placement parameters on decontamination effect, and the research results show that a reasonable layout of ultrasonic oscillator was conducive to the uniformity of sound pressure and could significantly improve the decontamination effect. Kim et al. [13] studied the influence of ultrasonic sound intensity on decontamination effect, obtained different ultrasonic sound intensity by adjusting ultrasonic decontamination process parameters, and the research results show that the decontamination efficiency increased with the increase of ultrasonic sound intensity.

Therefore, it will provide a theoretical basis for the removal of hot spots of radioactive pollutants by carrying out research on ultrasonic decontamination process parameters, to explore decontamination efficiency under different decontamination process parameters, to obtain the best decontamination process parameters.

2. Pollutant formation and ultrasonic decontamination mechanism

2.1 Pollutant formation mechanism

The formation mechanism of radioactive pollutants in the pipeline mainly with six forms including crystallization fouling, particulate fouling, chemical reaction fouling, corrosion fouling, biological fouling and solidification fouling [14], among which crystallization fouling, particulate fouling and corrosion fouling are the three most common forms. Crystallization fouling and particulate fouling are pollutants deposited in the medium on the pipeline surface, relative to corrosion fouling, its adhesion is relatively small, and it is easier to remove with ultrasonic decontamination technology. For corrosive dirt, the pipeline itself participates in the chemical reaction to generate corrosive products, ultrasonic decontamination technology can remove most of the corrosive products, for the remaining corrosive products with large adhesion, the cleaning technology of ultrasonic decontamination and chemical decontamination can be used to improve the decontamination efficiency. Therefore, for different kinds of radioactive pollutants, ultrasonic decontamination technology and its corresponding decontamination composite technology can obtain better decontamination effect.

2.2 Ultrasonic decontamination mechanism

Ultrasound generally refers to sound waves with a frequency above 20kHz, which show a certain regular sound field when propagating in the liquid. It uses an ultrasonic generator to excite the ultrasonic transducer and transmits the generated sound waves into the fluid inside the pipeline, thus, generating cavitation effect and mechanical effect in the medium [11, 15]. Through the interaction between ultrasound and the fluid medium, the cleaning of pollutants on the inner wall of the pipeline is achieved.

(1) Cavitation effect: when the ultrasonic propagates in the liquid, it will form a high-frequency and reciprocating cycle of positive and negative pressure, change the internal state of the medium, and produce many microbubbles in the medium. The microbubbles expand and break in the medium, resulting in a violent impact, making the pollutants crushed and refined, reducing their adhesion ability, and falling off to complete the removal of pollutants.

Cavitation effect is the main decontamination mechanism of ultrasonic decontamination. If cavitation effect is to occur in the pipeline, the sound pressure amplitude or sound intensity needs to exceed the cavitation threshold. Therefore, there needs to be negative pressure in the liquid, and the negative pressure should be greater than the structural strength of the liquid. The cavitation threshold P can be expressed by the following formula, P_0 is the static pressure of the liquid (Pa), P_v is the vapor pressure (Pa), σ is the surface tension coefficient of the liquid (N/m), R_0 is the initial radius of the cavitation nucleus (m).

$$P = P_0 - P_v + \frac{2}{3\sqrt{3}}\sqrt{\frac{\left(\frac{2\sigma}{R_0}\right)^3}{P_0 - P_v + \frac{2\sigma}{R_0}}} \quad (1)$$

The motion equation of the microbubbles produced by cavitation can be expressed by the Rayleigh-Plesset equation, R is the radius of the cavitation bubble (m), ρ is the density of the liquid (kg/m^3), $P(R)$ is the pressure in the cavitation microbubbles (Pa), μ is the dynamic viscosity of the liquid (Pa·s), P_w is the pressure at infinity (Pa).

$$R\frac{\mathrm{d}^2 R}{\mathrm{d}t^2} + \frac{3}{2}\left(\frac{\mathrm{d}R}{\mathrm{d}t}\right)^2 = \frac{1}{\rho}\left[P(R) - \frac{2\sigma}{R} - \frac{4\mu}{R}\frac{\mathrm{d}R}{\mathrm{d}t} - P_w\right] \quad (2)$$

(2) Mechanical effect: when the ultrasonic propagates in the liquid, a huge sound pressure will be generated in the medium, and the oscillation caused by the sound pressure will make the liquid produce a violent mechanical movement, which will impact the pollutants. The shear force generated at the contact surface of the pollutants and the pipe wall will make the pollutants broken and fall off, to achieve the purpose of pollutant removal.

2.3 Ultrasonic decontamination monitoring and evolution

Currently, the commonly used monitoring and evaluation methods include weighing method, thickness measurement method, temperature difference method and radioactivity monitoring. By detecting the removal number of pollutants in the process of ultrasonic decontamination, it is used to characterize the ultrasonic decontamination efficiency, to monitor the decontamination effect in the decontamination process, and to evaluate the decontamination effect after ultrasonic decontamination. The monitoring and evaluation of ultrasonic decontamination technology is conducive to timely optimization of decontamination program and decontamination process and obtain the best decontamination effect. In decontamination of radioactive hot spots, the engineering application effect of ultrasonic decontamination technology can be monitored and evaluated by monitoring the change of exposure dose rate and environmental dose rate during decontamination.

3. Experimental method

The dynamic ultrasonic decontamination test system is conducive to simulating the operating environment of the pipeline and realizing the efficient and accurate removal of pollutants in the pipeline. The pipeline decontamination test system consists of ultrasonic decontamination controller, ultrasonic decontamination transducer, fixture, sample pipe, circulating pipe, water temperature heater, control valve and circulating pump, as shown in Figure 1. The sample pipe is made by 304 stainless steels, and the water temperature heater, control valve and circulating pump, it can realize the flow rate and temperature adjustment of the medium in the pipeline, which is more suitable for the actual application scenario of the pipeline.

The ultrasonic decontamination system consists of an ultrasonic control system and the ultrasonic transducer system, which are independently developed and designed. Working parameters of ultrasonic decontamination system: power AC220V ± 10%, frequency 38.25kHz, and decontamination power is adjustable from 0 to 400W. This study achieves the optimal decontamination process parameters by adjusting the decontamination power and changing the decontamination time.

In order to characterize the decontamination efficiency qualitatively and quantitatively under different decontamination process parameters, and compare the decontamination effect, the decontamination effect was quantification characterized by metal film. By sticking the metal film on the inner surface of the pipeline, the ultrasonic transducer is fixed on the outer surface of the pipeline, and the liquid medium is water. The decontamination process parameters are set through the ultrasonic controller, and the decontamination test is carried out on the sample pipeline. After the decontamination test, the metal film is removed from the inside of the pipeline, surface scanning and numerical treatment are perform, and the ratio of the damaged area on the metal film surface to the decontamination area is calculated that is the decontamination efficiency ($\lambda = \frac{A_s}{A_t}$, λ is decontamination efficiency, A_s is the area of the damaged area, A_t is the total area of the metal film) of ultrasonic decontamination. In decontamination of radioactive hot spots, the engineering application effect of ultrasonic decontamination technology can be monitored and evaluated by monitoring the change of exposure dose rate and environmental dose rate during decontamination.

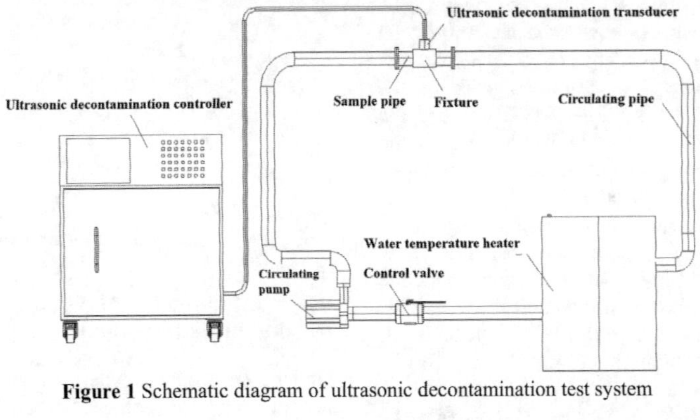

Figure 1 Schematic diagram of ultrasonic decontamination test system

4. Results and Discussion

4.1 Influence of power on ultrasonic decontamination effect

Ultrasonic power is the main factor to determine the strength of ultrasonic sound pressure, the cavitation effect of ultrasonic decontamination is closely related to the ultrasonic sound pressure intensity. Cavitation can occur only when the sound pressure intensity exceeds a certain value, which is the cavitation threshold value. Therefore, the study of decontamination effect under different decontamination power is of great help to improve the removal of pollutants in the pipeline. The test conditions are as follows: frequency 38.25kHz,

 https://doi.org/10.1051/e3sconf/202452802015

temperature 24°C, flow rate 20L/min, time 20min, decontamination power: 150W, 200W, 250W, 300W, 350W, 400W.

Figure 2 shows the morphology of the ultrasonic decontamination effect at different powers. The white area represents the metal film, and the black area represents the damage caused by ultrasonic cleaning to the metal film. As can be seen from the figure, with the increase of power, the damage area of the metal film becomes larger, indicating that the decontamination effect of ultrasonic is better with larger power. It is worth noting that many small and dense pores are distributed on the surface of the metal film, which is mainly caused by ultrasonic cavitation. Under the action of ultrasound, many microbubbles are generated inside the pipeline medium. The shock wave generated after the microbubbles expand and rupture, will impact the surface of the metal film, and then many micropores will leave on the surface of the metal film. Meanwhile, as the cleaning process progresses, the micropores continue to expand, forming larger pores.

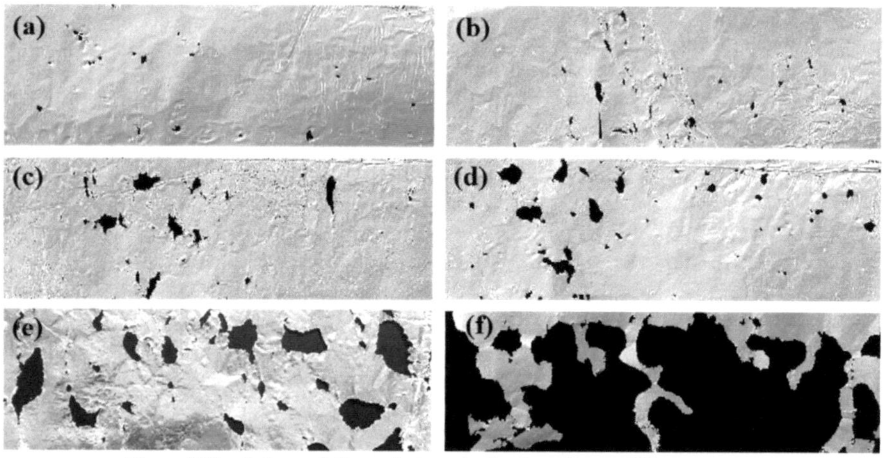

Figure 2 Morphology of ultrasonic decontamination effect at different powers, (a) 150W, (b) 200W, (c) 250W, (d) 300W, (e) 350W, (f) 400W

In order to characterize the ultrasonic decontamination efficiency quantitatively at different powers, the damage behavior of metal films at different powers was calculated, as shown in Figure 3. Figure 3 shows the variation curve of ultrasonic decontamination efficiency with power. It can be seen from the figure that the decontamination efficiency increases with the increase of power. When the power reaches 400W, the highest removal rate of pollutants is 62.69%. At low power, the decontamination effect is not significant, the decontamination efficiency is below 10%. Therefore, the ultrasonic decontamination efficiency increases, as the power increases, mainly because the mechanical effect and cavitation effect of ultrasound are significant with the increase of power, which is beneficial for the removal of pollutants in pipelines. Therefore, choosing a higher decontamination power without damaging pipeline material can achieve better decontamination effects. Further exploration is needed on the damage behavior of pipeline materials under different powers.

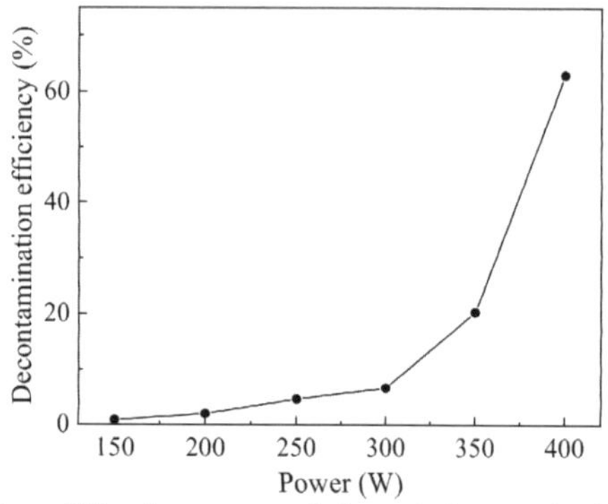

Figure 3 The change curve of ultrasonic decontamination efficiency with power

4.2 Influence of time on ultrasonic decontamination effect

Under the same experimental conditions, as the decontamination time increases, the ultrasonic decontamination efficiency will gradually reach saturation. Therefore, studying the ultrasonic decontamination efficiency at different times is of great guiding significance for improving the work efficiency during operations. Especially for decontamination of radiation hot spots inside the pipelines in service operation of nuclear power plants. The longer the operation time, the more radiation dose the operator is exposed to. Reasonable control of ultrasonic decontamination operation time can effectively reduce the collective dose. Therefore, stopping the operation at the maximum decontamination efficiency, is the optimal solution to ultrasonic decontamination, when the effect of decontamination operations cannot be directly observed, and it can also protect the safety of operators. The test conditions are decontamination power 400W, frequency 38.25kHz, temperature 24°C, flow rate 20L/min, decontamination time: 1min, 2min, 5min, 10min, 15min, 20min, 40min, 60min.

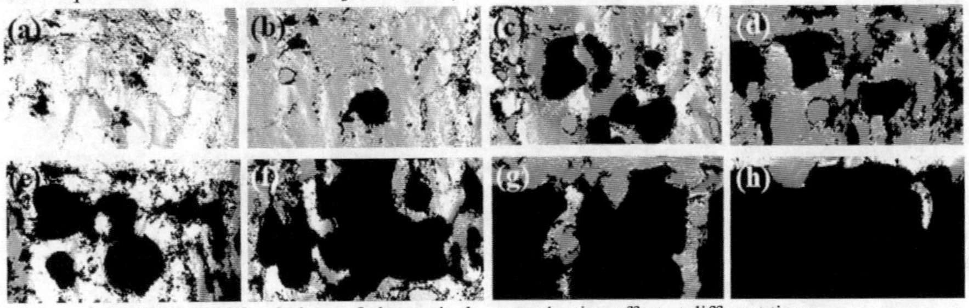

Figure 4 Morphology of ultrasonic decontamination effect at different times,
(a) 1min, (b) 2min, (c) 5min, (d) 10min, (e) 15min, (f) 20min, (g) 40min, (h) 60min

Figure 4 shows the morphology of ultrasonic decontamination effect at different times. It can be seen from the figure that with the increase of time, the damage area of the metal film is larger, indicating that the ultrasonic decontamination effect is better with the increase of time. When the decontamination time is 60min, the decontamination efficiency is above 90%, and the ultrasonic decontamination efficiency basically reaches saturation. If the decontamination operation continues, a more significant decontamination effect will not be achieved. This is consistent with the research results of Zhang et al. [16].

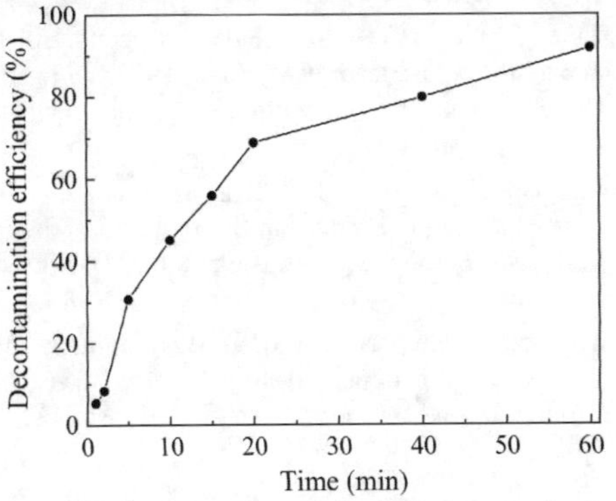

Figure 5 The change curve of ultrasonic decontamination efficiency with time

In order to intuitively display the decontamination effect under different decontamination time, the decontamination effect under different decontamination time was quantified, as shown in Figure 5. Figure 5 shows the variation curve of ultrasonic decontamination efficiency with time. As can be seen from the figure, under fixed power and frequency, decontamination efficiency increases with the increase of time. When the decontamination time is about 60min, decontamination efficiency has basically stabilized, which is the most optimized decontamination test process.

4.3 Engineering application

There are a lot of radioactive pollutants deposited in the pipeline of a nuclear power plant, which leads to a high dose rate around the pipeline, and the collective radiation dose level of the operators is increased during the maintenance of the working area near the pipeline, which seriously affects the health of the operators. Therefore, ultrasonic decontamination tests are carried out on hot spots to remove radioactive pollutants inside the pipeline, to reduce the environmental dose rate level near the pipeline hot spots, and to ensure the safety of operators and the environment. Table 1 shows the change of pipe hot spot exposure dose rate before and after decontamination in different ultrasonic decontamination zones, and Table 2 shows the change of environmental dose rate around pipe hot spot before and after decontamination in different ultrasonic decontamination zones.

Table 1 Exposure dose rate changes before and after ultrasonic decontamination

Test point	Before decontamination (mSv/h)	Time (min)	After decontamination (mSv/h)	Exposure dose reduction rate
A1	19.80	60	8.10	59.09%
A2	3.50	60	0.87	75.20%

Table 2 Environmental dose rate changes before and after ultrasonic decontamination

Test point	Before decontamination (mSv/h)	Time (min)	After decontamination (mSv/h)	Environmental dose reduction rate
A1	1.35	60	0.13	90.81%
A2	1.02	60	0.09	91.18%

As can be seen from Table 1 and Table 2, after the radioactive pollutants in the pipeline is ultrasonic decontaminated, the exposure dose rate at the pipeline hot spot is reduced by 75.20% at the highest, and the environmental dose rate is reduced by more than 90%, which greatly reduces the collective dose level of operators working and ensures the health of the operators and environmental safety.

5 Conclusion

In this paper, the ultrasonic decontamination efficiency under different decontamination process parameters was studied by building a dynamic test bench for ultrasonic decontamination, and the engineering application was carried out in the pipeline. The main conclusions are as follows:

(1) By adjusting different ultrasonic decontamination power, the decontamination efficiency was tested under different decontamination power. With the increase of ultrasonic decontamination power, the decontamination efficiency also gradually increased. When the power reached 400W, the highest pollutant removal rate was 62.69%. The main reason is that the mechanical effect and cavitation effect of ultrasonic are significant with the increase of power, which is conducive to the removal of pollutants in the pipeline. However, when the power increases to a certain value, it will cause damage to the pipeline material. Therefore, the subsequent damage effect of power on the material needs to be further studied.

(2) By setting different ultrasonic decontamination time, the decontamination efficiency was tested under different decontamination time. The decontamination efficiency increased with the increase of ultrasonic decontamination time. When the decontamination time was 60min, the ultrasonic decontamination efficiency was more than 90%.

(3) Through the ultrasonic decontamination application on the pipeline hot spots, after ultrasonic decontamination, the exposure dose rate at the pipeline is reduced by 75.20% at the highest, and the environmental dose rate is reduced by more than 90%, which greatly reduces the collective dose level of operators working and ensures the health and environmental safety of operators.

Acknowledgements

This work was financially supported by the by the Peak Planning Project of CGN (R-2023SZRE21TF)

References

1. Zhang X J, Wang R, Yang S Y (2020) Development Direction of Pipeline Corrosion Research on Oil Refining and Chemical Enterprises[J]. Materials Science Forum, 993: 1230-1234.
2. Xiao D (2019) Corrosion Factor Analysis and Corrosion Protection Strategy of Petroleum Pipeline[J]. Chemical Engineering Design Communications, 45: 45.
3. Huang J P (2021) Chemical cleaning in newly built units under pressure steam Applications on pipes. Cleaning World, 37: 22-23
4. Huang H J, Cui X, Zhang Y (2019) Analysis of the causes of tube and pipeline blocking in sulige gas field[J]. International conference on materical science and engineering technology.
5. Song Y L, Zhao Y P, Wang L, et al (2022) Preliminary Application of On-line Ultrasonic Decontamination of Pipelines at a NPP[J]. Radiation Protection bulletin, 42: 27-31.
6. Mahadik S, Balasubramanian V (2017) A Novel Engraving Planarization Innovation to Rectify Non-Consistency Post Chemical Mechanical Cleaning[J]. Indian Journal of Public Health Research and Development, 8: 129.
7. Nian S C, Sun H X, Cheng Y (2015) Mechanism based on pipe cleaning apparatus of high-pressure water jet[J]. 5th International Conference on Information Engineering for Mechanics and Materials.
8. Hosokawa H, Nagase M, Fuse M, et al (2013) Development of a suppression method for deposition of radioactive cobalt after chemical decontamination: application of a ferrite film formation process to an actual BWR plant[J]. Nuclear Science and Engineering, 175: 135-148.
9. Tsuyoshi I, Hosokawa H, Yanagisawa S, et al (2020) Carbon steel recontamination by Co-60 incorporation into iron (II) oxalate dihydrate formed during chemical decontamination[J]. Journal of Nuclear Science and Technology, 58:1-6.
10. Dai P (2016) Ultrasonic Decontamination Technology Application in Nuclear Power Plant[J]. Development & Innovation of Machinery & Electrical Products, 29: 46-48
11. Wang Q Z, Zhao X T, Cheng J M, et al (2006) Experiment and research on the scale inhibition and heat transfer enhaneement of ultrasonic[J]. Electrical Equipment, 12: 38-40.
12. Worapol T, Jatuporn T (2019) Development of an industrial ultrasonic cleaning tank based on harmonic response analysis[J]. Ultrasonics, 91: 68-76.
13. Kim W J, Park K H, Choi J, et al (2010) Visualization and minimization of disruptive bubble behavior in ultrasonic field. Ultrasonic, 50: 798-802.
14. Soudagar S R, Samant S D (1995) Seniquantitative characterization of ultrasonic cleaner using a novel piezoelctric pressure intensity measurement probe[J]. Ultransonics Sonochemistry, 2: 49-53.
15. Zheng Z B (2012) Ultrasonic sound field simulation and descaling experimental study[D]. Changsha University of Science & Technology.
16. Zhang H G, Guo C N (2005) Discussion on ultrasonic cleaning efficiency[J]. China Science and Techology Information, 24: 114.

E3S Web of Conferences **528**, 02016 (2024) https://doi.org/10.1051/e3sconf/202452802016
NET-LC 2024

Carbon-Energy Synergetic Energy Power System Economic Dispatch Method

Chong Zhiqiang[1], Song Hongyu[2], Li Zhenbin[1], Chen Liang[1], Yu Guangyao [1], Guo Haochen [3,*], Wu Zhijun [3], Dong Yitong [3]

[1] State Grid Tianjin Electric Power Research Institute, Tianjin 300384, China
[2] Baodi Power Supply Branch, State Grid Tianjin Electric Power Company, Tianjin 301899, China
[3] Key Laboratory of Smart Grid of Ministry of Education, Tianjin University, Tianjin 300072, China

Abstract. Low carbon is a key driver for the future development of power systems. This paper proposes a carbon-energy synergetic approach for economic dispatch in energy power systems. Firstly, by integrating multi-energy conversion models and carbon emission flow theories, a coupled interactive mechanism model is constructed. Building upon this foundation, an economic optimization dispatch model for comprehensive energy power systems is introduced, leveraging the synergistic value of carbon energy. This method supports economic dispatch in comprehensive energy power systems, considering both security and economic requirements while reducing carbon emissions. This integrated approach offers a promising solution for achieving low-carbon operational objectives and supporting the transition to sustainable energy infrastructure. Typical case studies in this paper validate the effectiveness of the proposed method.

1. Introduction

The challenge posed by carbon emissions represents a significant global concern, imposing constraints on societal development. In recent years, various nations have successively introduced “zero carbon” or “carbon neutrality” objectives, thereby raising the issue of carbon emissions to a level of paramount national strategic importance.[1] In 2020, China articulated its visionary goal: “China will independently enhance its national contributions by implementing more robust policies and measures, striving to attain peak carbon dioxide emissions before 2030, and ultimately aiming for carbon neutrality by 2060.” Integral to China’s pursuit of its “dual carbon” objectives is the development of a novel power system prominently featuring a substantial proportion of renewable energy sources.

The low-carbon economic dispatch of power systems is an indispensable prerequisite for ensuring the secure and stable operation of such systems. Extensive research endeavors have been undertaken in this domain. Li et al. [2] conducted an analysis delving into the comparative aspects between carbon emission flow and power flow calculations, alongside examining the factors influencing carbon emission flow within power systems. Zhang et al. [3] advanced a recursive algorithm tailored for computing carbon emission flow within power systems, drawing upon the adjacency properties inherent to node carbon potentials. Meanwhile, Guo et al. [4] delved into the distribution mechanism of carbon emission flow within the power grid, elucidating correlations among unit-node carbon flow, unit-branch carbon flow, and unit-load carbon flow. In a complementary vein, Lu et al. [5] devised a carbon flow tracking model founded upon conventional power flow tracing methodologies, thereby facilitating economic dispatch in power systems.

However, conventional operations of power systems predominantly pivot on electricity flow, neglecting the spatiotemporal distribution of carbon flow within the “source-grid-load” framework, along with the interplay between electricity and alternate energy forms such as gas and heat. Consequently, optimization and scheduling strategies crafted under this paradigm often fail to account for the direct coupling interactions among multiple energy flows, thereby posing challenges in concurrently addressing the system’s low-carbon assessment criteria. Furthermore, quantifying carbon emissions across diverse system components proves to be a challenging endeavor. Notably, considerable disparities manifest within the spatiotemporal distributions of carbon flow and electricity flow, compounded by intricate interactions, all of which are subject to concurrent constraints imposed by factors such as the carbon emissions intensity of source-side units, network topology, and load energy consumption. As such, characterizing the synergistic relationship between carbon and energy becomes a formidable task[6].

Operational scheduling within current frameworks predominantly prioritizes electricity flow, accentuating system economic viability and security, albeit at the expense of insufficient consideration for low-carbon operational objectives and overlooking the ramifications of “cold-heat-electricity-gas” synergy on carbon flow distribution within the system. Moreover, prevalent systems often rely on electricity generation volumes and associated carbon emissions intensities to compute carbon

*Corresponding author’s e-mail: ghctju@tju.edu.cn

emissions, thus exacerbating challenges in quantifying carbon emissions from various system components. In contrast, the optimization and scheduling of integrated energy systems necessitate comprehensive consideration of the distinctive characteristics precipitated by the coupling of multiple energy flows, which entails the seamless integration of spatiotemporal distribution analyses of carbon flow and nascent domains such as carbon trading markets, thereby facilitating the attainment of low-carbon operational objectives for power systems[7].

Therefore, this paper proposes a carbon-energy synergetic energy power system (EPS) economic dispatch method. Firstly, based on the multi-energy conversion relationship model of the system, combined with the theory of carbon emission flow, a coupled interactive mechanism model of carbon flow, power flow, and multi-energy flow in the power system is established. Building upon this foundation, the synergistic value of carbon-energy is fully leveraged to develop an economic dispatch model for comprehensive EPSs with carbon-energy synergy. This model ensures consideration of both the security and economic requirements of system operation while effectively reducing carbon emissions and supporting the construction of novel power systems.

2. Coupled interactive mechanism model of power systems

2.1 Multi-energy flow and carbon trading coupling model

The energy hub describes the functional relationship between multiple energy inputs and outputs in a multi-energy coupling system, as shown in Equation (1):

$$\underbrace{\begin{bmatrix} L_1 \\ L_2 \\ \vdots \\ L_n \end{bmatrix}}_{L} = \underbrace{\begin{bmatrix} c_{11} & c_{12} & \cdots & c_{1m} \\ c_{21} & c_{22} & \cdots & c_{2m} \\ \vdots & \vdots & \ddots & \vdots \\ c_{n1} & c_{n2} & \cdots & c_{nm} \end{bmatrix}}_{C} \underbrace{\begin{bmatrix} P_1 \\ P_2 \\ \vdots \\ P_m \end{bmatrix}}_{P} + \underbrace{\begin{bmatrix} k_{11} & k_{12} & \cdots & k_{1u} \\ k_{21} & k_{22} & \cdots & k_{2u} \\ \vdots & \vdots & \ddots & \vdots \\ k_{n1} & k_{n2} & \cdots & k_{nu} \end{bmatrix}}_{K} \underbrace{\begin{bmatrix} H_1 \\ H_2 \\ \vdots \\ H_u \end{bmatrix}}_{H} \tag{1}$$

The matrix form of Equation (1) is shown in Equation (2):

$$\boldsymbol{L} = [\boldsymbol{C} \quad \boldsymbol{K}] \begin{bmatrix} \boldsymbol{P} \\ \boldsymbol{H} \end{bmatrix} \tag{2}$$

where $L_i\left(i=1,2,\cdots,n\right)$ represents the output of the i-th energy form in the energy hub; $P_w\left(w=1,2,\cdots,m\right)$ represents the input of the w-th energy form in the energy hub; c_{iw} denotes the coupling factor, typically composed of the product of allocation coefficients and equipment efficiency coefficients; $H_\tau\left(\tau=1,2,\cdots,u\right)$ represents the charging and discharging power of the τ -th energy form; $k_{i\tau}$ denotes the energy storage coupling factor; $\boldsymbol{L}$ denotes the output power vector; $\boldsymbol{C}$ represents the coupling matrix of conversion devices; $\boldsymbol{P}$ represents the input power vector; $\boldsymbol{K}$ represents the coupling matrix of storage devices; $\boldsymbol{H}$ represents the power vector of energy storage charging and discharging.

The energy hub model embodies the processes of energy transmission, conversion, and storage, but it struggles to depict the coupling relationship between multiple energy flows and carbon trading volumes. Carbon trading volume refers to the difference between the carbon emissions of the system and the gratuitous carbon quota. Currently, in China, the calculation of carbon trading volume is determined by the difference between the carbon emission coefficients of various types of energy and the gratuitous carbon quota coefficients published by provincial and municipal governments. In this paper, we introduce carbon trading volume into Equation (1) to establish a coupled model between multiple energy flows and carbon trading volume, referred to as the multi-energy flow and carbon trading coupling model, as shown in Equation (3):

$$\begin{bmatrix} L_1 \\ L_2 \\ \vdots \\ L_n \\ E \end{bmatrix} = \begin{bmatrix} c_{11} & c_{12} & \cdots & c_{1m} \\ c_{21} & c_{22} & \cdots & c_{2m} \\ \vdots & \vdots & \ddots & \vdots \\ c_{n1} & c_{n2} & \cdots & c_{nm} \\ \alpha_1-\beta_1 & \alpha_2-\beta_2 & \cdots & \alpha_m-\beta_m \end{bmatrix} \begin{bmatrix} P_1 \\ P_2 \\ \vdots \\ P_m \end{bmatrix} + \begin{bmatrix} k_{11} & k_{12} & \cdots & k_{1u} \\ k_{21} & k_{22} & \cdots & k_{2u} \\ \vdots & \vdots & \ddots & \vdots \\ k_{n1} & k_{n2} & \cdots & k_{nu} \\ \varphi_1-\rho_1 & \varphi_2-\rho_2 & \cdots & \varphi_u-\rho_u \end{bmatrix} \begin{bmatrix} H_1 \\ H_2 \\ \vdots \\ H_u \end{bmatrix} \tag{3}$$

It can be abbreviated as:

$$\underbrace{\begin{bmatrix} \boldsymbol{L} \\ E \end{bmatrix}}_{M} = \underbrace{\begin{bmatrix} \boldsymbol{C} \\ \boldsymbol{A} \end{bmatrix}}_{D} \boldsymbol{P} + \underbrace{\begin{bmatrix} \boldsymbol{K} \\ \boldsymbol{B} \end{bmatrix}}_{U} \boldsymbol{H} \tag{4}$$

The matrix elements in the (n+1)-th row represent the difference between the carbon emission coefficient and the free carbon quota coefficient; $\alpha_w\left(w=1,2,\cdots,m\right)$ denotes the carbon emission coefficient of the w-th input energy source in the energy hub; $\beta_w\left(w=1,2,\cdots,m\right)$ represents the free carbon quota coefficient of the w-th input energy source in the energy hub; $\varphi_\tau\left(\tau=1,2,\cdots,u\right)$ stands for the carbon emission coefficient of the τ -th energy storage in the energy hub, while $\rho_\tau\left(\tau=1,2,\cdots,u\right)$ denotes the free carbon quota coefficient of the τ -th energy storage in the energy hub; E denotes the carbon trading volume of the energy hub; $\boldsymbol{A}$ is the $1\times m$ dimension carbon trading volume coefficient matrix for the energy supply side, $\boldsymbol{A}=[(\alpha_1-\beta_1),(\alpha_2-\beta_2),\cdots,(\alpha_m-\beta_m)]$; $\boldsymbol{B}$ is the $1\times u$

 https://doi.org/10.1051/e3sconf/202452802016

dimension carbon trading volume coefficient matrix for energy storage, $\boldsymbol{B}=[(\varphi_1-\rho_1),(\varphi_2-\rho_2),\cdots,(\varphi_u-\rho_u)]$; $\boldsymbol{M}$ represents the $(n+1)\times 1$ dimension carbon-energy output matrix; $\boldsymbol{D}$ is the $(n+1)\times m$ dimension combined matrix of carbon-energy coupling coefficients for the energy supply side, while $\boldsymbol{U}$ is the $(n+1)\times u$ dimension combined matrix of carbon-energy coupling coefficients for energy storage. The carbon emission coefficients α_w and free carbon quota coefficients of various input energy sources β_w, as well as the carbon emission coefficient φ_τ and free carbon quota coefficients ρ_τ of energy storage, can be obtained from the carbon emission quota allocation plans announced by provincial and municipal governments. Additionally, the elements in the coupling matrices $\boldsymbol{C}$ and $\boldsymbol{K}$ of conversion equipment and storage equipment, respectively, can be determined by the efficiency and energy distribution ratio coefficients of various types of energy conversion equipment in Equation (4). Hence, the elements in the proposed combined matrices $\boldsymbol{D}$ for the energy supply side and $\boldsymbol{U}$ for energy storage are known quantities. Given the energy input data, various energy output powers and their corresponding carbon trading volumes can be determined through the carbon-energy coupling coefficient matrix.

The carbon-energy coupling model established in this paper, Equation (4), integrates the multi-energy flow $\boldsymbol{L}$ and carbon trading volume $\boldsymbol{E}$ in matrix form, reflecting the coupling relationship between internal multi-energy flow and carbon trading volume within the system, as well as the connection with input-side energy sources. Through Equation (4), it is possible to obtain information on the system's multi-energy flow while simultaneously determining the carbon trading volume involved in the carbon market.

2.2 Coupled interactive mechanism model of power systems

The typical EPS structure, as shown in Figure 1, consists of combined heat and power (CHP) units, gas boilers (GB), electric chillers (EC), electric boilers (EB), battery energy storage systems (BESS), photovoltaic (PV) systems, and wind turbines (WT), among other equipment [8].

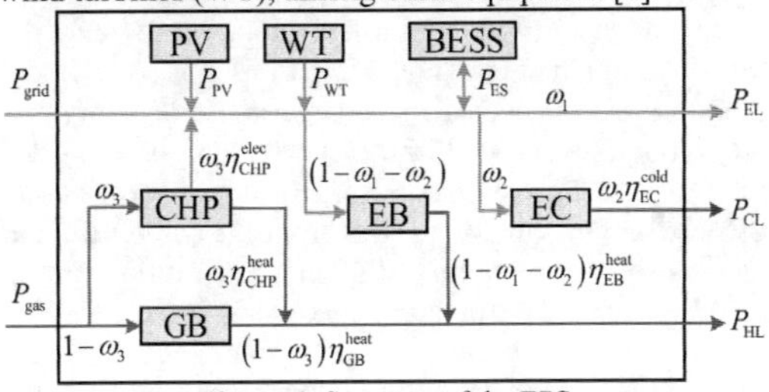

Figure 1. Structure of the EPS.

The EPS purchases natural gas and electricity from external sources and utilizes energy conversion devices along with electric energy storage to meet users' demands for electricity, heating, and cooling loads. Additionally, the EPS can participate in carbon trading to profit from surplus carbon emission allowances. The following is an analysis of the carbon emission sources of the EPS. The main sources of carbon emissions for the EPS include emissions from primary energy supply and energy storage devices, namely, purchased electricity from the grid, natural gas consumption, electricity generation from PV and WT systems, and carbon emissions from electric energy storage. Although PV, WT, and electric energy storage can achieve zero carbon emissions during operation, they still generate certain carbon emissions during the manufacturing and transportation processes. Therefore, adopting a life cycle analysis approach can normalize the carbon emission coefficients during the operation of PV, WT, and electric energy storage [9].

According to Equation (4), the coupling relationship between multi-energy flows and carbon trading in Figure 1 is modeled, as shown in Equation (5):

$$\begin{bmatrix} P_{EL} \\ P_{CL} \\ P_{HL} \\ E_{tr} \end{bmatrix} = \begin{bmatrix} \omega_1 & \omega_1\omega_3\eta_{CHP}^{elec} & \omega_1 & \omega_1 \\ \omega_2\eta_{EC}^{cold} & \omega_2\omega_3\eta_{CHP}^{elec}\eta_{EC}^{cold} & \omega_2\eta_{EC}^{cold} & \omega_2\eta_{EC}^{cold} \\ (1-\omega_1-\omega_2)\eta_{EB}^{heat} & (1-\omega_1-\omega_2)\omega_3\eta_{CHP}^{elec}\eta_{EB}^{heat}+\omega_3\eta_{CHP}^{heat}+(1-\omega_3)\eta_{GB}^{heat} & (1-\omega_1-\omega_2)\eta_{EB}^{heat} & (1-\omega_1-\omega_2)\eta_{EB}^{heat} \\ \alpha_{grid}-\beta_{grid} & \alpha_{gas}-\omega_3\eta_{CHP}^{elec}\beta_{CHP}-(1-\omega_3)\eta_{GB}^{heat}\beta_{GB} & \alpha_{PV}-\beta_{PV} & \alpha_{WT}-\beta_{WT} \end{bmatrix} \times \begin{bmatrix} P_{grid} \\ P_{gas} \\ P_{PV} \\ P_{WT} \end{bmatrix} - \begin{bmatrix} \omega_1 \\ \omega_2\eta_{EC}^{cold} \\ (1-\omega_1-\omega_2)\eta_{EB}^{heat} \\ \varphi_{ES}-\rho_{ES} \end{bmatrix} P_{ES} \tag{5}$$

where P_{grid}, P_{gas}, P_{PV}, and P_{WT} represent the purchased power from the grid, natural gas injection power, and power injection from PV and WT, respectively; P_{EL}, P_{CL}, P_{HL}, and E_{tr} represent the electric/cooling/heating loads and carbon trading volume, respectively; ω_1 denotes the power allocation coefficient for supplying electric loads; ω_2 represents the power allocation coefficient for supplying EB; ω_3 denotes the gas power allocation coefficient for supplying CHP; η_{CHP}^{elec}, η_{CHP}^{heat}, η_{GB}^{heat}, η_{EB}^{heat}, and η_{EC}^{cold} represent the electric efficiency, heat efficiency of CHP, heat efficiency of GB, heat efficiency of EB, and cooling efficiency of EC, respectively; P_{ES} represents the charging/discharging power of the battery energy storage system (BESS); when it is greater than 0, it indicates charging, and when it is less than 0, it indicates discharging; α_{grid}, α_{PV}, α_{WT}, α_{gas}, and φ_{ES} represent the unit power carbon emission coefficients for purchased power, PV, WT, natural gas, and the battery energy storage system, respectively; β_{grid},

β_{PV}, β_{WT}, β_{CHP}, β_{GB}, and ρ_{ES} represent the unit power-free carbon quota coefficients for purchased power, PV, WT, CHP, GB, and the battery energy storage system, respectively. In Equation (5), the carbon trading volume E_{tr} of EPS is the sum of the carbon trading volumes for purchased power, PV, WT, natural gas, and the battery energy storage system (if the carbon emission coefficient for any part exceeds the carbon quota coefficient, the carbon trading volume for that part is positive; if the carbon emission coefficient for any part is less than the carbon quota coefficient, the carbon trading volume for that part is negative). When it is positive, EPS needs to purchase carbon emission rights from the carbon trading market; when it is negative, EPS can sell surplus carbon emission rights in the carbon trading market.

Based on the carbon emission quota allocation plans published by provincial and municipal governments, the carbon emission coefficients and free carbon emission coefficients for purchased power, natural gas, renewable energy, and battery energy storage in Equation (5) can be obtained [10]. Meanwhile, based on the given efficiencies of various energy devices, the multi-energy flows inside the EPS shown in Figure 1 and the carbon trading volume participating in the carbon market can be calculated by using Equation (5).

It should be noted that ω_1, ω_2, and ω_3 are intermediate variables introduced to meet the modeling requirements of the energy hub. During model solving, these variables can be expressed by using the output power of each equipment unit and the efficiency of the equipment. The mathematical expressions for ω_1, ω_2, and ω_3 are given by Equations (6)-(8):

$$\omega_1 = 1 - \frac{P_{EC}^{elec} + P_{EB}^{elec}}{P_{grid} + \omega_3 \eta_{CHP}^{elec} P_{gas} + P_{PV} + P_{WT} + P_{ES}^{dis} - P_{ES}^{ch}} \quad (6)$$

$$\omega_2 = \frac{P_{EB}^{elec}}{P_{grid} + \omega_3 \eta_{CHP}^{elec} P_{gas} + P_{PV} + P_{WT} + P_{ES}^{dis} - P_{ES}^{ch}} \quad (7)$$

$$\omega_3 = \frac{G_{CHP}}{P_{gas}} \quad (8)$$

3. Carbon-energy synergetic energy power system economic dispatch model

The economic dispatch model of the EPS considering low-carbon optimization operation aims to minimize the total cost, including fuel cost, maintenance cost, and carbon trading cost. The objective function is formulated as Equation (9):

$$\min F = C_{\mathrm{Main}} + C_{\mathrm{Fuel}} + C_{\mathrm{CO_2}} \quad (9)$$

$$C_{\mathrm{Main}} = \sum_{t=1}^{T} P_{j,t} c_{\mathrm{main},j} \quad (10)$$

$$C_{\mathrm{Fuel}} = \sum_{t=1}^{T} \left(q_{t,\mathrm{gas}} \lambda_{\mathrm{gas}} + q_{t,\mathrm{elec}} \lambda_{\mathrm{elec}} \right) \Delta t \quad (11)$$

$$C_{\mathrm{CO_2}} = \lambda_{\mathrm{c}} E_{\mathrm{tr}} \quad (12)$$

where C_{Fuel} represents fuel costs; C_{Main} denotes maintenance costs; $C_{\mathrm{CO_2}}$ stands for carbon trading costs, when $C_{\mathrm{CO_2}}$ is negative, it indicates selling, resulting in profits for EPS; when $C_{\mathrm{CO_2}}$ is positive, it indicates buying, incurring costs for EPS; λ_{c} represents the carbon trading price; $c_{\mathrm{main},j}$ denotes the maintenance cost per unit power of the j-th device; $P_{j,t}$ represents the output power of the *j*-th device at time *t*; *T* represents the total operational time per day, which is 24 hours; $q_{t,\mathrm{gas}}$ and $q_{t,\mathrm{elec}}$ represent the natural gas purchase quantity and the purchased electricity from the external power system at time *t*, respectively; λ_{gas} and λ_{elec} represent the prices of natural gas and electricity, respectively; Δt denotes the time interval of system operation, which is set to be 1 hour.

The constraints of EPS include both equality constraints and inequality constraints. The equality constraints consist of the system power balance constraints established by using the carbon-energy coupling model, as shown in Equation (5). The inequality constraints consist of constraints on the upper and lower limits of equipment output during operation, as shown in Equations (13)-(14).

$$0 \le P_{j,t} \le Q_j \quad (13)$$

$$0 \le P_{\mathrm{grid}} \le \overline{P}_{\mathrm{grid}} \quad (14)$$

where Q_j represents the rated capacity of equipment *j*; $\overline{P}_{\mathrm{grid}}$ denotes the maximum transmission power of the interconnection line.

Additionally, the energy storage system needs to satisfy the energy storage relationship before and after charging and discharging, as well as the constraints on maximum charging and discharging power and energy storage capacity, as shown in Equations (15) to (19):

$$W_{\mathrm{ES}}^{t+1} = W_{\mathrm{ES}}^{t} + (P_{\mathrm{ES}}^{\mathrm{ch}} \eta_{\mathrm{ES}}^{\mathrm{ch}} - \frac{P_{\mathrm{ES}}^{\mathrm{dis}}}{\eta_{\mathrm{ES}}^{\mathrm{dis}}}) \Delta t \quad (15)$$

$$0 \le P_{\mathrm{ES}}^{\mathrm{ch}} \le u_{\mathrm{ES}} P_{\mathrm{ES}}^{\mathrm{ch,max}} \quad (16)$$

$$0 \le P_{\mathrm{ES}}^{\mathrm{dis}} \le \left(1 - u_{\mathrm{ES}}\right) P_{\mathrm{ES}}^{\mathrm{dis,max}} \quad (17)$$

$$W_{\mathrm{ES}}^{\mathrm{min}} \le W_{\mathrm{ES}} \le W_{\mathrm{ES}}^{\mathrm{max}} \quad (18)$$

$$W_{\mathrm{ES}}^{0} = W_{\mathrm{ES}}^{\mathrm{end}} \quad (19)$$

where W_{ES}^{t} represents the energy storage before charging or discharging; W_{ES}^{t+1} t represents the energy storage after charging or discharging; $P_{\mathrm{ES}}^{\mathrm{ch}}$ and $P_{\mathrm{ES}}^{\mathrm{dis}}$ represent the charging and discharging power of the energy storage system, respectively; $\eta_{\mathrm{ES}}^{\mathrm{ch}}$ and $\eta_{\mathrm{ES}}^{\mathrm{dis}}$ represent the charging and discharging efficiency of the energy storage system, respectively; $P_{\mathrm{ES}}^{\mathrm{ch,max}}$ and $P_{\mathrm{ES}}^{\mathrm{dis,max}}$ represent the maximum charging and discharging power of the energy storage system, respectively; u_{ES} is an introduced binary variable used to control whether the

https://doi.org/10.1051/e3sconf/202452802016

energy storage system can simultaneously charge and discharge; W_{ES}^{min} and W_{ES}^{max} represent the minimum and maximum energy storage capacity of the energy storage system, respectively; W_{ES}^{0} and W_{ES}^{end} represent the start and end times of the operating cycle of the energy storage system, respectively.

4. Case analysis

4.1 Case parameters

In this paper, the typical EPS structure shown in Figure 1 is used as a test example. The EPS adopts the 'grid-connected not online' mode. Taking winter as an example, the winter typical day data is selected to represent the operation of EPS. The electricity/heat/cooling load curve of typical winter days is shown in Figure 2 [11].

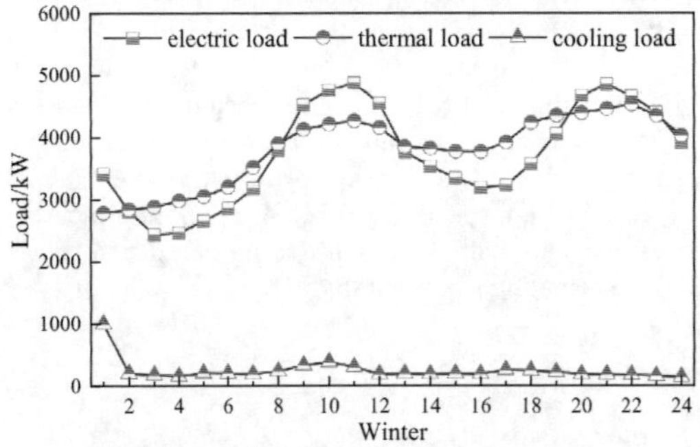

Figure 2. Typical daily load curve in winter.

The operating parameters of all devices in Figure 1 are shown in Table 1. Time-of-use electricity price, carbon trading price, and natural gas price are shown in Table 2. The carbon emission coefficient of the unit active power of different energy types and the free carbon quota coefficient allocated by the government are shown in Table 3 [10].

Table 1. EPS equipment parameters.

Equipment	Capacity/kW	Maintenance coefficient yuan/kW	Conversion efficiency
CHP	600	0.04	Electric: 0.30 Heat: 1.732
GB	3300	0.02	0.94
EB	3600	0.02	0.88
EC	2700	0.02	4.10
PV	5600	0.035	-
WT	4000	0.030	-
BESS	600 kWh	0.026	Charge:95% Discharge:95%

Table 2. Energy prices and carbon trading prices.

	Time	Price yuan/kWh
Electric price	Valley period (0:00-8:00)	0.4
	Peak period (8:00-12:00, 16:00-20:00)	1.6
	Ordinary period (12:00-16:00, 20:00-24:00)	0.8
Carbon trading prices	All day	0.35
Natural gas prices	All day	0.26

Table 3. Carbon emission coefficient and carbon quota coefficient.

Type of energy (g/kWh)	Grid	Wind	PV	BESS	Gas
Carbon emission coefficient	1303	43	25	91.33	564.7
Carbon quota coefficient	798	78	78	91.33	CHP: 424 GB: 152

4.2 Results analysis

To illustrate the correctness of the carbon-energy coupling model and the effectiveness of the carbon-energy collaborative optimization method, three scenarios are set up for comparative analysis.

Scenario I: traditional EPS economic dispatch without considering environmental protection;

Scenario II: EPS low-carbon scheduling to minimize carbon emissions without considering the economy;

Scenario III: considering the economy and environmental protection, the EPS carbon energy collaborative scheduling is carried out by using the method proposed in this paper.

Taking scenario III as an example, the correctness of the carbon-energy coupling model in this paper to describe the dependence relationship between EPS multi-energy flow and carbon trading volume is illustrated.

It can be seen from Equation (5) that the efficiency of each device in the carbon energy coupling coefficient matrix is used as a component element in the matrix to reflect the attachment relationship between the multi-energy flow and carbon trading volume within EPS. Therefore, this paper takes changing the power generation efficiency of CHP as an example to analyze the dependence relationship between multi-energy flow and carbon trading volume. The power generation efficiency of CHP under scenario I is increased from 0.3 to 0.35 and 0.4. The output power and purchase power of each device on typical days in winter and the comparison results of EPS carbon trading volume are shown in Figures 3 and 4.

It can be seen from Figure 3 that when the power generation efficiency of CHP increases from 0.3 to 0.35 and 0.4, the output power of each device in EPS changes accordingly. The output power of CHP increases, and the consumed natural gas power increases, which are 411.30 kW and 13833 kW, respectively. The output power of GB decreases, and the consumed natural gas power decreases, which are 411.65 kW and 14492.3 kW, respectively. The increased natural gas power of CHP is less than the reduced natural gas power of GB so the total natural gas power consumed by EPS decreases, which are 0.35 kW and 658.93 kW, respectively. In addition, the amount of electricity purchased from the external power grid is reduced, and the carbon emissions generated by EPS are reduced. At the same time, the increase in CHP electrical efficiency reduces carbon emissions due to energy non-conversion, which in turn leads to an increase in surplus carbon quotas available for sale, that is, an increase in carbon trading volume. As shown in Figure 4, the carbon energy coupling model Equation (5) in this paper can not only describe the coupling relationship between multi-energy flows but also describe the coupling relationship between multi-energy flows and carbon trading volume within EPS.

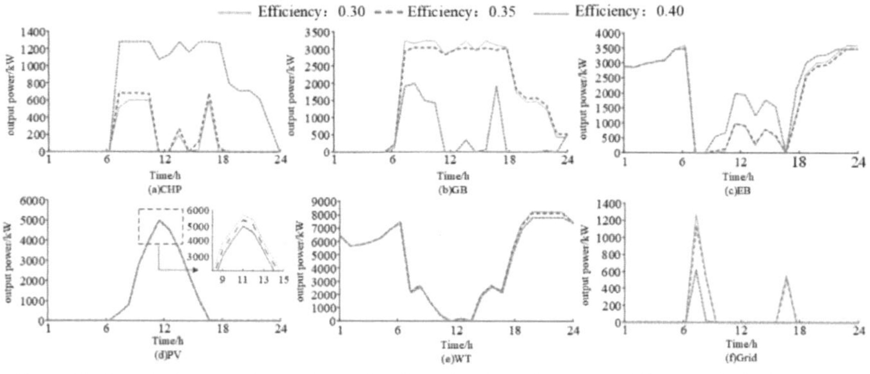

Figure 3. The output power of each piece of equipment and purchased electricity on typical winter days.

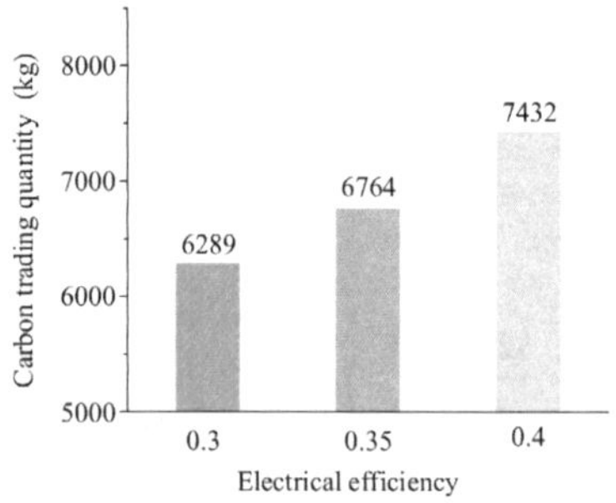

Figure 4. The carbon trading volume comparison chart of different CHP electrical efficiency.

The comparison results of the scheduling schemes in the three scenarios are shown in Table 4. The negative sign '-' of the numerical results of carbon trading volume and carbon trading cost in Table 4 indicates the carbon emissions sold and the carbon trading gains obtained by EPS participating in carbon trading.

Table 4. EPS dispatching schemes in three scenarios.

Scenarios	Fuel cost /yuan	Maintenance /yuan	Carbon trading cost /yuan	Total cost /yuan	Carbon emissions /kg
I	36268.22	4749.34	-543.45	40474.11	63684.68
II	40556.05	4859.21	-2987.21	42428.05	50948.25
III	36816.38	4854.31	-1901.30	39769.39	59948.26

Compared with scenario I and scenario II, the fuel cost, maintenance cost, and total cost of scenario II increased by 4287.83 yuan, 109.87 yuan, and 1953.91 yuan respectively, but the carbon transaction cost of scenario II decreased by 2443.76 yuan and the carbon emission decreased by 12736.43 kg. This shows that the use of EPS low-carbon scheduling can reduce the carbon emissions of the system without considering the economy of the system, but it needs to increase the total operating cost.

Compared with scenario I and scenario III, the fuel cost of scenario III increased by 548.16 yuan, the maintenance cost increased by 104.97 yuan, but the carbon transaction cost of scenario III decreased by 1357.85 yuan, the total cost decreased by 704.72 yuan, and the carbon emission decreased by 3736.42 kg. This shows that the carbon energy coordinated economic dispatch method proposed in this paper can consider the economy and environmental protection of the system, reduce the total operating cost of the system while reducing the carbon emission of the system, and realize the coordinated operation of EPS energy saving and carbon reduction.

5. Conclusion

This paper establishes a carbon-energy coupled interactive model for comprehensive energy power systems and proposes a carbon-energy synergetic economic dispatch method for energy power systems. The simulation verification yields the following conclusions:

1) Economic dispatch in energy power systems requires a thorough consideration of the "carbon-electricity" synergy, balancing system economy and safety while incorporating carbon emission targets;

2) The proposed method can effectively utilize complementary synergies among multiple energy flows, enhancing the low-carbon economic performance of system operation.

Future research will further consider the uncertainty of carbon markets and explore stochastic optimization dispatch methods.

Acknowledgments

This work was financially supported by the research project of State Grid Tianjin Electric Power Company (KJ-22-1-24).

References

1. Fahad lqbal, Anwar Shahzad Siddiqui (2017) Optimal configuration analysis for a campus microgrid—A case study. J. Protection and Control of Modern Power Systems., 2(3): 1-12.
2. Li Peng, Wang Jiahao and Wang Zixuan (2023) Collaborative optimal scheduling of the community integrated energy system based on multi-energy supply/demand-cost mapping analysis. J. Proceedings of the CSEE., 43: 4307-4319.
3. Zhang Shenxi, Wang Danyang and Cheng Haozhong (2022) Key technologies and challenges of low-carbon integrated energy system planning for dual carbon targets. J. Automation of Electric Power Systems., 46: 189-207.
4. Guo Jingrong, Xiang Yue and Wu Jiajie (2023) Low-carbon optimal scheduling of integrated electricity-gas energy systems considering CCUS-P2G technology and risk of carbon market. J. Proceedings of the CSEE., 43: 1290-1303.
5. Lu Zhigang, Yang Yu and Geng Lijun (2018) Low-carbon economic dispatch of the integrated electrical and heating systems based on benders decomposition. J. Proceedings of the CSEE., 38: 1922-1934.
6. Mahdi Nozarian, Alireza Fereidunian (2023) Masoud Barati. Reliability-Oriented Planning Framework for Smart Cities: From Interconnected Micro Energy Hubs to Macro Energy Hub Scale. J. IEEE Systems Journal., 17: 3798-3809.
7. Mazzoni Stefano, Ooi Sean, Nastasi Benedetto (2019) Energy storage technologies as techno-economic parameters for master-planning and optimal dispatch in smart multi energy systems. J. Apply Energy., 254: 113682.
8. Zhao Haipeng, Miao Shihong and Li Chao (2022) Research on optimal operation strategy for park-level integrated energy systems considering cold-heat-electric demand coupling response characteristics. J. Proceedings of the CSEE., 42: 573-589.
9. Tang Hao (2018) Capacity optimization of wind/PV hybrid power generation system based on life cycle cost/carbon emissions. D. Ningxia University.
10. Wang Zesen, Shi Yan and Tang Yanmei (2019) Low carbon economy operation and energy efficiency analysis of integrated energy systems considering LCA energy chain and carbon trading mechanism. J. Proceedings of the CSEE., 39: 1614-1626.
11. Mu Yunfei, Wang Congshan and Cao Yan (2022) A CVaR-based risk assessment method for park-level integrated energy system considering the uncertainties and correlation of energy prices. J. Energy., 247: 123549.

Advancements in Substation Inspection Robots: A Review of Research and Development

Xiaoyan Wei * [1,2] ,William P. Rey [1]

[1] School of Information Technology, Mapua University, Manila, Philippine
[2] Xi'an Siyuan University, Shaan, Xi'an,China

Abstract: The emergence of power station inspection robots signifies a significant leap forward in intelligent automation within the domain of substation maintenance. These robots are equipped with the capability to autonomously traverse through substations, conducting meticulous inspections of equipment and facilities. By replacing manual inspection methods, they bring forth a multitude of benefits including heightened efficiency, cost reduction, and enhanced personnel safety. As the realms of artificial intelligence and robotics continue to advance, the development of substation inspection robots has become a focal point in the power industry. This paper offers a comprehensive overview of the current landscape of inspection robots, offering insights from both domestic and international contexts. Furthermore, it delves deeply into the critical technologies that underpin substation inspection robots, showcasing the cutting-edge advancements driving their unparalleled performance. In addition to providing a comprehensive overview of existing technologies, this work offers a forward-looking perspective on the potential applications of these robots. It provides a glimpse into the future of substation maintenance practices, envisioning a landscape where these robots play an indispensable role in optimizing operational efficiency, ensuring regulatory compliance, and enhancing overall system reliability. Moreover, the paper outlines key areas for future technical development, paving the way for continued innovation and advancement in this dynamic and rapidly evolving field.

1. Introduction

Substations play a critical role in ensuring the consistent and efficient supply of power, making regular inspection of substation equipment imperative for their long-term, stable operation [1][2][3][4]. Currently, manual inspection predominates as the method for outdoor substation inspections. However, this approach is beset by challenges such as high labor intensity, poor accuracy, and timeliness issues. The emergence of substation inspection robots, as autonomous intelligent systems, has emerged as a focal point in addressing the need for unattended substation monitoring [5].

In contrast to traditional manual methods, substation inspection robots offer a solution that effectively mitigates the shortcomings associated with manual inspections [6][7][8]. These robots are capable of replacing on-duty personnel in executing routine inspection tasks for substation equipment, thereby significantly enhancing the efficiency and quality of substation operations. By promptly identifying faults and issuing early warnings, they contribute to reducing the risks of equipment failure and power outages. Furthermore, they alleviate the workload of inspection personnel, lower the operational and maintenance costs of substations, and elevate the overall intelligence level of substations.

Acknowledging the pivotal role that substation inspection robots play in facilitating regular inspections and their considerable potential for further development, it becomes evident that conducting a thorough review of research and development in this field is imperative [9][10]. This paper seeks to delve deeply into the multifaceted landscape of substation inspection robotics, aiming not only to provide researchers with an enhanced understanding but also to furnish valuable perspectives for the advancement of these technologies.

To achieve this objective, this paper aims to provide a comprehensive understanding of the current state of research and development in the field of substation inspection robots, while also outlining prospects and challenges. By doing so, it endeavors to stimulate widespread adoption and utilization of these robots across various operational contexts within the power industry. In detail, section II, this paper offers an extensive overview of the existing landscape of substation inspection robots, examining both domestic and international developments. Through an in-depth analysis of current practices and trends, readers will gain valuable insights into the global evolution of this technology, thus laying a solid foundation for subsequent discussions. Section III delves into the intricate details

*Correspondence: xwei@mymail.mapua.edu.ph; wprey@mapua.edu.ph

of the key technologies that serve as the backbone of substation inspection robots. By dissecting these technologies and elucidating their functionalities, this section provides researchers and practitioners with a deeper understanding of the technical underpinnings driving the effectiveness and efficiency of these robots. Finally, in Section IV, the paper adopts a forward-thinking approach by exploring the potential avenues for future advancements in substation inspection robots. By outlining potential applications and addressing anticipated technical challenges, this section aims to inspire innovation and collaboration within the research community, ultimately paving the way for the realization of more sophisticated and robust inspection solutions. Through this structured and comprehensive approach, this paper endeavors to not only inform but also inspire action, driving the widespread adoption and implementation of substation inspection robots to enhance the reliability, efficiency, and safety of power infrastructure worldwide.

2. Research status

The development of substation inspection robots has arisen from the challenging conditions prevalent in outdoor substations, characterized by adverse working environments, significant inspection burdens, and insufficient reliability. These robots, built on flexible mobile platforms, are equipped with advanced technology, including high-definition imaging systems, video acquisition devices, and a variety of sensors. This enables them to meticulously assess equipment operation and monitor the substation environment effectively. Globally, substation inspection robots have gained widespread popularity and adoption, consistently delivering satisfactory inspection outcomes [11].

2.1 Research status of substation inspection robots abroad

Since the 1980s, Japan Canada, and other countries have carried out research on transformer inspection robots. Japan's Mitsubishi Corporation and Tokyo Electric Power successfully developed a 500kV substation inspection robot in 1980. The visible light camera and infrared camera equipped on the inspection robot can monitor the operation of the power equipment in the substation and the temperature of the equipment in real-time, as shown in Figure 1.

In 2008, Hydro-Quebec in Canada developed a suspended cable inspection robot that runs smoothly on cables up to 735kV and 1000A through four pulleys [12] and can cross the obstacles at the cable connections, as shown in Figure 2.

Figure 1. Japan Mitsubishi inspection robot

Figure 2. Canada Quebec inspection robot

The University of Sao Paulo in Brazil developed a substation inspection robot with steel cable suspension. The inspection robot monitors the thermal changes of equipment in the substation in real-time through its infrared thermal camera [13]. In 2012, the New Zealand Power Grid Corporation cooperated with Massey University to develop an all-terrain substation inspection robot. The robot carries a high-definition visible light camera to acquire and return images and videos of equipment in the substation and also carries an ultrasonic sensor for environmental detection and collision prevention. However, the working mode of the inspection robot is remote control, which cannot realize autonomous operation. Later, with the advancement of technology, a Japanese university combined GPS positioning technology to develop an intelligent robot, capable of remote accurate positioning and navigation, as well as data interaction and real intelligent operation [14]. In 2014, Beaudry et al. developed a substation inspection robot equipped with a 3D laser, infrared thermal imager, and high-definition camera based on the Husky A200 wheeled mobile robot platform. The inspection robot can autonomously navigate and inspect the substation, read all kinds of dial data, and monitor the temperature of equipment in the station. In 2015, Sweatt and others from the Colorado School of Mines in the United States designed a WiFi-based communication and positioning inspection robot. This type of robot can realize its positioning through WiFi fingerprinting and can be used in dangerous working environments such as high temperatures and toxic gas.

The development of substation inspection robots has entered a period of rapid growth in recent years. More and more foreign technology companies, well-known universities, and scientific research institutions have researched relevant robot technologies, yielding a significant number of research results in robot navigation and positioning, sensor utilization, and motion control. The United States is also a leader in robot research, development, and manufacturing technology. As early as the early 1980s, the prototype with inspection robot

https://doi.org/10.1051/e3sconf/202452802017

functionality was developed in California, USA [15]. The line patrol robot can traverse overhead power lines, utilizing visual equipment to monitor insulators, line junction points, and corona losses.

2.2 Research status of substation inspection robot in China

Compared with developed countries in Europe and the United States, as well as neighboring Japan, the development level of substation inspection robots in China is still in its infancy, having only achieved preliminary intelligence. For instance, in 1999, Shandong Luneng Intelligent Company, a subsidiary of the Research Institute of Shandong Electric Power Company of the State Grid, initiated research on substation inspection robots and established the Robot Technology Laboratory of the State Grid to delve into electric robot technology. By 2004, the first robot prototype was successfully developed, followed by the development of a series of substation inspection robots under the support of the National 863 Program. The second-generation substation inspection robot was successfully trialed at the Changqing 500kV Substation in 2005. Presently, the fifth-generation substation inspection robot has been successfully developed and deployed.

In 2007, the Shandong Electric Power Research Institute proposed the system structure of a substation inspection robot, laying the groundwork for further research and development in this field. Detailed descriptions of the realization methods of image inspection, instrument recognition, and temperature detection on the inspection robot were also provided. In 2012, the Shenyang Institute of Automation of the Chinese Academy of Sciences successfully developed a substation inspection robot capable of traveling on a preset rail, enabling all-weather inspection under harsh natural conditions such as rain, snow, and wind. Additionally, Chongqing University developed a wheeled mobile substation inspection robot, which was successfully deployed at the 500kV Banan power station.

Figure 3. Substation patrol robot developed by Guozi

Zhejiang Guozi Robot Technology Co., Ltd. has successfully developed a LiDAR-based inspection robot, which utilizes trackless navigation technology and can conduct comprehensive inspections of substations under various weather and road conditions, as depicted in Figure 3. Currently, this inspection robot is being utilized by major power grid companies across China, spanning 27 provinces in the country. It facilitates the application of full coverage of substations ranging from 66kV to 1000kV, effectively assisting power workers in completing a multitude of repetitive and high-risk tasks. In 2018, Qin et al. from Wuhan University developed a cable inspection robot based on LiDAR data positioning, as illustrated in Figure 4. This robot autonomously performs positioning inspections using LiDAR data and detects cable faults using its built-in camera, thereby significantly reducing labor costs and enhancing fault detection accuracy. The involvement of numerous scientific and technological enterprises has notably accelerated the pace of domestic substation inspection robot productization.

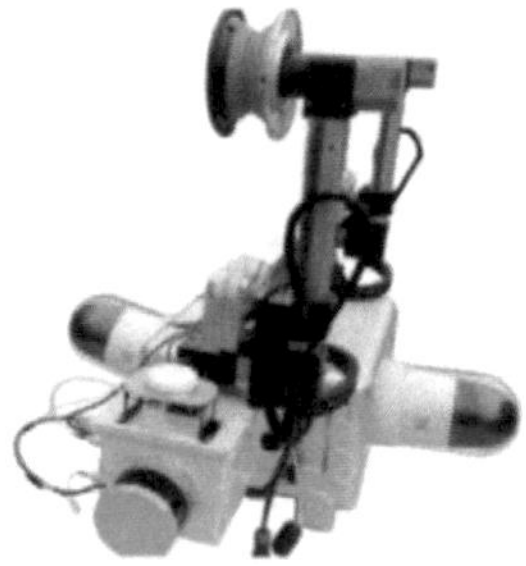

Figure 4. Cable inspection robot

Figure 5. Langchi Xinchuang

As depicted in Figure 5, the substation inspection robot developed by Shenzhen Longchi Xinchuang Technology Co., Ltd. can replace manual labor, enabling all-weather professional inspections of hazardous areas and critical equipment within substations. Through the analysis of equipment and component degradation trends based on big data, this robot establishes a maintenance foundation for equipment condition, ensuring the safety of power plant operations and maintenance. In 2019, China's Nanjing Yijiahe Company introduced a track inspection robot equipped with various sensors, offering both equipment inspection and indoor environment monitoring functions, as illustrated in Figure 6 [16]. Furthermore, in 2022, China's Zhejiang Autonomous Robot Company launched the MR200 intelligent orbital inspection robot, as showcased in Figure 7.

Figure 6. Nanjing Yijiahe track inspection robot

https://doi.org/10.1051/e3sconf/202452802017

Figure 7. MR200 intelligent orbital inspection robot

Overall, the increasing maturity of technologies such as LiDAR positioning, image recognition, and artificial intelligence has greatly enhanced the capabilities of intelligent and unmanned inspection robots. This advancement has generated heightened interest and exploration among research institutions, university laboratories, and enterprise teams. These entities are actively involved in studying and developing core technologies for substation inspection robots, thereby paving the way for new avenues of innovation in the realm of intelligent and unmanned inspection systems.

3. Substation inspection robot key technology

The effectiveness and efficiency of substation inspection robots rely heavily on several key technologies that enable accurate and comprehensive inspections. This section highlights the essential technological components that underpin the operation of these robots in the following subsections.

3.1 Visual recognition technology

Substation inspection robots require robust visual capabilities to accurately identify substation equipment and detect abnormal conditions [17]. Vision technology stands as a cornerstone in the realm of robotics. Leveraging sensors such as LiDAR and cameras, these robots acquire information about each piece of equipment in the substation, facilitating precise inspection. By employing image recognition algorithms and artificial intelligence technology, these robots gain the ability to autonomously identify equipment faults, enabling timely warnings and interventions. To further enhance the positioning accuracy of these robots, multi-sensor fusion can be employed. By integrating data from multiple sensors, the inspection robot's positioning accuracy can be significantly improved. Looking ahead, as technology continues to advance, robot positioning technology is poised to become even more accurate and intelligent.

3.2 Positioning and navigation technology

Substations typically encompass vast areas with densely distributed equipment, necessitating precise positioning and navigation capabilities for robots to autonomously carry out inspection tasks. Technologies such as GPS, inertial navigation systems, and laser ranging offer accurate positioning information for robots. Moreover, leveraging SLAM (Simultaneous Localization and Mapping) algorithms enables robots to autonomously navigate complex environments, evade obstacles, and plan inspection paths. In SLAM algorithms, localization, and mapping are closely intertwined and mutually dependent [18]. The establishment of an accurate map of an unknown environment relies on the precision of the robot's localization algorithm, while localization within an unknown environment map depends on the establishment of a local map. Over the past three decades, SLAM and related technologies have emerged as research hotspots in fields such as unmanned driving, robotics, and indoor navigation and positioning. To further enhance the positioning capabilities of robots, multi-sensor fusion can be employed, leading to improved positioning accuracy for inspection robots. Looking ahead, as technology continues to evolve, robot positioning technology is expected to become even more precise and intelligent.

3.3 Communication and data transmission

The substation inspection robot requires real-time communication with the monitoring center and the transmission of collected data to the background for analysis [19]. Stable and reliable communication and data transmission technology are essential to ensure the effectiveness of robot inspection. Wireless communication technologies such as WiFi, 4G/5G, etc., along with data compression and encryption technologies, and cloud platform storage technologies, all provide the necessary technical support for robot communication and data transmission.

3.4 Intelligent decision and control technology

Robots need to have intelligent decision-making and control capabilities and can make reasonable behavior choices according to the actual situation [20]. Through artificial intelligence technologies such as reinforcement learning and deep learning, robots can learn from historical data and make more adaptive decisions. The use of an advanced control algorithm can realize the precise operation of the robot and ensure the completion of the inspection task.

4. Prospect of substation intelligent inspection robot technology

The future of substation intelligent inspection robot technology holds immense promise, driven by ongoing advancements in key areas such as artificial intelligence, sensor integration, and communication systems. This section outlines several key prospects for the continued evolution and enhancement of substation inspection robots.

4.1 Real-time high-precision navigation and positioning

In order to ensure the stability of autonomous navigation and the high-quality acquisition of images, substation inspection robots have stringent requirements for the positioning accuracy of parking points during inspection tasks [21]. In practical applications, the slow speed of the robot leads to longer inspection times for the entire station. Therefore, it becomes essential to significantly enhance the walking speed of the vehicle body during autonomous inspection, ensuring smooth navigation and precise parking. Research on real-time and high-precision navigation technology suitable for dynamic environments has become an inevitable trend for the future.

4.2 Simplified environment awareness technology

Due to limitations in hardware performance and algorithm complexity, SLAM technology cannot achieve real-time simultaneous localization and mapping. To ensure environment perception in the scene, professionals must build the SLAM algorithm offline based on the actual scene conditions before enabling autonomous operation. This process provides preliminary information about the scene environment but reduces the ease of use of the inspection robot to some extent. Additionally, current laser SLAM algorithms still require various sensors to assist LiDAR in meeting accuracy requirements. However, due to cost, size, and the specialized environment of substations, many environmental perception sensors cannot be installed, thus diminishing the accuracy of the substation inspection robot's environmental perception. With the rapid improvement of hardware performance and the continuous innovation of SLAM algorithms, the realization of simplified SLAM algorithms is the inevitable development direction for inspection robots.

4.3 Inspection system communicates with the power information smoothly

At present, there is no standard scheme for robot systems to access the Intranet, and network security problems for robot systems to access the Intranet still exist. Therefore, under the premise of increasing efforts to solve the network security problem, it is also necessary to formulate a full set of schemes for the robot system to access the Intranet to ensure the interoperability of the robot system and the Intranet business, so that data fusion analysis can be carried out in combination with the power grid data, and the robot business can be deeply customized based on the characteristics of the power grid. Therefore, it is also an important research direction to solve the interconnection between robots and Intranet services in the future.

5. Discussion

The substation inspection robots represent a holistic application system that not only transforms traditional inspection methods but also aligns with the future trajectory of substation automation. By harnessing robotic technology, these robots not only alleviate the burden on human resources but also elevate inspection quality, resulting in more efficient and effective inspection practices. A pivotal aspect of the effectiveness of substation inspection robots lies in their ability to accurately navigate and plan inspection routes. Precise localization and path planning are essential for ensuring thorough inspections and optimizing resource utilization.

However, challenges such as achieving accurate positioning, dealing with complexities in path planning, and considering economic factors along the route highlight the necessity for continuous research and development efforts in this domain. This study emphasizes the importance of location and path planning for inspection robots, particularly in outdoor substation scenarios. Overcoming these challenges will not only improve the efficiency and reliability of substation inspections but also contribute to cost-effectiveness and sustainability in power infrastructure management.

6. Conclusion

In conclusion, the advancement of substation inspection robots represents a pivotal step toward enhancing the efficiency, reliability, and sustainability of substation maintenance practices. By addressing challenges related to localization and path planning, further research and development efforts can unlock the full potential of these robots, ultimately contributing to the advancement of power infrastructure management. Through ongoing innovation and refinement, substation inspection robots will continue to play a vital role in ensuring the safe and sustainable operation of substations, thereby supporting the broader goals of power infrastructure reliability and resilience.

References

1. Fan C., Zhu Z., Chen W., Qin K., Huang H., & Huang Z. (2024, January). Navigation of Substation Inspection Robot Based on Machine Vision. In ICSETPSD 2023: Proceedings of the First International Conference on Science, Engineering and Technology Practices for Sustainable Development, ICSETPSD 2023, 17th-18th November 2023, Coimbatore, Tamilnadu, India (p. 251). European Alliance for Innovation.
2. Shan C., Lu L., Zhou H., Su Z., Dai Y., Su T., & Pei J. (2023, December). Research on Multi-sensor Fusion Positioning and Navigation of Substation Inspection Robot Based on Beidou RTK and 3D Laser. In 2023 IEEE 4th China International Youth Conference on Electrical Engineering (CIYCEE) (pp. 1-6). IEEE.

https://doi.org/10.1051/e3sconf/202452802017

3. Jin M., Tang Y., & Wang Z. (2024, February). Comprehensive Analysis of Inspection Robot Technology and Application in Substation. In 2023 International Conference on Data Science, Advanced Algorithm and Intelligent Computing (DAI 2023) (pp. 136-147). Atlantis Press.
4. Chi Y., Li Y., Ye J., Li L., Xu K., & Liang X. (2022, April). Study on requirements for test and detection of substation inspection robot. In 2022 5th International Conference on Energy, Electrical and Power Engineering (CEEPE) (pp. 1062-1067). IEEE.
5. Song Y., Deng P., Cai M., & Chen F. (2015). The application of inspection robots in substation inspection. In Cloud Computing: 5th International Conference, CloudComp 2014, Guilin, China, October 19-21, 2014, Revised Selected Papers 5 (pp. 189-196). Springer International Publishing.
6. Li F., Ye S., & Han G. (2023, December). Research on formation control algorithm of multiple inspection robots in substations. In Third International Conference on Control and Intelligent Robotics (ICCIR 2023) (Vol. 12940, pp. 453-458). SPIE.
7. Cai T., Ren Y., Dong L., Wang X., Jia F., & Tian M. (2023, November). Positioning and Mapping Technology for Substation Inspection Robot Based on Gaussian Projection. In 2023 29th International Conference on Mechatronics and Machine Vision in Practice (M2VIP) (pp. 1-6). IEEE.
8. Lu S., Zhang Y., & Su J. (2017). Mobile robot for power substation inspection: A survey. IEEE/CAA Journal of Automatica Sinica, 4(4), 830-847.
9. Zhang J., Yang X., Wang W., Guan J., Ding L., & Lee V. C. (2023). Automated guided vehicles and autonomous mobile robots for recognition and tracking in civil engineering. Automation in Construction, 146, 104699.
10. Zhang H., Su B., & Meng H. (2017). Development and implementation of a robotic inspection system for power substations. Industrial Robot: An International Journal, 44(3), 333-342.
11. Huang W. (2019). Application of intelligent robot in a substation. Shang Dong Industrial Technology,07:206.
12. Takahashi H. (1989). Development of Patrolling Robot for Substation. Japan IERE Council, Special Document R-8903,1989,10:19.
13. Ren Z., Liu S., Long Y. (2011). Inspection robot suspended on power transmission lines based on expert system. Future Intelligent Information Systems, pp 349-357.
14. Allan J, Beaudry J. (2014). Robotic systems applied to power substations state-of-the-art survey. Proceedings of the 2014 3rd International Conference on Applied Robotics for the Power Industry. Foz do Iguacu, Brazil, p.1-6.
15. Zhang X. (2015). Research on the motion control system of substation inspection vehicles. Xi 'a Technological University.
16. Jin M, Tang Y, Wang Z. (2023). Comprehensive Analysis of Inspection Robot Technology and Application in Substation. Proceedings of the 2023 International Conference on Data Science. Advanced Algorithm and Intelligent Computing.
17. Yang L, Fan J, Liu Y, Peng J, Liang Z. (2020). A Review on State-of-the-Art Power Line Inspection Techniques. IEEE Transactions on Instrumentation and Measurement,9350-9365
18. Wang C, Yin L, Zhao Q, Wang W, Li CY, Luo B. (2020). An intelligent robot for indoor substation inspection. Industrial Robot: the international journal of robotics research and application.47:
19. Zhang T, Dai J. (2020). Electric Power Intelligent Inspection Robot: a Review. Journal of Physics: Conference Series, Volume 1750, 3rd International Conference on Mechanical, Electrical and Material Application (MEMA 2020) 6-8 November 2020, Chongqing, China.https://doi.org/10.1088/1742-6596/1750/1/012023.
20. Alhassan A, Zhang X, Shen H, Xu H. (2020). Power transmission line inspection robots: A review, trends and challenges for future research. Electrical Power and Energy Systems 118 (2020) 105862.https://doi.org/10.1016/j.ijepes.2020.105862.
21. Cao W., Chen, Z., Wu C., & Li T. (2024). A Method for Matching Information of Substation Secondary Screen Cabinet Terminal Block Based on Artificial Intelligence. Applied Sciences, 14(5), 1904.

https://doi.org/10.1051/e3sconf/202452802018

Research on evaluation methods for thermal bridge treatment measures for building envelopes

Qian Cai[1,*], Lei Chen[2], Changjun Wang[1], Jing Ren[1], Siqi Wang[1]

[1] Beijing Building Research Institute Co.,Ltd. of CSCEC, Beijing, China
[2] China Construction First Group Co., Ltd,Beijing, China

Abstract. Along with the promotion of low-carbon and zero carbon buildings such as ultra-low energy buildings, near zero energy buildings, and zero energy buildings, there is higher and higher demand to the thermal performance of building envelopes. Thermal bridges have a crucial impact on the thermal performance of building envelope structures. In order to achieve the goal of energy conservation and carbon dioxide reduction in building structures, it is necessary to conduct refined calculations and evaluations on thermal bridge treatment measures. In this paper, three thermal bridge treatment measures evaluation methods from different dimensions are presented: evaluation method based on reduction degree of thermal bridge effect, evaluation method based on carbon dioxide emissions from thermal bridges, and evaluation method based on technical economy of thermal bridges.

1. Introduction

The global energy consumption of buildings is constantly increasing. According to statistics from the US Energy Information Administration, the energy consumption of the construction industry has increased by approximately 1.3% annually and is expected to reach 22% by 2050 [1]. Of this, 35% is for heating and air conditioning, 11% is for lighting, 18% is for primary electrical equipment, and the remaining 36% is for other miscellaneous energy [2]. According to 2022 Research Report of China Building Energy Consumption and Carbon Emissions released by China Association of Building Energy Efficiency, the total energy consumption of the entire construction process (including the stage of building material production, construction, and operation) in China is 2.27 billion tons of Coal Equivalent (tce), accounting for 45.5% of China's total energy consumption, and the total carbon dioxide emissions of the entire construction process in China were 5.08 billion tCO2, accounting for 50.9% of China's total carbon dioxide emissions. China is gradually accelerating measures for building energy conservation and emission reduction. The promulgation and implementation of the national standard Technical Standards for Near Zero Energy Buildings (GB/T 51350-2019) means that China's ultra-low/near zero energy buildings have entered a stage of rapid development [3]. The impact of thermal bridges in ultra-low energy buildings far exceeds that of ordinary energy-saving buildings. Therefore, special designs must be carried out to eliminate or weaken thermal bridges in the design of ultra-low energy buildings, and the impact of thermal bridges on energy consumption should also be included in the building energy consumption calculation [4].

When designing thermal bridge for enclosure structures, thermal bridge simulation software is generally used to calculate the thermal bridge coefficients for different thermal bridge measures, which are used as the basis for evaluating and comparing the methods for the selection of thermal bridge treatment measure. At present, most scholars analyse ultra-low energy building thermal bridges mainly focus on the causes and control methods of thermal bridges [5], and propose design optimizations by simulating the thermal bridge coefficients of typical parts in different climate zones and working conditions [6]. However, there is no systematic study on the comprehensive evaluation of thermal bridges treatment measures from the perspective of carbon dioxide emissions throughout the entire life cycle of buildings.

This article comprehensively analyses and evaluates the current conventional thermal bridge treatment measures from three aspects: the reduction degree of thermal bridge effect, the carbon dioxide emissions per unit thermal bridge, and the technical economy of thermal bridge measures. For the latter two evaluation perspectives, this article proposes two new quantitative calculation coefficients and corresponding calculation formulas, namely the unit thermal bridge carbon dioxide emission coefficient and the unit thermal bridge technical economy coefficient. The evaluation methods proposed in this article provide a multi-dimensional and more intuitive quantitative evaluation method for building thermal bridges, assist designers to predict and evaluate before the implementation of the heat bridge measures, so as to achieve a truly low-carbon and cost-effective thermal

*Corresponding author's e-mail: chitty210@126.com

https://doi.org/10.1051/e3sconf/202452802018

bridge optimization design, and reflect the low-carbon effect and value of thermal bridge measures.

2. Evaluation method based on reduction degree of thermal bridge heat transfer coefficients effect

2.1. Evaluation principle

Usually, researchers use thermal bridge simulation software to calculate the heat transfer coefficients of thermal bridges processed by thermal bridge treatment measures, and compare the results with the heat transfer coefficient of unprocessed thermal bridges to reflect the reduction effect of thermal bridge treatment measures on thermal bridge effects. However, at this stage, the reduction effects of different thermal bridge treatment measures have only been characterized by simple thermal bridge heat transfer coefficients, and it is unable to visually display and compare the reduction degree and intensity of different thermal bridge treatment measures.

This article proposes a new quantification coefficient, namely reduction degree of thermal bridge effect, which can be used to evaluate the impact of thermal bridge treatment measures on weakening the thermal bridge effect. When comparing the thermal bridge effect reduction degree of different thermal bridge treatment measures using the evaluation method proposed in this section, it should be ensured that the construction of the treated thermal bridge nodes is consistent. The quantification coefficient proposed in this article is more intuitive and clearer than the method of only comparing the values and differences of thermal bridge heat transfer coefficients, and can better reflect thermal bridge effect reduction degree generated by different thermal bridge treatment measures.

The reduction degree of thermal bridge effect is characterized by the reduction degree rate of thermal bridge, which is shown as Equation (1):

$$RAR = \frac{UTC-PTC}{UTC} \times 100\% \qquad (1)$$

where RAR represents reduction degree rate of thermal bridge, UTC represents the heat transfer coefficient of the unprocessed thermal bridge. PTC represents the heat transfer coefficient of the thermal bridges processed by thermal bridge treatment measures.

2.2. Case Studies

This section takes the external balcony as the research object of the thermal bridge node and applies different thermal bridge treatment measures in order to evaluate the thermal bridge reduction degree caused by different treatment measures.

The exterior wall structure of the external balcony is made of 200mm thick reinforced concrete exterior wall and 250mm thick graphite polystyrene board insulation. The length of the external balcony board is 2m. The thermal bridge node of the exterior wall is the external balcony board. The thermal bridge treatment measures in this article are to lay 100mm thick extruded polystyrene boards of different lengths on the upper and lower sides of the external balcony board. The construction of the external balcony thermal bridge and the thermal bridge treatment measures with different laying lengths are shown in Figure 1. The laying lengths of the extruded polystyrene boards are 0.5m, 1m, 1.5m, and 2.0m, respectively. Comparisons of heat transfer coefficient reduction degrees of thermal bridge nodes under different treatment measures and without any treatment measures are made.

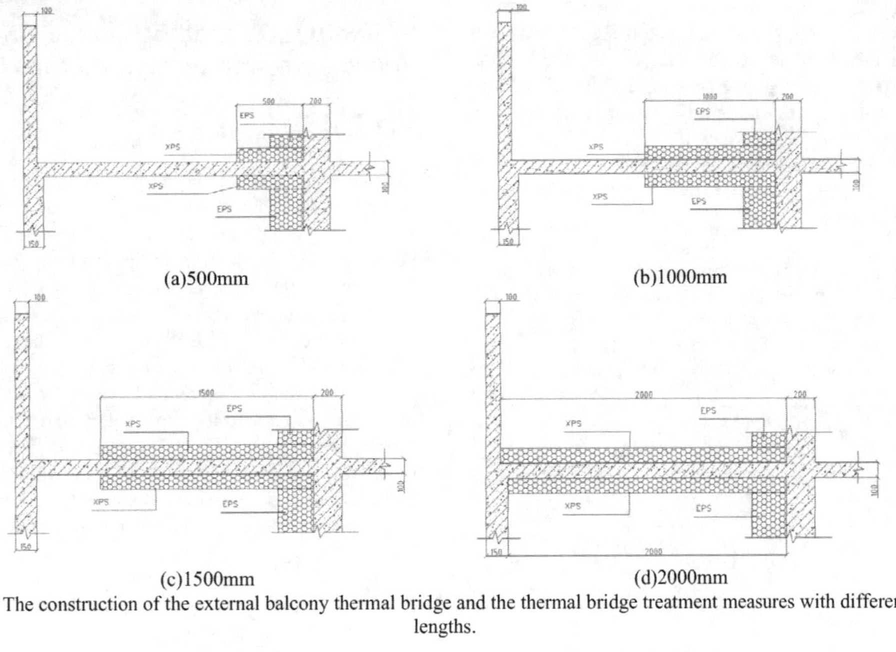

(a)500mm (b)1000mm

(c)1500mm (d)2000mm

Figure 1. The construction of the external balcony thermal bridge and the thermal bridge treatment measures with different laying lengths.

https://doi.org/10.1051/e3sconf/202452802018

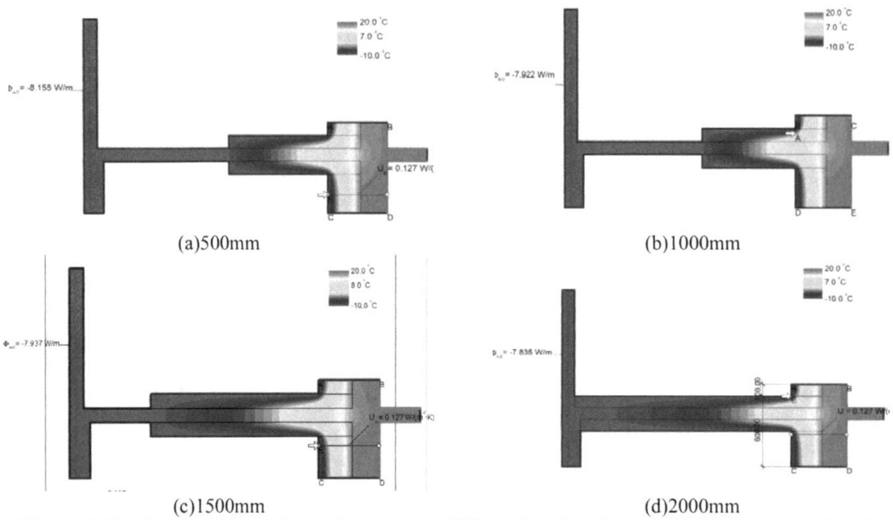

Figure 2. Simulation results of thermal bridge under different lengths of extruded polystyrene board paving.

2.3. Analysis Results

The thermal bridge analysis software Filixo is used to calculate the thermal bridge heat transfer coefficients of external balcony boards with extruded polystyrene boards of 0.5m, 1m, 1.5m, and 2.0m lengths respectively, as shown in Figure 2. Calculate the reduction degree rate of thermal bridge RAR of different thermal bridge treatment measures by using the equation proposed in Section 1.1. The results show that the RAR values of thermal bridge treatment measures for extruded polystyrene boards with laying lengths of 0.5m, 1m, 1.5m, and 2.0m are 34%, 51%, 54% and 55% respectively.

According to the calculation results of RAR, it can be seen that as the laying length of the extruded polystyrene board increases, the corresponding decrease in thermal bridge heat transfer coefficient gradually slows down. Based on this trend, it can be inferred that the thermal bridge effect reduction degree of the external balcony boards caused by extending the laying length of the extruded polystyrene board will not exceed 60%. The numerical values, reduction degrees and change trend of thermal bridge heat transfer coefficients are shown in Table 1 and Figure 3.

Table 1. The numerical values and reduction degrees of thermal bridge coefficients of external balcony boards with different lengths of extruded polystyrene boards.

Lengths of extruded polystyrene boards/ (m)	Thermal bridge coefficients/ (W/m.K)	Thermal bridge coefficient reduction degrees
0	0.38	/
0.5	0.25	34%
1	0.185	51%
1.5	0.176	54%
2	0.172	55%

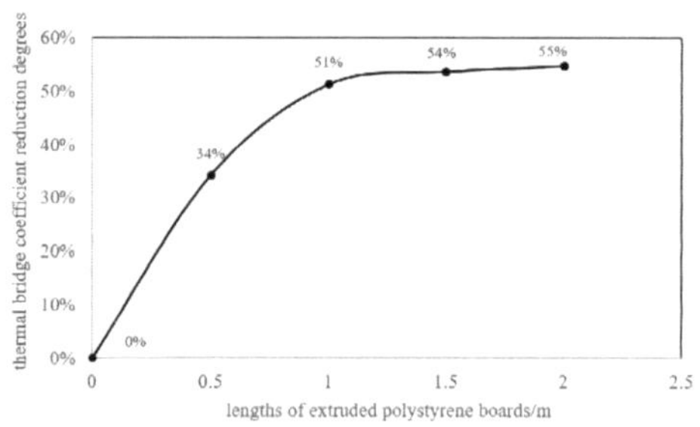

Figure 3. Change trend of thermal bridge heat transfer coefficients.

3. Evaluation method based on carbon dioxide emissions from thermal bridges

3.1. Evaluation principle

Generally speaking, thermal bridge treatment measures are compared based on the thermal bridge effect reduction effect, that is, the impact on the thermal performance of building envelopes and the building energy consumption. With the continuous development of green building concepts, building carbon dioxide emissions have become an important indicator for low-carbon building evaluation. The life cycle carbon dioxide emissions of buildings include the emission increment caused by the increase in material usage due to thermal bridge treatment measures. Therefore, it is necessary to evaluate thermal bridge treatment measures from the perspective of carbon dioxide emissions. When comparing thermal bridge treatment measures, not only should the impact of thermal bridge effect reduction caused by the measure on the energy consumption of the building operation, but also the increase in carbon dioxide emissions caused by the measure itself should be considered.

The unit thermal bridge carbon dioxide emission of the thermal bridge treatment measure can be calculated by using Equation (2):

$$UTE = \frac{CTM}{UTC-PTC} \quad (2)$$

where UTE represents the unit thermal bridge carbon dioxide emission of the thermal bridge treatment measure, CTM carbon dioxide emissions caused by thermal bridge treatment measures. UTC represents the heat transfer coefficient of the unprocessed thermal bridge. PTC represents the heat transfer coefficient of the thermal bridges processed by thermal bridge treatment measures.

3.2. Case Studies

In this section, the parapet wall is used as the thermal bridge model for heat transfer analysis. The arrangement of thermal insulation materials for the analysed parapet wall is as follows: the outer layer material is made of 250mm thick molded polystyrene board, the inner layer material is made of 250mm thick extruded polystyrene board, and the roof is made of 300mm thick extruded polystyrene board. The construction of the parapet wall is shown as Figure 4. The thermal bridge treatment measures are to install 100mm thick thermal insulation layers with different materials on the top of the parapet wall. The materials used for thermal insulation layers include extruded polystyrene board, rock wool board, and rigid foam polyurethane board.

In section 3.3, the competition of the carbon dioxide emissions caused by insulation layers with different materials and the competition of the ratio of carbon dioxide emissions caused by different thermal bridge treatment measures to the corresponding thermal bridge effect reduction degrees are conducted. The larger the ratio, the greater the carbon emission cost caused by the corresponding thermal bridge treatment measure.

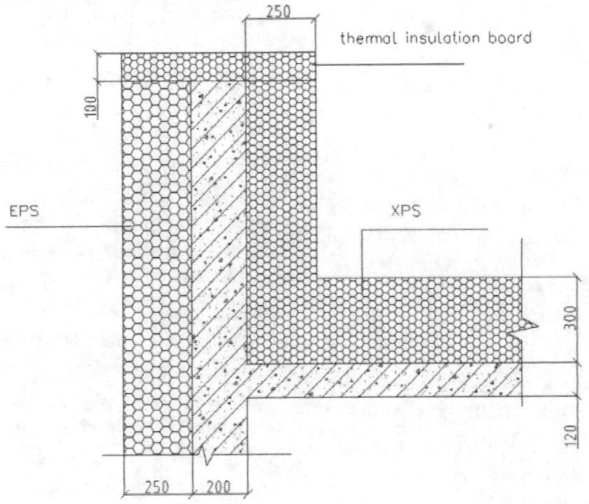

Figure 4. The construction of the parapet wall.

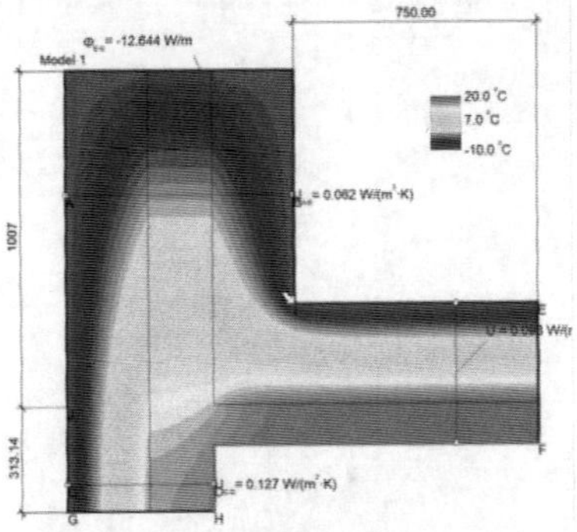

(a) Thermal bridge simulation without thermal bridge treatment measures

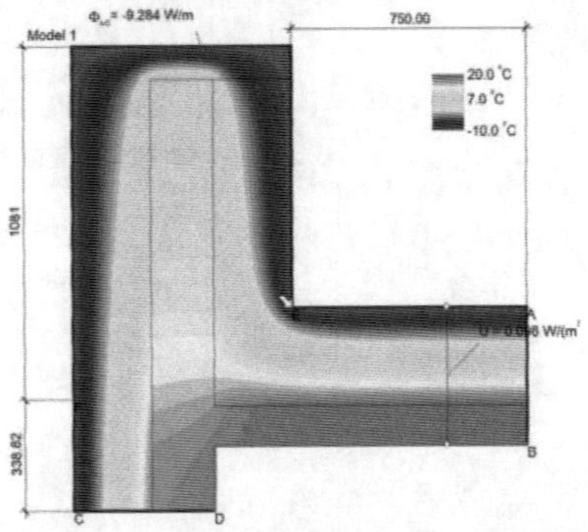

(b) Thermal bridge simulation of parapet wall with extruded polystyrene board laid on top

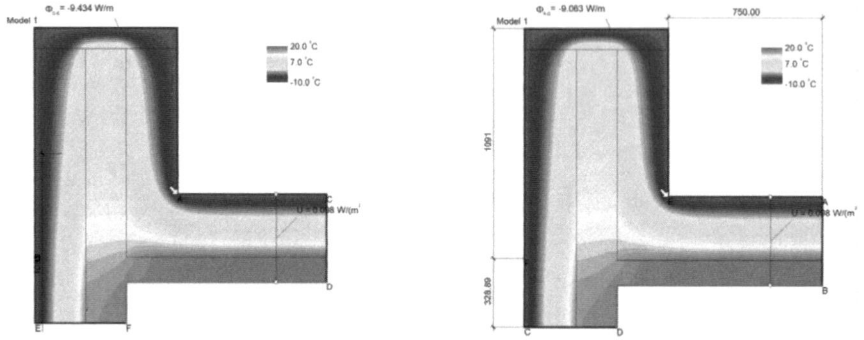

(c) Thermal bridge simulation of parapet wall with rock wool board laid on top (d) Thermal bridge simulation of parapet wall with rigid foam polyurethane board laid on top

Figure 5. Heat transfer simulations of thermal bridge models with different thermal insulation materials on top and without any thermal insulation material.

3.3. Analysis Results

The thermal bridge analysis software Filixo is used to calculate the thermal bridge heat transfer coefficients of parapet walls with extruded polystyrene board, rock wool board, and rigid polyurethane board laid on top, respectively. And Heat transfer simulations of thermal bridge models with and without thermal bridge treatment measures are shown in Figure 5. According to Appendix D of the Standard for Building Carbon Emission Calculation (GB/T51366-2019) [7], the carbon dioxide emission factors for extruded polystyrene board, rock wool board, and rigid foam polyurethane board are 5020kg CO_2 e/t, 1980kg CO_2 e/t, and 5020kg CO_2 e/t respectively. And the densities of extruded polystyrene board, rock wool board, and rigid foam polyurethane board are 30kg/m^3、140kg/m^3 and 35kg/m^3 respectively. According to calculations, the carbon dioxide emissions per unit length of 100mm thick and 700mm wide extruded polystyrene board, rock wool board, and rigid foam polyurethane board are 10.5kg CO_2 e/m, 19.4kg CO_2 e/m, and 12.8kg CO_2 e/m, respectively. According to the carbon emission calculation formula for thermal bridges, it can be concluded that when 100mm thick extruded polystyrene board, rock wool board, and rigid foam polyurethane board are used for the thermal bridge treatment measures of the parapet wall, the corresponding unit thermal bridge carbon dioxide emissions are 86.41kg CO_2 e·K/W, 167.28kg CO_2 e. K/W, and 100.70kg CO_2 e·K/W. The larger the unit thermal bridge carbon emissions, the greater the carbon emissions consumed by thermal bridge treatment measures, which is less conducive to carbon reduction throughout the entire life of the building.

From the calculation results as shown in Figure 6, it can be seen that the thermal bridge treatment measure using rock wool has the highest unit thermal bridge carbon emissions, while the thermal bridge treatment measure using extruded polystyrene board has the lowest unit thermal bridge carbon emissions. Therefore, from the perspective of carbon emissions throughout the entire life cycle of buildings, the thermal bridge treatment measure using extruded polystyrene board is the most desirable.

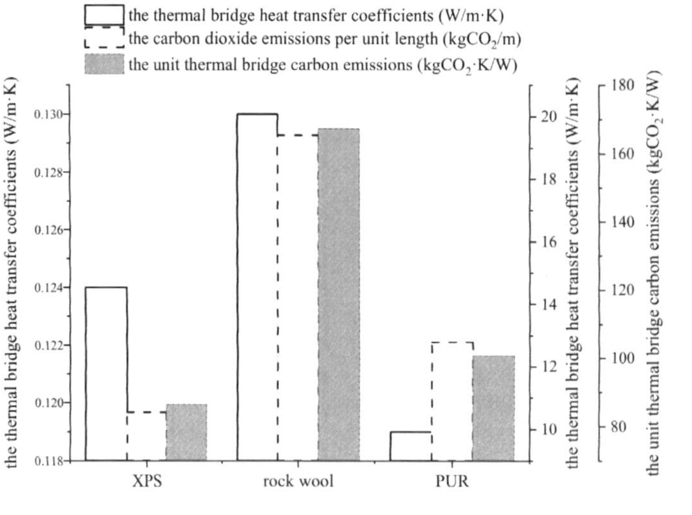

Figure 6. Trend of unit thermal bridge carbon emissions.

https://doi.org/10.1051/e3sconf/202452802018

4. Evaluation method based on technical economy of thermal bridges

4.1. Analysis Results

Cost control is an important part of construction project management. Therefore, thermal bridge treatment measures should be evaluated from the perspective of construction cost. This article proposes a new indicator, namely the unit thermal bridge technical economy coefficient, to evaluate the technical economy of thermal bridge treatment measures. The calculation formula for the thermal bridge technical economy coefficient is shown in equation (3). This indicator can be used to compare the economic costs generated by different thermal bridge treatment measures. Using this indicator to evaluate thermal bridge treatment measures can not only avoid using less effective treatment measures due to cost control, but also avoid selecting high-cost thermal bridge treatment measures based solely on the reduction effect of thermal bridges, which can hinder the promotion of thermal bridge treatment measures.

$$UEC = \frac{TBC}{UTC-PTC} \tag{3}$$

where UEC represents the unit thermal bridge technical economy coefficient, TBC represents the cost expenditure of the thermal bridge treatment measure, UTC represents the heat transfer coefficient of the unprocessed thermal bridge. PTC represents the heat transfer coefficient of the thermal bridges processed by thermal bridge treatment measures.

4.2. Case Studies

In this section, the connection between the ground and the exterior wall is used as the thermal bridge model for heat transfer analysis. At the connection, the thermal insulation material is 250mm thick exterior extruded polystyrene board for the wall, 100mm thick extruded polystyrene board for the upper side of the floor slab. The lower side of the floor slab is an unheated basement, which does not belong to the ultra-low energy consumption building area. Three thermal bridge treatment methods are provided as follows: (1) Lay 1m wide and 100mm thick rock wool boards along the inner side of the underground floor slab and the outer wall at the connection point, (2) Extend the external insulation material of the exterior wall downwards for 1 meter towards the direction of the underground exterior wall, (3) Combine the treatment measure (1) and treatment measure (2) together. The construction of the connection area and three thermal bridge treatment measures are shown in Figure 7.

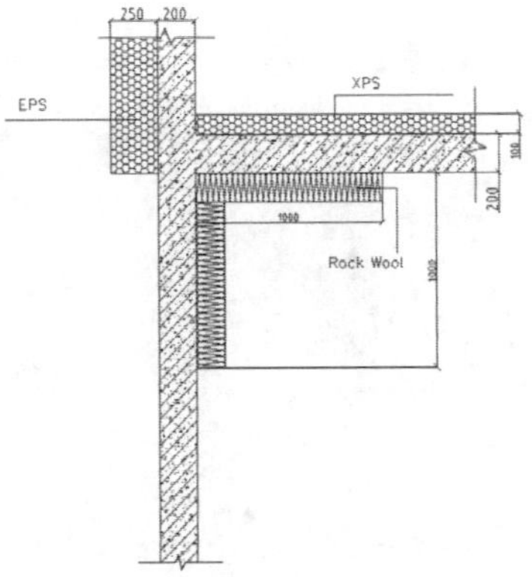

(a) Thermal bridge treatment measure (1)

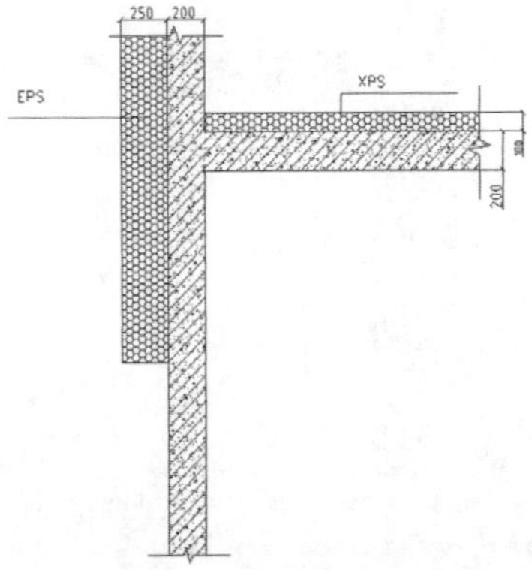

(b) Thermal bridge treatment measure (2)

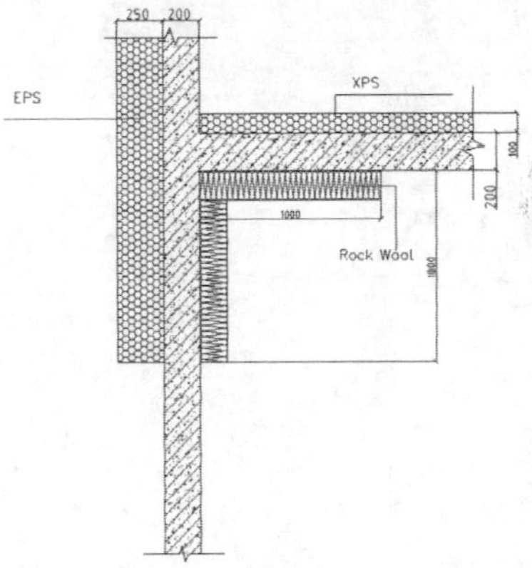

(c) Thermal bridge treatment measure (3)

Figure 7. The construction of the connection area and three thermal bridge treatment measures.

 https://doi.org/10.1051/e3sconf/202452802018

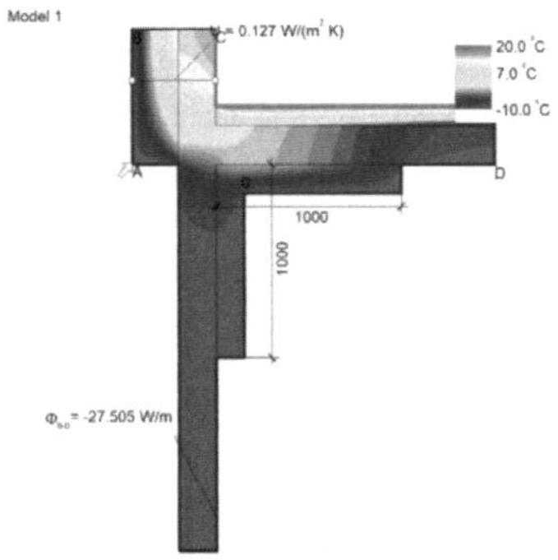

(a) Thermal bridge treatment measure (1)

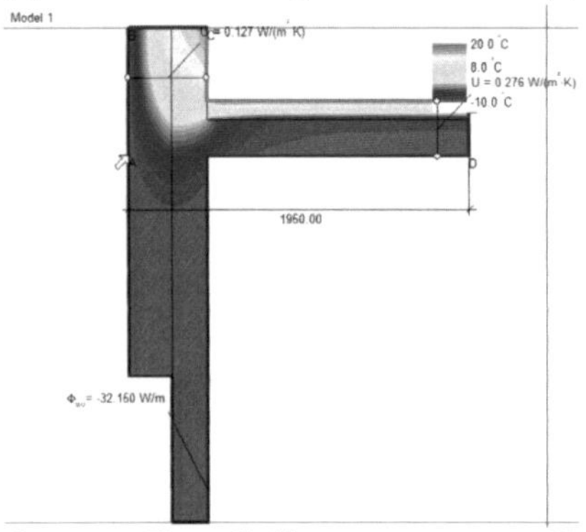

(b) Thermal bridge treatment measure (2)

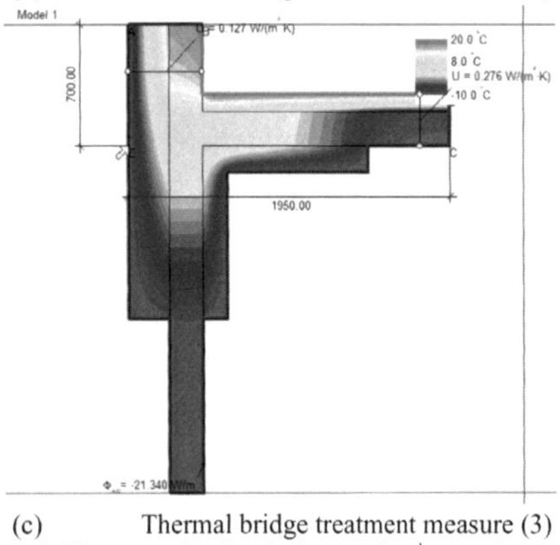

(c) Thermal bridge treatment measure (3)

Figure 8. Heat transfer simulations at the connection between the ground and the exterior wall under three thermal bridge treatment measures.

4.3. Analysis Results

The thermal bridge analysis software Filixo is used to calculate the thermal bridge heat transfer coefficients of the connections processed by three thermal bridge treatment measures. And Heat transfer simulations results of three measures are shown in Figure 8. The thermal bridge heat transfer coefficients of three measures are 0.289W/(m·K), 0.444W/(m·K) and 0.083W/(m·K) respectively. The cost of rock wool board is about 400 RMB/m^3, and the cost of extruded polystyrene board is about 620RMB/m^3. Therefore, the unit cost of the three thermal bridge treatment measures are 76 RMB/m, 155 RMB/m, and 231 RMB/m, respectively. According to equation (3), the unit thermal bridge technical economy coefficients of the three measures are : 434.28 RMB·K/W, 7750 RMB·K/W, and 606.30 RMB·K/W, respectively. The trends of technical economy coefficients are shown as figure 9.

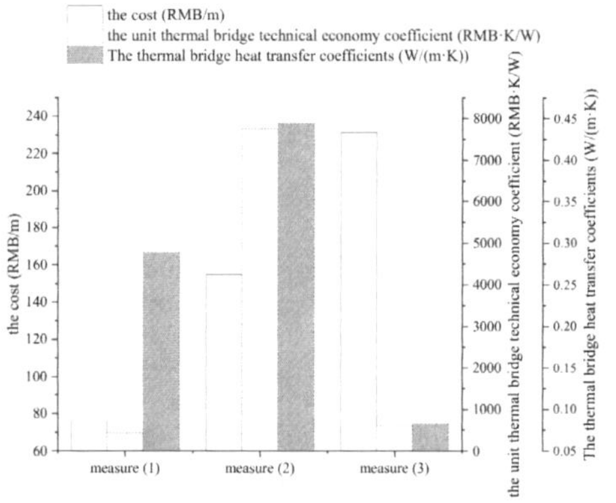

Figure 9. The trends of technical economy coefficients.

The lower the technical economy coefficient, the lower the cost of unit thermal bridge effect reduction, and the better the technical economy and cost-effectiveness of this thermal bridge treatment measure. The thermal bridge treatment measure (2) has the maximum technical economy coefficient, which means that its cost of unit thermal bridge effect reduction is the highest and the technical economy efficiency is the worst. And this type of thermal bridge treatment measure should be avoided. The thermal bridge treatment measure (1) has the minimum technical economy coefficient. However, based solely on the thermal bridge effect reduction degree, the thermal bridge treatment measure (3) has the best thermal bridge reduction effect, and its technical economy feasibility is moderate. Therefore, after multiple considerations, the thermal bridge treatment measure (3) is recommended.

5. Conclusion

This article presents three evaluation methods for thermal bridge treatment measures, namely (1)the evaluation method based on reduction degree of thermal bridge heat transfer coefficients effect, (2)the evaluation method based on carbon dioxide emissions from thermal bridges, and (3) the evaluation method based on technical economy of thermal bridges. Corresponding evaluation coefficients and calculation formulas have been proposed for each evaluation method.

Evaluation method (1) is suitable for comparing measures when only considering the reduction degree of thermal bridge effect. Evaluation method (2) is applicable for comparing measures taken from the perspective of reducing carbon emissions throughout the entire life cycle of buildings. Evaluation method (3) is suitable for evaluating the technical economy of measures from a comprehensive perspective of engineering cost and the degree of thermal bridge effect reduction. Three thermal

https://doi.org/10.1051/e3sconf/202452802018

bridge treatment measure evaluation methods can be used simultaneously or separately.

In order to maintain the advanced nature and applicability of the evaluation method, the evaluation method will be further studied and timely integrated into new technologies and materials, and regularly updated and revised, to adapt to the changing market demand and industry development trend.

Acknowledgments

This work was financially supported by the Research on Evaluation and Acceptance of Low Carbon Buildings. (PT-2022-07).

References

1. U.S. Energy Information Administration Office of Energy Analysis, 2019. International Energy Outlook 2019 with Projections 2050.
2. U.S. Department of Energy, 2015. Chapter 5: increasing efficiency of building systems and technologies. Quadrennial Technology Review, An Assessment of Energy Technologies and Research Opportunities 143e181. September.
3. Zhen Yu, Wei Liu . Development status and prospects of passive low energy buildings in China, 2018, Power Demand Side Management, 20(5), 2018: 1-4.
4. China Academy of Building Research. GB /T 51350—2019, Technical Standard for nearly zero energy buildings. China Architecture & Building Press, China, 2019
5. Zimin Wu, Hongliang Chu, Xiaochen Li, et al. Construction drawing optimization of ultra-low energy consumption building from the view point of real estate. New building materials, 2020 (9) : 15-19.
6. Li Peng, Yuliang Pan, Hongwei Shen, et al. Numerical Simulation of Thermal Bridge for Ultra Low Energy R esidential Buildings. BUILDING SCIENCE. 2022(6) : 197-205.
7. Ministry of Housing and Urban-Rural Development of the People's Republic of China. GB/T51366-2019, Standard for Building Carbon Emission Calculation. China Architecture & Building Press, China, 2019.

Overview of the faults and diagnosis methods of the hydraulic system of modern coal mining machines

Zhi Li[1,2]

[1]CCTEG Taiyuan Research Institute Co., Ltd. ,Taiyuan ,China
[2]Shanxi Tiandi Coal Machinery Equipment Co., Ltd. Inner Mongolia Branch, Ordos ,China

Abstract: The high power, high efficiency, low failure rate, and intelligence of coal mining equipment are important directions for the development of coal mining machines both domestically and internationally. The implementation of efficient and low failure coal mine equipment is an important prerequisite for ensuring the smooth deployment of intelligent mines. At present, coal mining equipment generally consists of three parts: mechanical, electrical, and hydraulic. However, hydraulic pressure is a necessary system for coal mining equipment to complete high-power and intelligent operations. This article will combine the author's years of work experience to summarize the current local faults and diagnostic methods of hydraulic systems in coal mine equipment.

1. Basic characteristics of hydraulic system of coal machine

The hydraulic system of the coal machine(hereinafter referred to as the "hydraulic system") is a mechanical equipment from the analysis of the main body of the coal machine,which is closely related to mechanical technology. Because the hydraulic system's own fluid properties are different from the mechanical structure, its working principle, working medium and structural process characteristics are different from the other two types of coal machine systems [1]. The hydraulic system has its own unique transmission mode, control mode and application mode.In the process of transforming coal mining equipment from highly mechanized to intelligent,the existing high-power,low-automatic hydraulic system setting technology can no longer meet the requirements.Therefore,part of the hydraulic system has completed the mutual information exchange and energy exchange with the electronic computer system.The current mainstream development direction of coal mining equipment is unmanned intelligence and remote operation.Therefore,it is necessary to realize the complete monitoring of all hydraulic systems of coal mining machinery and make corresponding fault diagnosis,and to give a reasonable maintenance plan is a necessary process for the current coal mining machinery discovery[2].This article hopes to realize the intelligentization of coal machine hydraulic system as soon as possible,and will explain some common coal machine hydraulic system faults and diagnosis methods below.

2. Overview of the faults of the hydraulic system of the coal machine and the fault characteristics

2.1. Overview of the failure of the coal machine hydraulic system

The failure of the hydraulic system is generally caused by the failure of the hydraulic components or hydraulic medium to reach the expected functional level,which causes the whole machine to not work normally.The failure of the hydraulic system of the coal machine is usually accompanied by vibration,noise and high temperature [3].However,most of these phenomena are caused by the inability of hydraulic equipment and external conditions to meet the requirements of normal operation. In most cases,the faults of the coal machine hydraulic system can be roughly divided into the following 6 attributes.

1) The faulty component is the component that has failed. Hydraulic components are the basic unit that constitutes hydraulic equipment. When the hydraulic components fail,the overall performance of the hydraulic equipment will decrease or fail.Therefore,only an in-depth and systematic understanding of the principle,structure,function and failure mechanism of hydraulic components can successfully analyze field failures and even eliminate them.

2) Fault parameters,that is,physical quantities that characterize the loss of or appearing function of the hydraulic device.Such as pressure,flow,leakage,speed,torque,action sequence,position,efficiency,vibration,noise and oil port

e-mail:1280026808@qq.com

vacuum,etc.When the above parameters exceed the design maximum value range,it means that the hydraulic system is malfunctioning.

3) Failure phenomenon,that is, when the key parameters of the hydraulic system are abnormal,the hydraulic system failure can be observed artificially.Such as obvious abnormal equipment noise,system weakness and high temperature.

4) Fault information, that is, characteristic information that reflects the internal damage of the hydraulic device.The fault phenomenon is only part of the fault information. Equipment alarm signals,system test analysis conclusions,equipment service life and operation records are also fault information. The fault information and the fault have a certain corresponding relationship,which is the starting point and basis for judging the fault.

5) The cause of the failure, that is, the initial cause of the failure.Mainly include oil pollution, mechanical wear and fracture failure,design and manufacturing problems, installation problems,inconsistent changes in link conditions, and human factors.

6) The scope of the fault,and the content of the fault design.Some failures are single, and some are comprehensive. The former is caused by an abnormality of individual factors,while the latter involves multiple links.For example, when the hydraulic oil is contaminated,it will cause more spool stuck in the hydraulic system,and the electromagnet will burn out.

2.2 Fault characteristics of coal machine hydraulic system

Because the working environment of coal mining machinery is worse than that of ground equipment,the fault investigation,fault diagnosis and fault resolution of the coal machine hydraulic system[4] have the following characteristics.

1) Concealment The damage and failure of hydraulic devices often occur in the deep interior.Due to the inconvenience of assembly and disassembly,the detection conditions on the site are limited and it is difficult to observe directly.In addition, the surface fault phenomenon of the hydraulic system is more difficult to obtain, such as the internal holes of the hydraulic main control valve of the continuous miner.Once collusion and blockage occur,The overall action of the continuous miner is out of adjustment, but the difficulty coefficient of the entire valve body to accurately determine the fault is too large.

For example,the schematic diagram of the hydraulic system of a coal machine is shown in Figure 1.The fault is that the working pressure is about 41MPa,and the unloading pressure is only about 1MPa lower than the working pressure. When the system runs for 10 minutes,the oil temperature of the hydraulic system rises above 50°C.After an on-site inspection, it was found that the main oil circuit and the relief valve unloading control oil circuit that caused this problem were abnormally colluding on the valve plate.In the figure 1 below,the collusion oil circuit is indicated by a dashed line. When the hydraulic pump circulates through the main oil circuit,a small part of the hydraulic oil will flow through the reversing valve through the orifice connected by the dotted line and then return to the hydraulic oil tank.The reversing valve has a small diameter and resistance,so the pressure of the hydraulic oil before the reversing valve is P1.This pressure makes the relief valve under the secondary pressure control, the relief valve spool cannot be fully opened,and the system pressure rises.The oil is heated by the throttle valve.At that time, the technicians did a detailed inspection of the hydraulic system components of the vehicle in the mine,and found no suspicious points.Finally,all components were subjected to a pressure test during the equipment overhaul,and it was found that the main overflow valve plate was caused by liquid leakage.

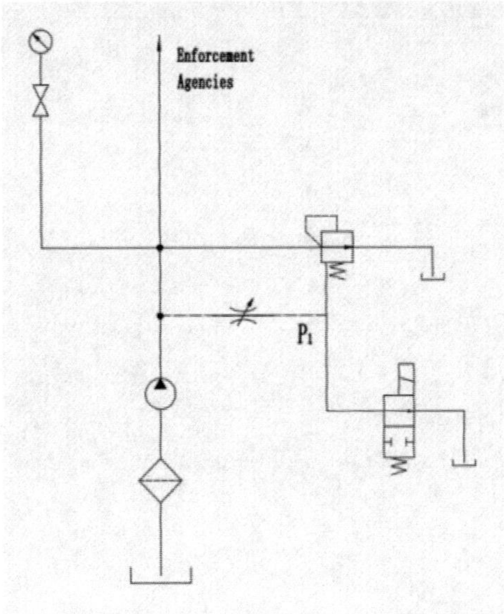

Figure 1. The hydraulic system with the main oil circuit and the unloading oil circuit in collusion

2) Staggeredness There are various overlaps and crossovers between symptoms and causes of hydraulic system failures.A symptom may have many causes.If the speed of the actuator is slow,the causes may include. excessive load,wear of the actuator itself,excessive guide rail error, leakage in the system, pressure regulating system failure, speed regulating system failure, pump failure, etc.A certain symptom may be caused by a combination of multiple reasons. For example.when the pump, reversing valve, and hydraulic cylinder are all in a worn state,the efficiency of the system is greatly reduced.When these parts are replaced one by one,the efficiency of the system will also increase.For comprehensive faults,the influencing factors should be fully considered,and the priority should be resolved one by one.

3) Randomness During the operation of the coal machine hydraulic system,it is affected by a variety of random factors, such as changes in underground voltage,changes in ambient temperature,and changes in the geology of machine excavated rock tunnels.The intrusion of pulverized coal during the mining process is also random.Due to the influence of random factors,the direction of the specific change of the fault is not certain,which will aggravate the difficulty of judgment and quantitative analysis.

4) Differences Due to differences in design,processing materials,and application environments,the wear and deterioration speed of hydraulic components varies greatly. General hydraulic component life standards cannot be applied in the field.Specific wear evaluation standards can only be determined for specific hydraulic equipment and hydraulic components.However,this process requires long-term operation data accumulation.

3. Diagnostic method for key parameters of hydraulic system of coal mining machine

The main work content of hydraulic fault diagnosis includes.

3.1 Formula calculation method

Some of the current hydraulic system parameters cannot be directly obtained and need to be obtained through mathematical models, as shown in the following formula:

$$q_v = \frac{v_\gamma \bullet v}{1000 \bullet \eta_v} \tag{1}$$

$$n = \frac{q_v \bullet 1000 \bullet \eta_v}{v_g} \tag{2}$$

flow: q_v*: unit:L/min*

displacement: v_g*: unit:* cm^3

Revolution: n: unit: rpm

Volumetric efficiency: η_v

3.2 Key parameter lookup table

For some key parameters that can be directly obtained or calculated through the 3.1 formula, they can be compared with the calibration values, and the comparison accuracy is shown in Table 1 below.

Table 1. Comparison accuracy of key parameters

Measurement parameters	Measurement accuracy level		
	A	B	C
Pressure (gauge pressure ＜ 0.2MPa)KPa	±1.0	±3.0	±5.0
Pressure (gauge pressure ＜ 0.2MPa)(%)	±0.5	±1.5	±2.5
flow(%)	±0.5	±1.5	±2.5
torque(%)	±0.5	±1.5	±2.5
torque(%)	±0.5	±1.0	±2.0
temperature(%)	±0.5	±1.0	±2.0

3.3 Simulation analysis comparison

For some key parameters that are difficult to directly measure in new hydraulic systems, simulation can be used to input measurable parameters and obtain other key explanations for fault analysis.The key parameter simulation of a certain system is shown in Figure 2.

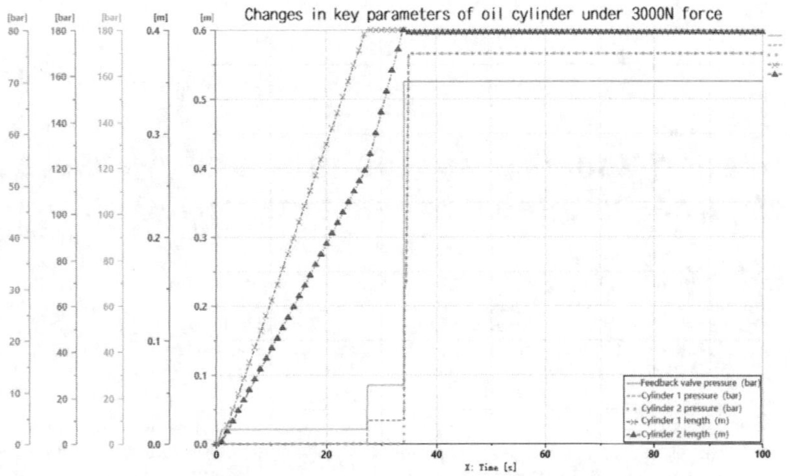

Figure 2.Simulation results of key parameters of a certain system

4. Conclusion

According to the current development trend of hydraulic system fault diagnosis technology both domestically and internationally, the hydraulic system fault detection methods currently being studied and preliminarily applied internationally include vibration diagnosis method, acoustic vibration method, thermodynamic diagnosis method, material optical analysis method, neural network[5], etc. However, the widely used methods mentioned above include formula calculation, table lookup, and model simulation and analysis. These widely used methods will have significant development and application in the future when combined with new technologies under research.

References

1. Liang Yufang, Chen Yongfeng. Application of field maintenance technology in fault diagnosis of coal mine machinery hydraulic system [J]. Coal Mine Machinery, 2010, 31(09): 236-238.
2. Ma Xianmin, Liang Lan, Zhang Yongqiang, Xie Leping. Research on D-S data fusion fault diagnosis of coal mine large electromechanical equipment based on GA-BP[J]. Coal Technology, 2016, 35(01): 218-221.
3. Wang Fei. Research on the failure of the hydraulic system of coal mining machinery[J]. Science and Technology and Innovation, 2020(14): 134-135.
4. Zhuraev Akbar Shavkatovich. Characteristics of packing gland seals in hydraulic systems of quarry excavators and results of comparative analysis of experimental tests,Journal | [J] Vibroengineering Procedia. Volume 54 , Issue . 2024. PP 252-257.
5. Park Chul Hong;Park Cheol Soo.Limitations and issues of conventional artificial neural network-based surrogate models for building energy retrofit,Journal | [J] Journal of Building Performance Simulation. Volume 17 , Issue 3 . 2024. PP 361-370.

Study on energy-food water footprint and pressure on available water resources in three northeastern provinces

Yanqiu Wang[1], Yuwei Jiang[1,*], Zhaoge Zhu[2], Jingzhu Wei[1]

[1]College of Economics and Management, Northeast Petroleum University,, Daqing 163318, China
[2]College of Mechanical Science and Engineering, Northeast Petroleum University, Daqing 163318, China

Abstract. The study measured the energy-food water footprint and food-food water footprint of 36 prefecture-level cities in the three northeastern provinces from 2000 to 2020 using food and energy-food water footprint evaluation models, and examined the pressurees on available water resources generated by the energy-food water footprint in the three northeastern provinces. The results show that the energy-food water footprints of the three northeastern provinces during the study period are spatially affected by differences in resource endowments, with more than 50 per cent of the regions experiencing high energy-food water pressurees, and the efficiency of water resources utilisation needs to be improved. In this regard, the paper makes recommendations to safeguard the sustainable development of the region and contribute to stabilising the energy-food-water chain balance.

1. Introduction

The three northeastern provinces (Heilongjiang, Jilin and Liaoning) are important food production bases and energy manufacturing bases in China, but there is a spatial mismatch in the distribution of the three resources in China [1]. Therefore, under the condition of limited total water resources, effective management of water resources to maximise the effect produced by water resources is an important way to ensure the sustainable development of the region [2]. In 2002, Dutch researchers Hoekstra et al. first formally presented the idea of water footprint, which refers to the amount of water required for all products and services in a given time and space, and is figuratively represented as the footprints that water treads in production and consumption [3]. *The Water Footprint Evaluation Manual* was first published in 2012, and ever since The concept of water footprinting has garnered global interest from academics and governments alike [4]. Nowadays, research on energy-food water footprint focuses on the efficient utilization of energy-water, food-water and energy-food-water systems. Guan Wei utilized an energy water footprint evaluation methodology to measure five different types of energy water footprints in China and examined the spatial correlation between China's energy-water footprints and its accessible water resources [5]. Mesfin M. Mekonnen et al. measured the water footprint of food and showed that the state of global food production will lead to non-sustainable use of blue water, which will also further put additional pressure on available water resources [6]. Qin Y conducted a systematic global analysis of seasonal competitive water consumption for irrigated agriculture and electricity production in major river basins around the world under potential climate warming scenarios [7]. More investigations are needed to fully understand the spatial pressure relationships between energy, food production and water availability. Therefore, using the concept of 'water footprint', this paper quantitatively analyses the water footprint of food and energy production and its pressure relationship with water resources in 36 prefecture-level cities in the three northeastern provinces. The aim is to create a benchmark for sustainable and efficient management of water resources.

2. Research methods and data sources

2.1. Research methods

2.1.1. Energy water footprint calculation model.

The direct water footprint, which represents the volume of water used during the manufacturing life cycle, and the indirect water footprint, which results from material input, are both included in the energy water footprint, the Water Footprint Energy Calculation The model is based on the following computation formulas:

$$WE = \sum_{p=1}^{n}(W_{p(blue,direct)} + W_{P(grey,direct)} + \sum_{q=1}^{n}(W_{q(blue,indirect)} + W_{q(grey,indirect)}) \quad (1)$$

*Corresponding author's e-mail: JYW5990714@126.com

https://doi.org/10.1051/e3sconf/202452802020

$$W_{grey,direct} = \frac{G}{C_{\max} - C_{nat}} = \frac{L \times V_p}{C_{\max} - C_{nat}} \quad (2)$$

where p denotes the pth production process, q denotes the qth type of energy, where W_E is the energy water footprint (10^8 m^3), W_{direct} and $W_{indirect}$ are the direct and indirect water footprints (10^8 m^3), $W_{blue,direct}$, $W_{grey,direct}$ are direct blue water footprint and direct gray water footprint (10^8 m^3), Indirect blue and gray water footprints are represented by $W_{blue,indirect}$ and $W_{grey,indirect}$, respectively(10^8 m^3). where L is the quantity of wastewater released for every energy produced (m^3/GJ), V_p is the pollution content of the wastewater that is released (mg/m^3), G is the quantity of pollutants released for every energy unit produced (mg/GJ), the maximum allowable concentration of pollutants in water is C_{max} (mg/m^3), The term "C_{nat}" describes the pollutant's initial concentration in the water body (mg/m^3).

2.1.2. Food water footprint calculation model.

The CROPWAT program can calculate the water quantity data for different crops under different meteorological conditions based on meteorological data, soil type, crop growth stage and other factors. The formula is as follows:

$$WF = W_f \times S \quad (3)$$

$$W_f = \frac{ET}{P} \times 10 \quad (4)$$

$$ET = \max(0, ET_c - P_{eff}) \quad (5)$$

$$ET_c = ET_o \times K_c \quad (6)$$

where WF represent the food water footprints, respectively, consumed during crop production (m^3). ET are water evapotranspiration (mm), respectively, W_f is the amount of water emitted per food unit (m^3/kg), P is the amount of food produced per area (t), S is the sown area of the crop (hm^2), 10 is a conversion factor to transform units of water depth into water per unit area (m^3/hm^2). Where P_{eff} is the effective rainfall amount (mm), ET_c is the crop water demand (mm), ET_o is crop evapotranspiration, K_c is the crop coefficient.

2.1.3. Energy-food water footprint pressure index.

In order to comparatively analyse the degree of pressure generated by the energy-food water footprint on available water resources in different regions, it is possible to refer to relevant studies and construct the energy-food water footprint pressure index (D_{ij}),, which is formulated as follows[8]:

$$D_{ij} = E_{ij} + F_{ij} \quad (7)$$

$$E_{ij} = \frac{WE_{ij}}{Q_{ij}}, F_{ij} = \frac{WF_{ij}}{Q_{ij}} \quad (8)$$

where D_{ij} is the region I's energy-food water footprint's pressure index in year j, where E_{ij} represents region I's energy-water footprint's pressure index in year j, while F_{ij} represents the region I's food water footprint's pressure index in year j. Q_{ij} is how much water resources (i.e., 40% of all water resources) region have access to in year j [9].

The study divided the energy-food water pressure levels into five levels based on the World Resources Institute's classification criteria for water pressure in China, as shown in the table:

Table 1. Classification of energy-food water pressure levels

pressure level	No pressure	Low pressure	Medium pressure	Higher pressure	High pressure
D_{ij}	<0.1	0.1-0.2	0.2-0.4	0.4-0.8	>0.8

2.2. Data sources and determination of parameters

2.2.1. Sources of information.

This paper selected pertinent data for the three northeastern provinces and examined the relationship between water resources available for energy-food production and the amount of water used for these purposes in 36 prefecture-level cities. The data required for the calculation come from the Statistical Yearbook and the Statistical bulletin. Determination of some parameters in the production and processing of raw coal, crude oil, natural gas and thermal power, and the reference coefficients for the conversion of energy to standard coal are based on national standards, guidelines and various statistical yearbooks. The National Meteorological Information Centre's China Meteorological Data Network provided the climate data.

2.2.2. Determination of parameters.

Crude coal, crude oil, natural gas and electricity are discounted to standard coal by factors of 0.7143 (kgce/kg), 1.4286 (kgce/kg), 1.2143 (kgce/kg) and 0.1229 (kgce/kW-h), respectively. The direct blue and grey water footprints of the energy unit production in the mining process of raw coal are 0.014 (m^3/GJ) and 0.137 (m^3/GJ), and the direct blue and grey water footprints of the energy unit production in the washing process are 0.007 (m^3/GJ) and 0.027 (m^3/GJ); and the direct blue and grey water footprints of the energy unit production in the mining process of crude oil are 0.167 (m^3/GJ) [10] and 0.016 (m^3/GJ), and 0.057 (m^3/GJ) [11] for the direct blue water footprint of energy unit production during crude oil processing; 0.077 (m^3/GJ) and 0.013 (m^3/GJ) for the direct blue and grey water footprints of energy unit production during natural gas extraction, and 0.006 (m^3/GJ) for the direct blue water footprint of energy unit production during natural gas processing and purification. value of 0.006 (m^3/GJ) [12]; direct blue and grey water footprint values of energy unit production of thermal power cooling system of 0.681 (m^3/GJ) and 0.083 (m^3/GJ) [13], and indirect blue and grey water footprint values of energy unit production of energy input process of thermal power of 0.063 (m^3/GJ) and 0.387 (m^3/GJ).

3. Results

3.1. Characterization of the energy-food water footprint's spatial variation

3.1.1. Characterization of the energy water footprint's spatial variation.

As illustrated in Figure 1, the water footprint of raw coal dominates the fossil energy water footprints of the areas in the three northeastern provinces; in 2000, this was the case for 26 regions. On the one hand, this is because the three northeastern provinces get their coal mostly from the aforementioned locations, which is rich in resources, and on the other hand, due to the rapid development of the country's economy, the demand for coal as one of the main sources of energy in the market also continues to increase, the water footprint of fossil energy was generally high in all three of the northeastern provinces' regions for raw coal, and it continued to rise as mining activities increased. Towards the end of the study period, there is a decreasing trend in the water footprint of raw coal in each region of the three northeastern provinces, and the areas dominated by it are mainly distributed in Jixi City, Hegang City, Shuangyashan City and other areas, while the areas dominated by the water footprint of crude oil are mainly concentrated in the oil and gas resource-rich areas such as Daqing City, Songyuan City, Panjin City and other areas. These areas are highly clustered and their distribution ranges are relatively stable. Areas with high thermal power water footprints are mainly located in Dalian, Changchun and Harbin, which are mostly areas with rapid urbanisation, high economic development and fossil energy-rich areas.

3.1.2. Characterization of the spatial variability of the water footprint of food.

As demonstrated in Figure 2, there are notable regional variations in the water footprint of food in each of the three northeastern provinces, with the Yichun city's total water footprint of food being the lowest (0.046 billion m^3), and the highest in the city of Qiqihar (2.782 billion m^3).). In certain regions, the water footprint of food decreased in 2005 as a result of heavy rainfall; in Hegang and Dandong, indicating that the total water footprint of food in the three northeastern provinces at the beginning of the study was low. From 2005 to 2020, there is an overall upward trend, and the high water footprint of food is mainly concentrated in Qiqihar, Harbin, Baicheng, Changchun, Jiamusi, and Suihua, which are suitable for large-scale mechanised farming due to a number of factors such as climate, topography, and geomorphology, yet the majority of the area is semi-arid and dry, with a serious shortage of precipitation resources and some areas suitable for growing. which consume more water. Consequently, agriculture has a comparatively large water footprint.

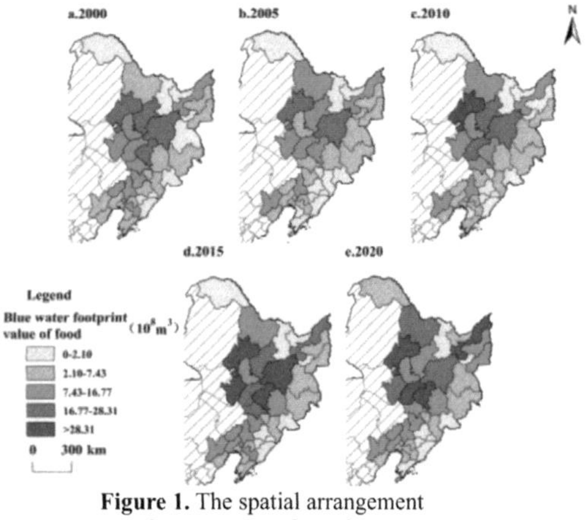

Figure 1. The spatial arrangement of energy water footprint

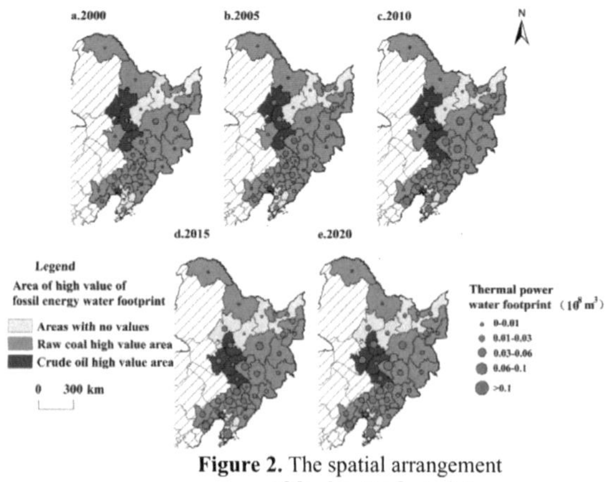

Figure 2. The spatial arrangement of food water footprint

3.2. Pressure of the energy-food water footprint on available water resources

The pressure on the amount of available water resources from the energy and food water footprints is constrained by the combination of the energy water footprint and the food water footprint. According to the classification of water pressure in Table 1, the spatial distribution diagram of energy-food water pressure index is drawn, as shown in Figure 3. It can be seen from Figure 3a, the regions with high energy water footprint pressure index are primarily found in Daqing City, Heilongjiang Province and Panjin City, Liaoning Province, while the rest of the regions are unpressureed areas. The reason for this is that Daqing and Panjin have large oil fields in China with abundant oil and gas resources. Daqing and Panjin have less available water resources than the other urban regions, which contributes to a higher energy water footprint pressure index in those places. The areas with high food water footprint pressure indices are mainly located in the Northeast Plain (temperate continental monsoon climate) and Songhua River Basin (north temperate monsoon climate) south of the Xiaoxinganling Mountains and west of the Changbaishan Mountains, which is in line with the food-growing areas and the climate suitable for agricultural development in the Northeast Provinces.

Comprehensive Figure 3a, 3b can be seen, the spatial map of the energy-food water footprint pressure on available water resources in the three northeastern provinces is basically the same as the spatial map of the food water footprint pressure on available water resources, with the food water footprint equivalent to 25 times that of the energy water footprint, and the high energy water footprint pressure area is also the high food water footprint pressure area, so the dominant factor in the degree of pressure on available water resources from the energy-food water footprint is the food water footprint, and in addition to the energy-food competition for water resources, the competition for water resources between various types of food is also very intense.

As can be seen from Figure 3b, the number of areas with high and higher pressure on water resources from the energy-food water footprint accounts for 52.78%, and their available water resources are relatively scarce. In the areas with high pressure and high pressure on energy and water resources, the water footprint of crude oil and blue water footprint of maize are larger in Daqing, and the water footprint of crude oil and blue water footprint of rice are larger in Panjin, indicating that the dominant factors that can negatively affect the pressure of the energy-food water footprint on water resources in Daqing are the spatial distribution of the water footprint of crude oil and the blue water footprint of maize, and that the contradiction between the production of crude oil and that of maize is the most prominent. The same reasoning can be obtained that the dominant factor that can negatively affect the pressure of Panjin City's energy-food blue water footprint on available water resources is the spatial distribution pattern of the water footprint of crude oil and the blue water footprint of rice, and the contradiction between crude oil production and rice production is the most prominent. Most of the rest of the pressure high value area of the main conflict is rice production and corn production.

The areas with medium water pressure are Heihe City, Shuangyashan City, Jixi City and Jilin City, which are mainly dominated by the water footprint of rice and soybeans, indicating that this type of area is still under a certain amount of pressure on water resources, and the demand for water resources in agricultural production is greater than that in energy production, which reflects the characteristics of the region's agricultural economy. The number of areas with low and no water pressure is 33%, and this type of area has small energy reserves, low food production, abundant water resources, and an overall low energy-food water footprint. Most of the regions in this category have a water footprint for rice and a water footprint for soybeans, and although soybeans and rice consume more water per unit of food, the cultivation of soybeans and rice, which consume more water, does not put a greater pressure on water resources due to the abundance of available water resources.

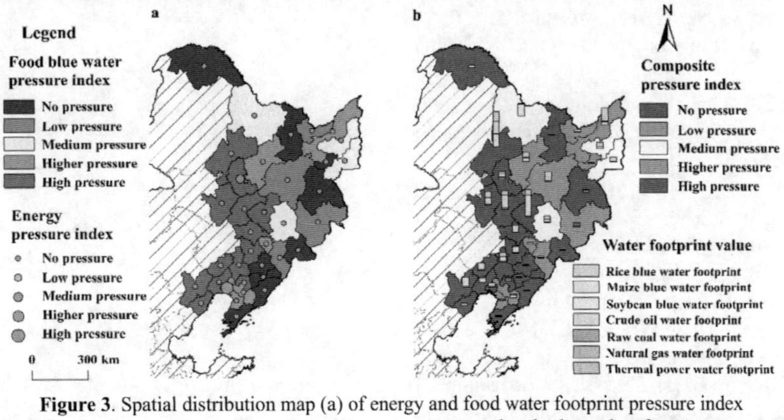

Figure 3. Spatial distribution map (a) of energy and food water footprint pressure index

4. Conclusions

(1) In terms of spatial pattern, the water footprint of energy and food is mainly affected by the difference in resource endowment, with the water footprint of energy mainly gathered in energy-rich areas such as Daqing and Panjin, and the blue water footprint of food mainly distributed in the Northeast Plain area south of the Xiaoxinganling Mountains and west of the Changbai Mountains. Water resources cooperation can be strengthened between different regions, taking into account the dual needs of water resources in energy and food production. Transregional water resource allocation can be achieved through water transfers, inter-basin water transfers and water trading.

(2) The three northeastern provinces as a whole are under clear pressure in terms of water use, and the regions clearly face challenges in water use, with the most prominent conflicts in crude oil, corn and rice production. Easing the pressure on water resources in the three northeastern provinces can be achieved mainly through agriculture, promoting the rational allocation and use of water resources on farmland, encouraging

agricultural restructuring, and reducing the cultivation of crops that have a high demand for water resources on the premise of guaranteeing food security.

Acknowledgements

This research has been supported by the Philosophy and Social Science Research Planning Project of Heilongjiang Province, Project No. 21JYB139.

References

1. Ma, L., Li, C., Hu, X., Wang, P., & Li, X. (2021). Synergetic change of water, energy and food in China: Quantitative description and challenges. Stochastic Environmental Research and Risk Assessment, 35(1), 43-68.
2. Cui S, Wu M, Wang X and Cao X. (2023). Optimization of planting structure of water-lifting irrigation system based on water footprint and water-energy-food linkage. Journal of Hydraulic Engineering, 54(8), 967-977.
3. Hung, A. H. P. (2002). Virtual water trade a quantification of virtual water flows between nations in relation to international crop trade.
4. Hoekstra, A. Y. (2011). The water footprint assessment manual: Setting the global standard. Routledge.
5. Guan W, Zhao X and Xu S. (2019). Spatiotemporal characteristics of China's energy and water footprint and its matching relationship with water resources. Resources Science, 41(11), 2008-2019.
6. Mekonnen, M. M. and Gerbens-Leenes, W. (2020). The water footprint of global food production. Water, 12(10), 2696.
7. Qin, Y. (2021). Global competing water uses for food and energy. Environmental Research Letters, 16(6), 064091.
8. Bai J and Zhang H. (2018). Spatial-temporal variation and driving force analysis of water-energy-food pressure in China. Chinese Journal of Geography, 38(10), 1653-1660.
9. GAO Tian, YANG X. (2021) Research on the decoupling relationship between food and energy water footprint pressure and green development in the Yellow River Basin[J].Water-saving Irrigation
10. LI He. (2003). Research on the index system of cleaner production in oil and gas extraction industry (heavy oil). Environmental Protection of Oil and Gas Fields, 13(1), 16-18.
11. WANG W. (2007). Ecological Analysis and Optimal Utilization of Water Resources in Petrochemical Industry (Doctoral dissertation, Shanghai: Tongji University).
12. Lei Y, Liu Q, Tang W, Qing S and Zhang J. (2011). The role of energy conservation and emission reduction in natural gas processing plants in promoting a low-carbon economy. Oil & Gas Chemicals, 40(4), 358-361.
13. Zhu F, Zhong L and Wang Q. (2009). The emission coefficient of the main pollutants in the thermal power industry. China Environmental Science Press.

https://doi.org/10.1051/e3sconf/202452802021

Simulation of natural ventilation for semi-enclosed buildings

Youwen Zhu[1], Xiaolin Sun[1,2,*], Jinfeng Wang[1,2], Haojie Hu[1], Zihao Liang[1]

[1]School of Food Science and Technology, Shanghai Ocean University, Shanghai, 201306, China
[2]Shanghai Professional Technology Service Platform on Cold Chain Equipment Performance and Energy Sav-ing Evaluation, Shanghai, 201306, China

Abstract. Energy-saving buildings have become a topical issue and widely used the energy-saving ventilation design, which aims to maximize the use of natural ventilation while reducing demand on mechanical ventilation. In this research, natural ventilation performance of two different semi-enclosed building layouts, "O" type and "L-L" type, are studied by numerical simulation. According to the simulation results, in the "O" type, the maximum indoor air velocity occurs in building A (orientation facing the inflow direction). The minimum indoor air velocity occurs in building C (orientation facing building A). In the "L-L" type, the maximum indoor air velocity occurs in building A (orientation facing the inflow direction). The minimum indoor air velocity occurs in buildings B and D (orientation parallel to the inflow direction). And air flow field simulation results show that, the "L-L" type exhibits better than the "O" type in natural ventilation. Furthermore, simulation results show that the natural ventilation performance in the room is influenced not only by the building orientation, but also by the window size.

1. Introduction

Energy-saving buildings has become a topical issue that highly valued by the government, and energy-saving ventilation design has been widely applied in that context. The energy-saving ventilation design aims to maximize the use of natural ventilation and, simultaneously, reduce reliance on mechanical ventilation through a well-designed building layout. This approach is intended to achieve energy savings[1].

In China, enclosed layout has been widely used in both residential and commercial buildings. The buildings along the street are often neatly arranged, creating a compact and unified building elevation. However, this layout can only ensure that some of the buildings have good orientation. Due to mutual occlusion, some buildings experience poor natural ventilation. The semi-enclosed buildings, which feature open areas in its layout, come in different types designed to enhance natural ventilation, including the "O" type, the "L-L" type, and others.

There are various factors that affect the natural ventilation of buildings. Chen Y et al.[2] focused on single bedrooms and how various factors, such as the window-wall ratio, influenced natural ventilation. Results revealed that under similar conditions, the volume of ventilation was positively associated with the window-wall ratio. Nomura M[3] selected 30 buildings and evaluated their air exchange rates, which ranged from 1 to 10 ac/h with no distinct peak. The results indicated that there was no clear association between air change rate and floor area, and natural ventilation primarily depended on the building design scheme. Zhang W et al.[4] used CFD to concentrate on hospital design and natural ventilation in buildings, aiming to enhance air quality. The results indicated that by setting reasonable types, locations, and heights of patios, the public space of a hospital could exhibit better natural ventilation, leading to improved comfort. Wu G. and Gan V.J[5,6] also used CFD to develop a model for enhancing the natural ventilation of high-rise buildings. Ma G et al.[7] used EnergyPlus to simulate natural ventilation of a railway station in Lianyungang City. Compared to mechanical refrigeration and air conditioner, using natural ventilation in conjunction with an air conditioner could reduce operation by 467h and save 13.1% of energy consumption. With the rise in air pollution, Ahmed T et al.[8] analyzed natural ventilation in warm climates. The results indicated that cross ventilation was effective in providing natural ventilation to lower indoor air temperatures. KazemiEsfeh M et al.[9] analyzed the natural ventilation of a curved roof at different wind angles and revealed that a semi-cylindrical curved roof could improve natural ventilation inside buildings. Bienvenido-Huertas D et al.[10] proposed the use of natural ventilation to de-crease cooling energy consumption during summer temperatures. Kyritsi E et al.[11] analyzed how window opening patterns impacted natural ventilation in office buildings. Gough H.L et al.[12] analyzed the impact of air flow direction on natural ventilation in an isolated building.

In this research, ANSYS is used to analyze the impact of window size and building orientation on natural ventilation for two different semi-enclosed building layouts, to provide suggestions for future building design.

*Corresponding author's e-mail: xlsun@shou.edu.cn

 https://doi.org/10.1051/e3sconf/202452802021

2. Simulation model and analysis methods

2.1. Building model

2.1.1. The "O" type

The four buildings are symmetrically arranged, looking like the letter "O". In this model, each building has four identical floors, and each floor contains four basic rooms. The doors of rooms are closed and the specific model is depicted in Figure 1.

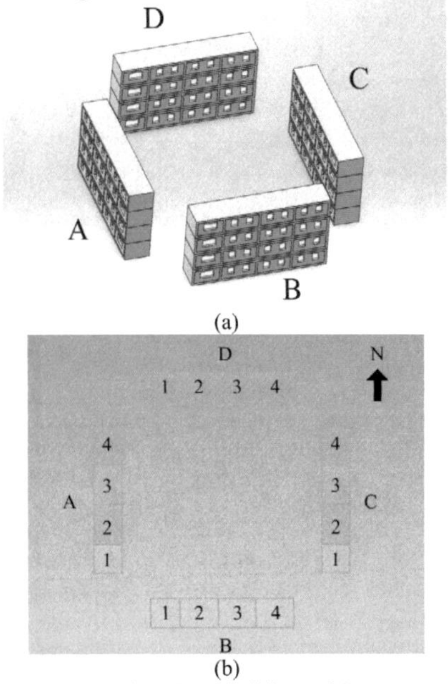

Figure 1. Semi-enclosed buildings of the "O" type.

Each floor has four rooms with lengths of 6m, 8m, 8m, and 8m, as shown in Figure 3. The thickness of wall is 0.4m. The depth of each room is 6m, and the height is 3m. Assuming that the air flow direction is from left to right. As shown in Figure 1, the building that directly faces the direction of the air flow is designated as building A. Other buildings, in an anticlock-wise orientation, are labeled as buildings B, C, and D. The distance between building A and building C is 42m, and the distance between building B and building D is also 42m. The four rooms on the third floor of each building are designated as room 1, room 2, room 3, and room 4 respectively, as shown in Figure 2. There is only one window in room 1, sizing 3m×1.5m. While for room 2, 3 and 4, there are two windows in each room, with the same window size 1.5m×1.5m, as shown in Figure 3.

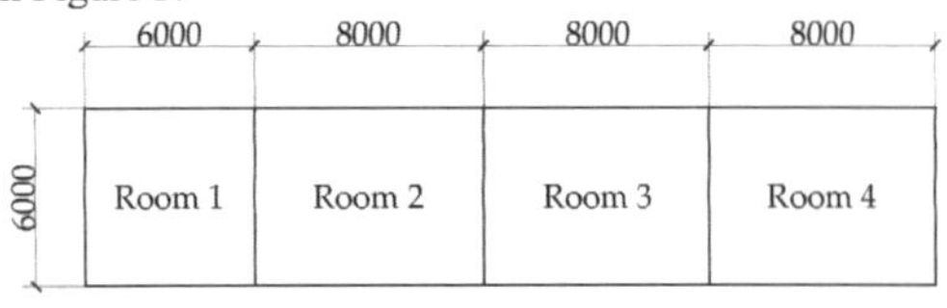

Figure 2. Floor plan.

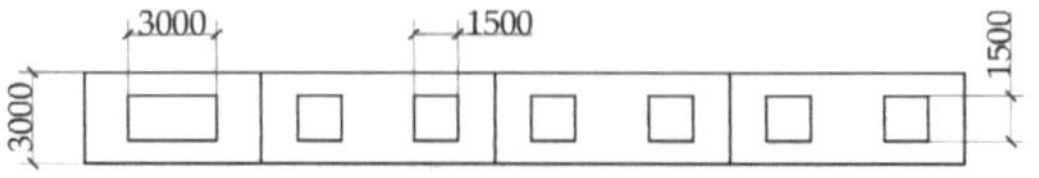

Figure 3. Floor elevation.

2.1.2. The "O" type

The two buildings are positioned in a way that looks like the letter "L". A pair of "L"s are placed symmetrically, forming an "L-L" shape. The doors of rooms are closed and the specific model is depicted in Figure 4. In this model, each building has four identical floors, and each floor contains four basic rooms, without considering the wall thickness. The room sizes are the same as the "O" type, as shown in Figure 3 and Figure 4. The distance between building A and building C is 42m, while the distance between building B and building D is 33m. Additionally, the distance between building A and building D, as well as building B and building C, is 6m.

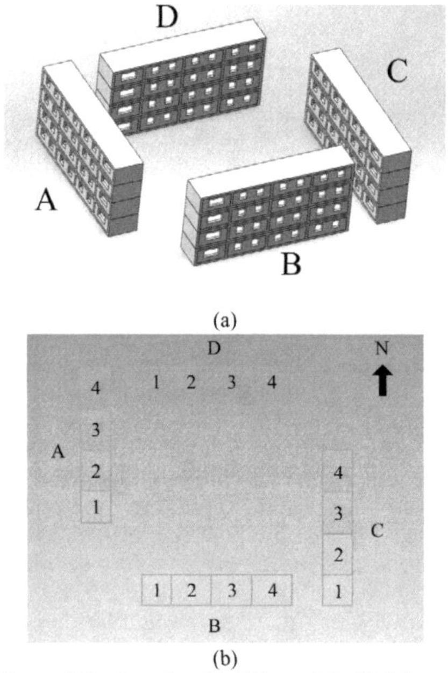

Figure 4. Semi-enclosed buildings of the "L-L" type.

2.2. Ventilation simulation and measurement method

2.2.1. Setting of air flow field

In this study, ANSYS Icepak is used to conduct numerical simulations of indoor natural ventilation. In the air flow field, the distance between the air intake boundary and building A is 82m. The distance between the air outlet boundary and building C is 164m. Additionally, the distance between the lateral air boundary and building B is 82m, as well as building D. The height of the air flow field is 97.2m, as shown in Figure 5 and Figure 6.

https://doi.org/10.1051/e3sconf/202452802021

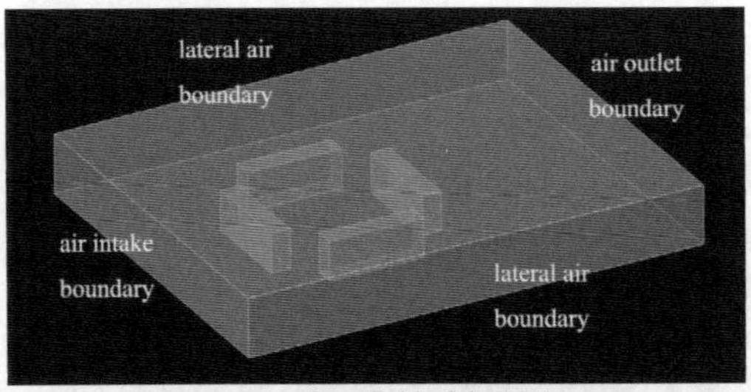

Figure 5. Air flow field of the "O" type.

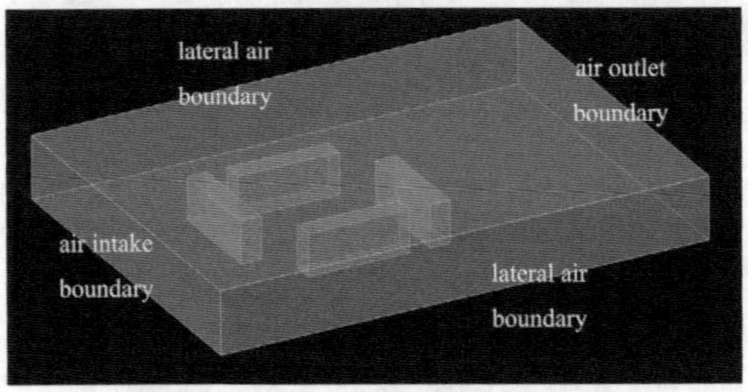

Figure 6. Air flow field of the "L-L" type.

2.2.2. Setting of air velocity and building design

According to the China Meteorological Data Set for Building Thermal Environment Analysis[13], the data for typical meteorological years in Shanghai is shown in Table 1. This research focuses on simple buildings in Shanghai, a representative area in the summer condition. The outdoor air velocity is set at 3.4m/s.

Table 1. Data of typical meteorological years in Shanghai.

Season	Dominant air flow direction	Frequency of air flow direction(%)	Average air flow velocity (m/s)
Spring	ESE(111.5°)	11.91	3.8
Summer	ESE(111.5°)	14.96	3.4
Autumn	NNE(22.5°)	14.41	3.8
Winter	NNE(22.5°)	14.24	3.5

Among basic structures, the school building is better suited to the requirements of semi-enclosed layout. Therefore, the building design is based on the school building in this research. According to GB 50099-2011: the Code for Design of School[14], the primary school main building should not exceed four floors, while the middle school main building should not exceed five floors. Consequently, in this research, a four floors building has been chosen as the standard research subject for modeling.

2.2.3. Assessing natural ventilation of the building

The natural ventilation performance of buildings can be evaluated on air velocity, ventilation volume, air change rate, air age and so on. And in this research, air velocity is adapted as the evaluation criterion.

2.3. Analysis method

According to the GB/T 50785-2012: Evaluation Standard for Indoor Thermal Environment in Civil Buildings[15], the sampling points should be arranged in places that are the most uncomfortable for people, including positions near windows and in-ner corners. The distance between the sampling points and adjacent walls should ex-ceed 0.5m. In this research, 5 sampling points are arranged in each room, and the av-erage air flow velocity of the 5 sampling points is taken as the mean air flow velocity of the room. Location of the sampling points are depicted in the Figure 7.

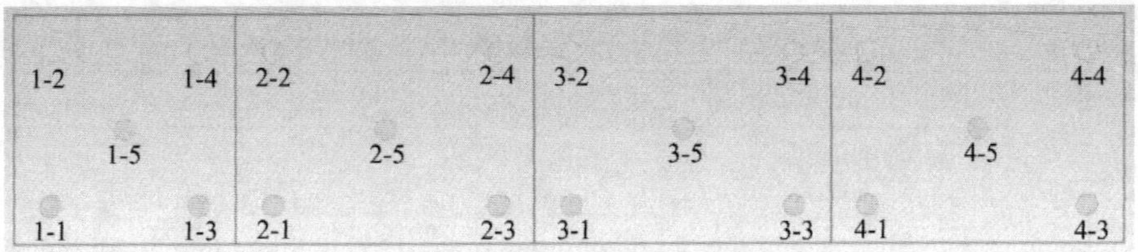

Figure 7. Location of the sampling points.

3. Simulation results

3.1. Ventilation simulation of the "O" type

The air flow is set from left to right with the velocity of 3.4m/s, and the calculation result of air flow field for the "O" type buildings is shown in Figure 8.

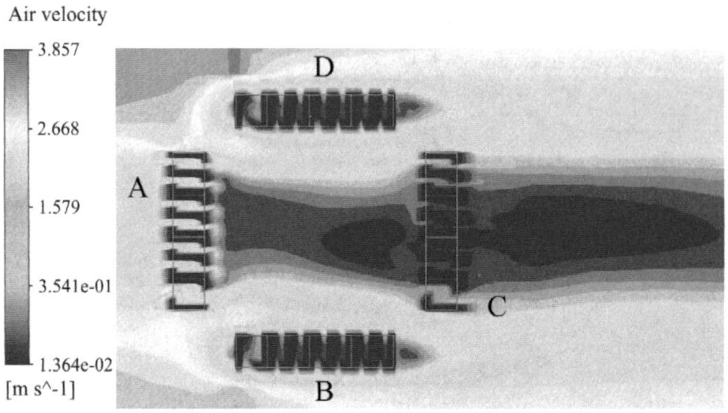

Figure 8. Air flow field of the "O" type

The color of air flow field in Figure 8 represents air flow velocity. In building A (orientation facing the inflow direction), the air flow velocity is highest. In building C (orientation facing building A), the air flow velocity in the other rooms is significantly lower. To assess the ventilation more accurately, this research quantitatively analyzes indoor air flow at sampling points located in the rooms on the 3rd floor.

In building A, according to Table 2, it could be observed that the maximum air flow velocity occurs in room 1, while the minimum air flow velocity occurs in the room 2. Room 1, with one large open window, performs significantly higher air flow velocity than the other 3 rooms that with two small open windows, indicating that the natural ventilation performance in the room will be reduced if one large window is divided into two small ones.

Table 2. Air flow velocity in building A of the "O" type.

Building A	Sampling point (m/s)					Average air flow velocity (m/s)
	1	2	3	4	5	
Room 1	0.33	0.21	0.21	0.38	1.20	0.446
Room 2	0.34	0.22	0.12	0.09	0.06	0.166
Room 3	0.36	0.21	0.19	0.18	0.10	0.208
Room 4	0.42	0.22	0.20	0.16	0.06	0.212

According to Table 3 and Figure 1, building B is parallel to the inflow direction, but the indoor air velocity of four rooms is still different. Room 1, with a large open window, exhibits higher air flow velocity than the other rooms. However, the average air flow velocity in building B is much lower than that in building A (orientation facing the inflow direction), indicating that the orientation of the building has significant influence on indoor natural ventilation in the "O" type.

Table 3. Air flow velocity in building B of the "O" type.

Building B	Sampling point (m/s)					Average air flow velocity (m/s)
	1	2	3	4	5	
Room 1	0.20	0.08	0.13	0.26	0.65	0.264
Room 2	0.12	0.08	0.10	0.17	0.06	0.106
Room 3	0.06	0.06	0.11	0.12	0.02	0.074
Room 4	0.07	0.07	0.10	0.14	0.06	0.088

According to Table 2 and Table 4 , the average air flow velocity in building C is significantly lower than that in building A. This indicates that building A (orientation facing the inflow direction) has significant influence on the indoor natural ventilation of building C (orientation facing building A) in the "O" type.

Table 4. Air flow velocity in building C of the "O" type.

Building C	Sampling point (m/s)					Average air flow velocity (m/s)
	1	2	3	4	5	
Room 1	0.12	0.10	0.08	0.04	0.32	0.132
Room 2	0.06	0.03	0.03	0.02	0.03	0.034
Room 3	0.12	0.03	0.08	0.04	0.06	0.066
Room 4	0.16	0.07	0.04	0.04	0.02	0.066

As shown in Figure 1, buildings B and D are symmetrically arranged. According to Table 3 and Table 5, the average air flow velocity in building D, is very close to that in building B that with the same orientations (orientation parallel to the inflow direction).

Table 5. Air flow velocity in building D of the "O" type.

Building D	Sampling point (m/s)					Average air flow velocity (m/s)
	1	2	3	4	5	
Room 1	0.19	0.09	0.14	0.28	0.72	0.284
Room 2	0.11	0.10	0.10	0.15	0.06	0.104
Room 3	0.10	0.07	0.11	0.16	0.06	0.100
Room 4	0.14	0.06	0.12	0.11	0.04	0.094

3.2. Ventilation simulation of the "O" type

The air flow is set from left to right with a velocity of 3.4m/s, and the calculation result of air flow field for the "L-L" type buildings is shown in Figure 9.

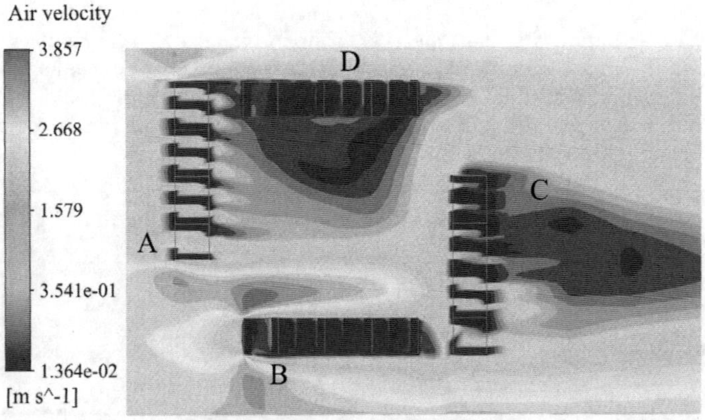

Figure 9. Air flow field of the "L-L" type.

The color of air flow field in Figure 9 represents air velocity. In building A (orientation facing the inflow direction), the air velocity is the highest. In buildings B and D (orientation parallel to the inflow direction), the air velocity of other rooms is significantly lower except for room 1 near building A. This research quantitatively analyzes indoor air flow velocity at sampling points located in four rooms on the 3rd floor.

In building A, according to Table 6, it could be observed that the maximum air flow velocity occurs in room 1, while the minimum air flow velocity occurs in room 2. Room 1, with one large open window, performs significantly higher air flow velocity than the other 3 rooms that with two small open windows, indicating that the natural ventilation performance in the room will be reduced if one large window is divided into two small ones.

Table 6. Air flow velocity in building A of the "L-L" type.

	Sampling point (m/s)					Average air flow velocity (m/s)
	1	2	3	4	5	
Room 1	0.36	0.08	0.19	0.42	1.40	0.490
Room 2	0.40	0.16	0.11	0.21	0.03	0.182
Room 3	0.42	0.18	0.10	0.20	0.02	0.184
Room 4	0.43	0.21	0.12	0.40	0.04	0.240

Building B is parallel to the inflow direction, as shown in Figure 4. According to Table 6 and Table 7, room 1, with a large open window, exhibits higher air velocity than the other rooms. However, the average air flow velocity in building B is much lower than that in building A (orientation facing the inflow direction), indicating that the orientation of the building also has significant influence on indoor natural ventilation in the "L-L" type.

Table 7. Air flow velocity in building B of the "L-L" type.

	Sampling point (m/s)					Average air flow velocity (m/s)
	1	2	3	4	5	
Room 1	0.12	0.02	0.12	0.21	0.36	0.166

https://doi.org/10.1051/e3sconf/202452802021

Room 2	0.09	0.04	0.08	0.12	0.04	0.074
Room 3	0.04	0.03	0.06	0.11	0.02	0.052
Room 4	0.05	0.03	0.09	0.13	0.05	0.070

According to Table 8 and Table 9, the average air flow velocity in building C is close to that in building A. This indicates that building A (orientation facing the inflow direction) has little influence on the indoor natural ventilation of building C (orientation facing building A) in the "L-L" type.

Table 8. Air flow velocity in building C of the "L-L" type.

Building C	Sampling point (m/s)					Average air flow velocity (m/s)
	1	2	3	4	5	
Room 1	0.28	0.08	0.12	0.26	1.20	0.388
Room 2	0.31	0.12	0.08	0.18	0.04	0.146
Room 3	0.25	0.12	0.05	0.18	0.08	0.136
Room 4	0.22	0.10	0.06	0.08	0.02	0.096

Building D is parallel to inflow direction and is shielded by building A, as shown in Figure 4. According to Table 7 and Table 9, the average air flow velocity in building D is close to that in building B. This suggests that in the "L-L" type, building A (orientation facing the inflow direction) has little influence on the natural indoor ventilation of adjacent building D (orientation parallel to the inflow direction), and that the orientation of the building is the primary influencing factor.

Table 9. Air flow velocity in building D of the "L-L" type.

Building D	Sampling point (m/s)					Average air flow velocity (m/s)
	1	2	3	4	5	
Room 1	0.11	0.06	0.04	0.12	0.32	0.130
Room 2	0.05	0.04	0.11	0.02	0.10	0.064
Room 3	0.32	0.03	0.09	0.04	0.04	0.104
Room 4	0.06	0.05	0.09	0.11	0.02	0.066

4. Comparison Results

To analyze the natural ventilation of two types, the average air velocity in four rooms of the "O" and "L-L" types is compared, as shown in Table 10 and Table 11.

Table 10. Air flow velocity in rooms of the "O" type.

"O" type	Room (m/s)				Average air flow velocity (m/s)
	1	2	3	4	
Building A	0.466	0.166	0.208	0.212	0.263
Building B	0.264	0.106	0.074	0.088	0.133
Building C	0.132	0.034	0.066	0.066	0.075
Building D	0.284	0.104	0.100	0.094	0.146

Table 11. Air flow velocity in rooms of the "L-L" type.

"L-L" type	Room (m/s)				Average air flow velocity (m/s)
	1	2	3	4	
Building A	0.490	0.182	0.184	0.240	0.274
Building B	0.166	0.074	0.052	0.070	0.090
Building C	0.388	0.146	0.136	0.096	0.191
Building D	0.130	0.064	0.104	0.066	0.091

Table 10 shows that, for the "O" type, maximum average air flow velocity occurs in building A, while minimum average air flow velocity occurs in building C. The average air velocity in building C is 0.075m/s, which is significantly lower than that of 0.263m/s in building A. The results indicate that building A (orientation facing the in-flow direction) significantly reduces the natural ventilation of building C (orientation facing building A) in the "O" type. The average air flow velocity in building B and D is significantly lower than that in building A. The natural ventilation of building B and D (orientation parallel to the inflow direction) is worse than that of building A in the "O" type.

Table 11 shows that, for the "L-L" type, maximum average air flow velocity occurs in building A, while minimum average air flow velocity occurs in building B. The average air velocity in buildings B and D is approximately 0.09m/s, which is significantly lower than the 0.274m/s in building A. The results indicate that the natural ventilation of building B and D (orientation parallel to the inflow direction) is worse than that of building A (orientation facing the inflow direction) in the "L-L" type.

Comparing Table 10 with Table 11, in the "O" type, building A (orientation facing the inflow direction) significantly reduces the natural ventilation of building C (orientation facing building A). In the "L-L" type, the natural ventilation of building C (orientation facing building A) is slightly influenced by building A (orientation facing the inflow direction), and building C still exhibits good ventilation performance. Additionally, the average air velocity of four buildings in the "O" type is lower than that in the "L-L" type, indicating that the "L-L" type exhibits better natural ventilation.

5. Conclusions

Based on the results and analysis above, conclusions are obtained as below:

1. In the semi-enclosed buildings, the "L-L" type exhibits better natural ventilation than the "O" type.

2. In the semi-enclosed buildings, orientation facing the inflow direction exhibits better natural ventilation compared to orientation parallel to the inflow direction. This indicates that the orientation of a building significantly influences on natural ventilation.

3. In the semi-enclosed buildings, using one large open window instead of two small open windows can enhance natural indoor ventilation.

In the semi-enclosed buildings, reasonable building orientation, well-designed building layout and larger window size can enhance natural ventilation and reduce the demand on mechanical ventilation, which will lead to energy saving.

Reference

1. Xu X. Research on design and reconstruction strategy of building ventilation system based on the concept of low carbon. Urbanism and Architecture. J. 2019, 16(28): 115-117. DOI:10.3969/j.issn.1673-0232.2019.28.024.
2. Chen Y, Wang X, Sun B et al. Wind tunnel experimental study on single room ventilation affected by miscellaneous factors. Journal of Zhejiang University (Engineering Science). J. 2012, 46(4): 658-664. DOI:10.3785/j.issn.1008-973X.2012.04.013.
3. Nomura M and Hiyama K. A review: Natural ventilation performance of office buildings in Japan. Renewable & Sustainable Energy Reviews. J. 2017, 74(7): 746-754. DOI:10.1016/j.rser.2017.02.083.
4. Zhang W, Zhang C and Wang Y. Study on natural ventilation simulation and design of double hallway inpatient building courtyard in hot and humid area. Journal of South China University of Technology (Social Science Edition). J. 2023, 51(4): 61-70. DOI:10.12141/j.issn.1000-565X.220361.
5. Wu G, Han D and Zhang J. Research on the space design of natural ventilation in high-rise office buildings based on ventilated corridors. Architectural Journal. J. 2023(2): 117-22.DOI:10.3969/j.issn.1671-6612.2023.01.012.
6. Gan V.J.L, Wang B, Chan C. M et al. Physics-based, data-driven approach for predicting natural ventilation of residential high-rise buildings. Building Simulation. J. 2021. DOI:10.1007/s12273-021-0784-9.
7. Ma G, Shen T and Yu T. Analysis on natural ventilation effect of large space buildings in coastal areas. Refrigeration&Airconditioning. J. 2023, 37(1): 88-91. DOI:10.3969/j.issn.1671-6612.2023.01.012.
8. Ahmed T, Kumar P and Mottet L. Natural ventilation in warm climates: The challenges of thermal comfort, heatwave resilience and indoor air quality. Renewable & Sustainable Energy Reviews. J. 2021, 138: 110669. DOI:10.1016/j.rser.2020.110669.
9. KazemiEsfeh M, Sohankar A, Shahsavari A.R et al. Experimental and numerical evaluation of wind-driven natural ventilation of a curved roof for various wind angles. Building and Environment. J. 2021: 108275. DOI:10.1016/j.buildenv.2021.108275.
10. Bienvenido-Huertas D, Sánchez-García D and Rubio-Bellido C. Analysing natural ventilation to reduce the cooling energy consumption and the fuel poverty of social dwellings in coastal zones. Applied Energy. J. 2020, 279: 115845.
DOI:10.1016/j.apenergy.2020.115845.
11. Kyritsi E and Michael A. An assessment of the impact of natural ventilation strategies and window opening patterns in office buildings in the mediterranean basin. Building and Environment. J. 2020, 175: 106384. DOI:10.1016/j.buildenv.2019.106384.
12. Gough H.L, Barlow J.F, Luo Z et al. Evaluating single-sided natural ventilation models against full-scale idealised measurements: Impact of wind direction and turbulence. Building and Environment. J. 2020, 170: 106556.
DOI:10.1016/j.buildenv.2019.106556
13. National Meteorological Information Center. *China Meteorological Data Set for Building Thermal Environment Analysis*. China Architecture & Building Press: The Beijing, China.
14. The Beijing: Standards Press of China. GB 50099-2011. Code for Design of School. https://www.nssi.org.cn/nssi/front/76841260.html
15. The Beijing: Standards Press of China. GB/T 50785-2012. Evaluation Standard for Indoor Thermal Environment in Civil Buildings. https://www.nssi.org.cn/nssi/front/78202043.html

https://doi.org/10.1051/e3sconf/202452802022

Study on coupling and coordinated development of green development of power grid and ecosystem—from the perspective of energy industry chains

Yanming Jin*[1], Xue Tan[1], Qian Zhao[1], Shixiong Jiang[2], Xi Li[2], Zhongqing Wang[2], Mengjie Xing[1]

[1]State Grid Corporation of China, Beijing, 102209, China
[2]State Grid Fujian Electric Power Science Research Institute, Fuzhou, 102209, China

Abstract. The advancement of green power grid technology, driven by the Tenth Five-Year Plan and the "Double Carbon" objective, is transforming the energy sector and promoting a greener economy. Understanding the link between green power grid development and the ecosystem is crucial. This study establishes an evaluation system to assess their coordination using a model. The analysis shows a progression towards coordinated development since 2006, with green initiatives playing a key role. This evolution highlights green development as a driving force behind power grid progress, rather than a constraint.

1. Introduction

The "Fourteenth Five-Year Plan" marks a pivotal phase for China's pursuit of high-quality economic growth and environmental conservation. China has set the ambitious target of "dual carbon," with the transition from fossil fuels to cleaner energy sources playing a central role in achieving carbon neutrality. In 2020, China's CO_2 emissions from energy combustion will be about 9.7 billion tons, and the power industry will be about 3.9 billion tons, accounting for 88% and 35% of the total CO_2 emissions [1]. In the future, a new energy-based system will be established. Seen from the perspective of the energy industry chain, the green transformation of the power grid not only impacts the restructuring of the energy power system but also plays a crucial role in shaping the development of environmentally friendly production methods and lifestyles in the economy and society.

Currently, the article on the green development of the power grid primarily concentrates on its internal growth and progress. Wu [2] proposed optimization suggestions for the environmental protection system of power grid enterprises in the context of the reform of the domestic ecological civilization system; Liu [3] developed an index system and evaluation model for multi-level power grid development, considering aspects such as safety. The green development of the power grid should extend beyond its internal environmental protection measures and consider its crucial position in the broader energy industry chain. By examining how the power grid positively influences energy transformation across the industry chain, a more comprehensive understanding of its importance can be gained. This study utilized national data from 2006 to 2019 to calculate the coupling coordination degree, drawing primarily from sources such as the China Energy Statistics Yearbook and the Environmental Statistics Yearbook. The analysis aimed to investigate the coupling and harmony association between the power grid's green improvement index system and the natural habitat evaluation index system. By emphasizing the power grid's role as a central hub for clean energy generation and consumption, this research presents a practical approach to fulfilling the goals outlined in the "14th Five-Year Plan" and the "double carbon" objectives, ultimately promoting high-quality green development.

2. Coupling evaluation index system

In terms of the energy industry chain, the interconnection between power grid advancement and the ecological environment manifests primarily in three dimensions [4-10]. Concerning the creation of a green power grid development evaluation index system, assessment criteria can be devised based on three key dimensions: collaboration with upstream green initiatives, achieving internal sustainability goals, and fostering downstream environmental progress [11-13]. The power grid collaborates with upstream entities to advance clean energy development by establishing a low-carbon new energy supply system, enabling multi-energy provision, facilitating inter-provincial and inter-regional clean energy transmission, ensuring supply-demand equilibrium in energy-rich regions with high demand, fostering peak shaving auxiliary service markets, and achieving source-grid coordination. In terms of power

*Corresponding author's e-mail:jinsezhuangxiu@sgeri.sgcc.com.cn

grid infrastructure, the environmental impact encompasses the entire lifecycle of construction, operation, and decommissioning, including issues like electromagnetic noise pollution, wastewater discharge, energy loss, and greenhouse gas emissions. Downstream efforts to promote clean energy consumption involve leveraging power substitution opportunities at the user level, increasing terminal electricity consumption rates, optimizing power resource allocation, reducing electricity management costs, and enhancing energy efficiency and emission reduction measures.

When developing the ecosystem evaluation index system, the impact of the power system on the environment can be thoroughly assessed by employing the Pressure-State-Response model [14-15]. Pressure indicators are used to gauge the environmental effects resulting from economic and social endeavors, with a primary emphasis on the release of atmospheric pollutants, greenhouse gases, and solid waste. State indicators track changes in the environmental conditions within a defined timeframe, with a particular emphasis on air pollutant emissions in the power sector and noise pollution from power infrastructure projects. Response index aims to reduce the adverse impact of human activities, focusing on waste disposal rates and investments in pollution control measures.

3. Research method

3.1. Data standardization processing

In the coupled system, the evaluation index system composed of multiple factors has different dimensions, orders of magnitude and index properties. Through data standardization, each index can be compared and the influence of dimension can be eliminated. In the green power grid development system, the future energy development strategy target or trend extrapolation method's predicted index value serves as the reference value. In the eco-environmental system, the method employed for normalization in the absence of a reference value.

3.2. Comprehensive evaluation function

Entropy weight method uses the intrinsic information of evaluation index to determine the weight and avoid the deficiency of subjective judgment. The green improvement of power grid and evaluation function of natural habitat are as follows.

$$P(X)=\sum_{j=1}^{n} w_j X_{ij} \tag{1}$$

$$E(Y)=\sum_{j=1}^{n} w_j Y_{ij} \tag{2}$$

$P(X)$ is the green development index of power grid, describing the level of green development within the power grid subsystem; $E(Y)$ signifies the ecological environment index, reflecting the overall development status of the ecological environment subsystem; w_j represents the weight of the j-th indicator; Y_{ij} denotes the standardized value of the j-th index for both power grid green development and ecological environment in the i-th year.

3.3. Coupling coordination model

Coupling is a term used to describe the interconnected and collaborative relationship between two or more systems, encompassing both the degree of coupling and the degree of coupling collaboration [14]. The level of interconnection indicates the extent of reciprocal impact between systems or internal components, whereas the level of interconnection collaboration measures the degree of synchronization among systems or internal components throughout the progression.

3.3.1. Coupling degree.

Let C indicates the degree of coupling [16-17]. Subsequently, the coupling model illustrating the interplay among multiple systems can be formulated as follows.

$$C=\left\{\frac{\prod(u_i,u_j,\cdots,u_n)}{\left[\frac{u_i+u_j+\cdots+u_n}{t}\right]}\right\}^{1/t} \tag{3}$$

u is the system performance function, utilized to compute the collective level of the subsystems; t is the number of systems and is equal to 2; i and j represents the i-th and j-th systems respectively. make $P(X)=u_1$, $P(X)=u_2$, substituting it into the subsystem overall level of power grid green improvement and ecological environment, and constructing a two-dimensional coupling model.

$$C=\frac{2\sqrt[2]{P(X)\times E(Y)}}{P(X)+E(Y)} \tag{4}$$

3.3.2. Degree of coupling collaborative.

While the coupling degree can signify the intensity of the correlation between systems, to accurately depict the practical level and condition of systems, it is essential to introduce a coupling coordination degree model, and let D represent coupling coordination degree [18]. α , β is the coefficient to be determined; $\alpha+\beta=1$; T represents the comprehensive assessment index of the two-dimensional system.

$$D=\sqrt{C\times T} \tag{5}$$

$$T=\alpha u+\beta u=\alpha P(X)+\beta E(Y) \tag{6}$$

This article emphasizes the sustainable improvement of the power grid, thus establishing $\alpha = 0.6, \beta = 0.4$. Based on the existing research results, according to the size of $P(X)$ and $E(Y)$, the coupling coordination degree is subdivided into three type, four subtype. Table 1 illustrates the correlation between the sustainable improvement of electric power network and the ecological system.

Table 1. Coupling coordination degree classification.

Type	Scope	Subtype	Third type	Implication
Advanced development	$0.8 < D_1 \leq 1$	Highly symbiotic	$P(X) > E(Y)$	Highly symbiotic- $E(y)$ lag behind
			$P(X) = E(Y)$	Highly symbiotic- synchronous development
			$E(Y) > P(X)$	Highly symbiotic- $P(x)$ lag behind
Transform development	$0.5 < D_1 \leq 0.8$	Coordinated development	$P(X) > E(Y)$	Coordinated development- $E(y)$ lag behind
			$P(X) = E(Y)$	Coordinated development-synchronous development
			$E(Y) > P(X)$	Coordinated development- $P(x)$ lag behind
Primary development	$0.2 < D_1 \leq 0.5$	Run-in development	$P(X) > E(Y)$	Run-in development- $E(y)$ lag behind
			$P(X) = E(Y)$	Run-in development- synchronous development
			$E(Y) > P(X)$	Run-in development- $P(x)$ lag behind
	$0 < D_1 \leq 0.2$	Lowly symbiosis	$P(X) > E(Y)$	Lowly symbiosis- $E(y)$ lag behind
			$P(X) = E(Y)$	Lowly symbiosis- synchronous development
			$E(Y) > P(X)$	Lowly symbiosis- $P(x)$ lag behind

4. Empirical research

4.1. Stage division

As the proficiency of the two systems enhances, the coupling coordination degree between them has progressively risen from 0.06 to 0.74, as shown in Figure 1.

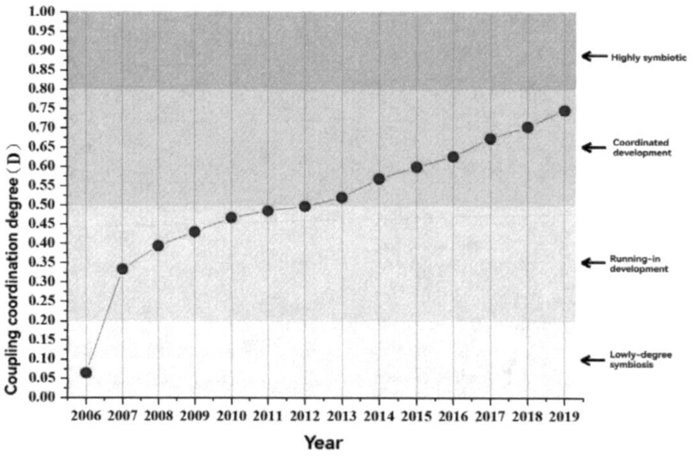

Figure 1. Coupling coordination degree.

4.1.1. Lowly symbiotic stage.

In 2006, it is a stage of low symbiosis, where where the improvement level of these two systems is relatively low, and the green development of the power grid is in a weak position. Currently, the power grid development prioritizes "generation-transmission-distribution" for economic gains, with ecological protection serving as a strict limit. The relationship between green power grid development and the ecosystem is characterized by mutual constraints and low symbiosis.

4.1.2. Run-in development stage.

In 2007-2011, it is a stage of running-in development, where the development level of the two systems is gradually increasing, and the development rate is equivalent, which restricts and promotes each other in the running-in. To address the conflict between power grid development and limited resource availability, the power grid system is emphasizing the achievement of low-carbon energy supply and the fulfillment of clean energy demand. Clean energy sources are progressively being integrated, and the utilization efficiency of renewable energy is continuously enhancing.

4.1.3. Coordinated development stage.

In 2012-2019, it is a stage of coordinated development, where the green development of the power grid and the ecological environment progress harmoniously and advance together. In 2015, the power grid achieved a higher level of green development than the ecological environment, signifying a pivotal moment in its evolution. The integration of green development principles across the production, transmission, and distribution processes has greatly boosted the power grid's environmental performance. Environmental protection now serves as a key driver of progress, continuously pushing for greener practices throughout the power grid's operations.

4.1.4. Highly symbiotic stage.

At the stage of high symbiosis, the levels of development of the two systems are highly coordinated, with the green development of the power grid continuing to play a positive role in improving the ecological environment system. During this stage, the focus of the green development of the power grid expands beyond the energy industry to include intelligent energy system as a platform, representing a new trend aimed at promoting green industry, sustainable transportation, eco-friendly buildings, and green lifestyles across society. In the future, the green development of the power grid and the ecological environment are expected to achieve a high degree of coexistence.

4.2. Sensitivity analysis

Considering that the assignment of weight α and β will affect the coupling degree between the green development of the power grid and the ecological environment. In this study, sensitivity analysis is carried out for three cases of α = 0.6; α = 0.7; α = 0.8. The research indicates that the higher the weighting of power grid green development, the more significant the degree of coupling coordination, and the green development of the power grid will facilitate the coordinated advancement of the ecological environment in the future,as shown in Figure 2.

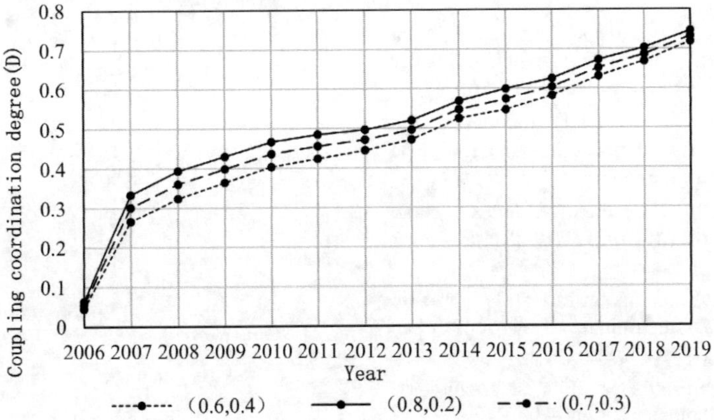

Figure 2. Sensitivity analysis of coupling coordination based on weight.

5. Conclusion

By analyzing the coupling mechanism between power grid green development and the ecosystem, the conclusions are obtained as below:

Between 2006 and 2019, the interplay between the green improvement of the power grid and the ecological environment progressed through stages of low symbiosis, transitional improvement, and collaborative improvement. The influence of the green advancement of the power grid on the ecological environment is increasing, suggesting that the adoption of strategies like electric energy substitution and clean substitution is vital for the integration and harmony between the power grid and the ecological environment system. This shift also indicates that green progress has evolved from impeding power grid development to propelling advancements within the power grid sector.

Acknowledgments

This work was financially supported by the State Grid Corporation's science and technology project "Evaluation technology and demonstration application of environmental protection ability of power grid enterprises". (project number: SGTYHT/21-JS-223).

References

1. Yang S, Ji Y and Zhang D (2018) Research and prospect of power system energy saving and emission reduction based on environmental effects. Journal of North China Electric Power University (Social Science Edition)., 4: 1–7.
2. Wu K and Zhao D (2020) Research on environmental protection management system of power grid enterprises facing green development. Value Engineering., 20: 12–13.
3. Liu W, Yu L, Zhang L, Yuan P, Wang Y and Zhao R (2020) Evaluation of power grid development level based on AHP and multilevel fuzzy comprehensive evaluation. Smart Power., 05: 80–85.
4. Fang Z (2017) On the impact of power grid construction on ecological environment and its prevention and control countermeasures. Resource Conservation and Environmental Protection., 2: 111–112.

5. Zhang L (2013) Study on Environmental Impact Assessment of Power Grid Construction Project. Inner Mongolia University.
6. Wang Z (2015) Study on Environmental Impact Assessment and Risk Prevention of UHV Power Grid Planning. North China Electric Power University.
7. Bai J (2012) Research on sustainability evaluation model and application of power grid construction projects. North China Electric Power University.
8. Liu Y (2010) Research on Sustainable Development of Electric Power Based on Ecological Carrying Capacity. North China Electric Power University (Beijing).
9. Song Z (2018) Research on DSM optimization under the background of air pollution prevention and control. North China Electric Power University (Beijing).
10. Zheng X and You W (2018) Innovation of price mechanism to promote green development of energy system. Price Theory and Practice., 4: 12–16.
11. Jiang Y (2017) Research on evaluation method and system of green development of Hainan electric power. North China Electric Power University.
12. Xin H (2016) Urban Green Power Index System and Evaluation Model. North China Electric Power University (Beijing).
13. Cheng C, Zhou Y and Yue K (2012) Strategic environmental assessment index system of urban green power. China Electric Power., 3: 76–80.
14. Ott W (1978) Environmental Indices: Theory and Practice. Science Publishers, Ann Arbor.
15. Engelbert S (1997) The index of sustainable economic welfare (ISEW) as an alternative to GDP in measuring economic welfare. Economic,. 1: 16–20.
16. Zhang X (2017) Study on the Coupling Relationship between Land Use and Ecological Environment in Beijing. Agricultural Science and Technology and Information., 08: 28–31.
17. Liu S, Zhang X, Hou Y, Chang W and Zheng Y (2018) Study on the Coupling Relationship between Urbanization and Ecological Environment in Beijing-Tianjin-Hebei Region. China Market., 19: 33–34.
18. Yang S, Qin Y, Deng W and Ji M (2017) Comparison of the coupling relationship between tourism economy and ecological environment of the top ten urban agglomerations in China. statistics and decision., 22: 131–134.

Study on coupling factors between green development of power grid and ecosystem—from the perspective of energy industry chains

Yanming Jin*[1], Xue Tan[1], Qian Zhao[1], Shixiong Jiang[2], Xi Li[2], Zhongqing Wang[2], Mengjie Xing[1]

[1]State Grid Corporation of China, Beijing, 102209, China
[2]State Grid Fujian Electric Power Science Research Institute, Fuzhou, 102209, China

Abstract. The green power grid development, with industry and regional characteristics, impacts energy systems and eco-friendly production. Ecosystem services are vital for sustainable development. Urgent research is needed to adapt to new situations. This study establishes index systems for power grid green development and ecology. Key factors affecting their coordination are identified, guiding the promotion of green power grid development. Focus on enhancing cross-regional clean energy allocation, waste disposal, noise monitoring, and increasing electric energy use in terminal consumption.

1. Introduction

Currently, China's ecological civilization construction includes the "double carbon" goal. Replacing fossil energy and reducing emissions in the energy sector are crucial for achieving carbon neutrality in China. The trend is towards constructing a new energy system centered on new energy sources[1]. The power grid will serve as the main hub for clean energy production and consumption, playing a key role in leading energy transformation and encouraging the advancement of carbon-reduced manufacturing and daily living. As the functional forms of the electrical network continue to evolve, its influence on both upstream and downstream sectors in terms of green development is also shifting. Regarding the sustainable advancement of the electrical grid proper, China's current environmental protection system has adopted a supervision mode characterized by "simplified approval procedures, enhanced oversight, and strict accountability", imposing stricter environmental protection requirements on the power grid. Ecosystem provides important ecological services for society, which is very important for supporting green production and maintaining the sustainable development of society [2]. In recent years, most studies tend to focus on the advancement of electrical grid proper. Wang Hua [3] established the indicators for the evaluation of electricity network energy conservation and environmental safeguarding concerning grid loss, energy consumption in operation, and environmental stewardship of the power network; Wang Wei [4] analyzed the evaluation system for the advancement of the eco-friendly power grid covering aspects such as safety, reliability, green initiatives, and economic sustainability. The sustainable advancement of the electrical grid should not only focus on its own environmental protection, but also take into account its crucial role. Therefore, it is necessary to research the development of both aspects to meet new challenges and determine the crucial elements impacting the eco-friendly advancement of the electrical grid. This study will design an index system for the green development of the electrical grid and the natural environment from the energy industry chain perspective, and analyze the interaction among the influencing factors of the electrical grid's green development and the natural environment using a coupling analysis method.

2. Index system for coupling evaluation

2.1. Index system for evaluating the eco-friendly advancement of electrical grids

Assessment criteria are set based on three perspectives [5], it includes three indicators. The primary indicators of integrated upstream green advancement of electrical grid (A1) include: the proportion of clean energy power generation to total power generation (B1), new energy usage ratio (B2), transmission capacity of trans-regional lines (B3) and energy storage on grid side scale (B4); The green advancement (A2) of electrical grid itself includes: assessment rate of the building environment project (B5), qualification rate of completed environment project (B6), sulfur hexafluoride reclamation rate (B7), integrated line loss ratio (B8), yearly electromagnetic interference monitoring extent at the substation (B9), treatment ratio of ordinary waste (B10), disposal ratio of used transformer oil. (B11) and disposal lead storage battery disposal rate (B12). The downstream advancement of electrical grid cooperation (A3) includes:

*Corresponding author's e-mail:jinsezhuangxiu@sgeri.sgcc.com.cn

https://doi.org/10.1051/e3sconf/202452802023

the substitution amount of electric energy (B13), the ratio of electric power to terminal energy application (B14) and the latest electric vehicle charging station (B15).

2.2. Index system for evaluating the ecosystem

The influence of electricity system on the natural environment is thoroughly assessed. The Pressure-State-Response model [6] serves as the foundational framework for constructing an ecosystem evaluation index system, comprising three primary indicators: pressure, state, and response. The pressure indicator reflects the environmental impact of human activities, encompassing effects like resource demand, material consumption, and emissions from industrial operations. Key metrics include per capita CO_2 emissions, individual SO_2 emissions, individual NOx emissions, individual particulate matter (dust) emissions, and individual industrial solid waste output. State indicators track environmental conditions at specific time points, evaluating ecosystem status, natural environment quality, and human life quality. Metrics selected include annual average concentrations of CO_2, SO_2, NO_2, $PM_{2.5}$, and regional acoustic environmental monitoring compliance rates. Response indicators focus on societal and individual actions to mitigate and rectify the adverse effects of human activities on the environment. Investment indicators for general industrial solid waste disposal, industrial hazardous waste disposal, and industrial pollution control are considered for remedial actions.

2.3. Index system for evaluating coupling

Coupling refers to the collaborative connection between multiple systems, including coupling degree and coupling fitting degree [6]. The level of coupling indicates the extent of interconnection between systems, while the degree of coupling collaboration reflects the level of synchronization between systems. In the context of carbon neutrality and the energy revolution, the coupling and collaborative advancement of green power grid development and ecological environment preservation involve a shift from conventional energy consumption and power system expansion models to align with the natural system's material circulation and energy flow patterns. This aims to optimize energy utilization, minimize waste emissions, achieve a balance between economic and ecological benefits, and foster a positive interaction between the power grid and the environment. This research concentrates on green power grid development, assigning a higher evaluation weight of 0.6 to the power grid's green development system compared to the ecological environment system. This choice enables an in-depth analysis of the coupling dynamics between these two systems.

3. Research method

3.1. Green advancement of electrical grid

The standardization method involves using the target value of strategic plan for energy advancement or the predicted projected trend-based indicator as a benchmark.

$$X_{ij}=\begin{cases}1-\dfrac{q_j-x_{ij}}{\max\left(q_j-\min_{1\le i\le n}\left(x_{ij}\right),\max_{1\le i\le n}\left(x_{ij}\right)-q_j\right)}, & x_{ij}<q_j\\ 1-\dfrac{x_{ij}-q_j}{\max\left(q_j-\min_{1\le i\le n}\left(x_{ij}\right),\max_{1\le i\le n}\left(x_{ij}\right)-q_j\right)}, & x_{ij}>q_j\\ 1 & ,x_{ij}=q_j\end{cases} \tag{1}$$

X_{ij} is the standardized value of the j-th indicator in the i-th evaluated year; q_j is the ideal value of the j-th indicator; x_{ij} is the value of the j-th indicator of the i-th evaluated year.

3.2. Ecosystem

Due to the indeterminacy of the reference value for the eco-environmental system, the range method is employed for normalization. equtions (2) and equtions (3) are positive and negative indicators respectively.

$$Y_{ij}=\frac{y_{ij}-\min(y_j)}{\max(y_j)-\min(y_j)} \tag{2}$$

$$Y_{ij}=\frac{\max(y_j)-y_{ij}}{\max(y_j)-\min(y_j)} \tag{3}$$

y_{ij} is the original value of the j-th index in the i-th year; Y_{ij} is the normalized value after the treatment of the j-th index in the i-th year; $\max(y_j)$ and $\min(y_j)$ is the peak and trough value of the j-th index in all years.

4. Empirical research

4.1. Data selection

The data selected for analysis spans from 2006 to 2019 and is primarily sourced from the China Energy Statistics Yearbook, Environmental Statistics Yearbook, Environmental Bulletin, and other industry-specific data sources. In the context of fostering sustainable advancement within the electrical grid sector, reference values are established based on the target metrics outlined in the company's 2030 strategic goals, such as the share of clean energy generation and the transmission capacity of cross-regional power lines, or through index values derived from trend extrapolation methods.

 https://doi.org/10.1051/e3sconf/202452802023

4.2. Assessment outcomes of green advancement of electrical grid

Since the initiation of the "Eleventh Five-Year Plan", the green development of the electrical grid has shown continuous improvement across its three dimensions. The self-assessment level of the electrical grid's overall green advancement has consistently ranked the highest, signifying a relatively high environmental protection standard in current power grid construction projects. Although the evaluation level of downstream power grids remains low, the rate of improvement is rapid. As the green advancement level of downstream electrical grids continues to enhance gradually, the overall green advancement of the electrical grid in the future is expected to further advance, as shown in Figure 1.

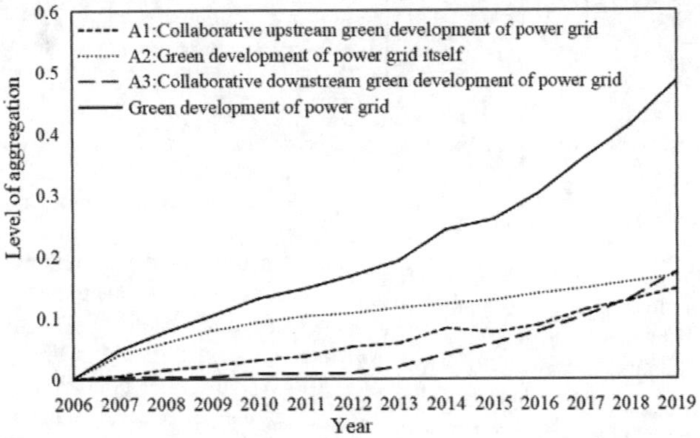

Figure 1. Comprehensive evaluation of green advancement level of electrical grid.

4.3. Assessment outcomes of eco-environmental system

Since the implementation of the "Eleventh Five-Year Plan", China's overall environmental ecosystem has been steadily enhancing. Specifically, while energy consumption in various sectors has been increasing along with economic development, the ecological pressure has been kept relatively stable, thanks to the macroeconomic policies implemented by the government. Except for a slight decline in 2012-2013, the ecological environment in other years increased year by year, and finally stabilized, indicating that the ecological environment in China is constantly improving; The level of eco-environmental response is also increasing year by year, which shows that under the national policies and measures, China's ability to deal with eco-environmental disturbances and protect the eco-environment is getting stronger and stronger, as shown in Figure 2.

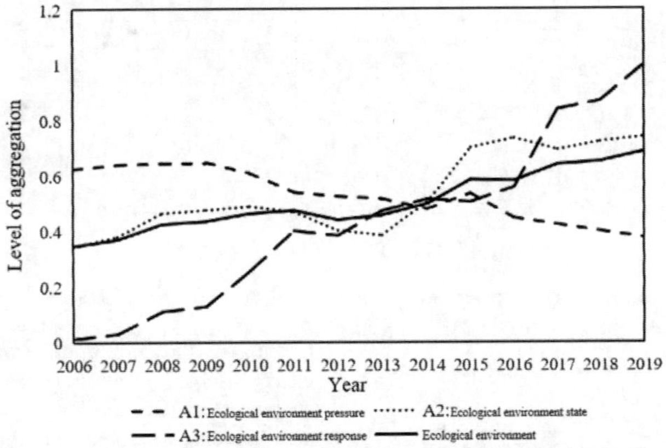

Figure 2. Comprehensive evaluation level of ecosystem.

4.4. Influencing factor

4.4.1. Collaborative upstream green advancement of electrical grid.

Within the context of promoting the coordinated green advancement of the electrical grid upstream, key positive indicators include the ratio of clean energy in overall power generation, the deployment rate of renewable energy, trans-regional transmission capacity, and grid-side energy storage capacity. As each of these indicators increases, the degree of coupling coordination also rises, demonstrating a strong correlation between positive indicators and the coordination degree of the two systems. This relationship has a facilitating effect and illustrates a consistent growth trend. Therefore, it is imperative to further advance the development of various clean energy sources, such as solar, wind, and hydro power, facilitate the regional distribution of new energy, enhance utilization efficiency, and maximize the green advancement benefits of the electrical system, as shown in Figure 3.

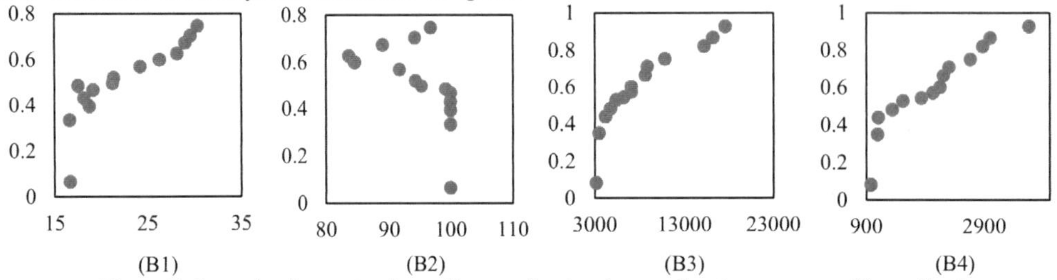

Figure 3. Green development and coupling coordination degree of upstream power grid coordination.

4.4.2. Green advancement of electrical grid itself.

Within the framework of the electrical grid's own green development, the overall line loss ratio stands as the sole negative indicator. As the line loss rate decreases and positive indicators increase, the degree of coupling coordination also grows. Notably, the ratio of environmental impact assessment for construction projects, the ratio of environmental approval for completed projects, the comprehensive line loss ratio, the annual monitoring coverage rate of electromagnetic noise in substations, and the general waste disposal ratio have a more significant impact on coupling coordination. This underscores the importance of intensifying environmental protection measures in power grid construction projects, emphasizing scrap recycling management throughout the entire lifecycle of electrical grid construction, operation, and decommissioning, and advancing efforts to reduce line loss rates through technical and managerial interventions, as shown in Figure 4.

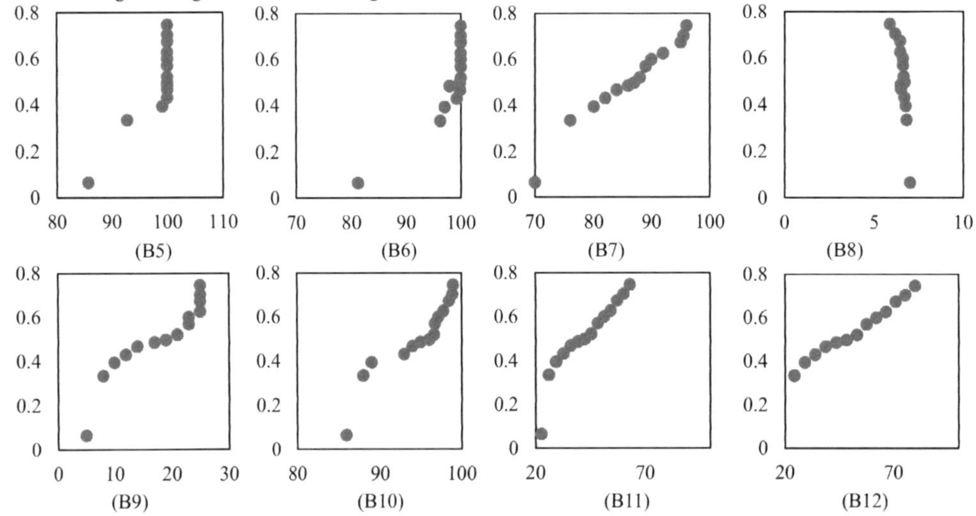

Figure 4. Green development and coupling coordination degree of power grid itself.

4.4.3. electrical grid coordinates downstream green advancement.

In the downstream green advancement coordination of the electrical grid, the electric energy proportion in terminal energy consumption significantly influences the ecological environment. Other indicators have a more substantial impact on coupling collaboration in the early stages of advancement, eventually stabilizing. This highlights the importance of implementing electric energy substitution strategies and enhancing overall societal electrification levels in downstream green power grid development. Promoting replacement of electric energy in critical sectors, restructuring final energy consumption, and leveraging the positive impact of downstream electrical grid cooperation on the natural environment are crucial steps, as shown in Figure 5.

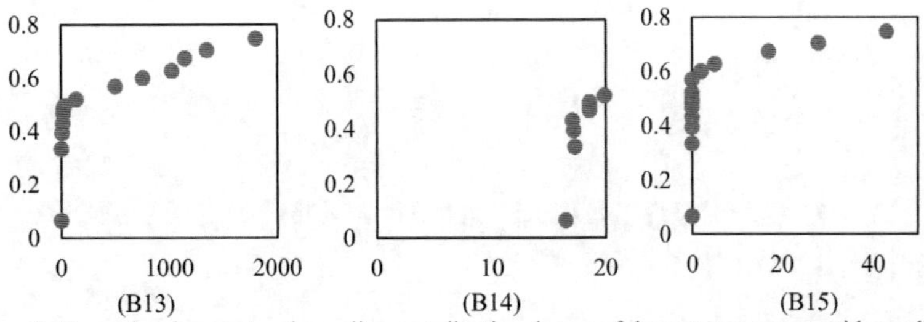

Figure 5. Green development and coupling coordination degree of downstream power grid coordination.

5. Conclusion

From the standpoint of the energy industry chain, we have established systems for green development in the electrical grid and ecosystem evaluation indices, and the following conclusions are obtained by digging back the key influencing factors through the indicators:

In the context of green advancement in the electrical grid, it is essential to prioritize boosting interregional transmission capacity and optimizing the distribution of clean energy in the upstream sector. To promote green advancement within the power grid, it is crucial to introduce enhanced management and technical strategies aimed at reducing overall line loss rates, expanding the monitoring in substations electromagnetic noise, and improving the efficiency of transformer oil and disposal of waste lead batteries. Furthermore, downstream green advancement efforts in the electrical grid should focus on promoting the adoption of electric energy and increasing the ratio of electric energy in final energy consumption.

Acknowledgments

This work was financially supported by the State Grid Corporation's science and technology project "Evaluation technology and demonstration application of environmental protection ability of power grid enterprises". (project number: SGTYHT/21-JS-223).

References

1. Zhu M, Nicholas S and Joseph E.Stiglitz (2023) Embracing the new paradigm of green development: a study on China's carbon-neutral policy framework. World Economy., 3: 3-30.
2. Tsujimoto M, Kajikawa Y and Tomita J (2018) A review of the ecosystem concept—Towards coherent ecosystem design. Technological forecasting and social change., 136: 49-58.
3. Wang H (2019) Research on Comprehensive Evaluation of Energy Conservation and Environmental Protection of Huizhou Power Grid Based on Analytic Hierarchy Process. Guangdong University of Technology.
4. Wang W (2015) Theoretical and Empirical Research on Green Power Grid Development. North China Electric Power University.
5. Jiang Y (2017) Research on evaluation method and system of green development of Hainan electric power. North China Electric Power University.
6. Ott W (1978) Environmental Indices: Theory and Practice. Science Publishers, Ann Arbor.

https://doi.org/10.1051/e3sconf/202452803001

An overview on carbon emission accounting systems and method at home and abroad

Hua Pan , Chang Liu, Yiqiao Pei, Rongshun Liu*

Central Southern China Electric Power Design Institute; Hubei Province,430071 China

Abstract. Carbon emission accounting is the basic premise for accurately grasping the change trend of carbon emissions, effectively carrying out various carbon emission reduction work, and promoting the green transformation of economy. It is also an important basis for different levels of entities to implement emission reduction tasks and evaluate the completion effect. At present, China attaches great importance to the construction of carbon emission accounting system, and has achieved certain results, but the existing accounting system has been difficult to adapt to the new requirements of China's carbon emission reduction work. Therefore, this paper starts from the carbon emission accounting methods, summarizes the current domestic and foreign carbon emission accounting systems, analyzes the shortcomings of the current carbon emission accounting system construction in China, and puts forward relevant suggestions for improving China's carbon emission accounting system.

1. Introduction

Greenhouse gas emissions have brought about a series of serious natural environmental problems such as global warming and biodiversity decline, and it has become a world consensus to deal with climate change and transition to a low-carbon economy [1-2]. In September 2020, China proposed that China will increase its independent contribution and adopt more powerful policies and measures. Carbon dioxide emissions strive to peak by 2030 and strive to achieve carbon neutrality by 2060[3].

Under the background of " double carbon, " exploring the application of carbon emission accounting system and accounting standards in line with China 's development characteristics, improving China 's carbon emission accounting system, and finding out the data and change trend of China's carbon emissions are the important basis and primary conditions for achieving the goal of " double carbon. " In recent years, China has attached great importance to the construction of carbon emission accounting system and has made some progress. However, due to the late start of carbon emission accounting in China and the lack of continuous basic data of carbon emission and the accumulation of accounting experience, there are some shortcomings in the practical application of carbon emission accounting [4-5].

In order to establish a unified and standardized statistical accounting system for carbon emissions, this paper summarizes the carbon emission accounting models at home and abroad, and compares them from three dimensions: accounting boundary, accounting model and emission factor. Finally, relevant suggestions are put forward for the construction of China's carbon emission accounting system.

2. Formatting of manuscript components

The main forms of carbon accounting can be divided into two categories : measurement-based and calculation-based. Specifically, from the perspective of existing greenhouse gas emission accounting methods, it can be summarized into three types : emission factor method, mass balance method, and measurement method [6].

2.1. Emission factor method

The basic idea of the emission factor method is to construct activity data and emission factors for each emission source according to the carbon emission inventory list, and take the product of activity data and emission factors as the carbon emission estimate value of the emission project. The calculation formula is as follows:

$$E=AD\times EF \tag{1}$$

Where: E is greenhouse gas emissions (such as carbon dioxide, methane, etc.); AD is the amount of production or consumption activities that contribute to greenhouse gas emissions; EF is the coefficient corresponding to the activity level data.

2.2. Mass balance method

The basic principle of mass balance method is the law of conservation of matter. This method calculates the greenhouse gas emissions by subtracting the carbon

*Corresponding author's e-mail: lrs6207@csepdi.com

content in the input material from the carbon content in the output material. The calculation formula is as follows:

$$E=(AD_i \times CC_i - AD_p \times CC_p - AD_w \times CC_w) \times 44/12 \quad (2)$$

Where, AD_i is the input amount of the i raw material; CC_i is the carbon content of the i raw material; AD_p is the output of the p product; CC_p is the carbon content of the p product; AD_w is the output of the w type of waste; CC_w is the carbon content of the w waste.

2.3. Measurement Method

The measurement method mainly measures the flow rate, flow rate and concentration of the emitted gas through monitoring means or continuous metering facilities recognized by the relevant state departments, and calculates the total emission of the gas with measurement data recognized by the environmental protection department.

3. Carbon emission accounting models at home and abroad

3.1. Foreign carbon emission accounting model

The Intergovernmental Panel on Climate Change (IPCC) of the United Nations recently released the *2019 Refinement to the 2006 IPCC Guidelines for National Greenhouse Gas Inventory*, which covers carbon emission projects in various aspects, including economic production, domestic consumption, waste treatment and land use [7].

From the perspective of accounting boundaries, in terms of accounting gas types, greenhouse gases include CO_2, CH_4, N_2O, hydrofluorocarbons, perfluorocarbons and other gases. In terms of accounting scope, the 2019 Refinement includes four sectors: Energy, Industrial Processes and Product Use (IPPU), Agriculture, Forestry and Other Land Use (AFOLU), and waste.

From the perspective of the accounting model, the emission factor method has been more mature in the accounting of emissions in the energy sector. The emission factor method is also a traditional and mainstream method used by the IPPU department to calculate emissions. However, considering the problems such as the change of emission factors caused by the updating of production processes in the actual situation, the IPCC also proposes to use the mass balance method for accounting accordingly. The specific method of AFOLU department is emission factor method. The waste department mainly uses the emission factor method.

3.2. Domestic carbon emission accounting model

At the national and provincial levels, China has formulated relevant measures to explain the statistical accounting methods for carbon emissions from energy activities at the national and provincial levels.

From the perspective of accounting boundary, carbon emissions in relevant methods refer to CO_2 emissions, and carbon emissions accounting boundary refers to energy activities and industrial production processes. The scope of national carbon emissions includes direct CO_2 emissions from fossil energy used as fuel, used as raw materials and part of the processing and conversion process in China, and carbon emissions from international fuel tanks are deducted. The scope of carbon emission accounting in provincial regions mainly includes the direct CO_2 emissions generated by fossil energy used as fuel, raw materials and part of the processing and conversion process in the administrative region of the province (region and city), and the indirect CO_2 emissions contained in the transfer of electricity between provinces.

From the perspective of the accounting model, in addition to the mass balance method used in the calculation of direct CO_2 emissions in the industrial production process in the national and provincial regions, the rest of the accounting methods used are emission factor methods.

In addition to the above accounting methods, there is also a guide to regional carbon emissions accounting.

From the perspective of the accounting boundary, the carbon dioxide emissions referred to in the guidelines include the direct carbon dioxide emissions (that is, carbon dioxide emissions from energy activities) generated by fossil energy consumption in the administrative area of the province (autonomous region, city), as well as the indirect emissions contained in electricity transfer. Among them, fossil energy consumption activities can be divided into: energy production and processing conversion, agriculture, industry and construction, transportation, service industry and other, residential life.

From the perspective of the accounting model, the method used in the accounting of each field is the emission factor method, and the energy consumption of each field is obtained from the energy balance table of the provinces (regions and cities). Among them, the energy consumption in the field of energy production and processing conversion is calculated by "processing conversion input" in the energy balance sheet; The energy consumption in the field of agriculture corresponds to "agriculture, forestry, animal husbandry and fishery" under "terminal consumption" in the energy balance sheet. Energy consumption in the industrial and construction sector corresponds to "industry" and "construction" under "final consumption" in the energy balance sheet; Energy consumption in the transport sector corresponds to "transport, storage and postal" under "terminal consumption" in the energy balance sheet; The energy consumption of services and other sectors corresponds to "wholesale, retail and accommodation and catering" and "other" under "final consumption" in the energy balance sheet; The energy consumption in the field of residential life corresponds to the "living consumption" under "terminal consumption" in the energy balance sheet.

At the enterprise level, take the "Guidelines for Accounting and Reporting of Greenhouse Gas Emissions for Enterprises" as an example.

From the perspective of accounting boundaries, the scope of accounting and reporting of greenhouse gas emissions from power generation facilities includes:

carbon dioxide emissions from fossil fuel combustion and carbon dioxide emissions from the purchase and use of electricity.

From the perspective of the accounting model, the emission of fossil fuel combustion is the sum of the carbon dioxide emissions generated by the combustion of various fossil fuels in power generation facilities during the statistical period, in which the carbon dioxide emissions generated by the combustion of fossil fuel i is calculated using the emission factor method, which is equal to the product of its consumption and fuel emission factors. CO_2 emissions from purchased electricity use can be obtained by multiplying purchased electricity use by the grid emission factor.

4. Comparison of carbon emission accounting models at home and abroad

4.1. Accounting boundary

From the perspective of accounting greenhouse gas types, in the international carbon emission accounting method, greenhouse gases include CO_2, CH_4, N_2O, hydrofluorocarbons, perfluorocarbons and other gases. At present, the accounting methods issued by China only consider the emissions of CO_2 greenhouse gases at the national, provincial, regional and city levels. At the enterprise level, most industries only consider CO_2 greenhouse gas emissions, and a small number of industries consider N_2O and SF_6 emissions.

From the perspective of domain division, foreign carbon emission accounting methods take the whole-society carbon emission sources as the basis for division, and divide the whole-society carbon emission sources into four major fields: energy, industrial process and product use, agriculture, forestry and other land use, and waste. Each field is refined accordingly, and the carbon emission activities contained in the field are divided in detail. China's carbon emission accounting method at the national level divides the sources of carbon emissions into major categories. Similar to the international method, the sources of carbon emissions are divided into direct emissions from fossil energy used as fuel, used as raw materials and part of the processing and conversion process. It is worth mentioning that the greenhouse gas accounting method of key industries introduced by the state does not cover the carbon emission sources of the whole society, and the accounting subject of the method itself is the industry-related enterprises, and the enterprise's own carbon emissions are comprehensively sorted out from the perspective of the enterprise, and the carbon emission sources of the whole industry are not considered from the perspective of the industry, and the carbon emission statistics are not considered from the perspective of the whole society.

4.2. Accounting model

From the perspective of basic accounting methods, the international carbon emission accounting model adopts emission factor method, and some industrial processes use mass balance method for accounting, and according to different activity level data, the adopted emission factor method can be further subdivided into two categories. One is fossil fuel combustion, whose accounting model can be expressed as the product of fossil fuel consumption and related emission factors. The other is industrial process emissions, whose accounting model can be expressed as product output and emission factor product. The national carbon emission accounting model mainly uses the emission factor method, and some industrial processes adopt the mass balance method. The accounting model is basically consistent with the international one, and can be divided into two categories: fossil fuel combustion and industrial process emissions. The expression form is basically consistent with the international, and is the product of consumption and related emission factors.

Due to the differences in the statistical standards of international, national and key provinces and cities, the specific parts contained in each carbon emission accounting model are inconsistent, but the accounting methods used in each part are basically emission factor method, which is expressed as the product of fuel consumption/product output and emission factor. For a small number of chemical production processes or processes that are difficult to classify and identify, mass balance method can be adopted for carbon emission accounting.

4.3. Emission factor

Foreign emission factors mainly come from the *2019 Refinement to the 2006 IPCC Guidelines for National Greenhouse Gas Inventory* issued by the IPCC, which contains default emission factors in four fields: energy, industrial process and product use, agriculture, forestry and other land use, and waste, involving very comprehensive industry categories and a very wide range of coverage.

Only the corresponding statistical accounting methods for carbon emissions from energy activities in national and provincial regions have been developed, and the default emission factors involved in the accounting methods have not been published. In addition, the industry default emission factors released by the state are not all directly given. For example, the default emission factors of fossil fuels are calculated by giving the default values of the parameters involved in the calculation formula, including carbon content per unit calorific value and fuel carbon oxidation rate. The default values of each parameter are derived from the *IPCC 2006 Guidelines for National Greenhouse Gas Inventories, the National Energy Statistical Yearbook, the Study of China's Greenhouse Gas Inventories, the Guidelines for the Preparation of Provincial Greenhouse Gas Inventories* and other relevant data. In addition, in addition to the subtle differences in the default emission factors of various industries, the default emission factors of different provinces and cities are also different, such as the emission factors of electricity and heat.

To sum up, the default value of emission factors is directly given in foreign countries, and the industry categories involved are the most comprehensive, providing a reference for the national and provincial and municipal industry default emission factors. In addition, the default emission factors of various industries released by the state vary slightly with different industries, and the default emission factors of different provinces and cities also have certain differences.

5. Conclusions

5.1. Existing deficiency

(1) The deviation of historical statistics of energy consumption is large.

Due to the inconsistency of national and provincial energy consumption statistics, there are great differences in historical data. According to the ' Energy Statistics Yearbook ' before 2014, the difference between the total energy consumption of each province and the total energy consumption of the country is 12% ~ 23% between 2005 and 2012. Although this difference narrowed to less than 3% after systematic adjustment in 2015, it expanded to more than 4% in 2016 and 2017, resulting in energy consumption becoming the main source of carbon emission accounting bias.

(2) The accounting model is relatively lagging

At the national and provincial levels, China's carbon emissions accounting methods are still based on the old IPCC guidelines, but have not been combined with the latest methods in the 2019 revision, and the results based on the lagged method can not accurately reflect the status quo of China's carbon emissions. At the enterprise level, compared with the measurement of carbon emissions based on emission factor method, the actual measurement method has the advantages of fewer intermediate links and higher accuracy. However, only a few enterprises use the actual measurement method to calculate carbon emissions, resulting in low accuracy of accounting data at the enterprise level.

(3) The measured values of emission factors are used less

In the current carbon emission accounting work, most of the emission factors use the default values in relevant standards, and the emission factors of relevant accounting projects are not measured. However, for the emission factors of fossil fuels, carbon content per unit calorific value of fuel is different in different regions of different enterprises, and there are also large differences in fuel combustion adequacy. If the default value is used for accounting, the result will be less accurate. For the emission factors in the production process, different enterprises in the same industry will also have obvious differences due to different production processes, and the use of relevant production process emission factors for accounting will lead to large errors.

5.2. Advices

(1) Establish a unified carbon emission accounting standard

A unified and scientific energy balance sheet will be issued for provinces and cities or key emission enterprises to visually record various data such as resource supply, processing and conversion, and terminal consumption of different energy sources. By calculating provincial/corporate carbon emissions based on the energy balance sheet, it can not only cover the energy consumption generated in the process of processing and conversion, but also avoid the double calculation of thermal power generation and heating.

(2) Strengthen the construction of basic databases

For emission factors, strengthen the calculation of emission factors for typical regions and typical energy products, use measured values instead of default values, develop more authoritative and scientific emission factors, and establish relevant dynamic databases. For other data related to carbon accounting, big data processing technologies such as machine learning and artificial intelligence can be relied on to realize the integration mining of existing data and the filling of historical missing values. At the same time, cross-verification of carbon emission data can be realized based on model construction.

References

1. Bjørn, A., Lloyd, S.M., Brander, M., et al. Renewable energy certificates threaten the integrity of corporate science-based targets. Nat. Clim. Change 12, 539-546. 2022.
2. Downie, J., Stubbs, W. Corporate carbon strategies and greenhouse gas emission assessments: the implications of scope 3 emission factor selection. Bus. Strat. Environ. 21, 412-422. 2012.
3. Myles R A, David J F, Chris H. Warming caused by cumulative carbon emissions towards the trillionth tonne [J]. Nature, vol. 458: 1163-1166.2009.
4. Chu S, Majumdar A. "Opportunities and challenges for a sustainable energy future," Nature, vol. 466, pp. 294-303, 2012.
5. J.Q. Li, J.F. Guo, S.J. Gao. "The current situation, problems and challenges of national carbon emission accounting work," Development Research, vol. 406, issue 6, pp. 9-14, 2020.
6. M.D. Liu, J.J. Meng, B.H. Liu. "Progress in the studies of carbon emission estimation," Tropical Geography, vol. 34, issue 2, pp. 248-258., 2014.
7. B.F. Cai, S.L. ZU, S.M. Yu. "The interpretation of 2019 refinement to the 2006 ipcc guidelines for national greenhouse gas inventory," Environmental Engineering, vol. 37, issue 8, pp. 1-11, 2019.

Problems and solutions in the development of new-energy vehicles in China

Amin Shao

Mianyang Teachers' College, Mianyang, Sichuan 621000, China

Abstract: Over decades of development, China has topped the world in energy generation and consumption. Despite the great achievements that China has made in the development of energy, there remain salient problems, such as energy shortage, unreasonable energy structure, and environmental pollution, making it imperative to develop new energies. This paper analyzes the development status-quo of new-energy vehicles (NEVs), unveils the problems in the development of the NEV sector, and puts forth solutions to these problems from both the microscopic and macroscopic angles, with a vision to find a path of development featuring higher production, increased prosperity, and ecological well-being.

1. Introduction

The advances in ecology and society have increased people's aspirations for a life of better quality. As a result, vehicles have become a necessary tool for travel in households, and the automobile industry has played a crucial role in economic and social progress. In the context of energy crises, China encourages energy conservation and emission reduction, making new-energy vehicles (NEVs) a new trend of development. NEVs not only alleviate energy crises but also reduce pollution, which is in line with the national endeavor of developing ecological civilization. Since the 18th National Congress of the Communist Party of China, the central government has made much effort to push forward the development of NEVs: providing policy support, investing labor, materials, and money, and enabling technical breakthroughs… Nonetheless, a range of constraints still exist to curb the progress of NEVs. As pointed out by President Xi, to move toward the new age of ecological civilization, we need to promote green development and low-carbon development, integrating ecological civilization into economic and social development [1]. While solving the problems in the development of NEVs and moving towards high-quality development, we need to pay attention to ecological civilization, so green development has become a mission needing prompt attention and actions.

2. Development Status-quo of NEVs

Over this decade, China has been holding up to the notion of green development, and on its way towards the goal of high-quality development, it has made enviable achievements in the development of NEVs. In July 2023, as reported in the Eco Forum Global Guiyang 2023, China has become the world's largest producer of NEVs for eight years straight, and the inventory takes up over half of the world's total, indicating that NEVs are gaining momentum across the nation.

2.1. Governmental support and favorable policies

As energy shortage and ecological civilization begin to gain public attention, the Chinese government has issued a range of policies to boost the development of NEVs. On Jan. 14th, 2009, the State Council passed the "Plan of Revitalization of the Automobile Industry" and proposed implementing the strategy of developing NEVs. On Feb. 17th, 2009, the central financial administration provided a one-time fixed amount of subsidy for pilot cities and demonstration organizations to buy or use energy-saving vehicles and NEVs, which marked the first time that the government promoted NEVs by way of direct financial subsidies. In 2011, the nation issued the need to initiate new pilot programs for the promotion of energy-saving or new-energy vehicles. In 2013, the policies that provide subsidies for NEVs were adopted across China. In November 2020, the "Development Plan for the NEVs Industry (2021-2035)" was released, which marked the start of the national policy to develop into a strong country of NEVs. The issuance of these policies encouraged more investment of automobile companies, and meanwhile boosted the progress of NEVs.

2.2. Technological innovation and progress

Technological progress is the key to the development of NEVs. For the time being, the NEV market is at its growing stage, though the trends differ among countries. The main NEVs on the market include the battery electric

Email: 494747824@qq.com

vehicle (BEV), the fuel cell electric vehicle (FCEV), and the plug-in hybrid electric vehicle (PHEV). In China, the BEV sector is well-developed, and the PHEV sector is gaining increasing momentum. The PLUS DM-i model of BYD, a typical model of PHEV, amassed the largest sales in 2022 among other PHEV models. Advances in NEV technologies, such as technologies regarding battery technology, electric control, and smart driving, have improved the performance and security of NEVs. China's NEV policies have resulted in the emergence of domestic companies with strong technological capabilities in the production of core components such as rechargeable batteries[2]. Consequently, NEVs are becoming the leading trend in the automobile industry, providing a cleaner and more efficient way of traveling, thus enjoying a bright future.

2.3. Expanded market

The market need is the very drive that pushes forward the NEV sector, and now the NEV sector is heralding its prime age. With the support of policies, the NEV sector is seeing increasing needs, driving the NEV industry forward. As reported by China Automotive Industry Association, China's NEV sector is undergoing exponential growth and now China ranks first in the world in terms of its NEV sales for eight years straight. In 2022, China produced about 7 million NEVs and sold 6.88 million NEVs, a year-on-year growth of 96.9% and 93.4%, respectively. The sales of NEVs in China took up 60% of the world's total sales. In the next few years, as the public awareness of environmental protection and energy conservation increases, the NEV sector is bound to witness substantially stronger growth.

3. Problems in the development of NEVs

3.1. Technological limits and bottlenecks

As President Xi pointed out in the 20th National Congress of CPC, there are still bottlenecks in the high-quality development of NEVs and the industry falls short of technological innovation. Despite the steady growth of the NEV sector, China is a latecomer in this sector and faces technological bottlenecks. Besides, in the domain of mainstream NEVs, such as FCEVs and hybrid electric vehicles (HEVs), China is short of the core technologies and has a large gap with the leading countries in this domain. In terms of technological innovation, China is at the stage of technological improvement, and the innovative strength differs considerably among the subjects in this sector; research institutes and universities are in want of stronger capacities for innovation. In particular, the national strength in innovation of battery technology is weak. Though China has invested much into research on electric energy, batteries, and electro-motor, it is still seeing a wide gap with developed countries in terms of battery endurance, battery power, charging accessibility and security, energy-saving performance, and the engine system of vehicles. To resolve the problems that confront China's NEV sector and grow into a world leader in the NEV domain, we need to overcome the technological bottlenecks before reaching healthy and sustained development.

3.2. Overdependence on governmental and market subsidies

The key drive behind the rise of NEV sales in China is policy support. However, over-reliance on policy support and governmental subsidies will put many Chinese self-developed NEV brands in an awkward situation. The average price of Chinese NEVs is around 200,000 RMB, and it is because of the governmental subsidies that the price reduces to a price slightly higher than 100,000 RMB. On one hand, cost control is challenging because of the high price of the batteries; on the other hand, market acceptance remains a concern. The price control of NEVs can only rely on the government. In 2023, the central government withdrew the subsidies for the NEV sector, posing a blow to the sales of NEVs.

3.3. Insufficient supportive facilities

At present, the supportive facilities for NEVs are insufficient, which hampers the promotion of NEVs on the market. Facilities like hydrogen refueling stations and gas stations are lacking; there are problems like insufficient fuel, crippled accessibility to fuel, and a shortage of maintenance stations that need to be solved. Allocation of charging piles is the foundation for the development of NEVs. Thus, to boost the healthy development of the NEV market, we need to rapidly increase the number of charging piles to remove the concerns about the development of NEVs.

4. Solutions for sustained development of NEVs

As President Xi has pointed out, we need to move steadily toward the carbon-neutrality goal and promote clean and efficient use of energy; in 2035, we will embrace a green lifestyle, secure reduction in carbon emissions after reaching the peak, and the ecology will improve substantially, reaching the goal of creating a beautiful China [3]. Developing NEVs is a crucial part of our initiatives to boost social and economic progress, and it is an indispensable section of our effort to reach the goal of creating a beautiful China.

4.1. Increasing governmental support and being market-oriented

4.1.1. Identifying the market position and making an effort on both the supply and demand sides

To create a high-level socialist market economy system, we need to make the market a decisive role in resource allocation while leveraging the power of the government.

https://doi.org/10.1051/e3sconf/202452803002

We need to identify the market position of the NEV industry, find the point to make breakthroughs, and by taking into account the technical and financial factors, determine which vehicle models should be the focus of development. On the supply side, we need to avoid the repetitive construction of homogenous models while boosting market needs on the demand side. In general, the technology for the PHEV is maturing, and thus this model of the vehicle should be the focus of promotion; the PEV enjoys great potential, and the lithium batteries it uses have good performance, so it should be the optimal model for public transport, given its endurance mileage. The FCVE, which is more expensive and has less mature technologies, is not suitable for broad promotion in China.

4.1.2. Improving policy support and increasing R&D investment

The development of NEVs requires full-scale policy support. In the NEV industry, the government encourages its market expansion by issuing supportive policies. Specifically, the government can provide subsidies for NEV manufacturers or provide consumption subsidies for consumers to boost production and sales of NEVs. They can also enlarge the adoption of NEVs on the market by providing incentives or cutting taxes, For example, the South Korean government provides subsidies and tax reductions in the new energy vehicle industry, enhancing its price competitiveness, attracting customers to purchase cars, and promoting large-scale production of cars[4]. Moreover, the government needs to invest more money into research on the batteries of NEVs, create a battery industry chain for NEVs, and promote the upgrading of NEV technologies to ensure the healthy development of the NEV industry.

4.1.3. Improving supportive facilities and creating a complete service chain

The development of NEVs requires a complete system of supportive facilities,

such as charging stations, charging piles, battery renting stations, and gas stations, and intensify the construction of charging facilities, which provide the most basic supporting facilities for the NEVs to displace traditional vehicles. It is necessary to provide an overall plan for allocation and the construction of maintenance service stations of NEV, and increase the accessibility of these facilities, thus solving the users' problems in finding charging piles or stations. Meanwhile, we need to leverage all resources available and direct more social capital to engage the private capital in the construction of supportive facilities. More supportive facilities need to be allocated in highly-populated areas like bus stations, parking lots, and shopping malls to play a demonstrative role in increasing the accessibility and availability of NEVs.

4.2. Breaking walls between industries and increasing inter-industry collaboration

4.2.1. Creating an ecosystem of healthy competition and adjusting the competition mode

The market competition can be fierce. In the bidding process, the integrators who hope the suppliers can win the bidding at a lower cost will lead the suppliers to bid by offering a lower price, which results in vicious competition. The suppliers should prioritize performance over the price, and overemphasis on the price is likely to cause problems like reduced quality and safety risks. Moreover, some companies compete by offering a price lower than its cost, which is suicidal. Companies need to adopt the strategy of differential competition and take a path that provides quality products at high prices instead of offering low-quality products at lower prices. Only in this way, can the NEV industry achieve healthy and sustained development.

4.2.2. Improving technological innovation and increasing corporate competitive force

In the new notion of development, innovation is defined as the primary drive of development. Innovation comes as the key that we can take to open the door for social and economic progress [5], and this is also true for NEVs. Invest more resources in the research and innovation of NEVs technology, especially achieving breakthroughs in battery technology, charging technology, and intelligent technology. Companies need to invest more in R&D of core technologies and extend the adoption of NEVs by technological innovation. For instance, innovation in the battery technology will increase the endurance mileage of NEVs and reduce the overall cost of the vehicle; innovation in the charging technology will enhance the efficiency and accessibility of charging; innovation in the smart driving technology will enhance the overall performance of the vehicle and provide better driving experience. More initiatives for technological innovation need to be performed to create a good environment of innovation. We also need to draw lessons from the developed countries, and in light of the national actualities, provide more assistance to the NEV industry and increase the competitive force of Chinese NEVs in the market.

4.2.3. Expanding global capacity cooperation and removing obstacles in the domestic-international dual cycle

As the NEV sector moves fast forward in China over these years, more and more investments have gone into this sector because of favorable policies and market directions. However, the global market of NEVs does not go well, leading the sector into overcapacity. Expanded global capacity cooperation is hence required to address the problem of overcapacity. First of all, we can transfer some overcapacity to developing countries in Southeast Asia, and Central and East Africa; we can also invest directly in the US or Europe to capture more shares in the local markets. Besides, we need to give more boost to the domestic market and create a greater market to mine more chances in the domestic market, building a domestic-international dual cycle with the domestic market as the focus. Be it in the cooperation or in the initiatives to boost circulation, research, and development of cutting-edge technologies of NEVs remain the kernel. Thus, we need to invest more in the NEV sector, build technological alliances[6], integrate capital and technological resources to create regional and professional research and development bases and realize open cooperation in innovation of core technologies by way of reasonable allocation of resources. Meanwhile, we also need to learn advanced technologies and experience in a targeted manner to increase the kernel competitive force of the NEV sector.

4.3. Increasing promotion and providing a good environment for development

4.3.1. Relying on governmental guidance and promotion to improve the environment of development

The development of ecological civilization is closely related to every one of us. The government needs to increase public awareness of environmental protection and enhance education about ecological civilization. We need to "increase the promotion of scientific and environmental protection of NEVs and provide more favorable policies for the NEV sector" [7], and allocate more supportive facilities to remove the concerns of citizens in green traveling. The NEV sector should invest more in the research and development of cutting-edge technologies to resolve technological bottlenecks, increase the public acceptance of NEVs, and enhance their willingness to buy and use NEVs.

4.3.2. Encouraging green production and consumption to promote a green and low-carbon lifestyle

China has been making an effort to increase the population's awareness of ecological civilization. In June 2023, the Third Convention of the Standing Committee of the 14th National Congress decided to make the 15th day of August the National Ecology Day, with a vision to create an atmosphere across China for low-carbon life, allow the consumers to feel the positive role that low-carbon traveling plays in environmental well-being, motivate the public to protect the environment and shape a set of low-carbon values for consumption. In reaching a carbon-neutral world, China's contribution to the global green and low-carbon transition is significant as the second-largest economy in the word[8].

5. Conclusion

The ecological environment is the basis for human survival and development[9], and ecological civilization is the foundation for the sustained development of mankind. China's new energy vehicle industry still has great potential for development. With the continuous progress of technology and the increasing maturity of the market, new energy vehicles will become one of the mainstream modes of transportation. We hope that in the near future, China can find a way to develop new energy vehicles with higher output, higher prosperity and better ecology, and make important contributions to global energy transformation and sustainable development.

References

1. Xi Jinping: President Xi's Comments on the Theory of Governance. 1st Vol., 2nd Ed. [M]. Beijing: Foreign Languages Press, 2018.01, p.121.
2. Maria A D ,Agnieszka M . Motives behind Sino-Japanese strategic alliances in the new energy vehicles sector in the age of the Belt and Road Initiative [J]. Asia Pacific Business Review, 2024, 30 (2): 274-299.
3. Xi Jinping: "Hold high the great banner of socialism with Chinese characteristics, strive in unity to build a modern socialist country in all aspects — speech by President Xi at the 20th National Congress of CPC. Beijing: People's Publishing House, 2022.
4. Euna L ,S. J M . Environmental Protection and Development of Technology-intensive Industries: The Case of New Energy Vehicle Industry in Korea [J]. Science, Technology and Society, 2021, 26 (3): 413-432.
5. Publicity Department of the CPC Central Committee: Scripts of important speeches by President Xi (2016 Edition) [M]. Beijing, People's Publishing House, 2016.04, p.133
6. Zheng Qingdong : Research Collection on Practising Xi Jinping's Economic Thought[M]. Beijing, People's Publishing House, 2023.02, p.80
7. Wang J., Zhang Y. R., Cheng S. S. Current conditions and problems in technological innovation of NEVs in China and solutions [J]. Enterprise Science and Technology & Development, 2022 (7): 20-22
8. Zhanar N ,Qingbin G ,Uzair A , et al. Can environmentally friendly technology help China to achieve a carbon neutrality target by 2060? An asymmetrical based study in China. [J]. Environmental science and pollution research international, 2023
9. Fan Wei ,etc: A hundred-year journey : The Communist Party of China sticks to the original intention [M]. Beijing, People's Publishing House, 2022.07, p.590

The Dilemmas and Countermeasures of China's Energy Development under the Background of Sustainable Development

Xiangchao Pan

Mianyang Teachers' College, Mianyang, SiChuan, China

Abstract: Energy is a vital resource for human survival as well as the material basis of all production and life. Environmental pollution and excessive resource consumption have become major issues in China due to the rapid consumption of energy driven by the deepening of reform and opening up and the acceleration of economic development. Problems and response strategies in the development and use of renewable energy have been analyzed from the perspective of sustainable environmental development in response to the existing situation of energy scarcity and environmental pollution in China. This analysis stresses the necessity of using renewable energy and its influence on environmentally sustainable development. It also provides prospects to achieve sustainable development in terms of energy utilization, the environment, and energy.

1. Introduction

With the rapid development of the economy after the reform and opening up, China's energy utilization has also encountered certain challenges. These problems are mainly seen in China's energy structure, where there is a very large amount of conventional energy and a relatively low proportion of renewable energy sources such as solar, hydropower, wind, biomass, etc. The sustainable development of China's environment and energy is greatly impacted by this situation. Currently, China is in its industrial development period. Despite its vast territory and wealth of resources, China faces several problems, including low per capita resource availability and a limited environmental carrying capacity as a result of its large population [1]. Consequently, we must prioritize the coordinated development of environmental protection and economic construction, as well as expand the development and application of renewable energy through legislation and technological innovation.

2. Problems in sustainable energy development in China

Patterson (1996) found that there is a positive correlation between Gross Domestic Product (GDP) and energy consumption, and the higher the value of energy productivity, the higher the energy productivity. [2] Energy is the engine of economic development, and the extraction and utilization of energy provide the assurances of rapid economic development. A prerequisite for attaining sustainable energy utilization is the development and use of renewable and sustainable energy, extending beyond the development and utilization of traditional energy products to associated fields of low-carbon renewable energy. Nowadays, there are still a lot of urgent issues with the sustainable development of energy, despite the rapid growth of the new energy product market in China.

2.1 Basic dependence of economic development on non-renewable resources

There is a close relationship between economic development and energy consumption. Rashe and Tatom (1977) first introduced energy into the Douglas production function, aiming to find the basic law between energy consumption and economic growth, which has attracted widespread attention [3]. Later, economists used econometric methods to study their relationship, and Hwang and Gum (1992) discovered a bidirectional Granger causality between energy and economic growth in Taiwan, China [4]. Pillar industries often have a strong dependence on resources because of the low level of resource development technology, which leads to waste and serious pollution during resource extraction. The sustainable development of the industry is neglected in the one-sided pursuit of output value. The extensive use of non-renewable energy resources, such as oil, coal, and natural gas, seriously harms the ecological balance of the environment. The original ecological environment on the surface of the earth has been destroyed by extensive exploitation of resources. On the other hand, pollution treatment and the technology for resource utilization have long lagged, making it impossible to recycle them for processing and further damaging the environment and ecology.

605081866@qq.com

2.2 Continued dominance of traditional production and consumption methods

2.2.1 The traditional production model mainly uses a large amount of resources first, "polluting first and then treating"

This process wastes resources and causes environmental pollution since it treats the pollution simply and roughly. The traditional production process follows the concept of "pollution first, treatment later". The majority of enterprises in our country apply a resource-dependent, pollution-first and treatment-second approach.

2.2.2 The traditional consumption method is that consumers primarily depend on conventional energy sources

Relevant data indicates that China's coal mining output will reach 4.45 billion tons in 2022, an 8% annual increase, making up over 50% of the world's total. The process of mining and burning coal releases a large amount of pollutants into the air, including carbon dioxide and sulfur dioxide, which are harmful to people's health [5]. Traditional production and consumption methods rely mostly on energy, which leads to enormous energy waste and requires an extensive investment of money, materials, and manpower to handle exhaust gas, wastewater, and waste. Besides, a large amount of national financial resources are spent on these measures. Traditional methods of production and consumption are no longer appropriate for meeting the demands of sustainable development, nor are they helpful for the long-term development of resource regeneration.

2.3 Low conversion rate of the renewable energy industry

2.3.1 The utilization of new energy accounts for a relatively low proportion

Although China has produced more new energy, its industrial utilization rate is quite low when compared to non-renewable energy. Despite having the most installed capacity of wind power in the world, there is little industrial use and inadequate large-scale operation. In terms of China's energy consumption structure, the energy production structure determines it. The National Bureau of Statistics of China reports that the consumption of coal accounted for 56.2% of the total energy consumption, while the consumption of oil accounted for 17.9% in 2022. The consumption of clean energy, which consists of natural gas, hydropower, nuclear power, wind power, and solar power, accounted for 25.9% of the total energy consumption [6]. Other renewable energy sources make up a small proportion of the total.

2.3.2 Expanding opportunities for innovation and breakthroughs in renewable energy technology

Wind power generation technology, biomass power generation technology, biomass formed fuel technology, solar energy utilization renewable energy technology, and crop regeneration product technology are all essentially at the inventive experimental stage of development. The demand for renewable energy cannot yet be satisfied by mature technologies, despite the country having invested a great deal of time and funds in research and development. Resource utilization and high technical requirements are essential in the development of renewable energy technologies. Encouraging the use of renewable energy in actual production and daily life involves not only long-term practical testing but also the majority of people's ability to practice and operate.

2.4 Incompatible policies and regulations for sustainable energy development

In the 11th Five-Year Plan, China has long advocated for changing its growth model and creating a resource-saving and environmentally friendly society. The cumulative energy consumption per unit of GDP decreased by 14% over the 13th Five-Year Plan period, saving 710 million tons of standard coal, which is an amount equal to the present annual energy consumption in the Yangtze River Delta region. Nevertheless, the local regulations for energy development and utilization are still unsound and inconsistent right now. Relying mainly on traditional energy for socioeconomic development will unavoidably be constrained by limited resources due to the sharp decline in resources. The upcoming environmental protection tax should play a part, but energy-related laws and regulations must be implemented at the point of use.

3. Basic requirements for sustainable energy development

Energy security can only be achieved through sustainable energy development. Sustainable energy development and utilization as well as energy-saving and emission-reduction measures for energy consumption are all included in the concept of energy security. Resolving the contradiction between energy supply and demand and attaining sustainable energy development through clean energy is the primary substance of sustainable energy development.

3.1 Saving the use of conventional energy

3.1.1 Transforming traditional production methods and industrial structure

The consumption of coal has started to be in a relatively stable trend in recent years as a result of the continuous improvement and implementation of environmental protection awareness, policies, and laws. Although the modernization of traditional industries is being driven by new technology, the share of tertiary industry in the national economy is steadily rising. The change in industrial structure results in an adjustment to the employment structure. The high-tech application of the first and second industries has drastically changed people's lifestyles and job opportunities, altering the basic reliance of traditional industries on traditional energy.

3.1.2 Green travel and low-carbon consumption

Green travel and low-carbon consumption are environmentally friendly lifestyles that contribute to the reduction of air pollution and carbon emissions, which in turn protects the environment. Sustainable development may be promoted by lowering the use of high-carbon items and choosing low-carbon modes of transportation like cycling, walking, or public transportation.

3.1.3 Further legalization and standardization of energy-saving management

China has created and implemented several energy-related laws and regulations to adjust activities that include energy development, utilization, and management, ensuring that these activities are legal, efficient, and rational. A great deal of effort has been made to assure energy security, efficiency, and sustainability. Laws and regulations, such as the Energy Conservation Law, Renewable Energy Law, and Regulations on Energy Conservation Management of Civil Buildings, will need to be formulated in the future to standardize and manage production, energy development, and utilization.

3.2 Increasing the scope of use of renewable clean energy industry

President Xi further elaborated on the specific arrangements and plans for carbon peaking and carbon neutrality goals at the Climate Ambition Summit on December 12, 2020. By 2030, China's CO_2 emissions per unit of GDP will have dropped by more than 65% from 2005. Non-fossil fuels will account for about 25% of primary energy consumption. Forest storage will have increased by 6 billion meters from 2005, and the total installed capacity of wind and solar power will surpass 1.2 billion kW [7]. Clean energy has been given a more significant historical mission as a key lever for reaching carbon reduction and carbon neutrality.

3.2.1 Expanding the use of wind energy

By the end of 2022, the total installed capacity of China's wind power generation industry reached 365 million kilowatts, representing an annual growth rate of 11.25%. The cumulative installed capacity of onshore and offshore wind power was 332.94 million kilowatts and 32.5 million kilowatts, respectively, with annual growth rates of 10.21% and 23.15%. Offshore wind power technology and equipment are gradually improving because China's main focus on resource utilization and development steadily moves towards the deep-sea field. It is possible to ensure the quality, stability, and safety performance of the wind power generation. According to the current development of wind power in China, wind power has operated on a sizable scale.

3.2.2 Enhancing the promotion and utilization of solar energy products

Currently, a substantial amount of research and development is being carried out on solar energy series products, including solar water heaters, photovoltaic cells, photovoltaic cables, and other solar power-generating products. However, the benefits of solar energy only contribute a small proportion to the development of the regional economy because they are seldom used in actual production and life, and there aren't many companies building and developing solar energy-related products or solar energy-related industries. In the future, it will be important to raise awareness of the environmental protection, energy conservation, and progressiveness of solar energy products. This will help the whole people develop new energy-saving habits, better promote the utilization of solar energy products, and increase the usage rate of solar energy.

4. Energy development strategy selection for sustainable green economy

4.1 Vigorously promoting the development of green and energy-saving industries

The economic and environmental development of a country is becoming more and more correlated with the choice of energy structure. The entire society must collaborate to achieve harmonious energy development and to create a society that conserves resources and the environment. With the rapid growth of China's green economy, the energy efficiency of industries has improved remarkably. Energy consumption increased by 2.8% annually on average from 2012 to 2019, which helped China's national economy and resulted in a notable improvement in energy utilization efficiency. According to the 14th Five-Year Plan, one of the main binding indicators for economic and social development is a 13.5% decrease in energy consumption per unit of GDP [8]. The material living standards of the people have also been greatly improved because of the rapid progress of the economy and technology as well as the ongoing increase in population. The consumption of energy continues to increase in this process. Large-scale consumption of fossil fuels has led to the emergence of environmental problems such as extreme weather, global warming, acid rain, etc. Moreover, the environment is deteriorating [9].

4.2 Accelerating the development of green and renewable energy

In contrast to conventional fossil fuels like coal, oil, and natural gas, renewable energy primarily refers to non-mineral energy that is continually available from nature and capable of regeneration. Renewable energy is also known as green energy because of its environmentally friendly characteristics in development and consumption. With the backdrop of achieving the "dual carbon" goal, new energy enterprises centered around solar and wind energy have ushered in promising development opportunities [10]. Currently, biomass, ocean, geothermal, solar, and wind energy are the main development areas.

4.2.1 Steadily promoting the development of solar energy.

Solar energy resources, a crucial component of clean, safe, renewable energy, are inexhaustible and have promising development prospects. With over two-thirds of its land area having more than 2,200 hours of sunshine annually, China is a country endowed with abundant solar energy resources. According to the latest resource assessment of the China Meteorological Administration, there might be 1.86 million GW of potential solar energy reserves on land. China's solar power generation reached 393 million kilowatts in 2022 [11]. However, high cost remains a significant barrier to the advancement of source technology and energy consumption methods when compared with conventional energy. It is necessary to eliminate the main barrier to the widespread application of photovoltaic power generation through policy support and technological progress.

4.2.2 Improving the economic efficiency of wind energy resource development

With the highest exploitable capacity and wind energy reserves in the world, China has abundant wind energy resources. Among them, 250 million kW of exploitable wind energy are found in the 10 m high onshore reserve, while 750 million kW are discovered in the offshore reserves, totaling 1 billion kW. According to statistics from the China Meteorological Administration, the annual average wind speed for wind energy resources in China in 2022 at a height of 10 meters is 0.82% lower than the average value in the previous 10 years. The wind energy resources in Northeast China, Inner Mongolia, Xinjiang, Gansu, and the Qinghai Tibet Plateau are commonly installed at a height of 70 meters for wind turbines, which is beneficial for wind power generation from the perspective of spatial distribution [12]. Owing to the substantial impact of seasons and climate, the proportion of wind power in the backbone power grid is generally controlled within 20% to maintain the stability and safety of the power grid.

4.2.3 Expediting the progress of biomass energy development

With the addition of 3.34 million kilowatts of installed capacity in 2022, China's biomass power generation now has a cumulative installed capacity of 41.32 million kilowatts. Among them, 2.57 million kilowatts of newly installed capacity of household waste incineration power generation make up the cumulative installed capacity of 23.86 million kilowatts. A total of 16.23 million kilowatts of installed capacity, including 650,000 kilowatts of newly installed biomass power generation from agricultural and forestry, are available. Biogas power generation has a cumulative installed capacity of 1.22 million kilowatts, of which 120,000 kilowatts are newly added. In this populous and agricultural country, dealing with household trash and agricultural waste like straw and rice husks is a dual challenge for environmental treatment and energy development [13]. These two issues are effectively connected by the development of biomass energy. China's biomass energy should have great development potential while addressing the economic, technological, and utilization aspects of biomass energy development.

4.3 Independent innovation of new energy technology research and development, establishment of industry university-research technology system

With the constantly growing scale of the new energy market, it is crucial to strengthen the independent innovation capability of the new energy industry and to set up an industry-university research technology system to support the sustainable development of the new energy industry. Independent innovation research and development as well as the establishment of an industry university research technology system are critical to the transformation of the new energy industry. On the one hand, independent innovation is a vital strategy to encourage the development of the new energy industry. On the other hand, the establishment of an industry-university research technology system can foster research collaboration between industry universities, strengthen the advancement and utilization of scientific and technological discoveries, and significantly bolster the fundamental competitiveness of the emerging energy industry. The new energy industry can only achieve a leap from scale expansion to quality improvement, environmentally friendly, and sustainable directions through continuous innovation, improving technological content, and enhancing market competitiveness.

4.4 Strengthening regulatory and policy support for the utilization of new energy

With an emphasis on safeguarding the sustainable development of the environment through policies and laws, China has progressively promulgated and implemented regulations and measures, such as the Renewable Energy Law and the Environmental Protection Law of the People's Republic of China. The new energy industry has unique challenges. Thus, the development and utilization of new energy require substantial financial support from the government, which might involve adopting tax incentives for enterprises located in new energy demonstration zones to lower their operating costs. The government should enhance its assistance in the new energy industry, implement more relevant policies and measures, expand the scale of the new energy market, and raise public understanding and acceptance of new energy to stimulate social investment enthusiasm. For instance, more people may become aware of the benefits and importance of new energy by carrying out various promotional activities, such as science lectures, TV commercials, online marketing, etc., which would stimulate market demand.

4.5 Strengthening international cooperation and actively participating in the formulation of international standards in the field of new energy

China must increase international cooperation and exchanges to develop new energy. There are two key strategies to consider. The first strategy involves enhancing global cooperation and exchange in new energy technologies and encouraging collaboration in China's research and development of important energy technologies, including, hydrogen, nuclear, and renewable energy. The second strategy involves strengthening international cooperation in the provision of raw materials for new energy. For example, it is necessary to import two-thirds of the uranium resources required for the production of nuclear power in China. Furthermore, China's voice and influence in the international new energy field can be strengthened via increasing international cooperation and actively contributing to the development of international standards in new energy. China can actively engage in the relevant work of international standard-setting organizations like the International Organization for Standardization (ISO) and the International Electrotechnical Commission (IEC) to contribute to the formulation of international standards in the field of new energy.

5. Conclusion

People's consumption structure have changed significantly as a result of the rapid development of China's economy and the continuous improvement of people's living standards. The main issues that China needs to address are resource scarcity and environmental pollution from the usage of non-clean energy like coal and oil. With the advancement of industry and technology, energy resources are becoming more limited, and various countries must deal with the unavoidable problem of sustainable development. According to its own resources and specific national conditions, China should formulate its energy development strategy to meet the increasingly expanding energy problem. This strategy should aim to utilize domestic resources while receiving energy from other sources and regions. Building a resource-efficient and environmentally friendly society and actively generating new energy is a common choice for Chinese society as well as a strategic strategy for sustainable economic and social development in the context of resource and energy constraints.

Acknowledgements

The work is supported by: 2022 Research Project on Ideological and Political Education Teachers in Higher Education Institutions by the Ministry of Education of the People's Republic of China: Research on the Integration of Excellent Traditional Chinese Culture into Ideological and Political Education Courses in Higher Education Institutions in the New Era (22JDSZK090);Key Research Base of Philosophy and Social Sciences of Sichuan Province, Sichuan University Ideological and Political Education Research Center 2020 General project: Celebrity Resources and Ideological and Political Education of University Students -- A Case study of Zhu De (CSZ20075);2020 Key project of Sichuan Libai Cultural Research Center: Research on the application value of celebrity resources in Ideological and Political education in local Universities (LB20-A01).

References

1. Li Lei. The Sustainable Development Principles and Strategies of Energy Economy in the New Era[J]. Trade Fair Economy, 2023(11):31-36.
2. Patterson M G. What is energy efficiency? [J]. Energy Policy, 1996, 24: 377-390.
3. Rashe R., Tatom J.Enegry Resouces and Potential GNP[J]. Federal Resevre Bank o f St Louis Review, 1977, 59 (6): 68-76.。
4. Hwang D.B.K, Gum B.. The causal relationship between energy and GNP: The case of Taiwan[J]. Journal of Energy and Development, 2000, 16: 219-226.
5. Wang Yuchao. Analysis of the Transformation Path of the New Energy Industry Based on the Concept of Sustainable Development [J]. *Industrial Innovation*, 2023.8:21-23
6. Statistical Bulletin on National Economic and Social Development of the People's Republic of China in2022[EB/OL][2023-02-28] https://www.stats.gov.cn/sj/zxfb/202302/t20230228_1919011.html?eqid=9ede71c2000017eb0000000464 2e38b4.
7. Shen Jun. Developing a Low Carbon Economy and Leading Green and Sustainable Development of Industries with New Energy [J]. *Cement Engineering,* 2021 (1): 1-6
8. During the 14th Five Year Plan period, China's energy consumption per unit of GDP will decrease by 13.5% - accelerating the formation of an energy saving society [N]. *People's Daily*, 2021-08-10(3).
9. Zheng Yujia. The significance of clean energy development and utilization for achieving sustainable development [J]. *Vitality*, 2023(5): 184-186.
10. Sun Jieling.Research on the Management Model of New Energy Enterprises Based on Sustainable Development [J]. *Enterprise Reform and Management,* 2023, (12): 52-53
11. Development of Renewable Energy in China in 2022 [EB/OL] [2023-02-15] https://www.ndrc.gov.cn/fggz/hjyzy/jnhnx/202302/t20230215_1348799.html.

12. In 2022, China's solar energy resources will be relatively large[EB/OL] [2023-04-06] https://www.cma.gov.cn/2011xwzx/2011xmtjj/202304/t20230406_5422838.html.
13. Chang Shiyan, Kang Liping. International policies for sustainable development of biomass energy and their implications for China [J]. *Journal of Agricultural Engineering,* 2017, (11): 1-10.

E3S Web of Conferences 528, 03004 (2024) https://doi.org/10.1051/e3sconf/202452803004
NET-LC 2024

Preliminary Exploration of Rational Settings for Sponge Airport Indicators in the Context of Low Carbon -- Taking Baiyun Airport as an Example

Lingling Zhang[1] , Daizhong Huang[2,3,*], Xu Pan[1] , Wangchun Chen[1] ,Shuaiyi Sun[1] , Zhenwen Xia[1]

[1] Civil Aviation Airport Planning and Design Research Institute Co., Ltd,beijing 100029,China
[2]Key Laboratory of the Pearl River Estuarine Dynamics and Associated Process Regulation, Guangzhou 510611, China
[3]Pearl River Water Conservancy Research Institute of the Pearl River Water Conservancy Commission, Guangzhou, Guangdong 510612, China

Abstract: Sponge city is an important means of urban stormwater management, and it is also the core concept of green and low-carbon ecological landscape construction.Currently, the rainfall and flooding problems of airports are becoming more and more obvious, and the construction of "sponge airports" is getting more and more attention. However, considering the operational needs of airports, the existing conventional sponge construction ideas cannot be directly applied to airport construction. How to set up a set of reasonable indicators that not only guarantee the safe operation of airports but also meet the requirements of the existing sponge city construction is one of the problems that need to be solved in the process of airport sponge city construction. Taking the third phase expansion project of Guangzhou Baiyun Airport as an example, the design principles of focusing on safety, adapting measures to local conditions, and meeting the overall standards, from the overall consideration of the airport, combined with the land use of each functional zone of the airport, reasonably set the sponge index, so as to provide a reference for the construction of "sponge airport", in order to promote the sustainable development of green and low-carbon construction.

1. Introduction

As a bridge between urban stormwater management and landscape design, sponge city is an important means to promote the construction of green ecological urban areas, low-carbon cities, and sustainable urban development[1],[2]. Combined with the current background of developing green aviation and promoting the construction of green airports, many new and expanded airports in China have proposed to build "ecological sponge-type airports" to achieve sustainable development of airports. The goal of sponge airport construction is to build an ecological green airport, which controls the total amount of runoff and runoff pollution outside the airport through comprehensive management and utilization of rainwater, ensures the safety of airport drainage and flood prevention, saves water resources, and builds an airport with natural water absorption, water storage, water purification, and water release capabilities[3].

The sponge city at the scale of the street block is an important tool and perceptual carrier for the construction of green and low-carbon cities. While integrating the requirements of sponge design, it is necessary to combine the differences in land use planning within the space and adapt to local conditions to construct a multi-dimensional green space suitable for sponge landing in the street block[4]. As an important part of urban development, airports are often considered as part of the urban construction area when preparing the special plan for sponge city. When setting sponge indicators, it is necessary to consider the needs of the city's own sponge construction, without considering the actual situation of airport operation. Therefore, there is a contradiction between the traditional sponge city concept and the drainage safety of Baiyun Airport.

Building an airport completely based on existing sponge planning indicators does not meet the requirements of airport operations or low-carbon city construction. Scholars at home and abroad have conducted a series of studies on sponge city construction in airports. The United States was the first to propose the concept of "low-impact development cities and green buildings." O'Hare International Airport achieved a non-discharge rate of 90% to 95% for precipitation through full roof greening[5]. Jaiyeola[6] analyzed that the collection and reuse of airport rainwater can not only conserve water resources but also prevent pollutants from entering the external environment. He also discussed the model methods for evaluating the potential of airport rainwater storage.. Hailin Zhang[7] proposed a low-impact development idea for airports in water-scarce

*Corresponding author's e-mail: 2418589582@qq.com

areas in the reconstruction project of Xi'an Xianyang International Airport. Yang Xinyu et al.[8] analyzed the application effects of different LID facilities in airport construction through case calculations. This series of research enriches the ideas of airport sponge construction, but there is no in-depth analysis of the contradiction between the planned sponge indicators and the functional requirements of each airport division. Sponge airport construction still has a long way to go. This article discusses the setting scheme of reasonable sponge indicators in the process of airport sponge construction through the setting of sponge indicators in the third phase expansion project of Guangzhou Baiyun International Airport, providing a reference for the subsequent "sponge airport" construction, with a view to promoting the sustainable development of green and low-carbon construction.

2. Study area

Baiyun Airport is located in the northern part of Guangzhou, at the junction of Renhe Town in Baiyun District and Xinhua Street in Huadu District, about 28 km in a straight line from the centre of Guangzhou, and was officially opened on 5 August 2004.[9] It was officially opened on 5 August 2004. In order to enhance the airport's security capacity and meet the continuously growing demand for air transport, Guangzhou Baiyun International Airport plans to implement a three-phase expansion project, which will include the construction of a new west second runway, east third runway and the related apron; T3 terminal building, T3 pre-floor integrated traffic centre and parking building and T3 work area; T2 terminal building, east four and west four finger corridors; as well as cargo, production and living auxiliary buildings and public utility support facilities. The proposed new construction land for the Phase III expansion is 15.6km^2 , and the satellite image of the airport is shown in Figure 1.

3. Research Methodology

3.1. Principles of sponge indicator design

(1) Safety First

To guarantee the safety of airport operations as a prerequisite, potential safety hazards must be eliminated and disaster prevention and mitigation capabilities must be enhanced.

(2) Adaptation to Local Conditions

By fully integrating the actual conditions of Baiyun Airport's land layout and operation scheduling, a reasonable sponge city construction target is set based on the relevant contents of the above plan.

(3) Overall Compliance

Based on the concept of "sponge," the overall planning and design of Baiyun Airport, infrastructure construction and operation, and the organic integration of sponge city construction aim to achieve overall sponge construction within the airport's red line that meets the standards.

3.2. Principles of sponge indicator design

Figure 1. Satellite map of the current situation of Baiyun Airpor

(1) Targeted Breakdown Programme

1) The total annual runoff control rate target breakdown is based on a parcel breakdown followed by review and adjustment.

2) Statistics on the area of each type of land, in the "Guangzhou City Construction Project Sponge City Construction Control Indicator Classification Guidelines" (Sui Shuihehu (2020) No. 7) in the different parcels of land on the basis of the total annual runoff control rate of the recommended value, combined with the area of various types of land in the area of the composition of a typical subsurface.

3) The total annual runoff control rate target is obtained by accounting for the control volume for each type of site area.

4) Account for the accessibility of the overall district goal based on the district-wide storage volume and adjust parcels if there is a shortfall.

5) Selection of control rates or consideration of the requirement to add controlled transfer volume on parcels with conditions, recalculated until requirements are met.

The specific decomposition process is shown in Figure 2.

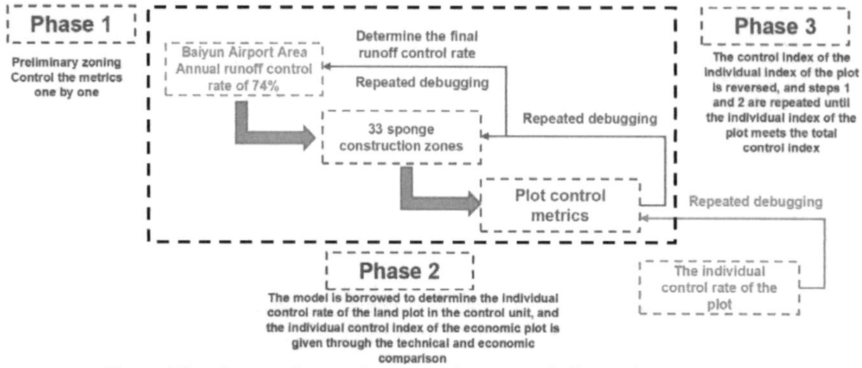

Figure 2. Step-by-step diagram for determining sponge indicators in sponge zones

(2) Method of calculating the size of the facility:

1) Volumetric method

For low-impact facilities where total runoff and runoff pollution are the control objectives, the design storage volume is generally calculated using the following formula:

$$V=10\times H\times \varphi\times F \quad (1)$$

Where: V - design storage volume, m^3 ;

H - Design rainfall, mm;

φ - Integrated rainfall runoff coefficient;

F - Area of the catchment, ha.

2) Flow method

Applies to the calculation of the size of transfer facilities such as transfer planting trenches, using the formula:

$$Q=\varphi\times q\times F \quad (2)$$

Where: Q - rainwater design flow, L / s;

φ - flow runoff coefficient;

F - catchment area, ha

3) Water balance method

Applicable to wet ponds, rainwater wetlands and other storage facilities scale calculation, facility volume first calculated in accordance with the volumetric method, and then according to the water balance calculation, taking into account the facility's water replenishment, external drainage, evaporation and other parameters, the design of the facility to adjust the volume of storage.

4. Overall indicator development

4.1. Planning sponge indicator requirements

In order to promote the sponge airport construction of Guangzhou Baiyun International Airport, according to the State Council's "Guiding Opinions on Promoting the Construction of Sponge Cities" (Document No. 75 of the State Council [2015]), and on the basis of fully connecting the "Guangzhou Sponge City Special Planning" and the "Special Planning for Sponge Cities in Guangzhou Airport Economic Zone (2019-2030)" indicator system taking values, combined with the functional positioning of Guangzhou Baiyun International Airport Based on the functional position of Guangzhou Baiyun International Airport and the characteristics of the region, the index system of sponge airport construction for the third phase expansion of Guangzhou Baiyun International Airport is determined from four aspects, such as water ecology, water environment, water resources, water safety, etc., as shown in the table 1.

Table 1. Planning sponge indicator requirements

form	serial number	Indicator name	metric	Nature of the indicator
ecohydrology	1	Total annual runoff control rate	≥74 per cent	restrictive
	2	Rate of water area	≥3.5 per cent	restrictive
	3	sunken green space	≥50 per cent	restrictive
	4	Ecological shoreline restoration rate	80 per cent	restrictive
water environment	5	Annual runoff pollution reduction rate	≥50 per cent (in SS)	restrictive
	6	Environmental quality of surface water	Class IV water	restrictive
	7	Percentage of rainwater and sewage diversion	100%	restrictive
	8	Sewage treatment rate	≥97 per cent	restrictive
water resources	9	Rainwater resource utilisation	≥3 per cent	restrictive
	10	Sewage reclamation utilisation rate	≥20 per cent	restrictive
water	11	Flood control standards	Effective response to a one-in-20-	restrictive

security			year rainstorm	
	12	Drainage system standards	5-year return period; 10-year return period for critical areas	restrictive
	13	flood protection standard	once in 100 years	encouraging

4.2. Analysis of the reasonableness of planning indicators

(1) Difficulty in meeting overall airport sponge standards

In order to meet the sponge acceptance requirements of Guangzhou, after the expansion of Baiyun Airport Phase III, the overall sponge index of the airport should meet the relevant planning requirements. The airport was put into use in August 2004, due to historical reasons, the existing T1, T2 airport scope can not be fully considered sponge city construction requirements, and this expansion for the scope of the airport involving the status quo. Therefore, it is difficult to rely only on the scope of Baiyun Airport Phase 3 expansion to set up sponge facilities to achieve the sponge standard of the whole airport.

(2) Difficulty in meeting some of the secondary indicators

Existing relevant plans require Baiyun Airport water area rate ≥ 3.5% and sunken green space ≥ 50%. However, considering that the total area of the red line of Baiyun Airport Phase III Expansion is 35.18km^2 and the green area is 12.14km^2 , in order to meet the above sponge index requirements, the water area of Baiyun Airport Phase III Expansion should be not less than 1.23km^2 and the area of the sunken green area should be not less than 6.07km^2 . This is not able to satisfy the demand of the airport land, nor can it satisfy the requirements of the airport's normal operation and safety, so it is very difficult to implement this kind of planning index. Therefore, it is very difficult to implement this kind of planning index.

(3)Airport's existing zoning sponges struggle to meet standards

The Guangzhou Airport Economic Zone Sponge City Special Plan (2019-2030) divides the Baiyun Airport Phase III Expansion Area into seven sub-areas, and for each sub-area, the annual total runoff control rate is required to be ≥74%, the annual runoff pollution reduction rate is required to be ≥50% (in terms of SS), and the storage volume per 10,000 square metres of hardened area is required to be ≥500m . ≥50% (in terms of SS), and ≥500m of storage volume per 10,000 square metres of hardened area[3] . Considering that Baiyun Airport contains different land types such as flight area, terminal area, cargo area and work area, some of the subareas are not able to set up storage facilities, and it is therefore difficult to require each subarea to meet the sponge index requirements independently.

4.3. Determination of overall airport indicators

Combined with the above analysis, it can be seen that the implementation of some of the planning indicators is more difficult, taking into account the actual situation of Baiyun Airport, "Guangzhou City Construction Project Sponge City Construction Control Indicator Classification Guidelines (Trial)" and "Guangzhou Airport Economic Zone Sponge City Special Planning (2019-2030)", under the premise of ensuring the safety of the airport flights, according to the first-level indicators, "should be done to do", the second-level indicators "can do" principle to determine the overall sponge indicator requirements are as follows. Under the premise of ensuring the flight safety of the airport, according to the principle of "should do as much as possible" for the first-level indicators and "can do as much as possible" for the second-level indicators, the overall sponge indicator requirements of Baiyun Airport are as table 2.

Table 2. Guangzhou Baiyun International Airport Phase III Expansion Project Sponge Construction Indicator Requirements

serial number	norm	Planning of Airports target requirement	this programme target requirement	form
1	Total annual runoff control rate	74 per cent	74 per cent	ecohydrology
2	Annual runoff pollution reduction rate	50%	50%	water environment
3	Percentage of rainwater and sewage diversion	100%	100%	
4	Rainwater resource utilisation	3 per cent	3 per cent	water resources
5	Sewage reclamation utilisation rate	≥20 per cent	≥20 per cent	
6	Flood control standards	Effective response to a one-in-20-year rainstorm	Effective response to a 100-year rainstorm	water security
7	Drainage system design criteria	5-year return period; 10-year return period for critical areas	The passenger terminal area and main arterial roads are designed for a return period of one in 10 years; the cargo area, aircraft maintenance area and other areas are designed for a return period of one in 5 years, and the design standard for the drainage system of important	

			underground facilities is not less than one in 50 years.
8	flood protection standard	once in 100 years	once in 100 years

4.4. Summary

Comprehensive analysis in this chapter shows that the existing sponge planning has not considered the actual situation of Baiyun Airport, and some of the planning indicators are more difficult to land in the airport design and construction. According to the requirements of the relevant technical documents, combining with the actual situation of the airport to set up realistic overall indicators, the next step to be done is to subdivide this series of overall indicators to the various sub-districts of the airport, in order to meet the needs of the airport design and later airport sponge acceptance. In this paper, the total annual runoff control rate is taken as a representative indicator, and through different decomposition methods, we explore the reasonable scheme of setting indicators for Baiyun Airport's sub-areas.

5. Zoning indicator development

5.1. Partitioning ideas and partitioning results

This zoning subdivided the airport's Phase III expansion area into three levels by taking into account the overall airport drainage zoning, the airport's internal drainage zoning and the airport's land use zoning:

(1) Level 1 zoning: the red line area of Baiyun Airport Phase III expansion is divided into Level 1 zoning according to relevant planning;

(2) Level 2 zoning: according to the airport drainage network design programme, the airport drainage zoning is treated as a level 2 zoning;

(3) Level 3 zoning: on the basis of level 2 zoning, further zoning is carried out by combining the land types of Baiyun Airport's flight area, terminal area, cargo area and working area.

The specific zoning plan is shown in Figure 3.

According to the above zoning principles, Baiyun Airport is divided into 1 level 1 zoning, 7 level 2 zoning and 33 level 3 zoning after the third phase expansion. The details are shown in Table 3 and Figure 4.

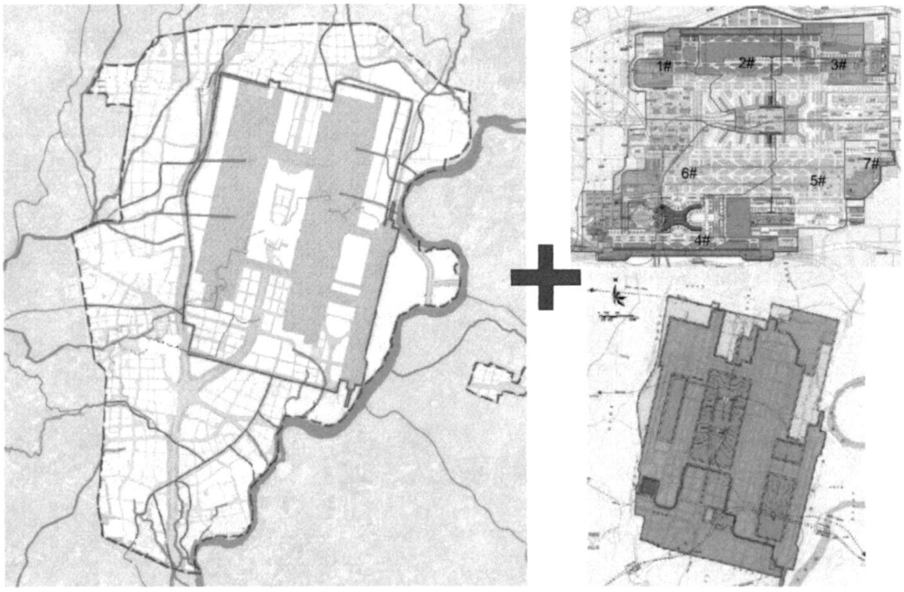

Figure 3. Sponge zoning basis map of Baiyun Airport

Table 3. Results of the sponge zoning of Baiyun Airport

Level 2 zoning	Level 3 zoning	instructions	Level 2 zoning	Level 3 zoning	instructions
1#	1-1#	Current built-up area	4#	4-1#	Expansion of the flight area
	1-2#	Current flight area		4-2#	Expansion of the terminal area
	1-3#	Current terminal area	5#	5-1#	Current flight area
	1-4#	Expansion of the flight area		5-2#	Expansion of the flight area
2#	2-1#	Expansion of the flight area		5-3#	Current terminal area
	2-2#	Extension Area Green Space		5-4#	Expansion of cargo area
	2-3#	Current flight area		5-5#	Expansion of the flight area
	2-4#	Current terminal area		5-6#	Current maintenance area
	2-5#	Forward Cargo Area		5-7#	Extension Area Green Space

3#	3-1#	Expansion of cargo area	6#	6-1#	Current flight area
	3-2#	Expansion of the flight area		6-2#	Expansion of the flight area
	3-3#	Current flight area		6-3#	Current terminal area
	3-4#	Status quo cargo area		6-4#	Current built-up area
	3-5#	Extension Area Green Space		6-5#	Expanded workspace
	3-6#	Current terminal area		6-6#	Expansion of the terminal area
	3-7#	Forward Cargo Area	7#	7-1#	Current flight area
				7-2#	Expansion of the flight area

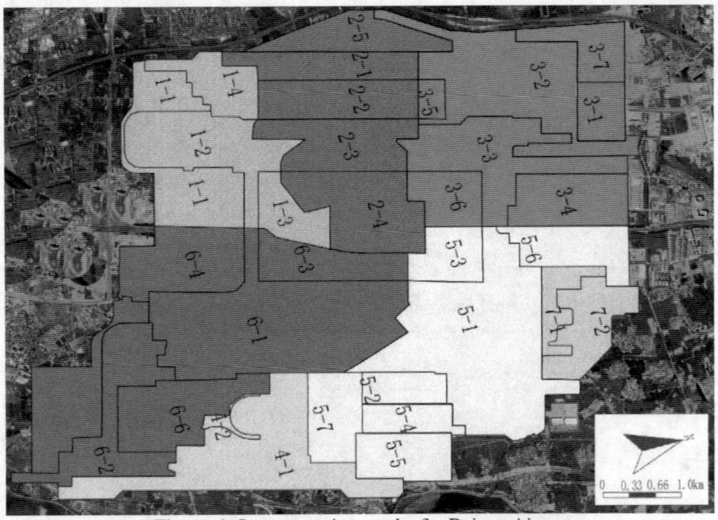

Figure 4. Sponge zoning results for Baiyun Airport

5.2. Completion of sponge indicator analyses independently by subdistricts

The Special Plan for Sponge City in Guangzhou Airport Economic Zone (2019-2030) proposes the same total annual runoff control rate indicators for the seven secondary sub-districts of Baiyun Airport, and if the sponge sub-districts set up a source control programme in combination with the actual situation of each district, and sub-districts consider the indicators of their own districts individually, the sponge indicators of the sub-districts are shown in the Table 4.

As can be seen from the table, taking into account the special characteristics of the airport, during the flood season, the flight area, terminal area, aircraft service area, cargo area and other functional areas of the airport should remove the rainwater from the hardened surfaces as soon as possible to ensure the safety of airport operation, and part of the area is unsuitable for the facilities related to "infiltration, stagnation, storage and purification". According to the underlay and sponge city construction background of each sub-area, the current airport area such as 1-3, 2-4, 3-6, 5-3 and 6-3 and other areas of the construction is more mature, basically there is no space for alteration, sponge city construction is more difficult. Therefore, it is not possible to meet the requirements of the relevant planning indicators by relying only on the independent resolution of rainfall runoff in each subarea.

Table 4. Summary of target indicators for sponge city construction in each sub-district

Level3 zoning	instructions	Runoff control rate	Level 3 zoning	instructions	Runoff control rate
1-1#	Current built-up area	99%	4-2#	Expansion of the terminal area	/
1-2#	Current flight area	33%	5-1#	Current flight area	/
1-3#	Current terminal area	/	5-2#	Expansion of the flight area	/
1-4#	Expansion of the flight area	/	5-3#	Current terminal area	/
2-1#	Expansion of the flight area	/	5-4#	Expansion of cargo area	/
2-2#	Extension Area Green Space	/	5-5#	Expansion of the flight area	99%
2-3#	Current flight area	/	5-6#	Current maintenance area	/
2-4#	Current terminal area	/	5-7#	Extension Area Green Space	/
2-5#	Forward Cargo Area	99%	6-1#	Current flight area	/
3-1#	Expansion of cargo area	/	6-2#	Expansion of the flight area	99%
3-2#	Expansion of the flight area	99%	6-3#	Current terminal area	/
3-3#	Current flight area	/	6-4#	Current built-up area	/
3-4#	Status quo cargo area	/	6-5#	Expanded workspace	99%
3-5#	Extension Area Green Space	/	6-6#	Expansion of the terminal area	/
3-6#	Current terminal area	/	7-1#	Current flight area	/

3-7#	Forward Cargo Area	/	7-2#	Expansion of the flight area	62%
4-1#	Expansion of the flight area	99%	average value		31%

5.3. Calculation and analysis of sponge indicators for drainage zoning in a comprehensive manner

In order to ensure the overall safety of the airport flood prevention and drainage after the expansion of Baiyun Airport Phase III, there are end storage tanks in each subzone of the airport. Storage pools in the control of their own sub-district rainfall at the same time also accept the upstream sponge sub-district rainfall water. Therefore, the end of the storage facilities are considered to use the upstream sponge sub-area rainfall control. According to the relevant design scheme: the maximum total storage volume of the airport construction storage facilities is 1,683,000m^3 . Since the rainwater from all sub-districts of the airport is collected into the storage ponds and then externally discharged, it is considered to make full use of the volume of the storage ponds to control the rainwater from the upstream sub-districts. New construction projects in the airport expansion area need to be developed in accordance with the concept and method of sponge city construction; to ensure that the sponge city construction in the expansion area meets the standard, and at the same time, try to ensure that the overall sponge city construction of the airport after the expansion meets the standard.The specific indicators for each zone are shown in Table 5.

Table 5. Summary of target indicators for sponge city construction in each sub-district

Level 2 zoning	Level 3 zoning	Runoff control rate	Level 2 zoning	Level 3 zoning	Runoff control rate
1#	1-1#	80.70	5#	5-1#	70.10
	1-2#			5-2#	
	1-3#			5-3#	
	1-4#			5-4#	
2#	2-1#	72.80		5-5#	
	2-2#			5-6#	
	2-3#			5-7#	
	2-4#		6#	6-1#	76.40
	2-5#			6-2#	
3#	3-1#	81.80		6-3#	
	3-2#			6-4#	
	3-3#			6-5#	
	3-4#			6-6#	
	3-5#		7#	7-1#	55.90
	3-6#			7-2#	
	3-7#		average value		74.00
4#	4-1#	63.10			
	4-2#				

5.4. Summary

Combined with the overall drainage zoning, internal drainage zoning and land zoning of the airport sponge construction zoning, respectively, consider each sub-district to complete the sponge indicators independently and secondary drainage zoning comprehensive consideration of sponge indicators of the two programmes to calculate the sub-district indicators. According to the calculation results, it can be seen that part of the sub-district is not suitable for setting up sponge-related facilities considering the operational functional requirements, and the sub-district can not independently complete the sponge indicator tasks, and the use of each secondary sub-district end of the end of the storage facilities for the upstream sponge sub-district rainfall regulation, in order to satisfy the airport sponge indicator requirements of the overall standard.

6. Conclusion

This article takes the third-phase expansion project of Guangzhou Baiyun Airport as an example to explore the reasonable setting scheme of sponge airport indicators in the context of low-carbon development. By considering the airport as a whole and integrating the land use situation of various functional zones within the airport, the rationality of the sponge airport setting scheme is discussed. The main conclusions and suggestions obtained are as follows:

(1)Considering low-carbon development, the existing sponge city planning indicators cannot be directly applied to airport sponge construction, and adjustments and optimization should be made based on the actual situation of the airport;

(2)Some of the secondary indicators such as the rate of water area and sunken green space in the sponge indicators of Baiyun Airport planning are difficult to implement, and should be adjusted in accordance with the principle of "doing everything that can be done";

(3)Baiyun Airport's three-level sub-districts can not independently complete the task of controlling the total amount of runoff within the sub-district, should be combined with the construction programme of the airport storage system, by the drainage sub-districts to carry out the overall control, to ensure the safety of airport operations at the same time, to meet the airport's overall sponge indicators to meet the standard requirements.

References

1. Chenhui Lin,Hao Wu, Exploring International Practices and Planning Carbon Reduction Technologies for Low-Carbon Neighborhoods [J], Human Habitat, 2022.
2. Yupei Ren,Jingli Yang,Xuefei Wu, From demand identification to simulation verification: Precise regulation of urban street green infrastructure on temperature [J]. Chinese Landscape Architecture, 2022, 38(11).
3. Peng Li,Research on the Technical Framework of Sponge Airport Rainwater System [D]. Civil Aviation University of China, 2018.
4. Feiyu Ouyang. Analysis of creating highlights of sponge city landscape at the scale of block under the background of low carbon [J]. Construction Technology, 2023, (15): 21-23.
5. Gene Peters. Vegetated Roof Installations at O9Hare and Midway International Airports[Jj,T& DI Congress,2011:977-986.
6. Jaiyeola. The management and treatment of airport rainwater in a water-scarce environment[J],International Journal of Environmental Science and Technology. 2017,2(14):421-434
7. Hailin Zhang,Xuewen Hao,Ying Xue. Thoughts on Sponge City Construction at Xi'an g YXianyang International Airport [J]. Civil Aviation Management, 2018(11): 57-60.
8. Xinxinang,Qianqian Wang,Jing Peng.Design and effect evaluation of low-impact development facilities for sponge airport [J]. Yellow River, 2022, 44(09): 126-130.
9. Xingxue Feng,Yongpeng Lv,Yuzhen Ling, et al. Exploring the construction of a "sponge airport" - taking Guangzhou Baiyun Airport Phase III as an example [J]. Urban Roads and Flood Control, 2022, (11): 110-114+17.

Fine calculation method for carbon emission of urban rail transit

Jingxuan Yang

School of Traffic and Transportation, Beijing Jiaotong University, Beijing 100044, China

Abstract. As an important part of modern integrated transportation system, urban rail transit generates carbon emissions mainly by consuming electric energy. Aiming at the problems such as single index and weak accuracy of the current carbon emissions accounting method for urban rail transit, this study builds a refined carbon emissions calculation model based on existing industry standards and norms, puts forward refined carbon emissions calculation factors under different conditions, and designs efficient algorithms to solve them. The research hopes to provide a basis for differentiated policy formulation for urban rail transit operators, so as to facilitate urban rail transit operators to obtain more say in carbon trading, complete carbon quotas more scientifically and rationally, and better help urban rail transit enterprises to complete the strategic goal of "carbon neutrality" and "carbon peak".

1. Introduction

Since 2006, China has been the world's largest emitter of carbon dioxide, and the transport sector, as one of the major sources of greenhouse gas emissions, occupies the third place among the emissions of all sectors, with an emissions share of 13.1% [1]. Urban rail transit includes a variety of standards, with high transportation efficiency, low environmental pollution properties, is the construction of digital transportation, green transportation, constitute an important part of modern comprehensive transportation system. In October 2021, The State Council issued the "Action Plan for Carbon Peaking before 2030", which includes the fifth point on "Green and low-carbon action for transport". To comply with this national policy, the China Urban Rail Transit Association released the "China Urban Rail Transit Green Urban Rail Development Action Plan" in August 2022. At present, some places like Shenzhen and Zhejiang Province have implemented carbon-inclusive methodologies and carbon emissions reduction policies [2]. However, at present, there is a lack of accurate carbon emission calculation methods, and the demand and requirements for accounting corporate greenhouse gas emissions with relatively uniform standards are rising. The accounting methods of corporate greenhouse gas emissions based on industry characteristics are being widely concerned. Extending to the enterprise value chain and focusing on the internal facilities of the enterprise are the two current trends.

The fine accounting of urban rail transit is an accurate calculation process of carbon emissions under different circumstances based on statistical data and equipment parameters combined with actual travel conditions after comprehensive consideration of regional, climate, equipment, passenger flow, operation and other conditions of urban rail transit system. Through the detailed accounting of carbon emissions under specific circumstances, urban rail transit operators can have a clearer and more accurate understanding of the carbon emissions of their operating lines, in order to better cooperate with the implementation of the national "dual carbon" strategy.

2. Influencing factors of carbon emissions in urban rail transit

2.1. Main methods of carbon emissions accounting

At present, the emission factor method is the most widely and most commonly used carbon accounting method. It is a carbon emissions estimation method proposed by the Intergovernmental Panel on Climate Change (IPCC) of the United Nations [3]. It can be simply understood as energy consumption plus an emission factor, which is a coefficient corresponding to energy consumption.

In the greenhouse gas emission accounting practice of enterprises, the most commonly used method for the calculation of carbon dioxide emissions generated by the purchase and use of electricity is also the emission factor estimation method, which means that the purchase and use of electricity is multiplied by the grid carbon emission factor to obtain the corresponding carbon emissions. Therefore, whether the grid carbon emission factor is used reasonably and its value is appropriate greatly affects the quality of greenhouse gas emission accounting, which is of great significance for the accurate assessment of carbon emissions (or carbon emissions reduction) of various regions, enterprises and projects, as well as the development of high-quality carbon peak and carbon neutral implementation plans. At present, the grid carbon emission factor is not updated in a timely manner, and the

Corresponding author's e-mail: yjx_xuaner@163.com

https://doi.org/10.1051/e3sconf/202452803005

spatial and temporal resolution is not enough. It is urgent to establish an objective, intuitive and accurate grid carbon emission factor system to provide scientific data reference for monitoring carbon emissions dynamics and implementing carbon reduction actions.

2.2 Analysis of influencing factors of urban rail transit energy consumption

Since the main form of energy used in the operation of urban rail transit is electric energy, carbon emissions generated by urban rail mainly come from purchased electricity. Discussing the influencing factors of carbon emissions of urban rail is equivalent to discussing the influencing factors of energy consumption of urban rail. Urban rail energy consumption refers to the energy consumption generated by urban rail passenger transport and related transportation auxiliary activities such as scheduling, locomotives and works. Taking metro as an example, its power consumption is mainly divided into two parts: train energy consumption and station energy consumption [4,5]. Some scholars have carried out qualitative analysis and identification of subway energy consumption and its influencing factors [6]. Some scholars have explored it, and some factors are listed in Table 1 below.

Table 1. Factors affecting energy consumption in subway operation.

Influencing factors of train energy consumption	Influencing factors of station energy consumption
Vehicle type, train formation, driving mode, ramp design, curve radius, station spacing, train weight, passenger volume, laying mode, season, motor type, station spacing, Locomotive traction characteristics, operation routing	Scale of the station and the number of entrances and exits, passenger flow, season, laying mode, mode of the station environmental control system, capacity and efficiency of the station and the section equipment.

2.3. Selection of influencing factors for fine calculation of carbon emission of urban rail transit

This study takes the metro lines that have been opened as the research object, and selects factors of carbon emission. Based on the analysis of the influencing factors of energy consumption and the actual situation, the station type, vehicle type, season factor and period factor are selected as the influencing factors for fine calculation, and the corresponding line conditions are listed.

2.3.1 Line conditions.

(1) Station type factor. The station type is related to the laying mode of the station. Under the same vehicle and passenger flow conditions, underground stations have more energy consumption in terms of ventilation, elevator, lighting, etc., so the underground station has greater energy consumption than the overground station. In this paper, the overground station type means that more than 50% of the stations in the line are overground stations, and the underground station type means that more than 50% of the stations in the line are underground stations.

(2) Vehicle type factor. Due to differences in vehicle width, self-weight, operating speed, and other technical indicators, the energy consumption of urban rail vehicles varies greatly between different types and models. This paper mainly discusses and analyses the two most widely used models: Type A and Type B.

2.3.2 Season and period conditions.

(1) Seasonal factor. The seasonal influence is mainly reflected in the power consumption of stations. Based on the data from Beijing Metro Line 5, the monthly energy consumption of stations equipped with an underground closed platform screen door system is significantly higher from June to October compared to other months. This paper discusses the impact of North China's hot summers and cold winters on station energy consumption due to noticeable changes in seasons. According to previous studies on temperature's effect on station energy consumption, November to May of the following year is classified as the cold season, while June to October is classified as the warm season.

(2) Period factor. The metro operates on different schedules during peak hours, low peak hours on weekdays, and rest days. The train departure frequency and intervals vary for each schedule, resulting in differences in train operation energy consumption. This paper distinguishes between these three situations.

Table 2. Influencing factors of subway operation energy consumption.

Line conditions		Season and period conditions	
Station type I	Vehicle Type J	Season K	Period L
Overground station	Type A	Cold season	Weekday peak hours
Underground station	Type B	Warm season	Weekday low peak hours
-	-	-	Rest day

Table 2 shows the influencing factors for the fine calculation of the four urban rail trasit operation energy consumption selected in this paper, where *I*, *J*, *K* and *L* represent the types number of each factor.

3. Fine calculation model of carbon emission

3.1. Carbon emission factor

This research model was established based on the relevant formulas and contents in *Shenzhen Low-carbon Public Travel Carbon Inclusive Methodology (Trial)* and *the White Paper of Urban Green Travel Index of Zhejiang Province*. Both of them propose to use the average passenger-kilometer emission factor to calculate personal carbon emissions of passengers in the project scenario, as shown below.

$$E_t = \left(EF_e \times AC_e\right) / Q_t \tag{1}$$

Where, E_t is the average passenger-kilometer emission factor of subway passengers in the project scenario, kgCO2/pkm. EF_e is the carbon emission factors of electricity, kgCO2/kWh. AC_e is the total annual electricity consumption of the subway in the project scenario, kWh. Q_t is the annual passenger turnover of subway in project scenario, pkm.

Among them, 'the White Paper of Urban Green Travel Index' of Zhejiang Province proposes to use correction factors to modify carbon emission factors, and the calculation formula of rail transit correction factors is as follows.

$$\alpha = f(k,t) \tag{2}$$

Where, k is the ridership rate of rail transit in a certain period of time (%). t is the weather corresponding to a certain area in a certain time period of rail transit.

The "White paper" does not provide a clear calculation method of correction factors. This paper refers to its carbon emission correction idea, proposes carbon emission factors of lines, classifies all rail transit line conditions, and the carbon emission factors of lines under each condition are the same.

Starting from the two categories of route conditions and season and time conditions given above, there are four sub-categories of factors. Among them, there are I types of stations, J types of vehicles, K types of seasonal factors and L types of time periods. In this study, I, J and K are all 2 and L =3. The number of each subclass can be multiplied to get the final subdivision result of all the influencing factors. Each result corresponds to a line carbon emission factor, for example, the first category is the overground station - type A - cold season - weekday peak hours.

3.2. Modeling

The carbon emission based on total power consumption E_a is calculated according to the internationally accepted emission factor method, the calculation formula is as follows.

$$E_a = \sum_{r=1}^{R} W_r \cdot EF_e \tag{3}$$

Where, E_a is the total annual carbon emissions from power consumption of all lines, kg. W_r is the total annual power consumption per line, kWh. EF_e is the national grid carbon emission factor, kgCO2/kWh. R stands for number of subway lines.

The basic average carbon emission factor was obtained according to the total energy consumption and mileage of rail transit operation. Its calculation formula is as follows.

$$e = \sum_{r=1}^{R} W_r \cdot EF_e / \sum_{r=1}^{R} D_r \tag{4}$$

Where, e is the base average carbon emission factor, kg/km. D_r is the total annual kilometers travelled by each line, km.

Carbon emissions generated by rail transit line operation are related to line conditions and line mileage, and the corresponding carbon emission factors of the same line under the same line conditions are the same. The total carbon emission E_b of the line is calculated as follows.

$$E_b = \sum_{i}^{I}\sum_{j}^{J}\sum_{k}^{K}\sum_{l}^{L} \alpha_{ijkl} \cdot d_{ijkl} \cdot e \tag{5}$$

Where, E_b is the sum of all line condition carbon emissions, kg. α_{ijkl} is the first line carbon emission allocation factor. d_{ijkl} is the kilometers of the line under the first condition, km.

The goal of the model is to allocate the carbon emission of electricity consumption to each line, so the carbon emission E_a based on the total power consumption and the total carbon emission E_b of the line should be consistent as far as possible, and the objective function can be written as follows.

$$\min C = |E_a - E_b| \tag{6}$$

Where, C is the difference between total carbon emission and total power consumption of each line, kg.

3.3. Constraint

The first type of constraint in this model is that all line carbon emission factors are positive and less than m .

$$0 < \sum_{i}^{I}\sum_{j}^{J}\sum_{k}^{K}\sum_{l}^{L} \alpha_{ijkl} < m \tag{7}$$

The second type of constraint is the plus, which is equivalent to all factors taking one without considering different line conditions, so all the carbon emission factors of the lines must add up to the total number of types of carbon emission factors of the lines.

$$\sum_{i}^{I}\sum_{j}^{J}\sum_{k}^{K}\sum_{l}^{L} \alpha_{ijkl} = ijkl \tag{8}$$

The third type of constraint is the station type carbon emission constraint, according to the existing literature, the annual energy consumption of underground stations in the northern region is more than that of overground stations, so the annual carbon emission of underground stations should be greater than that of overground stations.

$$\sum_{j}^{J}\sum_{k}^{K}\sum_{l}^{L} \alpha_{1jkl} d_{1jkl} e \geq \sum_{j}^{J}\sum_{k}^{K}\sum_{l}^{L} \alpha_{2jkl} d_{2jkl} e \tag{9}$$

The fourth constraint is the carbon emission constraint of vehicle type. The carbon emission of type A car is usually greater than that of type B car because of its wide body and large passenger capacity.

$$\sum_{i}^{I}\sum_{k}^{K}\sum_{l}^{L} \alpha_{i1kl} d_{i1kl} e \geq \sum_{i}^{I}\sum_{k}^{K}\sum_{l}^{L} \alpha_{i2kl} d_{i2kl} e \tag{10}$$

The fifth constraint is the seasonal carbon emission constraint. In northern cities, the total carbon emission of stations in summer is higher than that in winter.

$$\sum_{i}^{I}\sum_{j}^{J}\sum_{l}^{L} \alpha_{ij1l} d_{ij1l} e \geq \sum_{i}^{I}\sum_{j}^{J}\sum_{l}^{L} \alpha_{ij2l} d_{ij2l} e \tag{11}$$

The sixth type of constraint is carbon emission constraint in period. The train frequency is high in peak hours, and the number of trains is much higher than that in low peak hours and rest days, so the train operation carbon emission is high.

$$\sum_{i}^{I}\sum_{j}^{J}\sum_{k}^{K}\alpha_{ijk1}d_{ijk1}e \geq \sum_{i}^{I}\sum_{j}^{J}\sum_{k}^{K}\alpha_{ijk3}d_{ijk3}e \geq \sum_{i}^{I}\sum_{j}^{J}\sum_{k}^{K}\alpha_{ijk2}d_{ijk2}e \quad (12)$$

4. Example analysis

This study does not consider the climate differences between different regions, so it mainly takes northern cities with hot summer and cold winter as the research object. Based on some subways in Beijing, Taiyuan and Qingdao, the corresponding data are selected for example analysis, and genetic algorithm is adopted to solve the problem.

Among them, the running kilometers data of train refer to the Annual Report of Beijing Transportation Development and other documents, and the sub-periods running kilometers data refer to the subway timetable combined with the train operation mileage. The value of m is 5.

Table 3 shows the calculation result of the carbon emission factors of the lines, and the objective function value is 0.0000000000, which meets the requirements.

Table 3. The calculation result of the carbon emission factors of the lines.

Season and period conditions / Line factors	Cold season - weekday peak hours	Cold season -weekday low peak hours	Cold season - all other days	Warm Season- weekday peak hours	Warm Season - weekday low peak hours	Warm Season - all other days
Overground station - type A	1.3860	0.7702	0.3960	0.7427	0.5677	0.4178
Overground station - type B	0.7770	1.2641	0.2924	0.7524	0.1091	0.0382
Underground station - type A	3.1066	1.1255	2.2264	2.5896	1.2384	1.9277
Underground station - type B	0.6667	0.5529	0.5549	1.0291	0.6399	0.8286

The line carbon emission factors provided in this study can enable urban rail operators to make fine calculations according to the characteristics of the lines under their control, and can also obtain different carbon emissions in different periods and seasons, which is convenient for enterprises to analyze their own emission behavior and formulate emission measures more targeted based on the analysis results, so as to better participate in carbon trading. The fine calculation method of urban rail operation enterprises is as follows.

$$CE = W \cdot EF_e \cdot \alpha \quad (13)$$

Where, CE is the carbon emission of the line, kg. α is the carbon emission factor of the line under the condition.

Next, Jinan Metro Line 2 is taken as an example for example analysis. It belongs to the situation of underground station - type B, and the climate of Jinan city is consistent with the model situation in this paper. Table 4 shows the carbon emission calculation results of Jinan Metro Line 2.

Table 4. Carbon emissions calculation results of Jinan Metro Line 2.

	Cold season - weekday peak hours	Cold season -weekday low peak hours	Cold season - all other days	Warm Season - weekday peak hours	Warm Season - weekday low peak hours	Warm Season - all other days
Carbon emission factors of the line	0.6667	0.5529	0.5549	1.0291	0.6399	0.8286
Fine caculation of total annual carbon emissions (kg)	7115037.50	836744.54	1180725.00	21858593.08	10192434.77	17519193.08

5. Conclusion

The paper first analyzes the factors that affect the energy consumption of urban rail transit, which mainly include train energy consumption and station energy consumption. Then, the fine calculation model of urban rail transit carbon emissions is constructed, and the line carbon emission factors under 16 conditions are distinguished and solved. The carbon emission of different lines and different periods of urban rail transit are calculated using the emission factor method. This corrected the defects of imprecise calculation caused by the use of national unified grid carbon emission factors. Finally, it estimates the carbon emissions of Jinan Metro Line 2 using the fine calculation method and provides fine calculation results for different periods' carbon emissions. The method proposed in this paper accurately evaluates the carbon emissions of urban rail transit and contributes to developing more scientific and reasonable strategies for reducing carbon emissions by operating enterprises.

References

1. Friedlingstein P, O'Sullivan M, Jones M W, et al. (2023) Global Carbon Budget Earth. Syst Sci Data, 15(12): 5301-69.
2. Ecology and Environment Bureau of Shenzhen Municipality (2021) Shenzhen low-carbon pu-blic travel carbon inclusive methodology. https://www.sz.gov.cn/attachment/0/927/927134/9442918.pdf. (in Chinese)
3. Wang Z, Chen F, Fujiyama T (2015) Carbon emission from urban passenger transportation in Beijing. Transportation Research Part D: Transport and Environment, 41(DEC.): 217-27.
4. Casals M, Gangolells M, Forcada N, Macarulla M, Giretti A (2014) A breakdown of energy consumption in an underground station. Energy and Buildings, 78: 89-97.
5. Guan B W, Liu X H, Zhang T, Wang X K (2023) Hourly energy consumption characteristics of metro rail transit: Train traction versus station operation. Energy and Built Environment, 4(5): 568-575.
6. Feng Z B, Chen W Y, Liu Y, Chen H Y and Skibniewski M J (2023) Long-term equilibrium relationship analysis and energy-saving measures of metro energy consumption and its influencing factors based on cointegration theory and an ARDL model. Energy, 263.

https://doi.org/10.1051/e3sconf/202452803006

Comparative analysis on the carbon emissions of traditional wooden fishing raft and HDPE fishing raft

Kai Xu[1], Jinyi Chen[1] and Xin Wu[1,2,*]

[1]Fuzhou University, Fuzhou, 350108, China
[2]Jiangsu Province Engineering Research Center of Construction Carbon Neutral Technology, Suzhou University of Science and Technology, Suzhou, 215011, China

Abstract. In the age of "global boiling", to settle the problem of climate change, the carbon peaking and carbon neutrality goals has been proposed by China to control carbon emissions. Fishing raft is a kind of platform used by fishermen for aquaculture. Presently, numerous studies on carbon emission concentrate on buildings on land. Few studies focus on fishing rafts' carbon emissions, which is significative to be analyzed. Therefore, this study calculates the carbon emissions of the main parts of two kinds of fishing rafts during the production and transportation stage of building materials to compare the difference in carbon emissions between the two. The results show that at the stage of building materials production and transportation, the carbon emissions of the main part of the wooden fishing raft mainly comes from the production of EPS floats. Additionally, the carbon emissions of HDPE fishing raft is significantly higher than that of the wooden one. Based on these results, this study proposes that the monitoring of the illegal production of EPS floats should be strengthened, and that the plastic with lower carbon emissions should be selected as the construction material of fishing rafts.

1. Introduction

2023 has broken many climate records and the world is warming up at an unprecedented rate, according to the Provisional State of the Global Climate 2023. In order to deal with the problem of climate change and achieve sustainable development, China has put forward the carbon peaking and carbon neutrality goals to control carbon emissions from all kinds of production and development activities. Fujian is a province with a well-developed aquaculture industry. According to the data from the Fujian Provincial Bureau of Oceans and Fisheries, in 2022, the total amount of aquatic products in Fujian Province was 8.614 million tons, of which the output of mariculture was 5.478 million tons; the unit output of mariculture reached 33 tons per hectare, ranking first in China. Fishing raft is a kind of platform used by fishermen for aquaculture, which integrates production and residence. Traditional wooden fishing raft (hereinafter referred to as wooden fishing raft) is mostly composed of wooden walkway boards and EPS floats [1]. The EPS floats of the wooden fishing rafts will gradually disintegrate in the process of using, thus forming a lot of floating garbage, causing serious pollution to the ocean [2-4], which is not conducive to the sustainable development of the marine fishery. As a result, the product is banned. The walkway boards and floats of the new plastic fishing raft (hereafter referred to as HDPE fishing raft) is made of high density polyethylene (HDPE). Compared with wooden fishing rafts, HDPE fishing rafts form much less floating garbage on the sea. In order to complete the upgrading of marine facilities, Fujian Province "14th Five-Year" Fishery Development Special Plan proposed the target of transforming 440,000 traditional fishing rafts into HDPE fishing rafts. At present, numerous studies on the carbon emissions of buildings focus on buildings on land. Few scholars have paid attention to the carbon emissions of fishing rafts. Therefore, this study calculates the carbon emissions of a wooden fishing raft and an HDPE fishing raft respectively, and compares the difference between the two, and gives the related improvement suggestion, which will help to enhance the carbon reduction benefits during the construction of fishing rafts.

(a)

*Corresponding author's e-mail: wuxin@fzu.edu.cn

 https://doi.org/10.1051/e3sconf/202452803006

(b)

Figure 1. Traditional wooden fishing raft (a) and new plastic fishing raft (b).

2. Review

2.1. Literature collation related to the nature of fishing rafts

The nature of the fishing raft is defined in a special way, as they are used for both productive and residential purposes. Hainan Lingshui boat-dwelling fishing rafts are a typical cluster of residential buildings on the sea [5]. Almost all of the residents' production and living are carried out on the fishing rafts. They seldom go to the land. The Uniform Standard for Civil Building Design (GB 50352-2019) defines civil building as "a general term for buildings for people's residence and public activities" [6]. Based on this definition and the characteristics of fishing rafts, fishing rafts can be regarded as civil buildings. The Standard for Calculating Carbon Emissions from Buildings (GB/T51336-2019) applies to the calculation of the carbon emissions of civil buildings [7]. Therefore, carbon emissions of fishing rafts can be calculated on the basis of the methods proposed in the standard.

2.2. Studies on the carbon emissions of buildings

2.2.1. Studies on the carbon emissions of the buildings on land.

Nuri Cihat Onat (2020) conducts a global review of carbon footprint studies on global construction industry using the Scopus database. The study shows that about 60% of the studies are focused on the Chinese construction industry, which suggests that the carbon footprint of China's construction industry has a significant impact on the low-carbon development of the global construction industry [8]. With the help of CiteSpace visual analysis function, Yijin Chen (2024) analyzes 214 literatures on building carbon emission in China during the period of 2009-2023, and the result shows that the studies mainly focus on low-carbon buildings, energy conservation, carbon reduction and building energy efficiency [9]. The carbon emission of prefabricated buildings is a study hotspot. Tao Yin (2022) considers the comparison of the stage emissions from construction stage, building materials production stage and transportation stage between the fabricated construction and the traditional cast-in-place construction.The study concludes that the carbon emissions of the fabricated construction is lower than that of the traditional cast-in-place construction [10]. Jia Chen (2024) proposes a life cycle carbon emission evaluation method for prefabricated buildings based on IOWA and vector pinch cosine in order to determine the sensitivity indicators affecting the life cycle carbon emissions of prefabricated buildings. The results of the study point out four sensitivity indicators of prefabricated building life cycle carbon emissions including the prefabricated building construction area, policy support, prefabricated building technology level and construction organization design rationality [11]. Most of these studies focus on the carbon emissions of the buildings on land, but less attention is paid to carbon emissions of buildings on the water, such as fishing rafts.

2.2.2. Studies on the relationship between green building evaluation system and green building carbon emission.

At present, the evaluation system of green building has many kinds. Kanhong Wang (2021) conducts a comparative study on green building evaluation standards at home and abroad, and points out that although the evaluation indexes of green buildings in various countries are based on energy saving, there is no clear carbon emission index of buildings [12]. Therefore, the green building performance and the carbon reduction performance is not necessarily a positive correlation. In terms of the ability to protect the water environment, an HDPE fishing raft is obviously better than a wooden fishing raft. But this does not mean that the carbon emissions of the former is lower than that of the latter. Achille-B. Laurent (2018) assesses the life cycle carbon footprint of complete non-residential building structures. The results show that structures using wood as a building material can significantly reduce greenhouse gas emissionse [13], which indicates that wood has good low-carbon benefits.

3. Methodology

This study selects Qida Village, Ankai Township, Lianjiang County, Fuzhou City, Fujian Province, a coastal fishing village where mariculture is the main industry, as the simulated construction site for the projects. On the basis of the whole life cycle carbon emission calculation model of the building, this study adopts the method of comparative analysis. The study simulates the construction of a wooden fishing raft and an HDPE fishing raft with the same specifications respectively, calculating their carbon emissions and comparing the diffenrence. Generally speaking, the whole life cycle carbon emission of a building is calculated in five stages: production of building materials, transportation of building materials, construction, operation and dismantling. Since the carbon emissions of the two types of fishing rafts are more or less the same in the operation, construction and dismantling stages, and the main difference between the two types of fishing rafts lies in their walkway boards and floats, this study mainly considers the carbon emissions of the walkway boards and floats of fishing rafts (hereafter the walkway boards and floats are regarded as the main part of the fishing rafts) in the production and transportation stage of the building materials. The following formula is used to calculate the carbon emissions:

$$C_{JC} = C_{SC} + C_{YS} \quad (1)$$

Among them:

C_{JC}: Carbon emissions from the production and transportation stage/kg

C_{SC}: Carbon emissions from the production process/kg

C_{YS}: Carbon emissions from the transportation process/kg

4. Data Processing and Results

According to the measured data, the cross-section size of a wooden walkway board is adopted 500mm×100mm, and the cross-section size of an HDPE walkway board is adopted 400mm×75mm. According to the calculation results of scholars, the carbon emission factor (EF) of general-purpose wood is 178kgCO_2e/m^3. According to the values specified in the Standard for Building Carbon Emission Calculation (GB/T51336-2019), the EF of HDPE is 2620kgCO_2e/t. Most wooden fishing rafts, when arranging walkway boards, usually do not place a whole long section of board directly in the linear direction of one walkway board, but rather, several sections of boards are fixed with nails in order to form a long section of walkway board (as can be seen in Figure 1), which can make the walkway board more resistant to fracture. Therefore, in this study, two boards with a cross sectional size of 250mm×100mm and a length of 5m are first fixed with nails to form a spliced board with a cross sectional size of 500mm×100mm and a length of 5m. Then five sections of spliced boards are joined together with nails to form a whole long section walkway board. Each of the jointed boards is overlapped in the linear direction by an area of 500mm in length, thus the length of the whole long walkway board is 23m. Twelve sections of such long walkway boards are arranged horizontally and vertically and each intersection is secured with nails to form the square overall frame of the raft. Therefore, the square grid shape of the raft has a side length of 23m, with each net cage having a side length of 4.6m. The HDPE raft has the same overall size as the wooden raft, except that the HDPE boards do not overlap in the linear direction of the whole long section of the walkway board. The material of the floats used in the wooden fishing raft is expanded polystyrene (EPS). According to the calculations of scholars, the EF of EPS is 7860kgCO_2e/t. In this study, 15 floats are arranged in the linear direction of each whole long section walkway board, so 180 of them are needed for one fishing raft. The float material used for the plastic fishing raft is still HDPE, and the number of units used is the same as the number of units used for EPS floats. Carbon emission data for the building material production process are shown in Table 1.

In terms of transportation, the fuel structure change of trucks in the last 5 years shows a trend of diesel-based and diversified dispersion [14], which shows that diesel trucks account for the main part of trucks in China. Wooden walkway boards, HDPE walkway boards and HDPE floats are selected from production sites within Lianjiang County. EPS floats are currently banned and several foam product factories in Lianjiang County that used to produce them have been cancelled, so it is assumed in this study that they are produced from one of the foam product factories that are currently cancelled and transported to the project site. The author learnt from a local timber factory that the weight of timber used to make fishing raft walkway boards is around 800kg per cubic metre, so the density of the timber is estimated to be 800kg/m^3 in this study. The author learnt from a local plastic fishing raft manufacturing factory that the average weight of HDPE walkway boards is 12kg/m, and that the average weight of an HDPE float is 12.5kg. Since the production of EPS floats have been banned locally, the author learnt from an online shop that the average weight of an EPS float is 6kg. The wood, HDPE walkway boards and HDPE floats are transported in medium-sized diesel trucks (8t capacity) depending on the weight of the shipment. The low density of the EPS floats may result in the trucks being loaded to capacity when the weight of the goods is still low, which can lead to inefficiencies when transported in 2t trucks. Therefore, the EPS floats are also transported in medium sized diesel trucks (8t capacity). Carbon emission data for the building material transportation process are shown in Table 2.

Table 1. Carbon emissions from the production process.

	Construction material	EF(production)	Material usage	carbon emission
Wooden fishing raft	General wood	178kgCO_2e/m^3	15m^3	2670kgCO_2e
	EPS	7860kgCO_2e/t	1.08t	8488.8kgCO_2e
HDPE fishing raft	HDPE(walkway boards)	2620kgCO_2e/t	3.312t	8677.44kgCO_2e
	HDPE(floats)	2620kgCO_2e/t	2.25t	5895kgCO_2e

Table 2. Carbon emissions from the transportation process.

	Construction material	EF(transportation)	Material usage	transportation range	carbon emission
Wooden fishing raft	General wood	0.179kgCO_2e/(t·km)	12t	45km	96.66kgCO_2e
	EPS	0.179kgCO_2e/(t·km)	1.08t	25km	4.833kgCO_2e

HDPE fishing raft	HDPE(walkway boards)	0.179kgCO_2e/(t·km)	3.312t	45km	26.678kgCO_2e
	HDPE(floats)	0.179kgCO_2e/(t·km)	2.25t	45km	18.124kgCO_2e

According to the data processing results of the two tables, during the building materials production and transportation stage, the carbon emissions of the main part of the wooden fishing raft mainly comes from the production of EPS floats; the main part of a wooden fishing raft of this size emits about 11.3t CO_2, while the main part of an HDPE fishing raft of the same size emits about 14.6t CO_2 during this stage. The latter emits about 30% more carbon than the former, which is an obvious difference.

5. Proposals

Although the production of EPS floats has been explicitly banned, in recent years the relevant authorities have still found manufacturers producing such floats. Therefore, the monitoring of the situation related to the illegal production of EPS floats should be stepped up.

Although the HDPE fishing raft is more conducive to the protection of the marine environment than the traditional wooden fishing raft, this study shows that the carbon emissions of its main part in the building materials production and transportation stage are higher than those of the wooden fishing raft. Guanyi Hou's team (2022) has collated the data of non-renewable energy consumption and CO_2 emissions of different plastics over their full life cycle [15]. From their study, it is known that HDPE has relatively high non-renewable energy consumption and CO_2 emissions. Therefore, it is recommended that the plastic with lower carbon emissions should be selected for the product development of the main part of the fishing raft, on the premise of maintaining the good wind and waves resistance and the low marine pollution of the fishing raft.

6. Conclusion

Fishing rafts are an important carrier of the time-honoured way of living on the water. Attention to and studies on fishing rafts are not only conducive to their own sustainable development, but also can promote the protection, inheritance and development of water habitat. This study uses the comparative analysis on carbon emissions between the wooden and HDPE fishing rafts constructed in Qida Village, Fuzhou, in the stage of the production and the transportation of building materials. The study reaches the following results with a view to providing more references and lessons for the improvement of water habitats and mitigating the greenhouse effect:

Firstly, in the building materials production and transportation stage, the carbon emissions of the main part of the wooden fishing raft are mainly from the production of EPS floats.

Secondly, in the building materials production and transportation stage, the carbon emissions from the construction of the HDPE fishing raft are significantly higher than that of the wooden fishing raft.

Based on these results, this study makes the following proposals:

Firstly, the monitoring of the situation related to the illegal production of EPS floats should be increased.

Secondly, on the premise of maintaining good wind and waves resistance of fishing rafts and low marine pollution, plastics with lower carbon emissions should be selected for product development of the main part of the fishing raft.

Acknowledgments

This paper was supported by the foundation of Jiangsu Province Engineering Research Center of Construction Carbon Neutral Technology, NO.JZTZH2022-0101. Many thanks for the experts took parts in this study.

References

1. Wang F, Wu C and Pan Y (2023) Design of a Kind of Tubular Plastic Aquaculture Fish Raft. Fujian Agricultural Machinery, 44: 13-16.
2. Wu C and Li X (2022) Design of Plastic Fishing Raft Net Box in Reservoir Area. Mechanical & Electrical Technology, 46: 55-57.
3. Mindong Daily (2023) Reform, Wind from the Sea. http://szb.ndpic.cn/html/2023-09/25/content_43107_16571090.htm.
4. Bian J (2022) Returning the Sea to a Clear Blue. https://fjnews.fjsen.com/wap/2022-07/18/content_31086667.htm.
5. Wang Y and Huang Y (2023) Culture and Tourism Ensemble Playing Folklore Songs. http://news.hndaily.cn/html/2023-10/09/content_58486_16597026.htm.
6. Ministry of Housing and Urban-Rural Development of the People's Republic of China (2019) The Uniform Standard for Civil Building Design. https://www.mohurd.gov.cn/gongkai/zhengce/zhengcefilelib/201905/20190530_240715.html.
7. Ministry of Housing and Urban-Rural Development of the People's Republic of China (2019) The Standard for Calculating Carbon Emissions from Buildings. https://www.mohurd.gov.cn/gongkai/zhengce/zhengcefilelib/201905/20190530_240723.html.
8. Onat N C and Kucukvar M (2022) Carbon footprint of construction industry: A global review and supply chain analysis. Renewable and Sustainable Energy Reviews, 124: 109783.
9. Chen Y and Liu Y (2024) Review of Domestic Research on Building Carbon Emissions based on CiteSpace.

Journal of Heilongjiang Vocational Institute of Ecol ogical Engineering, 37: 11-16.

10. Yin T and Mei C (2022) Comparative Analysis of Carbon Emission from Different Construction Modes in Construction Field. Construction Quality, 40: 24-27.
11. Chen J, Li M and Liu J (2024) Carbon emission evaluation in life cycle of prefabricated buildings based on IOWA operator an d Vector Angle Cosine. http://kns.cnki.net/kcms/detail/42.1675.T.20231127. 1758.002.html.
12. Wang K and Ma Y (2021) Research on green building evaluation system based on building carbon emission analysis. Energy Conservation, 40: 75-77.
13. Laurent A B, Van Der Meer Y and Villeneuve C (2018) Comparative life cycle carbon footprint of a Non-residential steel and wooden Building structures. https://www.researchgate.net/publication/329781689_Comparative_Life_Cycle_Carbon_Footprint_of_a_Non-Residential_Steel_and_Wooden_Building_Structures.
14. Kong J, Xing Z and Pei H (2023) Prediction of fuel development of freight vehicles in China under the carbon peaking constrain t. International Petroleum Economics, 31: 72-78.
15. Hou G, Weng Y, Diao X, Song X, Zhou Y and Fu Y (2022) The Current Development Situation and Future Development of Biodegradable Plastics Industry. Materials China, 41: 52-67.

New Paradigm for Business Model Design and Innovation: An Insight of Product Lifetime Value Based on Electric Vehicle

Zang Xiaohui[1,2], Raja Nazim Abdullah[2,*], Liu Lihua[3,4], Liu Jing[1]

[1]Liuzhou Vocational & Technical College, China
[2]Universiti Pendidikan Sultan Idris, Malaysia
[3]Guangxi University of Science and Technology, China
[4]Universiti Putra Malaysia, Malaysia

Abstract. The business model is an effective and well-structured management tool that helps organizations acquire and hold onto a competitive edge over their competitors. This paper proposes a framework called "The Product Lifetime Value Business Model" for business model design and innovation of electric vehicle company. The framework consists of four layers, the innermost layer represents the business model, followed by the Product Lifetime Value layer and 6V model, the outermost layer comprises 17 additional dimensions that provide a more detailed and in-depth analysis. The concept of Product Lifetime Value is proposed and can be categorized into six distinct stages: planning, production, delivery, revenue generation, application, and recycling in this paper. In addition, a "6V" model is also proposed based on electric vehicle industry which contains value proposition, value creation, value delivery, value capture, value of after-sales services, and value of residual.

1. Introduction

The business model is a valuable and structured management tool that supports companies in gaining and maintaining a competitive advantage over their rivals [1-2]. Although the term "business model" was introduced in a scholarly publication in 1957, it gained significant attention only in the late 1990s [3]. Since the early 2000s, it has experienced substantial growth and widespread adoption across various business and management disciplines [1, 4]. To stay ahead in the dynamic corporate landscape, companies must proactively adapt or transform their existing business models through innovation [5]. Recently, there has been an increasing focus on business model innovation within the field of business model analysis [6].

The electric vehicle industry developed quickly in recent years and has steadily permeated various sectors, including individual mobility, public transportation, and logistics [7]. Electric vehicles are different from traditional gas-powered vehicles. New innovations brought numerous fresh companies to the vehicle industry, but within a few years of their development, a multitude of them declared bankruptcy [7]. Moreover, the rise of electric vehicles posed a significant challenge to traditional automotive brands. Some faced bankruptcy or underwent transformations, while surviving companies had to pursue innovation in both products and business models to ensure long-term sustainability.

As widely acknowledged, the strategic interplay between business models and innovation plays a pivotal role in effectively commercializing new technologies. This symbiotic relationship is crucial for fostering the sustainable growth of electric vehicle companies. To design and innovate business models, an electric vehicle brand company should consider various aspects. This article aims to develop a comprehensive framework for business model design from the perspective of product lifetime value. The proposed framework not only provides practical guidance for electric vehicle brand companies in their business model design or innovation but also contributes to advancing theoretical research on business models and innovation.

Section 2 of the article furnishes theoretical groundwork pertaining to business models, innovation, and Product Lifetime Value. The proposed Framework is expounded upon in Section 3, with a detailed discussion of its constituent elements in Section 4. A succinct summary is presented in Section 5, accompanied by suggestions for prospective research directions.

2. Underlying Concepts

2.1. Business Model

Business model is a conceptual framework that can assist in strategically taking into account the specifics of how the business operates [8]. It outlines a series of actions designed to meet the demands of the end-user and explains how the operating company plans to generate

*Corresponding author's e-mail: rajanazim@fpe.upsi.edu.my

https://doi.org/10.1051/e3sconf/202452803007

revenue through the value-added component of the business [9-10].

As Magretta argued that:A good business model answers Peter Drucker's age-old questions: Who is the customer? And what does the customer value? It also answers the fundamental questions every manager must ask: How do we make money in this business? What is the underlying economic logic that explains how we can deliver value to customers at an appropriate cost [11]?

To be more precise and in line with academic rigor, a business model should provide satisfactory responses to the following questions: 1) What offerings does a company provide that are perceived as valuable by customers? 2) How does a company's system create these values? 3) How are these created values effectively communicated and transmitted to customers? 4) How can these generated values be captured as revenues for the company? 5) How are these created values distributed internally within the organization and externally among stakeholders.

Business models are critical to a company, and a good business model can fully utilize the company's strengths for better firm performance. As Sohl and partners argued that for two companies that provide the similar products in the same industry can adopt two different business models would have very different firm performance [12]. Hence, there is a key managerial question is that how do different business models affect firm performance. Actually, in order to maintain well performance, business model should be constantly evaluated critically from a dynamic viewpoint, keeping in consideration that internal or external changes over time may necessitate business model innovation or evolution [10].

2.2. Business Model Innovation

Business model can be viewed as a tool for creating a competitive advantage and improving performance of company [2]. Business model innovation (BMI) is considered as "designed, novel, and nontrivial changes to the key components of a company's business model or the framework connecting these components" [5, 13]. Hence, business model innovation research can be considered in terms of the process and content of innovation [5].

The research on business model innovation process has increased [5], include the antecedents, moderator, and outcomes [13-14]. Timo Böttcher and Jörg Weking identified six categories of antecedents and seven potential outcomes of business model innovation [15]. Evers et al. identified antecedents and outcomes of BMI in firm internationalization though systematic literature review [16]. It can be found that most researcher take business model innovation as a whole or main body to research.

However, there is a very critical issue in the actual management. It is difficult to carry out the overall business model innovation in a short period of time for an incumbent firm. They need to know how many aspects of the business model innovation. Then they can decide which one should be started first, and gradually derived to all aspects. Darek. M. Haftor and Ricardo Climent Costa give five dimensions of BMI for industrial incumbent firm through multi-case exploration, they are exchangeable, actor, activity, transaction mechanism and governance [5]. Some scholars proposed BMI with the components of value proposition innovation, value creation, and value delivery innovation and value capture innovation [17-19]. Therefore, researchers have not formed a unified view for the specific aspects or components of business model innovation.

2.3. Product Lifetime Value

The concept of Product Lifetime Value (PLV) encompasses both the duration of a product's existence and its inherent worth. While the definition of value remains consistent across academic research, it will not be reiterated in this article. However, it is important to note that the interpretation of product lifetime in this study different from the conventional understanding and will be expounded upon subsequently.

The concept of Product Lifetime, also known as Product Life Cycle (PLC), is a well-researched and established theoretical framework in the field [20]. The PLC encompasses the four distinct stages - introduction, growth, maturity, and decline - that characterize the evolution of a product throughout its manufacturing and sales processes [21]. During the introduction stage, a new product is launched into the market; subsequently, in the growth stage, markets embrace and accept it [21]. As it progresses to the mature stage, sales and profits reach their peak before entering a declining phase where sales gradually decrease [21]. Therefore, the PLC primarily focuses on marketing aspects as an extensive long-term model. However, it does not elucidate how enterprises derive benefits from value-added processes or explain specific enterprise-specific product life cycle value addition.

Life Cycle Assessment (LCA) serves as a valuable tool for analyzing the potential environmental impacts of products and processes from a comprehensive life cycle perspective [22], encompassing the entire journey from raw material utilization to system recycling or disposal [23]. Verma and partners contend that an electric vehicle's life cycle can be divided into three primary stages, one of which is referred to as "Cradle to grave" LCA, signifying the complete lifespan of an electric vehicle starting from raw material processing (cradle) until its ultimate disposal stage (grave) [23]. Consequently, when evaluating business models or fostering innovation in the realm of electric vehicles, adopting an LCA approach that encompasses the entire PLC - spanning from raw materials through recycling or disposal - proves highly beneficial.

PLV can be regarded as the comprehensive process of delineating a product's lifecycle, encompassing its inception, raw material processing, and eventual disposal phase, all from the perspective of value. This article categorizes PLV into six distinct stages: planning, production, delivery, revenue generation, application, and recycling.

3. A proposed Framework- Product Lifetime Value Business Model

Business models delineate the manner in which companies generate economic benefits for themselves by providing value to their consumers [24]. Lüdeke-Freund et al. have further categorized business models into four dimensions, namely: (1) value proposition encompassing products and services; (2) value delivery involving target customers and processes of delivering value; (3) value creation comprising partners, stakeholders, and processes of creating value; and (4) value capture entailing revenues and costs [24]. However, it is imperative to acknowledge that repair, maintenance, re-exchange, and reuse also significantly influence customers' choices within the automobile industry. Consequently, business models for electric vehicles should incorporate not only those dimensions but also consider the values associated with after-sales services as well as residual worth. This comprehensive approach can be referred to as the "6V" model, they are value proposition to customer, value creation, value delivery, value capture, value of after-sales services, and value of residual.

Therefore, the PLVBM framework is proposed based on the concept of the business model from a value flow perspective to analyse business models and drive innovation of electric vehicles. The model consists of four layers, which are progressively subdivided from the innermost to the outermost layer. The innermost layer represents the business model, followed by the PLV layer. The third layer is the 6V model, and the outermost layer comprises 17 additional dimensions that provide a more detailed and in-depth analysis based on the 6V model. The framework form can be expressed as the figure 1.

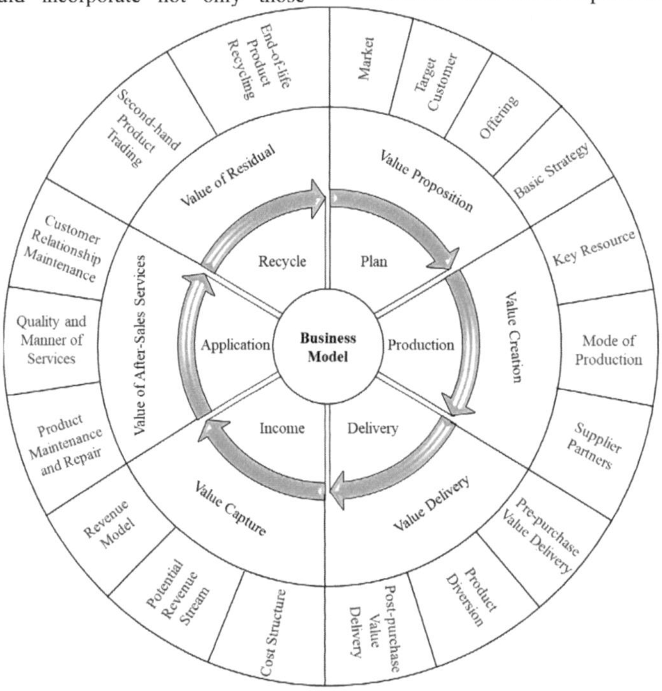

Figure 1. A Proposed Product Lifetime Value Business Model Framework.

Business model design and innovation of electric vehicle contain six aspects base on product lifetime value in this paper. It can be expressed as the formula 1.

$$\textit{Business model} = \textit{Plan} + \textit{Production} + \textit{Delivery} + \textit{Income} + \textit{Application} + \textit{Recycle} \quad (1)$$

Then, it can be further written base value perspective as the following formula 2.

$$\textit{Business model} = \textit{Value Proposition} + \textit{Value Creation} + \textit{Value Delivery} + \textit{Value Capture} + \textit{Value of After-Sales Services} + \textit{Value of Residual} \quad (2)$$

Both Formula 1 and Formula 2 indicate that electric vehicle company should consider the content involved six factors at the right of Formula when design and innovate its business model.

4. Discussion

4.1. Value Proposition to Customer

A value proposition refers to an ensemble of goods and services that are provided to customers, aimed at meeting their requirements and generating value to them, whether that value is symbolic or actual [25-26]. Value proposition to customer in PLVBM means what the firm will deliver to its customers, why customers will be willing to pay for it, and the firm's basic approach to competitive advantage [8], include: the offering, the target customer, and the basic strategy to win customers [8], as well as market.

Market in PLVBM means refers to market segment, represents a group of individuals or businesses who have

 https://doi.org/10.1051/e3sconf/202452803007

certain commonalities that differentiate them from other segments. Target customers are mainly come from market segment, who are most likely to find value in what the company provides. The offering in PLVBM is means what the company plans to offer to the target market and customer, such as offer product, or service, or product and service. Basic strategy in PLVBM refers to product or service price, quality, technology, brand, etc.

4.2. Value Creation

Value creation is frequently referred to as a critical component of the business model since it serves as the foundation for meeting customer interests [9]. It is the core of business model research, highlight the value generated for the company and its customers together with its underlying procedures and operations [25]. Value creation encompasses the method by which value is created both inside and outside, through collaboration with suppliers and consumers [27]. Value creation in PLVBM could answer the question as Thomas Clauss et al. said "What operational processes and resources will be necessary to create the services and products offered", mainly include key resource, mode of production, and supplier partners [28].

Key resources in PLVBM means are companies' critical assets, capabilities, and elements that are necessary for the company to operate, create value, and achieve its business objectives. Production mode refers to how company produce vehicles' main part such as body, battery, battery manage system, electric motor and so on, including self-made, cooperation, entrusting others, and others. Supplier partners refers to the cooperation with suppliers because there are lots of part in a vehicle and most of them are purchased by company. The suppliers are very important for an electric vehicle brand company.

4.3. Value Delivery

Value can be perceived in various forms, such as quality, price, convenience, innovation, and customer service. It is crucial to deliver value to customers. Customer value is the result of balancing the benefits and utilities a customer derives from a product (such as quality, advantages, worth) against any costs or losses incurred in acquiring and utilizing it [29].

Value delivery refers to the process of providing value to customers through the delivery of products, services, or solutions. It involves aligning business activities with customer needs and preferences to create a positive and memorable customer experience. Value delivery comprises the procedures and actions required to deliver the finished product or service [30].

Customer value is totally unrelated to price, which is a distinct advantage [31], especially industrial field. Good value delivery can enhance the cognition of consumers and promote the purchase of products or services by consumers. The customer's perception of value is not only at the delivery stage of a product or service, but throughout the entire stage of a company's operations. Hence, value delivery in PLVBM include: pre-purchase value delivery, product diversion, and post-purchase value delivery.

Pre-purchase value delivery refers to the processes and activities through which a company delivers its cultural and product values to potential consumers, enabling them to become aware of, recognize, and follow them in advance. Product diversion, like other scholars, encompasses the processes and activities required to deliver the final offering. Post-purchase value delivery means that the product and company values should continue to be delivered even after the consumer has made the purchase, thereby reinforced their decision, and enabled them to experience the additional value that the product brings.

4.4. Value Capture

Value capture, often known as monetization refers to the means, by which an organization gets paid by its customers and is able to keep operating commercially [32], explain how the organization getting its profit to cover the fees that have been paid [33]. Value capture, more specifically, is the process of securing revenues from value creation and distributing those gains across participating parties, including suppliers, clients, and partners [34].

Value capture in PLVBM include elements such as cost structure, potential revenue stream, and revenue model [28, 30]. Cost structure refers to the composition and breakdown of the costs that company incurs in its business operations, is importance for managers to assessed profitability, make pricing decisions, and other actions. Revenue stream refers to the various sources of income or revenue that a company generates as a result of its business activities, identify potential revenue stream is important for company's financial stability and growth. Revenue model describes how an organization makes money from its goods, services, and other offerings, put into another context, it describes how money enters the organization and should include strategies like direct sales, leasing, charging fees, or other revenue generated from value-added services in the electric vehicle industry.

4.5. Value of After-Sales Services

After-sales services, which are the actions provided by a manufacturer or dealer following the purchase of a product or service, with the objective of supporting customers during their utilization of the acquired product or service, are an essential component of any successful strategy to gain a competitive edge over peers [35]. The provision of after-sales services has grown in significance as a means of generating additional income [36]. In the automotive industry, customer emphasis is particularly pronounced on enhanced after-sales services as a determinant of value for money. It is widely acknowledged that after-sales services have evolved into a pivotal element of the marketing strategy for automobile companies, owing to the manifold benefits and enduring rewards they confer on the organization in both the immediate and extended time frames [35]. It is essential

https://doi.org/10.1051/e3sconf/202452803007

to change the after-sales business model paradigm in order to benefit from the additional revenue stream [37].

Value of After-Sales Services in PLVBM refers to the benefits and advantages that customers and companies gain from the services provided to customers after a product has been sold, include product maintenance and repair, quality and manner of services, and customer relationship maintenance. Product maintenance and repair refers to provide maintenance services, regular check-ups, and repairs to ensure the longevity and optimal performance of electric vehicles, plays an important role in after-sales services [35], has intrinsic value [38]. Quality and manner of services refers to the level of excellence, professionalism, and customer-centric approach that company demonstrates when delivering its services to customers, encompasses the overall experience that customers have while interacting with the company. Customer relationship maintenance refers to the ongoing efforts and strategies that company employs to nurture, sustain, and enhance its relationships with existing and potential customers, is crucial for building customer loyalty, maximizing customer lifetime value, and creating a positive brand perception.

4.6. Value of Residual

The value of residual for a vehicle is a projection of its future market value—indicating the anticipated price at which the user or another entity could sell the vehicle after a suitable period of use [39], it refers to its predicted worth at a certain point in the future [40]. Value of residual related to the depreciation of vehicle. The simplest way to understand depreciation is as the difference between a new vehicle's purchase price and its residual value after a certain period of time [41]. Given that the majority of new-car buyers do not retain ownership of their vehicles throughout the entirety of its lifetime, the depreciation of a vehicle emerges as a crucial factor influencing the overall cost of vehicle ownership and shaping consumer purchasing behavior [41].

For EV customers, they express concerns about diminished battery performance leading to a shorter electric range, the rapid evolution of technologies, and swift model upgrades—all of which contribute to the lower residual values associated with EVs [42]. Low residual values may be a critical market obstacle to the broad adoption of new EVs and should therefore be considered into the vehicle adoption evaluation to better represent the overall cost of vehicle ownership [42].

Value of residual in PLVBM refers to the anticipated value of an electric vehicle after a certain period of ownership, and be influenced by second-hand products trading and end-of-life product recycling. Second-hand products trading refers to the buying and selling of used or pre-owned EV in the marketplace, and involves transactions where individuals or businesses sell EV they no longer need or use to others who are interested in acquiring them at a lower cost compared to buying new items. End-of-life product recycling refers to the process of collecting, disassembling, and reprocessing EV that have reached the end of their usable life or have become obsolete, is an essential practice in promoting sustainability, resource conservation, and responsible waste management.

5. Conclusion

Most of article research on business model design and innovation focus on value proposition, value creation, value delivery, and value capture [18, 24]. However, there is limited research that considers business model design and innovation of after product sales. With the increasing consumer awareness of environmental protection and their growing emphasis on after-sales service quality, business model design and innovation after product sales have become increasingly important for companies, particularly in the electric vehicle industry. Therefore, this paper proposes a framework for business model design and innovation from the perspective of product lifetime value. This article contributes both theoretically and managerially.

5.1. Theoretical Contributions

The first contribution is proposed the conception of Product Lifetime Value (PLV). This article argue that Product Lifetime Value (PLV) is different from Product Life Cycle (PLC). Product Lifetime Value (PLV) can be seen as the process of defining a product from conception, raw material processing to the disposal phase from the perspective of value, and can be divided into six stages: plan, production, delivery, income, application, and recycle.

A second contribution of this article is considered the business model design and innovation after product sale in proposed framework. The business model design and innovation can be divided into 6 aspects from value chain perspective, contain value proposition, value creation, value delivery, value capture, value of after-sales services, and value of residual, this can be called "6V" model.

5.2. Managerial Implications

In addition to its theoretical contribution, this article has several important implications for companies in the electric vehicle industry regarding their business models and innovation.

First, for start-up electric vehicle brands, managers can utilize the Product Lifetime Value Business Model (PLVBM) as a tool to develop business models that are grounded in the concept of product lifetime value. This approach can help them create sustainable and customer-centric business models from the outset, considering the entire lifecycle of the product.

Second, for existing electric vehicle brands, managers can leverage the PLVBM framework to innovate their company's business model. By adopting this approach, they can identify opportunities to enhance their competitive advantage and improve their overall competitiveness in the market. This could involve rethinking value proposition, value creation, value delivery, and value capture strategies, with a particular focus on after-sales services and customer satisfaction.

Overall, the PLVBM framework provides valuable guidance for both start-ups and established companies in the electric vehicle industry, enabling them to develop and innovate their business models to thrive in a rapidly evolving market.

5.3. Future Research

There are several potential areas for future research based on the PLVBM.

First, researchers can analyze the business models of different enterprises across various industries using the PLVBM framework. This comparative analysis can shed light on how different industries approach and integrate the concept of product lifetime value into their business models. It can also provide insights into the effectiveness and applicability of the PLVBM across different contexts.

Second, empirical research can be conducted in different industries to validate and test the various elements and components of the PLVBM. This research can help to establish the validity and reliability of the framework and provide empirical evidence of its effectiveness in different industry settings.

Third, researchers can explore the relationship between business models based on the PLVBM and enterprise performance. This research can investigate how the adoption and implementation of the PLVBM framework impact key performance indicators such as profitability, market share, customer satisfaction, and sustainability. It can provide valuable insights into the potential benefits and outcomes of incorporating product lifetime value into business models.

By addressing these research areas, scholars can further advance our understanding of the PLVBM and its implications for business model design and innovation across industries.

Acknowledgments

This work was financially supported by the 2023 Annual project of Liuzhou Vocational & Technical College (2023SB01) (Research on Business Model Innovation of New Energy Vehicles from the Perspective of Value Chain).

References

1. Miller, K., McAdam, M., Spieth, P. and Brady, M (2021) Business models big and small: Review of conceptualisations and constructs and future directions for SME business model research. Journal of Business Research, 131: 619–626.
2. Teece, D. J (2010) Business Models, Business Strategy and Innovation. Long Range Planning, 43: 172–194.
3. Osterwalder, A., Pigneur, Y. and Tucci, C. L (2005) Clarifying Business Models: Origins, Present, and Future of the Concept. CAIS, 16.
4. Zott, C., Amit, R. and Massa, L (2011) The Business Model: Recent Developments and Future Research. Journal of Management, 37: 1019–1042.
5. Haftor, Darek. M. and Costa, R. C (2023) Five dimensions of business model innovation: A multi-case exploration of industrial incumbent firm's business model transformations. Journal of Business Research, 154: 113352.
6. Belussi, F., Orsi, L. and Savarese, M (2019) Mapping Business Model Research: A Document Bibliometric Analysis. Scandinavian Journal of Management, 35: 101048.
7. Ziegler, D. and Abdelkafi, N (2022) Business models for electric vehicles: Literature review and key insights. Journal of Cleaner Production, 330: 129803.
8. Richardson, J (2008) The business model: an integrative framework for strategy execution. Strategic Change, 17: 133–144.
9. Åström, J., Reim, W. and Parida, V (2022) Value creation and value capture for AI business model innovation: a three-phase process framework. Review of Managerial Science, 16: 2111–2133.
10. Wirtz, B. W., Pistoia, A., Ullrich, S. and Göttel, V (2016) Business Models: Origin, Development and Future Research Perspectives. Long Range Planning, 49: 36–54.
11. Magretta, J (2002) Why business models matter. Harvard Business School Boston, MA, USA.
12. Sohl, T., Vroom, G. and Fitza, M. A (2020) How Much Does Business Model Matter for Firm Performance? A Variance Decomposition Analysis. Academy of Management Discoveries, 6: 61–80.
13. Foss, N. J. and Saebi, T (2017) Fifteen Years of Research on Business Model Innovation: How Far Have We Come, and Where Should We Go? Journal of Management, 43: 200–227.
14. Zhang, H., Xiao, H., Wang, Y., Shareef, M. A., Akram, M. S. and Goraya, M. A. S (2021) An integration of antecedents and outcomes of business model innovation: A meta-analytic review. Journal of Business Research, 131: 803–814.
15. Böttcher, T. P. and Weking, J (2020) Identifying Antecedents and Outcomes of Digital Business Model Innovation. In ECIS.
16. Evers, N., Ojala, A., Sousa, C. M. P. and Criado-Rialp, A (2023) Unraveling business model innovation in firm internationalization: A systematic literature review and future research agenda. Journal of Business Research, 158: 113659.
17. Dai, Y., Wang, Y., Xu, B., Wu, Y. and Xian, J (2020) Research on image of enterprise after-sales service based on text sentiment analysis. International Journal of Computational Science and Engineering, 22: 346–354.
18. Pies, I. and Schultz, F. C (2023) The governance of sustainable business model innovation—An Ordonomic Approach. Scandinavian Journal of Management, 39: 101246.

19. Shakeel, J., Mardani, A., Chofreh, A. G., Goni, F. A. and Klemeš, J. J (2020) Anatomy of sustainable business model innovation. Journal of Cleaner Production, 261: 121201.
20. Mitchell, S.-L. and Clark, M (2019) Reconceptualising product life-cycle theory as stakeholder engagement with non-profit organisations. Journal of Marketing Management, 35: 13–39.
21. Zhang, Y. and Wei, F (2021) SMEs' charismatic leadership, product life cycle, environmental performance, and financial performance: A mediated moderation model. Journal of Cleaner Production, 306: 127147.
22. Chang, Y.-J., Neugebauer, S., Lehmann, A., Scheumann, R. and Finkbeiner, M (2017) Life Cycle Sustainability Assessment Approaches for Manufacturing. Sustainable Manufacturing, 221–237.
23. Verma, S., Dwivedi, G. and Verma, P (2022) Life cycle assessment of electric vehicles in comparison to combustion engine vehicles: A review. Materials Today: Proceedings, 49: 217–222.
24. Lüdeke-Freund, F., Gold, S. and Bocken, N. M (2019) A Review and Typology of Circular Economy Business Model Patterns. Journal of Industrial Ecology, 23: 36–61.
25. Freudenreich, B., Lüdeke-Freund, F. and Schaltegger, S (2020) A Stakeholder Theory Perspective on Business Models: Value Creation for Sustainability. Journal of Business Ethics, 166: 3–18.
26. Geissdoerfer, M., Pieroni, M. P., Pigosso, D. C. and Soufani, K (2020) Circular business models: a review. Journal of Cleaner Production, 277: 123741.
27. Clauss, T (2016) Measuring business model innovation: conceptualization, scale development, and proof of performance. R&D Management, 47: 385–403.
28. Thomas Clauss, Michael Abebe, Chanchai Tangpong, and Marianne Hock (2019) Strategic Agility, Business Model Innovation, and Firm Performance: An Empirical Investigation. IEEE Trans. Eng. Manage. 68: 767–784.
29. Bick, G., Brown, A. B. and Abratt, R (2004) Customer perceptions of the value delivered by retail banks in South Africa. International Journal of Bank Marketing, 22: 300–318.
30. Burström, T., Parida, V., Lahti, T. and Wincent, J (2021) AI-enabled business-model innovation and transformation in industrial ecosystems: A framework, model and outline for further research. Journal of Business Research, 127: 85–95.
31. Hinterhuber, A (2008) Value delivery and value-based pricing in industrial markets. Creating and managing superior customer value, 14:381-448.
32. Müller, J. M., Buliga, O. and Voigt, K.-I (2018) Fortune favors the prepared: How SMEs approach business model innovations in Industry 4.0. Technological Forecasting and Social Change, 132: 2–17.
33. Sinansari, P., Putri, A. A. and Lopatka, A (2022) Value of Sustainable Business Model in Mining Company: A Case Study. Procedia Computer Science, 207: 4142–4150.
34. David Sjödin, Vinit Parida, Marin Jovanovic, and Ivanka Visnjic (2020) Value Creation and Value Capture Alignment in Business Model Innovation: A Process View on Outcome-Based Business Models. Journal of Product Innovation Management, 37: 158–183.
35. Adusei, C. and Tweneboah-Koduah, I (2019) After-Sales Service and Customer Satisfaction in the Automobile Industry in an Emerging Economy. Open Access Library Journal, 6: 1–21.
36. Majava, J. and Isoherranen, V (2019) Business model evolution of customer care services. Journal of Industrial Engineering and Management (JIEM), 12: 1–12.
37. Rudnick, M., Riezebos, J., Powell, D. J. and Hauptvogel, A (2020) Effective after-sales services through the lean servitization canvas. International Journal of Lean Six Sigma, 11: 929–942.
38. Marais, K. B. and Saleh, J. H (2009) Beyond its cost, the value of maintenance: An analytical framework for capturing its net present value. Reliability Engineering & System Safety, 94: 644–657.
39. Wróblewski, P. and Lewicki, W (2021) A Method of Analyzing the Residual Values of Low-Emission Vehicles Based on a Selected Expert Method Taking into Account Stochastic Operational Parameters. Energies, 14: 6859.
40. Kim, M., Choi, J., Kim, J., Kim, W., Baek, Y., Bang, G., Son, K., Ryou, Y. and Kim, K.-E (2023) Trustworthy Residual Vehicle Value Prediction for Auto Finance. AI Magazine, 44: 394–405.
41. Rush, L., Zhou, Y. and Gohlke, D (2022) Vehicle Residual Value Analysis by Powertrain Type and Impacts on Total Cost of Ownership. Argonne National Lab. (ANL), Argonne, IL (United States).
42. Guo, Z. and Zhou, Y (2019) Residual value analysis of plug-in vehicles in the United States. Energy Policy, 125: 445–455.

Study on the predicament and sustainable development of the alumina smelting industry in China

Jingwu Yao[1], Huasha He[1,*], Xiaohui Liu[1,*]and Liwen Yao[2]

[1]Business school of Yulin Normal University,Yulin,Guangxi,537000,China
[2]Arts and Social Sciences School of Monash University Malaysia, Bandar Sunway, Selangor Darul Ehsan, 47500, Malaysia

Abstract: On the basis of a comprehensive review of the production and technology status of China's alumina industry, the main contradictions and challenges in the development of China's alumina industry in terms of resources, energy and environmental governance are analyzed, and the production cost is compared and analyzed worldwide. It is believed that the main factors restricting the sustainable development of alumina industry in China are the lack of sufficient high-quality bauxite resources in China, the production technology needs to be further optimized, and the generated solid and hazardous waste is difficult to realize harmless and resource utilization. In this regard, this paper puts forward the strategy for the sustainable development of China's alumina industry, that is, strictly control the disorderly and rapid expansion of production capacity and reasonable layout, optimize the resource and energy supply structure, implement the industrial development strategy of high quality, energy saving and low consumption to improve the core competitiveness, and accelerate the realization of the standard discharge and resource utilization of waste gas, wastewater and solid waste. The results of this paper are of great significance to the strategy of industrial structure adjustment and sustainable development of alumina industry in China.

1. Introduction

China's rapid economic development has promoted the strong demand for aluminum, which has strongly promoted the rapid development of China's alumina industry. Through the digestion and absorption of imported technologies and independent scientific and technological innovation, a pattern of common development of various forms of ownership has been achieved, a complete alumina industrial system has been formed, and China has become the world's largest aluminum production, consumption, import and export trading country. In 2022, China's alumina reached 81.862 million tons, accounting for 56.61% of the global alumina production. The technological innovation and scientific and technological progress of China's alumina oxide industry have also made remarkable achievements.

However, the huge scale of alumina industrial production has led to increasingly prominent resource, energy and environmental issues. Especially in recent years, due to the significant increase in domestic alumina production, environmental protection and other policies, domestic bauxite resources have been difficult to meet the demand, the proportion of imported bauxite resources has increased year by year, and the external dependence of bauxite has exceeded 60%[1]. The solid waste produced by alumina industry is increasing year by year, which has caused great pressure and impact on the ecological and environmental protection..

In view of the serious constraints on resources, energy and the environment, as well as the limits of the market development capacity, the development and survival of China's alumina industry has reached a very critical period. Carefully study the production and technology status of China's alumina oxide industry, analyze the main contradictions and challenges facing the development of China's alumina oxide industry, compare and analyze the production cost competitiveness on a global scale, and put forward the strategic development trend of China's alumina oxide industry production and technology. This is of great significance to the strategic research of industrial structure adjustment and high-quality sustainable development of China's alumina industry.

In this paper, the second part introduces the relevant research literature, the third part analyzes the production and resource status of China's alumina industry, the fourth part analyzes the technology and production cost composition of China's alumina industry, and the fifth part. The advantages and disadvantages of China's alumina industry are analyzed, the sixth part puts forward the relevant strategies and suggestions for the sustainable development of China's alumina smelting industry, and the seventh part puts forward the conclusion.

*Corresponding author's e-mail: 25861556@qq.com(H H), 70587999@qq.com (X L)

2. Research on the development of alumina smelting industry

2.1. Research on the alumina smelting technology

According to Yin Keting (1996), the production process of alumina in China is complex, the process is long, and the energy consumption is 1-3 times higher than that of foreign countries. Resource extraction is poorly managed and seriously wasted [2]. Liao Zhihua (2020) believes that the disorderly development of alumina industry will lead to overcapacity and vicious competition. The large proportion of bauxite imports and the high concentration of import countries will bring uncertain factors [3]. Ma Lifang (2019) believes that the transformation and management experience of cleaner production of alumina not only effectively solve the environmental protection problem, but also obtain better economic benefits [4]. Zhang Qingxuan (2018) believes that effectively reducing energy consumption in alumina production and realizing sustainable development of energy conservation and environmental protection are issues that alumina production enterprises must attach great importance to [5]. Li Zhuozhi (2017) proposed that the innovation of alumina smelting technology would play a positive role in improving red mud treatment technology and promote the healthy and sustainable development of China's economy [6]. Xue Shengguo et al. (2017) found that resource utilization and large-scale disposal of red mud in the solidification rate smelting process played a positive role in environmental risk prevention and control management and promoting the healthy development of alumina industry [7]. Yue Ling (2015) believes that China's alumina enterprises should actively explore overseas markets, develop and occupy overseas bauxite resources, and ensure the sustainable and stable development of China's alumina industry [8].

2.2. Research on the sustainable development of alumina industry.

Che Junwen et al. (2018) believe that the alumina industry is a resource, capital and technology-intensive raw material industry. Due to the large amount of tailings and red mud produced in the production process, a certain amount of alkali-containing wastewater is generated in the production process of alumina, resulting in very serious water pollution. Therefore, water pollution in the alumina industry has the most serious environmental impact [9]. Hao Haizheng (2019) believes that for the sustainable development of bauxite resources in Laos, along with the development of alumina industry, environmental protection has become the focus of the industry. Therefore, in the promotion of aluminum ore dressing technology, industry managers should recognize the value of ecological environmental protection and comprehensive utilization of resources. Implementation of green industrial development projects [10]. According to Xu Bin (2018), for resource-based enterprises, the upgrading of industrial pattern can improve the comprehensive utilization rate of resources and reduce production and manufacturing costs. Through systematic technological innovation and optimization, the comprehensive utilization rate of resources is constantly improved and the manufacturing cost of the whole industrial chain is reduced [11]. Zhou Faxing (2018) found that the most important measure to reduce energy consumption in alumina production is to update and reform production technology, seeking to meet the purpose of energy saving and consumption reduction while improving work efficiency. In addition, the quality of the raw materials used is strictly controlled, and the performance of the finished alumina products is committed to improving, so as to stand out in the same type of goods [12]. Wang Ning (2018) believes that alumina enterprises continue to improve their own production technology, reduce environmental pollution and excessive energy consumption, and achieve harmonious social and economic development [13].

Shao Ke (2023) believes that the improvement of resource utilization efficiency can not only reduce manufacturing costs and improve production efficiency, but also reduce the dependence on limited resources, reduce the impact on the environment, and promote the sustainable development of industrial economy [14]. Zhan Jianren (2023) believes that in the practice of industrial economic development, in order to further promote the economic development of the region, it is of significant value for regional development to think deeply about the sustainable development strategy of industrial economy under the new situation [15]. Huo Zhihong et al. (2022) believe that the standardization and mobility of low-carbon technology can promote the transfer and sharing of carbon reduction technology, and its upgrading process will promote the transformation of regional industrial economy to the low-carbon sustainable development path of "deindustrialization" and "modernization" [16].

2.3. Research on the influence of the new energy on the sustainable development of alumina industry.

Du Songqing (2019) believes that new energy plays an important role in structural transformation, and the development of new energy can drive regional economic development [17]. Jin Shuaiyu (2017) believes that China's energy development is faced with such problems as the increasingly exhausted exploitation of non-renewable resources, the slow development of renewable and clean energy, and the need to improve energy exploitation technology [18]. Xu Meiling et al. (2017) believe that under the environment of low-carbon economy, the research and development of new energy technologies effectively promotes the development of low-carbon economy [19]. Lin Yong (2023) believes that with the increasing market share of renewable energy, technology transfer has become the focus of attention of the international community [20]. Cao Li et al. (2017) believe that technology is the guarantee for the development of low-carbon economy, and the

 https://doi.org/10.1051/e3sconf/202452803008

development of new energy sources and research and development of new energy technologies are of great significance to promote the development of low-carbon economy [21].

3. Analysis of the production and resource status of alumina industry in China

3.1 Analysis of the production status of alumina industry in China.

Since 1954, China's alumina production began, but it was not until 2005 that China's alumina production increased rapidly, mainly because the high-temperature Bayer process technology for the treatment of diashite ore has been fully promoted and applied. See Figure.1.

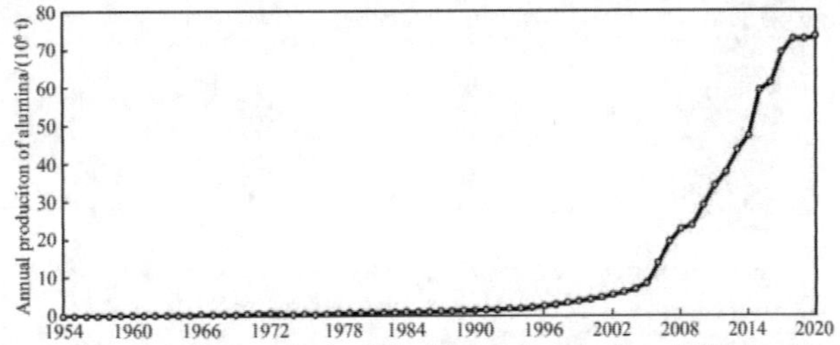

Figure 1. Changes in the China's annual production of alumina

Most alumina production enterprises are located in the regions of China with rich bauxite reserves, namely Shanxi, Henan, Guangxi and Guizhou. Alumina enterprises in Shandong Province, due to its proximity to the coast, mainly rely on imports of bauxite to produce alumina.

3.2. Analysis of the bauxite resources

According to the report on China's mineral resources released by the Ministry of Natural Resources in 2019 [22], in 2018, China's bauxite resources amounted to 5.17 billion tons, but its proven reserves were only 1 billion tons, accounting for only 3.3% of the world's reserves, which is far lower than Guinea, Australia, Brazil, Vietnam and Jamaica [23]. See Figure 2.

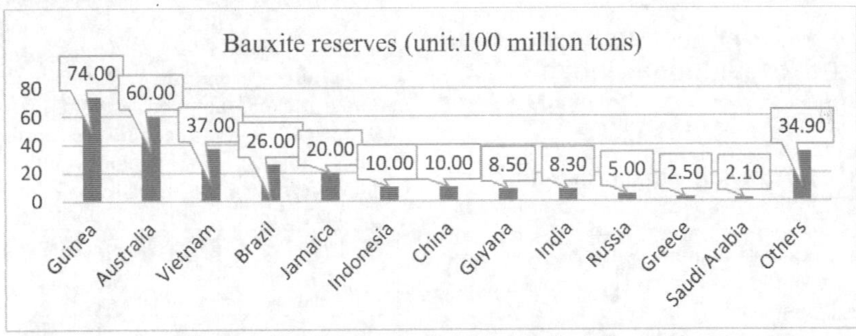

Figure 2. Bauxite reserves of major countries in the world

Therefore, in the world, China's bauxite reserves are relatively small, and the bauxite that can be economically utilized is all refractory diaspore ore, and the grade is low. Shanxi, Henan and other regions in northern China have a large number of large-scale aluminum oxide plants, and the distribution is concentrated. Due to decades of excessive mining, the local bauxite mining grade decline, serious depletion, and even depletion, the current supply of aluminum silicon ratio(A/S)is basically 3.5~5.5, resulting in high ore consumption and alkali consumption in alumina production and lack of competitiveness, has been difficult to support the local large-scale alumina industry. To this end, China's bauxite imports in recent years has shown an increasing trend year by year.

4. Analysis of the technology and cost of alumina industry in China

4.1. Analysis of the technical status of alumina industry in China

China's alumina industry has developed a series of major innovation achievements through decades of independent innovation in view of the resource characteristics of China's medium and low grade diaspore ore. Bayer process indirect heating enhanced dissolution technology and efficient enhanced Bayer process technology solve the key technical problems of using Bayer process to treat diaspore ore [24]; The Bayer process of mineral processing and the Bayer process of limehave solved the technical problems of economic treatment of medium and low grade diaspore ore to produce alumina [25].Through digestion, absorption and independent innovation,a series of large-scale energy-saving Bayer process key equipment has been developed, which has realized the large-scale and energy-saving of China's aluminum oxide industry.

4.2. Analysis of the alumina production cost ratio in China

Taking 2019 as an example, the comparison of alumina production and operation cost composition between China and other countries is shown in Figure 3.

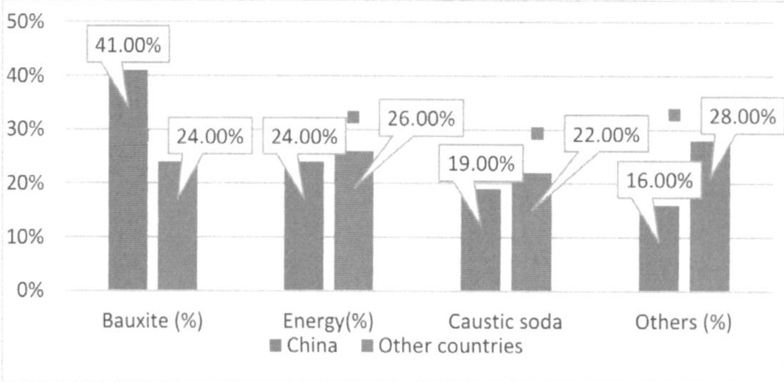

Figure 3.Comparison of alumina production cost distributions in 2019

It can be seen from Figure 3,The biggest difference in alumina production costs between China and other countries is the cost of bauxite. China's bauxite resources are poor and the ore price is high, so bauxite accounts for 41% of the production cost, while foreign countries only account for 24%. However, China's share of other alumina costs is relatively low, mainly because of China's lower labor and maintenance costs.

5. Analysis of the advantages and disadvantages of alumina industry in China.

5.1 Advantages of the alumina production in China

The advantages of China's alumina industry are: huge production capacity, buffer capacity to adjust production, lower prices of domestic mines in the south, a certain degree of competitiveness, lower labor and maintenance costs, coal and caustic soda prices are more reasonable.

It can be seen that China's alumina industry to improve the core competitiveness of the development of ideas lies in: as much as possible to exploit low-cost bauxite, bauxite resources in poor areas of the aluminum oxide plant transfer to the coastal areas to use high-quality imported ore; High efficiency and low consumption alumina production technology has been developed to achieve further energy saving and consumption reduction.

5.2 Disadvantages of the China's alumina industry.

5.2.1 Bauxite resources are in serious shortage.

China's overall bauxite resources lack, low grade, complex composition, difficult to meet the needs of the huge alumina industry, must import bauxite. According to Chinese customs data, in 2023, China imported a total of 141,565,900 tons of bauxite, a record high, an increase of 12.69% over 2022. A total of 99.256,800 tons of Guinean bauxite were imported in 2023, also a record high, an increase of 40.94% over 2022. In 2023, 34,6114 tons of Australian bauxite will be imported, an increase of 1.51% over 2022. Guinea and Australia's bauxite imports accounted for 94.56% of the total imports in 2023, occupying an absolute position. In 2023, China's total consumption of bauxite is 216.29 million tons, of which 80.24 million tons of domestic ore consumption, 136.05 million tons of imported ore consumption, imported ore consumption accounts for 62.90% of the total consumption. In addition, according to relevant experts, China's alumina smelting will demand about 220 million tons of bauxite in 2024, see Table 1.

Table 1. Projected demand for 2024(unit: 10000 tons)

Region(Province)	Imported ore	Domestic ore	total demand
Shandong	7700	0	7700
Shanxi	1770	3100	4870
Guangxi	1120	2280	3400
Henan	950	1290	2240
Guizhou	0	1200	1200
Hebei	1300	0	1300
Chongqing	1020	0	1020
Yunnan	0	320	320
Total	13860	8190	22050

It can be said that the resource guarantee problem of imported ore has become the soft underside of China's alumina industry. This has become a problem that China's alumina industry needs to pay great attention.

5.2.2. Alumina industry waste gas purification and standard discharge lack of the advanced technology.

The waste gas of alumina industry mainly includes the flue gas of alumina production, boiler flue gas and other gases containing a variety of pollution sources, among

https://doi.org/10.1051/e3sconf/202452803008

which the main pollutants are SO_2, NO_x, CO, CO_2, volatile organic compounds, etc., which cause harm to the environment and human body. For alumina industrial waste gas purification and standard emissions, China has no advanced technology, and there is still a certain gap compared with other countries in the world.

5.2.3 There is a lack of reasonable measures for the treatment of alumina industrial pollutants.

The pollutants in the alumina industry are mainly solid waste, of which the solid residue produced by the alumina industry is red mud, because the red mud contains 2 ~ 5g/L of attached alkali (in the attached water) and the mass fraction of 3% ~ 12% of the combined alkali (in the solid waste residue), causing pollution to the atmosphere, soil and groundwater. Because of the rapid increase of red mud accumulation in the storage yard and the danger of dam break, it is urgent to solve the problem of safe storage, harmless disposal and large-scale resource utilization of red mud. For the treatment of solid waste such as red mud, China has no corresponding advanced technology and lacks reasonable measures.

6. Strategies and suggestions for the sustainable development of alumina industry in China.

6.1 The disorderly and excessive expansion of alumina smelting capacity should be strictly controlled.

The scale of China's alumina industry is the world's first, the output accounts for more than 50% of the world's total output, the development to the present stage encountered serious resources, energy and environmental constraints, so we must strictly control the total scale of alumina production capacity, the focus of the work shifted to achieve high-quality operation, improve core competitiveness, promote energy conservation and emission reduction and realize resource recycling applications.

We must grasp the direction of capacity transfer. Alumina production capacity should be transferred from bauxite shortage or poor areas to coastal areas suitable for the construction of the aluminum oxide plants or foreign areas rich in bauxite resources, the use of foreign high-quality bauxite production of alumina, in order to reduce alumina production costs, or to energy rich and cheap areas to reduce smelting production costs and improve core competitiveness.

6.2 The poor bauxite resources need to be solved as soon as possible.

To solve the problem of shortage of bauxite resources in China as soon as possible, two approaches need to be taken, one is to speed up the development of efficient utilization technology of domestic complex components of bauxite, the other is to broaden the import channels of foreign high-quality bauxite and efficiently deal with it, or to encourage the construction of factories in the Belt and Road region rich in bauxite resources and energy and politically stable. That is, domestic and foreign resources together.

The domestic complex composition of the monohydrate duralumite mine needs to develop as soon as possible to economically treat the production of alumina technology, need to make a breakthrough in recent years, the focus is on the development of light burning desulfurization technology to treat high-sulfur bauxite, the development of the wet series technology to treat high-silicon bauxite, and try to produce alumina at a lower operating cost than the imported mine.

6.3 The problem of the unreasonable and short energy structure need to be Solved.

China's alumina industry needs to make full use of green energy such as abundant hydropower in southwest China, replace thermal power in northern China to produce alumina, and use coal and other resources as little as possible to reduce carbon dioxide emissions.

6.4 The problem of the deep purification and standard discharge of alumina industrial flue gas need to be solved.

The flue gas treatment of China's alumina industry mainly includes dust reduction, desulfurization and removal of the organic matter, which need to implement two methods: clean production and deep purification of emission flue gas. The technology is developed to strictly control the process conditions and reduce emissions in the production process, and for other harmful substances in the flue gas, the corresponding dust reduction and desulfurization technology is used to ensure that the exhaust gas is discharged to the standard.

6.5 The safe storage and resource reuse of the alumina industrial solid waste should be implemented.

It should be vigorously developed safe storage and large-scale low-cost resource utilization technology of red mud. Accelerate the development of technology that can use red mud in large quantities, and produce raw materials and auxiliary additives that can be used in building materials industry, metallurgical industry, chemical industry and environmental industry.

6.6 The deep recovery efficiency of the waste heat utilization in alumina production should be Increased.

Considering that in the aluminum oxide production process, the energy consumption of roasting accounts for about 25% to 30% of the comprehensive energy consumption of aluminum oxide, under normal circumstances, the flue gas temperature of the aluminum hydroxide baking furnace can reach up to 200°C, and the high-temperature baking furnace flue gas will inevitably take away part of the aluminum oxide. At the same time, due to the waste heat of baking furnace flue gas, a lot of heat and water loss, which has a very bad impact on the environment. Partial sensible heat of baking furnace flue gas is used to achieve deep recovery of the residual heat of flue gas. Direct heat exchange between low temperature water and baking furnace flue gas is adopted by spraying method, and the sensible heat of high temperature flue gas is directly transferred to low temperature water, and the latent heat of water steam is also guaranteed to enter the low temperature circulating water. The waste heat of roaster flue gas is used to pre-concentrate the evaporated liquid of the alumina plant, reduce the evaporation water of the evaporation station, thus reducing the evaporation steam consumption, and further achieve the purpose of reducing the energy consumption of alumina production.

7. Conclusion

According to the above analysis, the main challenges facing the development of China's alumina industry include excessive production capacity, unreasonable layout, poor quality of bauxite resources, energy conservation and consumption reduction still lag behind international advanced indicators, serious environmental pollution and lack of key treatment technologies.

In this regard, China's alumina smelting industry should strictly control the disorderly and rapid expansion of production capacity and reasonable layout, optimize the supply structure of resources and energy, implement high quality, energy saving, low consumption to improve core competitiveness, accelerate the realization of exhaust gas, wastewater and solid waste discharge and resource utilization, and realize the stable, healthy and green sustainable development of China's alumina industry.

Acknowledgments

This paper is the phased results of the following projects:

(1)Project name:Research on the coupling development of bauxite resource economy between China and Laos under the background of the Belt and Road Initiative; project principal: Jingwu Yao; Project number: G2020SK015.

(2)Project name: Research on the new context theory and the Development of the Cultural and Tourism Industry; project principal:Huasha He; Project number: G2019SK23.

References

1. Ma dongcan, Xu yamei, Guo chunqiao.Analysis on the current situation of using imported bauxite in domestic alumina plants. China Nonferrous Met, 2019(1):50-51.
2. Yin Keting. Current Situation, Problems and Development Countermeasures of Aluminum production in China [J]. Light Metals. 1996(12):9-14.
3. Liao Zhihua. Analysis on the influence of port alumina project on the development of aluminum industry in China [J]. Light Metals. 2020(05):1-4.
4. Ma Lifang.Process transformation and cleaner production of alumina industry [J]. Resource Conservation and Environmental Protection. 2019(11):28..
5. Zhang Qingxuan.Energy Consumption analysis and energy saving research of Alumina production [J]. World Nonferrous Metals. 2018(15):25+27.
6. Li Zhuozhi, LI Chen. Research on storage and harmless technology of red mud in alumina industry [J]. Technology and Market, 2017,24(12):37+40.
7. Xue Shengguo, Li Yubing, Guo Ying. Research progress on environmental impact of red mud in alumina industry [J]. Journal of University of Chinese Academy of Sciences, 2017,34(04):301-402.
8. Yue Ling. Opportunities and challenges of China's alumina industry under the new situation [J]. China High-tech Enterprises. 2015(02):7-8.
9. Che Junwen, Zhou Shenghe. Research on Environmental Audit of Alumina Industry in Economic activities [J]. Modern Marketing (Management Edition). 2018(02):47.
10. Hao Haizheng. Analysis on Promoting aluminum ore dressing Technology to Promote Alumina industry Development[J]. World Nonferrous Metals. 2019(24):206+208.
11. Xu Bin. Development strategy of bauxite industry in China under the new situation [J]. Light Metals. 2018(02):6-8.
12. .Zhou Faxing. Analysis of energy consumption in Alumina production [J]. World Nonferrous Metals. 2018(16):23+35.
13. Wang Ning. Development Status and Future trend of alumina production technology [J]. Metallurgy and Materials. 2018,38(04):90+92.
14. Shao Ke. Research on the relationship between the transformation and upgrading of manufacturing industry and the sustainable development of Industrial economy.] Shanghai Enterprises. 2023(09):36-38.
15. Zhan Jianren. Research on sustainable development countermeasures of industrial economy: A case study of Xiuyu District, Putian City, Fujian Province [J]. Small and medium-sized enterprise Management and technology. 2023(17):166-168.
16. Huo Zhihong, Huo Jingeng. Analysis on Sustainable Development of Industrial economy under Low-carbon Goal: A case study of Hebei Province [J]. China Market, 2022(35):67-69.

17. Du Songqing. Sustainable Development path of new energy in China [J]. Foreign Investment in China. 2019(13):80-81.
18. Jin Shuaiyu. Research on Energy Power Engineering and Sustainable Development of Energy [J]. China High-tech Zone. 2017(23):37.
19. Xu Meiling, Fan Dongliang, Zang Xiaoman. Development of new energy technology in low-carbon economy [J]. Chemical Engineering Design Bulletin, 2017,43(04):202.
20. Lin Yong. The strategic role of renewable energy technology transfer in energy transformation. Energy Conservation, 2019,42(03):87-89.
21. Cao Li, Zhang Jun.Analysis on the development of new energy technology under the environment of low-carbon economy [J]. China New Technology and New Products. 2017(06):112-113.
22. Ministryof Natural Resources,PRC.China Mineral Resources. Beijing: Geological Publishing House, 2019.
23. Lee Bray E. Minerals Yearbook 2017: Bauxite and Alumina (Advance Release). Virginia: U S Geological Survey, 2017.
24. Gu Songqing. Alumina production technology with high efficiency and low consumption from Chinese bauxite resource. Chin JNonferrous Met, 2004,14: 91-97.
25. Jiang Yuehua.Development of alumina production new process and new technology and its application. Light Met, 2007(2):11-13+21.

Analysis of various types of new energy storage revenue models in China

Lili Liu [1], Ying Zhang [2] and Yang Yu [3, *]

[1] China Energy Construction Group Liaoning Electric Power Survey and Design Institute Corporation, Shenyang, 110000, China
[2] China Power Engineering Consultant Group Northeast Electric Power Design Institute Corporation, Changchun, 130000, China
[3] State Grid Electric Power Research Institute Corporation, Nanjing, 210000, China

Abstract: In the current environment of China's vigorous development of energy storage, it is essential to carry out research on the benefits and economic evaluation of new energy storage. This paper establishes a framework for analyzing the revenue models of various types of energy storage under different scenarios. The framework complements the lack of previous studies on energy storage regulation under power generation systems such as wind power and coal power. In addition, a variety of scenarios were developed for the application of energy storage in the spot market, secondary service market, capacity market and user-side trading market. The results show that the case study energy storage plant has the highest revenue in the spot market, followed by the capacity market, and relatively low revenue in the secondary service market, while the leasing service can also bring a lot of revenue for the energy storage plant and thus become one of the more promising energy storage revenue models in the future.

1. Introduction

In 2017-2020, the power grid responded to the decision of the Energy Bureau and the Development and Reform Commission to reduce the rate of wind and light abandonment, and fully utilized the adaptive consumption of new energy resources in the power system resources, so as to reduce the rate of wind and light abandonment to 2%. At the same time, the pressure on the power grid is highlighted, some provinces began to require the power side configuration of energy storage. According to the national energy bureau statistics, from 2012 to 2022, the domestic new energy storage installed compound growth rate of up to 95%. 2022 domestic new energy storage installed capacity will reach 7.3GW, a year-on-year growth of nearly 200%. From the domestic energy storage installed type distribution, renewable energy distributed energy storage and independent energy storage installed proportion of 45% and 44%, respectively, the proportion of installed FM energy storage is only 1%, industrial and commercial household side energy storage installed proportion of 10%. China's current type of energy storage is still mostly used for supporting new energy installations in front of the meter energy storage installations.

Since 2021, the state has released a number of documents to promote the development of the energy storage industry, and the energy storage industry has thus ushered in a historic development opportunity. The Guiding Opinions proposed that by 2025, the installed capacity of clean energy storage will reach more than 30 million kilowatts, and at the same time, it is necessary to improve the incentive mechanism for new energy and energy storage projects, and to put forward policies to guarantee the grid scale consumption. The vast majority of provinces in China have strict construction requirements for energy storage facilities related to wind power and photovoltaic power stations. Under the stimulation of the mandatory energy storage policy, China's energy storage industry demand has surged and the industry is developing rapidly.

It can be seen that in the context of China's vigorous development of energy storage, it is crucial to strengthen the research on the benefits and economic evaluation of energy storage. Literature [1] proposes two kinds of revenue models and specific quantitative measurement methods for reducing power loss and participating in peaking auxiliary services. Literature [2] proposed the revenue models of "wind farm + energy storage" and "photovoltaic power station + energy storage", and used time series production simulation to calculate the power generation of "wind farm + energy storage" and "photovoltaic power station + energy storage", as well as analyzed the economy and operation mode of the new energy field station. Literature [2] proposed a modeling and evaluation method for the economic benefits of energy storage, taking into account the reduction of unit losses and deferred investment returns. Literature [3] proposed an energy storage capacity fixing method based on new energy side storage for grid peaking capacity. Literature [4] established an investment return analysis model to statistically analyze the cash flow of the whole life cycle of the PV energy storage system based on the investment of the PV energy storage system, the system power

*Corresponding author's e-mail: 1184034411@qq.com

generation, the enterprise electricity load curve, the electricity price and other revenue influencing factors. Literature [5] established a user multi-time demand response model and a comprehensive demand response model, and took the annual revenue and PV consumption rate as the optimization objective of collaborative energy scheduling of PV energy storage system. Literature [6,7] constructed a two-stage economic allocation model based on portfolio theory, which solved the problem of optimal allocation of user-side economic capacity, while considering the markets that user-side energy storage can participate in, the revenue, and the risks it may face. Literature [8] studied the current problems of high energy storage costs and unclear revenue models, respectively, and established an economic return model from the perspective of investors and comprehensive social returns, and capacity allocation with optimal economic indicators, and breakeven and sensitivity analysis of important projects affecting the economic benefits of energy storage.

In addition to conducting research related to the benefits of energy storage, specific studies addressing the functional role of energy storage in clean power generation systems and sorting out the various types of energy storage operation are also crucial. Literature [9] has studied and analyzed the application of lithium-ion batteries, flywheels, compressed air, and supercapacitors in gas-fired generating units. Literature [10] describes an energy management strategy that minimizes battery aging in electric vehicles through a combination of batteries and ultracapacitors. Literature [11] proposed a nonlinear coordinated power allocation strategy based on the state of charge (SOCsc) of supercapacitors. Literature [12] describes the functional role of energy storage in the power system and proposes an optimal allocation method of energy storage system (ESS) by combining with the actual demand of power grid..

Comprehensive the above literature, this paper analyzes the revenue model of various types of energy storage, and establishes the revenue model of different types of energy storage, selects the typical and reasonable basic data, and conducts specific measurements on the revenue of energy storage under different markets, so as to quantify the comprehensive revenue of various types of new energy storage in China, and to provide references for the future research on the economics of new energy storage and the formulation of development strategies.

2. The application of various types of energy storage scenarios and its revenue model analysis

2.1. Coal side distribution and storage and its revenue model

The growth of new coal-fired power projects nationwide is larger in 2022. 27.35 million kilowatts of installed coal-fired power generation capacity will be put into operation (including trial operation) throughout the year, mainly in Inner Mongolia, Shanxi, Anhui and Jiangxi provinces; 96.26 million kilowatts of installed coal-fired power generation capacity will get started, mainly in Inner Mongolia, Guangdong, Jiangsu and Hunan; the newly approved installed coal-fired power generation capacity will achieve 86.55 million kilowatts, 4.7 times of that in 2021, with Guangdong and Anhui being the provinces with the most newly approved Guangdong and Anhui are the provinces with the most newly approved capacity; 54.39 million kilowatts of in-stalled coal-fired power generation is in planning status, mainly in Hubei, Shaanxi and Jiangsu.

In 2022, the growth of new coal-fired power generation projects across the country will be large. In the whole year, 27.35 million kilowatts of coal-fired power generation capacity will be put into operation (including trial operation), mainly concentrated in Inner Mongolia, Shanxi, Anhui, Jiangxi and other provinces; 96.26 million kilowatts of new coal-fired power generation capacity will be built, mainly concentrated in Inner Mongolia, Guangdong, Jiangsu, Hunan and other provinces; the newly-approved installed capacity of coal-fired power generation will reach 86.55 million kilowatts, which will be 4.7 times of that of 2021, of which Guangdong and Anhui are the provinces with the most newly-approved installed capacity. Newly approved coal-fired power generation capacity will reach 86.55 million kilowatts, 4.7 times that of 2021, with Guangdong and Anhui being the provinces with the most newly approved installed capacity.

2.1.1.Benefit of saving unit spare capacity

Electricity loads vary at different times of the day, so coal power units need to take on a peaking role. The actual power generation of a coal power unit directly reflects the supply capacity of the peak load, while the lower flexibility of coal power units prevents them from reaching full capacity, which affects the economy of the overall unit operation. Energy storage can be used to reduce load peaks by charging during low load hours and discharging during peak hours. And the substitution effect of energy storage can release the capacity of coal power units, thus improving the utilization rate and economy of the units.

2.1.2.Benefits of load tracking to reduce unit creep

Load tracking is the dynamic adjustment of the slowly changing and continuously varying load to achieve real-time balance. The variable load can be subdivided into basic load and climbing load according to the actual situation of generator operation, and load tracking is mainly employed to climbing load. By adjusting the magnitude of energy storage output, the climbing rate of conventional energy units are able to be reduced, allowing them to transition to the dispatch command level as smoothly as possible. Load tracking has high requirements for discharge response time, requiring corresponding time at the minute level.

2.1.3. Benefits of participating in the frequency regulation market

Conventional units (thermal power and hydropower) usually respond to AGC signals to regulate the energy imbalance of the grid in a short period of time, but its volatility and randomness make the energy imbalance of the grid in a short period of time to aggravate, and the traditional energy is usually slower in frequency regulation, and there is a certain lag in responding to the instructions of the grid scheduling. Energy storage and fire power plants collectively participate in the AGC FM market offer, FM ancillary services market clearing in accordance with the "cost-effective priority, time priority, on-demand scheduling" principle, according to the market players FM service sorting price from low to high, until the winning party FM regulation rate total to meet the grid FM regulation of the winning party FM service maximum declared price as the FM service clearing price. The compensation for FM auxiliary services is measured and settled in units of units or energy storage facilities, and the compensation and shared costs are counted on a daily basis and settled on a monthly basis. The role of energy storage on the generation of each unit is shown in Table 1 below.

Table 1. The role of energy storage on power generation of each unit.

Unit Type	Climbing capacity of power generation equipment/ (%×min-1)	Short-time climbing capacity demand of the grid/ (MW×min-1)	Total power demand of the corresponding power generation equipment/MW	Energy Storage Power/MW	The influence of energy storage on the replacement of traditional power supply
Hydroelectric units	30	10	33.33	20	1.67
Gas units	20	10	50	20	2.50
Coal-fired units	2	10	5500	20	25.00

The regulation of the load component is generally AGC, which requires fast charging and discharging in a short time, and the use of electrochemical energy storage requires a large charging and discharging multiplier. The load component of the system frequency regulation changes in a cycle of minutes and seconds, requiring a high response speed, generally second response. FM energy storage is mainly installed on the thermal power plant side to assist the thermal power plant to provide secondary FM auxiliary services. When the thermal power units participating in the secondary FM are limited by the climbing rate and cannot accurately track the dispatching FM instruction, the AGC capability of the thermal power units can be improved through high-speed response energy storage, so as to obtain more AGC compensation benefits.

Thermal power energy storage equipment is typically 250 kW or more in power and generates revenue by increasing power generation or providing ancillary services. By adding electrochemical energy storage equipment, thermal power plants can improve the comprehensive FM performance indicators (measuring the FM frequency, FM time and FM accuracy) of the units, so that the units can undertake more FM tasks. The general configuration of FM service energy storage duration is only 0.5 hours, which reduces the investment in electrochemical energy storage, and energy storage power stations that entered the FM field earlier can recover the cost in about 2 years. Auxiliary service compensation funds mainly from the apportionment between power generation enterprises, the past annual national auxiliary service market size of 15 billion yuan, and AGC FM accounted for nearly 30%, about 4.5 billion yuan.

There are two joint energy storage FM plants in Shandong, namely Huadian International Laicheng Power Plant and Datang Linqing Thermal Power, equipped with 9MW/4.5MWh and 9MW/9MWh energy storage devices respectively. With the addition of energy storage, the combined value and regulation mileage of the thermal power units have increased significantly, which improves the FM compensation benefit of the thermal power units. After the commissioning of the energy storage power plant of Huadian International Laicheng Power Plant at the end of December 2021, the monthly FM compensation revenue of the power plant increased significantly, with an average monthly compensation revenue of RMB 7. 396 billion in January-June 2022; and after the commissioning of the energy storage power plant of Datang Linqing Thermal Power Plant in April 2022, the average monthly compensation revenue in May-June 2022 increased to RMB 5,233,000. From July 2021 to June 2022, the total monthly compensation in the FM auxiliary service market in Shandong Province increased from RMB26 million to RMB135 million per month.

2.2. Centralized scenery power side distribution and storage and its revenue model

At present, a number of domestic provinces require new energy enterprises to configure a certain percentage of energy storage, in order to improve the volatility of electricity, tracking power generation plans, peak shaving and valley filling, power storage to reduce the abandoned light and wind, while enhancing the new energy frequency regulation and voltage regulation capabilities.

And the energy storage system has a flexible four-quadrant operation capability, which can smooth the new energy output and maintain voltage stability, which is conducive to the friendly grid connection of new energy. At the same time, energy storage device and distributed photovoltaic system composed of user-side photovoltaic energy storage system, can be in the photovoltaic power generation peak period, storage of surplus on-line power, smoothing photovoltaic power generation fluctuations,

https://doi.org/10.1051/e3sconf/202452803009

and in the night time peak tariff hours to release the power, reduce the user peak tariff expenditures, increase the user's income, and its profit comes from the user's peak tariffs and Internet access to the difference in tariffs of the surplus power. Compared with wind power, photovoltaic and other intermittent energy sources, energy storage power generation costs are lower than thermal power, so it has a greater advantage in terms of operating income.

2.2.1. Cost analysis of scenery energy storage

The cost of wind and solar energy storage system is mostly composed of primary investment cost and O&M cost, while the primary investment cost is mainly made up of power cost and capacity cost, specific analysis of the cost components of the centralized wind and solar power side distribution and storage system are as follows:

(1) Investment cost

Assuming that the equipment replacement cost during the normal use period is not considered, the investment cost mainly contains battery equipment cost (mainly battery pack and battery pack management system) and non-battery equipment cost (including AC side transformer and circuit breaker rectifier/inverter system). The investment cost C_1 can be expressed as:

$$C_1 = \left(C_E + C_p\right) \times E \tag{1}$$

Where: C_E is the unit cost of battery facility, C_p is the unit cost of non-battery facility, E is the rated capacity.

(2) O&M cost

The annual O&M cost of is related to its operation condition, which includes the personnel salary, insurance, repair cost and other operation cost. The annual O&M cost C_2 can be expressed as:

$$C_2 = C_1 \times Q \times \left[1 + (n-1)i\right] \tag{2}$$

Where: Q denotes the O&M rate, n denotes the operating life, and i denotes the annual growth rate of O&M cost, which is generally 2%.

The cycle life of is another crucial parameter in its cost analysis. The cycle life is the number of thorough charge/discharge cycles of the battery when the nominal capacity is reduced to 80% of the initial rated capacity of the energy storage battery. The dominant factors that influence the cycle life include temperature extremes, overcharge and discharge, depth of charge and discharge (DOD) and charge and discharge rate. The cycle life is a function of the depth of charge and discharge when operating under benchmark charge and discharge current, voltage, and all kinds of condition. The life years n can be expressed as:

$$n = \frac{T_{life}}{L_{year}} \tag{3}$$

Where: T_{life} indicates the cycle life under the corresponding DOD; L_{year} indicates times of charge/discharge cycles per year.

Assuming that the equipment replacement cost during the normal service life is not considered, for the purpose of analysis, the linear method is used to cost share the total investment cost over its life time and superimpose it with the annual operation and maintenance cost to attain the annual value of the cost AC , which is calculated by the formula:

$$AC = \left(C_E + C_P\right) \times E \times \frac{1}{n} + C_1 \times Q \times \left[1 + (n-1)i\right] \tag{4}$$

2.2.2. Revenue analysis of scenery energy storage

The scenery energy storage system revenue mainly includes the revenue from wind and light abandonment, frequency regulation auxiliary services, the carbon market emission reduction, and the green power in the green certificate market.

(1) Abandoned wind and abandoned light revenue

Some provinces in China require new energy power plants to distribute an assured percentage of energy storage devices for diminishing the rate of wind and light abandonment. Take wind power station configuration energy storage as an example, wind power station can continue to generate electricity at night as the load is low, energy storage power station can be charged at low load, its charging cost can be regarded as zero, and discharged at peak load hours when the electricity price is higher. Therefore, the revenue generated to consume the abandoned wind and light R_1 can be expressed as follows:

$$R_1 = N \times E \times DOD \times P_d \tag{5}$$

Where: N is the number of days of operation of the energy storage plant, E is the rated capacity, DOD is the depth of discharge, P_d is the price of discharge.

(2) Frequency regulation auxiliary service revenue

The compensation method of FM auxiliary services in China's regional power grids largely refers to the U.S. PJM auxiliary services market. The revenue of FM service contains two parts: capacity compensation and mileage revenue, and the annual average revenue of FM service of energy storage power plant R_2 is:

$$R_2 = P \times \frac{t_{ct}}{t_{mt}} \times Y_c \times 12 + N \times \sum_{i=1}^{o} L_i Y_m K \tag{6}$$

Where: P denotes the rated power of the energy storage, t_{ct} denotes the time of the month when the energy storage is put into operation, t_{mt} denotes the time of the month, o represents the number of calls per day, L_i represents the FM mileage offered by the first call in the same direction as the command, K denotes the FM performance factor, Y_c denotes the unit capacity compensation of the monthly energy storage, and Y_m denotes the FM mileage compensation.

(3) Carbon market emission reduction revenue

After the configuration of the energy storage system, it'll supersede the conventional coal-fired units to provide electricity, reduce the emission of greenhouse gases and pollutants, attain carbon quotas, and participate in the carbon market to obtain emission reduction revenue. The annual emission reduction revenue obtained by the energy storage power plant participating in the carbon market R_3 is:

$$R_3 = N \times E \times DOD \times P_{CO_2} \times E_f \tag{7}$$

Where: P_{CO_2} is the carbon price in the carbon market and E_f is the carbon emission factor of coal power.

(4) Green certificate market green power revenue

China has been implementing green certificate trading since 2017, and scenic energy storage power plants can submit information and proof to the National Renewable Energy Management Center to apply for green certificates; then sell green certificates to other market players through the green certificate market to obtain green power revenue. The annual green power revenue obtained from the participation of energy storage power plants in the green certificate market R_4 is:

$$R_4 = \frac{N \times E \times DOD \times P_{ge}}{1000} \tag{8}$$

Where: P_{ge} is the green card price.

2.2.3. Economic analysis of scenic energy storage

The scenic energy storage system revenue mainly includes the revenue from wind and light abandonment, frequency regulation auxiliary services, the carbon market emission reduction, and the green power in the green certificate market.

(1) Basic data

The scenic energy storage set up in this section can be profited by participating in the power market, the FM accessory service market, the carbon trading market, and the green certificate market, with the following calculation conditions:

Energy storage capacity:

The installed capacity of photovoltaic and wind power at the site in 2023 is 22.06 million kilowatts. In this section of the economic calculation, the capacity of the configured energy storage system is 100MW/200MWh, the life span is set to 8 years, and the annual operating days are set to 350 days.

Spot market peak hour tariff:

The peak hours of the site are 9:00-11:00 and 15:00-17:00 every day. the spot market price change curve of a typical day is selected, and the average market price of the spot market peak hours is calculated to be 0.384 yuan/kWh, and the discharge price is taken as 0.384 yuan/kWh.

Construction cost of the energy storage system:

The energy storage system selected in this section uses lithium iron phosphate batteries that can be charged and discharged quickly, with high power and long life. The latest unit cost bid price of the energy storage system builder is between RMB 1600/kWh and RMB 2000/kWh, and the construction cost unit cost selected in this section is RMB 1800/kWh, of which: construction project cost is RMB 126/kWh, accounting for 7% of the static investment; battery equipment purchase cost is RMB 1404/kWh, accounting for 78% of the static investment; installation project cost is RMB 90/ kWh, accounting for 5% of the static investment; other costs are 180 RMB/kWh, accounting for 10% of the static investment.

O&M cost of energy storage system:

The O&M cost of different types of electrochemical energy storage equipment varies. The annual O&M cost of lithium-ion battery accounts for about 2% of its investment and construction cost. Among them, repair costs account for about 1.5% of its investment and construction costs, insurance costs account for about 0.25% of its investment and construction costs, and other operating costs account for about 0.25% of its investment and construction costs, all with an annual growth rate of 2%. The wage expense of the energy storage plant with rated capacity of 200MWh is set at 1.5 million yuan/year, and the annual growth rate is set at 5%.

Parameters affiliated to the intervention of energy storage systems in various markets:

In the auxiliary service market, energy storage mainly participates in AGC frequency adjustment service, assuming that its capacity compensation is 240 Yuan/MWh*month, and the compensation for calling mileage is 3 RMB/MW, assuming that it calls a full mileage within 15 minutes, which is 100MW, and the number of calls is 96 times per day. In the carbon market, the carbon price is set to RMB40/ton, and the carbon emission factor of coal power is 800g/kWh. In the green certificate market, the green certificate price is set to 175 Yuan/kWh.

The specific parameters involved in the scenic energy storage economics measurement are displayed in Table 2:

Table 2. Scenic energy storage economic analysis parameters.

Parameter	Numerical value	Unit	Parameter	Numerical value	Unit
Unit price of battery equipment	1404	Yuan/kWh	Rated power	100	MW
Unit price of non-battery equipment	396	Yuan/kWh	Capacity compensation	240	Yuan/MWh*month

Rated Capacity	200	MWh	Number of calls	96	times
Operation and maintenance rate	2	%	Single call mileage	100	MW
Discount rate	8	%	Mileage compensation	3	Yuan/MW
Annual growth rate of O&M costs	2	%	FM performance factor	1.8	
Maximum service life	8	Year	Carbon Price	40	Yuan/ton
Depth of charge and discharge	90	%	Carbon Emission Factor per kWh	800	g/kWh
Spot market peak hour tariff	0.384	Yuan/kWh	Green certificate unit price	175	Yuan/sheet

(2) Calculation example analysis

Based on the above parameters, the economic is measured. From the investor's perspective, as long as the capacity rate of the energy storage system is 1, the scenic energy storage gains from participating in the energy spot market to consume abandoned wind and light are 26.88 million yuan/year, the gains from participating in the FM auxiliary service market are 15.12 million yuan/year, the gains from participating in the carbon market are 2.016 million yuan/year, and the gains from participating in the green certificate market are 11.025 million yuan/year. The economic analysis of the whole life cycle is shown in Table 3:

Table 3. Scenic energy storage whole life cycle economic analysis (million yuan).

	Construction Period	Operation Period							
Year	1	2	3	4	5	6	7	8	9
Rated Capacity Rate	100%	100%	97.5%	95%	92.5%	90%	87.5%	85%	82.5%
Abandoned wind and abandoned light revenue		2419.2	2358.72	2298.24	2237.76	2177.28	2116.8	2056.32	1995.84
FM auxiliary service revenue		1814.4	1814.4	1814.4	1814.4	1814.4	1814.4	1814.4	1814.4
Carbon market emission reduction revenue		201.6	196.56	191.52	186.48	181.44	176.4	171.36	166.32
Emission reduction revenue from green certificate market		1102.5	1074.94	1047.38	1019.81	992.25	964.69	937.13	909.56
Operating income (tax included)		5537.7	5444.62	5351.54	5258.45	5165.37	5072.29	4979.21	4886.12
Operating income (excluding tax)		4900.62	4818.25	4735.87	4653.50	4571.12	4488.75	4406.38	4406.38
Government subsidies		2000	1800	1700	0	0	0	0	0
Cash inflow		6900.62	6618.25	6435.87	4653.50	4571.12	4488.75	4406.38	4406.38
Operating Costs		864.91	886.70	909.16	932.31	956.16	980.76	1006.11	1598.30
Value-added tax and surtax		0	0	0	0	0	0	454.60	584.98
Net cash flow before income tax	-36200	6035.71	5731.54	5526.71	3721.19	3614.96	3507.99	2945.66	8455.11
Accumulated net cash flows before income taxes	-36200	-30164.29	-24432.74	-18906.04	-15184.85	-11569.88	-8061.89	-5116.23	3338.88
Adjustment for income taxes	0	239.25	210.50	181.59	152.51	123.25	93.80	64.16	0
Net cash flow after income tax	-36200	5796.46	5521.04	5345.12	3568.68	3491.71	3414.19	2881.49	8455.11
Accumulated net cash flow after income tax	-36200	-30403.54	-24882.50	-19537.38	-15968.70	-12476.99	-9062.80	-6181.30	2273.81

In the economic evaluation, the internal rate of return, payback period, total investment return, and net capital profit margin are selected as the economic effect evaluation indexes. The economic effect evaluation is displayed in Table 4:

Table 4. Evaluation of the economic effect of scenic energy storage.

Economic evaluation indicators	**Numerical value**	**Economic evaluation indicators**	**Numerical value**
IRR (before tax)	2.03%	IRR (after tax)	1.38%
Payback period (before tax)	8.61 Years	Payback period (after tax)	8.73 Years
Total Return on Investment	1.30%	Net Capital Profit Margin	0.59%

As can be seen from Table 4, the scenic energy storage project has poor economics and the internal rate of return is lower than the standard discount rate, but the capital recovery can be completed within the operation period.

Further, the peak tariff of power spot market, mileage compensation unit price of AGC FM, carbon price and green certificate price are selected as the sensitivity analysis factors, and the after-tax internal rate of return is considered as the analysis index to carry out the sensitivity analysis of scenic energy storage, and the results are shown in Table 5 below:

Table 5. Sensitivity analysis table for scenic energy storage.

Uncertainties	**Rate of change (%)**	**After-tax internal rate of return (%)**	**Change in internal rate of return (%)**	**Sensitivity factor**
Basic Plan	0	1.38		

 https://doi.org/10.1051/e3sconf/202452803009

Peak Tariff	-10	0.81	-41.30	4.13
	-5	1.09	-21.01	
	5	1.66	21.01	
	10	1.95	41.30	
Mileage Compensation	-10	0.91	-34.06	3.41
	-5	1.14	-17.39	
	5	1.61	17.39	
	10	1.84	34.06	
Carbon Price	-10	1.33	-3.62	0.36
	-5	1.35	-2.17	
	5	1.40	2.17	
	10	1.43	3.62	
Green Card Price	-10	1.12	-18.84	1.88
	-5	1.25	-9.42	
	5	1.51	9.42	
	10	1.64	18.84	

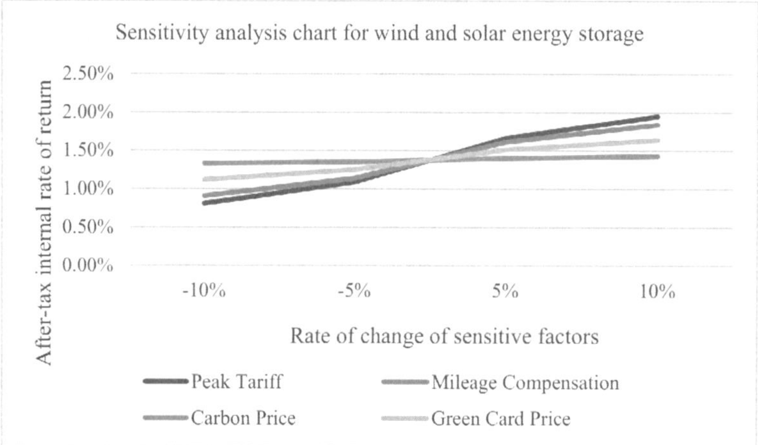

Figure 1. Sensitivity analysis chart for wind and solar energy storage.

As can be seen from Figure 1 above, The peak tariff and the unit price of AGC FM mileage compensation in the power spot market are sensitive factors for the economics and should be focused on, while the impact of carbon price and green certificate price on the economics of scenic energy storage is relatively small.

2.3. Independent shared energy storage and its revenue model

Shared energy storage is invested, operated and maintained by a third party or manufacturer, and the power and capacity of the storage system is leased to the target user, and the lessee is charged rent referring to the principle of "who benefits and who pays". Before 2015, the total cost of auxiliary services in China accounted for less than 1.5% of the total electricity bill, the current ratio has risen to 2.5%, and the power auxiliary services cost ratio will continue to rise.

In 2022, there are already 11 provinces with new energy storage power plants participating in the peaking accessory service market, with an installed capacity of 1.123 million kW, trading 170 million kilowatt hours of electricity, and an average clearing price of 0.42 yuan/kWh. 2025 and 2030, the whole society's electricity consumption is expected to reach 9.5 trillion kilowatt hours and 11 trillion kilowatt hours, and the national average sales price of 0.6 yuan/kW is calculated. Auxiliary service fee accounts for 3% of the cost of electricity consumption of the whole society, it is estimated that the auxiliary service market size will reach 171 billion yuan, 198 billion yuan.

Independent energy storage revenue model can be roughly divided into: shared leasing, spot arbitrage, auxiliary services, capacity tariff.

2.3.1. Electric energy spot market revenue

According to the national daily maximum load of 1.1 billion kilowatts, the minimum load of 800 million kilowatts calculation, the peak and valley difference is only 300 million kilowatts, if 150 million kilowatts of which will be filled with energy storage will lead to a decline in the peak and valley difference, the enterprise earns the price difference will also decline. Some provinces in the spot market construction of the initial period, the ceiling price control is more strict, although this move is conducive to ensuring the smooth operation of the market, but if the ceiling price is obviously low, will not be able to play the role of investment incentives for energy storage. 2022 In May, the General Office of the National Development and Reform Commission (NDRC)

https://doi.org/10.1051/e3sconf/202452803009

and the Comprehensive Department of the National Energy Administration (NEA) issued the Notice on Further Promoting the Participation of New Energy Storage in Electricity Markets and Dispatch Utilization, mentioning that the new type of energy storage can participate in the electricity market as an independent storage, and that an independent power station will send electricity to the grid, and its compliant charging power will not bear the transmission and distribution tariffs and governmental funds, and other additional costs, and the electricity tariff will be reduced by about 0.1-0.2 yuan/kWh.

In the case where energy storage participates in the spot market for peak-valley arbitrage, based on a sample of 67 consecutive days of spot trading in Shandong Province in 2022, it can be seen that the maximum peak-valley spread in the spot market spans a wide range, with a minimum value of less than RMB 100/MWh and a maximum value of more than RMB 1,500/MWh, with an average value of RMB 674/MWh. High electricity prices of more than RMB 1,000/MWh are characterized by a typical sharp spike, with a duration of less than a quarter to half an hour. The duration is less than a quarter hour to half an hour, while the duration of the high tariff period that reaches the hourly level corresponds to a tariff range of RMB 600/MWh, so the configuration of hourly peaking energy storage plants cannot be effectively discharged during the highest tariff period, which will greatly cut down on the revenue of new energy storage. The energy storage power station does not need to bear transmission and distribution tariffs and governmental funds and additional costs, can reduce the charging cost, and further expand the single energy storage intervene in the spot market revenue. With charging tariff of 0.1 yuan / (kWh), discharge tariff of 0.5 yuan / kWh, capacity charge of 0.02 yuan / kWh, charging and discharging efficiency of 85%, the annual number of 360 times of charging and discharging as the basis for calculating the 100MW/200MWh standalone energy storage power station annual spot market revenue can reach 21.96 million yuan.

Shandong power spot market as an example, Shandong May 1, 15 real-time power spot trading price is negative, May 2, 17 real-time power spot trading price is negative, two days within 48 hours of the real-time spot price of negative time period reached 32 hours. 20:00 May 1 to May 2, 17:00, the negative price lasted for 22 hours. May 1 all-day real-time power spot trading price, the average price is - 0.01302 yuan/kWh. For energy storage, the negative price means a significant reduction in charging costs, but ultimately whether it can be profitable, but also depends on the price difference. From the point of view of time-sharing tariffs, the highest price difference on May 1, 2022 was 0.42609 yuan/kWh, and the highest price difference on May 2 was 0.58408 yuan/kWh, and the price difference did not change significantly. Negative tariffs appear frequently, to the new energy generation has brought the impact, but to the photovoltaic + energy storage, distributed photovoltaic + energy storage to bring opportunities for development. 2022 Shanxi spot market peak and valley spread average of about 0.6 yuan / kwh, to 100,000 kilowatts of independent energy storage project for calculation, if 4 hours to recharge and discharge once, recharge and discharge once can earn 240,000 yuan. In addition to the maintenance time, a year by 250 days of operation days, a year can be 60 million yuan of spot market revenue.

2.3.2. Auxiliary service market revenue

According to the existing mechanisms at home and abroad, the electricity market is generally categorized into the energy market and the secondary services market. In the FM market, energy storage power stations must bid on the same platform as generating units. Common new energy storage auxiliary services in the form of new energy storage is mainly peak and frequency regulation (primary FM, secondary FM), the specific amount of revenue varies from province to province, mostly in the form of compensation according to the peak power charge, the price ranges from 0.15 yuan / kWh (Shandong) to 0.8 yuan / kWh (Ningxia). In Guangdong, for example, 3% of the peak load size is distributed over a single duration of no more than 4 hours, with no more than 30 occurrences throughout the year and a cumulative duration of no more than 50 hours. FM is mostly based on FM mileage compensation, and the price is RMB 0.1-15/MW according to the degree of the unit's response to AGC's FM instruction. some provinces have adopted a market-oriented approach to FM auxiliary service offerings and cost sharing. The FM market in Shandong, Shanxi, Fujian, Guangdong and other provinces has entered into operation or trial run. At present, the installed capacity of new energy in Shanxi has reached 40 million kilowatts, and the demand for FM can reach 800-1000 times per day, with a demand mileage of about 120 MW per FM. According to the existing FM market offer, if energy storage participates in the FM market, it can gain about 5.5 million yuan. In addition to one FM, energy storage also has spot market gains.

2.3.3. Capacity market revenue

There is great uncertainty in the utilization hours of energy storage, and it is difficult to maintain its economy by relying on electricity prices alone, so capacity tariffs are needed to stabilize the overall economy of energy storage.2022 In March 2012, Shandong Province's Notice on Capacity Compensation Tariffs for the Electricity Spot Market proposed that, before the operation of the capacity market, the cost of capacity compensation for the generating units participating in the electricity spot market would be collected from the user side, with a tariff standard of per kW 0.0991 yuan (including tax) per kW. The policy has been implemented since May 1, 2022, and is valid until December 31, 2026. Capacity tariff compensation is available for energy storage. It is estimated that the annual capacity compensation revenue of 100MW/200MWh independent energy storage power station can reach RMB43.41 million (calculated on 365 days).

2.3.4. Leasing or sharing of electricity savings proceeds

Revenues are shared between the energy storage developer and owner, with the customer paying a percentage or a certain amount of the energy savings to the owner of the storage asset. Typically, the lease payment is based on the developer's fixed investment costs and thus the amount is essentially fixed, whereas the shared revenue model is usually based on a percentage of the revenue that varies based on how much electricity is saved each month, and so it varies in amount. The shared revenue savings model is less common among individual residential customers and is mostly applied to commercial and industrial customers. When the shared revenue model is used on an individual basis, the contract term is typically 10 years or more. The leasing model, on the other hand, generally has a short-term trend, with a contract period of 2 to 4 years, and also supports monthly leases, so it is more flexible for users.

For example, a capacity of 100MW/200MWh independent energy storage power station, according to the investment of 420 million yuan, 30% of the project capital, internal rate of return of 8% of the indicators for calculation, the annual income needs to reach 64 million yuan to recover the cost; and independent energy storage power station can be leased to the capacity of wind power, photovoltaic enterprises, wind and photovoltaic power enterprises to obtain access to the indicators, so that the enterprise still owns the system of independent Therefore, the enterprise still owns the independent operation right and income right of the system. Independent energy storage capacity leasing costs using market bidding, set the maximum price and minimum guaranteed price, the general lease transaction price of about 330 yuan / kW. At present, the capacity leasing rate of independent energy storage power stations in Shandong Province is only about 20%. If all the capacity can be leased, the annual capacity leasing income of 100MW/200MWh independent energy storage power station can reach RMB 33 million, so the leasing transaction mode is also a more profitable option for independent energy storage power station.

Independent energy storage power plants can simultaneously carry out three modes of shared leasing, spot arbitrage and capacity tariff compensation. For example, for a 100MW/200MWh stand-alone energy storage plant in Shandong, it is estimated that the annual income from spot arbitrage will be about 20 million yuan, the income from shared leasing will be about 30 million yuan, and the income from capacity tariff will be about 6 million yuan. Based on a total investment of about 450 million yuan and a financing cost of 4.65%, the project can realize a return on capital of more than 8%.

In addition, Qinghai, Ningxia, Hubei, Shaanxi and Shanxi have successively introduced policies favorable to the application of shared energy storage power stations. The compensation standard for energy storage peaking in Qinghai Province is 0.5 yuan/kWh, and the number of annual utilization hours is not less than 540 hours; the compensation price for peaking service of the pilot energy storage project in Ningxia in 2022 and 2023 is 0.8 yuan/kWh, and the number of annual call-ups is not less than 300; Hubei and Shaanxi have committed that the energy storage leasing can be regarded as a new energy storage quota; Shanxi has made it clear that the shared energy storage power station can participate in the market of peak shifting and frequency regulation . Many of the above policies have also brought a strong development boost for shared energy storage power plants.

3. Conclusion

Considering the current development situation of new energy storage systems in China and the lack of research on revenue models, this study takes China's new energy storage systems as the main research object, practically analyzes the revenue models of multiple types of energy storage systems, and based on a large amount of reliable actual data, measures the revenues of energy storage systems under different markets. Through the above research we can get the following conclusions:

(1) Under the electricity spot market, the peaking tariff and the unit price of AGC FM mileage compensation are sensitive factors affecting the economics of energy storage systems, which should be focused on in future related studies, while the carbon price and the price of green certificates have a relatively small impact on the economics of new types of energy storage.

(2) The results show that energy storage plants have the highest returns in the spot market, followed by the capacity market, and relatively low returns in the ancillary services market, while leasing services can also bring a lot of revenue for energy storage plants, becoming a more promising one of the future revenue models for energy storage plants.

References

1. Liu Junhui, Yang Meng, Zhao Yang, Bai Hongkun, Zhao Wenjie and Zhang Yihan; the revenue calculation and economic analysis of new energy stations. Henan Electric Power 2022:8993. https://doi.org/10.19755/j.cnki.hnep.2022.s1.033.
2. Zhu Qing, Shen Chenshu, Li Muzi and Liu Kai; Research on the configuration and revenue model of large-scale centralized wind power energy storage system. Electrical technology 2022:48. https://doi.org/10.19768/j.cnki.dgjs.2022.17.002.
3. Chen Yong, Yang Biao, Li Zhibin and Long Tao; Research on the investment income analysis model of distributed photovoltaic energy storage system. China New technologies and New Products 2022:13941. https://doi.org/10.13612/j.cnki.cntp.2022.12.004.
4. Liu Jicheng Liu Yang, Yanyu and Li Yinghuan; photovoltaic energy storage optimal return model under investment and demand constraints. Computer Simulation 2022:130-133 + 139.
5. Zhou Lili, Xiang yue and Chen Lingtian; research on economic allocation of user-side energy storage capacity based on risk-benefit analysis. China Electric Power 2021:18797.

https://doi.org/10.1051/e3sconf/202452803009

6. Song Yuanjun, He Kai, Shi Jinyong and Ke Huimin; Research on the economic evaluation model of user-side energy storage scheme based on cost analysis. Electrical appliances and energy efficiency management technology 2020:8690. https://doi.org/10.16628/j.cnki.2095-8188.2020.05.014.
7. Luo Wen; Peng Wei; Zheng Jianqing; optimal configuration and economic analysis of photovoltaic side battery energy storage system. Automation Expo 2023:669.
8. Zhao Yuzhong, Sun Guobin, Chen Tong, Luo Qinghong and Zhou Yumei; Application analysis of new energy storage system in gas unit. Technology and Innovation 2023:1768. https://doi.org/10.15913/j.cnki.kjycx.2023.11.054.
9. Yang Jinlin; power distribution strategy of independent photovoltaic-hybrid energy storage system. Power supply technology 2023:5337.
10. B. Nguyen, R. German, J. P. F. Trovo and A. Bouscayrol, Real-time energy management of battery/supercapacitor electric vehicles based on an adaptation of Pontryagin's minimum principle, IEEE Trans. Veh. Technol., vol. 68, no. 1, pp. 203-212, Jan. 2019.
11. Qin Yuxiang, Zhang Weijun and Zhou Le; analysis of the role of energy storage device in the multi-energy flow system. Energy Engineering 2022:820. https://doi.org/10.16189/j.cnki.nygc.2022.04.002.
12. M.T. Tolmasquim, P.M.A. Senra, A.R. Gouvêa, A.O. Pereira, A.C. Alves, M. Moszkowicz Strategies of electricity distributors in the context of distributed energy resources diffusion Environ Impact Assess Rev, 84 (2020), Article 106429, 10.1016/j.eiar.2020.106429.

Carbon reduction effectiveness and efficiency of earth-berming design for speed skating ovals

Zheyuan Zhao[1,2], Peng Luo[1,2], Sihan Xia[1,2]* and Teng Fei[1,3]

[1] Harbin Institute of Technology, Harbin, Heilongjiang, 150001, China
[2] Key Laboratory of Cold Region Urban and Rural Human Settlement Environment Science and Technology, Ministry of Industry and Information Technology, Harbin, Heilongjiang, 150001, China
[3] The Architectural Design and Research Institute of HIT Co., Ltd., Harbin, Heilongjiang, 150001, China

Abstract. Earth-berming design is a common energy-saving practice for speed skating ovals. This study explores the impacts of different earth-berming conditions on the carbon emissions during the operation phase of the venue and evaluates the carbon reduction efficiency by taking into account the incremental carbon emissions caused by the excavation during the construction phase. In this study, three different climate zones were selected as the study environment to simulate the carbon reduction effect of two basic forms of mainstream-scaled speed skating ovals under various earth-berming conditions and to summarise the fitting equations of carbon reduction efficiency. The study found that the total annual carbon emissions (*CE*) of both forms are the lowest in the severe cold zone in general, and the decreased carbon (*dCE*) from the earth-berming design positively correlates with the excavated volume (*V*) with a maximum value of 98.9tCO_2e/y. Among all regions, the maximum *dCE* for all forms is highest in the severe cold zone, 5.1% and 9.8% more than that in the hot-summer-cold-winter zone, which has the worst performances. The carbon reduction efficiency of the earth-berming design was measured by critical gain time (*T*). The earth-berming covariates and *T* for both forms could be fitted by a quadratic equation, which showed that the larger the burial depth on the large-span space side of Form 1 compared to the auxiliary space side, the smaller the *T*-value, whereas the greater the *V* of Form 2, the smaller the *T*-value.

1. Introduction

Global warming and the consumption of non-renewable energy make carbon neutrality increasingly important. According to the International Energy Agency (IEA), building operations account for 30% of global final energy consumption [1]. In China, public buildings which are characterized by high carbon intensity, accounting for 38.6% of the carbon emissions from the construction sector [2]. It shows that carbon reduction is important throughout the entire lifecycle of buildings. With the advancement of China's ice sports promotion policies, the amount of speed skating ovals characterized by huge energy consumption will increase. Therefore, it is necessary to apply decarbonization techniques from the perspective of architectural design.

In the early design stage, earth-berming is considered an effective energy-saving method, and it has been applied in many completed projects. The Calgary Olympic Oval reduced the enclosure surface area exposed to the air for heat exchange by 30% [3]. The "Ice Ribbon" put 75% of the floor area underground to improve the overall energy efficiency of the building [4]. Among the 49 speed skating ovals in the world, nearly one-third were found to have adopted the earth-berming design. However, the increase in carbon emissions caused by earth-berming during the construction phase should not be ignored, and how to balance it with the carbon reduction in the operational phase is worth more consideration.

1.1. Energy saving of ice sports venues

Scholars have done a variety of investigations on energy conservation of ice sports venues, but most of the research target is ice hockey arena, which is on a smaller scale. The methods include applying residual temperature recirculating cavity envelope [5], using inflatable membrane structure [6], optimizing spatial scales and spatial combinations [7] and changing building forms [8].

1.2. Eath-berming and Energy-Saving

Many studies have proved the decarbonization effect of the earth-berming design. F Beigli and R Lenci (2016) found that earth-berming design positively impacts indoor thermal stability in Iran [9]. Liyan Song (2021) found the thermal stability of soil is a natural insulator and cold protector compared to conventional enclosures [10]. Boyuan Gong (2023) studied the total building energy consumption of a typical Turpan residence and found that the total building energy consumption decreased with the

*Corresponding author's e-mail: 570358132@qq.com

increase of buried depth [11]. C Cong *et al.* (2023) found a one-storey earth-bermed museum reduced the annual energy consumption by up to 488 MWh [12].

However, there is also a significant increase in energy consumption caused by excavation. Marco L Trani *et al.* (2016) studied a residential construction project in Milan with three underground floors, calculating that earthwork activities generated 28948.37 kg of CO_2 [13]. L Pinky Devi and Sivakumar Palaniappan (2017) found that the energy consumption from soil excavation and transportation generated energy consumption of 14-89 MJ/m³ and 19-135 MJ/m³, respectively through five cases[14].

1.3. Research Gap and Objective

Existing studies on energy-saving for ice sports venues focus on HVAC and material optimisation, which leads to very few studies on energy saving strategies in the early design stage. Moreover, most research objects are ice hockey venues rather than speed skating ovals. In addition, existing research on the decarbonization effects of earth-berming lacks a cross-life-cycle "input-benefit" correlation study.

In this study, we generate typical research models for mainstream-sized speed skating ovals; simulate the quantitative relationship between various earth-berming design options and decarbonisation effects; consider the carbon emission increment brought about by excavation in the construction phase and the carbon emission reduction in the operation phase and perform a fitting analysis to derive a benefit function. The results of this study reveal the impact of the earth-berming characteristic parameters on the carbon emissions during the operation phase of speed skating ovals and derive the most economical design scheme under the consideration of the additional carbon emissions increment from excavation.

2. Methodology

Figure 1 shows the research framework. Firstly, this study established the representative geometric model of mainstream-sized speed skating ovals and the simulated ground, and generated the energy consumption testing model further; secondly, the energy consumption simulation parameters were set according to spatial performance requirements and climate characteristics; then, the relationship between the testing models and the ground was continuously adjusted, and the performance of the energy consumption all testing models were recorded; finally, the results of the simulation were analysed [15].

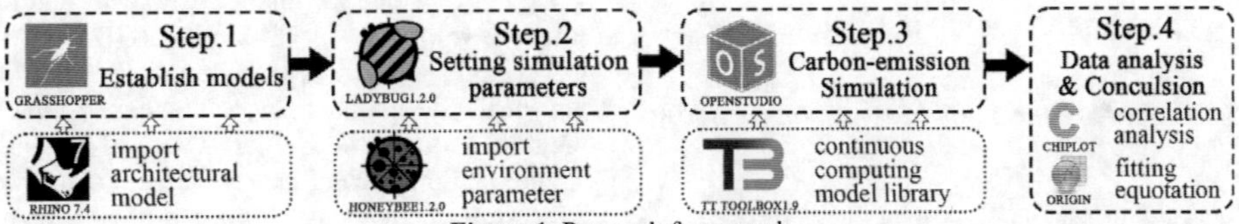

Figure 1. Research framework.

2.1. Study area

All the speed skating ovals in China are currently constructed in severe cold and cold climate zones as the popularity of ice skating varies from north to south. In the future, the hot-summer-cold-winter zones will also become potential speed skating oval construction sites. Therefore, this study takes the three zones above as the study area, selects the representative cities with similar topographic features in each zone - Harbin, Beijing and Wuhan - and derives their climate data for energy simulation.

2.2. Establishment of typical research model

In order to limit the research object and ensure the comparability of the testing models, this study has screened the scale, shape and spatial combination of the testing model. Currently, 10 out of 11 speed skating ovals operating in China and more than 60% of such venues around the world are under 6000 seats, most of which are characterised by a semi-capsule large-span space volume on the long side of a narrow rectangular auxiliary space volume with a unilateral stand inside. Considering that this type of speed skating oval is still mainstream in China in the future, it is simplified into typical research models.

As shown in Figure 2, the research models cover an area of 16000 m^2, with a total floor area of 17480 m^2. The sports space has a floor area of 13000 m^2, including a 400 m speed skating track that meets the International Skating Union (ISU) standards [16]. The internal space is divided into sports space, roof space, audience space and auxiliary space with different environmental characteristics, among which the sports space is subdivided into overhead space, racing space and rink space vertically. Two basic forms derive from the position of the sports field: Form 1 has the same elevation of the sports field as the floor slab of the auxiliary space, and in Form 2 it is the same as the first floor of the stand. The audience space has a 5-storey unilateral stand with the height of the first floor being 3m. The auxiliary space is 6 m on the first floor and 5 m on the second floor. The height of the roof is 13.7 m. The building is east-west oriented and the rectangular volume (including the audience and auxiliary space) is on the south side. The testing models for the two forms without earth-berming are denoted as *p1s* and *p2s*.

The two long edges of the bottom of the main body of the venue are extended to form two parallel lines, and both of them are used to form a combined surface to simulate the ground through extrusion and release operations. The surface is determined by the burying depths *d1* and *d2* of the two long edges of the venue's main body. In Form1, $0<d1<6$, $0<d2<6$; in Form2, $0<d1<3$, $0<d2<3$. The two forms take the values within the above ranges with a step size of 0.6m when generating the testing models, coming out of the testing model group 1 with 121 samples and group 2 with 36 samples, respectively.

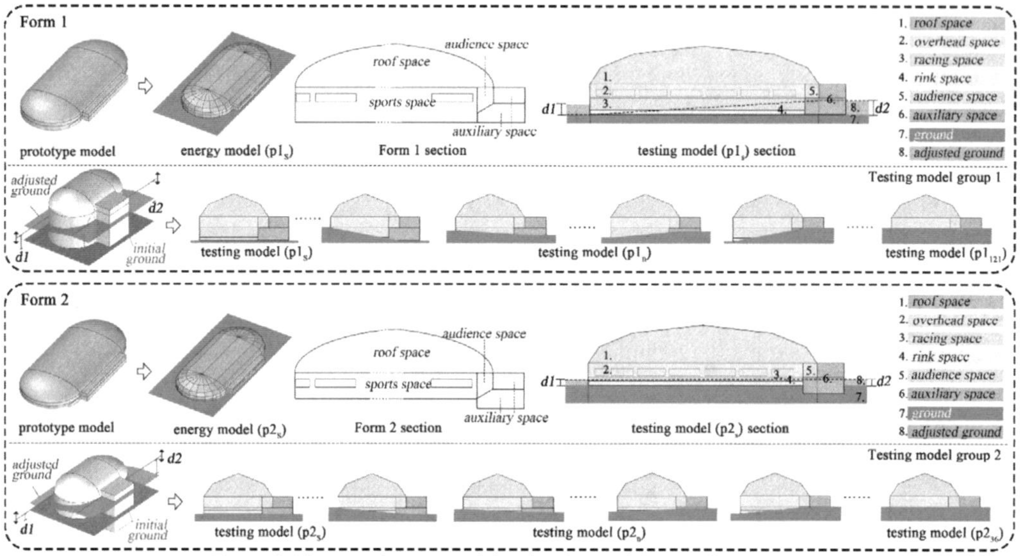

Figure 2. Diagram of energy model and testing model generation process.

2.3. Energy simulation parameter setting

2.3.1. Calculation method of operational energy consumption

In order to eliminate factors other than the earth-berming parameters, the energy simulation model operating parameters and running schedule are unified according to the "2022 Special Regulations Technical Rules Speed Skating. FINAL" issued by the International Skating Union (ISU), "Code for Green Building Performance Calculation for Civil Buildings" (JGJ/T 449-2018), and "Code for Design of Sports Buildings" (JGJ 31-2003). Since the focus of this study is on the early design stage of the speed skating oval, the parameters of the building envelope simulation in different climate zones do not take into account the specific construction, and the lowest value in the "Code for Energy Efficiency and Renewable Energy Utilization in Buildings" (GB 55015-2021) is uniformly adopted. There is no national specification for the special design parameters of speed skating tracks, so they are set according to the statistical results of earlier studies and the current mainstream operation parameters. All testing models have the same track design. Table 1 lists the set parameters for the building envelope. Table 2 lists the set operation parameters for the building energy simulation.

Table 1. Setting parameters for the building envelope.

Parameter	Roof U-value ($W/m^2 \cdot K$)	External wall U-value ($W/m^2 \cdot K$)	Ground floor R-value ($m^2 \cdot K/W$)	External window U-value ($W/m^2 \cdot K$)	SHGC
Harbin	0.25	0.4	1.1	2.5	0.4
Beijing	0.4	0.55	0.6	2.7	0.4
Wuhan	0.4	0.8	0.6	3.0	0.3

Table 2. Setting operation parameters for the internal environment.

Parameter		Heating (°C)	Cooling (°C)	Humidity (%)	Occupancy (people/ m^2)	Lighting (W/ m^2)
Roof space		——	——	30~65	——	——
Sports space	Overhead space	16	22	30~65	——	
	Racing space	12	18	30~65	——	
	Rink space	-9	-5	30~45	0.03	48
Audience space		18	22	40~65	0.85	15
Auxiliary space		18	26	40~65	0.40	8.8

The formula for calculating the annual carbon emissions of the building CE ($kgCO_2e/y$) is equation (1):

$$CE = EUI \times A \times OM \quad (1)$$

The annual carbon reduction dCE($kgCO_2e/y$) for each undercut design testing model is calculated as equation (2):

$$dCE = CEs - CEn \quad (2)$$

where *EUI* (kWh/(m^2·y)) is the energy use intensity, *A* (m^2) is the energy simulation area of the energy model, *CEs*(kgCO_2e/y) is the total annual carbon emission of the no-earth-berming scenario as the initial control model, *CEn*(kgCO_2e/y) is the total annual carbon emission of each testing model. *OM* is the average carbon emission factor of 0.5703 kgCO_2/kWh for the current Chinese power grid [17].

2.3.2. Calculation method of carbon emission of excavation

Carbon emissions caused by excavation vary under various construction environments and construction methods. According to the calculation result of Xia Wang (2012) [18] and J Chou *et al.* (2013) [19], the average energy consumption of excavation and removal of earth varies from 32 kWh/m³ to 208.2 MJ/ m³. In this paper, to make the research results more intuitive, the incremental energy consumption per cubic meter of excavation in China in 2012 is converted to current carbon emission efficiency, which is taken as $\triangle E$=30 kWh/m³ by comprehensively comparing with the results of other existing studies. The formula for calculating the annual carbon emission increment *eCE* (kgCO_2e/y) from the excavation for each testing model is equation (3) where *V* (m³) is the excavated volume:

$$eCE = \Delta E \times V \times OM \quad (3)$$

2.3.3. Calculation method of decarbonization efficiency

To characterise the decarbonisation efficiency of the earth-berming design more intuitively, this study introduces the critical gain time *T* (yr), which is calculated as equation (4):

$$T = \frac{eCE}{dCE} \quad (4)$$

The value of critical gain time *T* indicates the time when the *eCE* and *dCE* are offset, and the speed skating oval will enter the "pure gain period" after *T* years.

3. Result

3.1. Overall energy performance in each climate zone

The two groups of testing models were substituted into the meteorological data of the three study regions to test the *CE* performance, and the distribution characteristics of the statistical results were obtained, as shown in Figure 3. The distribution domains of the *CE* values of both forms were the lowest in the severe cold zone, the second lowest in the cold zone, and the highest in the hot-summer-cold-winter zone; their median values were respectively 4390.83tCO_2e/y, 4453.20 tCO_2e/y, 4467.99 tCO_2e/y and 4369.40 tCO_2e/y, 4433.50 tCO_2e/y, 4444.97 tCO_2e/y. The variance of *CE* thresholds was significantly higher in the severe cold zone than in other zones, indicating that in the severe cold region, the effect of decarbonisation is more sensitive to the change in the way of earth-berming.

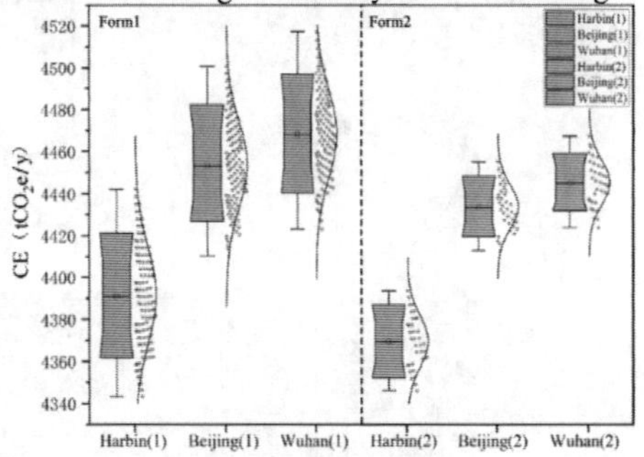

Figure 3. Threshold state maps of *CE* distribution for the two forms across three climate zones.

3.2. Effect of earth-berming design on carbon emission benefits

Figure 4 shows the correlation analyses of the excavation covariates *d1* and *d2, d1-d2* characterising the slope of the ground, the total excavated volume (*V*), the annual carbon reduction (*dCE*) and the critical benefit time (*T*) for each form in each climate zone. The results indicates that both groups of testing models have the strongest correlation with *V* and a weaker correlation with the *d1-d2* difference. Figure 5 further demonstrates that the relationship between *dCE* and *V* is a fluctuating positive correlation state. Generally, the *dCE* increases aligned with *V*, but the *dCE* caused by the slope of the ground varies under a similar volume. The maximum *dCE* of Form 1 and Form 2 among the three zone is 2.2% and 1.1% of total *CE*, respectively, the amount can be up to 98900.11 kgCO_2e/y and 47684.41 kgCO_2e/y, offsetting the *CE* from all the non-sports spaces.

For the critical gain time *T*, both *V* and *dCE* of Form1 have a low correlation with it, while *d1-d2* characterizing the ground slope has a strong correlation, and the correlation strength decreases from the severe cold, cold and then hot-summer-cold-winter zones in order. In contrast, in the case of Form 2, the ground slope is less influential on *T* than in Form 1, but the negative correlation of *dCE* and *V* with *T* values is significantly stronger. The influence of *V* is strongest in hot-summer-cold-winter zones, while the influence of *dCE* is strongest in severe cold zones.

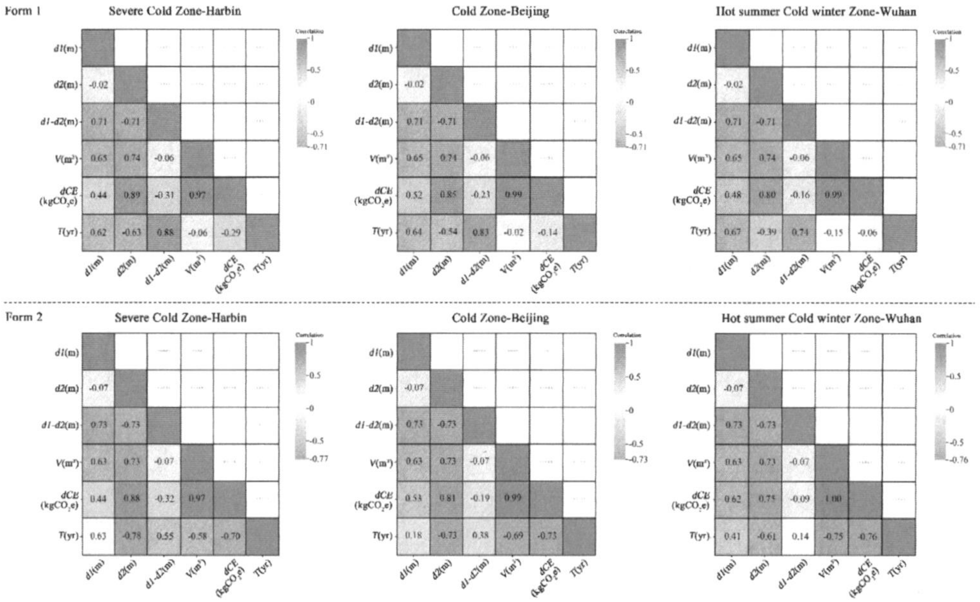

Figure 4. Correlation analysis matrix of the earth-berming design parameters with decarbonization performance.

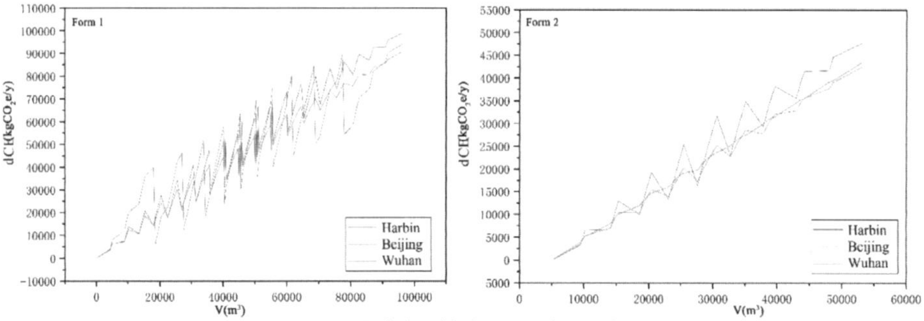

Figure 5. Relationship between *dCE* and *V*.

3.3. Prediction of decarbonization efficiency for earth-berming

In order to further explore the complex relationship between earth-berming and *T* to provide an estimation tool, the covariates *d1*, *d2* and *T* are fit in equation (5), which directly characterise the relationship between the building and the ground, and the results are shown in Figure 6.

$$T = a \times d1^2 + b \times d2^2 + c \times d1 \times d2 + e \times d1 + f \times d2 + k \quad (5)$$

The values of the coefficients *a, b, c, e, f,* and *k* in each case are shown in Table 3. The adjusted R^2 of this fitting equation reaches 0.811~0.917, which shows that it has a high degree of fit and good accuracy in practical application. For Form 1, the top 10 shortest *T*-values in all climatic zones are kept with *d1* =0 m and *d2* taking values within 0.6-6 m. The minimum value of *T* is 12.41 yr from the severe cold zone. When the largest carbon reduction is adopted, i.e., *d1* and *d2* both obtain the maximum value of 6, the *T*-values of the three zones are 16.61yr, 18.13yr, and 17.46yr, slightly lower than the median. Meanwhile, for similar *V*, the *T* values increase with the increase of *d1-d2*. For Form 2, the *T* reach the lowest values of 16.49yr and 20.71yr, respectively, in the severe cold and cold zones, when *d1*=0m and *d2*=3m. When the *d1* and *d2* take the maximum values of 3m, the *T*-values are 19.05yr and 21.41yr, higher than the median in each group. In the hot-summer-cold-winter zone, the scenario with the lowest *T*-value is when *d1* and *d2* reach the maximum value of 3m, respectively, when the *T*-value is 20.92yr.

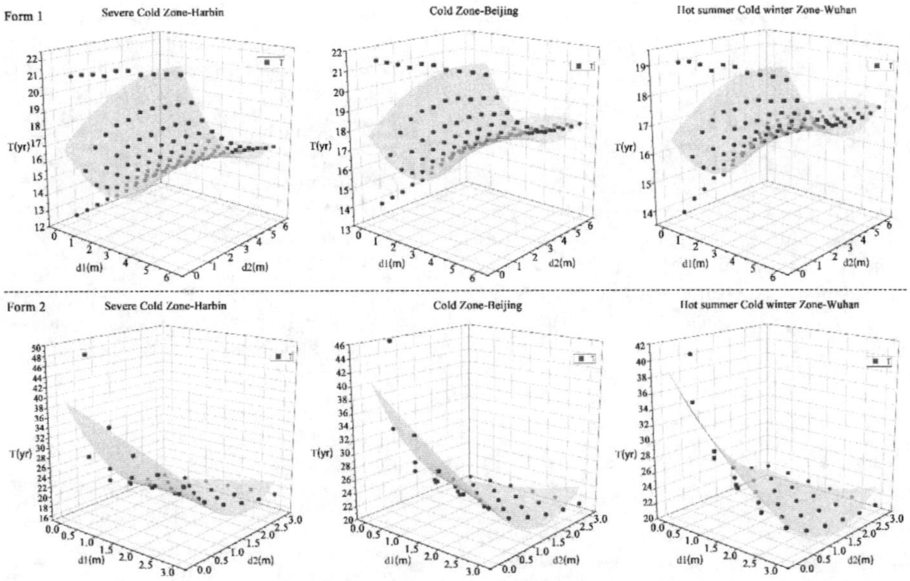

Figure 6. Fitting relationship diagram of *d1*, *d2* and *T*.

Table 3. Coefficients of fitting equation.

Coefficients	Harbin	Beijing	Wuhan	Harbin	Beijing	Wuhan
	Form 1	Form 1	Form 1	Form 2	Form 2	Form 2
a	-0.11048	-0.10473	-0.07621	-0.1063	0.27983	0.60485
b	0.27159	0.24028	0.16618	3.06066	2.21957	1.58476
c	-0.06245	-0.04633	-0.03817	1.71378	1.90878	1.79981
e	1.65134	1.38784	0.96749	-2.73729	-5.19171	-6.64441
f	-2.25792	-1.83075	-1.10832	-17.02155	-13.62064	-10.35118
k	16.80659	17.70851	16.49139	39.03791	40.85425	38.45486
R^2	0.91666	0.89094	0.85675	0.81111	0.81363	0.82489

4. Discussion

4.1. Interpretation of main findings and study implications

This study addresses the need for research on the impacts of earth-berming design for speed skating ovals on carbon emissions and carbon reduction benefits, which has reference significance for future practices in China. First, it was found that the annual total *CE* decrease with increasing *V* in all conditions. and the *CE* of Form 1 is overall higher than that of Form 2. Regardless of the forms, the *CE* is the lowest in the severe cold zone. Second, the carbon reduction effect from earth-berming is more pronounced in the severe cold zone, and the change in the *dCE* due to the ground slope is more pronounced than in other climate zones. In addition, in Form1, the greater the burial depth of the non-auxiliary space side relative to the other side under similar *V*, the shorter the time of critical gain. In Form 2, the smaller the *V*, the longer the *T*, and this phenomenon is especially obvious in hot-summer-cold-winter zone.

The results of this study are practical to the design practice of speed skating ovals. Firstly, it can guide the low-carbon design in different climate zones in China and help architects make choices in site selection. Secondly, this study also provides a formula for evaluating the carbon reduction benefits so that architects can estimate the total carbon emissions and decarbonisation efficiency by substituting architecturally operational variables and avoiding over-designing based on empiricism. Finally, the technical framework of this study is also applicable to similar sports building designs to assess the impact of earth-berming design on energy saving and decarbonisation.

4.2. Limitations and future work

In order to make the conclusions more intuitive, this study uses a constant to estimate the carbon emissions per unit volume increment brought about by excavation. In future research, the impact of different construction techniques, environment and building structure on the carbon emission increment will be considered to obtain the critical gain time, and self-generated methods will carry out the optimisation of low carbon emission construction solutions, substituting the results of which into the current research framework will further improve the accuracy and practicability of the prediction model.

5. Conclusions

This study summarises two mainstream-sized speed skating oval forms, uses ladybug + honeybee plug-ins to calculate the impact of earth-berming on operational carbon emissions and excavation carbon increments, and compares the carbon reduction effect and efficiency of the two forms under different climatic conditions. The main conclusions are as follows:

(1) From 3.1, the *CE* decreases with increasing *V* and the *CE* of Form 1 is generally higher than that of Form 2. The *CE* of both forms is the lowest in the severe cold zone and the highest in the hot-summer-cold-winter zone, indicating the sformr is the best construction site for speed skating ovals.

(2) From 3.2, the maximum *dCE* of Form 1 in the severe cold region is 5.1% more than the hot-summer-cold-winter zone, and in Form 2, the value is 9.8% To shorten the *T*, Form 1 should bury the side of the large space deeper relative to the other side, and Form 2 only needs to increase the volume of excavation to achieve a better carbon reduction effect and efficiency.

(3) From 3.3, a quadratic equation can express the carbon reduction efficiency of both forms in all climate zones. the difference in the *T*-values in Form 1 is much smaller than Form 2. The minimum *T*-value of Form 2 reaches more than 19 yr, significantly higher than that of most scenarios of Form 1. However, some low-level venues often adopt Form 2, if we want to obtain carbon reduction benefits, we should maximise the operation time after completion rather than being easily abandoned.

Acknowledgments

This work was financially supported by the Key Research and Development Plan of Heilongjiang Province "Research and development of an intelligent cloud platform technology for monitoring carbon emissions from urban buildings in cold regions based on digital technology" (2022ZX01A33) and National Natural Science Foundation of China (NSFC) project "Design Theory and Method of Sustainable Intelligent Lunar Research Station" (52238002).

References

1. International Energy Agency. (2022). Tracking. https://www.iea.org/energy-system/buildings#tracking
2. Li, X., Li, Y., Zhou, H., Fu, Z., Cheng, X., & Zhang, W. (2023). Research on the Carbon Emission Baselines for Different Types of Public Buildings in a Northern Cold Areas City of China. *Buildings*, 13(5), 1108. [https://doi.org/10.3390/buildings13051108]
3. Elliott, J. (2007). Green vs. grey: A comparative energy analysis of two olympic speed-skating ovals. *Journal of Architectural and Planning Research*, 160-177. [http://www.jstor.org/stable/43030798]
4. Li, J., Zheng, F., Luo, G., Li, S., Wang, Z., Chen, L.& Shao, B., (2023) Key Technology on Green Construction of National Speed Skating Oval (Ice Ribbon). *Construction Science and Technology*.11,34-36+40. [doi:10.16116/j.cnki.jskj.2023.11.005.] (in Chinese)
5. Sun, X., & Mei, H. (2014). A Study on the system of residual cold heat recirculation cavity wall in ice sports building. *Architectural Journal*, S1, 158-160. (in Chinese)
6. Lin, W., Liu, X., Li, S., & Zhang, T. (2022). Investigation on thermal environment and heat transfer characteristics in ice rinks with different envelopes. *Building and Environment*, 219, 109250. [https://doi.org/10.1016/j.buildenv.2022.109250]
7. Ma, X., & Fei, T., (2023). Research on energy efficiency design of spatial layout of ice sports arenas in cold regions based on Ladybug + Honeybee. *Contemporary Architecture*, S1, 22-27. (in Chinese)
8. Sutton, J. (2023). Thermodynamics of Ice Arena Geometry, Low-Emissivity of Building Materials, and Natural Ventilation: An Energy Efficiency Study of an Ice Arena in Rochester, NY. Rochester Institute of Technology ProQuest Dissertations Publishing, 30819389. [https://repository.rit.edu/cgi/viewcontent.cgi?article=12794&context=theses]
9. Beigli, F., & Lenci, R. (2016). Underground and semi underground passive cooling strategies in hot climate of Iran. *Journal of Environmental Science*, 5(3), 1-12.
10. Song, L., (2022). Study on climate adaptability design of semi-underground space in cold region, (MA thesis, Jilin Jianzhu University). [DOI: 10.27714/d.cnki.gjljs.2021.000153]:38-39. (in Chinese)
11. Gong, B. (2023). Research on the design of new residential houses in Turpan based on comprehensive thermal storage performance improvement, (MA thesis, Xi'an University of Architecture and Technology). [DOI: doi/10.27393/d.cnki.gxazu.2023.000812] (in Chinese)
12. Cong, C., *et al.* (2015). Energy analysis of relics museum buildings. *Energy Procedia*, 75, 1809-1818. [https://doi.org/10.1016/j.egypro.2015.07.150]
13. Trani, M. L., Bossi, B., Gangolells, M., & Casals, M. (2016). Predicting fuel energy consumption during earthworks. *Journal of cleaner production*, 112, 3798-3809. [https://doi.org/10.1016/j.jclepro.2015.08.027]
14. Devi, L. P., & Palaniappan, S. (2017). A study on energy use for excavation and transport of soil during building construction. *Journal of cleaner production*, 164, 543-556. [https://doi.org/10.1016/j.jclepro.2017.06.208]

15. Ji, X., Tang, J., & Zhang, J. (2022). Effects of Salt Stress on the Morphology, Growth and Physiological Parameters of Juglans microcarpa L. Seedlings. *Plants*, 11(18), 2381. [https://doi.org/10.3390/plants11182381]
16. International Skating Union. (2022). Special Regulations Technical Rules Speed Skating. FINAL, 2022. https://www.isu.org/speed-skating/rules/ssk-regulations-rules/file
17. Ministry of ecology and environment the people's republic of China. (2023). Notice on Doing a Good Job in 2023-2025 Reporting and Management of Greenhouse Gas Emissions of Power Generation Enterprises. [https://sthjj.xm.gov.cn/zwgk/tzgg/202303/P020230315587633934087.pdf] (in Chinese)
18. Wang, X., (2012). Life Cycle Assessment for Carbon Emission of Residential Building (MA thesis, Tianjin University). [https://kns.cnki.net/kcms2/article/abstract?v=Fhes7GDiHN2MwBcvKDIRv3znxAtlWKBoVEKPAyPQK7GcH7wtBR-Sbrc79KgAhEUztHKd3snZDDO3iUR1OFGEQ4o-U5FDQjWnWTML8uAQRP7VcHNFrGpNkjl-9Idiu4ghRxP6RUn6OMHBhfWGmeWg6A==uniplatform=NZKPTlanguage=CHS] (in Chinese)
19. Chou, J., & Yeh, K., (2013). Life-cycle carbon emission simulation and environmental cost analysis for building construction: a comparison of prefabrication and cast-in-place work, *Construction Management Association of The Republic of China*, 93, 41-63. [https://doi.org/10.6505/CMJ-2012-93-41-23]

Research on the Measurement of the Relationship between Carbon Emissions and Economic Development in China ——Based on the Tapio Decoupling theory and the Environmental Kuznets Theory Analysis

Yuxin Chen[1] and Yan Wang[1,2*]

[1]Information Engineering College, Hubei University of Chinese Medicine, Wuhan, China
[2]School of Electrical Engineering and Automation, Wuhan University, Wuhan, China

Abstract: In this paper, based on the panel data of China's carbon dioxide emissions, economic and energy development from 2000 to 2022, per capita GDP and carbon emissions are used as analysis indicators.The environmental Kuznets theory serves as the basis for an analysis of the correlation between economic progress and carbon dioxide emissions.The findings indicate that China's GDP and carbon emissions increased between 2000 and 2022. From 7,912 yuan in 2000 to 85,698 yuan in 2022, the per capita GDP grew by over ten times. Although per capita carbon emissions generally increased, they declined between 2014 and 2017, It might be connected to the nation's strong initiatives to support green development and the restructuring of the energy structure.For specific explanations,please refer to 4.1 *Temporal evolution characteristics of carbon emissions and GDP*. China is still in a transitional phase when it comes to environmental preservation and economic development, as seen by the N-shaped curve connecting the country's carbon dioxide emissions and economic growth rate. Based on this theory, the Tapio decoupling theory provides an understanding of the decoupling relationship between carbon emissions and economic growth based on carbon accounting. The irrational use of coal, oil, natural gas and other fossil fuels as the main energy has caused a series of environmental problems and seriously restricted economic development. Therefore, Stronger innovation and promotion of low-carbon and green technologies are required, as is the reduction of carbon emissions from the industrial structure and method of economic progress.

1. Introduction

The continuous rise in global carbon dioxide emissions has caused widespread concern, making carbon peak and carbon neutrality an urgent need to drive comprehensive and profound changes in the social system. To address this global challenge, we must minimize our carbon footprint by incorporating carbon peaking and carbon neutrality into the overall plan for ecological progress. China's efforts in this regard, including increasing resource input and promoting the wide application of renewable energy, have not only had a profound impact on domestic sustainable development, but also provided valuable experience and examples for global sustainable development.However, these measures alone will not be enough to fully achieve the twin goals of carbon reduction and sustainable development. Therefore, while promoting economic development, we need to deeply study the relationship between economic development and carbon emissions, and formulate corresponding policies and strategies accordingly. Such studies not only broaden our understanding of how economic activity affects carbon emissions, but also provides an important basis for formulating more scientific and effective carbon emission reduction measures.In general, researching the connection between economic development and carbon emissions is very important. It will provide critical guidance for achieving our carbon peak and carbon neutrality goals, while also laying a solid foundation to protect the planet's ecological environment and create a better future for future generations.

In recent years, the evaluation and measurement of high-quality economic development has become a research hotspot. Chen,et al [1] innovatively constructed an index system of high-quality economic progress from five dimensions of innovation, balance, stability, sustainability and equity by using China's provincial panel data from 2011 to 2019, and measured it with entropy weight method. Chen ,et al[2] used the principal component analysis method to select the total GDP to reflect the level of economic progress. With the growing awareness of climate change worldwide, the measurement and analysis of carbon emissions become particularly important. Wang, et al. [3] took Guangxi as an example, combined with its specific development background, and made use of the data from 2005 to

* Corresponding author's e-mail: shuxuewy2016@stmail.hbtcm.edu.cn

2020. By using the extended STIRPAT model, they explored in depth the impact of four factors, namely economy, population, structure and technology, on carbon emissions.The nexus between economic development and carbon emissions has always attracted the attention of environmental economics research. Grossman, et al. [4] conducted a pioneering empirical analysis exploring the correlation between environmental quality and per capita income, and proposed that "pollution increases with the increase of per capita GDP at the low income level, and decreases with the increase of GDP at the high income level". Subsequently, Panayotou, et al. [5]named this relationship the Environmental Kuznets Curve (EKC). However, with further research, Friedl, et al.[6] analyzed the Austrian data and found that the actual situation was more in line with the "N" shaped curve. T.Luzzati , et al. [7]conducted an extensive study on the relationship between energy consumption and per capita GDP in 113 countries, and found that the relationship in most countries was not a classic inverted U shape, but a monotony positive correlation, which challenged the EKC hypothesis to some extent. In the context of China, Chen,et al. [8]using data from 2000 to 2018, combined with EKC model and decoupling theory, they identified that the association between per capita GDP and per capita carbon emissions exhibits a distinctive "inverted N-type" EKC. Using the Tapio index, Zhuang,et al. [9] examined the decoupling features of 20 nations that release greenhouse gases over time, including China. The analysis focused on the decoupling relationship between economic growth and carbon emissions .Wang,et al[10] (2023) employed the Tapio decoupling model to examine the disconnection between economic growth and carbon emissions.Xiao,et al. [11] (2013) used the Tapio decoupling model to study the decoupling elasticity between industrial carbon emissions and industrial economic growth during the 15th and 11th Five Year Plans.Li,et al[12]used the OECD decoupling index and Tapio decoupling index to measure the connection in Shanxi Province between the decoupling between industrial economic expansion and carbon dioxide.

Through the review of the existing literature, we can find that although domestic and foreign scholars have a wealth of studies on the connection between the economic growth and carbon emissions, most of the research still focuses on a specific region or province. There exists a paucity of systematic investigation into the specific structure of the correlation curve between carbon emissions and national economic development, as well as the temporal evolution characteristics of carbon emissions.Therefore, this paper aims to construct comprehensive economic indicators and carbon emission indicators for the whole country, analyze the dynamic characteristics of the data from 2000 to 2022, and carry on the thorough discussion using the environmental Kuznets theory to analyze the correlation between the growth of economic and carbon dioxide emissions,with a view to providing scientific basis for relevant policy formulation.

2. Research Approach

The research approach of this article is shown in Figure 1.

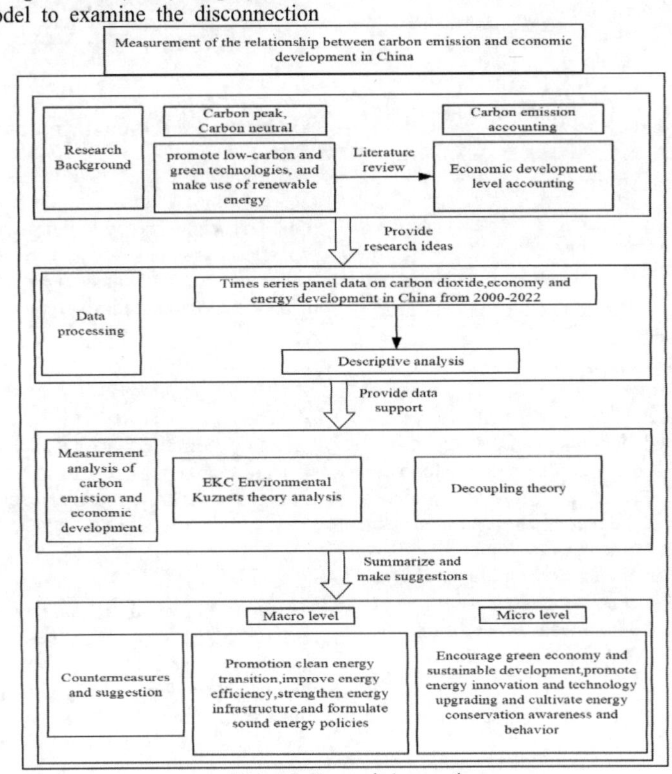

Figure 1. Research Approach

3. Research methods and data sources

This paper uses the EKC theory to study the correlation between the stage of economic development level and carbon emission.

3.1 EKC model.

The hypothesis that the "inverted U-shaped" environmental Kuznets curve exists between the environment and the economy was first proposed by American economist Grossman&Kruegar in 1991.GDP per capita can be a good indicator of the economic growth of a country or region.Its cubic polynomial model is as follows:

$$y = \beta_0 + \beta_1 X_{it} + \beta_2 X_{it}^2 + \beta_3 X_{it}^3 + \beta_4 Z_{it} + \varepsilon_{it} \quad (1)$$

Where :Y represents the level of environmental quality variables, which refers to per capita carbon dioxide emissions; X stands for the variable of the level of economic development, which in this paper refers to the per capita GDP; Z represents other factors affecting the environment (random error term).And the relationship between EKC model parameters and curves is shown in Table 1.

Table 1. Relationship between EKC model parameters and curves

Parameter value	Curvilinear relation	Curve shape
$\beta_1 = \beta_2 = \beta_3 = 0$	curvelessness	
$\beta_1 \geq 0$, β_2、$\beta_3 = 0$	primary curve relation	straight line
$\beta_1 \leq 0$, β_2、$\beta_3 = 0$	primary curve relation	straight line
$\beta_1 \geq 0$, $\beta_2 \leq 0$、$\beta_3 = 0$	conic relation	Inverted U-shaped curve
$\beta_1 \leq 0$, $\beta_2 \geq 0$、$\beta_3 = 0$	conic relation	U-shaped curve
$\beta_1 \geq 0$, $\beta_2 \leq 0$、$\beta_3 \geq 0$	cubic curve relation	N-shaped curve
$\beta_1 \leq 0$, $\beta_2 \geq 0$、$\beta_3 \leq 0$	cubic curve relation	inverted N-shaped curve

3.2 Tapio Decoupling theory.

The decoupling theory explores the correlation between environmental contamination and economic development, and analyzes the degree of closeness between the two. The decoupling status of economic development and ecological environment is an important tool for measuring sustainable development. Many foreign researchers have applied decoupling theory to research on reducing greenhouse gas emissions. This article mainly uses the Tapio model to determine the decoupling status from 2000 to 2022, and constructs the following decoupling model:

$$\varepsilon = \frac{\frac{\Delta C}{C}}{\frac{\Delta GDP}{GDP}} \quad (2)$$

In the formula: ε :The decoupling elasticity coefficient,

C :represents the per capita carbon emissions,

ΔC : represents the change in per capita carbon emissions,

GDP:represents the per capita GDP output value,

ΔGDP : represents the change in per capita GDP output value

According to the decoupling theory, Table 2 lists the divisions of different decoupling states.

Table 2. Division of decoupling states

Decoupling states		$\frac{\Delta C}{C}$	$\frac{\Delta GDP}{GDP}$	ε
decoupling	Strong decoupling	-	+	$(-\infty, 0)$
	Weak decoupling	+	+	$[0, 0.8]$
	Recessive decoupling	-	-	$(1.2, +\infty)$
Coupling	Recessive coupling	-	-	$[0.8, 1.2]$
	Expansive coupling	+	+	$[0.8, 1.2]$
Negative decoupling	Strong negative decoupling	+	-	$(-\infty, 0)$
	Expansion negative decoupling	+	+	$(1.2, +\infty)$
	Weak negative decoupling	-	-	$[0, 0.8]$

3.3 Methods of measuring carbon emissions.

According to the carbon emission calculation guide released by China Carbon Accounting Database (CEADs), eight energy sources, coal, crude oil, natural gas, diesel, gasoline, fuel, kerosene and coke, carbon emissions are estimated using the index estimation method. The equation is expressed as follows:

$$C = \sum_{i=1}^{8} E_i * C_i \quad (3)$$

Where: c :carbon emissions, unit: 10,000 tons; E_i :energy consumption, calculated by 10,000 tons of standard coal; C_i:carbon emission coefficient of energy, is the type of energy, take 8 categories.

3.4 Data sources

This study utilizes data spanning from 2000 to 2022 to investigate the correlation between China's economic development and carbon emissions.The data on per capita GDP and per capita carbon emissions in China from 2000 to 2022 are shown in Table 3.Per capita GDP serves as an economic proxy indicator, while per capita carbon emissions serve as an environmental proxy

https://doi.org/10.1051/e3sconf/202452803011

indicator. The data utilized in this research are sourced from the China Statistical Yearbook, China Energy Statistical Yearbook, and pertinent datasets provided by the National Bureau of Statistics.

Table 3. 2000-2022 China's per capita GDP and per capita carbon emissions

Particular year	Real GDP (trillion)	GDP per capita (yuan)	Carbon emissions (10,000 tons)	Carbon emissions per capita (tons)
2000	10.028	7912	370334	2.922
2001	11.086	8686	387832	3.039
2002	12.172	9476	418318	3.257
2003	13.742	10634	485291	3.755
2004	16.184	12450	559997	4.308
2005	18.732	14326	633844	4.848
2006	21.944	16694	704039	5.356
2007	27.009	20442	767368	5.808
2008	31.924	24039	788919	5.941
2009	34.852	26116	848996	6.362
2010	41.212	30734	929436	6.931
2011	48.794	36166	1015548	7.527
2012	53.858	39624	1048458	7.714
2013	59.296	43369	1078059	7.885
2014	64.356	46755	1089841	7.918
2015	68.886	49800	1077182	7.787
2016	74.640	53608	1079429	7.753
2017	83.204	59426	1103673	7.883
2018	91.928	65410	1149929	8.182
2019	98.652	69962	1177107	8.348
2020	101.357	71776	1194812	8.461
2021	113.324	80224	1246632	8.825
2022	121.021	85698	1266742	8.973

4. Results

4.1 Temporal evolution characteristics of carbon emissions and GDP

Based on the time evolution characteristics of carbon emissions and GDP from 2000 to 2022, sketchup model was used to draw the column charts of carbon emissions, total GDP, per capita carbon emissions and per capita GDP respectively.The horizontal coordinate was divided into time years, and the three charts all described the change characteristics over time. The carbon emissions, real GDP, per capita GDP, and per capita carbon emissions are displayed in Figures 2, 3, and 4 below.

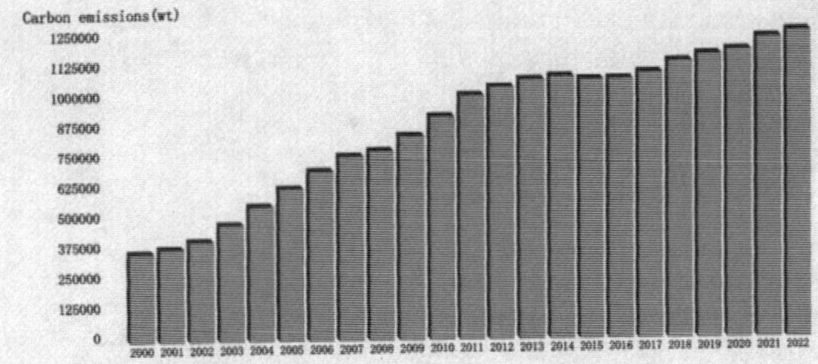

Figure 2. 2000-2022 total carbon emission

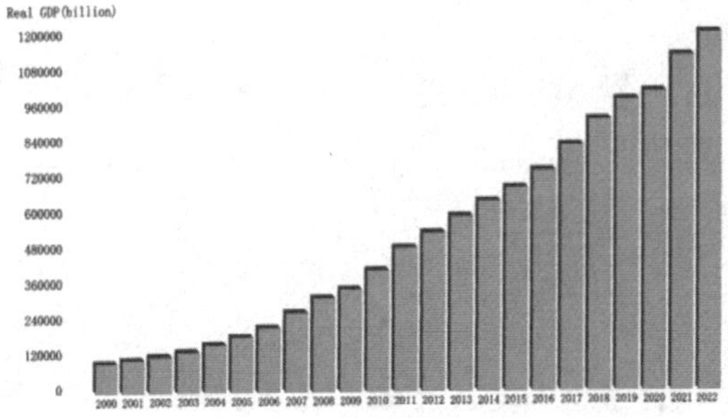

Figure 3. Real GDP Data from 2000 to 2022

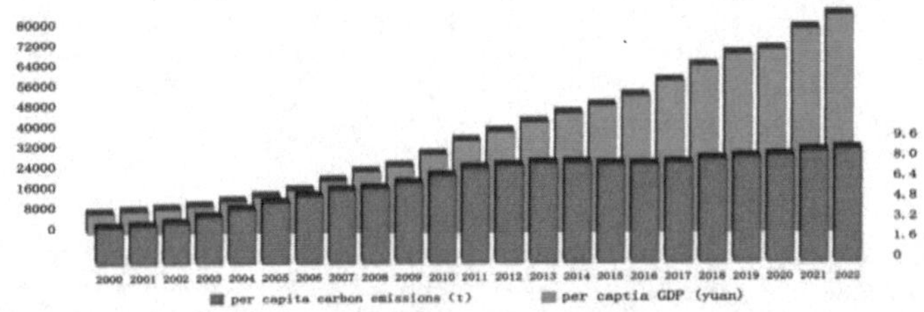

Figure 4. Carbon emissions per capita and GDP per capita from 2000 to 2022

As can be seen from Figure 4, Both per capita GDP and per capita carbon emissions are typically on upward trajectory,per capita GDP shows an upward trend year by year, from 7,912 yuan in 2000 to 85,698 yuan in 2022, an increase of nearly 10 times. The trend of per capita carbon emissions has been increasing generally,but have declined in recent years, which may be related to the country's vigorous promotion of energy structure transformation and green development. From 2014 to 2017, China's energy system continuously developed around the strategy of "energy production and consumption revolution". The reform of state-owned energy enterprises in September 2014, the reform of the power system in March 2015, the structural reform of the energy supply side in November 2015, the implementation of the "Internet plus" smart energy demonstration project organized by the National Energy Administration in February 2016, the oil and gas system reform in May 2017, and the establishment of the national carbon market in December. These measures to promote the transformation of energy structure have led to a decrease in per capita carbon emissions between 2014 and 2017.

4.2 EKC model analysis

Based on the statistical data from 2000 to 2022, the regression curve equation of environmental development level and economic development level was obtained using Matlab software programming, and the regression curve diagram is drawn as shown in Figure 5:

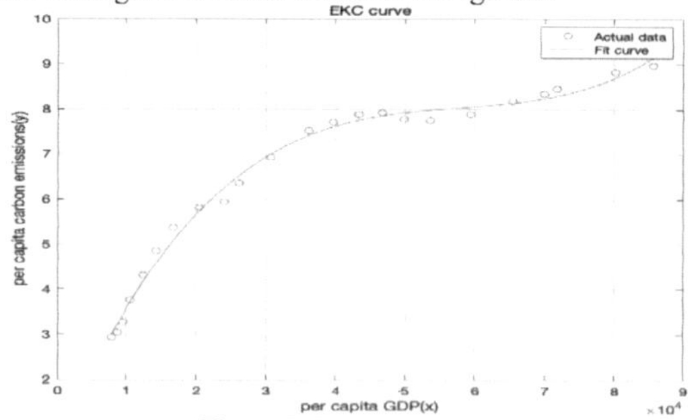

Figure 5. Regression curve

Conclusion: The model equation can be obtained from the data in the table as follows:

$$y = 3.5961e\text{-}14x^3 - 6.2284e\text{-}09x^2 + 0.00037x + 0.4332$$

$$R^2 = 0.99288$$

Combined with the data in the table, it can be concluded that from 2000 to 2022, China's total carbon emission and GDP generally change in the same direction, and with the increase of GDP, the total carbon emission will also increase. The characteristics described by this model are basically consistent with the actual situation. The R^2 coefficient of determination of this model is 0.99288, which is greater than 0.9, indicating that there is a strong correlation between per capita GDP and per capita carbon emissions. The actual and predicted trends of per capita carbon emissions and per capita GDP are roughly the same, indicating that the EKC model has a good simulation of experimental data, but the prediction of the relationship between future carbon emissions and GDP may be distorted. The equation conforms to an "N" type curve. Therefore, it can be proved that the relationship between China's economic growth and carbon emissions during 2000-2022 is an "N" type environmental Kuznets curve. The relationship between economic development and environmental development is not a simple linear relationship. In the first turning point, per capita co_2 emissions first rise and then fall; At the second turning point, when GDP per capita increased, co_2 emissions per capita gradually increased. At present, China is at the left end of the second turning point, and carbon emission reduction will constrain the expansion of China's economic growth scale.

Therefore, the real relationship between China's total carbon emissions and GDP needs to be verified on the basis of data obtained from China's continuous economic development in the coming period of time. Therefore, this paper chooses the cubic "N-type" curve to describe the relationship between China's carbon emission and GDP.

4.3 Tapio Decoupling theory analysis

Based on the above Tapio decoupling model and the status quo of carbon emissions and economic development in China, the decoupling status of carbon emissions and economic growth in China during 2000-2022 is calculated, as demonstrated in Table 4. Among these, a strong decoupling state denotes simultaneous economic development and reduction in carbon emissions; Weak decoupling means that between economic growth and carbon emissions, carbon emissions increase more slowly than economic growth; Expansive coupling refers to a scenario where both economic development and carbon emissions increase concurrently, with carbon emissions rising at a pace largely parallel to that of economic development. Expansion negative decoupling refers to a situation where carbon emissions outpace the rate of economic growth.The domestic decoupling index is transformed between relative decoupling and absolute decoupling.

Table 4. The decoupling status of carbon emissions and economic growth in China from 2000 to 2022

year	GDP(trillion)	$\Delta GDP/GDP$	Carbon emissions(trillion)	$\Delta CO_2/CO_2$	$T(CO_2,GDP)$	Decoupling state
2000	10.028	-	370334.02	-	-	-
2001	11.086	0.10	387831.51	0.045	0.47	Weak decoupling
2002	12.172	0.09	418317.77	0.073	0.82	Expansive coupling
2003	13.742	0.11	485291.18	0.138	1.21	Expansion negative decoupling
2004	16.184	0.15	559996.72	0.133	0.88	Expansive coupling
2005	18.732	0.14	633843.85	0.117	0.86	Expansive coupling
2006	21.944	0.15	704038.82	0.100	0.68	Weak decoupling
2007	27.009	0.19	767368.05	0.083	0.44	Weak decoupling
2008	31.924	0.15	788919.10	0.027	0.18	Weak decoupling
2009	34.852	0.08	848995.85	0.071	0.84	Expansive coupling
2010	41.212	0.15	929436.10	0.087	0.56	Weak decoupling
2011	48.794	0.16	1015548.12	0.085	0.55	Weak decoupling
2012	53.858	0.09	1048457.67	0.031	0.33	Weak decoupling
2013	59.296	0.09	1078058.64	0.027	0.30	Weak decoupling
2014	64.356	0.08	1089841.41	0.011	0.14	Weak decoupling
2015	68.886	0.07	1077181.74	-0.012	-0.18	Strong decoupling
2016	74.640	0.08	1079428.62	0.002	0.03	Weak decoupling
2017	83.204	0.10	1103673.46	0.022	0.21	Weak decoupling
2018	91.928	0.09	1149929.22	0.040	0.42	Weak decoupling
2019	98.652	0.07	1177107.25	0.023	0.34	Weak decoupling
2020	101.357	0.03	1194811.96	0.015	0.56	Weak decoupling
2021	113.324	0.11	1246631.60	0.042	0.39	Weak decoupling
2022	121.021	0.06	1266742	0.016	0.27	Weak decoupling

According to the data in Table 4, the following conclusions can be inferred:

1)In 2015, there was a notable decoupling observed between China's carbon emissions and economic growth, with carbon emissions decreasing despite the country's continued rapid economic expansion. In 2015, China accomplished emission reduction development, aligning with the action objectives of carbon peak goals, and the decoupling state exhibited relatively favorable outcomes.

2)In 2003, there was an expansion negative decoupling between China's carbon emissions and economic growth, indicating that the growth rate of carbon emissions far exceeded economic growth. China's carbon emissions have not been effectively controlled yet.

3) 2002, 2004, 2005, and 2009 were all expansive coupling, indicating that the economic growth rate and carbon emission growth rate were the same, and

development was more balanced; The remaining years of China's the growth in the economy and carbon emissions show a weak decoupling state, indicating that China's carbon emissions decoupling elasticity is good, and the current energy-saving and emission reduction measures are feasible,But there is still room for improvement.

4) China's actual GDP has shown a trend of increasing year by year, from 10.028 trillion yuan in 2000 to 121.021 trillion yuan in 2022, an increase of nearly 12 times. Simultaneously, both total carbon emissions and per capita carbon emissions have exhibited a consistent upward trend year by year,especially with rapid growth since 2004. In 2013 marks the zenith of total carbon emissions and per capita carbon emissions,which have since decreased.

5) The decoupling model in this article specifically explores the correlation between China's economic development and carbon emissions, explores the correlation between economic development and environmental pollution, analyzes the degree of closeness between the two, measures the decoupling situation between economic development and ecological environment, constructs a decoupling index, quantifies the correlation between economic development and environmental pollution, and the decoupling indices are transformed between relative decoupling and absolute decoupling. A strong decoupling state indicates economic development and reduced carbon emissions; A weak decoupling state indicates economic development and an increase in carbon emissions. However, the elastic decoupling index only describes the nonlinear relationship between environmental pollution and economic development. At the same time, the spatial and temporal development of economic development is unbalanced, and there may be a mismatch between decoupling development and economic development. These factors all constrain the construction of green decoupling countries.

5. Conclusion

In this study, the Kuznets environmental hypothesis is employed to examine the association between China's carbon emissions and economic development spanning from 2000 to 2022.The research results of this paper show that the relationship between China's economic growth and carbon emissions during 2000-2022 is an "N" type environmental Kuznets curve. The relationship between economic development and environmental development is not a simple linear relationship. Per capita GDP and per capita carbon emissions are generally on the rise, and per capita GDP is on the rise year by year, from 7,912 yuan in 2000 to 85,698 yuan in 2022, the per capita GDP grew by over ten times. There has been a general upward trend in per capita carbon emissions, but have declined in recent years, which may be related to the country's vigorous promotion of energy structure transformation and green development. In 2015, China's carbon emissions and economic growth showed a significant decoupling state, this indicates that despite the rapid growth of the country's economy, its carbon emissions declined.China achieved emission reduction development, which met the action requirements of carbon peak goals and the decoupling state was relatively ideal. In 2003, the decoupling between China's carbon emissions and economic growth exhibited a state of expansion negative decoupling. The growth rate of carbon emissions was much higher than economic growth, and China's carbon emissions had not been effectively controlled. In 2002, 2004, 2005, and 2009, expansive coupling was observed, indicating a simultaneous rise in both economic growth rate and carbon emission growth rate, thereby fostering a more balanced development. In the subsequent years, China's economic growth and carbon emissions displayed a state of weak decoupling, indicating that China's carbon emissions decoupling elasticity is good.The current energy-saving and emission reduction measures are feasible,but there is still room for improvement.

References

1. Chen Hailong, Li yang. Spatial effect analysis of carbon emission intensity on high-quality economic development[J]. *TJYJC*,**2022(23)**:117-120.
2. Chen Chong, Lin Qin. Study on the relationship between carbon dioxide emission, economy and energy developmentin China[J]. *Journal of Mudanjiang Normal University*,**2014(05)**:16-19.
3. Wang,A.N.,Bei,L.Analysis of influencing factors of carbon emission in Guangxi based on STIRPAT model[J].*Environmental.Protection Science*,**2023(07)**
4. GROSSMAN G M,KRUEGER A B.Environemtal impacts of a North American Free Agreement[R].*NBER working paper* **No.3914,1991**
5. Panayotou T.Empirical Tests and Policy.Analysis of Environmental Degradation at Different Stages of Economic development[C].*Technology and Employment programme,International Labor Office,Geneva,Working Paper* **WP238，1993**
6. Friedl B ,Getzner M.Determinants of co_2 emissions in a small open economy[J].*Ecological Economics*,**2003,45(1)**:133-148
7. T. Luzzati, M. Orsini. Investigating the energy-environmental Kuznets curve[J].*Energy*, **2009,(34)**:291-300
8. Chen Huiting,Yi Qiguo.Analysis of the relationship between China's economic development and carbon emission[C].*Proceedings of 2021 Annual Science and Technology Conference of Chinese Society of Environmental Sciences*,**2021(10)**:54-59.
9. Zhuang Guiyang. Low-carbon Economy: China's Development Path in the context of Climate Change. *Beijing: China Meteorological Press.*
10. Wang Y. The decoupling relationship between carbon emission and economic development in Northern China and its influencing factors.**2023 (3)**
11. Xiao Hongwei, YI Danhui, Zhou Mingyong. Research on the relationship between Industrial

https://doi.org/10.1051/e3sconf/202452803011

Economic Development and carbon Emission Decoupling in China [J]. *Fujian Forum (Humanities and Social Sciences Edition)*.**2013(3)**

12. Li Zhongmin, Qing Dongrui.A Study on the Decoupling Between Economic Growth and Carbon Dioxide:A Case on the Shanxi Province[J]. *Fujian Tribune (the Humanities & Social Sciences monthly)* ,**2010(2)** :67 - 72

Research on the distribution of harmful substances in automotive products based on chemical quantitative

Tian Li[1,*], Meng Hou[1], Rujie Yu[1] and Qian Sha[2]

[1]Automotive Data of China(Tian Jin） Co., LTD., Tianjin, China
[2]China Electronics Engineering Design Institute Co., Ltd., Beijing, China

Abstract. At present, China's automotive industry lacks standardized standards and testing methods for conducting harmful substance testing and evaluation of products.This article finely disassembled a passenger car model and conducted detection and analysis of harmful substances. More than 320000 materials were tested, involving 85 assembly components. This article studies six harmful substances in automotive materials, including lead, mercury, cadmium, hexavalent chromium, polybrominated biphenyls, and polybrominated biphenyl ethers. The content of harmful substances and their distribution in different materials and components are analyzed. This article provides technical reference for conducting hazardous substance assessments in the automotive industry through data analysis. The harmful substance mainly contained in the studied automotive product is lead, which is mainly distributed in materials such as copper alloys, aluminum, steel, solder, plastics, etc. The content of mercury, cadmium, hexavalent chromium, polybrominated biphenyls, and polybrominated biphenyl ethers in automobiles is generally low.

1. Introduction

Since 2000, the European Union has successively issued“ELV Directive”“RRR Directive”and other mandatory regulations on the use of harmful substances in automotive products[1,2].In 2015, China issued the “Management Requirements for Harmful Substances and Recyclability of Automobile”, which requires domestic passenger cars to comply with relevant standards for lead, mercury, cadmium, hexavalent chromium, polybrominated biphenyls, and polybrominated biphenyl ethers.With the release and implementation of policies, Chinese automotive companies have gradually begun to attach importance to harmful substance substitution technology and research and development work, and the content of harmful substances in products has been significantly reduced.

2. Research methods

This article provides a detailed disassembly and inspection of a vehicle model. The entire vehicle was disassembled to form 385 primary assembly components, covering more than 20000 material nodes. In addition, harmful substance content testing was conducted on each material node, and the quality was tested. Among them, the disassembly requirements shall be carried out in accordance with the GB/T 33460 standard. The testing method shall be carried out in accordance with the standards of QC/T 941, QC/T 942, QC/T 943, QC/T 944, GB/T 23263, and QC/T 1131. The analysis of test results is based on GB/T 30512.

3. Current situation of hazardous substance management in domestic and foreign automobiles

3.1. Current situation of EU management

The European Commission issued a directive on scrapped vehicles in 2000, requiring the implementation of hazardous substance and recycling management for M1 and N1 category vehicles, as well as three wheeled motor vehicles and their components and materials, except for three wheeled motorcycles. In terms of harmful substances, the EU stipulates that member states should control and reduce the use of four harmful substances in cars, including lead, mercury, cadmium, hexavalent chromium, etc., and should promote recycling and avoid discarding hazardous waste to prevent such substances from being released into the environment .

3.2. Current management situation in South Korea

The Resource Recycling Agency of the South Korean Ministry of Environment formulated and promulgated the Resource Recycling and Utilization Act for Electrical and Electronic Products and Vehicles (referred to as the Recycling Act) on April 27, 2007, and officially implemented it on January 1, 2008.This bill is the first

*Corresponding author’s e-mail: litian@catarc.ac.com

law in South Korea specifically aimed at the recycling and reuse of electronic and electrical products and automotive products. It draws on the advanced aspects of the European Union's WEEE, RoHS, and ELV directives and is the main law in South Korea for implementing the management of hazardous substances and recycling in automobiles.The bill strictly limits the content of four harmful substances, including lead, mercury, cadmium, and hexavalent chromium, in automotive products, and proposes that automobile manufacturers/importers should disclose whether they comply with the restrictions on the use of harmful substances.

3.3. Current Management Situation in China

In 2015, China issued the "Management Requirements for Harmful Substances and Recyclability of Automobile", which requires domestic passenger cars to comply with relevant standards for lead, mercury, cadmium, hexavalent chromium, polybrominated biphenyls, and polybrominated biphenyl ethers.With the continuous implementation of ELV management in our country, the industry has established an ELV control mechanism of "enterprise declaration compliance verification result disclosure", and has issued thirteen batches of "List of Compliance with the Management Requirements for Harmful Substances and Recyclable Utilization of Vehicles" for With the implementation of ELV management policies, the use of six controlled harmful substances, including lead, mercury, cadmium, hexavalent chromium, polybrominated biphenyls, and polybrominated biphenyl ethers, has been significantly reduced. Among them, the average usage of lead (excluding lead-acid batteries) has been reduced by about half from 220 grams per vehicle (2015), and the cumulative reduction in lead usage has reached more than 10000 tons.

4. Results and Analysis

4.1. Classification analysis of automotive product materials

The materials used in automobiles can be roughly divided into two categories: metallic materials and non-metallic materials.Metal materials include steel plate, cast iron and other heavy metal materials, aluminum, magnesium, titanium and other light metals and their alloy materials, foam metal and so on. Non metallic materials include polymers, glass, plastics, rubber, etc.

The quality comparison of different materials in the automotive products analyzed in this article is shown in Figure 1.Among them, the metal mass is the largest, at 1034.14kg; Next are plastic and polymer materials, with masses of 164.38kg and 150.75kg respectively; Rubber and glass materials are relatively few, with masses of 33.66kg and 32.65kg respectively; The total weight of other materials is 4.68kg.

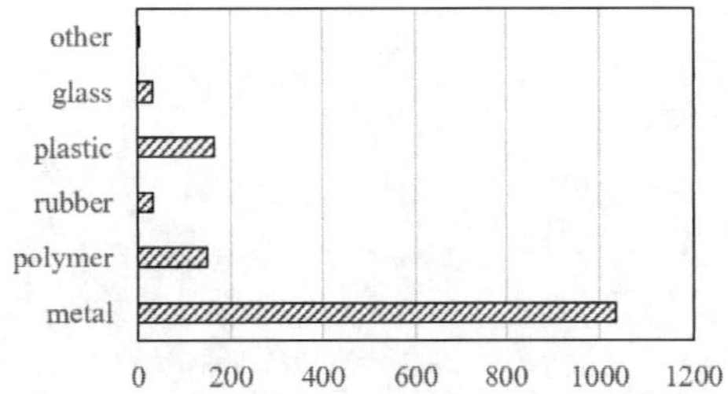

Figure 1. Classification analysis diagram of automotive materials.

4.2. Analysis of lead distribution in automotive products

After lead and its compounds enter the human body, they may deposit in the body, causing damage to the nervous, hematopoietic, digestive, renal, cardiovascular, endocrine and other systems, leading to lead poisoning[3].Lead in automotive products is often used in lead-acid batteries, cable sheaths, etc.

4.2.1 Distribution of lead in the entire vehicle

The distribution of lead in the entire vehicle is shown in Figure 2.Out of more than 20000 testing materials in the entire vehicle, there are a total of 813 points containing lead materials.The proportion of sites with lead content exceeding 10000mg/kg is 15.74%, mainly composed of copper alloys, aluminum materials, solder materials, etc., mainly distributed in battery assemblies and electronic and electrical equipment (ceramics), all of which are exempted components and materials.Among them, the aluminum lead content in the battery electrode plate is the highest, with multiple points exceeding 900000mg/kg.The sites with lead content ranging from 1000 to 5000mg/kg have the highest proportion, accounting for 56.58%(See Figure 3 for details).Mainly distributed in steel and electronic and electrical equipment (ceramics).Most of them are exempted materials and components, but after chemical quantitative analysis, the lead content of three plastic parts exceeds 1000mg/k, which may not meet relevant standards.

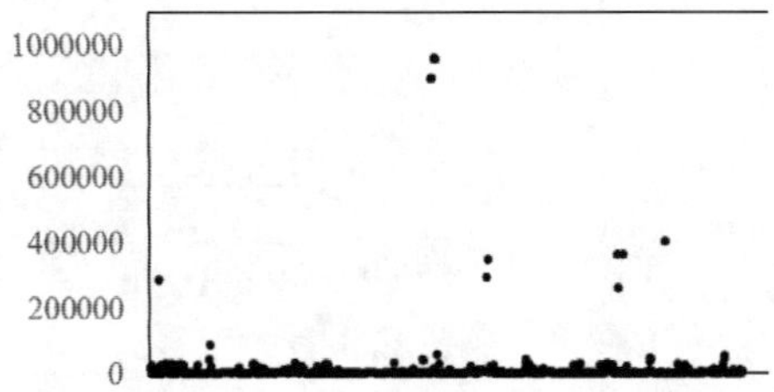

Figure 2. Distribution diagram of lead in the entire vehicle

https://doi.org/10.1051/e3sconf/202452803012

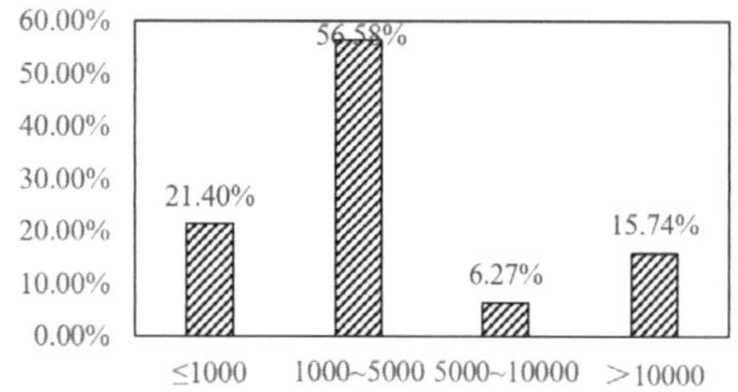

Figure 3. Diagram of the proportion of materials with different lead contents

4.2.2 Distribution of lead in different types of materials

From the distribution of lead in different materials,as shown in Figure 4. Plastic contains the least amount of lead, with only 4 locations detecting the presence of lead; Next is metallic materials, with a total of 244 lead containing sites. There are 565 lead containing sites in other materials. Other materials include electrolytic paper, electronic and electrical equipment, etc.

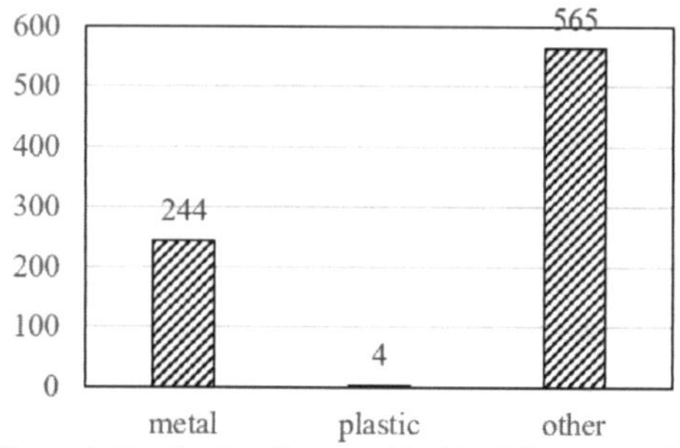

Figure 4. Distribution diagram of lead in different materials

By analyzing the lead content of different types of materials,as shown in Table 1. The main lead containing materials in the entire vehicle are copper alloy, aluminum, coating, steel, solder, and plastic. From the quantity of materials containing lead, copper alloy has the highest quantity, at 99; Next are aluminum, coatings, and steel, with the least amount of plastic containing lead. From the perspective of lead content, solder has the highest lead content, with a maximum value of 91965mg/kg and an average value of 74024mg/kg; Next is copper alloy, with a maximum value of 91965mg/kg, an average value of 33290mg/kg, and an average value of 18922mg/kg; The lowest lead content is in the coating, with a maximum value of 628mg/kg and an average value of 168mg/kg.

Table 1 Distribution of lead in different materials

Material type	Quantity of lead containing materials (pieces)	Maximum value (mg/kg)	Minimum value (mg/kg)	Average value (mg/kg)
Soldering	12	91965	67576	74024
Coating	34	628	13	168
Steels	31	3456	16	356
Aluminum	79	2467	17	180
Plastic	4	2600	320	1532
Copper Alloy	99	33290	7313	18922

By analyzing the proportion of materials with different lead contents in different intervals, as shown in Figure 5. It can be seen that the lead content of solder is generally higher, all higher than 10000mg/kg; The lead content in the coating is relatively low, all below 1000mg/kg. The lead content of most copper alloys ranges from 5000 to 10000mg/kg; The lead content of most plastics ranges from 1000 to 5000mg/kg; Most aluminum materials are below 1000mg/kg, while a few materials are between 1000-5000mg/kg.

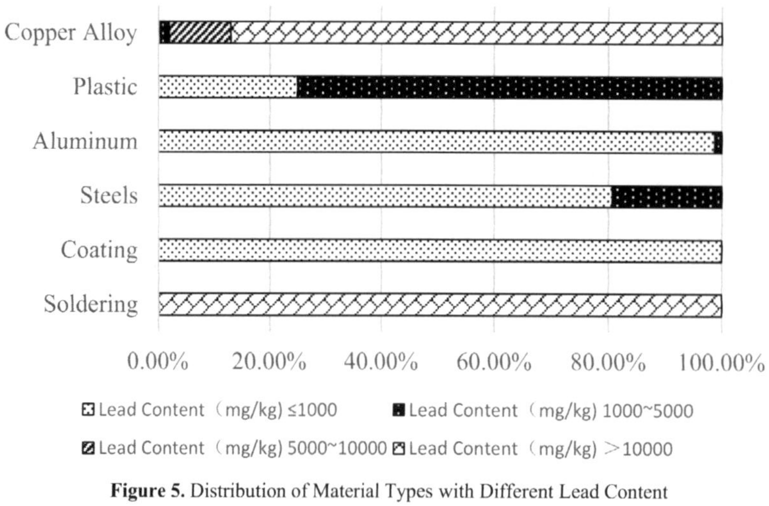

Figure 5. Distribution of Material Types with Different Lead Content

4.3. Analysis of mercury distribution in automotive products

Oral, inhalation, or exposure to mercury can cause brain and liver damage. Mercury can accumulate in living organisms and is easily absorbed by the skin, respiratory tract, and digestive tract[4].Most harmful substances such as mercury exist in the form of compounds in the interior of vehicles, especially in seats, gearshift handles, instrument panels, and enamel materials. Some car paint on the surface of the car also contains mercury.

According to statistics on the distribution of mercury in the entire vehicle, As shown in Figure 6. Only three points were detected for the presence of mercury, with concentrations of 1620mg/kg, 2381mg/kg, and 2726mg/kg, respectively.Mercury containing materials are mainly distributed on the tungsten wire in license plate lamps. Its material type belongs to electronic and electrical equipment and is an exempted component.

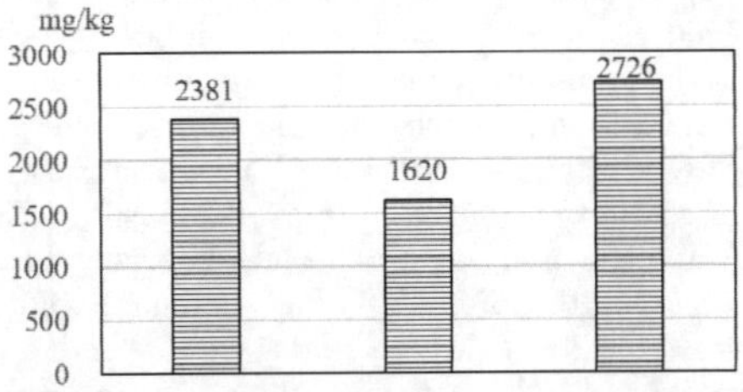

Figure 6. Distribution of Mercury in Different Materials

4.4. Analysis of cadmium distribution in automotive products

When the environment is contaminated with cadmium, cadmium can accumulate in organisms and enter the human body through the food chain, causing chronic poisoning[5].Cadmium coatings can enhance the corrosion resistance of metals such as steel, brass, and aluminum, and can be used as corrosion protection coatings for screws, fastening devices, vehicle chassis, and other structural components in industries such as electronics and automotive manufacturing.

According to the analysis of the distribution of cadmium in the entire vehicle, the results are shown in Figure 7. In the entire vehicle, a total of 13 material testing points detected the presence of cadmium.The cadmium content in the entire vehicle is generally low.The material with the highest cadmium content is electrolytic paper, which is 45mg/kg. Cadmium is widely distributed in metal materials, such as copper alloys, aluminum materials, coatings, etc. The main components involved include electrodes, metal washers, and camera shells.

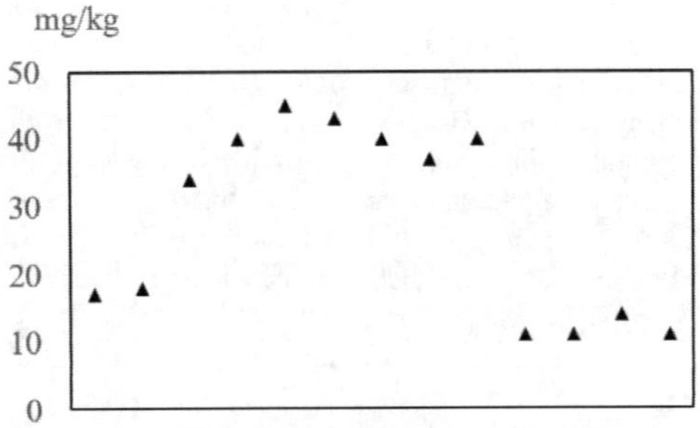

Figure 7. Distribution diagram of cadmium in different materials

4.5. Analysis of the Distribution of Hexavalent Chromium in Automotive Products

Hexavalent chromium may cause skin irritation, tissue and cell damage, and even digestive system damage to the human body[6,7].The industrial uses of hexavalent chromium compounds include chromate pigments in dyes, paints, inks, and plastics, chromates added as preservatives to paints, primers, and other surface coatings, and chromates electroplated on metal components to provide decorative or protective coatings.

The distribution of total chromium in automotive products was analyzed using XRF instruments, and the results are shown in Figure 8.The entire vehicle has 20000 points containing chromium, with a focus on materials such as metal, plastic, and rubber.But these chromium elements contain trivalent chromium and hexavalent chromium. Through further chemical quantitative analysis, the results showed that the entire vehicle did not contain hexavalent chromium.

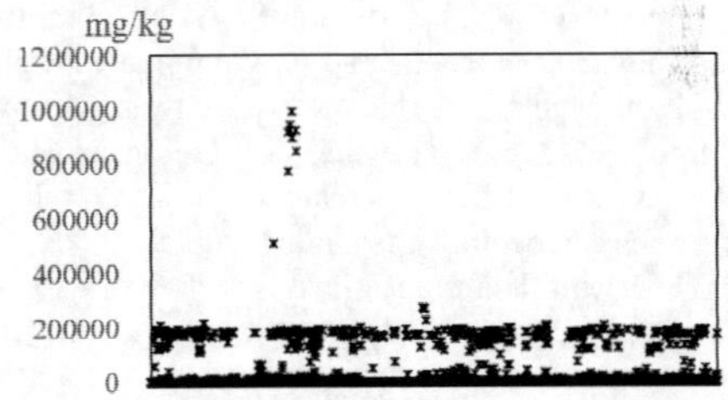

Figure 8. Distribution of Total Chromium in Different Materials

4.6. Analysis of the distribution of brominated flame retardants in automotive products

Polybrominated diphenyl ethers and polybrominated biphenyls, their toxicity mainly affects the human endocrine system and fetal growth. When plastics containing such flame retardants are overheated or incinerated, they will produce bromodiphenyldioxins or furans, which are carcinogens and can cause serious and widespread pollution, including soil, water, and air.Brominated flame retardants can be added to composite materials during the manufacturing process of automotive products to improve their fire resistance.

Through chemical quantitative detection and analysis, it was found that the entire vehicle does not contain polybrominated biphenyls (PBBs), but a small portion of materials contain polybrominated biphenyl ethers (PBDEs), as shown in Table 2.The polybrominated diphenyl ethers (PBDEs) contained in the entire vehicle are mainly nine brominated diphenyl ethers and ten brominated diphenyl ethers, and are mainly distributed in materials such as plastics and rubber, involving components such as shock absorbers and plastic shells.

Table 2 Distribution of Polybrominated Diphenyl Ethers in Different Materials

No.	Component Name	PBDEs（mg/kg）	Material type
1	Shock absorber rubber ring	Nine bromodiphenyl ether: 768 Decabromodiphenyl ether: 17098 (exempted)	rubber
2	plastic shell	258	plastics
3	plastic shell	255	plastics
4	plastic shell	256	plastics
5	rubber block	Nine bromodiphenyl ether: 840 Decabromodiphenyl ether: 15847 (exempted)	rubber

5. conclusion

This article provides a detailed disassembly and inspection of a vehicle model, involving 385 primary assembly components and over 20000 materials. This study first conducted initial screening using XRF instruments, and then determined the accurate content of lead, mercury, cadmium, hexavalent chromium, polybrominated biphenyls, and polybrominated biphenyl ethers through chemical quantitative analysis. Meanwhile, this article also analyzed the distribution of harmful substances in different materials. The following conclusions can be drawn from this article:

1)The harmful substance lead is widely distributed in automotive materials. A total of 813 materials in the vehicle contain lead, which is mainly distributed in components or materials such as batteries, electronic and electrical equipment, steel, and aluminum.

2)Lead in automotive materials is mainly distributed in electronic and electrical equipment. Metal materials have the second highest lead content. Lead content is relatively low in plastics.

3)Mercury and cadmium are less distributed in automotive materials. The automotive products in this study do not contain hexavalent chromium.

4)In this inspection, no polybrominated biphenyls (PBBs) were found in the car, and only a small portion of the materials contained polybrominated biphenyl ethers (PBDEs), mainly nine brominated biphenyl ethers and ten brominated biphenyl ethers..

In summary, the automotive products studied and tested in this article generally contain harmful substances such as lead, mercury, and cadmium, but all of them meet the standard requirements. It can be seen that with the release and implementation of policies related to the management of harmful substances in automotive products in China, the harmful substances in automotive products have been effectively controlled.

References

1. Is European end-of-life vehicle legislation living up to expectations? Assessing the impact of the ELV Directive on 'green' innovation and vehicle recovery[J]. Jason Gerrard;;Milind Kandlikar.Journal of Cleaner Production,2005.
2. The End-of-life Vehicle Ordinance in the German automotive industry – corporate sense making illustrated[J]. Tillmann Blume;;Michael Walther.Journal of Cleaner Production,2012.
3. Yoo;;Bong-Soo Lim;;Jin-Ho Park.Waste Management,2004(6)
4. Alpha-Lipoic Acid Attenuates Cadmium- and Lead-Induced Neurotoxicity by Inhibiting Both Endoplasmic-Reticulum Stress and Activation of Fas/FasL and Mitochondrial Apoptotic Pathways in Rat Cerebral Cortex[J]. Yan Yuan;Shi Wen Zhao;Shuang Quan Wen;Qiao Ping Zhu;Li Wang;Hui Zou;Jian Hong Gu;Xue Zhong Liu;Jian Chun Bian;Zong Ping Liu.Neurotoxicity Research,2021.
5. Cadmium exposure modulates the gut-liver axis in an Alzheimer's disease mouse model[J]. Zhang Angela;Matsushita Megumi;Zhang Liang;Wang Hao;Shi Xiaojian;Gu Haiwei;Xia Zhengui;Cui Julia Yue.Communications Biology,2021
6. Status and environmental management of soil mercury pollution in China: A review[J]. Liu Sijia;Wang Xuedong;Guo Guanlin;Yan Zengguang.Journal of Environmental Management,2021.
7. Hexavalent chromium-reducing plant growth-promoting rhizobacteria are utilized to bio-fortify trivalent chromium in fenugreek by promoting plant development and decreasing the toxicity of hexavalent chromium in the soil[J]. Soni Sumit K.;Kumar Govind;Bajpai Anju;Singh Rakshapal;Bajapi Yashi;Laxmi ;Tiwari Sudeep.Journal of Trace Elements in Medicine and Biology,2023.

https://doi.org/10.1051/e3sconf/202452803013

Analysis of the impact of differentiated carbon tax policies on shore power technology selection by port and shipping enterprise

Xudong Liu*, Shuyin Zhang

Institute of Logistics Science and Engineering, Shanghai Maritime University, Shanghai,201306, China

Abstract. Reducing carbon emissions has become a common goal in today's world for the protection of the environment, and the shipping industry, as an important force in the development of global trade, will be unavoidably involved in this conversion. This research paper develops an evolutionary game model between port and shipping enterprise under the differentiated carbon tax policy, analyses the evolutionary stabilization strategies between port and shipping enterprise, and explores the influence of relevant influencing factors on the evolutionary path. The results show that the differentiated carbon tax policy can effectively incentive port and shipping enterprise to adopt shore power technology, and the greater the difference between carbon taxes the more obvious the promotion effect of shore power. In addition, lowering the construction and operation costs for port and shipping enterprise to use shore power technology, and increasing other costs of not using shore power technology can contribute effectively to the willingness of shipping companies to utilize shore power.

1. Introduction

With shipping accounting for about 80% of global trade transportation and 2.89% of global greenhouse gas emissions, there is an urgent need to reduce carbon emissions from the shipping industry[1].Lately, the International Maritime Organization (IMO) has been considering the possibility of introducing a global shipping carbon tax, which would be levied based on a ship's carbon emissions, with the proceeds supporting the development of low-carbon technologies[2].But a harmonized carbon tax policy could have a negative impact on smaller shipping enterprises that are less resilient to external risks, meanwhile potentially reducing the overall capacity of the global shipping market[3].In contrast, a differentiated carbon tax policy could target different ports and shipping enterprises, thereby reducing greenhouse gas emissions more effectively, meanwhile taking into account economic efficiency, fairness and industry competitiveness[4], [5].In practice, some countries and regions have begun to explore and implement differentiated carbon tax policies. For example, the use of low-carbon or carbon-free fuels is encouraged through the imposition of carbon taxes at different rates on ships using different types of fuels; or tax incentives or subsidies are provided for ships and ports adopting energy-efficient technologies and operational optimization to reward their efforts to reduce emissions[6].For the purpose of reducing the pollution to the environment caused by power generation from moored ships, more and more experts and scholars have begun to advocate shutting down the auxiliary engines of the ships and connecting the ships with the shore power grid to get power while moored, i.e., utilizing the shore power to generate electricity[7], [8].However, some ports and shipping enterprises are reluctant to adopt shore power due to its high construction and operating costs. Therefore, how to promote the use of shore power in ports and shipping enterprises has become a hot topic of concern today.

This paper focuses on the decision behavior of port and shipping enterprise on whether to utilize shore power and its interaction under the guidance of differentiated carbon tax policy, and analyses the dynamic evolution process and steady state of the two parties under different combinations of strategies through the perspective of evolutionary game theory.

2. Evolutionary game model construction and solving

2.1. Specimens results

The participants of the game are ports and ship enterprises, both of which are finite rational beings and will pursue the maximization of their own utility. Among them, the probability that the port "uses shore power" in the strategy of using shore power is x ($0 \le x \le 1$), then $1 - x$ is the probability of "not using shore power". Meanwhile, assuming that the probability of ship enterprises "using shore power" is y ($0 \le y \le 1$), then $1 - y$ is the probability of "not using shore power".

When the port chooses to carry out the "use of shore power" strategy, it needs to pay the corresponding

*Corresponding author's e-mail: 2396735705@qq.com

construction costs C_{p1}, in addition, the construction of shore power, after the adoption of shore power technology, the use of shore power for the port of the incremental revenue for the port of I_1, but the port needs to bear a certain degree of shore power operation and maintenance costs C_{p2}.

Similarly, when the port chooses to carry out the "use of shore power" strategy, meanwhile, the ship enterprise chooses the "use of shore power equipment" strategy, at this time, the ship enterprise needs to bear the cost of transforming the shore power system as C_{s2}, the incremental revenue of the ship after adopting the shore power technology as I_2. When ports or ship enterprises use shore power technology, they need to pay carbon tax as T_1; when they do not use shore power technology, they need to pay carbon tax as T_2,T_1<T_2.

2.2. Construction of the model and its solution

According to the previous hypotheses, the payoff matrix of game between port and shipping enterprise is established (see Table 1).

Table 1. Payoff matrix for port and ship enterprise

		ship enterprise	
		Accepting shore power	No shore power accepted
port	Implementation of shore power	$R_{po}-T_1Q_2-C_{p1}-C_{p2}+I_1$, $R_{so}-T_1Q_4-C_{so}-C_{s1}-C_{s2}+I_2$	$R_{po}-T_1Q_2-C_{p1}-C_{p2}$, $R_{so}-T_1Q_4-C_{so}-C_{s1}-C_{s2}$
	No implementation of shore power	$R_{po}-T_2Q_1$, $R_{so}-T_2Q_3-C_{s3}$	$R_{po}-T_2Q_1$, $R_{so}-T_2Q_3-C_{s3}$

2.2.1. *An Analysis of the Evolutionary Stability of Port Strategies.*

For ports, the expected benefits of "implementing shore power" are:

$$U_{11}=y(R_{po}-T_1Q_2-C_{p1}-C_{p2}+I_1)+(1-y)(R_{po}-T_1Q_2-C_{p1}-C_{p2}) \tag{1}$$

The expected benefits of "no shore power implementation" are:

$$U_{12}=y(R_{po}-T_2Q_1)+(1-y)(R_{po}-T_2Q_1) \tag{2}$$

Then the average return is:

$$U_1=xU_{11}+(1-x)U_{12} \tag{3}$$

The dynamic equation for the replication of the port is:

$$F(x)=x(x-1)(C_{p1}+C_{p2}-T_2Q_1+T_1Q_2-I_1y) \tag{4}$$

2.2.2. *Analysis of the Evolutionary Stability of Ship Enterprise Strategies.*

For shipping enterprises, the expected benefits of "accepting shore power" are:

$$U_{21}=x(R_{s0}-T_1Q_4-C_{s0}-C_{s1}-C_{s2}+I_2)+(1-x)(R_{s0}-T_1Q_4-C_{s0}-C_{s1}-C_{s2}) \tag{5}$$

The expected benefit of "not accepting shore power" is:

$$U_{22}=x(R_{s0}-T_2Q_3-C_{s3})+(1-x)(R_{s0}-T_2Q_3-C_{s3}) \tag{6}$$

Then the average return is:

$$U_2=yU_{21}+(1-y)U_{22} \tag{7}$$

The replicated dynamic equation for the shipping enterprises is:

$$F(y)=y(y-1)(C_{s0}+C_{s1}+C_{s2}-C_{s3}-T_2Q_3+T_1Q_4-I_2x) \tag{8}$$

2.3. Equilibrium strategy analysis for evolutionary game models

Setting the dynamic equations for replication equalized to zero results in four pure strategy equilibrium points, obtain the 5 equilibrium points of this game matrix:O(0,0),A(1,0),B(0,1),C(1,1),D[$(C_{s0}+C_{s1}+C_{s2}-C_{s3}-T_2Q_3+T_1Q_4)/I_2$, $(C_{p1}+C_{p2}-T_2Q_1+T_1Q_2)/I_1$)]The Jacobi matrices calculated from the dynamic equations for the replication of harbour and shipping companies are as follows:

$$J=\begin{bmatrix}(2x-1)(C_{p1}+C_{p2}-T_2Q_1+T_1Q_2-I_1y) & -I_1x(x-1)\\ -I_2y(y-1) & dF(y)/dy\end{bmatrix} \tag{9}$$

$$dF(y)/dy=c_{22}=(2y-1)(C_{s0}+C_{s1}+C_{s2}-C_{s3}-T_2Q_3+T_1Q_4-I_2x)$$

Taking the equilibrium points, and bringing them into and , respectively, results are obtained about the system equilibrium points corresponding to the following table 2.

Table 2. Different cases of eigenvalues of Jacobi matrices

Equilibrium points	Det(J)	Tr(J)	Stability
O（0，0）	+	-	ESS
A（1，0）	+	+	Unstable
B（0，1）	+	+	Unstable

C (1, 1)	+	-	ESS
D (x*, y*)	-	x	Saddle point

Based on local stability analysis, the equilibrium point is an ESS stable point when >0 and <0. In this paper, O(0,0) and C(1,1) are analyzed:

(1) O(0,0)

When $Q_1T_2 - C_{p1} - C_{p2} - Q_2T_1 < 0$, $C_{s3} - C_{s0} - C_{s1} - C_{s2} + Q_3T_2 - Q_4T_1 < 0$,the evolutionary stabilization strategy for port and shipping enterprise is (not implemented, not used). If $Q_1T_2 - C_{p1} - C_{p2} - Q_2T_1 < 0$, the port enterprise to implement shore power investment costs (construction costs, technology costs, labour costs, etc.) and the cost of operation and maintenance is large, the port enterprise tends not to implement; if $C_{s3} - C_{s0} - C_{s1} - C_{s2} + Q_3T_2 - Q_4T_1 < 0$, the cost of using shore power for the shipping enterprise is greater than the cost of not using shore power, and the final stabilization point falls at (no implementation, no use), and the evolution path diagram is shown in Figure 1(a).

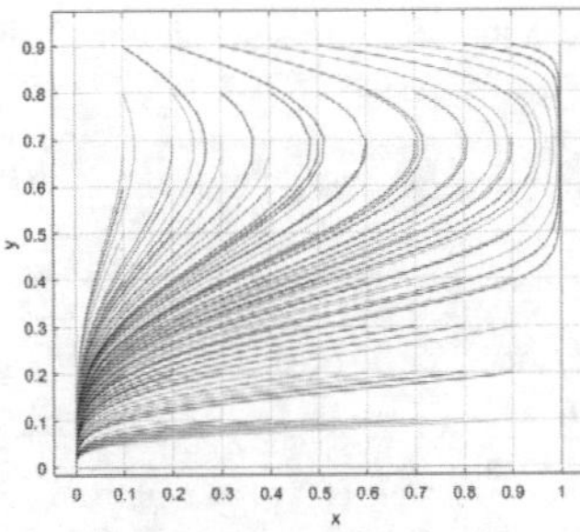

(a)

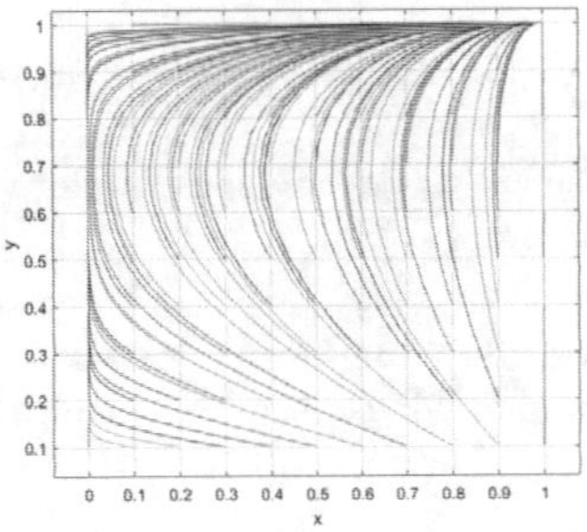

(b)

Figure 1. Evolutionary path diagram

(2) C(1,1)

When, $C_{p1} + C_{p2} - I_1 - Q_1T_2 + Q_2T_1 < 0$, $C_{s0} + C_{s1} + C_{s2} - C_{s3} - I_2 - Q_3T_2 + Q_4T_1 < 0$, the evolutionary stabilization strategy for port and shipping enterprises is (implement, use). If $C_{p1} + C_{p2} - I_1 - Q_1T_2 + Q_2T_1 < 0$, the port implementation of shore power investment costs (construction costs, technology costs, labour costs, etc.) and operation and maintenance costs are low, the incremental benefits for the port from the use of shore power by ships after the adoption of shore power technology is larger, the port tends to implement; if $C_{s0} + C_{s1} + C_{s2} - C_{s3} - I_2 - Q_3T_2 + Q_4T_1 < 0$, the cost of the use of shore power by the shipping enterprise is lower than the cost of not using shore power, the incremental benefits for the ship after the adoption of shore power technology is higher, and ultimately stabilized. point falls on (implementation, use), and the evolution path diagram is shown in Figure 1(b).

According to the evolutionary stability point O(0,0) in Table 2, i.e., the port does not put the shore power system into operation, the shipping enterprise does not receive shore power and the port puts the shore power system into operation and the shipping enterprise receives shore power C(1,1), the gaming behaviour reaches stability.A(1,0) and B(0,1), meanwhile, exhibit evolutionary instability, and D$((C_{s0} + C_{s1} + C_{s2} - C_{s3} - Q_3T_2 + Q_4T_1)/I_2\,,\,(C_{p1} + C_{p2} - Q_1T_2 + Q_2T_1)/I_1)$ is a saddle point.The evolution process of port and shipping enterprise's implementation of shore power is shown in Figure 2. Combining this figure with the above analysis, it can be seen that O(0,0) and C(1,1) are the evolutionary stable strategies in the evolutionary game system of port and shipping enterprise. And point D is used as the saddle point, although the specific strategy choice of both parties' decision-making cannot be determined, the dynamic evolution process of the game between port and shipping enterprise can be obtained through the movement of the saddle point.

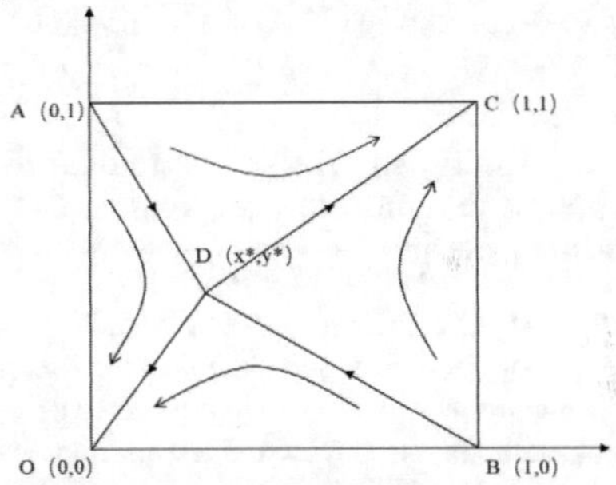

Figure 2. Dynamic evolutionary processes.

O(0,0) and C(1,1) are the evolutionary stabilization strategies in the evolutionary gaming system of ports and shipping enterprises. The final evolutionary outcome depends on the magnitude of the probabilities represented by the area of S_{ADBC} and S_{AOBD}.If $S_{ADBC} > S_{AOBD}$, the probability that the system converges to (0,0) is less than the probability that (1,1); if $S_{ADBC} < S_{AOBD}$, the probability that the system converges to (1,1) is less than the probability that (0,0).In order for both ports and shipping enterprises to implement shore power, measures should be taken to increase the area of S_{AOBD}. Further analysis of the factors affecting the area of S_{ADBC} and S_{AOBD} gives the main factors affecting the evolution path of both sides.

$$S_{ADBC} = S_{ADC} + S_{BDC} = 1 - \frac{1}{2}((C_{s0} + C_{s1} + C_{s2} - C_{s3} - Q_3T_2 + Q_4T_1)/I_2 + (C_{p1} + C_{p2} - Q_1T_2 + Q_2T_1)/I_1) \quad (10)$$

https://doi.org/10.1051/e3sconf/202452803013

3. Impact of different influences on evolutionary pathways

3.1. The effect of construction cost C_{s0} on the evolutionary path to shore power use by shipping companies:

Proposition 1. Holding the rest of the factors constant, the larger the construction cost of a shipping enterprise using shore power, the smaller the probability of cooperation between the two parties.

Proof: the more S_{ADBC} is a partial derivative of C_{s0} , $\partial S_{ADBC}/\partial C_{s0} = -(2I_2)^{-1} < 0$,and S_{ADBC} is a decreasing function of C_{s0}, the greater the probability that the system converges to (0,0).

3.2. Shipping enterprises are required to bear the cost of shore power usage during berthing C_{s1} on the evolutionary path:

Proposition 2. Holding the rest of the factors constant, the greater the cost of shore power usage in the berthing process that the shipping enterprise has to bear, the lower the probability of co-operation between the two parties.

Proof: the more S_{ADBC} is a partial derivative of C_{s1}, $\partial S_{ADBC}/\partial C_{s1} = -(2I_2)^{-1} < 0$, and S_{ADBC} is a decreasing function of C_{s1}, the greater the probability that the system converges to (0,0).

3.3. The shipping enterprise has to bear the cost of modifying the shore power system C_{s2} on the evolutionary path:

Proposition 3: The greater the cost that the shipping enterprise has to bear to modify the shore power system, the lower the probability of co-operation between the two parties, when the rest of the factors are constant.

Proof: the more S_{ADBC} is a partial derivative of C_{s2}, $\partial S_{ADBC}/\partial C_{s2} = -(2I_2)^{-1} < 0$, and S_{ADBC} is a decreasing function of C_{s2}, the greater the probability that the system converges to (0,0).

3.4. The effect of cost C_{s3} of not using shore power in shipping enterprises on the evolutionary path:

Proposition 4: The greater the cost to the shipping enterprise of not using shore power, the greater the probability of co-operation between the two parties, holding the rest of the factors constant.

Proof: the more S_{ADBC} is a partial derivative of C_{s3}, $\partial S_{ADBC}/\partial C_{s3} = (2I_2)^{-1} > 0$, and S_{ADBC} is an increasing function of C_{s3}, the greater the probability that the system converges to (1,1).

3.5. The impact on the evolutionary path of the carbon tax T_1 to be paid by port or shipping enterprises adopting shore power :

Proposition 5: Holding the rest of the factors constant, the smaller the carbon tax to be paid by the port or shipping enterprise that adopts shore power technology, the higher the probability of cooperation between the two parties.

Proof: the more S_{ADBC} is a partial derivative of T_1 , $\partial S_{ADBC}/\partial T_1 = -2^{-1}(Q_4 I_2{}^{-1} + Q_2 I_1{}^{-1}) < 0$, and S_{ADBC} is an increasing function of T_1, the higher the probability that the system converges to (0,0).

3.6. The impact on the evolutionary path of the carbon tax T_2 to be paid by port or shipping enterprises that do not adopt shore power:

Proposition 6: Holding the rest of the factors constant, the larger the carbon tax to be paid by the port or shipping enterprise that has not adopted shore power technology, the higher the probability of cooperation between the two parties.

Proof: the more S_{ADBC} is a partial derivative of T_2,$\partial S_{ADBC}/\partial T_2 = 2^{-1}(Q_3 I_2{}^{-1} + Q_1 I_1{}^{-1}) > 0$, and S_{ADBC} is an increasing function of T_2, the higher the probability that the system converges to (1,1).

3.7. Impact of construction cost C_{p1} on the evolutionary path of inputs for the use of shore power in ports:

Proposition 7: Holding the rest of the factors constant, the greater the construction cost invested in the use of shore power in the port, the lower the probability of co-operation between the two parties.

Proof: the more S_{ADBC} is a partial derivative of C_{p1} , $\partial S_{ADBC}/\partial C_{p1} = -(2I_1)^{-1} < 0$, and S_{ADBC} is a decreasing function of C_{p1}, the greater the probability that the system converges to (0,0).

3.8. Impact of operating costs C_{p2} on evolutionary pathways for the use of shore power in ports:

Proposition 8: The greater the operating cost of using shore power in the port, the lower the probability of co-operation between the two parties, holding the rest of the factors constant.

Proof: the more S_{ADBC} is a partial derivative of C_{p2} , $\partial S_{ADBC}/\partial C_{p2} = -(2I_1)^{-1} < 0$, and S_{ADBC} is a decreasing function of C_{p2}, the greater the probability that the system converges to (0,0).

4. Discussion

This study has two theoretical significance. On the basis of domestic and international literature on differentiated carbon tax and port and shipping shore power emission reduction research, this paper reveals the decision-making logic and behavioral motivation of all parties between

 https://doi.org/10.1051/e3sconf/202452803013

economic interests and environmental protection under the influence of the differentiated carbon tax policy, which can effectively incentive port and shipping enterprise to take emission reduction measures. Secondly, this study provides an in-depth analysis of the economic effects of the differentiated carbon tax policy, and explores the specific manifestations of the policy's incentive effects on ports and shipping enterprises and its impact on the competitiveness of the shipping industry as a whole. The results of the study show that through the enforcement of differentiated carbon tax policies, not only can we enhance the environmental performance of the port and shipping industry and promote the green transformation of the industry, but we can also strengthen the competitiveness of the port and shipping industry in the international arena and promote sustainable development.

The paper has also provided the government with practical policy insights. This paper proposes an innovative strategy to incentive port and shipping enterprises to reduce emissions through the policy of differentiated carbon tax.Firstly, the results of the study can provide guidance for the government to optimize the design of carbon emission reduction policies. By understanding the gaming behaviors of ports and shipping enterprises, the government can adjust the carbon tax rate in a targeted manner to achieve optimal emission reduction. Meanwhile, through the introduction of shore power technology for emission reduction, port and shipping enterprises can reduce carbon emissions and environmental risks, and based on the results of the study, the government can also formulate corresponding motivational policies that encourage port and shipping enterprise to use shore power more often to reduce emissions.

5. Conclusion

This paper analyses whether port and shipping enterprise make use of shore power or not under the consideration of differentiated carbon tax policy, constructs a two-party evolutionary game model of port-shipping enterprise under the consideration of differentiated carbon tax policy, and analyses the influencing factors of the strategy choices of each participant subject in the promotional process of port and shipping enterprise regarding the use of shore power for emission reduction under the differentiated carbon tax policy. The following conclusions are drawn:

(1) The construction and operation cost of shore power technology is negatively correlated with the willingness of port and shipping enterprise to use shore power, i.e., the higher the cost invested in shore power technology, the lower the probability of port and shipping enterprise to use shore power, and vice versa. China's port shore power there are many unfavourable factors for the promotion of shore power technology, for example, compared with the traditional diesel power generation, the use of shore power involves power stability, adaptability and other technical issues, which makes the use of shore power cost is higher. Therefore, the state should increase the research and extension of technology for shore power, reduce the cost of using shore power technology, and give certain subsidies to incentive port and shipping enterprise to use shore power technology.

(2) The higher the cost to the shipping enterprise of not using shore power, the higher the probability that the shipping enterprise will use shore power.The use of shore power can be promoted by adjusting the costs to shipping enterprise of not using shore power. For example, measures such as increasing the cost of bunker fuel for ships and increasing the environmental costs of using bunker fuel after entering a port could be used to induce shipping enterprise to use shore power.

(3) Differentiated carbon tax policy can effectively promote the use of shore power in ports and shipping enterprise, and the greater the difference between carbon taxes the more obvious the effect of the promotion of shore power.It is found that the higher the carbon tax for port and shipping enterprise not using shore power and the lower the carbon tax for using shore power will finally tend to both sides meanwhile using shore power. Therefore, the differentiated carbon tax policy can effectively promote the use of shore power and enhance the emission reduction effect of port and shipping enterprise. The implementation of this policy will create positive incentives for port and shipping enterprise to promote sustainable development and mitigate climate change.

References

1. Kim H, Koo KY and Joung T-H (2020) A study on the necessity of integrated evaluation of alternative marine fuels. Journal of International Maritime Safety, Environmental Affairs, and Shipping 4:26–31.
2. Meng L, Liu K, He J, Han C and Liu P (2022) Carbon emission reduction behavior strategies in the shipping industry under government regulation: A tripartite evolutionary game analysis. Journal of Cleaner Production 378:134556.
3. Lagouvardou S, Psaraftis HN and Zis T (2022) Impacts of a bunker levy on decarbonizing shipping: A tanker case study. Transportation Research Part D: Transport and Environment 106:103257
4. Tan Z, Miu C, Xu S and Wang Y (2021) Performance evaluation of remanufacturing strategies under uniform carbon tax and discriminated carbon tax. Journal of Systems Engineering 36:102–119.
5. Ma N, Yin G, Li H, Sun W, Wang Z, Liu G and Xie D (2022) The optimal industrial carbon tax for China under carbon intensity constraints: a dynamic input–output optimization model. Environ Sci Pollut Res 29:53191–53211.
6. Avetisyan M,(2018) Impacts of global carbon pricing on international trade, modal choice and emissions from international transport. Energy Economics,76:532–548
7. Yu J, Voß S and Tang G (2019) Strategy development for retrofitting ships for implementing shore side electricity. Transportation Research Part D: Transport and Environment 74:201–213.

8. Yin M, Wang Y and Zhang Q (2020) Policy implementation barriers and economic analysis of shore power promotion in China. Transportation Research Part D: Transport and Environment 87:102506.

https://doi.org/10.1051/e3sconf/202452803014

How do the construction of urban high-speed railways influence green total factor energy efficiency?

Ting Yu

School of Public Affairs, Zhejiang Gongshang University, Hangzhou, Zhejiang, 310018, China

Abstract. As a typical measure for reducing carbon footprint, the opening of high-speed railways (HSRs) is a powerful lever for attaining the strategic purpose of carbon peak and carbon neutrality. Derived from the data of 265 prefecture-level cities in China from 2001 to 2018, this study adopts the Dynamic Network Slacks-Based Measure (DNSBM) model and introduces link variables and carry-over variables to conduct a dynamic assessment of green total factor energy efficiency (GTFEE). The study finds that the opening of HSRs significantly improved the GTFEE and this conclusion is robust under multiple conditions such as PSM-DID, counterfactual test and exclusion of other policy interference. Heterogeneity analysis indicates that the construction of HSRs has a more significant positive influence on GTFEE in smart city pilot cities, non-old industrial base cities middle and lower reaches of the Yangtze River Economic Belt (YREB). Mechanism testing indicates that economy concentration effect, structural optimization effect, and innovation driven effects are the transmission channels of HSRs in promoting low-carbon transition.

1. Introduction

Since the reform, China has adopted a development model characterized by high energy consumption and contamination. This model has made it the second biggest economy in the word. Nonetheless, this extensive model of economic development has given rise to corresponding environmental challenges, leading to China becoming the world's largest energy consumer. Within the framework of green and sustainable development, inefficient utilization of energy and extensive development models are inherently unsustainable. The enhancement of GTFEE involves a dynamic process of improving the energy structure, facilitating the shift to higher-value industries, and realizing a low-carbon transition. Therefore, promoting GTFEE has become a crucial measure for achieving high-quality economic development.

To attain sustainable development and improve energy efficiency through green transition, the opening of HSRs and the construction of an HSRs network should play crucial roles. In July 2016, China's State Council approved the *Medium and Long Term Railway Network Plan*, which aims to achieve regional interconnection and coordinated development throughout China by 2030. These developments prompt the topic of discussion in this study: could the ongoing improvement of the HSR connections effectively promote the enhancement of GTFEE? If so, what mechanism underlies this relationship? This will provide empirical support and valuable references for leading efforts in realizing the strategic objectives of peaking carbon emissions and attaining carbon neutrality.

Diverging from prior research, this study introduces these novel ideas: Firstly, from the perspective of the establishment of HSRs, this paper utilizes the staggered difference-in-differences (DID) model to identify the net policy effects of HSRs construction on GTFEE, and elaborating on the transmission channels. Second, based on the DNSBM model, the dynamic assessment of the GTFEE of 265 cities in China from 2001 to 2018 through link variables and inter-temporal variables.

2. Literature review

As an essential transportation infrastructure, the inauguration and building of HSRs markedly influence the economy, social development, and low-carbon transition of cities along the route. Firstly, in terms of economic benefits, Knaap and Oosterhaven (2011) found that HSRs opening possessed a notably beneficial effect regarding the per capita GDP of cities along the route [1]. Scholars also pay attention to the "siphon effect" caused by HSRs opening (Schloetzer, 2012) [2]. In terms of environmental benefits, Yue et al. (2015) demonstrated that HSR exhibits clear characteristics of low carbon emissions and high environmental protection [3]. Guo (2021) found that cities with higher centrality in HSRs networks tend to experience fewer industrial water pollution incidents [4].

Currently, there are abundant achievements in investigating the factors that affect energy efficiency. Scholars generally believe that the introduction of comprehensive environmental regulations will positively impact urban energy efficiency (Cheng et al., 2019) [5]. Additionally, technological progress (Chunbo and David,

Corresponding author's e-mail: yut200209@163.com

https://doi.org/10.1051/e3sconf/202452803014

2008) [6], infrastructure development (Chen, 2021) [7], and urban environmental legislation (Zhang et al., 2022) [8] all constitute significant factors influencing GTFEE.

Upon examining the current body of literature, it can be found that there are abundant research results on the spin-off effects of HSRs opening and the influencing factors of GTFEE. However, there are still some shortcomings: firstly, while existing studies have extensively discussed HSRs opening and energy efficiency improvement, few studies have combined the two, resulting in an insufficient comprehension of the mechanism of action. Secondly, most studies use the static DEA method containing non-expected output for energy efficiency measurement, with the non-radial DNSBM model being rarely used.

3. Theoretical analysis and research hypotheses

The opening of HSRs not only leads to a direct reduction in carbon emissions through improved transportation modes but also exerts a considerable indirect influence on energy efficiency improvement. The opening of HSRs not only reduces the use of other energy-intensive vehicles, but also reduces the carbon emissions caused by road traffic congestion. On the other hand, it increases the likelihood of logically allocating innovation elements among different regions, accelerates knowledge spillover and promotes producers to achieve higher industrial specialization with less resource consumption. Drawing from this, the paper suggests the subsequent hypothesis:

Hypothesis 1: The opening of HSRs is beneficial for enhancing GTFEE.

Viewing through the lens of economy concentration effect, the inauguration of HSRs is expected to foster the economic concentration of cities along this trajectory, thereby impacting their carbon emission intensity and improving energy efficiency. The opening of HSRs can stimulate the flow and redistribution of capital, population, technology, and other factors, and attract more enterprise investment and population concentration, thereby improving the resource efficiency required for economic development per unit of output. From the perspective of structural optimization effect, the opening of HSRs will be accompanied by a higher passenger volume, and then advance the consolidation and growth of the tertiary industry with strong mobility of production factors, and effectively promote the transition of low-carbon economy. Compared with the first and second industries, the service industry takes knowledge and technology as the core, has the traits of low pollution intensity, and holds the potential to drive the industry's evolution towards intelligence, environmental conservation, and enhanced value. Viewing from the angle of innovation driven effects, the opening of HSRs provides an innovative platform and impetus for enhancing GTFEE. not only facilitates the rapid flow and sharing of innovative elements but also generates knowledge spillover effects. Drawing from this, the paper suggests the subsequent hypothesis:

Hypothesis 2: The opening of HSRs can improve GTFEE through economy concentration effect, structural optimization effect and innovation driven effect.

4. Model construction, variable selection and data source

4.1. Model construction

This study, grounded in theoretical examination, employs the staggered DID to assess the overall policy effect of HSRs' opening on GTFEE. Establish the following equation (1).

$$GTFEE_{it} = \alpha + \beta HSR_{it} + \eta X_{it} + \mu_i + \varphi_t + \varepsilon_{it} \quad (1)$$

Within this framework, $GTFEE_{it}$ symbolizes the aggregate energy efficiency factor of city i in the year t; HSR_{it} is the independent variable of the construction of HSRs. The coefficient β reflects the net policy effect of HSRs construction on GTFEE. X_{it} symbolizes the collection of control variables, μ_i symbolizes the city-fixed effect, φ_t symbolizes the year-fixed effect, ε_{it} is a random error term.

4.2. Selection and measurement of variables

4.2.1. Dependent variable: green total factor energy efficiency.

The DNSBM model facilitates the gathering and examination of pertinent input and output variables, along with carry-over and free link variables, to derive GTFEE. Within the production sector, the input factor is energy consumption, while the output factor is GDP. Within the realm of research and development, the input factor represents the city's total expenditure on science and technology, while the output factor is the aggregate tally of its patents. Labor force can be used as the free link variable amidst the departments of production and R&D, quantified through the aggregate employment count in each city throughout the years. In this paper, carry-over variables are classified into two categories. One is to set the city capital stock as a free intertemporal variable and calculate it by adopting the perpetual inventory approach. The other is the bad carry-over variables, which mainly include some environmental pollutants, such as CO2, PM2.5, etc.

4.2.2. Independent variable: the construction of high-speed railway.

Within this study, HSRs' opening is taken as the independent variable. If city i opens its first HSR in year t, HSR_{it}=1 is set from year t and subsequent years. Otherwise, HSR_{it}=0.

4.2.3. Control variables.

They mainly include: urban economic development level (*gdp*): the measurement of this study is determined by the logarithmic value of the per capita GDP in urban areas; urbanization rate (*ur*): this paper gauges the percentage of non-agricultural population in the overall population as the year concluded; foreign direct investment (*fdi*): this article measures the logarithmic value of the utilization of foreign investment in cities; government financial expenditure (*gov*): evaluated using the logarithm of local government's general public budget expenditure; city productivity (*pro*): determined by comparing the total urban product with the employee count per unit at the period's conclusion.

4.3. Data sources and statistical analysis

The data from this study falls into three distinct groups: First, the data related to the opening of HSRs over the years are obtained from the National Railway Administration. Additionally, macroeconomic figures like urban GDP and fiscal spending originate from the *China Urban Statistical Yearbook* spanning various years. Thirdly, data on energy consumption at both provincial and industrial tiers originate from the *China Energy Statistical Yearbook*. Table 1 exhibits the detailed statistical information pertaining to the variables in this article.

Table 1. Descriptive statistics of variables.

Variable	Obs.	Mean	St. Dev.	Min.	Max.
GTFEE	5035	0.2759	0.4470	0	1
gdp	5035	10.0730	0.8464	4.5951	13.0557
ur	5035	0.4131	0.2255	0.0381	1
fdi	5035	11.3165	2.0918	2.4932	16.8742
gov	5035	13.4190	1.2480	8.3603	18.2405
pro	5035	12.3044	0.7215	8.7073	14.4988

5. Analysis of empirical results

5.1. Benchmark regression results

This paper empirically tests the net effect of HSRs opening on GTFEE by constructing a two-way fixed effect model of city and year. Table 2 displays the standard regression outcomes. The columns (1) through (6) are listed as all control variables including city economic growth level, urbanization rate, etc., while regression results of controlling for fixed effects of years and cities. One can discover that the coefficient of HSR variable is notably positive. The benchmark regression results in Table 2 indicate that the construction of HSRs exhibits a beneficial effect on enhancing GTFEE.

Table 2. Benchmark regression results[a].

	GTFEE					
	(1)	(2)	(3)	(4)	(5)	(6)
HSR	0.0119*	0.0120*	0.0116*	0.0115*	0.0149**	0.0154**
	(0.0066)	(0.0066)	(0.0065)	(0.0065)	(0.0061)	(0.0062)
gdp		0.0004	-0.0002	0.0006	0.0088	0.0052
		(0.0081)	(0.0082)	(0.0082)	(0.0085)	(0.0083)
ur			0.0181	0.0183	0.0212	0.0244
			(0.0203)	(0.0203)	(0.0199)	(0.0198)
fdi				-0.0013	0.0003	0.0001
				(0.0015)	(0.0014)	(0.0014)
gov					-0.0535***	-0.0541***
					(0.0055)	(0.0054)
pro						0.0098
						(0.0069)
_cons	0.1402***	0.1364*	0.1403***	0.1417*	0.7562***	0.6807***
	(0.0018)	(0.0816)	(0.0018)	(0.0819)	(0.1043)	(0.1230)
City FE	Y	Y	Y	Y	Y	Y
Year FE	Y	Y	Y	Y	Y	Y
N	5035	5035	4970	5035	5035	5035
R^2	0.539	0.539	0.539	0.539	0.570	0.570

[a] Notes: standard errors clustered at the urban level are in parentheses. *, **, and *** respectively significant at 10%, 5%, and 1%. The following empirical tables encompass every control variable along with fixed effects.

5.2. Parallel trend test

The study employs the event study approach to further confirm the parallel trend that the opening of HSRs influences the shift towards a low-carbon urban environment. The first year before the opening of HSRs is taken as the d_1 period, and so on to the d_4 period; the first period after the implementation is taken as d1 period, and so on to d5 period, and the first year before the opening of the HSRs is taken as the base period (current period). This is evident from Figure 1, the regression coefficients before the base period have a small fluctuation range and fail the significance test, suggesting the absence of any notable disparity in GTFEE level between the experimental and the control group before the construction of HSRs, which passes the parallel trend test.

https://doi.org/10.1051/e3sconf/202452803014

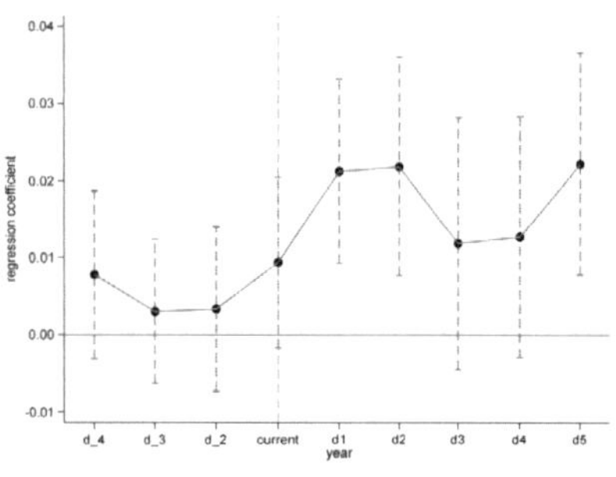

Figure 1. Results of the parallel trend test.

5.3. Robustness test

5.3.1. PSM-DID test.

This study further uses the PSM-DID technique for validation. Specifically, the article embraces the nearest neighbour matching method (the matching ratio is set at 1:1 and 1:3, respectively), and the nuclear radius matching method to estimate samples that were paired. The empirical results are displayed in columns (1)-(3) of Table 3. The coefficients of HSR are consistently and markedly positive, indicating strong robustness of the conclusion.

Table 3. Robustness test.

	GTFEE							
	(1)	(2)	(3)	(4)	(5)	(6)	(7)	(8)
HSR	0.0172***	0.0156**	0.0158**	0.0005	-0.0050	-0.0045	0.0153**	0.0153**
	(0.0064)	(0.0062)	(0.0061)	(0.0061)	(0.0056)	(0.0053)	(0.0062)	(0.0062)
N	4743	4970	4683	5035	5035	5035	5035	5035
R^2	0.571	0.570	0.562	0.569	0.569	0.569	0.571	0.570

5.3.2. Counterfactual test.

This article also conducts a counterfactual check by constructing a false opening time for HSRs. Specifically, the opening time of HSRs in cities is advanced by 1 year, 2 years and 3 years respectively, which continues to be multiplied by the grouping dummy variable, and this interaction term is included in the benchmark regression model for estimation again. The columns (4) to (6) of Table 3 are listed as the test results of constructing false legislation time. It can be found that the coefficients of HSR dummy variables fail to succeed in the test of significance, which is contrary to the conclusion of benchmark regression. The results of the counterfactual test once again show that the opening of HSRs is beneficial for enhancing the GTFEE.

5.3.3. Controlling other policy interference.

The establishment of national high-tech development zones, which promote close cooperation between enterprises and scientific research institutions, may greatly advance the growth of low-carbon economy. Executing pilot policies for cities with low carbon emissions could enhance the subjective initiative of low-carbon technology in pilot cities, and then encourage a shift towards a low-carbon lifestyle. This paper sets dummy variables for the above pilot policies respectively. The regression outcomes are displayed in columns (7) - (8) of Table 3, the positive impact of the opening of HSRs on GTFEE is valid.

5.4. Heterogeneity analysis

5.4.1. Heterogeneity policy pilot cities.

The pilot policy for smart cities in China is designed to significantly enhance the transition towards eco-friendly environments by investing more energy in promoting and researching technologies. The study categorizes the sample into smart city pilot cities and non-pilot cities to explore the varied effects of HSRs inauguration on GTFEE. The estimated results are displayed in columns (1) - (2) of Table 4. Observation reveals that the opening of HSRs has a more significant beneficial impact on enhancing energy efficiency in the pilot area of smart city policy. As a new urban development concept and management model, smart city plays an indispensable role during the achievement of the "dual carbon" objective and the low-carbon transition evolution of cities.

Table 4. Heterogeneity analysis.

	GTFEE					
	(1)	(2)	(3)	(4)	(5)	(6)
HSR	0.0203**	0.0087	0.0223**	0.0099	0.0237*	0.0020
	(0.0079)	(0.0102)	(0.0102)	(0.0072)	(0.0124)	(0.0130)
N	2869	2166	1957	3078	1425	532
R^2	0.602	0.505	0.577	0.598	0.607	0.547

5.4.2. Heterogeneous urban geographic location.

The growth of China's economy is greatly dependent on the YREB. Table 4, columns (3) - (4) display the estimated outcomes of categorizing all sampled cities into urban groups within and outside the YREB. It is evident that the inauguration of urban HSRs markedly enhances the low-carbon conversion of urban areas within the YREB. Additionally, the study categorizes the YREB into upstream and middle and downstream cities, the empirical results in column (5) – (6) of Table 4 reveal that initiating HSRs markedly enhances GTFEE in urban areas along the Yangtze River's middle and lower segments. The central and lower segments of the YREB boast robust urban innovation skills, the ease of transport will enhance urban collaboration. Consequently, launching HSRs in the central and lower segments of the Yangtze River plays a more significant role in enhancing the GTFEE.

5.5. Transmission mechanism test

By computing the proportion of the extra value of non-agricultural output to the region designated for urban administration, the degree of urban economy concentration can be objectively measured. Enhancing the industrial framework entails altering the balance among primary, secondary, and tertiary industries, which is measured by the percentage of the tertiary industry in the study. The urban innovation level index used in this paper is derived from the Urban innovation index proposed by the Industrial Development Research Center of Fudan University in the *China Urban and Industrial Innovation Capacity Report 2017*. Table 5 shows the mechanism test results of the impact of HSRs opening on GTFEE. Column (1) – (2) show that the construction of HSRs is conducive to promoting urban economy concentration, and the economic clusters will promote the improvement of GTFEE. The estimated results in column (3) - (4) indicate that the opening of HSRs can advance the enhancement of city industrial frameworks, and the industrial structure upgrading can promote GTFEE. The estimated results in columns (5) - (6) indicate that HSRs' opening can promote low-carbon transition by improving the overall innovation level.

Table 5. Mechanism test.

	scale	*GTFEE*	*Ind*	*GTFEE*	*innov*	*GTFEE*
	(1)	(2)	(3)	(4)	(5)	(6)
HSR	0.3840*** (0.0934)		0.3822*** (0.0998)		0.3168*** (0.0463)	
scale		0.0088*** (0.0027)				
Ind				0.0001*** (0.0000)		
innov						0.0206*** (0.0061)
N	5033	5033	5033	5033	4240	4240
R^2	0.661	0.574	0.384	0.569	0.898	0.625

6. Conclusions

Initially, this study elucidates the theoretical mechanism and transmission channel of the opening of HSRs affecting GTFEE; secondly, using DNSBM model method, the energy efficiency level of 265 cities from 2001 to 2018 is identified and dynamically evaluated, and the DID is used for empirical test. The research results indicated that: The opening of HSRs has markedly facilitated the process of urban low-carbon transition, and this conclusion remains robust under multiple circumstances such as PSM-DID, counterfactual test and exclusion of other policy interference. Furthermore, introducing HSRs in urban areas significantly enhances GTFEE in pilot cities of smart cities, and the central and lower segments of the YREB. The test of mechanisms reveals economy concentration effect, structural optimization effect, and innovation driven effects are the transmission channels for the opening of HSR to facilitate low-carbon shift.

The above conclusions have important policy implications to adhere to the opening and improvement of urban HSRs network, improve GTFEE, and foster energy sustainability and emission control goals. First, continuously improve and carry out unique HSRs opening policies conforming to local situations, comprehensively use a variety of scientific policy tools, and stimulate the endogenous driving force of GTFEE improvement. Second, take the construction of HSR as an important driving force, induce enterprises to settle in and thus generate economy concentration, and concurrently, fully utilize the potential of innovation platform to accelerate the spillover effect of green technology. Third, strengthen regional coordination and cooperation to increase HSR construction. Break the barriers of inter-regional flow of various factors, optimize the layout of the existing HSRs network, maximize the economic and ecological advantages resulting from the construction of HSRs, and further promote the improvement of GTFEE.

References

1. Knaap, T., Oosterhaven, J. (2011) Measuring the welfare effects of infrastructure: a simple spatial equilibrium evaluation of Dutch railway proposals. J. Res. Transp. Econ., 31(1):19-28.
2. Schloetzer J D (2012) Process integration and information sharing in supply chains. J. Account. Rev., 87(3):1005-1032.
3. Yue Y, Wang T, Liang S, et al (2015) Life cycle assessment of high speed rail in China. J. Transport. Res. D-Tr. E., 41: 367-376.
4. Guo H, Chen C, Dong X and Jiang C (2021) The evolution of transport networks and the regional water environment: the case of Chinese high-speed rail. J. Reg. Stud., (6):1084-1110.

5. Cheng J, Yi J, Dai S, et al (2019) Can low-carbon city construction facilitate green growth? Evidence from China's pilot low-carbon city initiative. J. J. Clean. Prod., 239:1158-1170.
6. Chunbo M and David I S (2008) China's changing energy intensity trend: a decomposition analysis. Energ. Econ., 3:1037-1053.
7. Chen Y and Lin B (2021) Understanding the green total factor energy efficiency gap between regional manufacturing-insight from infrastructure development. J. Energy., 237(3):121553.
8. Zhang B B, Yu L and Sun C W (2022) How does urban environmental legislation guide the green transition of enterprises? Based on the perspective of enterprises' green total factor productivity. J. Energ. Econ., 110:106032.

E3S Web of Conferences 528, 03015 (2024)
NET-LC 2024 https://doi.org/10.1051/e3sconf/202452803015

Diffusion of ESG Behavior Improvement Strategy for New Energy Enterprises under Weighted Benefit Risk Mechanism ——Based on Complex Network Evolution Game

Dan Lei[1], Qin Liu[2,*]

[1]School of Management, Wuhan University of Technology, Wuhan, Hubei, 430070, China
[2]School of Entrepreneurship, Wuhan University of Technology, Wuhan, Hubei, 430070, China

Abstract. In order to explore the impact of the enterprise's right, responsibility and profit risk mechanism on the enterprise's ESG behavior improvement strategy in the context of carbon emission reduction, this study uses Matlab to build a new energy enterprise cooperation game network model and establish the NW small world initial network. According to the process and result of multi-agent complex network game, this paper analyzes the strategic choice and development factors of ESG behavior in new energy enterprises. It is found that when the awareness of executives increases, the profits of enterprises will increase, but the spread of ESG behavior improvement is not significant; When the competitive intensity of enterprises increases, the profits of enterprises with ESG behavior improvement will decrease, and the diffusion of ESG behavior improvement strategy will be slow. When the price of carbon trading increases, the income of enterprises adopting ESG behavior will increase, but the sensitivity of carbon trading price to the expansion of enterprise network is not significant.

1. Introduction

The global climate issue makes the world further reflect on the relationship between human beings and nature, and the ESG value management concept of pursuing sustainable economic development has gradually become the consensus of all mankind. The scope of enterprise value creation extends from economic value to external environment and social value. The ESG concept requires enterprise managers to take into account economic, social and environmental benefits, and provide social and environmental responsibility and performance information that meets the needs of stakeholders. ESG evaluation system is an important means to promote enterprises to implement low-carbon transformation and achieve sustainable development. In recent years, with China's economy gradually shifting from high-speed growth to high-quality development stage, enterprises need to constantly strengthen the attention to external environment, social ethics and internal governance, and properly coordinate the relationship between human and nature, human and society, so as to achieve high-quality development of enterprises[1].

Therefore, how to spread the ESG behavior improvement strategy of China's new energy enterprises needs to be further explored. This paper starts with constructing a more perfect ESG evaluation system for new energy industry, and evaluates the ESG sustainable development ability of enterprises based on the enterprise right, responsibility, profit and risk mechanism.

2. Research design

This study determines the attributes of the firm body and studies the influence of individual firms' game decisions on themselves and their partners. The study design is as Figure 1：

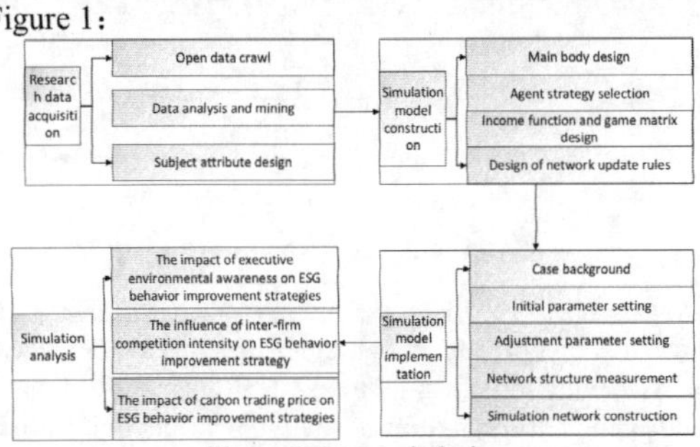

Figure 1. Research design

The research objects are new energy enterprises and government. The role of policy instruments is complex, it needs to consider the interactions between multiple agents, and system dynamics do not apply well to micro, individual decision changes. At the same time, the policy simulation process of this study focuses on discussing the effect of different right, responsibility, profit and risk mechanisms and their combinations on enterprise cooperation network. This is a dynamic adjustment process, so this study is more suitable for multi-agent simulation methods[2].

*Corresponding author's e-mail : D13714557850@outlook.com

3. Game model design

3.1 Game Model

3.1.1 Construction of small-world network model

With the rapid development of the Internet society, the relationship between enterprises is getting closer and closer, with the characteristics of a small world. In addition, the enterprises in this paper are universal, that is, there is no serious heterogeneity among enterprises. Businesses are connected through small-world networks, which reflects the impact of preference-attachment effects on business ecosystems. Firms are assumed to be profit maximizers, and by interacting with their immediate neighbors through the rules of evolutionary games, firms are able to provide profitable products[3]. The small-world network composed of enterprises is represented by G= (V, E), where V={V_i} represents the enterprise node in the network, and the element represents the edge V_i and enterprise V_j between enterprises in E, assuming that all connections in the network are undirected, if there is a connection between enterprise I and enterprise j, then (V_i, V_j) =1, otherwise (V_i, V_j) =0.

By constructing a nearest-coupled network with N nodes, they form a ring in which each node is connected to its adjacent K/2 nodes, where K is even.

An edge is added to a pair of randomly selected nodes with probability P. There is at most one edge between any two different nodes, and no node can connect to itself.

3.1.2 Game Model under Market Mechanism

Based on the research framework of Huang et al, we studied how subjects such as new energy enterprises make appropriate strategic choices over time under government supervision. These decisions change the system environment, which in turn affects the policy choices of related agents. Their interaction continued until the end of the simulation[4]. Other types of parameters are shown in the Table 1:

Table 1. Variable definition

Variable	Defined
T1	Normal income of enterprise
T2	Corporate Extra Revenue
C1	Business normal input
C2	Improving Corporate Input
Q1	Enterprise carbon emissions
Q2	Carbon emissions after enterprise ESG behavior improvement
F	Penalties for companies that do not reduce carbon emissions
a	Executive environmental awareness
b	Competitive intensity
c	Carbon trading price

Potential adopters and adopted companies are participants in the diffusion network of ESG behavior policies. Each entity has two kinds of policies: adopting ESG behavior and not adopting ESG behavior. The revenue function and game matrix are shown in the Table 2:

Table 2. Enterprise game matrix under market mechanism

		New Energy Enterprise 1	
		adopt	reject
New Energy Enterprise 2	adopt	$T_1+T_2-C_2$; $T_1+T_2-C_2$	$T_1+T_2-C_2$; T_1-C_1
	reject	T_1-C_1; $T_1+T_2-C_2$	T_1-C_1; T_1-C_1

3.1.3 Game Model under Government Regulation

It is necessary for the government to take macro-control measures to motivate and guide enterprises to adopt ESG behavior improvement strategies. The government's environmental regulation mainly restricts enterprises through relevant policies, and the environmental awareness of executives, competition intensity among enterprises and carbon trading price will affect the change of income function. Set a company environmental awareness of a competition intensity of b, carbon trading price of c. In addition, enterprises reject ESG behavior improvement strategy when carbon emissions are Q, ESG behavior improvement strategy when carbon emissions are Q ', Q' < Q, the government limits enterprises that reject ESG behavior improvement to F. Therefore, based on the game model under the market mechanism, it is improved to obtain the game model of whether enterprises adopt ESG behavior improvement under government regulation. The revenue function and game matrix are as Table 3:

Table 3. Enterprise game matrix under government mechanism

		New Energy Enterprise 1	
		Adopted	Reject
New Energy Enterprise 2	Adopted	$T_1+T_2*(1-qd)-C_2-Q'c$;$T_1+T_2*(1-qd)-C_2-Q'*c$	$T_1+T_2*(1-qd)-C_2-Q'*c$; $T_1-C_1-Q*c-F$
	Reject	T_1-C_1-Qc-F; $T_1+T_2*(1-qd)-C_2-Q'c$	$T_1-C_1-Q*c-F$; $T_1-C_1-Q*c-F$

3.2 Evolutionary Rules

In each generation game, all nodes play against each of its neighbors once, and the proceeds accumulate. According to Fermi's rule, we get the following formula:

$$P_{(i\leftarrow j)} = \frac{1}{1+e^{\frac{U_i-U_j}{k}}}$$

Where, U_i and U_j are the average returns of the two nodes respectively, node i is the node with the highest average returns among the neighbors of node j, and k is the noise. Ui is the current round of revenue i, U_j is the current round of revenue j. This function indicates that when the return of individual i is lower than that of j, it is easy for i to accept j's strategy. However, if the return of i is higher than that of j, i will still adopt j's strategy with a weak probability. This irrational choice of individual i is described by k, which describes the noise factor of the environment and reflects the uncertainty of the individual when the strategy is updated. The closer the value of k is to 0, it means that the irrational choice of the individual is approaching zero, and the strategy update is certain. Will choose to learn, otherwise will stick to their original strategy; When the value of k approaches infinity, it means that an individual is in a noisy environment, unable to make rational decisions, and can only randomly update his own strategy[5].

4. Simulation results analysis

4.1 Parameter Assignment

According to this paper, the initial probability (x) of new energy enterprises choosing production improvement ESG behavior strategy is 0.1[6].The Table 4 describes the initial parameters:

Table 4. Parameter setting

Parameter	Value taken	units
T1	0.04	Million
T2	0.04	Million
C1	0.025	Million
C2	0.035	Million
Q1	0.00002	Million
Q2	0.000012	Million
F	0.0003	Million
a	0.4	
b	0.3	
c	60	

4.2 Simulation case

4.2.1 Diffusion of Improving Strategies for ESG Behavior by Executive Environmental Awareness Levels

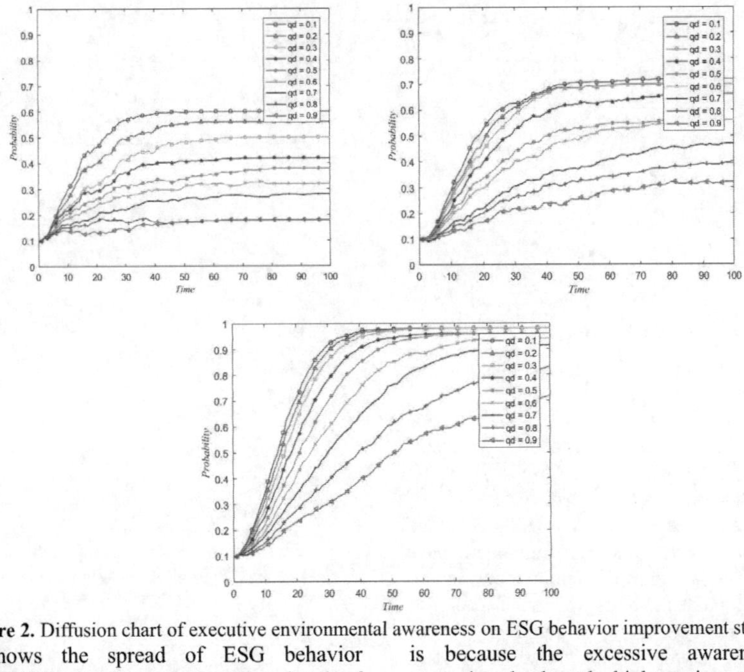

Figure 2. Diffusion chart of executive environmental awareness on ESG behavior improvement strategies

Figure 2 shows the spread of ESG behavior improvement strategies at the corresponding level of awareness level of executives of new energy enterprises. It can be seen that the efforts of enterprises to improve the environmental awareness of executives can effectively improve the earnings of enterprises.

In particular, when executives have too high awareness of environmental protection, the improvement degree of ESG behavior of enterprises will decrease. This is because the excessive awareness of corporate executives leads to the high earnings of enterprises, which does not need to share with similar enterprises, making enterprises an isolated individual. Therefore, increasing the level of environmental awareness among executives alone will not achieve the goal of significantly increasing the proliferation of ESG behavior improvement strategies in enterprises.

4.2.2 Effect of Competition Intensity on ESG Behavior Improvement Strategies

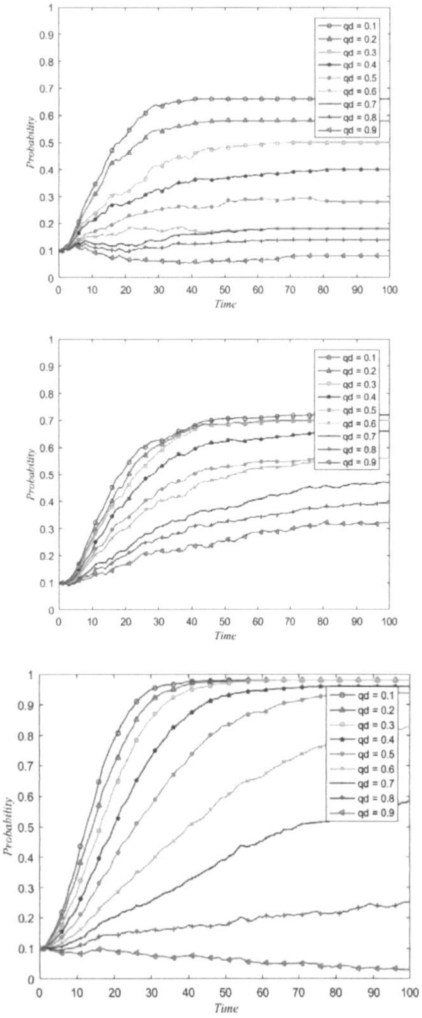

Figure 3. Competition intensity on ESG behavior improvement strategy diffusion graph

Figure 3 shows the improvement of ESG behavior under the corresponding degree of competition intensity b[0.1, 0.9] among new energy enterprises. With the increase of competition intensity in enterprise networks of different sizes, the improvement degree of ESG behavior decreases correspondingly. It can be seen that the increase of inter-firm competition intensity cannot improve the spread of inter-firm ESG behavior improvement strategies. This is because the increase in competition intensity will reduce the additional income of new energy enterprises adopting ESG behavior and increase the cost under the same income. When the benefits of new energy enterprises participating in the improvement of ESG behavior are greater than those that do not improve ESG behavior, more enterprises will be encouraged to adopt ESG behavior, thus increasing the intensity of competition. In addition, the larger the network size, the more sensitive the response to competition intensity. In other words, as the number of enterprises in the network increases, the diffusion of ESG behavior improvement strategies under the same competitive intensive increases.

4.2.3 Effect of Carbon Trading Price on the Diffusion of ESG Behavior Improvement Strategies

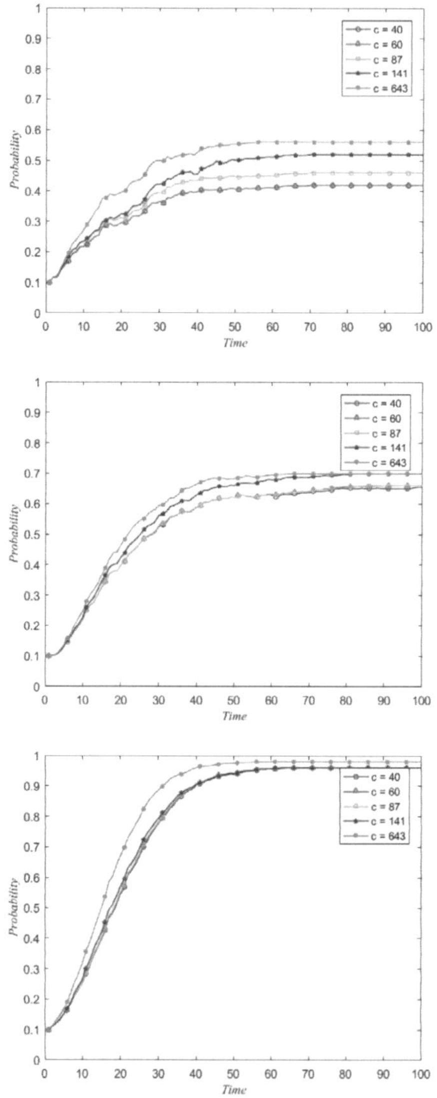

Figure 4. Carbon trading price on ESG behavior improvement strategy diffusion chart

Figure 4 shows the diffusion of ESG behavior improvement strategies at low (40 yuan/ton), medium (60 yuan/ton) and high (87 yuan/ton) levels in China, as well as Australia (141 yuan/ton) and the EU (653 yuan/ton) under corresponding carbon trading prices. Generally speaking, with the increase of carbon trading price in the network of different scales, the carbon emission trading between enterprises decreases, and the degree of diffusion increases accordingly. It can be seen that the increase in the price of carbon trading among enterprises is conducive to improving the diffusion level of ESG behavior improvement strategies. This is because an increase in the price of carbon trading will reduce the profits of new energy companies. When the carbon emission income is higher than the emission reduction income, it encourages more new energy enterprises to adopt ESG behavior to improve and reduce carbon emission. Moreover, the larger network is even less sensitive to the price of carbon trading. Although with the increase of the number of new energy enterprises in the network, the diffusion degree of ESG behavior improvement strategy will increase correspondingly under the same carbon trading price, due to the increase of network scale, the increase of the number of new energy enterprises and the increase of carbon trading frequency among enterprises, carbon trading price will not be paid too much attention. On the other hand, the increase of new energy enterprises in the network increases the number of neighbors, thus expanding the cumulative benefits of participants. Therefore, the strategy of new energy enterprises with resource advantages will be emulated by more enterprises. On the other hand, with the increase of the number of optional learning objects, the learning and updating rate of enterprises will be more efficient, so enterprises will choose to improve ESG behavior, thus making ESG behavior improvement strategies spread. This paper discusses the choice of ESG behavior improvement strategy from the perspective of network characteristics, and reveals that strengthening the connection between enterprises in the industry is conducive to the dissemination of ESG behavior improvement strategy.

5 Comparison and summary of simulation cases

Comparing the game results of each scheme, it can be seen that enterprises should adopt ESG behavior and establish a model conforming to the standard to ensure the normal operation of the game network. When enterprises operate under the government mechanism, the network of evolutionary game can exist stably. In the meantime. For the network, there are some obvious different game results due to the different combination of enterprise's right, responsibility, profit and risk mechanism.

At present, the environmental awareness of corporate executives can bring additional benefits to enterprises, but it will not promote the spread of ESG behavior improvement strategies among enterprises. Meanwhile, with the increase of the scale of enterprises in the network, the degree of choice diffusion for ESG behavior improvement also increases.

Compared with the environmental awareness of corporate executives, the improvement of ESG behavior in the intensity of competition among enterprises has a better diffusion effect. The increase in the intensity of competition among enterprises will reduce the additional income of enterprises conducting ESG behavior. At the present stage, the small number of enterprises conducting ESG behavior will bring higher income to enterprises[7].

In the context of complex network evolution game, neither the carbon trading price for new energy enterprises nor the intensity of competition among enterprises can fully improve ESG behavior. Therefore, comprehensive ESG behavior improvement requires the integration of multiple measures.

Acknowledgment

This work is supported by the National Social Science Foundation of China, and National innovation and entrepreneurship training program for college students.

Reference

1. Wang Dan, Liu Jun. (2022) National rejuvenation and Sustainable development: ESG management Innovation towards World-class enterprises [J]. Tsinghua Management Review,2022(10):96-105.
2. Huang X, (2021) Agent-based modelling for market acceptance of electric vehicles: Evidence from China, Sustainable Production and Consumption, 28:206-217
3. Shi Y, Wei Z, Muhammad Shahbaz, Zeng Y, (2021) Exploring the dynamics of low-carbon technology diffusion among enterprises: An evolutionary game model on a two-level heterogeneous social network, Energy Economics, 101
4. Huang X.(2022) Electric vehicle charging station diffusion: An agent-based evolutionary game model in complex networks, Energy, 257.
5. KM Ariful Kabir(2020) The impact of information spreading on epidemic vaccination game dynamics in a heterogeneous complex network- A theoretical approach, Chaos, Solitons & Fractals, Volume 132
6. Liu P, Li X., Li J, (2024) Low carbon technology diffusion of competitive firms under cap and trade mechanism --evolutionary game analysis based on complex network, Systems Engineering-Theory & Practice, [2024-03-10]:1-28.
7. Ibrahim Alnafrah,(2024) ESG practices mitigating geopolitical risks: Implications for sustainable environmental management, Journal of Environmental Management, Volume 358

Influencing factors on carbon emission in China transport industry based on PVAR model: A supply-demand perspective

Tianyi Yang*

College of Civil Engineering, Nanjing Forestry University, Nanjing, Jiangsu Province, 210000, China

Abstract: The control of carbon emission is of great significance to environmental protection. Transport accounts for about 10% of total carbon emissions. Therefore, energy conservation and emission reduction in the transportation sector have become critically important. This study takes 30 provinces in China as examples and uses a panel VAR (PVAR) model to study the impact of various factors on carbon emissions from the aspects of supply and demand.In terms of supply, transportation structure, environmental investment, and road density are key factors, and their impacts are dynamic.On the demand side, industrial structure, per capita GDP, and urbanization rate are major influencers. Specifically, urbanization and openness levels drive transportation carbon emissions in the east, per capita GDP in the central region, and industrial structure in the west. This paper will provide concrete and feasible policy suggestions for the characteristic provinces on transportation carbon emission.

1. Introduction

Statistical analysis shows that globally, energy-related CO2 emissions from the transport sector account for more than 20% of total emissions and are expected to double by 2050. China is believed to be the world's largest carbon emitter. In the context of rapid urbanization and global challenges, the issue of carbon emissions from China's transportation sector has attracted increasing attention. Therefore, it has always been the focus of continuous attention to study the influencing factors of transportation carbon emissions and reduce emissions[1]. Most studies tend to classify influencing factors only from the perspective of supply or demand[2][3][4] .pply and demand are the two main sources that influence carbon emissions, and considering one in isolation is not conducive to identifying key influencing factors. Therefore, this study constructs a framework to analyze the factors affecting transport carbon emissions from both supply and demand perspectives. The effects of these influences vary over time. At present, there is no clear discussion on whether the factors affecting transport carbon emissions have changed. Therefore, the panel vector autoregressive (PVAR) model is used in this study to simulate the time-varying effects of these factors on carbon emissions. China is a vast country with significant differences in geographical location, economic scale, scientific and technological level and resource endowment among different regions. Due to the different regional characteristics, the impact effect of the influencing factors may vary. Therefore, this study divided the region into eastern, central and western regions based on geographical location to explore the differences in the effects of these factors [5]. A panel vector autoregressive (PVAR) model was then used to simulate the dynamic effects of these factors on 30 provinces in China from 2005 to 2019. The study then divided the regions into eastern, central and western regions based on geographical location to explore differences in the effects of these factors. This study is helpful to provide specific, feasible, real-time and effective transport carbon emission policy recommendations for provinces with different characteristics

2. Methodology

2.1 TRANSPORTATION CARBON EMISSION

In this paper, ipcc carbon emission estimation method is adopted[6], and the total amount is given by equation (1) :

$$Emission = \sum_{i} F_i \times EF_i \qquad (1)$$

Where, i is the type of fuel consumed by urban traffic I n a year, such as anthracite and coking coal. Natural gas, liquefied petroleum gas, etc. F_i is the consumption of urban transportation fuel, x is its carbon emission factor, and EM is the total carbon emission in one year. Various carbon emission factors are shown in Table 1.

Table 1. Carbon emission coefficient of fuel

Fuel type	Rawcoal	gasoline	kerosene	diesel	fuel oil
Carbonemission factor (kgc/kg)	0.4855	0.8602	0.8273	0.8443	0.8647

*Corresponding author:2302229266@qq.com

https://doi.org/10.1051/e3sconf/202452803016

2.2 RELATED VARIABLES OF CARBON EMISSIONS

This study will explore the impact of relevant factors on carbon emissions from the aspects of supply and demand(Table 2). The supply-side variables include energy structure, transportation energy intensity, highway network density, transportation structure, technological innovation, and environmental protection investment. Demand-side variables include urbanization rate, GDP per capita, population density, industrial structure, openness to the outside world, and average service area per school and average number of people per medical bed[6].

Table 2. Carbon emission variables

Symbol	Variable	Definition
A	Energy structure	Coal consumption/total energy consumption
B	Transportation energy consumption	Energy consumption Carbon emission coefficient Sum of energy consumption
C	Roads per 10,000 people	Total miles per 10,000 people
D	Traffic structure	Rail freight volume/Road freight volume
E	Number of patent applications	Number of patent applications by province and year
F	investment in environmental protection	The government invests GDP in environmental protection
G	urbanization rate	Urban population/total population
H	Gross regional product per capita	Regional GDP/ resident population
I	Population density	Total population/built-up area
J	Degree of opening up	Foreign investment GDP
K	Average service area per school	Area/Area Total Number of schools by type
L	Average number of persons served per medical bed	Total population/total number of beds in health facilities
M	Transportation energy intensity	Transportation Energy consumption Transportation industry
N	Industrial structure	Output value of the tertiary industry/total output value

2.3 PVAR MODEL

The PVAR model established by Love and Zicchino[7] has been applied to cross-field research. PVAR model inherits the advantages of traditional VAR model, all variables are set as endogenous variables, and the orthogonal impulse response function can truly reflect the dynamic relationship between variables. At the same time, panel data can observe individual heterogeneity and explain economic problems without requiring the entire economic structure. In the analysis of interdependent economic problems, cross-section scale can make up for the deficiency of time series scale. The PVAR model is shown in equation (2).

$$Y_{it} = A_0 + \sum_{k=1}^{n} A_k Y_{t-k} + \mu_t + \lambda_i + \varepsilon_{it} \tag{2}$$

Where: i represents region, t represents time; k stands for lag period; n represents the lag order and A_0 represents the intercept term vector. A_k represents the coefficient matrix; μ_i and λ_i represent individual effects and time effects, respectively, and ε_{it} represents the random error term. A stable PVAR model can estimate the impulse response function (IRF). In this study, STATA 16 was used for PVAR model and IRF analysis. LLC, HT and IPS were used to select the optimal lag order of PVAR model. Then Granger causality test is used to test the causal relationship between variables.

PVAR model can more effectively depict the transmission mechanism between variables and meets the theoretical model's requirements for the characteristics of endogenous variables. This model is suitable for data with a wide cross-section and a short time series. With the help of PVAR model, the relationship between influencing factors and transportation carbon emissions can be analyzed more accurately. Meanwhile, the endogeneity of these variables can be avoided during the analysis process, and the transmission mechanisms of various influences can be clearly described.

3.Data

The data on carbon emissions and their influencing factors in transportation mainly comes from the China Energy Statistical Yearbook, China Urban Statistical Yearbook, and Provincial Statistical Yearbook. This study divides 30 provinces into eastern, central, and western regions for research, as shown in Table 3.

Table 3. Geographical division

Region	Province
East	Jiangsu, Beijing, Tianjin, Fujian, Liaoning, Shanghai, Shandong, Zhejiang, Hebei, Hainan, Guangdong
Central	Heilongjiang, Jilin, Hunan, Hubei, Jiangxi, Anhui, Shanxi, Henan
West	Ningxia, Qinghai, Chongqing, Inner Mongolia, Sichuan, Guizhou, Xinjiang, Guangxi, Yunnan, Shaanxi

4. Result

4.1 IMPULSE RESPONSE ANALYSIS FOR ALL SAMPLE PROVINCES

In this paper, LLC test, HT test and IPS test are used. If two of the three tests pass the test, the data is considered stable; if they fail, the first-order difference is used to continue the test; if they fail, the second-order difference is used again. The results are shown in Table 4.

Table 4. Unit- root test

Variables	LLC tests p-value	HT tests p-values	IPS tests p-values
D	0.0000**	0.0000***	0.0000***
F	0.0000***	0.0000***	0.0000*
I	0.0000***	0.9795	0.0000***
K	0.0000*	0.0000***	0.0000***
L	0.0000***	0.6477	0.0000***
M	0.0000***	0.0000***	0.4784
dA	0.0000***	0.0000***	0.0000***
dB	0.0000***	0.0000***	0.0000***
dC	0.0000***	0.0000***	0.0000***
dG	0.0000***	0.0000***	0.0000***
dH	0.9996	0.0000***	1.0000
dN	0.0000***	0.0000***	0.0000***
ddH	1.0000	0.0000***	0.0000***

In this paper, AIC, BIC and HQIC are calculated to determine the optimal lag order, and the lag order is selected as 1 order. It is concluded that the overall data variable A, C, D, E, F, G, H, I, J, K, M, N, (Table 5), the eastern data variable A, C, D, E, G, H, I, M, N ,the central data variable and the western data variable A, C, E, G, H, I, K, M, N pass the Granger causality test and are numerically correlated with the total carbon emission.

Table 5. Granger test results of overall regional data

Null hypothesis	*P*-value
A does no Granger-cause carbon emission	0.000***
C does no Granger-cause carbon emission	0.000***
D does no Granger-cause carbon emission	0.000***
E does no Granger-cause carbon emission	0.000***
F does no Granger-cause carbon emission	0.000***
G does no Granger-cause carbon emission	0.000***
H does no Granger-cause carbon emission	0.000***
I does no Granger-cause carbon emission	0.000***
J does no Granger-cause carbon emission	0.000***
K does no Granger-cause carbon emission	0.000***
M does no Granger-cause carbon emission	0.070*
N does no Granger-cause carbon emission	0.000***

Note: *$p < 0.10$, **$p < 0.05$, ***$p < 0.01$.

For all samples, from the supply side perspective, the effect of C is negative,which shows that the higher the density of highway network, the lower the carbon emission.

This may be attributed to higher road network density, better urban transportation facilities, and a greater inclination of residents towards low-carbon public transportation. The effect of D is negative, indicating that the larger the railway freight volume, the less carbon emission, and the carbon emission from road transportation is more environmentally friendly than that from railway. The effect of F is negative. It shows that increasing investment in environmental protection facilities can effectively reduce energy consumption. The influence degree of supply side factors can be generally ranked as the transportation structure, followed by environmental protection investment, and finally the density of highway network and patent application are the same.

On the demand side, the effect of G is positive.With the increase of urban population, carbon emissions from industrial transportation and living will also increase. the effect of H is first positive then negative. the effect of I is negative. Public transportation is developed in densely populated cities, and some private vehicles are restricted to travel, reducing energy consumption. The effect of N is negative. The definition of industrial structure is the proportion of the tertiary industry in the total, and the carbon emission of cities with large industrial proportion is larger, and the transportation of raw materials will increase carbon emission. the effect of J is positive. Foreign investment is mainly in industry, accelerating industrialization and leading to increased carbon emissions.

Demand-side factors shock degree analysis: industrial structure and per capita GDP first, second urbanization rate, followed by foreign investment, and finally per capita medical beds and population density.

4.2. IMPULSE RESPONSE ANALYSIS FOR SUBSAMPLES

4.2.1 EASTERN REGION

The effect of A is positive (Figure.1). As the share of coal in the energy mix increases, transport carbon emissions are likely to be affected and increase The effect of D is positive. Perhaps due to the industrial development in the east, the proportion of patents used to reduce emissions is not very large, resulting in increased emissions. The effect of G is positive, because the more urban population in eastern China, the more complex, rich and diverse production activities, so carbon emissions will naturally increase. The effect of H is negative. It may be that the transportation facilities in the developed eastern cities with high gdp are more developed, and people can travel easily and low-carbon, which reduces the carbon emissions of transportation.

Analysis of the shock degree of demand-side factors in the East: urbanization rate and foreign investment first, followed by per capita GDP, and finally population density.

https://doi.org/10.1051/e3sconf/202452803016

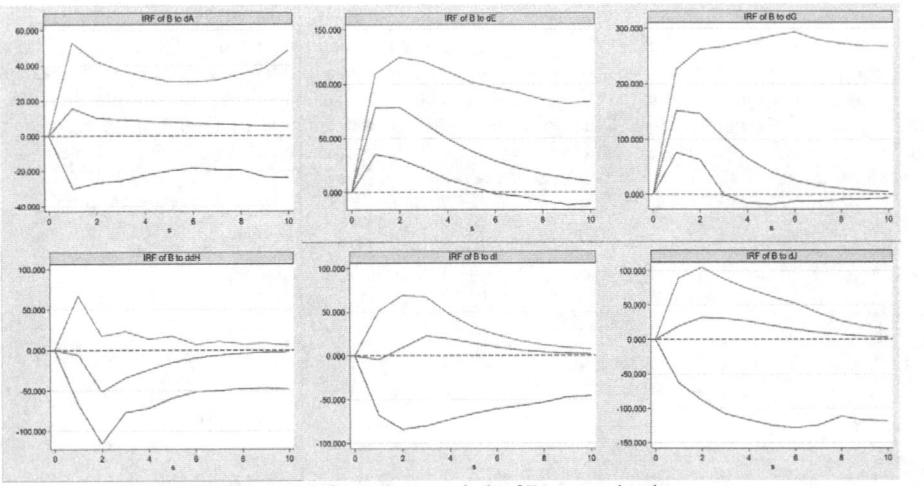

Figure 1. Pulse response analysis of Eastern region data

4.2.2 CENTRAL REGION

The effect of A is positive. Many cities in central China are energy-oriented and consume a lot of coal (Figure.2). The effect of D is negative. The effect of F is negative. The effect of G is positive. With the urbanization of central China, the government can take some measures to improve the traffic system. Such as promoting public transport and development. This will reduce the dependence on private cars and thus reduce carbon emissions. The effect of M is positive. It shows that the greater the transportation energy consumption, the greater the carbon emission. The central region is more affected than the general and western regions. The government can adopt electric or new energy vehicles to reduce carbon emissions. The effect of N is negative. The effect of H is negative. It is also possible that the central region is a place with high gdp and relatively developed transportation facilities, convenient for people to travel and low carbon. The effect of I is negative.

Analysis of the impact degree of demand-side factors in central China: the first is per capita GDP, followed by industrial structure, urbanization rate, foreign investment, per capita medical beds, and population density.

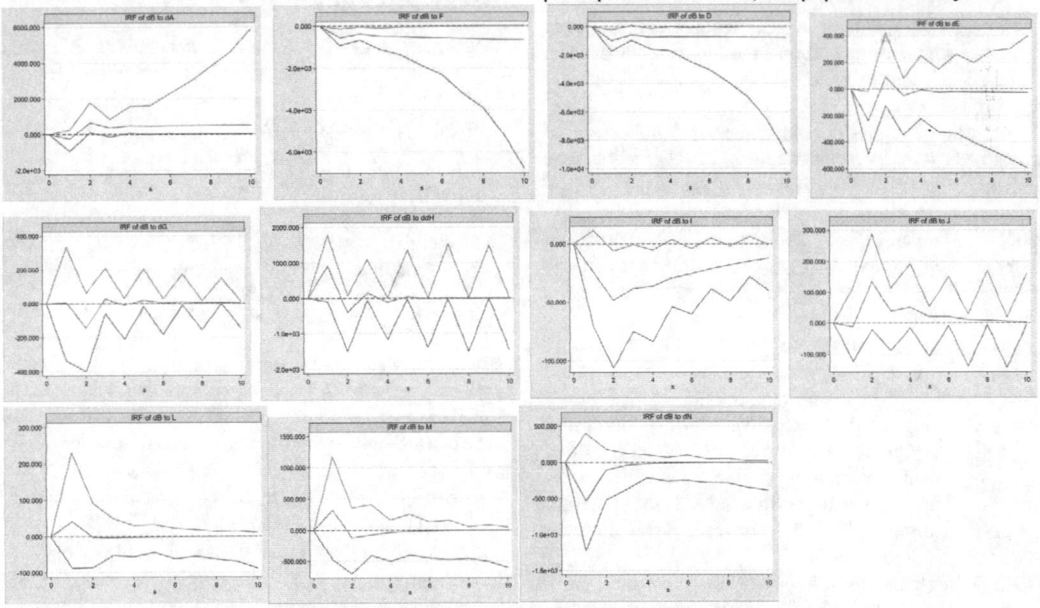

Figure 2. Pulse response analysis of Central region data

4.2.3 WESTERN REGION

In terms of the data of western China, The effect of C is positive, indicating that the greater the density of road network, the greater the carbon emission (Figure.3). Public transportation in western China is still relatively backward, and private cars are still the main ones, and the increase of roads will inevitably result in increased carbon emissions. The effect of E is positive. Although the central and western regions are both negative, the impact of the central region is greater. The effect of F is positive., and the impact of environmental investment is the same in the

Midwest. The effect of H is first positive and then negative.Western facilities lag behind, while people's purchasing power, people's production activities. The demand for this comparison is relatively large, which is why carbon emissions are high. The west should follow the example of the east to satisfy people's needs and develop public transportation. To satisfy people's needs and thus reduce carbon emissions. The effect of I is first negative and then positive. The effect of N is negative.

Analysis of the impact degree of demand-side factors in western China: the first is industrial structure, followed by urbanization rate, foreign investment, per capita GDP, per capita medical beds, and population density.

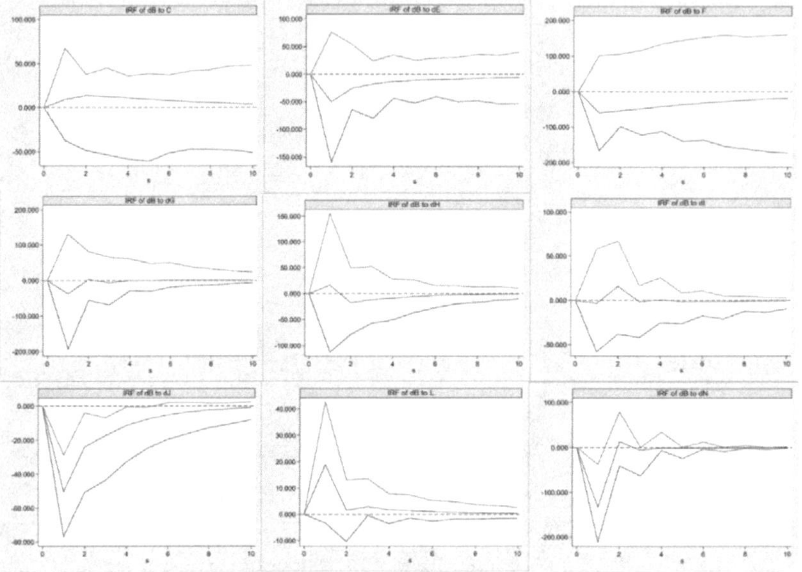

Figure 3. Pulse response analysis of Western region data

5. Conclusion

Based on pvar model, this paper studies the influence degree of each factor on carbon emission from the aspects of supply and demand, and gives reasonable adjustment suggestions.

The eastern region should reduce the proportion of coal and adjust the energy structure. Actively increase the number of patent applications for reducing energy consumption, appropriately reduce the urbanization rate, rationally plan cities, and avoid too dense population. Moderate cooperation with foreign companies to avoid the increase of carbon emissions caused by excessive industrialization.

The central region should also improve its energy structure. Promote public transport and development, reduce dependence on private cars, and thus reduce carbon emissions caused by excessive traffic intensity. Rationalize the distribution of health care facilities and avoid long-distance commuting to obtain medical services. Encourage non-motor vehicles, subway and other forms of public transportation. We will increase investment in environmental protection.

The western should follow the example of the east to meet people's needs while developing public transport, increasing the density of the road network and reducing car dependence. To meet People's Daily needs and thus reduce carbon emissions.

Acknowledgments

The author would like to thank the teachers and classmates in College of Civil Engineering,Nanjing Forestry University for their contributions to the method selection and the acquisition of data.

References

1. Wei Li , Ni Hao , Can Lu, "China's carbon emissions in conjunction with economic growth and clean energy use
", 119484(2023)
2. Dehu Yang a, Changming Zhu a,* , Jianguo Li a , Yating Li a , Xin Zhang b , Cunjian Yang c , Shuai Chu a, "Exploring the supply and demand imbalance of carbon and carbon-related ecosystem services for dual-carbon goal ecological management in the Huaihe River Ecological Economic Belt," Science of the Total Environment 912,169169(2024)
3. Manzhi Liu , Jinfeng Wang, Jixin Wen, Gang He, Jixin Wu, Huayang Chen, Xiaotao Yang, "Carbon Emission and Structure Analysis of Transport Industry Based on Input-output Method: China as an Example," Sustainable Production and Consumption 33,168–188(2022)
4. IPCC, 2023: Climate Change 2023: Synthesis Report. Contribution of Working Groups I, II and III to the Sixth Assessment Report of the

Intergovernmental Panel on Climate Change [Core Writing Team, H. Lee and J. Romero (eds.)]. IPCC, Geneva, Switzerland, pp. 35-115, doi: 10.59327/IPCC/AR6-9789291691647.

5. Yingji Xia, Chenlei Liao, Xiqun (Michael) Chen, Zheng Zhu, Xiaorui Chen, Lixing Wang, Rui Jiang, Marc E. J. Stettler, Panagiotis Angeloudis & Ziyou Gao, "Future reductions of China's transport emissions impacted by changing driving behaviour", 1228–1236 (2023)
6. Yuan Chen, Ahmed Bouferguene, Hong Xian Li, Hexu Liu ,Yinghua Shen,Mohamed Al-Hussein, "Spatial gaps in urban public transport supply and demand from the perspective of sustainability," Journal of Cleaner Production 195, 1237e1248(2018)
7. Valery I.Khabarov*,Olga G Khabarova, "Effciency of Labor Resources use and Issues of Transport Supply and Demand Development in Large Cities and Agglomerations," Transportation Research Procedia 61,308–313(2022)

https://doi.org/10.1051/e3sconf/202452803017

Assessing China's Path to Carbon Neutrality: Analyzing Historical and Projected Electricity-Related Carbon Emissions in the Context of the Dual Carbon Plan

Xiaorui Xu

Nanjing Agricultural University, Nanjing, China

Abstract. Carbon emissions have become a hot topic in the energy sector. China has proposed the "Dual Carbon Plan," aiming to peak its carbon dioxide emissions by 2030 and achieve carbon neutrality by 2060, whereby carbon emissions will be completely offset or absorbed, resulting in a net zero or negative carbon emission balance. However, China's heavy reliance on coal for electricity generation, coupled with issues of overcapacity and relatively backward technology, not only leads to significant carbon emissions but also poses challenges for a rapid transition to clean energy. This paper analyses historical data on China's total carbon emissions and carbon emissions from electricity generation. Additionally, it predicts future electricity-related carbon emissions based on policy and technological scenarios proposed by existing scholars, discussing the feasibility of achieving the goals set forth in the Dual Carbon Plan.

1. Introduction

1.1. China's energy composition and structure

China's energy structure has long been dominated by coal, followed by hydroelectric power. Since the reform and opening up, energy demand has continuously increased, gradually transitioning towards high-efficiency, low-pollution forms of oil and gas power generation. New energy sources represented by wind power, nuclear power, and solar energy are growing rapidly. Specific data on the power generation structure and the utilization rate of power generation equipment in the past few years are shown in Table 1 and Table 2, respectively.[1]

Table 1. China's power generation structure table (unit: GW)

Indicator name	2017	2018	2019	2020	2021	2022
Total power generation	1777	1900	2010.66	2200.58	2376.92	2564.05
hydroelectric power	341	352	356.4	370.16	390.92	413.50
Thermal power generation	1106	1144	1190.55	1245.17	1296.78	1332.39
Nuclear power generation	35.82	44.66	48.74	49.89	53.26	55.53
Grid-connected wind power	163.67	184.26	210.05	281.53	328.48	365.44
Grid connected photovoltaic power generation	130.25	174.63	204.68	253.43	306.56	392.61

Table 2. China power generation equipment utilization table (unit: hour)

Indicator name	2015	2016	2017	2018	2019	2020	2021
Total average for power plants 6 MW and above	3969	3785	3786	3862	3825	3758	3817
hydroelectric power	3621	3621	3579	3613	3726	3827	3622
Thermal power generation	4329	4165	4209	4361	4293	4216	4448
Nuclear power generation	7350	7042	7108	7499	7394	7453	7802
Grid-connected wind power	4091	3429	2700	2721	2082	2073	1281

1.2 The problem and causes of China's production capacity

Currently, the biggest issue concerning China's production capacity is overcapacity, predominantly observed in the manufacturing sector. The term was initially introduced in "Monopolistic Competition Theory." In 1947, Chamberlin

9211310114@stu.njau.edu.cn

systematically elaborated on the concept of "overcapacity" from the perspective of microeconomics, indicating that under conditions of non-perfect competition equilibrium, overcapacity represents an inefficient state where economic organizations deviate from "full capacity." Presently, foreign scholars primarily analyze the causes of overcapacity from the perspective of enterprises and at the micro-level, considering it a rational strategic behavior for firms to self-adjust to market competition, with relatively less discussion on addressing the overcapacity issue. Given the longstanding problem of overcapacity in China, domestic scholars have conducted in-depth explorations from various perspectives, including market mechanisms, technological advancement, and top-level governance within enterprises, through diverse theoretical frameworks and empirical studies. [2]

1.3 The international carbon situation and China's carbon-related policies

Recently, global climate change has become a focal point of international attention, as uncontrolled climate change could lead to severe consequences. Given the close correlation between climate change and greenhouse gas emissions, there has been a global consensus to strive for the goal of achieving "carbon neutrality." On December 12, 2015, nearly 200 parties under the United Nations Framework Convention on Climate Change (UNFCCC) reached the Paris Agreement during the Paris Climate Change Conference. This agreement, following the Kyoto Protocol, became the second legally binding climate agreement and provided a framework for global action on climate change. On April 25, 2023, the European Council adopted the Carbon Border Adjustment Mechanism, signaling the completion of the entire legislative process for this mechanism.. [3]On September 22, 2020, General Secretary Xi Jinping announced during the general debate of the 75th United Nations General Assembly that China will strive to peak carbon dioxide emissions before 2030 and work hard to achieve carbon neutrality before 2060. Over the past three years, China has actively and steadily promoted the "dual carbon" initiative, vigorously advancing energy revolution, promoting the transformation of industries towards green and low-carbon development, advocating for green lifestyles, and promoting comprehensive green transformation of the economy and society. The Central Committee of the Communist Party of China and the State Council issued the "Opinions on Fully, Accurately, and Comprehensively Implementing the New Development Concept and Doing a Good Job in Carbon Peak and Carbon Neutrality," and the State Council released the "Action Plan for Peak Carbon Emissions by 2030." Various relevant departments have issued 12 implementation plans for key areas and key industries as well as 11 supporting guarantee plans. Additionally, 31 provinces, regions, and municipalities have formulated local implementation plans for peak carbon emissions, forming a "1+N" policy system for peak carbon emissions and carbon neutrality with numerous measures and plans. [4]

1.4 Research content and directions

To achieve the goals of China's dual-carbon plan and propose reasonable and effective solutions to the issue of overcapacity, this paper first organizes historical carbon emissions data. Starting from the changes in the proportion of various carbon emission projects each year over the past three years, it analyzes possible reasons. It then focuses on the electricity sector, which has the largest proportion, and makes appropriate speculations and analyses on the reasons for the changes in values throughout the year, combined with actual emissions from different seasons of life and production.

After a certain background analysis, this paper also proposes a series of hypotheses based on existing production policies. This includes scenarios such as the development situation if other carbon-related policies are not implemented based on the existing 14th Five-Year Plan, and the adoption of carbon capture and storage technologies at three different rates. The analysis of these scenarios is based on reasonable calculations and speculations derived from previous analyses of historical data. Additionally, the paper incorporates other scholars' research on the changes in the proportion of different power generation methods as an additional supplement.

2. Carbon emission status analysis

2.1. The composition and changing pattern of China's carbon emissions in previous years

Table 3. Composition of China's carbon emissions in 2020, 2021 and 2022

Types of carbon emissions	2020	2021	2022
Power generation and heating	41%	46%	51%
Manufacturing, Smelting and Construction	39%	36%	28%
Transportation	11%	7%	10%
other industry	9%	11%	11%

 https://doi.org/10.1051/e3sconf/202452803017

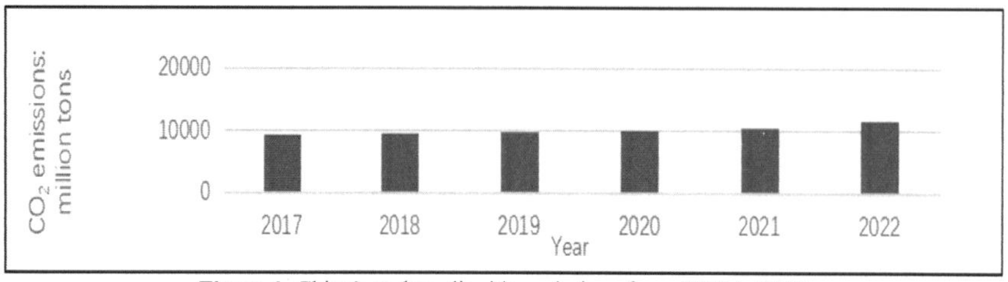

Figure 1. China's carbon dioxide emissions from 2017 to 2022

According to Table 3, the power generation and heating industry consistently accounts for around 50% of the annual maximum carbon emissions. This is because electricity generation is one of the fundamental infrastructures in modern society, essential for various industries and daily life activities. The combustion of fossil fuels is one of the primary methods of electricity generation globally, leading to significant emissions of carbon dioxide. Additionally, inevitable energy losses occur during the transmission of electricity, further exacerbating the already substantial emissions. Following the power generation and heating industry, the manufacturing, smelting, and construction industries have a decreasing trend in their proportion of emissions year by year. The proportions of emissions from the transportation and other industries are relatively low and show minor fluctuations, remaining at around 10%. Meanwhile, as shown in Figure 1, China's total carbon emissions have increased annually from 2017 to 2022. In 2017, China's carbon emissions were 9462.7 million tons (Mt), which grew to 10523 Mt in 2021 and reached 11480 Mt in 2022, with a growth rate of 9.1%.[5][6][7][8] Looking at the timeline, this increase is likely due to China's economic recovery after the pandemic, which involved a significant increase in energy consumption. Moreover, the pandemic hindered the progress of transitioning to clean energy, contributing to the peak in carbon emissions growth from 2021 to 2022.

2.2. Carbon emissions from China's power industry

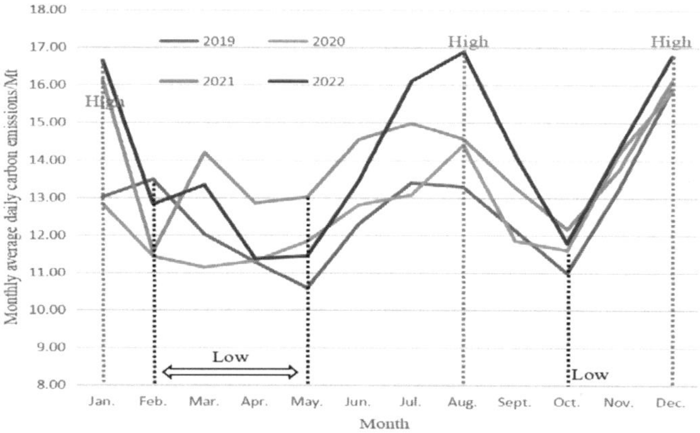

Figure 2. Average monthly carbon emissions of China's power industry from 2019 to 2023

Figure 2 contains the daily average carbon emissions data of China's electricity industry from 2019 to 2022, obtained from the Global Real-time Carbon Data website. The chart is based on the average values calculated from the data on the 10th, 20th, and 30th of each month (taking the last day of February for February). Generally, carbon emissions are higher in January, August, and December, while they are lower from February to May and in October. Taking 2022 as an example, the monthly average carbon emissions for January, August, and December were 16.64 Mt, 16.89 Mt, and 16.77 Mt, respectively. In contrast, the values for February, March, April, May, and October were only 12.82 Mt, 13.33 Mt, 11.37 Mt, 11.44 Mt, and 11.79 Mt, respectively. [8]Different electricity demand patterns in different months may be related to factors such as temperature variations, production demands, and holiday commercial electricity usage. For instance, the electricity consumption required for heating or cooling equipment and commercial electricity usage during holidays could influence the electricity demand in different months.

3. Numerical prediction of China's future power energy carbon emissions

3.1. Forecast of future electricity carbon emissions based on population development, urbanization process and changes in economic structure

Du Xiaohu, Zhou Xingbo, and Zhou Jianping conducted a relevant analysis of China's electricity carbon emissions level against the backdrop of carbon neutrality. They used economic growth as a foundation to determine energy and electricity growth processes and stage targets that are compatible with it. They also considered factors such as urbanization and predicted the future development of electricity: Total electricity consumption in society continues to grow, with the growth rate showing a trend of "high in the beginning, low in the middle, and stable in the end." From 2022 to 2035, electricity maintains an annual average growth rate of 3% to 4%, reaching 15.5 trillion kWh by 2035. In the following 15 years, the growth rate slows down, with an annual average growth rate of 0.2% to 2%, reaching 18.5 trillion kWh by 2060. After 2060, electricity essentially stabilizes. This paper considers the trend of electricity consumption as the equivalent trend of carbon emissions in the electricity sector and does not consider differences in carbon emissions among different power generation methods when predicting future electricity-related carbon emissions. It takes the growth rate of 3% before 2035 and the growth rate of 0.2% from 2035 to 2060 as the lower limit; and the growth rate of 4% before 2035 and the growth rate of 2% from 2035 to 2060 as the upper limit, and takes the arithmetic average of the two stages as the average growth rate. [9]

By combining the predicted growth rates mentioned above with China's annual electricity carbon emissions data from the Global Real-time Carbon Data, the predicted graph in Figure 3 can be obtained.

The reason for setting 2035 as the boundary between the two stages is because it falls at the end of the "14th Five-Year Plan." According to the current "carbon peak" target for 2030, after 2035, the amount of coal-fired electricity generation will decrease significantly, decreasing by 1.0 to 1.5 trillion kWh every five years. By meeting this requirement, the change in electricity carbon emissions as shown in Figure 3 can be realized.

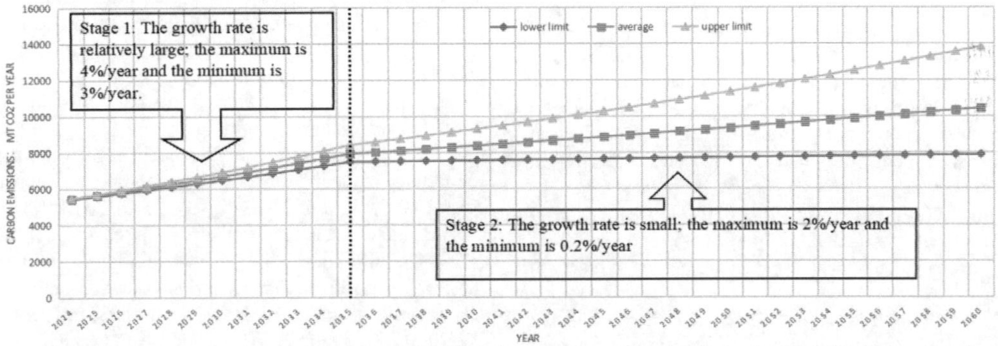

Figure 3. Forecast of carbon emission data under normal economic and population development from 2024 to 2060

3.2. Prediction of carbon emission values in the case of CCS technology intervention

Carbon Capture and Storage (CCS) is the process of capturing carbon dioxide gas directly from coal-fired power plants or other industrial processes. Its main goal is to prevent carbon dioxide from entering the Earth's atmosphere, thereby mitigating the exacerbation of the impacts of excess greenhouse gases. Captured carbon dioxide is transported and stored underground in geological formations. The advantages of CCS technology are significant: it can reduce emissions at the source. According to statistics from the International Energy Agency, CCS could potentially remove up to 20% of carbon dioxide emissions from industrial and energy production facilities. Additionally, due to the high concentration of carbon dioxide in CCS processes, carbon dioxide is more easily separated from adsorbents, reducing the social cost of carbon. However, its disadvantages cannot be ignored. The foremost is the cost of CCS itself. Implementing CCS technology could increase electricity costs by 50% to 80%. Additionally, the uncertainty and safety of long-term carbon dioxide storage capacity also pose significant challenges.

Tian Peining, Liang Xiao, Guan Yujie, Zhao Yixin, Mao Baohua, Xue Ting, and others have conducted relevant research on the impact of CCS (carbon capture and storage) technology on carbon emissions. In their studies, they point out that, under the condition of unchanged power generation structure, as the efficiency of power generation, energy utilization efficiency, and CCS technology level improve, the lifecycle carbon emission factors of various power generation methods will decrease, thereby having a beneficial impact on reducing the intensity of carbon emissions from electricity generation. Three scenarios are set in the research: rapid transformation, baseline transformation, and slow transformation. The baseline transformation scenario refers to the development trend that continues the current situation, with changes in various parameters following the current trend of natural development, gradually transitioning to clean and green power generation modes, with plans to scale up coal-fired power CCS technology in 2028; rapid and slow transformations involve the large-scale deployment of relevant technologies in 26 and 30 years, respectively. The lifecycle carbon emissions of the three scenarios will peak in 2025, 2027, and 2030, respectively, with peaks of 5.139 billion, 5.205 billion, and 5.338 billion tons. By 2060, they will decrease to 1.182 billion, 1.578 billion, and 2.421 billion tons, respectively, representing a carbon reduction of 74%, 66%, and 48% compared to 2021. Using these rates as the rate of carbon emission reduction in the prediction process, the growth rate before the use of CCS technology in the three scenarios is replaced by the average growth rate in Section 3.1, yielding the rate of carbon emissions change in the three scenarios. From this, Figure 4 can be plotted. [10]

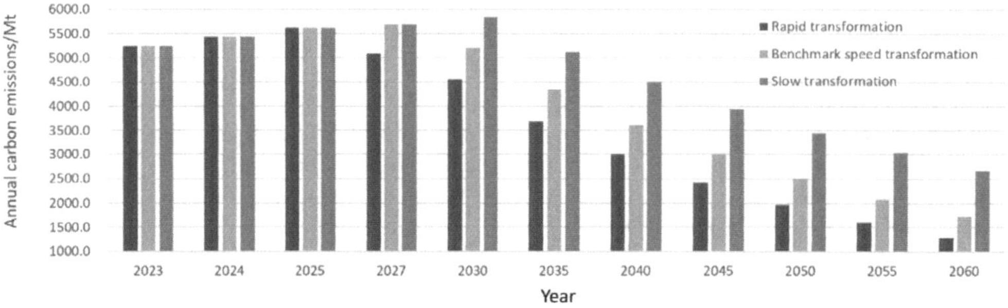

Figure 4. Numerical changes in electricity carbon emissions under three scenarios

3.3. The impact of changes in the proportion of various power generation methods on carbon emissions in the future

The two main scenarios regarding the composition of power generation methods are the Existing Policy Scenario and the Clean Energy Scenario. Under the Existing Policy Scenario, annual deployment of wind and solar power generation is limited to the current targets set by the government, with wind and solar power generation capacities meeting the Chinese government's target of reaching 1.2 billion kilowatts by 2030. The proportion of non-fossil energy generation meets the government's targets of reaching 50% by 2030 and expected to reach 60% by 2035.

Under the Clean Energy Scenario, by 2035, the proportion of non-fossil energy generation in China increases to 80%. Wind and solar power generation capacities approach the current 2030 targets by 2025 and reach 1.994 billion kilowatts by 2030 and 3.069 billion kilowatts by 2035. The proportion of non-fossil energy generation increases to 65% by 2030 and reaches 80% by 2035.In the Existing Policy Scenario, the intensity of carbon dioxide, sulfur dioxide, and nitrogen oxide emissions (per kilowatt-hour of electricity generated) decreases significantly, but due to the growth in electricity demand, absolute emissions increase in the short term, decreasing only by 12% by 2035 compared to 2020.

In the Clean Energy Scenario, by 2035, carbon dioxide emission intensity decreases by 70% compared to 2020, dropping to 0.16 tons of carbon dioxide per megawatt-hour, and absolute emissions decrease by 50% (reducing by 1.66 billion tons of carbon dioxide). In the Clean Energy Scenario, sulfur dioxide and nitrogen oxide emission intensity are also lower. [11]

4. Conclusion

Based on the information provided, a certain assessment of China's current electricity carbon emissions status can be made: In 2022, China entered the phase of economic recovery after the pandemic, during which energy consumption increased significantly. Simultaneously, the progress of clean energy transition was hindered due to the pandemic, resulting in a peak in carbon emissions growth rate from 2021 to 2022. Additionally, different months may have varying electricity demand due to factors such as temperature and production needs, often resulting in higher electricity carbon emissions in January, August, and December each year.

If we only estimate China's future carbon emissions levels based on population development, urbanization, and changes in economic structure, we would anticipate a continuous increase in electricity carbon emissions until 2060. However, there's a turning point in 2035, as it marks the end of the "14th Five-Year Plan" and significant reductions in coal-fired power generation are expected after 2035, with a decrease of 1.0~1.5 trillion kWh every 5 years, in accordance with the "carbon peak" goal set for 2030. The introduction of CCS technology leads to three different transformation scenarios—slow, baseline, and rapid—with peaks in 2025, 2027, and 2030 respectively. By 2060, carbon emissions are projected to decrease to 11.82 billion, 15.78 billion, and 24.21 billion tons respectively, representing a reduction of 74%, 66%, and 48% compared to 2021. Assuming these rates as the rate of carbon emissions reduction during the forecasting process, and using the average growth rate of 3.1 before the use of CCS technology, we can estimate the approximate values of the three scenarios in the future.

In addition, clean energy plays a significant role in emissions reduction. With proper regulation and support from government policies, technological advancements, and societal demand, China is expected to achieve a green and low-carbon transformation in the electricity industry, making positive contributions to sustainable development and global climate governance.

References

1. Wikipedia contributors. Energy Structure of the People's Republic of China [G/OL]. Wikipedia, 2024(20240102)[2024-01-02]. https://zh.wikipedia.org/w/index.php?title=%E4%B8%AD%E5%8D%8E%E4%BA%BA%E6%B0%91%E5%85%B1%E5%92%8C%E5%9B%BD%E8%83%BD%E6%BA%90%E7%BB%93%E6%9E%84&oldid=80332884.
2. Xueyang Chen, Luofu Zhang. Has the Belt and Road Initiative alleviated China's overcapacity? —— Sample calculations from 281 urban agglomerations

[J/OL]. Journal of Sichuan Administration College,1-18.

3. Shen Liang, Fan Tao, Mu Qun, Zhou Chunlei, XU Zhongping. Study on evaluation standardization system of carbon emission accounting methodology based on "Electricity-Carbon EmissionCalculation Model "[J]. China Soft Science, 2024, (S1): 318-333+340.
4. Chen Weiwei. Continue to promote green and low-carbon transformation - the "double carbon" goal has achieved positive results in the past three years [EB/OL](2023-9-22)[2024-03-08]. http://m.news.cn/2023-09/22/c_1129877403.htm
5. Shifair. China's carbon peak and carbon neutrality timetable and roadmap [EB/OL](2022-06-10)[2024-01-14] https://www.shifair.com/informationDetails/16147.html
6. Hongyuan Futures. "Carbon Neutrality" Series Special Topic Four: China's "Dual Carbon" Policy Review and Carbon Emissions Trading Tracking [EB/OL](2022-09-19)[2024-01-14] https://www.fxbaogao.com/detail/3384053
7. Gongyn. Analysis of China's carbon dioxide emissions, emission structure and transaction volume in 2022 [EB/OL](2023-04-07)[2024-03-08]. https://www.gonyn.com/industry/1398873.html
8. Carbon Monitor. Data Center - China[EB/OL](2022)[2024-02-01]https://carbonmonitor.org.cn/user/data.php?by=China
9. Du Xiaohu, Zhou Xingbo, Zhou Jianping. Analysis of China's Electricity Carbon Emission Level under the Background of Carbon Neutrality [J/OL]. Journal of Hydroelectric Engineering:1-11[2024-03-08].http://kns.cnki.net/kcms/detail/11.2241.TV.20231215.1626.002.html.
10. Pei-Ning T, Xiao L, Yu-Jie G, et al. Whole life cycle carbon emission and power generation structure transformation pathway planning of China's power[J]. Advances in Climate Change Research, 2024, 20(1): 97.
11. Nikit Abhyankar, Jiang Lin, Fredrich Kahrl, Shengfei Yin, Umed Paliwal, Xu Liu, Nina Khanna, Amol A Phadke, Qian Luo, David Wooley, Mike O'Boyle, Olivia Ashmoore, Robbie Orvis, Michelle Solomon. Achieving an 80% Carbon Free Electricity System in China by 2035[R]. Energy Technologies Area, Berkeley Lab2022:7

 https://doi.org/10.1051/e3sconf/202452803018

Research on the whole life cycle energy consumption and pollution emission of new energy vehicles and traditional fuel vehicles based on the perspective of single vehicle currency of the same vehicle model

Yelai Zhu *

School of mathematics and statistics,Chongqing University,Chongqing College Town, Shapingba District, Chongqing, China

Abstract: To construct an evaluation framework based on whole-life energy consumption and pollution emissions, the control variable method is used when selecting vehicles for comparison. By assessing a new energy vehicle (the Audi Q2L e-tron) and a traditional fuel vehicle (the Audi Q2L) of the same model, differences in emission and energy consumption are solely attributed to power forms. Results are monetized to facilitate analysis of the societal benefits of new energy vehicles, reflecting environmental impact in monetary terms. The results of the paper show that although the replacement of traditional fuel vehicles by new energy vehicles can achieve carbon emission reduction and considerable energy savings in the energy cycle, due to the fact that new energy vehicles produce more sulphur dioxide (SO_2) and nitrogen oxides (NO_x) in the production phase, the full replacement of traditional fuel vehicles by electric vehicles at this stage will not be able to achieve an improvement in the overall welfare of the society. Therefore, under the current conditions, the improvement of new energy vehicles should start with finding cleaner and more environmentally friendly ways in the manufacturing stage of automobiles, as well as modifying the way of producing electricity.

1. INTRODUCTION

As environmental and resource problems such as climate change, emission pollution, and the shortage of traditional fossil fuels become more and more serious, the State has gradually attached importance to energy conservation and environmental protection. In order to support the development of new energy vehicles, the State has listed new energy vehicles as one of the strategic emerging industries in the "Outline of the Fourteenth Five-Year Plan and Vision 2035 for the National Economic and Social Development of the People's Republic of China". Additionally, it has proposed the "New Energy Vehicle Industry Development Plan (2021-2035)" which aims to ensure that by 2025, the sales volume of new energy vehicles will reach the total sales volume of new automobiles. In the "Development Plan for New Energy Vehicle Industry (2021-2035)", it is proposed that the sales volume of new energy vehicles will reach about 20% of the total sales volume of new vehicles by 2025.

However, are new energy vehicles truly more low-carbon and environmentally friendly compared to traditional fuel vehicles, and is the cost of their utilization lower? Presently, China lacks a comprehensive assessment system for evaluating emissions reductions in new energy vehicles. This deficiency impedes China's new energy vehicles from progressing towards the objectives of low pollution and high developmental returns. Based on this, this paper through the development of China's new energy vehicles to systematically grasp the status quo, control the car with the same brand and the same model to ensure that the vehicle appearance, vehicle manufacturing process, automotive material composition, the type and number of parts, the overall quality, wind resistance coefficient, and other aspects of the data are the same, which is measured by the different emissions and energy consumption results are entirely caused by the different forms of power to compare a single new energy vehicle with a traditional fuel car. Comparing the pollution emission and energy cost of a single new energy vehicle with that of a conventional fuel vehicle over the whole life cycle, and subsequently monetising the data obtained, the final results can be more intuitively seen whether the replacement of fuel vehicles with new energy vehicles has significantly improved the social benefits, thus revealing the social impacts of replacing conventional fuel vehicles with new energy vehicles, and the social changes that may be triggered by the rapid emergence of the new energy automobile industry in China. Furthermore, this paper furnishes valuable insights for

*1780921441@qq.com; phone 86 13858586716

discussions on pertinent issues and aids in formulating policy proposals.

2. RESEARCH METHOD

This paper uses the life cycle assessment (LCA) methodology.The life cycle[1] refers to the entire process from production to use to end-of-life recycling, and the LCA methodology can be used to derive the environmental impacts of products at each stage. In the case of automobiles, this paper divides the life cycle of automobiles into the automobile manufacturing stage, the production of power batteries, the use of automobiles, and the end-of-life stage of automobiles. Otherwise, the impacts on the environment have been categorized into energy consumption and pollution emissions.

In this paper, the Audi Q2L and Audi Q2L e-tron have been chosen as the respective models for comparison between fuel-powered and electric vehicles. The aim is to control variables effectively, with both vehicles being sourced from the MQB platform of the Volkswagen Group in Germany. The only discernible difference between them is the type of energy source. Other key aspects such as vehicle exterior, interior, construction, body structure, materials, type and quantity of components, overall mass, and wind resistance exhibit nearly identical characteristics. This uniformity ensures that any variations in measured emissions and energy consumption are exclusively attributed to the differences in power forms, rather than the numerous other factors mentioned above. Notably, the paper does not encompass plug-in hybrid models. This omission is deliberate due to the fact that the emissions and energy consumption of such vehicles are contingent upon the owner's choice of using either fuel or electricity. Given the resultant variability in emissions based on usage patterns, plug-in hybrid vehicles are considered unsuitable for comparative analysis.

In terms of data collection, this paper argues that most academic articles use the Argonne National Laboratory's GREET vehicle life cycle database for life cycle calculations and that the databases for pollution and energy consumption at the vehicle manufacturing stage, vehicle use stage, and vehicle end-of-life stage originate from this model, which makes the data lack of relevance. In comparing the energy consumption and emissions of fuel-electric vehicles with those of new energy-electric vehicles, we have selected the Audi Q2L and Audi Q2L e-tron. However, the GREET model is essentially a weighted average of all the mainstream vehicles currently on the market. Equating this average to the Audi Q2L and Audi Q2L e-tron is inherently inaccurate. Therefore, we used the following method to calculate the lifecycle energy consumption and pollution emissions of fuel and electric vehicles.

3. ENERGY AND POLLUTION LAWS OVER THE LIFE CYCLE OF FUEL AND ELECTRIC VEHICLES

3.1 Energy consumption and pollutant emissions during the manufacturing phase of vehicle production (excluding batteries)

The Audi Q2L and Audi Q2L e-tron are equipped with a start-stop battery, while the Audi Q2L e-tron is also equipped with a power cell. When calculating the energy consumption and pollutant emissions of the vehicle's manufacturing phase, the start-stop battery is included in this case without the power cell.

The material of the main part of a car can be simplified to steel, cast iron, aluminium, copper, plastic, glass, rubber, and others.

Based on the data built into the GREET model in the Argonne National Laboratory's LCA report[2], it is possible to derive the percentage of each material excluding the power battery of an electric vehicle but including the start-stop battery of both. Combined with the energy and pollution coefficients of the various raw materials used in the manufacturing process (derived from the GREET model data for the entire life cycle of the vehicle), the emissions and energy consumption of both are shown in the following Table 1:

Table 1. Pollutant Emissions and Energy Consumption in Vehicle Production (excluding batteries)

Type of vehicle	Energy consumption (kJ)	Pollutant emissions (kg)						
		CO_2	CO	SO_2	NO_X	dust	CH_4	VOC
Audi Q2L	9×10^7	9623.99	166.47	87.6	24.47	159.9	21.56	0.62
Audi Q2Le-tron	8.68×10^7	9689.25	160.05	123.97	25.01	154.47	20.95	0.53

3.2 Energy Consumption and Pollution Emissions in the Production of Power Battery Stage

Although electric vehicles do not directly emit polluting gases during their use. However, their lithium-ion power batteries still generate energy consumption and pollution during the manufacturing process. As the mass of the lithium-ion power battery of an electric vehicle is very large, it will far exceed the energy consumption and pollution generated by the lead battery of a fuel vehicle during the manufacturing process. The data referenced at this stage is mainly based on the data provided by the GREET model, see Table 2:

Table 2. Pollution emissions and energy consumption in the production of automotive batteries

Type of vehicle	Energy consumption (kJ)	Pollutant emissions (kg)						
		CO_2	CO	SO_2	NO_X	dust	CH_4	VOC
Audi Q2L	1.58×10^5	8.74	0.0093	0.262	0.019	0.058	0.08	0.014
Audi Q2L e-tron	3.55×10^7	2925.6	1.1	15.16	4.5	5.51	11.02	0.58

3.3 Stages of vehicle use

3.3.1 Calculation methods and formulae in this paper

The GREET model data lacks specificity as it provides a weighted average of mainstream cars, failing to capture the unique characteristics of vehicles like the Audi Q2L and Audi Q2L e-tron. However, the reason we previously used GREET model data for the production phase of the vehicle was that it was difficult to accurately obtain data on energy consumption and pollution emissions for the Audi Q2L and Audi Q2L e-tron during the production phase. In contrast, data on fuel consumption, electricity consumption, and harmful gas emissions are relatively easy to obtain during the vehicle's driving phase, so we decided to discard the LCA research methodology[3].

For the fuel car Audi Q2L, it is assumed that its unit fuel consumption is C L/km, the total mileage during its life cycle is S km, the density of petrol is ρ kg/L, the calorific value of petrol is Q kJ/kg, and the total energy consumption is EC kJ. its total energy consumption during its life cycle is calculated by Equation (1) as follows:

$$EC = C \cdot S \cdot \rho \cdot Q \quad (1)$$

Currently, the National Stage VI Motor Vehicle Pollutant Emission Standard specifies the limit value of Likg/km for pollutants emitted per unit km from fuel vehicles, so the total mass M of pollutants emitted over the life cycle of the Audi Q2L can be calculated by Equation (2):

$$M = Li \cdot S \quad (2)$$

For the electric vehicle Audi Q2L e-tron, it is assumed that the unit power consumption is C kWh/km, the total driving range over the life cycle is S km, and the total energy consumption is EC kJ. According to the unit conversion, 1kWh = 3600kJ. Its total energy consumption during its life cycle is calculated by Equation (3) as follows:

$$EC = C \cdot S \cdot 3600 \quad (3)$$

Pollution from electric vehicles comes mainly from the electric energy production stage. The energy consumption calculated in the above equation is the actual energy consumed by the vehicle. However, electricity is lost in the production stage, transmission stage, and charging stage. Therefore, the efficiency of electricity generation is not 100 percent. These efficiency ratios are available in official ONS data and the relevant literature. Setting the efficiency as η and the raw energy consumption as EC0 , Equation (4):

$$EC_0 = \eta \times EC \quad (4)$$

By reviewing the literature, it can be learnt that the mass of pollutants to be emitted per 1 kWh of electricity produced is m. Then the total mass of pollutants emitted during electricity production, M, can be calculated by Equation (5):

$$M = m \cdot EC_0 \quad (5)$$

3.3.2 Calculation of energy consumption in the use phase of the vehicle

Determining the mileage of a car during its operational phase is a critical consideration. There are significant differences in the usage habits of different car owners, so there is no uniform standard for the mileage of each car during its usage phase. To ensure precision, reliance on official data is essential. According to the "Regulations on Compulsory Motor Vehicle Retirement Standards" issued by the State, both vehicles are classified as small and micro non-operating passenger vehicles, and their service life is recommended to be retired after 600,000 km. Therefore, we have established the mileage of these vehicles to be 600,000 kilometres.

Let the total mileage driven by the Audi Q2L during the use phase be S_1 , the fuel consumption per unit mile be c_1 , the density of petrol be ρ, the calorific value of petrol be q, and the total energy consumed during the use phase be Q_1 . The total energy consumed by the Audi Q2L is given by:

$$Q_1 = S_1 \cdot c_1 \cdot \rho \cdot q \quad (6)$$

According to the official data, S_1 = 600,000km, c_1 = 6.1L/100km, ρ = 0.73kg/L, q = 46,000kJ/kg. The calculation gives the total energy consumed by the Audi Q2L over its life cycle as Q_1 = 1.23 x 10 9kJ

Data from the National Bureau of Statistics shows that China's total refining efficiency over the last decade has been 95%, indicating that the efficiency of refining crude oil into petrol is 95%. Therefore, the actual energy consumption of the Audi Q2L during its lifetime is 1.23×10 9 ÷ 95% = 1.3×10 9kJ.

Let the total distance travelled by the Audi Q2L e-tron during the use phase be S_2 , the electricity consumption per unit mile be c_2 , and the total energy consumed during the use phase be Q_2 . The formula for calculating the total energy consumed is:

$$Q_2 = S_2 \cdot c_2 \quad (7)$$

According to the official data, S_2 = 600,000km, c_2 = 13.9kWh/100km, 1kWh = 3,600kJ. The calculation gives the total energy consumed by the Audi Q2L e-tron over its life cycle as Q_2 = 83,400kWh = 3×10 8kJ

Taking the above data into account, we can conclude that, all other indicators being equal, the total energy consumption of a purely electric vehicle is only 23 per cent of that of a purely fuel-efficient vehicle.

As evident from the previous figures, the Audi Q2L has a total mass of 1,405 kilograms and the Audi Q2L e-tron has a total mass of 1,610 kilograms. This is due to the fact that the mass of the powertrain of an electric car is much greater than that of a fuel vehicle, and the battery in particular is significantly greater than that of fuel. Despite the theoretical expectation that electric cars should consume more energy, the opposite holds true. This is because the energy conversion efficiency of an electric motor is much higher than that of an internal combustion engine[4].

3.3.3 Calculation of pollution emissions during the use phase of the vehicle

The Audi Q2L complies with the most stringent China 6B emission standards, including carbon monoxide (CO) at 500 mg/km, non-methane hydrocarbons (NMHC) at 35 mg/km, nitrogen oxides (NOX) at 35 mg/km, and particulate matter (PM) at 3 mg/km. Through calculations, the pollutants emitted by the Audi Q2L over its entire 600,000-kilometer usage cycle are shown in the Table 3:

Table 3. Emissions of pollutants from the Audi Q2L over its lifetime

Type of pollutant	CO	NMHC	NO_X	PM	CO_2
Pollutant mass (kg)	300	21	21	1.8	81830

3.3.4 Calculation of energy consumption and pollutant emissions during the electricity production phase

The Audi Q2L e-tron does not emit pollutants directly into the atmosphere during its life cycle. However, as the country's electricity generation relies mainly on thermal power generation, the environmental pollution caused by electric vehicles is mainly concentrated in the electricity production phase. According to the National Bureau of Statistics, around 45 per cent of the thermal energy in the process is converted into electricity, while the remaining 55 per cent is wasted in the form of heat.

Based on the emission standards for power plants, the mass of pollutants emitted per 1kWh of production, and taking into account the wasted heat, the total pollutant emissions during the electricity production phase were calculated as: 2.7 kg of dust, 39,685 kg of carbon dioxide, 18.6 kg of sulphur dioxide, and 26.7 kg of nitrogen oxides.

The combined calculations give us the energy consumption and pollution emissions of the Audi Q2L and Q2L e-tron in the use phase, as detailed in Table 4.

Table 4. Energy consumption and pollution emissions of the Audi Q2L and Q2L e-tron in the use phase and in the electricity production phase

Type of vehicle	Pollutant type and emission (kg)							Energy consumption (kJ)
	CO	CO_2	SO_2	NO_x	NMHC	PM (dust)	PN (particulate matter)	
Audi Q2L	300	81830	-	21	1.8	1.8	3.6×10^{11} pcs.	1.3×10^9
Audi Q2L e-tron	-	39685	18.6	26.7	-	2.7	-	5.71×10^8

Upon comparing the calculated results, we observe that electric vehicles exhibit approximately 50% lower greenhouse gas emissions in terms of CO2 and energy consumption during the usage stage, compared to conventional fuel vehicles. However, there are differences in the types of pollutants emitted by each vehicle. Both types emit NO_X and PM. Conventional fuel vehicles emit additional pollutants such as CO, NMHC, and PN, while electric vehicles emit additional SO_2, although the difference in emission quantities is not significant. Thus, from this perspective, the advantage of electric vehicles over conventional fuel vehicles in mitigating environmental pollution is not particularly pronounced. Their most significant advantage lies in reducing energy consumption and mitigating the trend of climate change.

3.4 Calculation of energy consumption and pollutant emissions at the end-of-life stage of a vehicle

China has not yet established a standardised vehicle scrapping standard. In order to ensure the fairness of the data, we chose to use the data in the GREET model.

Yan Huajun *et al.* (2018) selected Santana, a fuel vehicle weighing 1100 kg, and BYD E6, an electric vehicle weighing 2295 kg, as their research subjects. They calculated their pollutant emissions and energy consumption at the end-of-life stage by using the GREET model data. However, the pollutant emission coefficients of these two vehicles at the end-of-life stage are not provided in the paper. Nevertheless, it is possible to extrapolate the pollutant emissions of the Audi Q2L and Audi Q2L e-tron from their calculations.

Table 5. Pollutant emissions and energy consumption at end-of-life for the Audi Q2L and Audi Q2L e-tron

Type of vehicle	Energy consumption (kJ)	Pollutant emissions (kg)						
		CO_2	CO	SO_2	NO_x	dust	CH_4	VOC
Audi Q2L	2.54 x 106	218.41	0.04	0.85	0.33	0.17	0.77	0.02
Audi Q2L e-tron	3.82 x 107	3283.14	0.53	12.77	4.98	2.5	11.58	0.29

It can be deduced from the data in the Table 5 that electric vehicles emit significantly more pollutants and consume significantly more energy at the end-of-life stage than fuel vehicles. This phenomenon is attributed to the fact[5] that the power batteries of electric vehicles contain more complex metal ions, which makes the end-of-life and recycling process more complicated.

3.5 Comparative analysis of life cycle energy consumption and pollution emissions

Summing the energy consumption and pollutant emissions of the four phases of vehicle manufacturing,

 https://doi.org/10.1051/e3sconf/202452803018

fuel and electricity production, vehicle use, and vehicle end-of-life, table 6 is obtained.

Table 6. Comparison of life cycle energy consumption and pollutant emissions between fuel and electric vehicles

Type of vehicle	Energy consumption (kJ)	Pollutant emissions (kg)						
		CO_2	CO	SO_2	NO_x	dust	CH_4	VOC
Audi Q2L	1.39 x 10^9	91681.14	466.51	88.71	45.82	161.92	22.41	0.65
Audi Q2L e-tron	8.78 x 10^8	55582.99	161.68	170.5	61.19	165.19	43.55	1.4

3.6 Calculation of environmental impact potential over the life cycle

The potential environmental impacts of the Audi Q2L and Audi Q2L e-tron are shown in Table 7:

Table 7. Potential environmental impacts over the life cycle of fuel and electric vehicles

Type of vehicle	global warming (PRC usage)	acidification	photochemical smog	Smoke and dust
Audi Q2L	92151.79 kg	120.78 kg	14.26kg	161.92kg
Audi Q2L e-tron	56497.50 kg	213.33 kg	6.86kg	165.19kg

It's evident that fuel cars exert a greater impact on global warming and photochemical smog, while electric cars have a greater impact on acidification. The reason for this is that fuel cars emit more CO_2 and CO, and electric cars emit more SO_2 and NO_X in the process of electricity production.

4. ANALYSIS OF RESULTS

4.1 Analysis of energy consumption

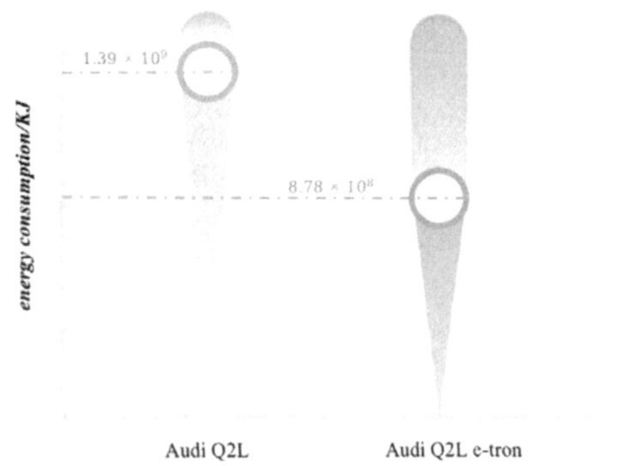

Figure 1. Full life cycle energy consumption of the Audi Q2L and Audi Q2L e-tron

From Figure 1 we get the full life cycle energy consumption of Audi Q2L is 1.39×109 KJ while the full life cycle energy consumption of Audi Q2L e-tron is 8.78×108 KJ.

Firstly, the relatively high energy consumption of the Audi Q2L (Fuel Vehicle) is mainly related to the fact that Fuel Vehicles rely on fossil fuels in their use. Thermal power generation is usually one of the main sources of fossil fuels, and the energy costs of fuel vehicles can be affected by the price and stability of fuel supply[6]. In addition, certain energy losses in thermal power generation reduce the energy efficiency of the whole system.

In comparison, the Audi Q2L e-tron (New Energy Electric Vehicle) is superior in terms of energy consumption. Firstly, electric vehicles use batteries as energy storage devices, and the manufacturing process of batteries may involve some energy consumption. However, EVs are more energy efficient in the use phase, as electric motors are typically more efficient at conversion than internal combustion engines. In addition, the overall environmental impact of electric vehicles will be lower if the electricity comes from renewable sources, rather than relying on high-emission energy sources such as thermal power generation.

Taking energy costs, thermal power generation and energy efficiency into account, the Audi Q2L e-tron offers advantages in terms of energy consumption and environmental impact over the entire life cycle. This underlines the potential of electric vehicles as a sustainable mode of mobility, particularly in the context of the drive to use renewable energy sources.

4.2 Analysis of pollutant emissions

4.2.1 Emissions perspective

Table 8. Total Pollutant Gas Emissions by Life Cycle Stage for Audi Q2L and Audi Q2L e-tron

Type of vehicle	Total pollutant gas emissions			
	The automotive production stage	Battery production stage	Utilisation stage and electricity production stage	end-of-life stage
Audi Q2L	10084.61	9.1823	82152.8	220.59
Audi Q2L e-tron	10174.23	2963.47	39733	3315.79

Table 9. Audi Q2L and Audi Q2L e-tron life cycle stages CO_2 emissions

Type of vehicle	CO_2				
	The automotive	Battery production	Utilisation stage and	end-of-life stage	entire life cycle

https://doi.org/10.1051/e3sconf/202452803018

	production stage	stage	electricity production stage		
Audi Q2L	9623.99	8.74	81830	218.41	91681.14
Audi Q2L e-tron	9689.25	2925.6	39685	3283.14	55582.99

The emissions of the Audi Q2L and Q2L e-tron in the use phase and the electricity production phase account for the majority of the life cycle emissions, and the most important of these emissions is CO_2.as shown by Table 8 and Table 9.

In the use phase of the Audi Q2L, the emission of 81,830 kg of CO2 raises concerns, significantly surpassing the 39,685 kg of CO2 emitted during the electric power production phase of the Audi Q2L e-tron. During the use phase of a fuel vehicle, CO_2 from the combustion of petroleum is the main source of pollution. This results in fuel cars contributing a significant amount of greenhouse gas emissions during use, which have a negative impact on the climate. The release of these greenhouse gases negatively impacts global warming, triggering extreme weather events and affecting the ecological balance. In terms of human health, extreme temperatures and meteorological events caused by climate change pose a threat to public health. As for electric vehicles, CO_2 emissions mainly come from the electricity production stage. At this stage, the way electricity is produced directly affects the overall environmental performance of electric vehicles. If the electricity comes from coal-fired or other high-carbon energy sources, CO_2 emissions will be relatively high. However, if renewable energy sources are used, CO_2 emissions will be significantly lower.New energy electric vehicles have made outstanding contributions in three aspects: environmental protection, energy protection, and energy security.

From the perspective of CO_2 emissions and the greenhouse effect, new energy electric vehicles have relatively low CO_2 emissions during the electricity production phase, especially if the electricity comes from renewable sources. This results in electric vehicles contributing less greenhouse gases over their lifecycle and a smaller burden on global climate change, which contributes to climate change mitigation and reduces the impact of global warming.

From the perspective of energy diversification and renewable resources, electrical energy can be produced from a wide range of renewable energy sources, such as solar and wind, thereby reducing dependence on finite fossil fuels, such as oil. This can help to promote energy diversification and reduce the overexploitation of finite resources[7].

From the perspective of improving energy security, the promotion of new energy electric vehicles has a positive impact on the country's energy security, as the country is able to reduce its dependence on imported oil, mitigate geopolitical risks, and build a more localised and sustainable energy system, thereby improving overall energy security. This is important for the country's economic stability and sustainable development.

4.2.2 Money volume perspective

Here we use monetary values to quantify greenhouse gas and air pollutant emissions. Regarding the cost of environmental pollution, there are many studies at home and abroad, and the commonly used methods to measure the cost of environmental pollution include the productivity change method, preventive expenditure method, replacement cost method, opportunity cost method[8], disease cost method and human capital method, alternative market method, etc. (Xu Songling, 1995; Zhou Yihong, 2004).

In this paper, the total social cost of pollutants is calculated by corresponding to the unit cost of the pollutant and the emission of the pollutant (Meng Xianchun, 2007), and the specific formula is as follows.

$$C = \sum e_i \times d_i \tag{8}$$

Where: C is the total cost of pollutant emissions; ei is the amount of pollutant emissions; di is the unit cost of emissions of the corresponding pollutant.

The calculation of pollutant emission costs in this paper refers to Meng Xianchun (2007)[9], China's corresponding pollutant unit emission costs di .

$$d_i = d'_i \times \frac{\text{China's CPI in 2016}}{\text{China's CPI in 2000}} \times R \tag{9}$$

Where: R is the annual average exchange rate of the euro to the renminbi in 2000; d'i is the unit emission cost of the corresponding pollutant in Europe in 2000, EUR/kg, from External Cost Values for EE SUT Framework (IER, 2010). d'i is the unit cost of emissions of the corresponding pollutant in Europe in 2000 in EUR/kg from the External Cost Values for EE SUT Framework (IER, 2010). d'i is obtained by adjusting d' using the Consumer Price Index (CPI) at constant 1978 prices and the annual average exchange rate of the euro to the renminbi for the year 2000 (1 euro = 7.63 yuan). The results are shown in Table 10.

Table 10. Unit costs for individual pollutant gases

emission	Unit cost/(€/t)	Unit cost/(Yuan/kg)
CO_2	21	0. 227
CH_4	480	5. 191
NO_x	6 700	72. 465
SO_x	6 500	70. 301
CO	62	0. 671
VOC	870	9. 410
dust	63 000	681. 384

Accordingly, we calculated the monetary value of pollution emissions at each stage.

https://doi.org/10.1051/e3sconf/202452803018

Table 11. Monetisation of pollutant emissions over the life cycle of fuel and electric vehicles

Type of vehicle	Monetary value of pollutant emissions at each stage of the life cycle (Yuan)				
	The automotive production stage	Battery production stage	Utilisation stage and electricity production stage	end-of-life stage	entire life cycle
Audi Q2L	119298.987	61.721	21524.966	253.295	141133.539
Audi Q2L e-tron	118201.543	5873.793	14090.645	3770.548	141943.344

From Table 11, it's evident that the monetary value of the car production process constitutes the bulk of the total life cycle monetary value. Conversely, pollutant emissions dominate the total life cycle emissions during both the use and electricity production phases. The underlying reason for this phenomenon is that the dust generated in the automotive production process is of a much larger order of magnitude than in the use and electricity production phases (see Figure 2), and the unit cost of dust emissions is much higher than that of other pollutants (see Figure 3).

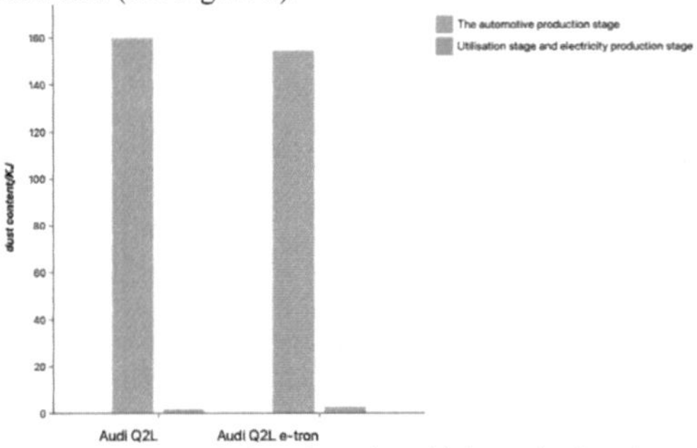

Figure 2. Dust emissions from the vehicle production phase versus the use phase and the electricity production phase

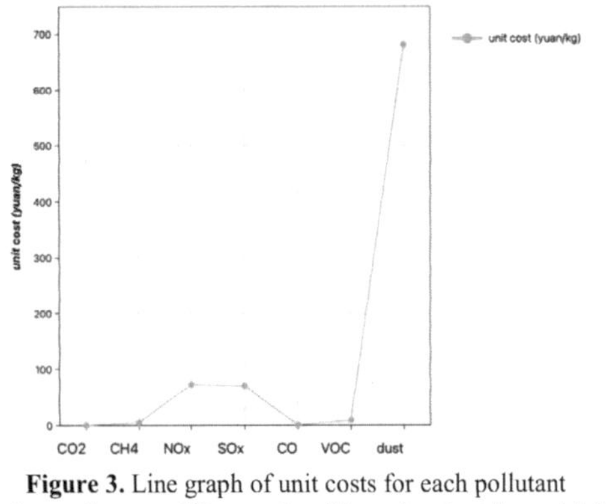

Figure 3. Line graph of unit costs for each pollutant

According to Figure 4, although electric vehicles emit significantly less greenhouse gases (CO_2) than fuel vehicles over their entire life cycle, the total greenhouse gas emissions are orders of magnitude greater than other pollutant gases. Theoretically, the social benefits of electric vehicles should be better than those of fuel vehicles, however, by calculating the monetary value, we find that the cost of pollutant gas emissions over the life cycle of electric vehicles is not lower than that of fuel vehicles. In fact, it is slightly higher, as depicted in Figure 5.

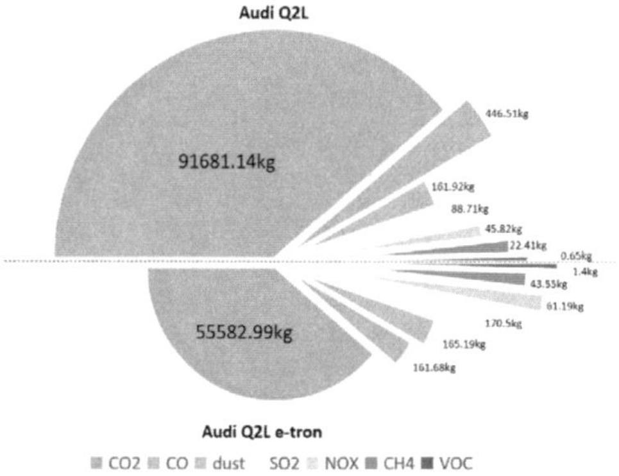

Figure 4. Emissions of individual pollutant gases over the entire life cycle of the Audi Q2L and Audi Q2L e-tron

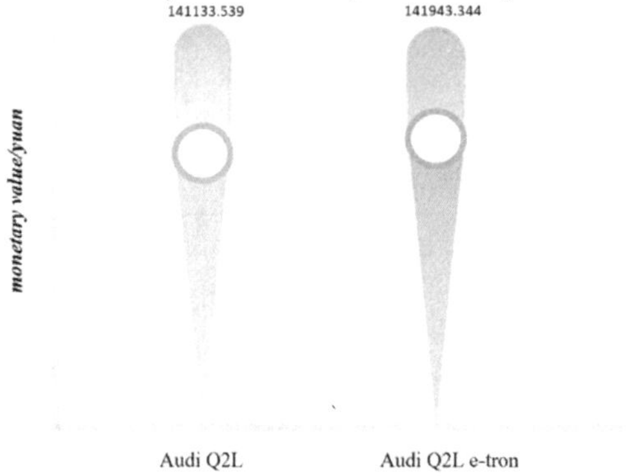

Figure 5. Monetary values over the entire life cycle of the Audi Q2L and Audi Q2L e-tron

The reason for this phenomenon is that although electric vehicles emit fewer greenhouse gases than gasoline vehicles throughout their entire lifecycle, the content of acid rain pollutants (SO_2) and nitrogen oxides (NO_X) is higher than that of gasoline vehicles (see Figure 6). Although the order of magnitude is smaller than greenhouse gases, the unit emission cost of acid rain pollutants and nitrogen oxides (NO_X: 72.465 yuan/kg, SO_x: 70.301 yuan/kg) The order of magnitude of CO_2 emissions is also much greater than the unit emission cost of greenhouse gases (CO_2: 0.227 yuan/kg). Therefore, when we separately calculate the emission costs of CO_2, SO_2, and NO_X throughout the entire lifecycle of fuel vehicles and electric vehicles, we found that although the CO_2 emission cost of fuel vehicles is higher than that of electric vehicles, the emission cost of SO_2 and NO_X of electric vehicles is also higher than that of fuel vehicles, and the sum of the emission costs of the three gases is similar for the two types of vehicles (see Figure 7).

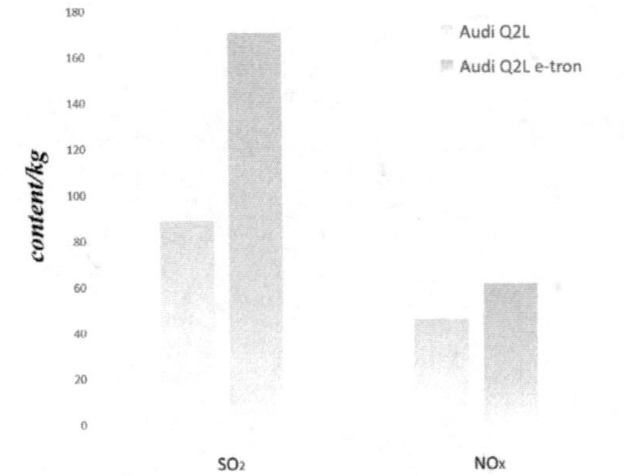

Figure 6. Throughout the life cycle of the Audi Q2L and Audi Q2L e-tron SO_2 and NO_X content histograms

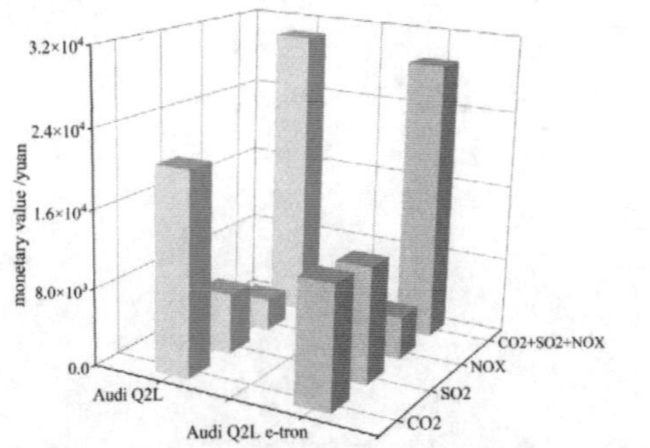

Figure 7. Audi Q2L and Audi Q2L e-tron throughout their life cycle CO_2 , SO_2 , NO_X Currency value histogram

Therefore, compared with fuel vehicles, although new energy electric vehicles greatly reduce greenhouse gas emissions throughout the life cycle, and play a role in addressing the problems of increasing climate change, serious emissions pollution, and shortage of traditional fossil fuels, as well as zero tailpipe emissions during driving and higher energy efficiency, they also increase sulfur dioxide as well as nitrogen oxides emissions, which aggravate the formation of acid rain, the destruction of the ozone layer, and the damage to plants and water. However, it also increases the emission of sulphur dioxide and nitrogen oxides, aggravating the formation of acid rain, the destruction of the ozone layer and the damage to plants and water bodies.From a monetary perspective, the social benefits of replacing fuel vehicles with new energy electric vehicles are not substantial when considering the monetization of pollutant gas emissions. Despite their environmental advantages, the overall impact on pollutant emissions does not translate into significant economic benefits.

5. CONCLUSIONS

This paper calculates the whole-life energy consumption and pollution emissions of the Audi Q2L and Audi Q2L e-tron, and presents the obtained results in monetary form, making the results more intuitively reflect the social benefits of the two vehicles.

The advantages of new energy electric vehicles in terms of total pollutant gas emissions remain significant, notably reducing carbon emissions and mitigating global warming. Therefore, the promotion of new energy electric vehicles is still imperative, but not directly in the national scope of the unified promotion, but should be first in the higher climate and health benefits and energy saving potential of the regional promotion, while the government incentives and subsidies in the policy should be maintained for a long time, in order to change the current electric vehicles and fuel vehicles in the competition, the status quo is still in a disadvantageous position.

However,due to the fact that new energy vehicles produce more sulphur dioxide (SO2) and nitrogen oxides (NOx) in the production phase, the full replacement of traditional fuel vehicles by electric vehicles at this stage will not be able to achieve an improvement in the overall welfare of the society. So under the existing conditions, the improvement of new energy vehicles should prioritize the adoption of cleaner and more environmentally friendly ways of producing vehicles at the manufacturing stage as well as reforming the way of producing electricity to improve the cleanliness of electricity[10]. At the same time, efforts should be made to find easier and cheaper ways to deal with nitrogen oxides and sulphur dioxide.

REFERENCES

1. Tagliaferri, C., Evangelisti, S., Acconcia, F., Domenech, T., Ekins, P., Barletta, D., & Lettieri, P. (2016). Life cycle assessment of future electric and hybrid vehicles: A cradle-to-grave systems engineering approach. Chemical Engineering Research and Design, 112, 298-309.
2. Mei, X. (2022). Comparative Study on the Environmental Impact of Electric Vehicles and Fuel Vehicles. Tianjin University of Science and Technology. Dissertation.
3. Wang, P. (2018). Research on the whole life cycle environmental impact assessment of electric vehicles and natural gas vehicles. Chang'an University. Dissertation.
4. Cohn, R. S. (1994). Vehicle Life Cycle Analysis. Cambridge, MA: MIT.
5. Jin, L., Lu, Y., Xie, J., et al. (2019). Environmental and Economic Benefit Analysis of New Energy Vehicle Full Life Cycle Based on GREET Model. Resources and Industries. vol. 21, no. 5, pp. 1-8. DOI:10.13776/j.cnki.resourcesindustries.20190904.001.
6. Hu, Y., Jin, Y., Zhang, S. (2021). Climate and Health Impacts and Social Costs of Electric Vehicle Replacement of Fuel Vehicles - A Cost-Benefit Analysis Based on a Single Vehicle Perspective. Journal of Peking University (Natural Science Edition), vol. 57, no. 5, pp. 916-926. DOI:10.13209/j.0479-8023.2021.068.

7. He, W., Hao, X., Chen, F. (2022). Carbon footprint evaluation of new energy vehicles based on life cycle. Journal of Northeast University of Finance and Economics, no. 2, pp. 29-41. DOI:10.19653/j.cnki.dbcjdxxb.2022.02.003.
8. Huo, H., Zhang, Q., Liu, F., & He, K. (2013). Climate and environmental effects of electric vehicles versus compressed natural gas vehicles in China: a life-cycle analysis at provincial level. Environmental science & technology, 47(3), 1711-1718.
9. Meng, X. (2007). Cost difference analysis of two kinds of buses based on full life cycle theory. Hunan University. Dissertation.
10. Arrow, K. J., Cropper, M. L., Eads, G. C., Hahn, R. W., Lave, L. B., Noll, R. G., ... & Stavins, R. (1996). Benefit-cost analysis in environmental, health, and safety regulation. Washington, DC: American Enterprise Institute, 1-17.

https://doi.org/10.1051/e3sconf/202452803019

Carbon emission monitoring method and empirical study of expressway construction period

Jun Wang[1], Xuejuan Yu[1], Xi Hai[2], Zhu Ming[3],and Yousen Yang[3,*]

[1]Jiangsu Traffic Engineering Construction Bureau, Jiangsu, Nanjing, 210004, China
[2]Jiangsu Middle Road Engineering Technology Research Institute Co., LTD, Jiangsu, Nanjing, 211800, China
[3]Middle road traffic testing technology Co., LTD, Jiangsu, Nanjing, 211800, China

Abstract. In order to improve the ability of carbon emission management in expressway construction, the precision management of carbon emission is realized. In this paper, the construction period carbon emission monitoring method is proposed, the carbon emission monitoring scope is defined, and the carbon emission monitoring content and method are determined. On this basis, taking the Zhangjinggao Yangtze River Bridge project as an example, analyze the monitoring data of construction carbon emissions. The results of the study showed that from July to December, the total carbon emission from construction was 668.30t. Among them, the carbon emissions of electricity, machinery and equipment, labor and construction water accounted for 47.02%, 40.68%, 11.99% and 0.31% of the total carbon emissions, respectively. In addition, the unit time carbon emission of the excavator is 12.12kg/h; The unit time carbon emission of the truck crane is 1.35kg/h.

1. Introduction

In recent years, global warming has been a serious challenge facing humanity. As an important promoter of the global climate governance process, China has made it clear that it will strive to reach the peak of carbon dioxide emissions by 2030 and achieve carbon neutrality by 2060[1].The transport sector is the third largest source of carbon emissions after industry and buildings, accounting for about a quarter of the world's total carbon emissions[2]. According to LIU Y's research, the fastest growing area of carbon emissions in the world is the transportation sector. From 2005 to 2020, the carbon emissions of the transportation industry increased by 149%, and the average annual growth efficiency reached 5.8%[3]. As the most common and important transportation infrastructure, expressway is a crucial carrier to promote the development of expressway transportation. During the construction process of expressway infrastructure, a large amount of carbon emissions will be generated due to material loss and the use of mechanical equipment. According to relevant studies, carbon emissions from road infrastructure construction account for 10% to 20% of total carbon emissions from road transportation[4]. Although the proportion of carbon emissions from expressway infrastructure construction is relatively low, the amount of carbon emissions is huge. Therefore, conducting research on carbon emission monitoring during construction is of great significance.

At present, domestic and foreign scholars have studied relatively little on carbon emission monitoring during the construction period, mainly focusing on the carbon emission accounting research of expressways. For example, Stripple et al. [5] were the first to use the LCA method to evaluate the CO2 of the entire road construction project, including raw material production and construction. Even including the maintenance and operation of roads, as well as the disposal or reuse of roads at the end of their life cycle. Baran dica et al. [6] evaluated greenhouse gases in road construction from a full lifecycle perspective and analyzed four construction projects in Spain.Specifically, the full life cycle assessment of road surfaces has always been a focus of attention for many researchers [7-9]. In addition, UCPRC has been committed to recommending common practices of LCA for pavement, and has proposed road life cycle assessment guidelines for practitioners, including a road life cycle assessment framework, and some recommended data and models that have been used in California and other states in the United States [10. At the same time, Abolhasani et al. [11] investigated the fuel use and emissions of non-road construction vehicles with real data. Frey et al. [12] summarized the actual data activity, fuel usage, and emission characteristics of selected automotive classifiers using petroleum diesel and bio-diesel. Although some scholars have studied expressway carbon emission monitoring, the monitoring is limited to operational vehicles, and no theoretical monitoring method has been formed for expressway construction carbon emission monitoring. Therefore, this article analyzes the structural characteristics of

*Corresponding author's e-mail: 2641247340@qq.com

construction energy consumption, proposes the scope of carbon emission monitoring during the construction period of expressways, clarifies the content and methods of carbon emission monitoring, and analyzes carbon emission monitoring data based on actual expressway engineering projects.

2. Construction energy consumption structure characteristics

In this paper, the characteristics of expressway construction energy consumption are investigated and analyzed. In the construction process, energy consumption comes from electricity, water, fuel and labor in living area, office area and construction site. Among them, the main equipment for fuel consumption is construction machinery, transportation machinery and processing machinery.

3. Carbon emission monitoring methods

3.1. Monitoring range

The whole life cycle of expressway is generally divided into 5 stages, which are material production stage, material transportation stage, site construction stage, operation and maintenance stage and demolition and recycling stage. Due to the fact that material production falls within the carbon emission range of production enterprises, it should not be included in the carbon emission monitoring scope during the construction phase. Therefore, this article divides the monitoring scope of carbon emissions generated at construction sites from the perspectives of time and space boundaries, as shown in Figure 1. As can be seen from Figure 1, the time boundary is the process of expressway construction from scratch, that is, from the start of the project to the completion and acceptance of the project. The space boundary refers to all production activities carried out on the construction site during the construction process, including material transportation, material processing, site construction operations and construction production management (living area and office area).

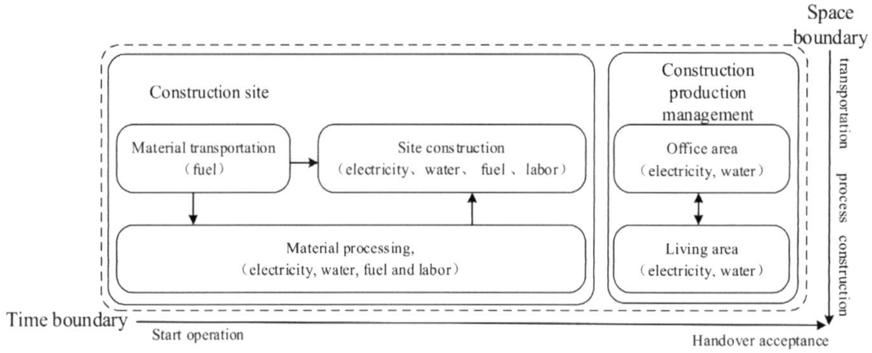

Figure 1. Carbon emission monitoring scope.

3.2. Monitoring content

The monitoring objects of construction carbon emissions mainly include two dimensions. The first dimension is the monitoring project, which mainly includes: temporary engineering, roadbed engineering, pavement engineering, bridge and culvert engineering, tunnel engineering, intersection engineering, traffic engineering and facilities along the line, greening and environmental protection engineering, other engineering, and construction site construction. The second dimension is the monitoring area, which mainly includes living areas, office areas, and construction sites.

According to the previous research and analysis, the main monitoring objects are fuel (diesel and gasoline), electricity, water, and labor. According to different monitoring objects, different monitoring methods are determined to obtain carbon emission monitoring data, as shown in Table 1.

Table 1. Data collection methods for construction carbon emissions.

Monitoring area	***Energy structure***	***Monitoring methods***
Office areas	water	Automatic collection
	electricity	Automatic collection
Living areas	water	Automatic collection
	electricity	Automatic collection
Construction sites	water	Automatic collection
	electricity	Automatic collection
	fuel (diesel and gasoline)	Statistics/Automatic collection
	labor	Statistics

3.3. Monitoring methods

This article focuses on carbon emission monitoring content, mainly using manual statistical methods and equipment automatic collection methods. Among them, for carbon emission data that cannot be collected through automatic collection by manual or some mechanical equipment, manual statistics should be used to fill out carbon emission reports, and they should be summarized and reported at the end of each month. The automatic collection method is set to collect carbon emission data every 5 minutes.

4. Empirical research

4.1. Project overview

This article takes the anchorage construction project of Section A3 of the Zhangjinggao Yangtze River Bridge as an example to analyze the carbon emission monitoring data from July to December 2023. The total length of the bridge project route is about 29.849km, which is divided into three parts:cross-river bridge, north connection and south connection. The entire line is constructed according to the standards of expressways, with a two-way eight lane cross river section and a design speed of 100km/h; The south and north connections are two-way six lane, with a design speed of 120km/h.

4.2. Analysis of total carbon emissions

According to the carbon emission monitoring data, during the construction process from July to December, the total carbon emission generated by machinery, equipment, electricity, water and labor consumption was 668.30 tons. Among them, the carbon emissions of machinery and equipment were 271 tons, accounting for 40.68% of the total carbon emissions; The carbon emission of electricity was 314.23 tons, accounting for 47.02% of the total carbon emission. The carbon emission of water was the least, 2.07 tons, accounting for only 0.31% of the total carbon emission. Artificial carbon emissions are 80.13 tons, accounting for 11.99% of the total carbon emissions, as shown in Figure 2. As can be seen from Figure 3, the carbon emissions in July, August and November reached 373.58 tons, accounting for 55.90% of the total carbon emissions.

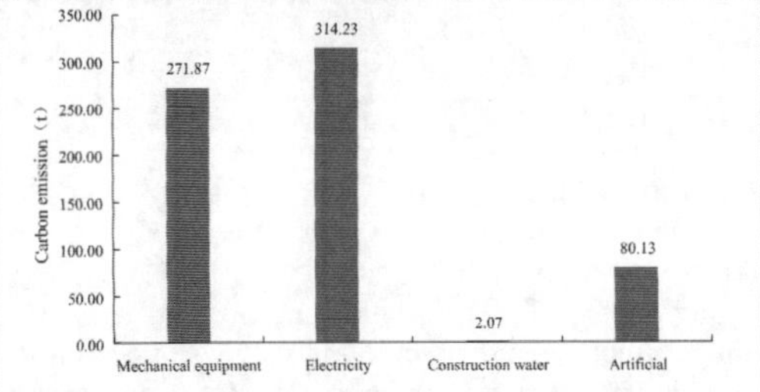

Figure 2. Carbon emissions from different energy sources.

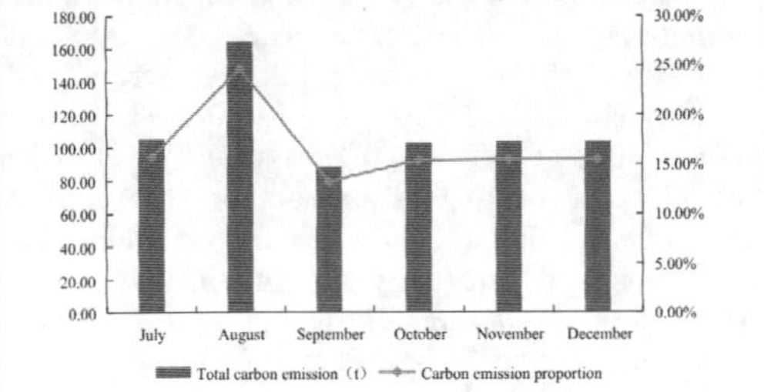

Figure 3. Monthly construction carbon emissions and proportion.

4.3. Carbon emissions from mechanical equipment

As can be seen from Figure 4, the total carbon emission of mechanical equipment is 271.87 tons. Among them, the highest carbon emissions of machinery and equipment in August was 85.46 tons, accounting for 31.44% of the total carbon emissions of machinery and equipment. The highest carbon emission of crawler crane is 196.79 tons, accounting for 55.95% of the total carbon emissions of mechanical equipment, but the carbon emission per unit time is 4.07kg/h; Automobile crane has the smallest carbon emission, 0.65 tons, accounting for 0.18% of the total carbon emission of mechanical equipment, but the unit time carbon emission is the smallest, 1.35kg/h, as shown in Figure 5.

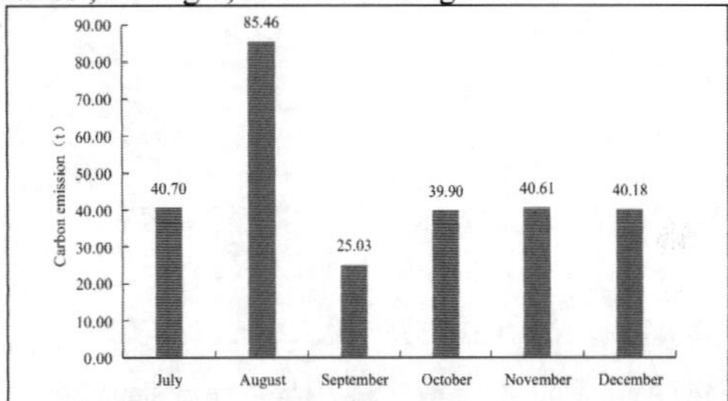

Figure 4. Carbon emissions from mechanical equipment.

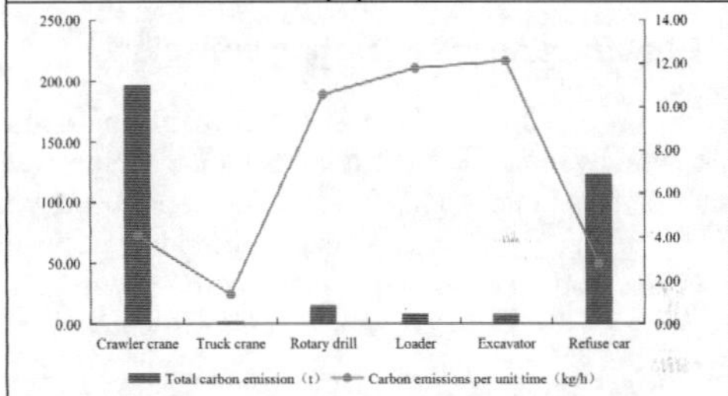

Figure 5. Carbon emission intensity per unit of different types of mechanical equipment.

4.4. Electricity carbon emissions

According to Figure 6, the carbon emissions from electricity are 314.23 tons. Among them, August used the most electricity, accounting for 20.02% of the total carbon emissions from electricity; The fluctuation range of carbon emissions from electricity in July, September, November, and December is not significant, with carbon emissions remaining between 49.5 tons and 51 tons; During the six months from July to December, the average monthly carbon emissions from electricity were 52.37 tons; In October, the carbon emissions from electricity were the lowest, accounting for only 15.77% of the total carbon emissions from electricity, at 49.56 tons.

In addition, the electricity carbon emission of the living area in August was 16.44 tons, accounting for 5.23% of the total electricity carbon emission. The average monthly carbon emissions from electricity in the living area are 13.93 tons. The electricity carbon emissions of the office area in August were 46.47 tons, accounting for

14.79 of the total electricity carbon emissions. The average monthly electricity carbon emissions of the office district are 39.85 tons, as shown in Figure 7.

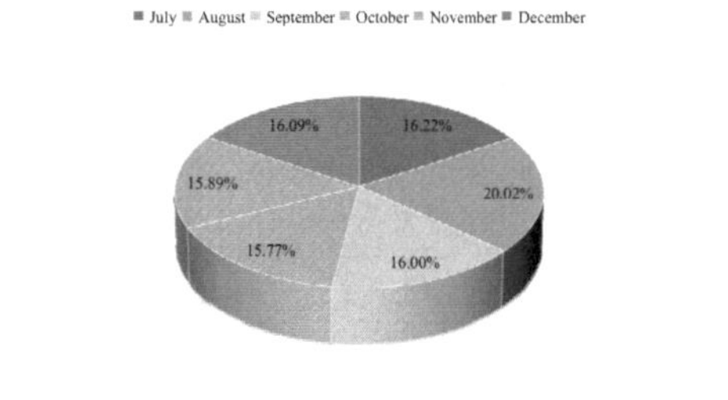

Figure 6. The proportion of carbon emissions from electricity.

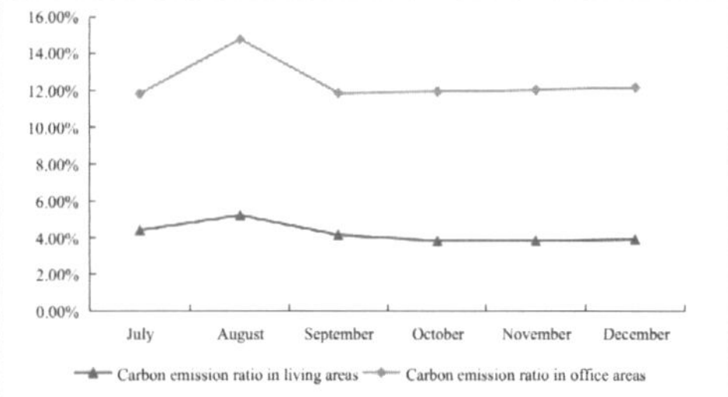

Figure 7. Monthly proportion of carbon emissions from electricity.

4.5. Construction water carbon emission

As can be seen from Figure 8, the carbon emission of construction water is 2.07 tons. Among them, water carbon emissions in August were 0.411 tons, occupying 19.90% of the total water carbon emissions. July and December had the lowest water carbon emissions, occupying 14.38% and 15.31% of the total water carbon emissions.

In addition, living areas had the highest water carbon emissions in August, accounting for 14.81% of the total water carbon emissions. The average monthly carbon emissions from water used in living areas is 0.238 tons. The office area had the highest water carbon emissions in December, accounting for 6.57% of the total water carbon emissions. The average monthly water carbon emission of the office area is 0.106 tons, as shown in Figure 9.

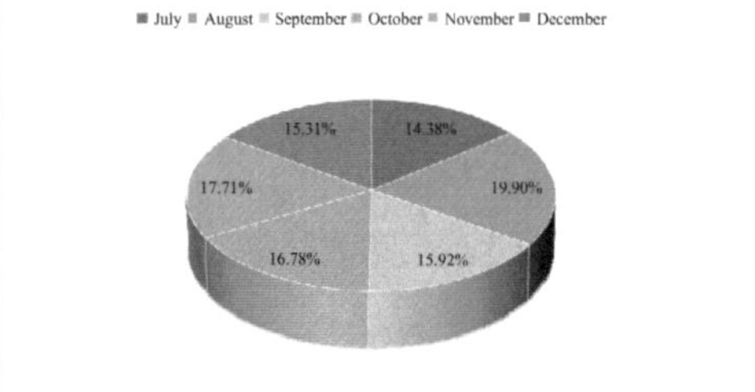

Figure 8. Monthly construction water carbon emission ratio

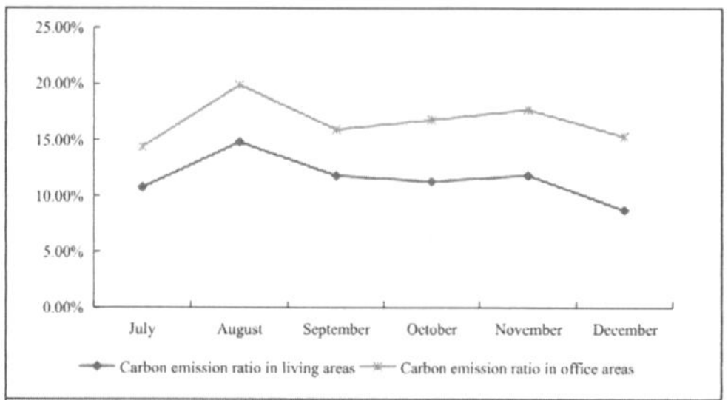

Figure 9. Monthly proportion of carbon emissions from water use.

4.6. Artificial carbon emission

As can be seen from Figure 10, the total artificial carbon emission is 80.13 tons. Among them, the highest labor consumption in August was 1798.5 working hours, generating 16.14 tons of carbon emissions, accounting for 20.14% of the total artificial carbon emissions. In July, the artificial carbon emission was 13.07 tons, accounting for 16.31% of the total artificial carbon emission. Artificial carbon emissions for the remaining four months were essentially the same, fluctuating between 12.40 tonnes and 13.00 tonnes; The average artificial carbon emissions for six months was about 13.35 tons.

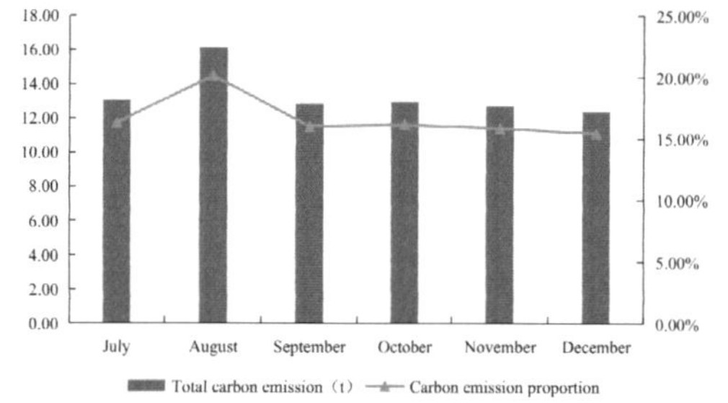

Figure 10. Artificial carbon emission and carbon emission proportion.

5. Conclusion

This article analyzes the characteristics of energy consumption structure during the construction period, proposes a construction carbon emission monitoring method, and takes the Zhangjinggao Yangtze River Bridge project as an example for analysis. According to research, the total carbon emissions from construction from July to December were 668.30t. Among them, electricity has the highest carbon emissions, accounting for 47.02% of the total carbon emissions; Secondly, the carbon emissions from mechanical equipment are 271 tons, accounting for 40.68% of the total carbon emissions; Artificial carbon emissions account for 11.99% of the total carbon emissions; The carbon emissions from construction water are the lowest, accounting for only 0.31% of the total carbon emissions. In addition, excavators have the highest carbon emissions per unit time, at 12.12kg/h; The carbon emissions per unit time of a truck crane are the smallest, at 1.35kg/h.

References

1. IPCC. Climate Change 2021: The Physical Science Basis[R].Cambridge and New York: Cambridge University Press,2021.
2. Yang Qi, Zhu Ronghui, Zhao Xiaoqiang. Carbon emission scenario prediction model for China's transportation industry[J]. Journal of Chang'an University(Natural Science Edition), 2014,34(5): 77-83.
3. LIU Y，WANG Y，LI D. Estimation and uncertainty analysis on carbon dioxide emissions from construction phase of real highway projects in China[J]. Journal of Cleaner Production, 2017, 144:337-346.
4. NIU Kai, ZHU Xiaodong, ZHANG Xingyu. Summary of Carbon Emission Accounting Methods and Emission Reduction Technologies in Road Construction Period[J]. Transport Energy Conservation & Environmental Protection, 2023, 19 (01): 23-26.
5. Stripple, H. Life Cycle Assessment of Road. A Pilot Study for Inventory Analysis, second ed. The IVL Swedish Environmental Research Institute, Stockholm, Swedish. 2001.
6. Kim, B., Lee, H., Park, H., Kim, H. Framework for estimating greenhouse gas emissions due to asphalt pavement construction. J. Constr. Eng. Manag.2012,138 (11), 1312-1321.
7. Liu, M., Qian, F., Zhan, X. Calculation model for energy carbon emission of building material transportation. E-Product E-service and E-entertainment (ICEEE), 2010 International Conference on. IEEE, 2010, 1-3.
8. Li, X., Liu, J., Xu, H., Zhong, P., Zheng, X., Wang, Z., 2011. Calculation of endogenous carbon dioxide emission during highway tunnel construction: a case study. In: Water Resource and Environmental Protection (ISWREP), 2011 International Symposium on, 2011, 2260-2264.
9. Harvey, J., Kendall, A., Lee, I.S., Santero, N., Van Dam, T., Wang, T. Pavement Life Cycle Assessment Workshop: Discussion Summary and Guidelines. Technical memorandum: UCPRC-TM-2010-03. University of California, Davis, California.2010.
10. Abolhasani, S., Frey, H.C., Kim, K., Rasdorf, W., Lewis, P., Pang, S. Real-world in-use activity, fuel use, and emissions for non-road construction vehicles: a case study for excavators. J. Air Waste Manag. Assoc.2012, 58 (8), 1033e1046.
11. Barandica, J.M., Fernandez-Sanchez, G., Berzosa, A. Applying life cycle thinking to reduce greenhouse gas emissions from road projects. J. Clean. Prod. 2013, 79-91.
12. Frey, H.C., Kim, K., Pang, S., Rasdorf, W.J., Lewis, P. Characterization of real-world activity, fuel use, and emissions for selected motor graders fueled with petroleum diesel and B20 biodiesel. J. Air Waste Manag. Assoc. 2012, 58 (10), 1274-1287.

 https://doi.org/10.1051/e3sconf/202452803020

Design and optimization of intelligent ventilation system in coal mine

Yong Wan[1,2,*]

[1]China Coal Technology and Engineering Group Chongqing Research Institute,Chongqing 400039, China
[2]State Key Laboratory of Coal Mine Disaster Prevention and Control, Chongqing 400037, China

Abstract: With the continuous development of the coal industry, coal mine safety and environmental protection have gradually become the focus of important attention. The design and optimization of mine ventilation systems is a key factor in ensuring air quality and safety in the working area. In order to improve the application level of automation and intelligent equipment of mine ventilation system, under the existing conditions of mine, the design significance and goal of mine intelligent ventilation system are taken as the starting point, and the design and application effects are analyzed. By optimizing the layout of ventilation routes, increasing the number of sensors, building an intelligent ventilation system, optimizing the ventilation scheme in the main production scenarios such as coal face, driving face and opening up main roadway in the daily production process of the mine, realizing rapid wind regulation and wind counteraction when disasters occur, and using air flow short circuit to form protective measures, effectively avoiding the expansion of disasters and accidents. At the same time, remote automatic control of ventilation facilities and equipment is carried out to effectively improve the optimal utilization of air supply, avoid the phenomenon of insufficient air supply and waste of air volume, and improve the intelligent level of mine ventilation system.

1. Introduction

Coal mine is a dangerous working environment, one of the most important dangers is the poisonous gas and coal dust produced in the mine. Through reasonable design and optimization of mine ventilation system, the harm of toxic and harmful gases and coal dust can be effectively reduced or removed, and the physical and mental health and life safety of relevant personnel in coal mining enterprises can be guaranteed. Most of China's coal mines are underground mining, and the occurrence conditions of coal seams are complex. With the increase of mining depth, the requirements for ventilation safety are becoming more and more stringent, and the research and design of mine intelligent ventilation system has become an important guarantee for coal mine safety production [1-3]. With the construction of intelligent mines and the wide application of 5G technology, coal mines are constantly developing in the direction of automation and intelligence on the basis of improving the mechanization level of modern mines [4-7]. The premise of mine safety production is to have a reliable and stable ventilation system, including perfect ventilation design, ventilation mode, high-power fan and ventilation route. Compared with traditional ventilation methods, the current ventilation system mostly adopts artificial fixed-point wind measuring station detection, and adopts ventilation facilities (such as air doors, wind barriers, wind Bridges, etc.) to adjust air volume and transform wind paths. With the increase in the scale of mine mining and the growth of ventilation routes, relying on artificial wind measurement can no longer meet the needs of long-term and continuous wind measurement operations in special locations, especially in places with toxic and harmful gases, and personnel can not timely detect air volume and wind flow. Based on this, the construction of highly automated intelligent ventilation monitoring system has important practical significance.

Intelligent ventilation system is a new type of mine intelligent ventilation system with science and technology as the core, and is the focus of smart mine construction, catering to the concepts of energy conservation, emission reduction, green and efficient [8]. At the same time, intelligent ventilation is the key to disaster emergency wind control and an effective measure for disaster prevention and reduction. The design and construction of intelligent mine ventilation system can guarantee the stability and safety of mine ventilation, and effectively make up for the shortcomings of manual management, real-time monitoring and perception of mine ventilation, intelligent fault diagnosis and early warning of disasters, active control and response to risks, improve the intelligent, information and automation level of mine ventilation, and meet the needs of intelligent mines.

*Corresponding author's e-mail:280210058@qq.com

2. Analysis of working principle and key elements of ventilation system in coal mine

Coal mine ventilation system is one of the key systems designed to ensure the air quality and safety in the mine. Its working principle is through the continuous operation of the main fan of the surface air shaft to form a circulation of air, the ground fresh air is constantly introduced into the mine to increase the oxygen concentration, to meet the needs of personnel for oxygen; Dilution and removal of toxic and harmful gases and diluted dust concentration control to ensure safe production; Maintain suitable climate conditions in mine and create good working environment. Mine ventilation can be divided into two types: natural ventilation and mechanical ventilation. (1) Natural ventilation: The ventilation method that uses the ventilation power generated by natural air pressure to cause the air to flow in the underground roadway is called natural ventilation. Natural wind pressure is generally relatively small and unstable, so the "coal mine safety regulations" stipulate that every mine must use mechanical ventilation. (2) Mechanical ventilation: The ventilation method that uses the ventilation power generated by the operation of the main fan to cause the air to flow in the underground roadway is called mechanical ventilation. In mines with mechanical ventilation, natural wind pressure also exists all the time, and affects the ventilation work of mines in various periods. Full attention should be paid to the ventilation management work, especially in high-gas mines.

3. Overall scheme of mine intelligent ventilation system

3.1. System composition.

Mine intelligent ventilation system is composed of underground monitoring equipment and ground control center. The architecture of mine intelligent ventilation system is shown in Figure 1. Underground monitoring equipment mainly includes sensors, cameras and industrial Ethernet ring network [9]. Specifically, it includes: 1) Gas sensor, damper switch sensor, wind speed sensor, wind speed and direction sensor, fan on/off sensor and other ventilation-related sensors are installed in the mine. The sensor monitors the underground ventilation environment in real time and collects data, which is collected by the communication sub-station to the underground switch; 2) The AI intelligent intrinsic safety camera is installed in the mine, and the corresponding intrinsic safety video analysis edge computing box is equipped. The camera monitors the ventilation facilities and equipment such as the damper, wind window and fan in the mine in real time, and collects video image data. The artificial intelligence recognition algorithm built into the video analysis edge computing box can understand and analyze video data in real time at the edge end; 3) The sensor data gathered to the switch and the video data identified by the edge computing box are transmitted to the ground control center through the underground industrial Ethernet ring network.

The ground control center is responsible for data analysis and storage, management and display of application systems. The application server is used to deploy the mine intelligent ventilation application system, and realize linkage control with the mine safety monitoring system, the underground personnel positioning system, and the broadcasting system. The data server stores large ventilation monitoring data collected by various sensors; The AI video analysis server deploies artificial intelligence algorithms to identify and analyze the video image content of the mine, and find the abnormal behavior of personnel and the abnormal state of facilities and equipment; The large screen in the dispatching room is used to display information such as numbers, graphics, and videos.

3.2. Research content.

The intelligent mine ventilation system is driven by dynamic monitoring big data, and a 3D ventilation simulation and hidden danger intelligent identification platform is established to realize real-time dynamic analysis of mine ventilation data, 3D visual display, and effective safety early warning and auxiliary decision-making. The main research contents include: 1) Dynamic monitoring large database. Integrate gas, CO, wind speed, wind direction, fan negative pressure, damper, duct, main ventilator, local ventilator and other sensor data to establish a large database of ventilation system. 2) 3D ventilation simulation and hidden danger intelligent analysis platform. The sensor monitoring data, camera monitoring video images, artificial intelligence image recognition algorithm, three-dimensional visualization ventilation simulation, ventilation report design, early warning alarm and auxiliary decision, BI large screen display, historical data query and other modules are integrated on a software platform to form a closed ventilation management system. The functional structure of mine intelligent ventilation system is shown in Figure 1.

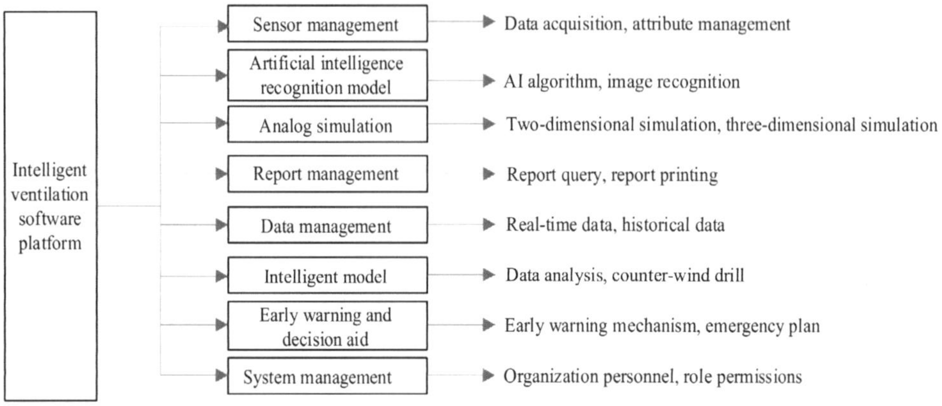

Fig.1 Functional structure of mine intelligent ventilation System

Through the construction of the system, the formation of an integrated comprehensive management platform, improve and improve the informatization level of coal mine ventilation management, timely grasp the operating status of the ventilation system, early detection of safety hazards, do a good job in prevention, and provide auxiliary decision-making basis for the dispatch and command of the mine ventilation system.

4. Function analysis of intelligent ventilation system

4.1. Intelligent optimized wind speed acquisition technology

Due to the objective phenomena such as the different size of the wind cross section and the different wind resistance caused by the roughness of the roadway surface at various points in the mine, precision errors often occur in wind speed monitoring [10-11]. Therefore, by changing the way of distribution, the step grid method is adopted to achieve dense distribution, the original section is cut and subdivided, and then the distribution law of air and wind speed in the section is analyzed, so as to obtain the actual position of relatively accurate average wind speed in the section. After optimization, the wind speed sensor is installed at a fixed point to improve the accuracy of wind speed acquisition. The distribution law of wind speed analyzed by ladder grid method is shown in Figure 2, and the measured wind speed distribution in rectangular section is shown in Figure 3.

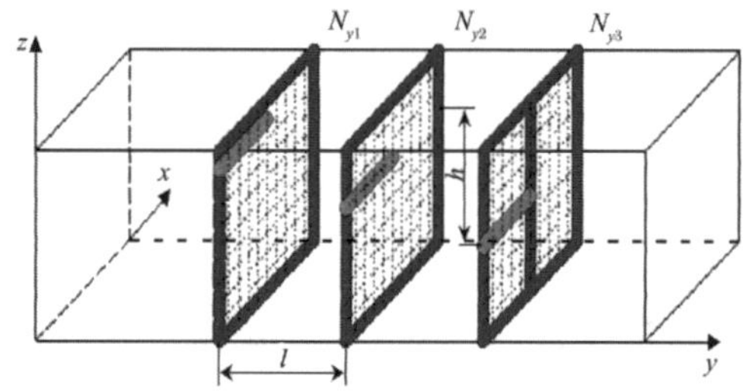

Fig. 2 Analyzing wind speed distribution by using a stepwise grid method

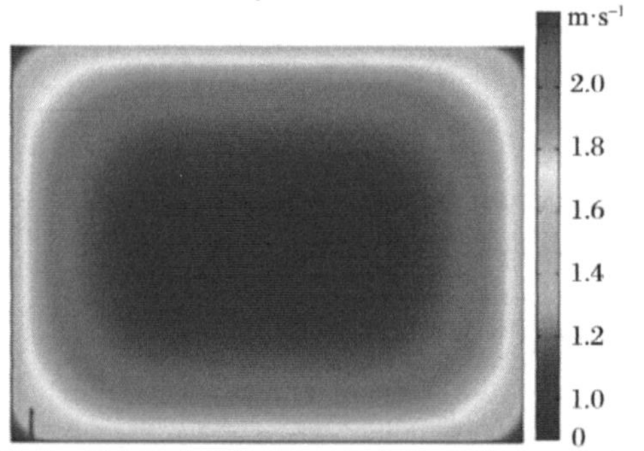

Fig. 3 Distribution of measured wind speed in rectangular section

4.2. Sensor layout optimization

In order to monitor the normal and stable operation of the mine ventilation system, it is necessary to install wind speed and wind pressure sensors on each mining face, and adhere to the principle of installing less equipment and wide monitoring range, so as to meet the purpose of multi-purpose equipment to the maximum extent [12]. Through the analysis of air flow route, ventilation mode, section shape and area, and wind supply purpose, the wind pressure sensor (FY) and wind speed sensor (FS) are arranged according to the route shown in Figure 4 to give full play to the maximum function of their monitoring points.

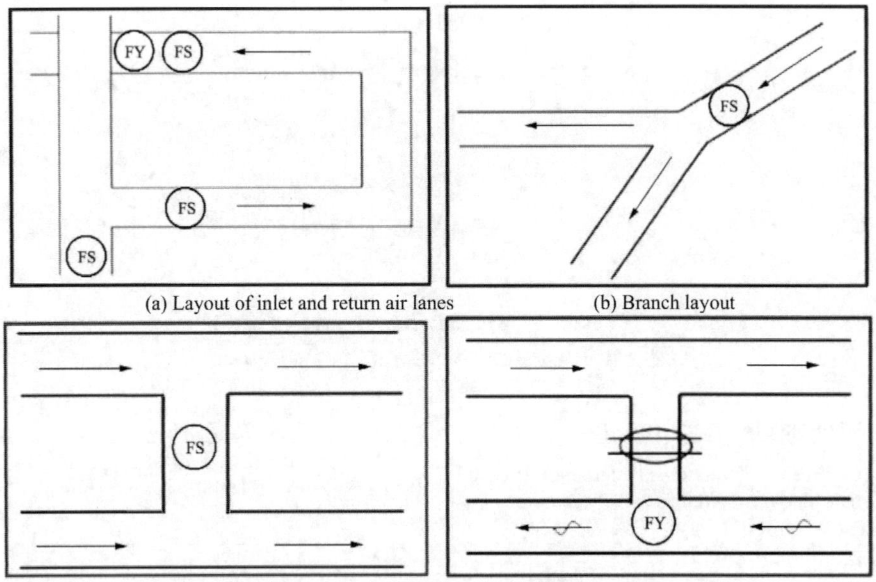

Fig. 4 Sensor layout of different ventilation routes

It is necessary to install sensors to monitor the different operation contents and ventilation methods at different points in the underground, especially at the detour yard, branch roadway and liaison roadway. In addition, the air flow instability is easy to occur on the return air flow side of the coal face with Y - and U-shaped ventilation modes. Before and after the air door for adjusting the air window is set in the main transportation roadway and the driving face, the air door is often opened in transportation, which is easy to cause a large influence of wind disturbance, resulting in local wind speed and wind pressure changes. These locations need to be monitored by sensors.

4.3. Intelligent decision and control function

Before building the intelligent ventilation monitoring system, the ventilation parameters and harmful gases of all mine laneway are measured and collected, and the upper warning value of safety index can be set in advance in the warning system according to the classification of monitoring content and functional needs. When the monitoring data exceeds the warning index, the program operation is carried out according to the automatic control instructions, and intelligent operations such as automatic air reverse, opening and closing the damper, adjusting and adjusting the Angle of the wind sash blade, fire information monitoring and alarm are issued at different levels, so as to realize automatic and unmanned supervision of dangerous work locations.

5. Software system design

5.1. System workflow

The stable and safe operation of the system is inseparable from the close cooperation between the underground and the surface equipment. The sub-station data processing of each underground safety monitoring system is particularly critical. After the data processing process of the sub-station starts, the sub-station starts to initialize the sensor of each node by relying on the CAN bus, and monitors the data, and judges and analyzes the information fed back by the ground monitoring system. If the system does not receive the data of the industrial Ethernet network, it needs to judge and analyze the CAN bus sensor data. Moreover, based on the actual states of methane sensor, wind speed sensor and wind pressure sensor, the operating state of the fan is judged and analyzed [13]. When it is determined that there is an abnormal operation of the fan, it is necessary to effectively control the equipment such as frequency converter and relay, and promptly report and dispose of the abnormal state; On the contrary, when the fan is in the normal state, the data can be transmitted by Ethernet. If the system receives Ethernet data during operation, it should be disposed of based on instructions and determine its status so that the system can interact. The data processing flow of the system sub-station is shown in Figure 5.

https://doi.org/10.1051/e3sconf/202452803020

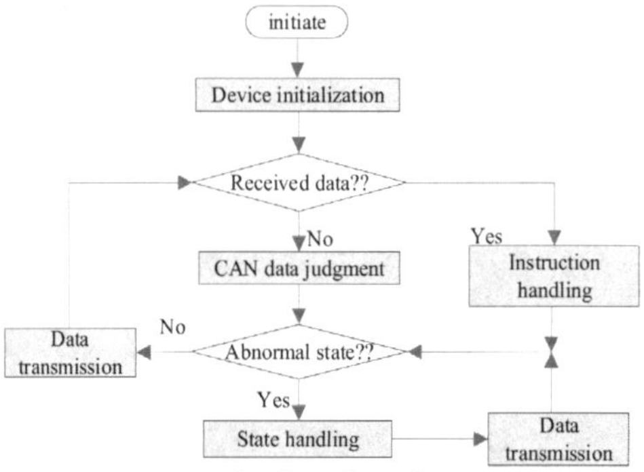

Fig. 5 Data processing flow chart of system sub-stations

5.2. System Display interface Design

The display interface of the intelligent ventilation system software in coal mine can comprehensively monitor and visually present the operating status of each node of the entire ventilation system, and remote control of related equipment in the system can be carried out with the help of the upper computer software. The display interface of the intelligent ventilation system is shown in Figure 6.

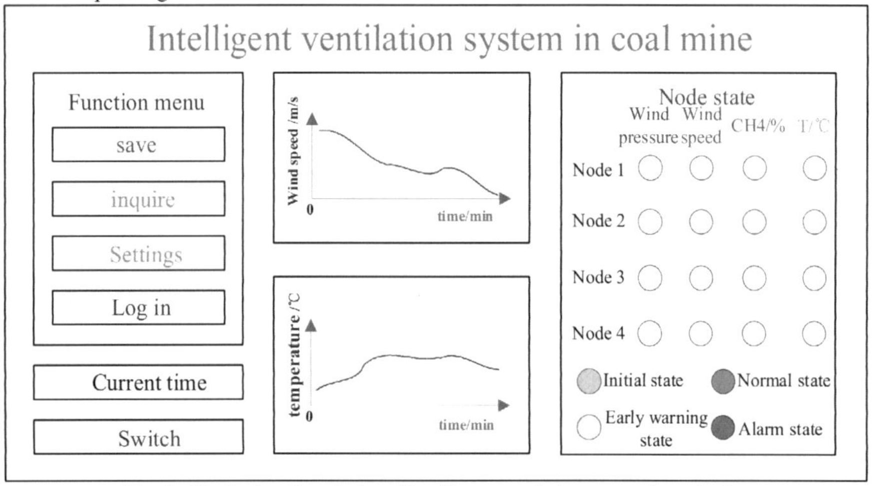

Fig. 6 Display interface design of intelligent ventilation system in coal mine

The whole interface contains three important departments: function menu, data curve and node status. The function menu can save and query node logo, time and other data. At the same time, it also has transmission parameter setting function and background system maintenance account login function, etc., which can fully present the status of time and device function partition. The data curve fully reflects the changes of temperature, humidity and wind speed at specific time nodes; According to the node status, you can understand the status of each node, such as the initial state, normal state and alarm state, etc. The setting function menu button can scientifically set the change of each state and the category of closed value, so as to coordinate with the change of node parameters, dynamically control the key fan parameters such as wind pressure, CH4, wind speed and temperature, and provide real-time and effective monitoring. For the monitoring system in the ground control room, the RGB high-definition large-screen electronic display system is installed, and the backup UPS power supply is continuous and stable, and the field data is continuously presented.

5.3. System Software Functions

5.3.1 Environmental monitoring function.

The system has related functions such as safety monitoring system and gas sensing equipment, which can dynamically and real-time monitor various harmful gases, connect the volume fraction with the intelligent ventilation platform, accurately analyze and control the underground environment, timely warn the situation of exceeding the standard, and propose wind control strategies for coping with disasters.

5.3.2 Intelligent analysis function.

Scientific methods such as neural network, machine learning and big data analysis can be selected to compare the equipment operation monitoring data, environmental monitoring data and ventilation perception data with the early warning threshold determined by the platform. If there is insufficient air volume, short circuit and inverter of air flow, it should be the first time to warn and make intelligent decisions quickly.

5.3.3 Intelligent control function.

The function can adjust and control the damper, window and other conditions intelligently with one key, and generate surveillance video data. At the same time, the operation of the ventilation equipment can be controlled remotely, and the switching quantity can be simulated to provide a basis for the ground command.

5.3.4 Intelligent decision-making function.

Real-time monitoring of ventilation network, and dynamic calculation of its air volume, the use of machine learning, artificial intelligence and other methods, simulation and decision-making of underground air flow, early warning of ventilation hazards and abnormal changes, rapid research and judgment of disaster, the development of emergency control measures, joint ventilation emergency, regulation and other equipment, the construction of a plan library, and put forward suggestions. At the same time, the optimal air flow control plan is determined intelligently and efficiently according to the conditions of the coal face, the air flow is adjusted by braking, and the adjustment results and suggestions are uploaded to the intelligent ventilation platform.

5.3.5 Mobile APP function.

Install and use mobile APP to provide timely feedback and comprehensive display of equipment operation status, intelligent control results, early warning and alarm information.

5.3.6 Camera AI learning control function.

According to the situation of pedestrians, vehicles and damper switches, accurate video recognition can be implemented to understand the actual situation of the scene and timely find the abnormal situation of the vehicle top impact damper and damper device, and carry out targeted early warning processing of these abnormal situations. At the same time, it also has the function of automatic locking of intelligent control damper, which can timely understand the trend and situation of underground disasters.

6. Field test and application effect

After the design of the intelligent ventilation system in the coal mine, the hardware template can be installed according to the design scheme through field test and application, and the related program of the control system can be arranged in the coal mine. The monitoring interface can fully display the air volume, wind speed and temperature of the underground ventilation system, and automatically control it, so as to intelligently supervise the entire coal mine ventilation system. Table 1 shows the field measurements of this ventilation system.

Table 1. Historical data of online fault monitoring of intelligent ventilation system

Serial number	Fault characteristics	Fault identification	Date of failure
1	The retarder vibrates badly	Gear lubrication is poor	2023.5.21
2	Local high temperature of machine	The bearings are badly worn	2023.5.24
3	High temperature of blade root	Fatigue damage occurred to the structure	2023.5.26
4	Rotor position high temperature	Bearing failure	2023.6.1
5	Motor vibration	Bearings are poorly lubricated	2023.6.12

The analysis of the above data can accurately judge the fault of the ventilation system and warn the accident in advance. The response time of intelligent fault identification of the intelligent ventilation system is only 0.63S, and the misjudgment rate does not exceed 0.25%, which saves the on-site manpower input, and the efficiency of the ventilation system has also been significantly improved, which effectively and timely prevents the launch of ventilation safety accidents in coal mines, enhances the safety management ability, and is conducive to the sustainable and stable development of coal mines.

7. Conclusion

Through the evaluation of the current situation of the ventilation system in the mine, it is pointed out that there are drawbacks such as low efficiency, short duration, easy to produce inspection blind area, and possible missing detection, which reflects the urgent need for upgrading and reforming the ventilation system.

Intelligent ventilation system can not only strengthen the intelligent degree of mine, but also effectively guarantee the safety of production and improve the level of ventilation construction. Mine leaders should pay attention to the optimization of ventilation system design, based on production, actively introduce the latest technology, constantly upgrade and optimize the mine ventilation system, improve the intelligence level, realize remote and real-time detection and regulation of the underground ventilation system operating status, more intuitive analysis can reduce energy consumption, ensure the normal and stable operation of the ventilation system, and contribute to the safety and efficient operation of mine ventilation. It can effectively avoid the occurrence of fire and other accidents, and lay a foundation for realizing high yield and high efficiency and essential safety of mine. For coal enterprises to build green environmental protection mine, to achieve the balance between sustainable economic development and environmental protection, has a broad market prospect.

Acknowledgments

Science and Technology Innovation Venture Capital Project of Tiandi Technology Co., LTD. (2023-TD-ZD001-005)

About the Author

Wan Yong (1980-), male, bachelor degree, senior engineer, has obtained 4 national authorized invention patents, mainly engaged in digital mining, monitoring and monitoring, intelligent instruments and other aspects of research and technology promotion, E-mail:280210058@qq.com.

References

1. CHENG Xiaozhi, WANG Kai, HAO Haiqing, et al.Research on intelligent regulation and control system and key technology of mine local ventilation ［J］. Industryand Mine Automation, 2021, 47（9）: 18-24.
2. TAN Zhanglu, WU Qi, XIAO Yixuan. Study on visualizationoptions of coal mine ventilation safety informa —tion［J］. Safety in Coal Mines, 2018, 49（12）: 86-89.
3. WEI Lianjiang, LI Sheng, WEI Zongkang, et al. Researchstatus and prospects of the influence of coal andgas outburst on ventilation system［J］.Coal science and technology，2021，49（3）：100-105.
4. [WANG Pengfei. Application of intelligent ventilation system and intelligent ventilation structure in mine [J]. Shandong Coal Science and Technology, 2023,41 (6) : 148-150.
5. Wallace Keith; Prosser Brian; Stinnette J.Daniel. The practice of mine ventilation engineering[J].International Journal of Mining Science and Technology,2015,25(02):165-169.
6. De Souza Euler. Application of ventilation management programs for improved mine safety[J].International Journal of Mining Science and Technology,2017,27(04):647-650.
7. WANG Yue. Research and Design of intelligent Monitoring System for mine fan [J]. Journal of Automation and Applications, 2019, 64(15) : 74-76.
8. LIANG Xiulong. Research on mine Intelligent ventilation System based on real-time monitoring system [J]. Western Exploration Engineering, 2023,35 (03) : 98-100+104...
9. ZHANG Fukai; SUN Yiran; SUN Junding; YUAN Guan; ZHANG Haiyan. Research on key technologies of mine Intelligent Ventilation System [J]. Safety in Coal Mine,2023,54(02):46-53.
10. LI Gang, CHENG Rongtao, WANG Kai. Research on local intelligent ventilation System for long driving face [J]. Mining Machinery, 2019, 51(10) : 69-73.
11. S.Sjöström; E.Klintenäs; P.Johansson; J.Nyqvist. Optimized model-based control of main mine ventilation air flows with minimized energy consumption[J]..International Journal of Mining Science and Technology,2020,30(04):533-539.
12. ZHONG Bin; ZHANG Xiaogang; WANG Lei; WANGYaoqiang. Intelligent Transformation of Mine Ventilation System and Its Application [J]. Mining Machinery,2024,52(03):56-60.
13. YAN Chuan; TAN Xiao. Design and application of intelligent ventilation system in coal mine [J]. Inner Mongolia Coal Economy,2024(01):124-126.

https://doi.org/10.1051/e3sconf/202452803021

Optimization of Initial Carbon Emission Allowance Allocation in Industrial Industries: A Case Study of Zhejiang Province, China

Kaiheng Hu[a], Qifen Li[b], Yongwen Yang[c], Junkai Deng[d]

Shanghai University of Electric Power, Energy and Environment Engineering Institute, Shanghai, China

Abstract—The construction of carbon market is one of the important means for China to reach the goal of carbon peaking and carbon neutrality, and the allocation of initial carbon emission right quota is the key link in carbon market transaction. At present, the historical method is the commonly used allocation method of initial carbon emission rights in China, and the use of this method is unfair and inefficient. Therefore, this paper proposes an allocation optimization model, selects a number of indicators, and optimizes the allocation of initial carbon emission rights based on the hierarchical analysis method and entropy weight method. Example analysis shows that the allocation optimization model helps to share the carbon emission responsibility and improve the allocation efficiency, realizes the optimal allocation of initial carbon emission rights among industrial sectors, and provides support for the emission reduction of industrial sectors.

1. Introduction

With the double growth of the world's economy and population, human activities have led to the emission of large amounts of greenhouse gases, and global warming is increasing [1]. In order to cope with the global climate change problem, so far some countries have co-signed conventions [2,3] to provide legally binding for emission reduction, and carbon trading mechanisms have been proposed, and each country has begun to develop its own carbon emission right trading market. For the free allocation of initial carbon emission rights, each industry or region has different economic development status, it is not possible to do equal treatment, but should bear "common but differentiated responsibilities [4]", which is also the basic guiding principle of carbon emission rights allocation. How to rationally and effectively allocate carbon emission allowances has become a hot issue in carbon trading.

Currently scholars' studies on carbon emission allowance allocation fall into two categories. The first category is allocation principles, and domestic scholars [5,6] mainly start from the principle of energy fairness to analyze the future distribution of carbon emission quotas and trends after China enters a new stage of economic development. The second category is allocation methods. Scholars [7,8] mainly based on the historical method and benchmarking method to equitably allocate carbon emissions at the provincial level in China.

Based on the above literature, this paper proposes an optimization model of initial carbon emission right allocation that takes efficiency and fairness into account. It constructs an inter-industry carbon emission right allocation index system and comprehensively evaluates the weights of the indexes based on the hierarchical analysis method and entropy weight method to realize the optimized allocation of the initial carbon emission right of each industrial industry in Zhejiang Province, and puts forward emission reduction suggestions for the direction of emission reduction of industrial industries in Zhejiang Province.

2. Distribution optimization model

2.1 Analytic hierarchy process

Analytic Hierarchy Process, referred to as AHP, refers to the decision method that decomposes the elements always related to decision into levels such as goals, criteria and schemes, and carries out qualitative and quantitative analysis on this basis.

The process of subjective evaluation using analytic hierarchy process is as follows:

Construct a judgment matrix for each indicator, and score and rank its importance by means of expert scores or questionnaire scores. The judgment matrix is shown in Table 1.

Table 1 Judgment matrix

B	***A1***	***A2***	**…**	***Am***
A1	A11	A12	…	A1m
A2	A12	A22	…	A2m
…	…	…	…	…
Am	Am1	Am2	…	Amm

[a]1031169209@qq.com, [b]liqifen73@163.com, [c]yangyongwen@vip.163.com, [d]dengjunkai826@163.com

In the judgment matrix constructed in Table 1, the relation in Eq.(1) must be satisfied.

$$\begin{cases} A_{ij} > 0 \\ A_{ij} = 1/A_{ji} \\ A_{ii} = 1 \\ i \neq j \\ i, j = 1,2,\cdots,m \end{cases} \quad (1)$$

The scale of the judgment matrix is shown in Table 2. In the comparison process, 1-9 is used to assign importance.

Table 2 Scale table

Serial Number	Factor i versus factor j	A_{ij} assignment
1	Equally important	1
2	Slightly important	3
3	Relatively important	5
4	Strongly important	7
5	Extremely important	9
6	The median of two adjacent judgments	2,4,6,8

By comparing each element index of the same level according to the importance degree, obtain a contrast degree contrast matrix .

$$B = \begin{bmatrix} a_{11} & \cdots & a_{1m} \\ \vdots & \ddots & \vdots \\ a_{m1} & \cdots & a_{mm} \end{bmatrix} \quad (2)$$

Verify the consistency of judgment matrix B. The consistency verification of judgment matrix is the key and premise to improve the accuracy of index determination. Its fundamental purpose is to test whether the constructed judgment matrix is scientific and logical, so as to make the evaluation results obtained by the judgment matrix more reliable. Whether the matrix passes the consistency verification depends on the consistency ratio CR, as shown in Eq.(3).

$$CR = \frac{CI}{RI} \quad (3)$$

It is necessary to calculate CI and judge RI before calculating CR. The calculation process of CI is shown in Eq.(4).

$$CI = \frac{\lambda_{\max} - n}{n-1} \quad (4)$$

In order to measure the size of CI, introduce a random consistency index RI. The size of RI is related to the order of the judgment matrix. In general, the larger the order of the judgment matrix, the more likely random deviation of consistency is to occur. RI can be found from Table 3.

Table 3 Standard value of random consistency index

n	1	2	3	4	5	6	7	8	9	10
RI	0	0	0.58	0.90	1.12	1.24	1.32	1.41	1.45	1.49

Then calculate the relative weights of each layer. In this paper, the eigenvalue weighting method is used to calculate the corresponding weights. First, normalize all factors by column.

$$H_{ij} = \frac{a_{ij}}{\sum_{i}^{m} a_{ij}} \qquad i, j = 1,2,\cdots,m \quad (5)$$

Next, add the matrixby row.

$$h_j = \sum_{i}^{m} h_{ij} \quad (6)$$

Finally, calculate the weight vector.

$$w_j^{(a)} = \frac{h_j}{\sum_{j=1}^{n} h_j} \quad (7)$$

2.2 Entropy weight method

Entropy was originally a function of the state of matter in thermodynamics, created by the German physicist Clausius in 1865 while studying the heat cycle. In thermodynamics, the greater the entropy, the greater the disorder of the molecular system, and the more random the molecular motion, so we can understand that the entropy is a measure of the disorder in the system. So entropy can be extended to other research fields to represent the degree of uncertainty of a certain index.

The entropy weight method (EWM) is"an o'jective evaluation method based on information entropy that can reduce the subjectivity of the weighting process. The higher the order of the system, the lower the entropy, and the greater the entropy. The entropy weight method shows the importance of the index according to the characteristics of different characteristics of the data itself, that is, the weight size.

The process of using entropy weight method for objective evaluation is as follows:

Select the index value to construct the decision matrix X.

$$X = \begin{bmatrix} x_{11} & \cdots & x_{1n} \\ \vdots & x_{ij} & \vdots \\ x_{m1} & \cdots & x_{mn} \end{bmatrix} \quad (8)$$

After collecting the index data, it is necessary to process the data to some extent. This paper adopts the linear normalization method. After standardized processing, the original data is transformed into dimensionless index values, so that each index value is at the same dimensional level, and can avoid the problem that some values are too large and occupy too much weight. After data processing is completed, the evaluation and analysis of index weights can be carried out.

In the allocation index system, there are positive indicators and negative indicators, and the higher the value, the more CEA can be allocated. The standardized processing formula for positive indicators is shown in Eq.(9).

$$x_{ij}^{'} = \frac{x_{ij} - \min(x_{1j}, x_{2j}, \cdots, x_{mj})}{\max(x_{1j}, x_{2j}, \cdots, x_{mj}) - \min(x_{1j}, x_{2j}, \cdots, x_{mj})} \quad (9)$$

A higher number of negative indicators means that fewer CEA are allocated. The standardized treatment formula for negative indicators is shown in Eq.(10).

$$x_{ij}^{'} = \frac{\max(x_{1j}, x_{2j}, \cdots, x_{mj}) - x_{ij}}{\max(x_{1j}, x_{2j}, \cdots, x_{mj}) - \min(x_{1j}, x_{2j}, \cdots, x_{mj})} \quad (10)$$

$$X^{'} = \begin{bmatrix} x_{11}^{'} & \cdots & x_{1n}^{'} \\ \cdots & \ddots & \vdots \\ x_{m1}^{'} & \cdots & x_{mn}^{'} \end{bmatrix} \quad (11)$$

Calculate the ratio of each indicator under each industry.

$$p_{ij} = \frac{x_{ij}^{'}}{\sum_{i=1}^{m} x_{ij}^{'}} \quad (12)$$

Calculate the entropy and difference coefficient in the matrix.

$$e_j = -\frac{1}{\ln m} \sum_{i-1}^{m} p_{ij} \ln p_{ij} \quad (13)$$

$$d_j = 1 - e_j \quad (14)$$

Calculate the weight of the corresponding indicator.

$$w_j^{(e)} = \frac{d_j}{\sum_{j=1}^{n} d_j} \quad (15)$$

2.3 Determining comprehensive weight

The comprehensive weight is expressed through the following equation:

$$w_i = \alpha \cdot w_i^{(\alpha)} + (1 - \alpha) \cdot w_i^{(e)} \quad (16)$$

3. Case analysis

The weights of each index and the comprehensive weights obtained through the analytic hierarchy process and entropy weight method are shown in Table 4.Through comparison, it can be found that under the analytic hierarchy process, the weight is more focused on historical carbon emissions and carbon intensity, while under the entropy weight method, the weight is more focused on carbon intensity and energy intensity.However, the weight of emission reduction responsibility under the comprehensive weight is 0.68. The weight of emission reduction responsibility is higher than the emission reduction potential and profitability, and the emission reduction responsibility will play a larger role in the allocation of CEA, which indicates that if the industry emits too much carbon dioxide, resulting in too much carbon intensity or energy intensity, then the industry should bear more emission reduction responsibility.The proportion of emission reduction potential and emission reduction capacity is relatively small, which indicates that an industry with a good energy structure and rich economic income has less impact on the allocation of CEA, and will have less pressure to reduce emissions.

Table 4 Weight results obtained by analytic hierarchy process and entropy weight method

Index	**Emission reduction responsibility**			**Emission reduction potential**		**Emission reduction capacity**	
	Historical carbon emissions	***Carbon intensity***	***Energy intensity***	***Energy structure***	***Scientific and technological input***	***Per-capita output value***	***Profitability***
The weight obtained by analytic hierarchy process	0.34	0.22	0.11	0.10	0.09	0.07	0.05
The weight obtained by entropy weight method	0.18	0.28	0.22	0.11	0.12	0.04	0.05
Comprehensive weight	0.26	0.25	0.17	0.11	0.11	0.06	0.05
	0.68			0.22		0.11	

The optimized allocation results of CEA are shown in Figure 1. It can be seen that after the adjustment of analytic hierarchy process and entropy weight method, each industry has different degrees of adjustment.Industries with relatively high historical carbon emissions, such as A17, A18, A22, A23, A24, A35, etc., have obtained fewer CEA compared with the allocation results of the historical method, which indicates that industries with higher historical carbon emissions have more responsibility to reduce emissions. Among them, A37 industry has the largest reduction, reducing 2017.75 million tons of quotas. As the industry with the largest carbon emission and energy consumption in the industrial field of Zhejiang Province, A37 has a great direct consumption of fossil energy and a great task of emission reduction.However, industries with low carbon emission have obtained more carbon emission permits by their excellent emission reduction potential and capacity, among which the A33 industry has increased the most, with an increase of 4,417,600 tons of quotas. A33 has a huge consumption of electricity, and energy consumption is mainly concentrated in equipment operation, but the fierce market competition has promoted the continuous improvement of energy utilization efficiency and continuous progress of technology. It performs well in the two indicators of emission reduction potential and emission reduction capacity, which helps to obtain more CEA.This optimized distribution result reflects not only the whipping of high-carbon emission industries, but also the reward of low-carbon emission green industries, which contributes to the improvement of fairness.

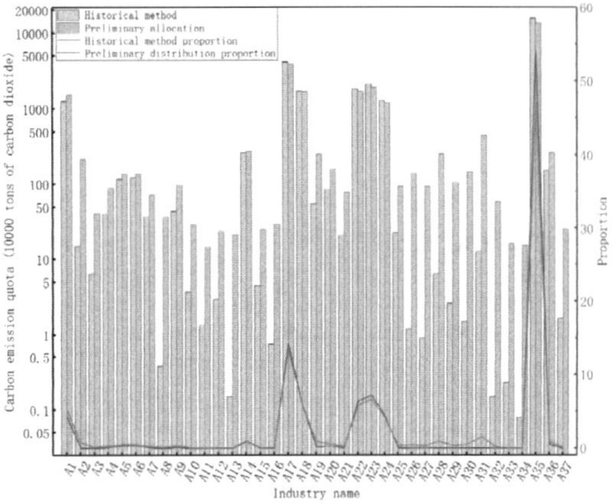

Figure 1 Preliminary allocation results of carbon emission permit quota

4. Conclusions

In this paper, the allocation optimization model is used to optimize the allocation of carbon emission rights among industries in the industrial field of Zhejiang Province. The results show that: first, the subjective and objective evaluation method combining the hierarchical analysis method and entropy weighting method can realize the reasonable allocation of index weights, and the optimal allocation under the subjective and objective evaluation method takes into account various factors, which reduces the role of the historical carbon emissions in the allocation to a certain extent; second, the allocation optimization model improves the fairness and efficiency compared with the commonly used historical method, and reduces the carbon emission right allocation gap among the industries. allocation gap between industries. The optimization method in this paper provides a realistic and feasible reference for the allocation of carbon emission rights among industries in Zhejiang Province.

References

1. G Akhmat, K Zaman, T Shukui, F Sajjad, MA Khan, MZ Khan. "The challenges of reducing greenhouse gas emissions and air pollution through energy sources: evidence from a panel of developed countries," J . Environ Sci Pollut Res Int, vol. 21(12), pp. 7425-7435, 2014.
2. H Thomas, W Oscar, L Markus, P Philipp. "The United Nations Framework Convention on Climate Change Secretariat as an orchestrator in global climate policymaking," J . International Review of Administrative Sciences,vol. 87(1), pp. 21-38, 2021.
3. K Yoomi, T Katsuya, M Shunji. "Environmental and economic effectiveness of the Kyoto Protocol," J . PLoS ONE, vol. 15(7), pp. e0236299, 2020.
4. C Shoibal, C Ananth, C Heleen, P Stephen, S Robert, T Massimo. "Sharing global CO2 emission reductions among one billion high emitters," J . Proceedings of the National Academy of Sciences of the United States of America, vol. 106(29), pp. 11884-11888, 2009.
5. A Guo, J Yang, F Zhong. "Carbon emission quotas and a reduction incentive scheme integrating carbon sinks for China's provinces: An equity perspective," EB/OL . Sustainable Production and Consumption, vol. 41, pp. 213-227, 2023.
6. G Fang, M Liu, L Tian, M Fu, Y Zhang. "Optimization analysis of carbon emission rights allocation based on energy justice—The case of China," J . Journal of Cleaner Production, vol. 202, pp. 748-758, 2018.
7. J Li, S Guo, K Wang. "Initial Quota Allocation Scheme Based on Energy Consumption and Carbon Emission Rights," J . 2022 9th International Forum on Electrical Engineering and Automation (IFEEA), pp. 740-745, 2022.
8. JG Zhou, YS Li, XJ Huo, XL Xu. "How to Allocate Carbon Emission Permits Among China's Industrial Sectors Under the Constraint of Carbon Intensity?" J . Sustainability, vol. 11(3), pp. 914, 2019.

https://doi.org/10.1051/e3sconf/202452803022

Study on the Impact of Digital Transformation on Carbon Emissions of Oil and Gas Companies

Liping Li[1], Qianyi Liu[1,*], Xinyu Xing[2], Mingxin Liang[1]

[1]College of Economics and Management, Northeast Petroleum University,, Daqing 163318, China
[2]College of Mechanical Science and Engineering, Northeast Petroleum University, Daqing 163318, China

Abstract. Based on the data of 84 listed oil and gas companies in China from 2012 to 2021, we empirically analyzed the impact of the digital transformation level of China's oil and gas companies on the level of carbon emissions through fixed panel model and mediation effect model.The results of the study show that the development of digital transformation significantly reduces the carbon emissions of oil and gas enterprises, and that digital transformation reduces the carbon emissions of enterprises mainly by being able to influence the optimization of enterprise structure and accelerate technological innovation. There is a moderating effect in the influence of digital transformation and carbon emissions for oil and gas enterprises. These conclusions hold after robustness tests and consideration of endogeneity issues.

1. Introduction

In order to cope with global climate change, China has made it clear at the UN General Assembly in 2020 that it will strive to reach the peak of carbon dioxide emissions by 2030, and strive to achieve carbon neutrality by 2060; against this backdrop, how to achieve the "dual-carbon" goal has become an important task for China nowadays. Under this background, how to take the "dual-carbon" goal as the lead to comprehensively promote the green transformation of the economy has become an important task facing China now.At present, enterprises are experiencing a new round of technological change, and digitalization has become a new production factor, which provides new possibilities for low-carbon development and green transformation of enterprises.It provides new possibilities for low-carbon development and green transformation of enterprises.Due to the ecological damage, depletion of oil and gas resources, and especially carbon emissions brought about by the development of oil and gas enterprises, this is obviously contrary to the green and low-carbon development under the leadership of the dual-carbon target.Therefore, whether digital transformation can help Chinese oil and gas enterprises reduce carbon emissions is a key issue related to the "dual-carbon" goal at the macro level and requires further study.

2. Literature Review and Theoretical Hypotheses

2.1. Impact of digital transformation of oil and gas companies on carbon emission

The oil and gas industry is an energy-intensive industry, and for energy-intensive industrial sectors, increasing the investment in digital technology can greatly reduce the level of its energy consumption, improve the efficiency of energy use, and reduce the generation of pollutants. [1] The green development capability of China's oil and gas enterprises needs to be cultivated urgently, and greenhouse gas emissions are an important factor affecting the green development capability of oil and gas enterprises, [2] Therefore, this paper puts forward the following hypotheses: H1: Digital transformation can reduce the carbon emissions of oil and gas enterprises.

2.2. Mediating effects of technological innovation and structural optimization in oil and gas enterprises

Digital transformation in the process of enterprise production and operation may promote enterprise technological innovation. First, through the digital transformation of the "ABCD" and other technologies into the enterprise manufacturing, organization and management of various businesses, will directly improve the enterprise's production technology innovation capacity [3].Secondly, digital communication technology can be complemented with

*Corresponding author's e-mail: 3211195565@qq.com

other production factors to promote the enhancement of the technological innovation ability of enterprises .

Based on digital technology has the attribute of technological progress, enterprises will apply it in the production and management process will promote the redistribution of various factors of production, so as to further optimize the organizational structure and production system within the enterprise, and effectively improve the allocation efficiency of resources .Specifically, oil and gas enterprises can apply digital technology to the entire life cycle from exploration and extraction, processing and refining, to product distribution, sales and even use. The digital transformation of oil and gas enterprises has also been extended from the manufacturing stage to the overall system including production, supply, sales and upstream and downstream supply chains [4].It can be seen that the digital transformation of enterprises helps to promote the production process of enterprises to the development of flexible, networked, platform, and reconfigure the production process of enterprises through the establishment of the industrial Internet of Things, which contributes to the optimization of the production process, improves the efficiency of resource use, and achieves the effect of reducing carbon emissions in the production and manufacturing process. Thus, this paper puts forward the following hypotheses:H2a Digital transformation is conducive to technological innovation of oil and gas enterprises, which in turn reduces the carbon emissions of oil and gas enterprises.H2b:Digital transformation is conducive to the structural optimization of oil and gas enterprises, which in turn reduces the carbon emissions of oil and gas enterprises.

2.3. The Moderating Effects of Environmental Regulation and Market Competition in Oil and Gas Firms

For oil and gas enterprises, the degree of importance attached to environmental protection by local governments will affect their motivation for carbon emission reduction. Therefore, in regions with strong environmental regulations, the effect of digital transformation on reducing carbon emissions of oil and gas enterprises is more significant..

As a market subject, the survival and development of enterprises are inevitably faced with the fundamental problem of market competition, according to which it is believed that the difference in the degree of market competition will change many behavioral choices of enterprises, including the carbon emission reduction effect of their digital transformation. Based on the above analysis, this paper puts forward the following hypotheses: H3a: Environmental regulation has a moderating effect on the impact of digital transformation on carbon emissions of oil and gas enterprises; H3b: Market competition has a moderating effect on the impact of digital transformation on carbon emissions of oil and gas enterprises; H3b: Market competition has a moderating effect on the impact of digital transformation on carbon emissions of oil and gas enterprises.

3. Research design

3.1. Variable Definition

3.1.1. Explanatory variable.

Corporate carbon emission levels: the use of single pollutant emissions does not provide a comprehensive measure of corporate carbon emission level,In this paper, we refer to the research method of Zhong Tingyong and Ma Fuqi to measure the carbon emission intensity of oil and gas enterprises, which is measured by the ratio of the enterprise's carbon dioxide emissions to the ratio with the enterprise's business revenue.Where, enterprise carbon emissions = (total energy consumption in the industry * carbon dioxide conversion coefficient * enterprise operating costs) / enterprise operating costs in the industry.

3.1.2. Core explanatory variables.

Digital transformation of oil and gas companies： Drawing on the method of Chenyu Zhao et al [5], the frequency of words related to digitization in the annual reports of listed companies is counted and the frequency of each keyword is used to form the final summed word frequency as a proxy indicator for the digital transformation of enterprises.

3.1.3. Control variable.

Referring to Xie Dongming[6] and others, this paper adds other control variables affecting carbon emission reduction of enterprises.These include the number of independent directors (Board), total asset turnover (ATO), Tobin's Q (Tobin Q), capital employed by major shareholders (Occupy), management expense ratio (Mfee), the company's age of incorporation (Firmage), the proportion of shares held by institutional investors (Inst), the net profit margin (Pro), and the cashflow ratio (Cashflow).

3.1.4. Intermediary variable.

(1)The number of enterprise patents (Patent): This paper introduces the total number of patent applications in the three categories of invention patents, utility model patents and design patents as a mediating variable to be tested. (2)Operating cost rate (Cost): Borrowing from Liu Hui's research [7], the operating cost ratio is used to measure the proportion of the enterprise's operating costs to revenues, reflecting the enterprise's overall operating cost structure. This paper argues that digital transformation can achieve the effect of reducing the level of carbon emissions by promoting the structural optimization of the production and operation process. Since the optimization of enterprise structure can be reflected in the reduction of the operating cost ratio, this paper takes the operating cost ratio, measured by the ratio of operating costs to operating revenues, as the mediating variable.

3.1.5. Moderator variable.

(1) Environmental regulation intensity (EV), (1) In regions with more stringent environmental regulations, companies tend to increase their investments in technologies to reduce carbon emissions in the course of digital transformation. However, in regions with relatively lax environmental regulations, oil and gas companies tend to be less motivated to reduce emissions due to the limited impact of environmental policies. Therefore, differences in the intensity of environmental regulations have a significant impact on whether firms increase their investment in emission reduction technologies during the digital transformation process. Based on this, this paper takes environmental regulation intensity as a moderating variable, measures the environmental regulation intensity (EV) of each region with the environmental regulation index of each city compiled by Liu Rongzeng et al. (2020) [8], and incorporates the interaction term between digital transformation and environmental regulation intensity into the model.

(2) Degree of market competition (Comp), Different degrees of market competition faced by enterprises will directly affect the results of technological innovation and structural optimization in the process of digital transformation, which in turn will lead to different results in reducing carbon emissions. In order to more accurately measure the degree of market competition faced by enterprises, this paper adopts the Herfindahl index of the industry in which the enterprise is located as an indicator. Specifically, when the Herfindahl index of an industry is lower than the overall median, we believe that the enterprise is facing relatively fierce market competition, and at this time we set the market competition dummy variable Comp=1; on the contrary, if the Herfindahl index is higher than or equal to the median, then it indicates that the market competition is relatively weak, and at this time, Comp=0. In this way, we are able to analyze the degree of competition in the market more accurately. on the digital transformation of enterprises and its effect on reducing the level of carbon emissions. Therefore, this paper takes the degree of market competition as a moderating variable and incorporates the digital transformation and market competition dummy variable (Digital × Comp) into the model.

3.2. Model setup

Based on the existing research, this paper constructs the following regression model to validate the research hypotheses proposed in this paper according to the mechanism framework proposed in the previous paper and the research hypotheses:

The direct impact of digital transformation on carbon emissions is shown in equation (1)..

$$\mathrm{Carin}_{it} = a_0 + a_1\,\mathrm{Digital}_{it} + \lambda\Theta_{it} + \sum\mathrm{Firm} + \sum\mathrm{Year} + \varepsilon_{it} \quad (1)$$

Referring to the study of Wen Zhonglin [9], the validation model of the mediation effect is shown in equations (2) and (3).

$$\mathrm{Mediator}_{it} = \gamma_0 + \gamma_1\,\mathrm{Digital}_{it} + \lambda\Theta_{it} + \sum\mathrm{Firm} + \sum\mathrm{Year} + \varepsilon_{it} \quad (2)$$

$$\mathrm{Carin}_{it} = \eta_0 + \eta_1\,\mathrm{Digital}_{it} + \eta_2\,\mathrm{Mediator}_{it} + \lambda\Theta_{it} + \sum\mathrm{Firm} + \sum\mathrm{Year} + \varepsilon_{it} \quad (3)$$

Referring to the study of Wen Zhonglin et al [9], the moderating effect validation model is shown in Equation (4):

$$\mathrm{Carin}_{it} = a_0 + a_1\,Digital_{it} + a_2\,Digital_{it} \times Moderator_{it} + a_3\,Moderator_{it} + \lambda\Theta_{it} + \sum Firm + \sum Year + \varepsilon_{it} \quad (4)$$

Included among these, Carin_{it} is the explanatory variable, denotes the carbon emission intensity of firm i in year t;The core explanatory variables are $\mathrm{Digital}_{it}$, Represents the degree of digital transformation of the enterprise in year i; θ_{it} is the set of control variables; $\sum$Firm and $\sum$Year represent time fixed effects and firm fixed effects, respectively; ε_{it} is the residual, Mediator was chosen to represent the mediating variable,Moderator is the moderating variable.

3.3. Sample selection and data sources

Oil and gas enterprises are a comprehensive concept covering all types of enterprises involved in the entire industrial chain of oil and natural gas from exploration, development, storage and transportation to processing and final sales. These enterprises include, but are not limited to, those specializing in oil and natural gas exploration and development, processing enterprises that carry out in-depth processing and refining of oil and gas products, chemical enterprises that produce all kinds of chemical raw materials and products, and sales enterprises that are responsible for bringing oil and gas products to the market.Combined with the availability of relevant index data, ST, *ST enterprises and oil and gas enterprises with incomplete data are excluded, and finally 84 listed oil and gas enterprises in domestic A-share stocks are selected as research samples, and the years from 2012 to 2021 are chosen as the research scope of this paper.Data from Cathay Pacific database (CSMAR), China Energy Statistical Yearbook, Wanderlust (Wind) database and corporate annual reports.

 https://doi.org/10.1051/e3sconf/202452803022

4. Empirical Tests and Analysis of Results

4.1. Base regression analysis

The VIF values for each explanatory variable are less than 10, indicating that there is no serious problem of multicollinearity. Table 1 Column (1) controls for time and individual fixed effects and column (2) adds control variables. The results show that digitization is significant at the 1% level, thus confirming hypothesis H1 of the paper.

4.2. Mechanism analysis

The regression results for the mediating role of technological innovation are shown in columns (3) and (4) of Table 1. The results show that digital transformation is an important driver of technological innovation in oil and gas firms, which in turn facilitates the process of innovation and transformation. This innovation significantly improves the efficiency of resource utilization and effectively reduces carbon emissions. This confirms the hypothesis H2a of this paper.

For the mediating role of structural optimization regression results are shown in columns (5) and (6) of Table 1. According to the results, in column (5), the Digital variable is significantly negative at the 5% level, indicating that digital transformation can drive firms to achieve structural optimization. Meanwhile, in column (6), the operating cost rate (Cost) is significantly positive as a mediating variable, indicating that the optimization of corporate structure reduces the increase in the effect of carbon emission level in oil and gas companies. It confirms the hypothesis H2b of this paper.

According to the regression results in column (7) of Table 1, "digital×EV" is significantly negative at the 5% significance level, indicating that the stronger environmental regulations have a positive impact on reducing carbon emissions of oil and gas enterprises, thus verifying the hypothesis H3a of this paper. The regression results in column (8) show that "Digital×Comp" is significantly negative at the 10% significance level, indicating that there is a moderating effect of market competition. This validates the hypothesis H3b of the paper.

Table 1. Regression analysis results

Variable name	(1)	(2)	(3)	(4)	(5)	(6)	(7)	(8)
	carin	carin	Patent	carin	Cost	carin	carin	carin
digital	-0.0017***	-0.0019***	0.2832**	-0.0019***	-0.0002**	-0.0013*		
	(0.001)	(0.001)	(0.131)	(0.001)	(0.000)	(0.001)		
Patent				0.0047				
				(0.031)				
Cost						2.3277***		
						(0.490)		
digital×ev							-0.9301**	
							(0.423)	
digital×comp								-0.0056*
								(0.003)
_cons	1.7408***	2.5100**	2.3929	3.1977***	0.7670***	0.7248	2.2571***	2.5084***
r2_a	0.9293	0.9292	0.9452	0.9288	0.7488	0.9328	0.9685	0.8812
id FE	Yes	Yes	Yes	Yes	Yes	Yes	Yes	Yes
year FE	Yes	Yes	Yes	Yes	Yes	Yes	Yes	Yes

Note. *,**,*** denote p<0.1, p<0.05, p<0.01 respectively

4.3. Robustness check

4.3.1. Replacement of explanatory variables.

"E" in the Bloomberg ESG scoring system is the score of environmental performance indicators such as carbon emission intensity, energy consumption intensity, water utilization rate and soil pollution of enterprises, and the higher the score is, the better the environmental performance of the enterprise, and the better the effect of carbon emission reduction. Accordingly, this paper takes the environmental score E in enterprise ESG as an explanatory variable, which is labeled as ESG_EP.

4.3.2. Excluding the effects of strategic corporate behavior.

Digital transformation has been a hot issue for investors in recent years, and the expectation of managers in oil and gas enterprises to take advantage of the widespread attention effect brought by digitization may result in the frequency of digitization-related terms being artificially inflated in the text to enhance the reputation and image of oil and gas enterprises, which in turn will be favored by the government. For the strategic hype behavior of managers in oil and gas enterprises, it may distort the authenticity of digital transformation indicators.First, some firms are engaged indigital transformation but are not presented in textual form in their annual reports, so this paper conducts regressions after removing the sample with zero digital transformation.Second, In the stock market crash of 2015, the stock prices of listed companies suffered unprecedentedly heavy losses. In order to cope with this predicament and boost stock prices, compared with other years, digital transformation, as an important means to enhance the competitiveness of companies, was noticed by a wide range of investors, and increased disclosure of related information helped to attract the attention of investors, thus stabilizing or boosting the stock price, and companies may have disclosed digital transformation-related terms more actively in that year, which has an impact on the conclusions of this paper.As shown in Table 2, the results of the regression between the degree of digital transformation and the level of carbon emissions of the oil and gas companies examined are consistent with the previous section, and the results are robust.

Table 2. Robustness test regression results

Variable name	(1) ESGEH	(3) carin	(4) carin
digital	0.0255***(0.006)	-0.0019***(0.001)	-0.0012**(0.001)
_cons	29.9330***(9.767)	2.3302**(1.137)	1.8864** (0.955)
r2	0.7290	0.9406	0.9407
r2_a	0.6881	0.9305	0.9311
id FE	Yes	Yes	Yes
year FE	Yes	Yes	Yes

4.4. Instrumental variables method.

There is a bidirectional causal relationship between the level of digital transformation of oil and gas enterprises and the level of carbon emission reduction of oil and gas enterprises, i.e., accelerating the digital transformation can reduce the level of carbon emission of oil and gas enterprises, and at the same time, the reduction of the level of carbon emission of oil and gas enterprises may lead to the enterprises' more motivation and demand for a high level of digital transformation, so this paper adopts an instrumental variable method to solve the endogeneity problem. In the selection of instrumental variables, drawing on the research of Zhu Shujin [10] and others, the 2SLS method is chosen, and the first-order lagged variables and second-order lagged variables of digital transformation are used as instrumental variables. The regression results, shown in Table 3, indicate that the instrumental variables selected in this paper are reasonable and reliable. It also proves to be consistent with the previous findings.

Table 3. Endogeneity treatment regression results

Variable name	Instrumental variables method Phase I(1) digital	Phase II(2) carin
digital		-0.0121***(0.086)
L.digital	0.6564***(0.042)	
L2.digital	0.1898***(0.044)	
control variable	Yes	Yes
id FE	Yes	Yes
year FE	Yes	Yes
Cons	3.2978(10.529)	-2.0904*(1.217)
Kleibergen-Paap rk LM		525.125
Cragg-Donald Walid F		657.814

5. Conclusions

This paper empirically analyzes the impact of digital transformation on carbon emissions of Chinese oil and gas companies using data from 2012-2021 of Chinese A-share listed oil and gas companies as a sample.The results show that digital transformation can significantly reduce the carbon emission effects of Chinese oil and gas companies, and this conclusion still holds after various robustness tests such as endogeneity treatment and substitution of important variables.Technological innovation and structural optimization serve as mediating factors in the impact of digital transformation on carbon emissions in oil and gas enterprises. On the other hand, environmental regulations and market competition play a modulating role in the relationship between digital transformation and carbon emissions in these enterprises.

Acknowledgements

This research has been supported by the Northeast Petroleum University School Fund Project, Project No. 2022QYDL.

References

1. Goralski M A, Tan T K.(2020). Artificial intelligence and sustainable development. The International Journal of Management Education, 18(1):100330.
2. S Zhang, J Li and Mengxue Li.(2023). Evaluation of green development capability of Chinese oil and gas enterprises under carbon emission constraint. Oil and Gas Storage and Transportation, 42 (02): 121-130.
3. Hilty L M, Aebischer B.(2015) ICT for sustainability: An emerging research field[A]. Hilty L M, Aebischer B. ICT innovations for sustainability. Cham: Springer.
4. May G, Stahl B, Taisch M,et al..(2017). Energy management in manufacturing: From literature review to a conceptual framework. Journal of Cleaner Production,167:1464–1489.
5. C Zhao, W Wang and X LI.(2021). How digital transformation affects enterprise total factor productivity[J]. Finance and Trade Economics, (7):114-129.
6. D Xie.(2020). Local Regulation, Vertical

https://doi.org/10.1051/e3sconf/202452803022

Regulation and Corporate Environmental Protection Investment: An Empirical Study Based on Listed A-share Heavy Pollution Enterprises. Accounting Research,(11).

7. H Liu, B Cong.(2022). Has Digital Transformation Promoted Energy Saving and Emission Reduction in Chinese Enterprises? Journal of Shanghai University of Finance and Economics, 24 (05): 19-32.
8. R Liu, C He.(2021). Research on the threshold effect of environmental regulation on urban residents' income inequality [J]. China Soft Science, (08): 41-52.
9. Z Wen, B Ye.(2014). Mediation effects analysis:Methods and model development. Advances in Psychological Science, 22 (05): 731-745.
10. S Zhu, Y Luo and W Duan.(2021). Service-oriented manufacturing, markup rate distribution and resource allocation efficiency. China Industrial Economy,(4):62-80.

https://doi.org/10.1051/e3sconf/202452803023

Evaluation of Carbon Emission Efficiency of Chinese Oil and Gas Enterprises

Yanqiu Wang, Mingxin Liang*

College of Economics and Management, Northeast Petroleum University, Daqing 163318, China

Abstract. This paper constructs a carbon emission efficiency evaluation model using the non-expected output SBM model and the Global-Malmquist-Luenberger model to statically and dynamically evaluate the carbon emission efficiency of 69 oil and gas enterprises from 2016 to 2021. The results found that the overall carbon emission efficiency of China's oil and gas enterprises is at a relatively low level, but with an upward trend. The results also show that there are differences in carbon emission efficiency among oil and gas enterprises, and that the carbon emission efficiency of most oil and gas enterprises improves thanks to the improvement of technical efficiency.

1. Introduction

Excessive carbon dioxide emissions are a major challenge of our time, prompting China to propose a "carbon capping and carbon neutral goal". Since the "carbon capping and carbon neutral" strategy was proposed, carbon emission efficiency has become the focus of attention. It includes both single-factor indicators such as carbon productivity, carbon index, and carbon emission intensity [1-3], as well as full-factor indicators that consider both economic benefits and environmental impacts [4]. In this paper, we select oil and gas companies, which are important players in China's efforts to achieve its climate goals. By assessing their carbon emission efficiency, it aims to facilitate their low-carbon transition.

2. Research methods and data sources

2.1. Research methods

2.1.1. Non-expected output SBM model.

The DEA method is widely used in the efficiency evaluation of different fields, and the traditional DEA models include CCR, BCC, etc., but the unintended outputs are not taken into account in the practical application, so the methods such as the unintended output SBM model are introduced [5], as shown in (1).

$$\min \rho = \frac{1 - \frac{1}{m}\sum_{i=1}^{m} s_i^- / x_{ik}}{1 + \frac{1}{n+z}\left(\sum_{j=1}^{n} s_j^+ / y_{jk} + \sum_{r=1}^{z} s_r^{b-} / b_{rk}\right)} \tag{1}$$

$$s.t. \begin{cases} X\lambda + s^- = x_k \\ Y\lambda - s^+ = y_k \\ B\lambda + s^{b-} = b_k \\ \lambda, s^+, s^- \geq 0 \end{cases}$$

Where, x_{ik} denotes the evaluated decision unit inputs, respectively, y_{jk} and b_{rk} denotes the values of expected and non-expected outputs, respectively, s denotes the slack variable for inputs and outputs, respectively, s^+ and s^- indicate excess and shortfall in output, respectively, m , n , z denote the number of decision units with redundant inputs, expected outputs and non-expected outputs, respectively, ρ denotes the efficiency value, which takes the range of (0,1], when $\rho = 1$, it indicates that the decision unit is efficient, i.e., there is an efficient relationship between inputs and outputs.

2.1.2. Global-Malmquist-Luenberger model.

The actual situation of enterprises is dynamic, Malmquist-Luenberger (ML) index can only be used to compare the changes of production efficiency in the short term, therefore, this paper adopts the Global-Malmquist-Luenberger (GML) index model, which includes all the observations in all the time periods when constructing the production frontier, and it can reflect the long-term trend more comprehensively [6], and the calculation formula is shown in (2). The GML index is greater than 1 indicates that carbon efficiency increases, less than 1 indicates that carbon efficiency decreases, and equal to 1 indicates no change.

$$GML_t^{t+1} = \frac{1 + \vec{D}^G(x^t, y^t, b^t; y^t, -b^t)}{1 + \vec{D}^G(x^{t+1}, y^{t+1}, b^{t+1}; y^{t+1}, -b^{t+1})} \tag{2}$$

*Corresponding author's e-mail: lmx76540@163.com

Under constant returns to scale, the GML index can be decomposed into an index of technical efficiency change (EFFCH) and an index of the rate of technical progress (TECH), as shown in (3).

$$GML_t^{t+1} = GMLEFFCH_t^{t+1} \times GMLTECH_t^{t+1} = \left[\frac{1+\vec{D}^t(x^t,y^t,b^t;y^t,-b^t)}{1+\vec{D}^{t+1}(x^{t+1},y^{t+1},b^{t+1};y^{t+1},-b^{t+1})}\right] \times \left[\frac{1+\vec{D}^G(x^t,y^t,b^t;y^t,-b^t)}{1+D^t(x^t,y^t,b^t;y^t,-b^t)} \times \frac{1+\vec{D}^{t+1}(x^{t+1},y^{t+1},b^{t+1};y^{t+1},-b^{t+1})}{1+\vec{D}^G(x^{t+1},y^{t+1},b^{t+1};y^{t+1},-b^{t+1})}\right] \quad (3)$$

2.2. Data sources and determination of parameters

2.2.1. Sources of information.

China's oil and gas industry has a high degree of monopoly, most leading oil and gas companies from PetroChina, CNOOC and Sinopec, the three state-owned group companies. In this paper, 96 enterprises were initially censored, excluding enterprises that were forced to be delisted, treated as ST, ST*, and did not disclose relevant data during the study year, and 69 enterprises were finally identified as sample enterprises. The required data were obtained from the Cathay Pacific database, the website of the National Bureau of Statistics, corporate annual reports, social responsibility reports, and environmental reports.

2.2.2. Determination of parameters.

According to the relevant literature and data availability [7-8], the evaluation index system selected in this paper is shown in the Table 1.

Table 1. Carbon Emission Efficiency Evaluation Indicator System for Oil and Gas Enterprises

Type of indicator	Indicators	Unit	Meaning of Indicator
Input indicators	Energy Consumption	Tons of standard coal	Enterprise Energy Consumption
	Capital stock	Million Yuan	Total fixed assets of enterprises
	Labor force	million	Number of enterprise employees
Desired output indicators	Main business income	million yuan	Enterprise main business income
Non-desired output indicators	Carbon Emission	tons	Total carbon dioxide emissions

3. Results

3.1. Static evaluation of carbon emission efficiency of oil and gas enterprises

The results of the static measurement of carbon emission efficiency are shown in Table 2.

Table 2. Annual Average of Carbon Emission Efficiency of Oil and Gas Enterprises

Stock Short Name	Annual Average Value	Stock Short Name	Annual Average Value	Stock Short Name	Annual Average Value
Shenzhen Energy	0.2797	Qixiang Tengda	0.5234	Changchun Gas	0.1466
Nanjing Public Utilities	0.2459	Jiu Li Special Material	0.3295	PetroChina Engineering	0.3870
Jinzhou Pipeline	0.3480	Yongtai Energy	0.3128	Taihe New Material	0.3017
Shandong Expressway	0.2743	National New Energy	0.2680	Hengli Petrochemical	0.3177
Huajin	0.3799	Changbao	0.3384	Shengli	0.2743
Delonghui Energy	0.2255	Shandong Molong	0.2580	Tianfu Energr	0.1989
Shenyang Chemical	0.3031	Tongyuan Petroleum	0.3107	Yueyang Xingchang	0.4122
Heng Yi Petrochemical	0.3999	Haiyang Oil Engineering	0.2889	Sinochem International	0.3938
Meida	0.3290	RENZHI	0.4462	Baomo	0.2369
Shandong Haihua	0.3086	Longzhou	0.2772	Potential Hengxin	0.2921
Rongsheng Petrochemical	0.4428	Petrochemical Machinery	0.3216	Petrochemical Oil Service	0.2580
Satellite Chemical	0.4169	Beacon Energy	0.2434	Huafeng Chemical	0.3019
Baichuan Energy	0.3254	Xinchao Energy	0.7337	Baoli International	0.3822

Blue Flame Holdings	0.1707	Haimo Technology	0.2773	China Merchants Shipbuilding	0.2929
Daqing Huake	0.3988	Yanggu Huatai	0.3308	Chongqing Gas	0.3096
Aoyang Health	0.2841	Hengtai Aipu	0.2802	Shenzhen Gas	0.2984
Hailide	0.3044	Huibopu	0.3610	Tongkun	0.3364
JUNIOR OIL	0.2377	Sifangda	0.2507	Lanco Hi-Tech	0.1894
Shanghai Petrochemical	0.5292	China National Petroleum Corporation	0.3841	China National Offshore Oil Service	0.2238
Shaanxi Natural Gas	0.4031	Oriental Electric Heat	0.2994	Guochuang High-Tech	0.4749
Shenkai	0.2930	Sinopec	0.5147	Xinao	0.3617
Jereh	0.3344	Ancai Gaoke	0.3180	Bomike	0.2863
Hengtong	0.4983	Guanghui Energy	0.2597	Compton	0.3816

3.1.1. Analysis of the trend of changes in the overall carbon emission efficiency of oil and gas enterprises.

Figure 1 illustrates the results of a comprehensive examination of the overall trends in carbon emission efficiency of 69 oil and gas firms from 2016 to 2021.A carbon emission efficiency value of nearly one "signifies that businesses have significantly reduced their carbon emissions." According to the results, the majority of businesses always have carbon emission efficiency values lower than 1, and very few businesses have ever achieved a situation where their efficiency values equal one in a given year. Thus indicating that businesses generally have low carbon emission efficiency. A calculation based on the annual average value, the sample oil and gas enterprises' average efficiency in releasing carbon from 2016 to 2021 is consistently low, exhibiting a pattern of erratic development. Nevertheless, there is potential to enhance the overall carbon emission efficiency of these enterprises.

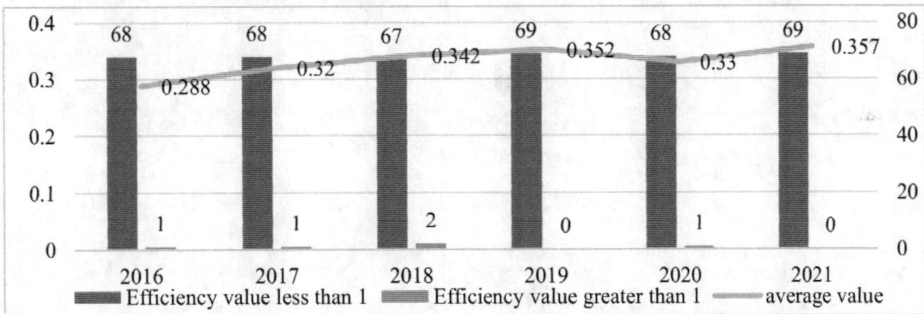

Figure 1. Trends in Overall Carbon Emission Efficiency of Oil and Gas enterprises

3.1.2. Analysis of the average annual carbon emission efficiency value of each oil and gas enterprises.

The carbon emission efficiency levels of various oil and gas companies range considerably from one another. Additionally, the organizations' carbon emission efficiencies exhibit an uneven distribution and significant modifications between calendar years. According to the annual average value of carbon emission efficiency, 69 enterprises are categorized into low, medium and high levels, and the annual average efficiency values are 0-0.3, 0.3-0.6 and 0.6-1. The percentage of oil and gas companies with low, medium, and high carbon emission efficiency is found to be 28, 40, and 1, respectively. This suggests that the majority of oil and gas enterprises have carbon emission efficiency levels that are in the middle range, and the variations that result could be influenced by any number of factors.

3.2. Dynamic Evaluation of Carbon Emission Efficiency of Oil and Gas Enterprises

3.2.1. Comprehensive analysis of dynamic efficiency values.

Table 3 presents the results obtained from the application of the GML model, which serves as a metric for assessing carbon emission efficiency.

Table 3. Carbon emission efficiency of oil and gas enterprises and its decomposition

YEAR	EFFCH	TECH	GML
2016-2017	1.038	1.055	1.095
2017-2018	1.165	0.912	1.062
2018-2019	1.125	0.917	1.032
2019-2020	0.911	1.033	0.941
2020-2021	1.184	0.938	1.111
Average Value	1.085	0.971	1.048

The GML of oil and gas enterprises has shown a slight fluctuating trend from 2016 to 2021. In terms of the mean value, the overall GML index of 69 oil and gas enterprises has a mean value greater than 1. This indicates that the level of carbon emission efficiency has been on an upward trend among oil and gas enterprises during this study period. The mean value of the technical efficiency change index is 1.085, indicating that oil and gas enterprises have achieved a better level of technical efficiency as a whole, while the technical progress index with a mean value of 0.971 indicates that there has been a decline in technological progress among oil and gas enterprises. In terms of time, the dynamic values of carbon emission efficiency measured by 69 oil and gas enterprises from 2016-2021 are 1.095, 1.062, 1.032, 0.941, and 1.111, respectively, and the GML indexes are all greater than 1 except for 2019-2020, which indicates that the overall oil and gas enterprises' carbon emission efficiency is developing in a favorable manner.

3.2.2. Analysis of carbon emission efficiency index and decomposition index of each enterprise.

The results of the annual average GML index and decomposition index are shown in Table 4.

Table 4. Average annual GML index and decomposition index for each oil and gas enterprises

Stock Short Name	GML	Stock Short Name	GML	Stock Short Name	GML
Huajin	1.022	Sifangda	1.095	Changbao	1.021
Shengli	1.002	Hengtai Aipu	0.972	Baomo	0.995
Tongyuan Petroleum	1.038	Shandong Molong	1.116	Qixiang Tengda	1.125
Delonghui Energy	1.080	Baoli International	1.065	Jinzhou Pipeline	1.031
Potential Hengxin	1.154	Haimo Technology	1.037	Shenkai	1.086
Xinao	1.033	Huibopu	1.049	Longzhou	1.030
Hailide	1.044	Ancai Gaoke	1.065	Bomike	0.828
China Merchants Shipbuilding	1.069	China National Offshore Oil Service	1.094	China National Petroleum Corporation	1.055
Heng Yi Petrochemical	1.124	Oriental Electric Heat	1.074	Sinochem International	1.080
Meida	1.025	Sinopec	1.020	Compton	0.968
Yueyang Xingchang	1.041	Yongtai Energy	1.070	Satellite Chemical	1.058
Shandong Haihua	1.040	Guanghui Energy	1.151	Changchun Gas	1.045
Jiu Li Special Material	1.012	Hengli Petrochemical	1.038	Blue Flame Holdings	0.980
Shenzhen Gas	1.066	RENZHI	1.015	Hengtong	1.179
Daqing Huake	1.041	PetroChina Engineering	1.145	Shanghai Petrochemical	1.071
Huafeng Chemical	1.107	Lanco Hi-Tech	1.122	Chongqing Gas	1.032
Aoyang Health	0.968	Yanggu Huatai	1.018	Shenzhen Energy	1.007
JUNIOR OIL	0.992	Jereh	1.126	Tongkun	0.999
Xinchao Energy	1.161	Tianfu Energy	1.112	Taihe New Material	1.079
Nanjing Public Utilities	0.942	Shandong Expressway	1.059	Guochuang High-Tech	1.099
Shenyang Chemical	0.999	Baichuan Energy	1.001	Beacon Energy	1.013
Shaanxi Natural Gas	0.996	Haiyang Oil Engineering	1.059	National New Energy	1.005
Petrochemical Machinery	0.995	Petrochemical Oil Service	1.036	Rongsheng Petrochemical	1.036

The changes in carbon emission efficiency of each oil and gas enterprises can be categorized into two scenarios for detailed explanation. (1) Among the12 enterprises analyzed, a declining trend in carbon emission efficiency is observed. Among these, technological advancements are identified as a limiting factor for improving carbon emission efficiency in 8 companies, while 3 companies are deemed to require the adoption of more sophisticated technology to enhance efficiency. Furthermore, the technical efficiency change index and the technical progress index of Bomaco are both below 1, implying the necessity for technological enhancements.(2) The trend in carbon emission efficiency among 57 oil and gas enterprises is observed to be rising. Within this group, 43 enterprises experienced enhancements in technical efficiency, while the technical progress of 5 enterprises contributed positively to the improvement of carbon emission efficiency. Among the 9 enterprises, both the change index of technical efficiency and the index of technological progress exceed 1, signifying that their advancements in technical efficiency and technological progress have positively impacted the enhancement of carbon emission efficiency.

4. Conclusions

The main conclusions are as follows:

(1) The carbon emission efficiency of the oil and gas enterprises in the sample is generally low, albeit exhibiting an upward trajectory over time.

(2) The enhancement in carbon emission efficiency among the majority of oil and gas enterprises can be attributed to advancements in technical efficacy, signifying progress in carbon emission control within the industry.

References

1. Kaya Y and Yokobori K. (1993) Global environment, energy, and economic development. Held at the United Nations University, Tokyo, 1993: 25-27.
2. Mielnik O and Goldemberg J. (1999) Communication the evolution of the Carbon Index in developing countries. Energy Policy, 27(5): 307-318.
3. Ang B W. Is the energy intensity a less useful indicator than the carbon factor in the study of climate change [J]. Energy Policy, 1999, 27(15): 943-946.
4. Sun H, Kporsu A K and Taghizadeh-Hesary F. (2020) Estimating environmental efficiency and convergence: 1980 to 2016. Energy, 208: 118224.
5. Tone K. (2001) A slacks-based measure of efficiency in data envelopment analysis. European Journal of Operational Research, 130(3): 498-509.
6. Oh D.A. (2010) Metafrontier approach for measuring an environmentally sensitive productivity growth index. Energy Economics, 32(1): 146-157.
7. Jadoon A K, Sarwar A and Javaid M F. (2023) Estimating environmental efficiency of the selected Asian countries: does convergence exist? Environmental Science and Pollution Research, 30(19): 55024-55033.
8. Moutinho V and Madaleno M. (2021) Assessing eco-efficiency in Asian and African countries using stochastic frontier analysis. Energies, 14(4): 1168.

Research on the calculation method of indirect carbon emissions of enterprises based on market transaction data

Pengwei Cong[1,*], Zheng Zhao[1], Xue Yu[2], Qingkai Sun[1], Xiaoxuan Zhang[1], Yu Wang[1]

[1]State Grid Energy Research Institute Co., Ltd, Beijing 102209, China
[2]State Grid Xinjiang Economic and Technological Research Institute, Xinjiang 830026, China

Abstract: Accurate accounting of indirect carbon emissions from electricity is conducive to motivating market entities to proactively and voluntarily reduce emissions, fairly promoting regional emissions reductions, and accelerating the formation of green and low-carbon consumption models. Based on the analysis of the problems faced by China's power carbon emission accounting, this paper considers different transaction types such as market-based electricity consumption, priority power purchase, and power grid agency power purchase, and proposes an indirect carbon emission accounting method for enterprise power sources based on power transaction data. The effectiveness of the proposed model and method is verified through case calculations for high-energy-consuming enterprises. The method proposed in this article can accurately describe the power consumption structure of enterprises and effectively encourage market entities to proactively and voluntarily reduce emissions.

1. Introduction

As the single industry with the largest proportion of carbon emissions in my country, electric power is crucial to achieving carbon peak and carbon neutrality goals [1]. Due to the characteristics of large cross-regional transmission of electricity, high emissions, and good data base, it is an international practice to control greenhouse gas emissions from both ends of electricity production and consumption. The United States, Australia, Canada, the United Kingdom, New Zealand and other countries are all in regional or enterprise Measures have been taken to control indirect carbon emissions from electricity. my country's seven carbon emissions trading pilots all include indirect carbon emissions from electricity consumption, requiring managed enterprises to report indirect emissions from power purchased within and outside the boundaries of enterprises and key facilities. The national carbon market is about to expand, and the next step will be to include industries such as electrolytic aluminum and cement, which will put forward higher requirements for indirect carbon emissions from electricity use by companies in these industries [2-3].

Currently, indirect carbon emission accounting for enterprise electricity use mainly uses the carbon emission flow tracking method. Literature [4-5] proposed the carbon emission flow theory based on the principle of proportional sharing. In view of the fact that the power flow allocated by the proportional sharing principle does not match the actual power flow, literature [6] starts from the basic circuit theory and calculates the load carbon flow rate based on the power transfer relationship between the source and the load. Literature [7] is also based on basic circuit theory. By defining four types of node admittance matrices, the complex power distribution relationship between each unit, each load and network loss in the system is obtained, thereby accounting for the carbon emissions of the entire system. In literature [8], the detailed model of carbon emissions in a hybrid power system with strong time-variability such as wind, water and solar energy is comprehended through quantifying the temporal changes in carbon intensity.Literature [9] examines the impact of distributed and demand-side response on power flow and carbon emissions once integrated into the power system. Literature [10] is based on the carbon flow tracking algorithm, supports real-time power flow analysis and carbon flow tracking of the power grid, and realizes real-time online calculation and analysis of carbon emissions in all aspects of the power grid. However, this method has low data transparency and requires grid operating mode data, topology data, generator operation data, generator carbon emission intensity data, load curve data, etc., making data verification difficult. On the other hand, this method is difficult to make a detailed distinction between the carbon emissions of corporate electricity loads. At present, the main network power flow calculation model is generally a node network of 220 kV and above. The node load is the aggregation of various power loads in the power supply area of the node. Carbon emissions are tracked and calculated to the node. It is difficult to indirectly calculate the carbon emissions of different enterprises under the node. Emissions are further differentiated.

With the improvement of China's carbon market and electricity market, all thermal power base quantities and

*Corresponding author's e-mail: imagecong@126.com

electricity prices have been liberalized, and industrial and commercial users have fully entered the market [11]-[12]. The characteristics of bilateral direct transactions in the electricity market have become increasingly obvious. Except for a small amount of priority power generation allocated to power grid agents for power purchase, other Both planned and market power are matched, forming a one-to-one economic and balanced relationship, which is more conducive to tracking carbon emissions according to "transaction flow". This study relies on the bilateral attributes of power transaction data, distinguishes different types of users, and proposes an indirect carbon emission accounting model for enterprise electricity consumption based on market transaction data based on market correspondence, electricity consumption of different transaction power supply types, and corresponding carbon emission factors. The proposed method relies on mature conditions such as the existing power transaction measurement system and transaction settlement basis to track the carbon emission flow according to "transaction flow". It does not require complex network power flow calculations and can effectively encourage market entities to proactively and voluntarily reduce emissions.

2. Indirect carbon emission accounting method for enterprise electricity consumption based on market transaction data

In the electricity market, transaction contracts formed through various types of transactions between different electricity consumers and power generation entities also have problems with the identification of their electricity consumption at the settlement level. It is necessary to agree on the various types of contracted electricity and deviation electricity on the electricity side. The settlement method realizes the decoupling of electricity at the settlement level. Therefore, according to the division of each enterprise's electricity consumption in market settlement at each time period, the virtual "power transaction flow" can also be determined, thereby achieving tracking of its corresponding power generation sources.

2.1. Market-oriented partial electricity

2.1.1. Carbon emission accounting on the power generation side.

$$Q_{Electric\ energy} = Q_{Inter-provincial\ contracts} + Q_{Inter-province\ day-ahead} + Q_{inter-province\ intra-day} + Q_{Intra-provincial\ contracts} + Q_{Intra-province\ day-ahead} + Q_{province's\ real-time\ electricity} + Q_{Inter-provincial\ contract\ deviations} + Q_{The\ leveling\ power} \quad (1)$$

Inter-provincial contracts and intra-provincial contracts are for medium and long-term electricity, and the relationship between the buyers and sellers is clear. The inter-provincial contract's power purchaser is the power grid company of other provinces; the intra-provincial contract is mainly a direct transaction within the province, and the power purchaser is a market entity within the province.

Inter-province day-ahead and inter-province intra-day are the inter-provincial spot market (if participating) electricity. This part of electricity can clarify the transaction path relationship between the purchase and sale of provinces, but cannot clarify the relationship between the sending-end power generation enterprise and the receiving-end individual user.

Intra-province day-ahead and province's real-time electricity are the provincial spot market electricity (if participating). This part of the electricity is cleared electricity energy through centralized transactions, and it is difficult to clearly define the relationship between power generation and user electricity. Referring to the spot market settlement principle, the power generation side is settled according to the node power price on the power generation side, and the user side is settled according to the weighted average power price. Similarly, for carbon emissions, the carbon emissions on the power generation side can be calculated based on the average carbon emission intensity of the power generation side nodes, and the user side can be calculated based on the weighted average average carbon emission intensity.

The leveling power is the difference between the power generation company's monthly grid-connected power and the total grid-connected power by time period.

If the average carbon emission intensity of power generation enterprise market entities is $E_{Market\ entities}$, then its total carbon emissions are:

$$E_{Market\ entities} = EF_{Market\ entities} \times Q_{Electric\ energy} \quad (2)$$

2.1.2. User-side indirect carbon emission calculation

For wholesale market users' electric energy charges, they generally include contract electric energy charges, day-ahead market electric energy charges, real-time market electric energy charges, leveling charges, and user-side price difference adjustment charges. Its contract power, day-ahead and real-time spot power, and leveling power are decoupled from each other and can be superimposed. The price difference adjustment on the user side does not involve electricity issues. It is mainly because when the electricity price at the unified settlement point on the user side changes, it has an impact on the settled electricity price. Therefore, the user's power consumption is composed as follows.

$$Q_{Electric\ energy} = Q_{contract} + Q_{day-ahead} + Q_{real\ time} + Q_{Leveling\ electricity} \quad (3)$$

Contract power is mid- to long-term trading power, and green power trading is also a type of mid- to long-term trading and belongs to contract power.

$$Q_{contract} = Q_{contract,\ Non-green\ electricity} + Q_{contract,green\ electricity} \quad (4)$$

A market-oriented user may sign medium- and long-term contracts with multiple power plants, and the carbon emission intensity of each power plant may be different.

The carbon emission measurement method for the contract part is as follows, in which the green power contract part has zero emissions.

$$E_{contract} = \sum EF_{i,\ \text{Non-green electricity}} \times Q_{i,\ \text{Contract non-green electricity}} \quad (5)$$

Spot market power. The difference between the day-ahead market electricity time-sharing reported electricity volume and the contract breakdown electricity volume. The difference between the actual time-of-use electricity consumption of users in the real-time market and the time-of-use electricity reported in the day-ahead market. It is difficult to trace the transaction relationship between users and power generation for spot electricity, and its carbon emissions can be calculated by weighted average carbon emission intensity on the power generation side.

$$E_{Spot\ power} = (Q_{Day-ahead\ power\ balance} + Q_{Real-time\ power\ balance}) \times EF_{Weighted\ average\ emission\ intensity\ of\ spot\ power\ generation} \quad (6)$$

$$EF_{Weighted\ average\ emission\ intensity\ of\ spot\ power\ generation} = (\sum EF_{i,\ Thermal\ power} \times Q_{i,\ Thermal\ power\ clears\ spot\ power}) / Q_{Total\ spot\ power} \quad (7)$$

$$Q_{Total\ spot\ power} = Q_{Thermal\ power} + Q_{New\ energy\ power} \quad (8)$$

Level the power on the power side. The difference between the user's monthly market-based transaction settlement electricity (including power consumption) and the accumulated time-sharing actual electricity consumption. Under the spot mode, low-voltage users who choose to directly participate in market transactions but do not meet the time-of-use measurement conditions (including the trading units of electricity sales companies acting as agents for such users) can only choose typical curves to sign contracts with power plants when participating in medium- and long-term transactions. For long-term contracts, the curve shape of later contracts cannot be adjusted. This type of user's electricity is measured and counted on a monthly basis. Positive deviations of electricity are settled at a certain percentage (for example, 1.1 times) above the mid- to long-term contract price. Negative deviations of electricity are generally settled according to the contract price. Since this part of the electricity is settled monthly, the corresponding relationship between time-of-use power generation and users cannot be determined, so it is calculated based on the weighted average carbon emission intensity of the monthly power generation side.

$$E_{Leveling\ power} = Q_{Leveling\ power} \times EF_{Monthly\ power\ weighted\ average\ emissions\ intensity} \quad (9)$$

$$EF_{Monthly\ power\ weighted\ average\ emissions\ intensity} = (\sum EF_{i,\ Thermal\ power} \times Q_{i,\ Actual\ monthly\ thermal\ power\ capacity}) / Q_{Total\ monthly\ thermal\ power} \quad (10)$$

2.2. Government pricing of electricity

2.2.1. Priority power generation and purchase scope and matching principles

Prioritize power generation. On the premise of ensuring the safe operation of the power grid and reliable power supply, meeting the electricity needs of premium purchasers, and not increasing the cross-subsidy burden on power grid enterprises, we must adhere to the boundaries between provinces and promote the power sources within the province to participate in the market as much as possible to achieve the desired release. Let go. At this stage, hydropower, nuclear power, and grid-to-grid and point-to-grid supporting power supplies with lower prices among external power, whose on-grid prices are significantly lower than the average power purchase price in the region, are included in the power grid to ensure safe supply and support, and are purchased and sold by power grid enterprises. Do not participate in market transactions.

Priority in purchasing electricity. Priority power purchase users are mainly residents (including public welfare electricity users who are subject to residential electricity prices) and agricultural electricity prices. Power grid enterprises guarantee power supply in accordance with government pricing. This part of the electricity accounts for 27% of the company's power supply (including line loss). It is recommended that priority should be given to ensuring safe supply of power (including imported coal and electricity) to match priority power purchases (accounting for 24% of the total power supply). The insufficient part should be supplemented by local high (affordable) price priority power generation, and the remaining balance should be unified by the power grid enterprises in the market. purchase.

2.2.2. Carbon emission calculation method

(1)Prioritize power generation

Prioritize power generation carbon emission intensity

$$EF_{Prioritize\ power} = \frac{EF_{exotic\ coal} \times Q_{exotic\ coal} + EF_{gas} \times Q_{gas} + EF_{Necessary\ for\ people's\ livelihood} \times Q_{Necessary\ for\ people's\ livelihood}}{Q_{External\ power} + Q_{Hydropower} + Q_{Nuclear\ power} + Q_{New\ energy\ power} + Q_{Other\ necessary\ power}} \quad (11)$$

Prioritize power generation and actual grid electricity

$$Q_{Prioritize\ power} = Q_{External\ power} + Q_{Hydropower} + Q_{Nuclear\ power} + Q_{New\ energy\ power} + Q_{Other\ necessary\ power} \quad (12)$$

Prioritize direct carbon emissions from power generation

$$E_{Prioritize\ power} = EF_{Prioritize\ power} \times Q_{Prioritize\ power} \quad (13)$$

(2)Priority purchase of electricity

For residents, agriculture and other priority electricity purchase users, their actual electricity consumption

$$Q_{Priority\ purchase\ of\ electricity} = Q_{resident} + Q_{agriculture} \quad (14)$$

Corresponding indirect carbon emissions

$$EF_{Priority\ purchase\ of\ electricity} = EF_{Prioritize\ power} \times Q_{Priority\ purchase\ of\ electricity} \quad (15)$$

2.2.3. Power grid agent purchases electricity

When power grid enterprises participate in medium- and long-term transactions on behalf of industrial and commercial users, the transaction curve is formed by referring to the typical curve of non-market electricity consumption.

The grid purchasing part is provided by partial priority generation and partial marketization. There are currently two distribution methods: "bundling" and "categorization and proportioning". “Bundling” refers to evenly distributing the electricity purchased by the power grid company (premium generation + marketization) to premium purchase and agency users. "Classification matching" means that the optimal power generation will be matched to optimal purchasing users first, and any remaining power will be distributed to agency users.

For some priority power generation and some market-oriented electricity, the weighted average carbon emission intensity is calculated, and then multiplied by the corresponding part of the electricity purchased by the grid to obtain the indirect carbon emissions of the electricity purchased by the grid. The formula is as follows

$$Q_{\text{Power grid purchasing power}} = Q_{\text{Some priority power generation}} + Q_{\text{Partial market power}} \quad (16)$$

$$EF_{\text{Partial priority power}} = (\sum EF_{\text{Discharge power}} \times Q_{\text{Corresponding power}}) / Q_{\text{Partial priority power}} \quad (17)$$

$$EF_{\text{Partial market power}} = (\sum EF_{\text{Discharge power}} \times Q_{\text{Corresponding power}}) / Q_{Partial\ market\ power} \quad (18)$$

The indirect carbon emissions from electricity purchased by the power grid

$$E_{Power\ grid\ purchasing\ power} = EF_{Partial\ priority\ power} \times Q_{Partial\ priority\ power} + EF_{Partial\ market\ power} \times Q_{Partial\ market\ power} \quad (19)$$

3. Example analysis

Based on the PLEXOS power market simulation software, a typical spot pilot is simulated. The power generation side reports the volume and quotes, while the user side reports the volume but does not quote. Analyze the indirect carbon emissions of the user side in the spot market model to verify the method model proposed in this article. Among them, the medium- and long-term transactions of a high-energy-consuming enterprise include green power transactions, medium- and long-term transactions of thermal power plant A, and medium- and long-term transactions of thermal power plant B. The enterprise declares the electricity consumption curve in the spot market based on its own electricity demand, and after clearing, the spot is formed. Positive and negative deviation power. It is assumed that the actual electricity consumption curve is consistent with the final settlement curve. The user's total electricity consumption that day was 10,852 MWh, including 1,438 MWh of green power in the medium and long term, 5,143 MWh in the medium and long term of thermal power plant A, 3,429 MWh in the medium and long term of thermal power plant B, and 842 MWh of spot power. . The composition of electricity consumption is shown in Figure 1:

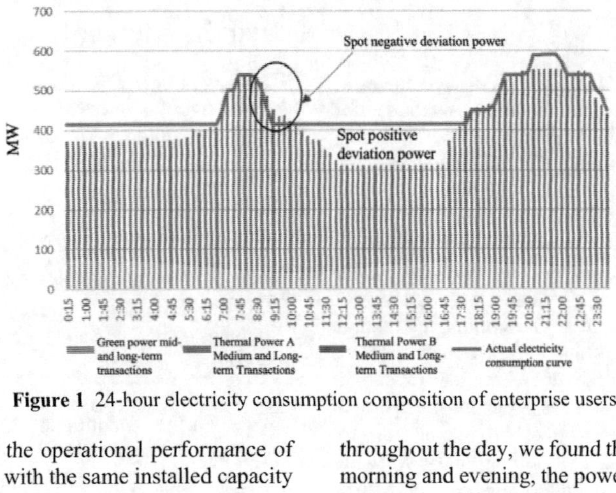

Figure 1 24-hour electricity consumption composition of enterprise users

Then, we delved into the operational performance of two thermal power plants with the same installed capacity of 1000 MW - A and B - within 24 hours, as shown in Figure 2. Specifically, Factory A included indirect carbon emission costs in its operational decisions, while Factory B did not include this cost in its calculations. By simulating the fluctuations in electricity demand throughout the day, we found that during peak hours in the morning and evening, the power generation of both plants increased, while during periods of reduced demand at night, the power generation correspondingly decreased. Specifically, the carbon emission cost of Plant A is 0.05 yuan per kilowatt hour, which is reflected in the revenue calculation, resulting in its revenue generally being lower than that of Plant B within 24 hours. For example, during peak morning hours (6:00-9:00), the revenue of Factory A is about 5% lower than that of Factory B, while during nighttime hours (such as 1:00-6:00 in the morning), this gap may widen to over 10%. This significant difference in benefits highlights the importance of environmental compliance costs in the financial analysis of power companies, and reveals that for thermal power plants participating in carbon emissions trading, effective management of carbon emissions costs is crucial for maintaining financial health and market competitiveness.

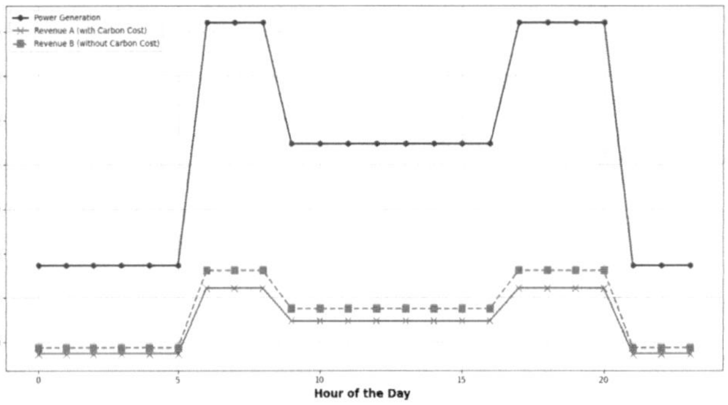

Figure 2 24-Hour power generation and revenue variation

The calculation results of the company's indirect carbon emissions from electricity use are shown in Table 1:

Table 1 Calculation results of indirect carbon emissions from enterprise electricity use

	Carbon emission intensity per unit of comprehensive power generation (tCO_2/MWh)	Indirect carbon emissions from electricity use(tCO_2)	total emissions(tCO_2)	The carbon emission intensity of the company's indirect electricity consumption on the day (tCO_2/MWh)
Thermal power plant A	0.78408	4032.5		
Thermal power plant B	0.81254	2786.2		
spot market	0.796	670.2		
High energy consumption enterprises			7489	0.69

4. Conclusion

This article analyzes the advantages and disadvantages of two indirect carbon emission tracking methods for enterprise power loads, based on "power flow" tracking and "power transaction flow tracking", taking into account different transaction types such as market-based power consumption, priority power purchase, and power grid agency power purchase. This paper proposes an indirect carbon emission accounting method for enterprise power based on power transaction data and different power sources. This method can effectively promote the linkage between the carbon market and the electricity market, guide the electricity side to actively search for low-carbon or even zero-carbon power sources in the market, promote the electricity side to reduce its own carbon emissions by participating in green electricity trading, and reflect the environmental value of clean energy. . At the same time, it will form positive incentives for new energy and low-carbon power sources to promote the development of clean energy.

Acknowledgments

This project is funded by SGCC Science and Technology Project "Research on the calculation method and key technology of carbon emissions in new power system" (SGSDDK00YJJS2100284).

Reference

1. Liu Mingda, Meng Jijun and Liu Bihan (2014) Research progress on carbon emission accounting methods at home and abroad [J]. Tropical Geography, 34(02): 248-258.
2. Chen Wan (2023) The national carbon market expansion is mature and one is included one by one [J]. Environmental Economics, (21): 48-49.
3. Qiu Zhongtao, Jin Yanming, Xu Chenzhi (2023) Analysis of the impact of average power emission factor selection on high energy-consuming industries under the expansion of the national carbon market [J]. China Electric Power, 56(12): 1-7.

4. KANG Chongqing, ZHOU Tianrui, CHEN Qixin, et al (2012) Carbon emission flow in networks[J]. Scientific Reports, 2.
5. KANG Chongqing, ZHOU Tianrui, CHEN Qixin, et al. (2015) Carbon emission flow from generation to demand: a network-based model[J]. IEEE Transactions on Smart Grid, 6(5): 2386-2394.
6. LI Yansong, LIU Qizhi, ZHANG Zhenbo, et al (2017) Algorithm of carbon emission flow based on power distribution[J]. Power System Technology, 41(3): 840-844.
7. BI Hanwen, FAN Xiaozhou, XIAO Hai, et al. (2023) A node admittance matrix algorithm to support the carbon emission tracing model of whole power system[J]. Proceedings of the CSEE, 43(20): 7881-7891.
8. Imran Khan,Michael W. Jack and Janet Stephenson.(2018)Analysis of greenhouse gas emissions in electricity systems using time-varying carbon intensity [J].Journal of Cleaner Production,Volume 184 , Issue . 2018. PP 1091-1101.
9. Farraokh Rahimi,Ali Ipakchi and Fred Fletcher(2016)The Changing Electrical Landscape: End-to-End Power System Operation Under the Transactive Energy Paradigm[J]. IEEE Power and Energy Magazine.
10. LI Yaowang, ZHANG Ning, DU Ershun, et al. (2022) Mechanism study and benefit analysis on power system low carbon demand response based on carbon emission flow[J]. Proceedings of the CSEE, 42(8): 2830-2842.
11. YUAN Hefei (2022) On the problems and countermeasures in carbon emission costing of thermal power generation enterprises[J]. Business Culture, (3): 106-108.
12. Wang Yirong, Chen Haolin, Lin Lishen, et al (2024) National carbon price forecast considering carbon emissions in the power industry [J/OL]. China Electric Power, 1-9.

Study on the cognitive system and key issues of green and low-carbon production mode and lifestyle

Jing Hu[1], Hu Yan[1] and Yuxin Miao[2*]

[1]State Grid Energy Research Institute Co., Ltd., Beijing 102209, China
[2] North China Electric Power University, Beijing 102206, China

Abstract. To promote the construction of ecological civilization and fundamentally solve ecological and environmental problems, it is necessary to form green and low-carbon production and lifestyle. This article focuses on the cognitive system of green and low-carbon production methods and lifestyles, analyzes the key elements of green and low-carbon production methods and lifestyle systems, and proposes typical scenarios and models of energy and electricity promoting terminal green energy consumption from an energy perspective. Based on economic and social perspectives, it constructs green cost guidance mechanisms and green behavior training mechanisms to comprehensively support the development of green and low-carbon production methods and lifestyles, and accelerate the construction of a green background for ecological civilization.

1. Introduction

Green development has become a global consensus, but there is currently no clear definition of green and low-carbon production and lifestyle. New types of industrial parks, represented by zero carbon industrial parks, are a key development direction for green and low-carbon industries. Scholars have conducted relevant research on the energy system planning and evaluation system of zero carbon parks [1-4]. The current focus of promoting green buildings includes the application of renewable energy in buildings, substitution of building electricity, and flexible interaction between electricity supply and building energy consumption. Building materials are the main source of carbon emissions. In commercial office building materials, concrete structures have advantages over steel structures in energy consumption, carbon emissions, and carbon emission costs [5]. The building photovoltaic storage direct flexible system, which integrates photovoltaic power generation, energy storage, DC distribution, and flexible electricity consumption, is a key technology for promoting building electrification and increasing the penetration rate of renewable energy [6].The focus of green development in agriculture and rural areas is to promote resource recycling, the development of distributed power sources in rural areas, and the improvement of agricultural machinery energy efficiency. There is a lot of research on rural energy development models and biomass utilization [7-9]. In summary, the current focus on green and low-carbon production and lifestyle is more on a specific field or scenario, but there is a lack of systematic research on constructing path mechanisms.

This article first proposes the concept and connotation of green and low-carbon production and lifestyle, constructs a "2344" cognitive system for green and low-carbon production and lifestyle, and then systematically proposes the path mechanism for building green and low-carbon production and lifestyle from the perspectives of energy, economy, and society. Finally, a summary and outlook are provided.

2. The connotation and cognitive system of green and low-carbon production and lifestyle

2.1. The connotation of green and low-carbon production and lifestyle

Building a green and low-carbon production and lifestyle is a systematic project for comprehensive economic and social development, including changes in development concepts, policy deployments, technological means, and other aspects. It is a new transformation of traditional production and lifestyle. According to national policy requirements and global practice, this article defines green and low-carbon production and lifestyle as an economic and social development mode and behavior concept characterized by the core goal of forming energy-saving and environmentally friendly production, simple and moderate living, high-level protection of the ecological environment, efficient and efficient recycling of resources, fair distribution of development results, and sustainable economic and social development.

Ecological and environmental issues require a change in development methods to be fundamentally resolved,

*Corresponding author's e-mail: 876602356@qq.com

https://doi.org/10.1051/e3sconf/202452803025

forming a green and low-carbon development mode, further forming a green and low-carbon production and lifestyle, and accelerating the construction of ecological civilization. Green development emphasizes the harmonious coexistence between humans and nature, and adhering to green development is a comprehensive and revolutionary change in production methods, lifestyles, ways of thinking, and values.

2.2. The connotation of green and low-carbon production and lifestyle

Building a green and low-carbon production and lifestyle requires coordinated efforts from all industries in society. This article proposes a cognitive system for green and low-carbon production and lifestyle based on content composition, cognitive perspective, core elements, and key areas, as shown in Figure 1.

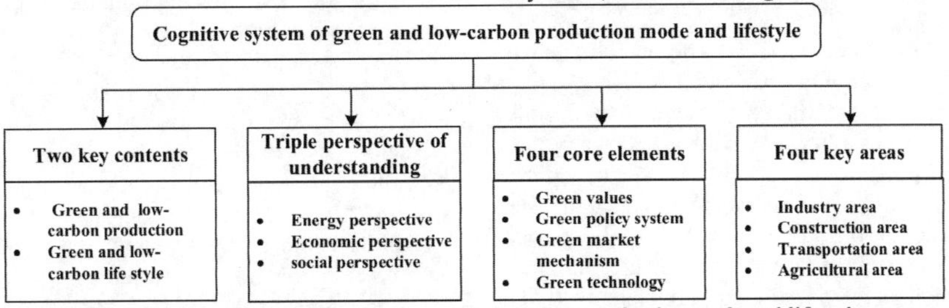

Figure 1. Cognitive system of green and low-carbon production mode and lifestyle.

(1) Two key contents

Green production mode is a productive force characterized by ecological environment protection, resource conservation and utilization, as well as corresponding production relations including politics, economy, culture, etc. Green lifestyle refers to the consumption and social interaction patterns and behavioral characteristics of residents based on the concepts of ecological environment protection, resource conservation and utilization, and harmonious coexistence between humans and nature.

(2) Triple cognitive perspective

From an energy perspective, it is necessary to pay attention to how to coordinate energy safety and supply guarantee with the green and low-carbon upgrading of terminal energy consumption methods. From an economic perspective, attention needs to be paid to how to coordinate green development with affordable and sustainable economic and social development. From a social perspective, it is necessary to pay attention to how to cultivate public green consumption habits.

(3) Four core elements

Green values are an important guide for building a green production and lifestyle. The green policy system is the fundamental guarantee for building a green production and lifestyle. The green market mechanism is a necessary condition for building a green production and lifestyle. Green technology is the internal driving force for building a green production and lifestyle.

(4) Four key areas

Industry, construction, and transportation are the three major energy consuming sectors, while urban and rural areas are the two major regions that form green and low-carbon production and lifestyle. Urban areas are the main battlefield for the green transformation of the industrial, construction, and transportation sectors, while rural areas drive the green development of agriculture and carry the mission of common green development among rural industrial, construction, and transportation sectors in the process of urbanization. Compared to urban areas, rural areas are an important growth pole for building green and low-carbon production and lifestyle.

3. Energy perspective: creating typical scenarios and models for green energy use

This section focuses on the four main energy consumption fields of industry, construction, transportation, and agriculture. Based on the analysis of the energy consumption characteristics and trends in each field, typical scenarios and patterns are proposed.

3.1. Industrial area

The low-carbon energy supply in industrial parks includes two typical modes: low-carbon/zero carbon energy self supply mode and large grid green power supply mode.

The low-carbon/zero carbon energy self-sufficiency model mainly relies on comprehensive energy in the park, supplemented by "green electricity trading + diversified carbon sinks" to achieve low-carbon/zero carbon energy supply, which is more suitable for areas with good new energy resource endowments.

The green power supply model of the large power grid is mainly based on "green power trading + diversified carbon sinks", supplemented by comprehensive energy in the park, to achieve low-carbon/zero carbon energy supply, which is suitable for areas with high maturity in the electricity market.

The integrated energy system of the park in two typical modes is centered on electricity, achieving integrated supply of multiple energy sources in the park through "green power from the large power grid + distributed new energy + energy storage + refrigeration and heating". According to the differences in high-temperature heat demand in the park, there are mainly three typical system structures, including "distributed new energy + energy storage + electricity for cooling and heating", "distributed new energy + energy storage + electricity for hydrogen production + electricity/hydrogen for cooling and heating", and "distributed new energy + energy storage + electricity/coal/gas for cooling and heating".

3.2. Construction area

The total demand for electricity in public buildings and the available space area are relatively large, and the property rights are relatively clear, making it the main scenario suitable for building flexible electricity systems. The construction field focuses on creating typical scenarios for low-carbon and efficient power supply and flexible interaction in public buildings, including 2 categories and 4 typical modes.

Price based flexible interactive mode, including two modes: time of use electricity price and fixed load.

The market-oriented flexible interactive mode includes two modes: user behavior adaptive response and point-to-point interaction.

The time of use electricity pricing model and user behavior adaptive response model are led by the power grid, providing customized interactive strategies that can maximize the initiative of building flexible electricity systems to participate in power demand response. The other two modes are dominated by the building electricity system and require a large amount of energy storage and advanced control technology equipment as support, while also requiring matching interaction and trading mechanisms, making it difficult to achieve in the short term.

3.3. Transportation area

Vehicle network interaction is the main typical mode of friendly interaction between the power grid and green transportation facilities. It is preliminarily estimated that the theoretical potential that can contribute to peak shaving capacity by 2030 is about 30-60 million kilowatts.

Orderly charging in car network interaction is gradually entering the stage of large-scale promotion, and V2G does not yet have the conditions for promotion in the short term. In scenarios such as dedicated charging stations, personal charging stations in residential communities, and public charging stations, dedicated charging stations mainly serve operational vehicles and have great potential for orderly charging, making them suitable as the main focus of recent orderly charging. Whether operational vehicles can be guided to the grid for valley charging depends on factors such as the difference in peak and valley electricity prices, whether the charging stations connected to the vehicles have automatic wake-up function, and the ratio of the number of charging stations to the number of operational vehicles.

3.4. Agricultural area

The development and utilization of rural new energy mainly includes three typical models: new energy+industry, comprehensive utilization of biomass resources, and distributed low-carbon comprehensive energy system.

The new energy+industry model is suitable for the central and eastern regions with abundant distributed photovoltaic and wind power resources, including "photovoltaic+modern agriculture", "new energy+rural landscape tourism", "new energy+ecological restoration", etc. It is recommended to promote it with good economic benefits.

The comprehensive utilization model of biomass resources is suitable for most rural areas, mainly focusing on the comprehensive utilization of biomass energy resources. The products include biogas, biogas, electricity, heat, organic fertilizer, etc. The collection, storage and transportation costs are high, and the current economic efficiency is not high.

The distributed low-carbon comprehensive energy system model mainly uses wind, light, and biomass, supplemented by energy storage and natural gas, to form a highly self-sufficient comprehensive energy system, providing various energy varieties such as cold, hot, and electrical. It is suitable for county-level industrial parks, agricultural industrial parks, large public buildings, etc. It has great potential in promoting new urbanization construction at the county level, and the current focus is on promoting demonstration applications.

4. Economic and social perspectives: Building a green cost diversion mechanism and a green behavior cultivation mechanism

4.1. Green cost diversion mechanism based on transmission and allocation of green costs in the industrial chain

This article proposes a green cost sharing mechanism based on the industrial chain, as shown in Figure 2. The core idea is to achieve cost and benefit balance among various stakeholders through the transmission and allocation of green costs among upstream and downstream enterprises in the industrial chain. At the same time, the government takes necessary policies and measures to ensure the implementation of the mechanism, including realizing green benefits, transmission and allocation of green costs (as shown in Figure 3), and developing green finance.

Green cost allocation method: Green costs consist of supply chain green costs and production chain green costs. The green costs shared by supply enterprises are calculated based on the proportion of their products and services provided to green investment costs and operating costs. The green costs shared by green production enterprises are the total green costs minus the green costs borne by the supply enterprises.

Green revenue sharing method: Green revenue is shared among upstream and downstream enterprises based on the proportion of green cost sharing. In order to enhance the enthusiasm of supply chain enterprises to participate in investing in green technology, allocation is made between supply and production enterprises based on the principles of cost sharing and benefit sharing.

By sharing green costs and benefits, we can promote the coordinated development of the green industry chain and supply chain, fully leverage the role of various market entities in the industry chain, vigorously improve the level of green supply chain management, and reduce subsidies from the government for green technology investment.

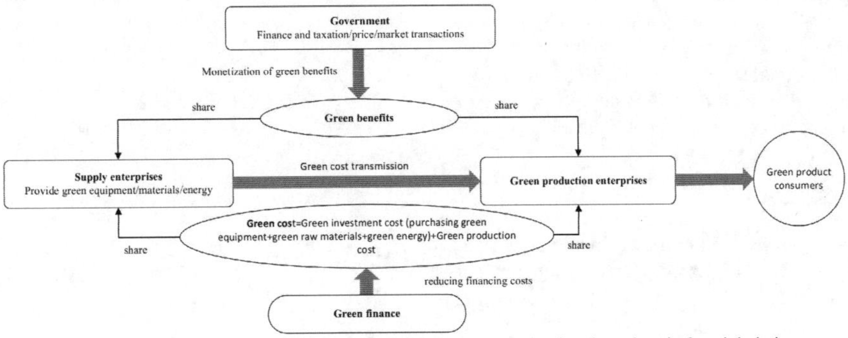

Figure 2. Green cost transmission and apportionment mechanism based on industrial chain.

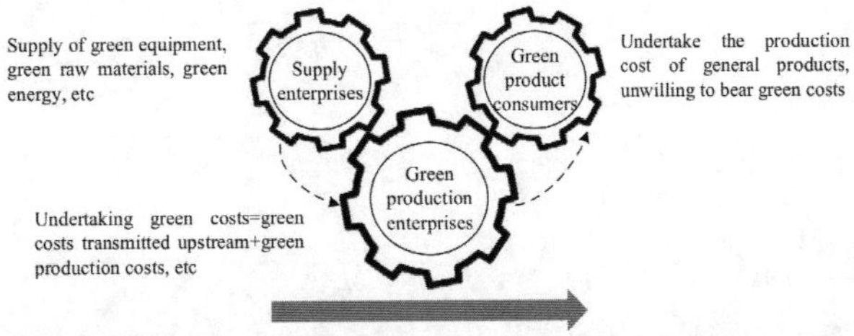

Figure 3. Green cost transmission apportionment process.

4.2. A Green Behavior Cultivation Mechanism Based on Maslow's Hierarchy of Needs Model

This article proposes a green behavior cultivation mechanism based on Maslow's hierarchy of needs model, which includes four aspects: focusing on habit cultivation, behavior recognition, green certification, and green incentives, as shown in Figure 4.

Habit cultivation is the precursor to practicing green behavior, focusing on four aspects: publicity, education, environmental protection volunteering, and supervision and reporting

The effective identification of green behavior is an important foundation for cultivating and guiding public green behavior. Based on green databases, quantitative analysis of the degree of greening of various life behaviors is conducted to enable the public to clearly and simply perceive green behavior.

Green certification is an important basis for obtaining green incentives for green behavior. A green behavior evaluation system is formed by measuring the green behavior of consumer subjects and measuring the green indicators of the entire life cycle of consumer objects. Currently, the green certification standards have not been unified, and there are significant differences in the integration mechanisms on different platforms, which affects the understanding of green behavior. It is necessary to build a unified green behavior certification mechanism for consumer subjects and consumer objects, and continuously update the green certification standards by combining market conditions and social responsibility factors.

Green incentives are the most effective means of directly guiding green behavior, with a focus on establishing effective incentive mechanisms in both active and passive aspects.

Figure 4. Green behavior training mechanism.

5. Conclusion

Building a green and low-carbon production and lifestyle is the main carrier for implementing the concept of green development, and it is also a theoretical innovation in the implementation process of sustainable development. The research significance of this article is as follows:

(1) We have constructed a cognitive system for green and low-carbon production and lifestyle, and proposed 4 typical scenarios and 10 typical green energy consumption modes;

(2) A green cost diversion mechanism based on the transmission and allocation of green costs in the industrial chain and a green behavior cultivation mechanism based on the Maslow hierarchy of needs model have been proposed, forming a support system for building a green and low-carbon production and lifestyle from the perspectives of conceptual cognition, path implementation, and mechanism improvement.

Acknowledgments

This work was supported by the Management Science and Technology Project of the State Grid Corporation of China, the project name: Research on Key Technologies for Optimizing and Enhancing Rural Power Supply Services to Support Agricultural and Rural Modernization (1400-202416298A-1-1-ZN).

References

1. Abdollahi HSG.Application of the Multi-objective optimization and riskanalysis for the sizing of a residential small-scale CCHP system[J]. Energy and buildings.2013, 60(May): 330-44.
2. SONG Zhuoran, CHENG Mengzeng, NIU Wei, et al. Key Technologies and Development Trends of Zero-Carbon Park Optimization Planning for Energy Internet[J]. Electric Power Construction, 2022, 43(12):15–26.
3. QIU Di, LIU Dong, GAO Fei, et al. Multi-scenario Modeling and Day-ahead Optimization of Park Multiple Energy System Based on Discrete Hybrid Automaton[J]. Proceedings of the CSEE, 2023, 3(01):135–147.
4. Wang Jingjing, Li Qifen, Yang Yongwen． Comprehensive evaluation of nearly zero carbon emission parks based on variable weight theory[J].Science Technology and Engineering, 2021, 21 (01):334–340.
5. KIM S,MOON ,SHIN Y,et al.Life comparaive analysis of energy consumption and cO, emisions of diferent building structural frame tyoesil The Scientific World Journal,2013,2013:175702.D01: 10.1155/2013/175702.
6. CHENG Hongbo, GUO Yuanxi, SHANG Zixua, et al. Study of Vehicle-to-Grid Interaction Income Considering Scenario Coefficient. Electric Power,2022,55(03):28–36.
7. FENG Kaihui, YAN Hu, DAI Wuzhen,etal. Optimal Operation of Rural Energy System with Biomass Biogas Power Generation Considering Energy Utilization Efficiency[J]. Electric Power, 2022, 55(07):172–178.
8. FENG Kaihui, LI Qionghui, HUANG Bibin,etal. Key Issues on Rural Energy Development in China[J].Electric Power,2022,55(06):1–8.
9. HONG Bowen,FENG Kaihui, MU Yunfei, etal. Discussion of Distributed Energy Resources Utility Mode and Its Application in Rural Areas of China[J].Electric Power,2020,53(02):99–104.

Energy Flow Modeling and Low Carbon Analysis of Process-oriented Production Line

Guohua Gao, Shiyue Pan, Zihua Zhang* and Zhenjiang Zhu

Beijing University of Technology,Beijing,100124,China

Abstract. Nowadays, energy and environmental problems are becoming more and more prominent, and the energy-based manufacturing industry also needs to face huge energy consumption problems. Therefore, production processes need to be investigated from an energy efficiency perspective at an early design stage so that maximum benefits can be obtained at low cost, low time and low risk. This process industry carbon emissions system does not calculate the consumption of raw materials and electricity in a generalized way at the boundaries of the plant. Rather, we delve into each step of the process and perform mathematical modeling of the process through a graphical approach, incorporating each link and factor that constitutes and influences the change in carbon emissions, and performing an aggregation of carbon emissions for the entire production line. It facilitates the subsequent improvement and optimization of the entire process-based production line, providing the basis for implementing and establishing the best energy-saving system.

1. Introduction

Rising energy prices, increasingly competitive markets, stringent environmental legislation, concerns about global climate change and interactions with customers are motivating manufacturing companies to improve production efficiency and minimize adverse environmental impacts. Fossil energy is still the main source of energy in the world[1]. A range of energy and environmental and potential problems have been plaguing governments. Global industrial energy consumption is expected to increase by 18 percent between 2015 and 2040. The process industry has been the image of high energy consumption and high carbon emissions. Therefore, the concept of "low carbon sustainability" should be adopted throughout the system lifecycle of process-based production lines, adding value while emphasizing energy and material savings. This means that the work of modeling the energy flow of process-oriented production lines is important to facilitate the low-carbon optimization of production lines.

Jan Peter Andersen conducted a study on the utilization and conversion of materials and energy in the steel industry, and developed a standardized energy consumption and process model for materials and energy.[2]. Liru Liu et al.[3] analyzed how changes in logistics during the production process would affect the overall production energy consumption based on a basic logistics diagram. Goutam Dutta has developed an optimization model for energy consumption to optimize the study of energy and greenhouse gases. [4].Professor Chen gave a study on process energy consumption, main production system energy consumption and enterprise energy consumption, respectively.[5] respectively. A neural network model and process energy consumption analysis model were built using both empirical and theoretical methods to analyze and study the process energy consumption. Academician Yin Ruiyu derived a comprehensive analysis flow chart for steel enterprises, which provides a better basis for studying the material flow, energy flow, and information flow in the overall process [6]. Wang et al. divided the energy in the system by source, destination and role, and further built an energy flow model of the steel enterprise using the heat balance method [7].

Taking a BOPP film production line as an example, this paper investigates the use of continuous Petri Net(PN) to achieve the modeling of energy flow in various types of continuous production systems, based on the study of energy flow models, for characteristics such as the graphical representation of PN in production systems that generally contain discrete, continuous and mixed production systems. And the information such as material flow, information flow, and energy flow, which are prevalent in the production system, are all reflected in the model.

2. Methodology

2.1. Introduction of Petri Net Theory

Petri nets (PN) [8] were first proposed in 1962 in the PhD thesis of German scholar Carl A. Petri, "Communication by automata theory", which was initially used to depict communication systems and has become a popular research topic for many scholars. With the continuous

*Corresponding author's e-mail: zhangzihua@bjut.edu.cn

research of scholars from all walks of life, PN has been greatly developed in both theory and application.PN has two kinds of nodes: location and variation. The position is represented by a circle and the variation is represented by a short line, and they are connected by arcs (directed arcs). In PN, directed arcs can point from position to variation or from variation to position, no independent position or variation is allowed.

The modeling and analysis of the continuous part of a production system can be achieved by continuous PN [9]. The continuous PN was first proposed by Rene David and summarized by an infinite division derivation of a marker in the basic PN, which can well describe the continuous part of the production process. The markers in the continuous PN position and the weights on the directed arc can be non-negative real numbers, and their positions and transitions are called continuous positions and continuous transitions. In modeling, the discrete position, continuous position, discrete variation, and continuous variation are denoted by the symbols in Fig.1. in order to distinguish them from the basic PN. A PN model is said to be hybrid PN if it has both discrete positions and discrete transitions as well as continuous positions. odeling of hybrid systems can be well implemented using hybrid PN [10].

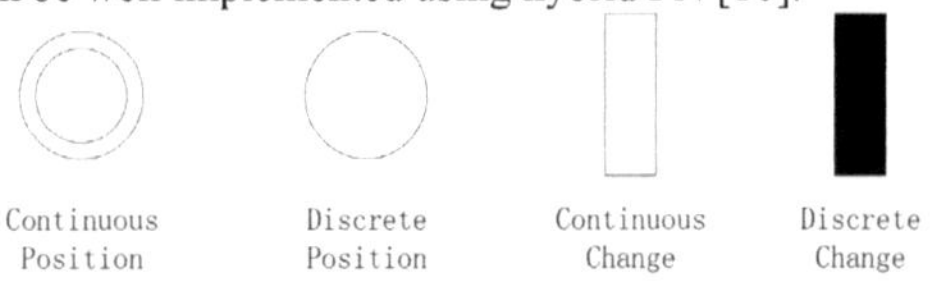

Figure 1. The sign of discrete and continuous nodes in PN.

3. Case study

3.1. Petri net energy flow model of discrete production system

In this section, the process of stretch film leaving the tractor and entering the film winder is analytically modeled. The film needs to be surface dusted before it enters the winder to ensure that the film breaks during the winding process due to surface foreign matter. The film is then wrapped on the core under the operation of the winder for subsequent cutting and rewinding. The process is a typical discrete production system with energy consumption. The PN energy flow is shown in Fig. 2. The specific meanings are shown in Tab. 1.

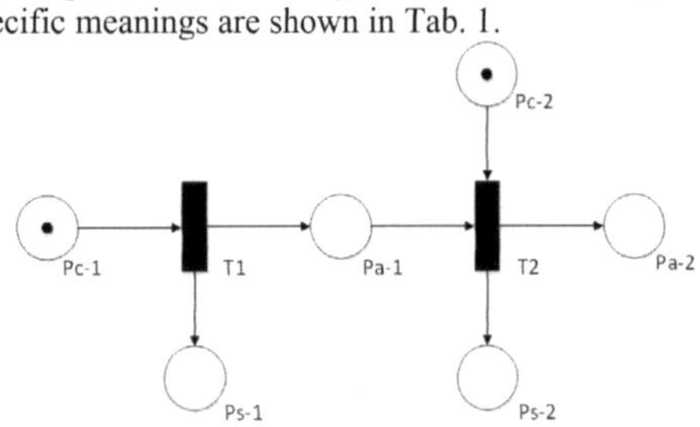

Figure 2. PN model of film winder operation.

Table 1. The position and change meaning of PN model in film winder operation.

Position/Change	Meaning
Ti	i=1--Dust removal process, i=2--Winding process
Pc-i	i=1--Stretched film, i=2--Roll core
Pa-i	i=1--Clean film, i=2--Master roll(Semi-finished products)
Ps-i	i=1--Energy consumed during dust removal process, i=2--Energy consumed in Winding process

3.2. Petri net energy flow model for continuous production system

The energy flow in a continuous production system is analyzed as an example of the first step in film manufacturing - batching and mixing. The process is to prepare and mix various raw materials in a certain ratio to facilitate the subsequent melting of the raw materials. As shown in Fig.3, the energy flow of the process is simulated with a continuous PN. The meaning of each position and change in the diagram is shown in the Tab.2.

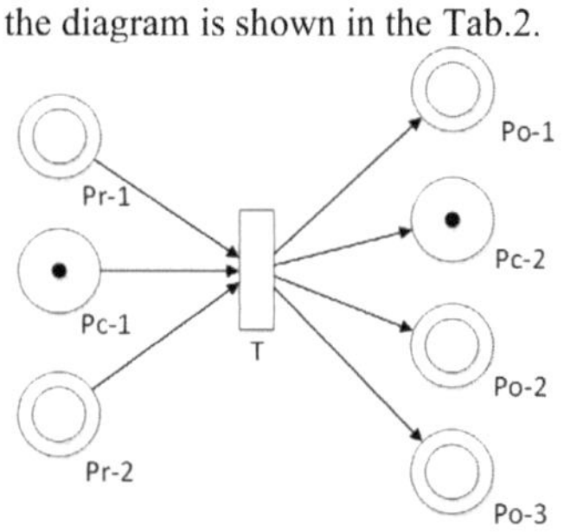

Figure 3. PN model of Batching and material mixing process.

Table 2. The position and change meaning of PN model in Batching and material mixing process.

Position/Change	Meaning
Ti	Mixer
Pc-i	i=1--Electric power, i=2--Mixture
Po-i	i=1--Energy used in the mixture, i=2--Heat energy dissipated, i=3--Unused electrical energy
Pr-i	i=1,2--Various raw materials

3.3. Petri net energy flow model for hybrid production system

3.3.1. Introduction of production process of automatic film production line

Biaxially Oriented Plastics Film (BOPF) is a film made of polymer melt or solution first made into sheet or thick film through a narrow head, then stretched in two vertical directions (longitudinal and transverse) in steps at a certain temperature and set speed in a special stretching machine, and made by appropriate cooling or heat

 https://doi.org/10.1051/e3sconf/202452803026

treatment or special processing (such as corona, coating, etc.).In paper, we focus on this part of the extrusion-casting system as an example to carry out the Petri net energy flow model of the hybrid production system for this process.

Film production is a typical process-type industry, which is not only accompanied by a continuous process of energy flow and material flow, but also driven by discrete events and the influence of the time factor, the BOPP film production process is long and complex, and there are problems of high energy consumption, low energy efficiency and low recycling rate of auxiliary materials in the processing and recycling process, which leads to the high cost of product processing, and serious environmental emissions. Production line energy source optimization, while the current production line carbon emission characteristics of the lack of effective research methods, restricting the further development of the film industry.

In the manufacturing process, the production process can be roughly divided into batching and mixing, melt extrusion, casting, longitudinal stretching (MDO), transverse stretching (TDO), haul-off winding, aging static, slitting, packaging process. According to the above eight process flows we analyze the energy, material input and output of each step of the whole production process, which provides the basis for the subsequent establishment of the energy network. The process flow is shown in Fig. 4.

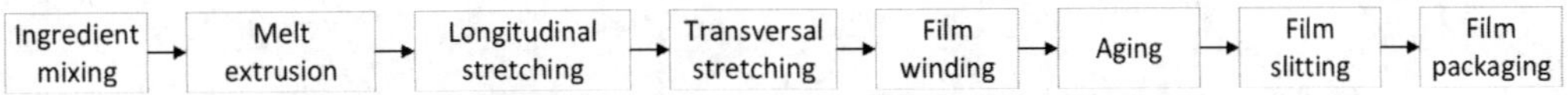

Figure 4. Flow chart of manufacturing process

3.3.2. Hybrid Petri net modeling of energy flow in automatic thin film production line

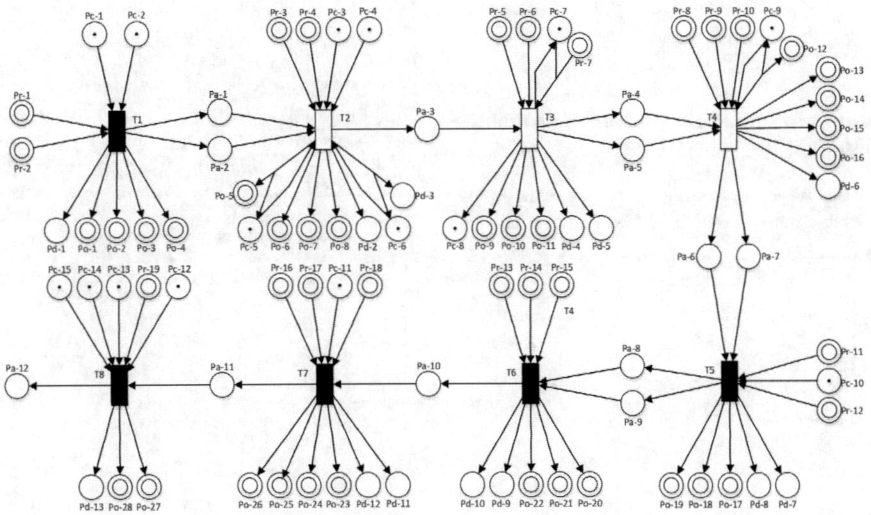

Figure 5. PN model of manufacturing process

Table 3. The position and change meaning of PN model in manufacturing process

Position/Change	Meaning
Ti	i=1--Ingredient mixing, i=2--Melt extrusion, i=3--Longitudinal stretching, i=4--Transversal stretching, i=5--Film winding, i=6--Aging, i=7--Film slitting, i=8--Film packaging
Pc-i	i=1--Polymer polypropylene particles, i=2--T1 process labor, i=3--Powder knotted sintered, i=4--Stainless steel filter discs, i=5--Water, i=6--Air, i=7--Oil, i=8--Water vapor generated by T3 process, i=9--T4 process air and water, i=10--Core roll, i=11--Blade, i=12--T8 process labor, i=13--Bag, i=14--Tape, i=15--Wrapping cardboard
Pr-i	i=1--Thermal energy of the heating system, i=2--Electrical energy of the mixer, i=3--Electrical energy of the power supply system, i=4--Thermal energy of the heating system, i=5--Electrical energy of the longitudinal drawer rollers, i=6--Electrical energy of cooling and sizing, i=7--Thermal energy of the oil belt, i=8--Electrical energy of the cross-drawing machine, i=9--Electrical energy of the cross-drawing machine rollers, i=10--Electrical energy consumed by the processing of the film characteristics, i=11--Electrical energy of the detector, i=12--Winder electrical energy, i=13--Crane energy consumption, i=14--Transportation trolley energy consumption, i=15--Air circulator electrical energy, i=16--Loading and unloading trolley energy consumption, i=17--Slitter energy, i=18--Tension control roller energy,

	i=19--Packer energy
Pa-i	i=1--Melt, i=2-- Melt thermal energy, i=3--Thick film, i=4--Semi-finished film, i=5--Semi-finished film thermal energy, i=6--Finished film, i=7--Finished film thermal energy, i=8--Rolled film, i=9--Rolled film thermal energy, i=10--Finished film after aging, i=11--Segmented rolled film, i=12--Finished rolled film
Pd-i	i=1--Electrical energy used by T1, i=2--Kinetic energy of ejected air, i=3--Kinetic energy of machine movement, i=4--Thermal energy for heating the metal rollers, i=5--Electrical energy for the operation of the longitudinal drawing machine, i=6--Kinetic energy for stretching the film, i=7--Energy of detection, i=8--Kinetic energy of winding, i=9--Energy consumed by movement of film roll, i=10--Kinetic energy of circulator, i=11--Control roll consumes kinetic energy, i=12--Loading and unloading roll kinetic energy, i=13--Packaging roll kinetic energy
Po-i	i=1--Granule bags, i=2, 7--Impurity waste, i=3, 10--Lost thermal energy, i=4, 6, 11, 14, 17, 21, 25, 28--Standby electrical energy, i=5--Waste filter, i=8--Waste filter, i=9, 13--Film breaking, i=12--Heat energy taken away by air and water, i=15--Heat energy emitted from cooling zone, i=16--Heat energy lost from rollers, i=18--Heat energy lost from cooling zone, i=18--Heat energy emitted by rollers, i=19, 22--Heat energy emitted, i=20--Heat energy dissipated from machine, i=23--Heat energy emitted from cutting rolls, i=24--Waste blades, i=26--Waste rolls of film, i=27--Packaging waste

Fig. 5 shows the PN energy flow of the thin film production process and Tab. 3 shows the meaning of the symbols in the figureWhere Pc denotes input material, Pr denotes input energy, Pa denotes flowing material and energy, Pd denotes energy used, and Po denotes emitted energy and material. Based on this part of the analysis and the steps of building the hybrid PN, we can simulate the model using Matlab/Simulink.

3.4. Simulation and analysis of Petri nets using Matlab/Simulink

The stateflow interaction with simulink can be realized by defining inputs and outputs. The inputs and outputs can be in the form of data or events. Each stateflow module belongs to a sub-package of simulink module. stateflow can be directly embedded into simulink simulation model, and in the initial stage of simulation, simulink will transform the logic graph drawn by stateflow into C language S function through compiler program, so that the two are organically combined together. All the attributes of the variables such as data form or event form can be set in the stateflow variable manager. Fig. 6 and Fig. 7 show example simulation diagrams.With the subsequent input of energy flow formulas and values for each process in the production of BOPP film, the specific energy direction can be simulated. Provide a basis for subsequent low-carbon improvements.

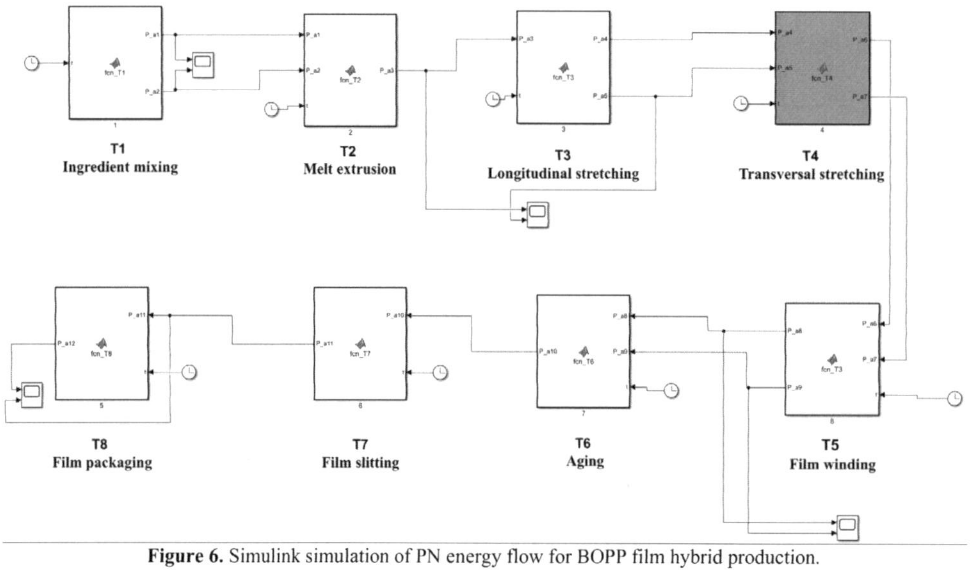

Figure 6. Simulink simulation of PN energy flow for BOPP film hybrid production.

 https://doi.org/10.1051/e3sconf/202452803026

```
%%   calcu
P_r8=Q_hlm*G2/A;
P_r9=Q_hlmk*G2/A;

P_r10= Q_spe*G2/A;


P_o13=m_pm/A ;
P_o12=c_m*v_m*t_m*d*rio_m*dT5 ;
P_o14=Q_hlmr ;
P_o15= v_hls*rio_a*c_a*dT6;
P_o16= v_hkt*rio_hkt*num_hkt ;

%%   out
t_r=min([t/3,1]) ;
P_a6=( P_a4-P_o13 )*t_r;

 P_a7=( P_r8 + P_r9 + P_r10 - P_o12 -P_o14 - P_o15 -P_o16 )*t_r;
end
```

Figure 7. Entering formulas related to transverse stretching in the Matlab editor

4. Conclusion and future work

In this paper, the energy network model of the production process of BOPP film production line is established by using PN network, and the common discrete, continuous and hybrid production systems in the actual production system are modeled and described by using ordinary PN, continuous PN and hybrid PN to realize the energy flow of the production process. The method of simulation and analysis of Petri nets using Matlab/Simulink software is also pointed out. This paper mainly provides a theoretical method for the establishment of a methodological framework for the low-carbon analysis of process-type production lines, and does not bring specific data into the simulation for the time being. In the future, through the optimization of the simulation model and the construction of formulas, the optimization points will be derived from the combination of specific data, so as to achieve the lowest carbon emissions from the process.

Acknowledgments

This project was supported by the National Key R&D Program of China (Grant No. 2023YFD2001100), Beijing Science and Technology Planning Project (Grant No. Z221100006422007), and R&D Program of Beijing Municipal Education Commission (Grant No. KM202310005034).

References

1. Zeng Chen. Assessment of energy saving and emission reduction efficiency and its coordination with economic benefits by regions in China[D]. Changsha: Hunan University, 2014.
2. Andersen J P, Hyman B. Energy and material flow models for the US steel industry[J]. Energy, 2001,26(2):137-159.
3. Liu L, Lu A, Lu Z, et al. Effect of material flows on energy intensity in process industries[J]. Energy, 2006, 31(12):1870-1882.
4. Dutta G, Sinha G P, Roy PN, et al. A Linear Programming Model for Distribution of Electrical Energy in a Steel Plant[J]. International Transactions in Operational Research, 1994, 1(1):17-29.
5. Chen Guang. Energy modeling of steel enterprises and dynamic simulation of their oxygen systems [D]. Shenyang: Northeastern University, 2002.
6. Yin Ruiyu. Analysis and integration of steel manufacturing processes[J]. Journal of Metals, 2000,36(10):1077-1084.
7. Wang J J, Cai Jiu ju, Zhang Q et al. Modeling energy flow in iron and steel enterprises[J]. China Metallurgy, 2006, 16(5):48-52.
8. Alla H, David R. Continuous and hybrid Petri nets. Cricuits systems and computers,1998,8(1):159~188.
9. David R, Alla H. On hybrid Petri nets. Discrete event dynamic system theory and applications,2001,11(1):9-40.
10. Yan-Jun L U, Dai H P. Hybrid Petri net modeling for urban traffic network[J]. Journal of Zhejiang University, 2007.

AUTHOR INDEX